Commemorating 150 Years of Justus von Liebig's Legacy

Commemorating 150 Years of Justus von Liebig's Legacy

Editors

Christoph Janiak
Sascha Rohn
Georg Manolikakes

Basel • Beijing • Wuhan • Barcelona • Belgrade • Novi Sad • Cluj • Manchester

Editors

Christoph Janiak
Institut für Anorganische
Chemie und Strukturchemie
Heinrich-Heine-Universität
Düsseldorf
Germany

Sascha Rohn
Department of Food
Chemistry and Analysis
Technische Universität Berlin
Berlin
Germany

Georg Manolikakes
Department of Chemistry
RPTU Kaiserslautern-Landau
Kaiserslautern
Germany

Editorial Office
MDPI
St. Alban-Anlage 66
4052 Basel, Switzerland

This is a reprint of articles from the Special Issue published online in the open access journal *Chemistry* (ISSN 2624-8549) (available at: www.mdpi.com/journal/chemistry/special_issues/VCET69OZN0).

For citation purposes, cite each article independently as indicated on the article page online and as indicated below:

Lastname, A.A.; Lastname, B.B. Article Title. *Journal Name* **Year**, *Volume Number*, Page Range.

ISBN 978-3-7258-0440-5 (Hbk)
ISBN 978-3-7258-0439-9 (PDF)
doi.org/10.3390/books978-3-7258-0439-9

Cover image courtesy of Public domain, via Wikimedia Commons

© 2024 by the authors. Articles in this book are Open Access and distributed under the Creative Commons Attribution (CC BY) license. The book as a whole is distributed by MDPI under the terms and conditions of the Creative Commons Attribution-NonCommercial-NoDerivs (CC BY-NC-ND) license.

Contents

About the Editors vii

Preface ix

Siegfried Schindler
Justus von Liebig
Reprinted from: *Chemistry* 2023, 5, 71, doi:10.3390/chemistry5020071 1

Robin Piani, Björn B. Beele, Jörg Rust, Christian W. Lehmann and Fabian Mohr
Coinage Metal Complexes Containing Perfluorinated Carboxylates
Reprinted from: *Chemistry* 2023, 5, 58, doi:10.3390/chemistry5020058 15

Florian Nährig, Yu Sun and Werner R. Thiel
Phosphine Functionalized Cp^C Ligands and Their Metal Complexes
Reprinted from: *Chemistry* 2023, 5, 62, doi:10.3390/chemistry5020062 36

Felipe Dornelles da Silva, Maximilian Roca Jungfer, Adelheid Hagenbach, Ernesto Schulz Lang and Ulrich Abram
Stabilization of 2-Pyridyltellurium(II) Derivatives by Oxidorhenium(V) Complexes
Reprinted from: *Chemistry* 2023, 5, 63, doi:10.3390/chemistry5020063 58

Hanns M. Weinert, Christoph Wölper and Stephan Schulz
Synthesis of 5-Metalla-Spiro[4.5]Heterodecenes by [1,4]-Cycloaddition Reaction of Group 13 Diyls with 1,2-Diketones
Reprinted from: *Chemistry* 2023, 5, 64, doi:10.3390/chemistry5020064 72

Aimée E. L. Cammiade, Laura Straub, David van Gerven, Mathias S. Wickleder and Uwe Ruschewitz
Synthesis, Structure, and Spectroscopic Properties of Luminescent Coordination Polymers Based on the 2,5-Dimethoxyterephthalate Linker
Reprinted from: *Chemistry* 2023, 5, 65, doi:10.3390/chemistry5020065 89

Heba Youssef, Jonathan Becker, Clemens Pietzonka, Ilya V. Taydakov, Florian Kraus and Klaus Müller-Buschbaum
Divalent Europium, NIR and Variable Emission of Trivalent Tm, Ho, Pr, Er, Nd, and Ce in 3D Frameworks and 2D Networks of Ln–Pyridylpyrazolates
Reprinted from: *Chemistry* 2023, 5, 69, doi:10.3390/chemistry5020069 102

Tobias Severin, Viktoriia Karabtsova, Martin Börner, Hendrik Weiske, Agnieszka Kuc and Berthold Kersting
Synthesis, Structures and Photophysical Properties of Tetra- and Hexanuclear Zinc Complexes Supported by Tridentate Schiff Base Ligands
Reprinted from: *Chemistry* 2023, 5, 70, doi:10.3390/chemistry5020070 124

Kathrin Kostka and Matthias Epple
Surface Functionalization of Calcium Phosphate Nanoparticles via Click Chemistry: Covalent Attachment of Proteins and Ultrasmall Gold Nanoparticles
Reprinted from: *Chemistry* 2023, 5, 72, doi:10.3390/chemistry5020072 142

Rana Ahmed, Inga Block, Fabian Otte, Christina Günter, Alysson Duarte-Rodrigues, Peter Hesemann, et al.
Activated Carbon from Sugarcane Bagasse: A Low-Cost Approach towards Cr(VI) Removal from Wastewater
Reprinted from: *Chemistry* 2023, 5, 77, doi:10.3390/chemistry5020077 159

Sergey P. Verevkin and Aleksandra A. Zhabina
Platform Chemicals from Ethylene Glycol and Isobutene: Thermodynamics "Pays" for Biomass Valorisation and Acquires "Cashback"
Reprinted from: *Chemistry* **2023**, *5*, 79, doi:10.3390/chemistry5020079 173

Lo'ay Ahmed Al-Momani, Heinrich Lang and Steffen Lüdeke
A Pathway for Aldol Additions Catalyzed by L-Hydroxyproline-Peptides via a β-Hydroxyketone Hemiaminal Intermediate
Reprinted from: *Chemistry* **2023**, *5*, 81, doi:10.3390/chemistry5020081 192

Julika Schlosser, Julian F. M. Hebborn, Daria V. Berdnikova and Heiko Ihmels
Selective Fluorimetric Detection of Pyrimidine Nucleotides in Neutral Aqueous Solution with a Styrylpyridine-Based Cyclophane
Reprinted from: *Chemistry* **2023**, *5*, 82, doi:10.3390/chemistry5020082 209

Stefan Buss, María Victoria Cappellari, Alexander Hepp, Jutta Kösters and Cristian A. Strassert
Modification of the Bridging Unit in Luminescent Pt(II) Complexes Bearing C^N*N and C^N*N^C Ligands
Reprinted from: *Chemistry* **2023**, *5*, 84, doi:10.3390/chemistry5020084 222

Nils Pardemann, Alexander Villinger and Wolfram W. Seidel
Improved Synthesis and Coordination Behavior of 1H-1,2,3-Triazole-4,5-dithiolates (tazdt^{2-}) with Ni^{II}, Pd^{II}, Pt^{II} and Co^{III}
Reprinted from: *Chemistry* **2023**, *5*, 86, doi:10.3390/chemistry5020086 235

Stephanie L. Faber, Nesrin I. Dilmen and Sabine Becker
On the Redox Equilibrium of TPP/TPPO Containing Cu(I) and Cu(II) Complexes
Reprinted from: *Chemistry* **2023**, *5*, 87, doi:10.3390/chemistry5020087 252

Sameera Shah, Tobias Pietsch, Maria Annette Herz, Franziska Jach and Michael Ruck
Reactivity of Rare-Earth Oxides in Anhydrous Imidazolium Acetate Ionic Liquids
Reprinted from: *Chemistry* **2023**, *5*, 94, doi:10.3390/chemistry5020094 266

Sebastian Derra, Luca Schlotte and Frank Hahn
Suzuki–Miyaura Reaction in the Presence of N-Acetylcysteamine Thioesters Enables Rapid Synthesis of Biomimetic Polyketide Thioester Surrogates for Biosynthetic Studies
Reprinted from: *Chemistry* **2023**, *5*, 96, doi:10.3390/chemistry5020096 283

Elaheh Bayat, Markus Ströbele and Hans-Jürgen Meyer
Unraveling the Synthesis of SbCl($C_3N_6H_4$): A Metal-Melaminate Obtained through Deprotonation of Melamine with Antimony(III)Chloride
Reprinted from: *Chemistry* **2023**, *5*, 99, doi:10.3390/chemistry5020099 295

Michel Stephan, Max Völker, Matthias Schreyer and Peter Burger
Syntheses, Crystal and Electronic Structures of Rhodium and Iridium Pyridine Di-Imine Complexes with O- and S-Donor Ligands: (Hydroxido, Methoxido and Thiolato)
Reprinted from: *Chemistry* **2023**, *5*, 133, doi:10.3390/chemistry5030133 307

About the Editors

Christoph Janiak

Christoph Janiak studied Chemistry at the Technical University Berlin (TUB) and the University of Oklahoma. He obtained his PhD at TUB in 1987, followed by postdoctoral stays at Cornell University and at BASF AG, Ludwigshafen in the polyolefin division. From 1991 to 1995, he carried out his Habilitation at TUB. Following a non-tenured professor position at the University of Freiburg from 1996 to 1998, he obtained tenure in 1998 as Associate Professor for Inorganic and Analytical Chemistry. In November 2010, he moved to the University of Düsseldorf as full professor (Chair) for Nanoporous and Nanoscaled Materials. His research interests include the properties and utilizations of metal–organic frameworks, covalent triazine frameworks and metal nanoparticles. Christoph Janiak co-authored over 700 research papers, book contributions and patents, with an h-index of 94 (Google Scholar).

Christoph received a Fonds of the Chemical Industry fellowship and award (1985–1987, 1988); the Heinz Maier-Leibnitz award (1991); the ADUC award for Habilitands (1996); a Heisenberg fellowship award (1997) and was a visiting professor at the University of Angers, France, and at Wuhan University of Technology, China. Currently, he is guest professor at the Hoffmann Institute of Advanced Materials at Shenzhen Polytechnic in China. He is a Fellow of the Royal Society of Chemistry (FRSC). He teaches General Chemistry, Coordination Chemistry, Main-group Chemistry and Analytical Chemistry and has co-authored several text books on Inorganic Chemistry (in German).

Sascha Rohn

Sascha Rohn, born 1973, is a full professor for Food Chemistry at the Technische Universität Berlin (Germany). He graduated from the University of Frankfurt/Main, Germany, with the first and second state examination in Food Chemistry, 1999. In 2002, he obtained his Ph.D. in Food Chemistry from the Institute of Nutritional Science, University of Potsdam, Germany, working on interactions of polyphenols with food proteins. After two years as a postdoc, he left Potsdam for Berlin, where he did a habilitation at the Institute of Food Technology and Food Chemistry of the Technische Universität Berlin. From October 2009 to October 2020, he was a full professor at the Hamburg School of Food Science, Institute of Food Chemistry, University of Hamburg, Germany. His group is dealing with the analysis of bioactive food compounds. More specifically, they are characterizing the reactivity and stability of bioactive compounds. The aim is to identify degradation products that serve as quality parameters, as process markers during food/feed processing or as biomarkers in nutritional physiology. Results of their work have been presented in more than 250 publications so far (Scopus h index is 60 and more than 10,000 citations). More than 30 well-known scientific journals ask Prof. Rohn regularly to review scientific manuscripts. From 2006 to 2012, he was chairman of the Northeastern branch of the German Food Chemical Society (LChG). From 2015 to 2023, he was also heading the Institute of Food & Environmental Research Bad Belzig, Germany, dealing with applied research in the fields of new natural raw materials for new food/feed/non-food products. He is actually member of the board of the German Nutrition Society (DGE) and chair of the scientific advisory board of the Max Rubner-Institut (MRI), Germany.

Georg Manolikakes

Georg Manolikakes studied chemistry at the Ludwig-Maximilians-Universität (LMU) München (Germany), the University of Oxford (UK) and the Universite Paris-Sud/Orsay (France). He obtained his PhD at LMU in 2009 under the supervision of Prof. Paul Knochel. After a postdoctoral stay with Prof. Phil S. Baran (Scripps Research Institute La Jolla/USA), he started his independent career as group leader at the Goethe University Frankfurt (Germany) in 2010. Since 2017, he has been a professor of Organic Chemistry at the RPTU Kaiserslautern-Landau (Germany). His research activities in the Manolikakes group focus on the development of novel efficient and sustainable methods for a modular and stereoselective synthesis of bioactive compounds using metal catalysis, photo- or electrochemistry as well application of these compounds in medicinal chemistry projects. Georg Manolikakes received a Liebig Fellowship from the Fonds der Chemischen Industrie (FCI), the Dr. Otto-Röhm Memorial Foundation Award, a Momentum Grant from the Volkswagen Foundation and the Rhineland-Palatinate State Teaching Award.

Preface

Justus Freiherr (Baron) von Liebig (12 May 1803–18 April 1873) was a renowned German chemist who made significant contributions to inorganic, organic, and agricultural chemistry. Liebig's work at the University of Giessen in Germany led to the development of modern teaching methods based on experimental laboratory-oriented work.

Liebig's contributions to inorganic chemistry are primarily seen in his work on plant nutrition and the development of modern fertilizers. He recognized that plants absorb essential inorganic nutrients in the form of salts. Liebig developed modern methods for the analysis of inorganic substances. He also popularized an earlier invention for condensing vapors, which came to be known as the Liebig condenser. In modern usage, it is still essential in many chemical laboratories.

Liebig made significant contributions to organic chemistry and is considered one of the founders of organic chemistry. Along with Friedrich Wöhler and Auguste Laurent, he pioneered the radical theory in the 1830s. Liebig's Kaliapparat simplified the determination of the carbon and the hydrogen content of organic compounds through combustion analysis. Thus, his work laid the foundation for the systematic analysis of organic substances, contributing significantly to the identification of numerous organic compounds. Liebig and Wöhler jointly realized that cyanic acid (HOCN) and fulminic acid (HCNO) represented two different compounds that had the same composition. This concept, later termed "isomerism", was a groundbreaking discovery in chemistry.

Liebig is often referred to as the "father of the fertilizer industry", as he realized that nitrogen and trace minerals (containing phosphorus and potassium) are essential plant nutrients. His work laid the foundation for the systematic use of fertilizers in agriculture. Liebig formulated the law of the minimum, which described how plant growth relied on the scarcest nutrient resource, rather than the total amount of resources available. This concept revolutionized the understanding of plant nutrition and has had a profound impact on agricultural practices, helping to improve crop yields and preventing food shortages. Liebig developed a manufacturing process for beef extracts, thereby contributing to food chemistry. A company, called Liebig Extract of Meat Company, introduced the Oxo brand beef bouillon cube. This contributed to the development of nutrient-rich food products, which nowadays would be called convenience food.

Liebig's reputation both as scientist and as teacher attracted many talented students; notable among them are August Kekulé, August Wilhelm von Hofmann, Adolphe Wurz and John Stenhouse.

A Special Issue of *Chemistry* was dedicated in 2023 to commemorate 150 Years of Justus von Liebig's Legacy. We thank the authors who contributed to this Special Issue, now collected in this Special Issue reprint.

Christoph Janiak, Sascha Rohn, and Georg Manolikakes
Editors

Editorial

Justus von Liebig

Siegfried Schindler

Institute of Inorganic and Analytical Chemistry, Justus Liebig University, 35392 Gießen, Germany; siegfried.schindler@anorg.chemie.uni-giessen.de

Abstract: This is a short overview of the life and achievements of Justus von Liebig. Clearly, this can only be an incomplete and somewhat personal view of the author, who has been a professor of inorganic chemistry at Justus Liebig University since 2002. Having already been interested in the work of Liebig for many years, and with a strong connection to the Liebig Museum in Giessen, the author hopes to provide some useful information about this great chemist, one of the founders of modern chemistry. The reader should find many interesting, probably new, facts about Liebig's major impact on chemistry, agriculture, nutrition, and pharmacology.

Keywords: Justus Liebig; history of chemistry; begin of modern chemistry

Citation: Schindler, S. Justus von Liebig. *Chemistry* **2023**, *5*, 1046–1059. https://doi.org/10.3390/chemistry5020071

Received: 20 April 2023
Revised: 1 May 2023
Accepted: 4 May 2023
Published: 6 May 2023

Copyright: © 2023 by the author. Licensee MDPI, Basel, Switzerland. This article is an open access article distributed under the terms and conditions of the Creative Commons Attribution (CC BY) license (https:// creativecommons.org/licenses/by/ 4.0/).

1. It all Started with a Bang

Justus Liebig was born on 12 May 1803, in Darmstadt (Hesse, Germany). His father owned a business, what today, most likely, you would call a drugstore, selling/preparing paints, and different materials, including chemicals. Early on, Justus had to help his father in the store, which he much preferred to school, where he did not enjoy learning old languages. Several times he had to fetch books for his father about chemistry from the library close by, which, from the age of 13, he read from beginning to end. Already at this early age, Justus wanted to become a chemist. An anecdote demonstrates this, as it was reported that his teacher and his classmates laughed at him when he gave this as an answer being asked in school what he would like to do later in life [1]. Justus enjoyed going to the fair in Darmstadt, where the marketeers (often charlatans/quacksalvers) performed all kinds of chemical tricks to attract people and to sell their wares, e.g., "universal medicine" that was declared to heal nearly every disease. To entertain the audience, they performed small explosions with the silver or mercury salt of fulminic acid (Figure 14). Some readers might be aware of the mercury salt used in the first season of the TV show Breaking Bad (not very realistic). There is an old book (published) for children about this time "Knallsilber—Geschichten aus dem Leben von Justus von Liebigs" (Figure 1a) [2], that describes how Justus watched at the fair how to create these explosions (Figure 10) and how he impressed his father, and especially his classmates, by preparing them himself. Most likely his teachers much less enjoyed this knowledge because all the boys then had easy access to these explosions. Silver fulminate is so unstable that the only application for this compound is in the bang snaps that are still sold today (for children!). Mainly they contain some sand wrapped together in paper with a tiny amount of silver fulminate (Figure 1b), and they explode when thrown on the floor or against a wall.

Due to his problems in school, Liebig did not finish high school. He left school when he was 14 and started an apprenticeship training in a pharmacy in Heppenheim (close to Darmstadt). However, here he also "failed" and left after only a few months. It was reported that he caused an explosion (or a fire) in the pharmacy when performing chemistry experiments (Figure 10) and was thrown out for that reason, but it is more likely that he did not like the work there at all. Back in Darmstadt, he worked again in his father's shop, produced bang snaps for sale, and even wrote his first publication about their chemistry in a journal. A chemist, Karl Wilhelm Gottlob Kastner, who became a professor in Bonn, was impressed and even wrote an introduction to this publication.

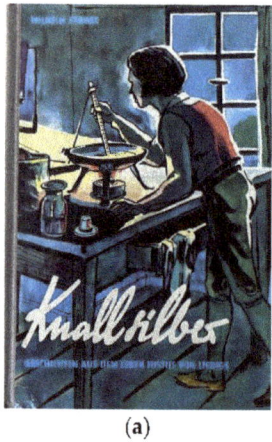
(a)

(b)

Figure 1. (a) The children book "Knallsilber" (Title in English: Silver fulminate—Stories of the life of Justus von Liebig); (b) Commercially sold bang snaps today kept in sawdust.

Liebig received the chance to join Kastner in Bonn and started studying chemistry there, despite not having finished high school. He followed Kastner to Erlangen in 1821, and at only 18 years old (Figure 2), he got into some serious trouble due to some student activities. Consequently, he had to escape a police investigation and flee back to Darmstadt. Despite these activities, he obtained a scholarship from the government of the state of Hesse to go to Paris to proceed with his studies. According to Liebig, France was the country where chemistry bloomed then, while Germany was left far behind.

Figure 2. Drawing of Liebig as a young student (photograph of a drawing at the Liebig Museum).

2. Paris

From October 1822, Liebig stayed in Paris for 17 months (his original stipend for 6 months was extended several times), and he made the acquaintance of some of the most famous chemists of the time, including L. N. Vauquelin, L. J. Thenard, and J. L. Gay-Lussac [1]. Liebig knew some French and studied it further because, in his opinion, it was necessary for every educated scientist. He was extremely impressed by the chemistry in France. He wrote to his very dear friend, the poet August von Platen [3]: *"It is a great shame how greatly the reputation of Germans has declined in physics, chemistry, and the other natural sciences. There is scarcely a shadow left, and they fight over this shadow like mad dogs. Contemporary German chemists presume to play at philosophy, thereby losing all their effectiveness.... The French and English proceed in the exact opposite fashion: here science is simply a mechanical stonework, the quasi-mathematical style of treating it allows no play of the mind whatever; but [this method] is at the moment very good, it has led recently to the most magnificent discoveries and is particularly useful for [practical] life"*. He admired the lectures by Gay-Lussac and the other professors, which were accompanied by experiments, and he was inspired to apply this later in his own teaching. His research focused again on fulminates, the explosives that had fascinated him as a boy. A short paper on that topic was printed in *Annales de chimie* and attracted significant attention. When the paper was presented by Gay-Lussac at the Académie des Sciences, Alexander von Humboldt was present and highly impressed by the young Liebig. Therefore, Humboldt recommended Liebig to Grand Duke Ludwig I. for a professorship at Giessen.

3. Giessen

Liebig was appointed in Giessen in May 1824, two weeks after his 21st birthday, as a (Ausserordentlicher) professor and became a full (Ordentlicher) professor in 1825; however, he was not welcomed at Giessen University because he had been put there by the Grand Duke and had not been chosen by the university/the professors [1]. Chemistry at that time was also not an established area of study but was regarded, rather, as a support for pharmacy or medicine. Thus, Liebig started to teach pharmacology, and his institute initially was pharmaceutical chemistry. Furthermore, when Liebig came to Giessen, there was already a full chemistry professor, Ludwig Wilhelm Zimmermann. They soon fell into a competition for students; however, it seemed that students preferred the teaching of Liebig, which must have been very lively and was combined with experimental demonstrations; many of these experiments are still presented in our first-semester chemistry lectures. In Germany, the word for a lecture is "Vorlesung" (a reading) that goes along with a reader for a lecturer. At that time, it was meant literally because professors mainly read from their books in a lecture. Liebig was different and thus attracted many students. The full circumstances are unclear, however, Zimmermann drowned in the river Lahn in 1825 (suicide suspected), and, thus, Liebig became a full professor of chemistry. Looking back, it is interesting to note that Liebig became a professor at 21 without having finished high school, with doubts about a proper doctoral dissertation (he received his doctoral title of Erlangen "in absentia), and having not completed a Habilitation [1]. Pretty amazing!

In Giessen, Liebig received a former guard house of a military barracks where he could set up his laboratory on the first floor and where he lived on the second floor together with his family. Over time, and sometimes pushing hard for money, the building was extended, and he obtained more space for research and teaching. Figure 3 shows the old laboratory with the stove and the front of the building (the door was opened for releasing toxic fumes). There was no heating system and no running water. Water for cooling a distillation was lifted in buckets attached to some tubing. In the newer part of the lab, hoods, quite similar to those used today, were later installed. The modern part of the laboratory and the way of running has been copied many times worldwide, e.g., in Russia or France (see below).

Figure 3. The stove of the old laboratory; front view of the Liebig Museum (door leading to the old laboratory; distillation setup; one of the hoods in the newer laboratory.

4. Students and Researchers in Giessen

The laboratory of Liebig became famous over time and attracted students and researchers from all over the world (Figure 10). Very early on, international exchange played an important role in the development of chemistry. Unfortunately, at that time, women could not yet participate. It is interesting to see who studied and/or worked with Liebig. Aside from the Germans, there were people from Belgium (1), Denmark (2), Great Britain (88), France (30), the Netherlands (3), Italy (2), Luxembourg (2), Mexico (1), Norway (2), Austria (11), Russia (21), Switzerland (43), Spain (1), Hungary (4), and United States (17) [4]. Some of the important chemists are shown on a genealogical tree in Figure 4, including Fresenius, Erlenmeyer, and Strecker, just to mention a few well-known names in chemistry. From there, it is obvious how chemistry developed further due to the influence of Liebig's laboratory.

Figure 4. Photograph of the genealogical tree at the Liebig Museum. Chemists in Figure 4 (from bottom to top and from left to right) Nobel prize winners are assigned with a star and the date of receiving the prize: Trunk: Justus von Liebig; 1. Branch: C. R. Fresenius, H. von Fehling, E. Erlenmeyer, A. Strecker, M. von Pettenkofer, A. Kekulé, H. Will, A. Sobrero, N. Zinin, J. Volhard, Th. Anderson, A. W. Hofmann, Ch. A. Wurtz, C. Schmidt; 2. Branch starting from Kekulé: A. von Baeyer* (1905), W. Körner, A. Ladenburg, Ch. G. Williams, J. Dewar, Th. Zincke, J. H.van't Hoff* (1901); Branch starting from Volhard: J. Thiele, D. Vorländer, H. Wieland* (1927), H. Staudinger* (1953), T. Reichstein* (1950), L. Ruzicka* (1939); Branch starting from Hofmann: W. von Miller, O. Wallach* (1910), R. Zsigmondy* (1925); Branch starting from Schmidt: W. Ostwald* (1909), S. Arrhenius* (1903), Th. W. Richards* (1914), W. Nernst* (1920), C. N. Lewis, H. von Euler* (1929), R. Abbeg, L. Ruzicka* (1939), F. Bergius* (1931), J. Langmuir* (1932), W. F. Giauque* (1949); Branch starting from von Baeyer: R. Willstätter* (1915), A. von Weinberg, E. Bamberger, H. Rupe, O. Fischer, E. Buchner* (1907), E. Fischer* (1902), R. Kuhn* (1938), P. Ehrlich* (1908), K. Landsteiner* (1930), P. H. Müller* (1948), A. Harden* (1929), O. Diels* (1950), O. H. Warburg* (1931), A. Windaus* (1928), K. Alder* (1950), A. Butenandt* (1939); Branch starting from Zincke: O. Hahn* (1944), H. Fischer* (1930).

A very important chemist from the Liebig laboratory was August Wilhelm von Hofmann, who later became president of the Chemical Society in London (now the Royal Society of Chemistry). After he accepted a position as a professor in Berlin, he was co-founder of the Deutsche Chemische Gesellschaft and became their president. He is well known, especially for his research on aniline (Figure 5).

Nikolay Zinin (Figure 5), a Russian chemist who later became the first president of the Russian Physical and Chemical Society, worked on the so-called benzoin condensation (discovered by Liebig) in Giessen. He later became a professor at the University of Kazan (afterward in St. Petersburg). He discovered that nitrobenzene could be reduced to aniline (general: the Zinin reaction, the reduction of nitro aromates to the corresponding amines). He played an important role in identifying aniline, which was finally confirmed by Hofmann [5]. The laboratory in Kazan was built according to the setup of Liebig's laboratory in Giessen, and it is now a museum (Museum of Kazan School of Chemistry). Some original samples of aniline prepared by Zinin are shown in the museum [5].

Figure 5. Memorial (copy; the original is in the Liebig Museum) of August Wilhelm von Hofmann (Text on the memorial: born in Giessen 8 April 1818; died in Berlin 5 May 1892; "A master and excellent teacher in chemistry; a successful researcher of aniline and aniline dyes") in Giessen (Frankfurter Strasse); a plaster bust of Nikolay Zinin at the museum of Kazan School of Chemistry.

Ascanio Sobrero, a student from Italy, who later became a professor at the University of Turin, was the discoverer of nitroglycerine in 1847. Alfred Nobel (who early on learned chemistry from Zinin in St. Petersburg) became familiar with nitroglycerine from Sobrero in 1850 when he was in Paris studying with Théophile-Jules Pelouze at the age of 17! Sobrero received his Ph.D. 1832 and described his time in Giessen well: *"In my last year in Giessen the number of students was 45. There were students of all nations, some with the goal to become a teacher, others with an interest in pharmacology, and many who wanted to do something with pure chemistry. The motivation to learn was very high for all students competing to perform well. However, there was no envy; all were like members of a large family, living under the same roof with the guidance of Liebig as a father"* (translation by the author) [4].

Friedrich August Kekulé, as Liebig was born in Darmstadt, started studying architecture in Giessen before switching to chemistry after attending some lectures of Liebig. In 1852, he finished his Ph.D. and later became a professor at the University of Bonn. The lecture notes of Kekulé on experimental chemistry have survived and (while hard to read) give an idea of the content of Liebig's lectures (Figure 6).

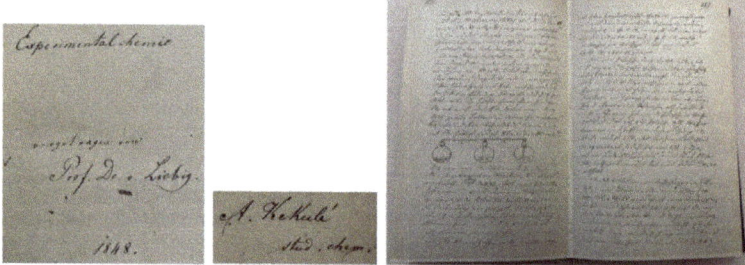

Figure 6. Photographs of the facsimile of the lecture notes of Kekulé.

Adolphe Wurtz spent one year in Liebig's laboratory after obtaining his Ph.D. in Strasbourg and was excited about the chemistry developments in Germany. Several French laboratories were later constructed based on the institute in Giessen. Wurtz became the first professor of organic chemistry at the Sorbonne in 1875. Wurtz described the work in Giessen similarly to Sobrero: *"What made life in Giessen so pleasant was the inner bond that prevailed among the chemists. United by the same scientific interest, suffused with the same devotion to the teacher, we worked the whole day together and supported each other at every instant with word and deed*" [3]. The book "Nationalizing Science—Adolphe Wurtz and the Battle for French Chemistry" by Alan J. Rocke is worth reading concerning chemistry at Liebig's time [3]. We could learn a lot from history to avoid repeatedly making mistakes.

5. Munich

After 28 years in Giessen, Liebig accepted an offer to move to Munich in 1852. Here, he obtained a new laboratory and had much fewer teaching duties. In Munich he even was in contact with the family of the King of Bavaria, who enjoyed attending spectacular chemistry presentations performed by Liebig. Unfortunately, during one of these lectures an accident occurred. Liebig was performing an experiment called the "Barking Dog" where NO is ignited in the presence of CS_2 in a long glass tube. The experiment's name is a consequence of the sound produced when the flame accelerates through the tube. The experiment was highly appreciated and should therefore be repeated. Then, a mistake occurred and instead of NO, oxygen was introduced into the system that caused an explosion when it was ignited. Despite being injured, Liebig was terrified to see blood on the faces of Queen Therese and Prince Luitpold. Luckily, none of the injuries were serious and there were no consequences. When Liebig realized that aging started to cause health problems he did not complain but instead he wrote to his sister in 1870 [1]: "If one has reached an age of 67 without ever been seriously sick one should be thankful, thank God for it and be content" (translated by the author). Liebig died on April 18, 1873; his grave is at the cemetery in Munich (Figure 7).

Figure 7. The grave of Liebig in Munich.

6. Achievements

Only a few examples of the many achievements Liebig accomplished can be presented here.

Aside from trying to understand the basics of chemistry, Liebig was always very interested in practical applications. For example, in Liebig's time, mirrors were made with mercury, a method that seemed to have been developed in Venice (Figure 8) and their production was dangerous due to the toxic mercury fumes. In Germany, especially around Fürth (close to Erlangen) there were several companies producing mirrors based on mercury. Even today, puddles of mercury are still sometimes found from that time when digging at construction sites in the areas of former mirror factories. Liebig developed a method to produce mirrors with silver; however, this business was unsuccessful because people preferred mercury-based mirrors. In contrast to the silver mirrors, these showed people much paler, which was preferred at that time. Only after the application of mercury was prohibited by law did silver mirrors start to be used.

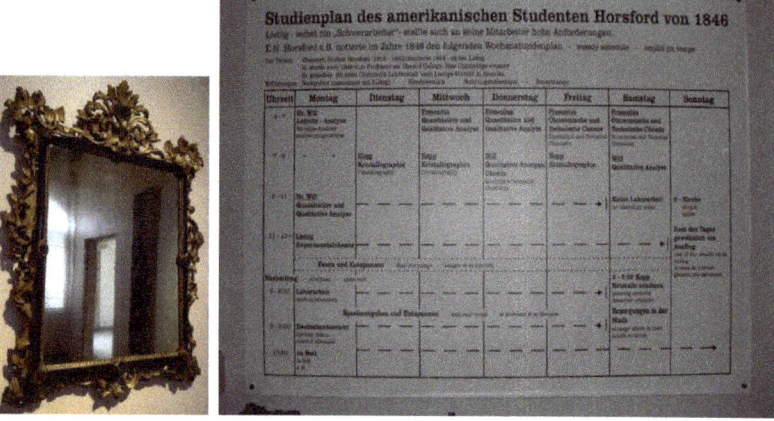

Figure 8. Photograph of a mirror made with mercury at a museum in Venice; timetable of Horsford.

Furthermore, Liebig invented baking powder that could be used in the place of yeast. However, again, his discovery did not lead to a successful business because people in Germany still preferred to use yeast. Instead, one of his students, the American Eben Norton Horsford, reformulated it later and became rich with it in the United States. After his studies with Liebig, Horsford was appointed as a professor at Harvard (Rumford Chair) where he was an early supporter of higher education for women. It is more than interesting to have a look at his timetable while he was studying with Liebig (Figure 8). Practical work or lectures started at 6 o'clock in the morning!

While Liebig was strongly involved in inorganic chemistry, he and Wöhler are especially important founders of organic chemistry. They introduced the concept of radicals in chemistry, which had nothing to do with molecules that we define today as radicals. The two chemists realized that while some molecule parts could be changed, other parts could be kept constant, for example, the benzoyl group in benzoic acid, benzaldehyde, etc. This finally became the extremely useful concept of functional groups. If cyanide is considered an organic molecule, Liebig also reported the first organocatalysis with the benzoin addition (often wrongly described as benzoin condensation). From the many organic compounds that Liebig investigated, only two more related ones should be mentioned here: chloroform and chloral hydrate. Independently from each other, chloroform was prepared in 1831 for the first time by the American Samuel Guthrie, the French Eugène Subeiran, and Liebig. Several years later, it became extremely useful as an anesthetic. Liebig prepared chloral hydrate around the same time, but its sedative properties were only recognized much later. Initially, it was wrongly assumed that chloral hydrate decomposes in the body to formic acid/formate and chloroform, which would cause the hypnotic effect. It is frightening to note that, according to Wikipedia, in the first 18 months of introducing it commercially in England, 17 million single doses were sold [6]!

Liebig can also be regarded as a pioneer in food chemistry, nutrition, and biochemistry. Many experiments were performed in his laboratory to gain better knowledge of these topics. A significant area of his research was in regard to milk for babies. At the time of

Liebig, a baby was in trouble if its mother could not provide milk for feeding. In that case, a nurse had to be found for breastfeeding. Cow milk alone is not suitable for babies. Liebig discovered a way to treat cow milk to make it suitable for infants and, thus, improved the chances of life and life generally for many. Furthermore, while in Munich, a friend's daughter was very ill with typhus and at the point of dying. Liebig managed to prepare a special soup for her and, thus, kept her alive.

While working on food, Liebig became aware of a problem in South America. Large numbers of cows were slaughtered there without using the meat because the primary interest was in harvesting their skin to produce leather. Without the technologies of today for freezing meat, it was impossible to keep it from going bad while trying to transport it over long distances. Liebig worked out a method to extract the meat as a solid paste that could be kept for a long time without the problem of rotting. He originally thought that he thus extracted the essence of the meat and that it would be a kind of a "superfood". This was not the case because the extract mainly consists of minerals and can be used to change the taste of your food, like, e.g., Marmite sold in England. However, the extract called "Liebigs Fleischextrakt" became very popular and is still sold today (Figure 9).

Figure 9. Liebigs Fleischextrakt; a guard book with collector cards inherited from the grandmother of the author's wife.

What was quite remarkable was how it was sold. Collector cards on many different topics were distributed together with each container of the Fleischextrakt and could be kept in a guard book (Figure 9). Four collector cards are presented in Figure 10. Thousands of these collector cards in many different languages were created.

Figure 10. Collector cards showing Liebig as a boy at the fair observing a marketer hitting a bang snap; young Liebig causing a fire in the pharmacy in Heppenheim; the laboratory of Liebig (the person on the right with the cylinder is Hofmann); and cows in South America.

Probably the most important research that Liebig completed was related to agriculture. For many different reasons, hunger had become a big problem for the population in Europe. The fields were exhausted and plant growth was only partially understood with many misconceptions. Liebig studied this in detail and realized that fertilizer had to be applied that provided all nutrients for the plants. He realized that if one important component was missing, an excess of all other nutrients will not help. He came up with a very good demonstration for this law of the minimum by showing a barrel that had planks of different heights (Figure 11). The plank that is the lowest would be the minimum.

 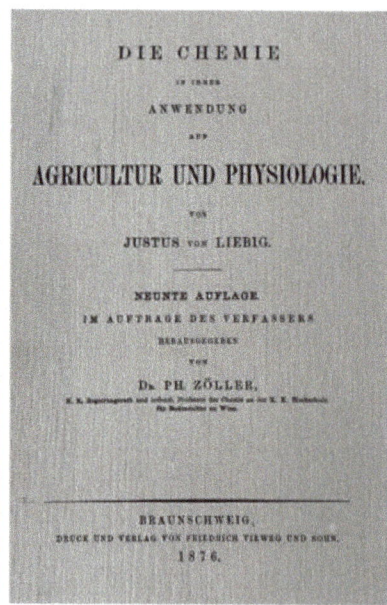

Figure 11. The barrel of Liebig shows the law of the minimum. Filling it up with water would lead to a leak at the plank with potassium ("Kali"). So, to improve the plant's growth, a potassium compounda needs to be added. A reprint of the book *Agricultur und Physiologie* by Liebig.

One can use this law of the minimum for our own nutrition. We could eat very healthily, however, if one component is too low (or missing), e.g., vitamin C, we still would be unhealthy and become sick over time. With this knowledge that he acquired over time (after many experiments and several failures, see below), Liebig finally could supply different fertilizers that helped to overcome the problem of starvation for many and, thus, enabled the feeding of many more people. Much of this knowledge was published in a book by Liebig "*Die Chemie in ihrer Anwendung auf Agricultur und Physiologie*" (Figure 11).

Finally, Liebig can be regarded as a pioneer in analytical chemistry. He realized very early on that exact analytical measurements are essential for understanding chemical reactions. In Giessen, he had a carpenter who made the set of scales for him. A photograph of the most accurate balance he used is shown in Figure 12. In contrast to today, such a set of scales took a long time to stabilize and was extremely sensitive to vibration. It was reported that, once, Liebig was smoking a cigar while waiting because that would be about the appropriate time for the set of scales to settle. To avoid any kind of vibration, the measurement result was read by observing the balance through a telescope from the door (Figure 12).

Figure 12. The best balance of Liebig with an accuracy of 0.3 mg and a maximum of 100 g; the telescope used to read the balance.

At the time of Liebig, there were only very few methods to analyze a compound. Chemists mainly had to rely on elemental analyses. Liebig had learned to perform these analyses well in Gay-Lussac's laboratory; however, the method was often inaccurate and took a lot of time. Therefore, he improved this method by introducing the Kaliapparat, which allowed a fast and accurate measurement of an organic compound's carbon and hydrogen content (Figure 13). The compound was burned, the water absorbed by the calcium chloride and the carbon dioxide by a potassium hydroxide solution in the Kaliapparat (the construction of the 5 bulbs ensured that all carbon dioxide was absorbed, and the solution was not pushed out). Afterward, both the tube with the calcium chloride and the Kaliapparat could be weighed, and it was possible to estimate the amount of carbon and hydrogen accurately. If the compound contained oxygen, the amount was just calculated as a difference. Thus, obtaining a molecular formula for glucose with CH_2O ("carbon with water, $C \times H_2O$") was possible, leading to the general name for many sugars being carbohydrates. The exact measurement of nitrogen was still a big problem and is discussed in great detail in the book *Nationalizing Science*, already mentioned above [3].

Figure 13. The Kaliapparat behind the U-tube with calcium chloride (the setup where the sample is burned is not shown); the author's membership card with the ACS logo.

This development was extremely important for the advancement of chemistry, now even allowing students to perform elemental analyses fast and accurately [7]. Before using the Kaliapparat this was time-consuming, and an expert was needed to perform the analysis correctly. An American student, Smith, asked Liebig for permission to use the Kaliapparat as a logo for the newly founded American Chemical Society. Many chemists in the US do not know what this symbol means; however, they have been reminded previously in an article published in the membership magazine of the ACS, C&E news [8]. Furthermore, it is not an eagle in the logo but a phoenix rising from the fire/ashes, fitting perfectly well with the idea of elemental analysis (Figure 13).

The result of an elemental analysis also led to a verbal fight between Liebig and Wöhler. Still working on the explosive silver fulminate, Liebig learned that Wöhler reported the same molecular formula for a supposedly stable compound. Both accused each other of not being capable of performing an elemental analysis correctly; however, as it turned out, they both were correct. With this dispute, they discovered the phenomenon of isomerism. Readers who have read or watched Harry Potter are fully aware of a so-called anagram ordering letters differently. Thus, in chemistry, it could make a big difference if the compound is written as HCNO or HOCN. The former is Liebig's fulminic acid and explosive, while the latter is cyanic acid, the compound Wöhler investigated (Figure 14). However, afterwards, Liebig and Wöhler became very good friends.

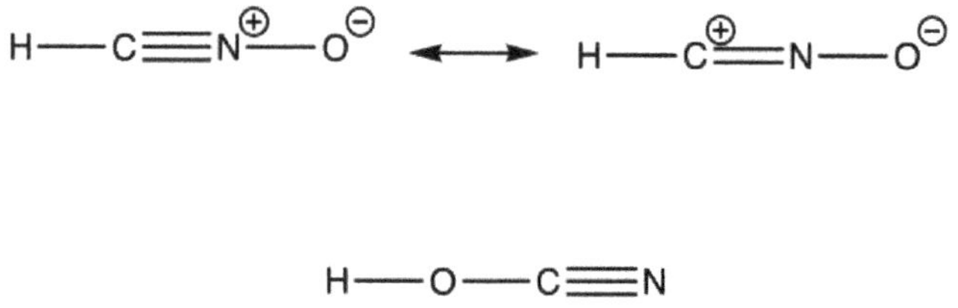

Figure 14. Isomerism of fulminic acid (**top**) and cyanic acid (**bottom**).

7. Failures

Finally, I think it is important to point out that even Liebig, who was an excellent and outstanding chemist, could make mistakes or fail in his work. This is how we all, and Liebig was no different, learn and improve our knowledge. For example, he missed discovering bromine as a new element when he was investigating the mineral waters of different fountains. He was so close but did not recognize it. Only quite late did he finally accept being wrong (as discussed above) that the Fleischextrakt was no superfood. While working on fertilizers he also made many mistakes that he learned to correct over time. Initially, he believed that plants receive nitrogen from the air and did not add nitrogen-containing compounds to the fertilizer. Furthermore, he thought fertilizer should be less soluble to avoid being washed out by rain. Due to these beliefs, there were disastrous results when his first fertilizers were applied. However, this is science: having an idea, testing it with experiments, and learning from the results. Liebig was very good at this. Additionally, Liebig failed to acknowledge the work of Carl Sprengel who recognized the importance of minerals in fertilizer before Liebig.

8. Conclusions

Liebig was an outstanding chemist/scientist. It is amazing what one person can achieve during a lifetime. Not only the many discoveries he made and the achievements for the advancement of chemistry/science, but also the aspects of teaching and the influence he had on other researchers. He also published a great deal of work in scientific journals for other chemists, but he also published in newspapers for the general public as well (Chemische Briefe—Chemistry Letters). It is remarkable that Liebig was already fully aware of the importance of international exchange, which is still relevant today (Figure 15).

Figure 15. International chemistry students (from Japan, Italy, Russia, and Kosovo) at the Justus Liebig University in the lecture room of Liebig 2021.

To really appreciate the accomplishments of Liebig, the reader is invited to visit the Liebig Museum in Giessen [4]. Furthermore, it is possible to learn more about Liebig (and the development in chemistry) at the museum (Deutsches Museum) in Munich. The museum in Giessen has existed since 1920. Currently, reconstruction is taking place due to a fire in December 2022 (Figure 16). Luckily, it only damaged the lecture room, and rebuilding it to its original state will be possible. On 29 March, a ceremony awarded the Laboratory of Justus Liebig with the "Historical Landmarks Award" of the European Chemical Society (Figure 16). Furthermore, an application process has been started to have it assigned as a UNESCO world cultural heritage site.

Figure 16. The lecture room at the Liebig Museum after the fire (already cleaned up); Historical Landmarks Award for the Laboratory of Justus Liebig (Prof. Dr. Angela Agostiano, President-elect of the European Chemical Society and Prof. Dr. Gerd Hamscher, Chair of the Justus Liebig Society).

Acknowledgments: The author would like to thank Bernd Commerscheidt (kommisarischer Kurator des Liebig-Museums) for checking the manuscript for mistakes concerning historical facts.

Conflicts of Interest: The author declares no conflict of interest.

References

1. *Ausstellungskatalog: Justus Liebig (1803–1873) Seine Zeit und Unsere Zeit*; Der Präsident der Justus-Liebig-Universität Gießen: Gießen, Germany, 2003.
2. Strube, W. *Knallsilber—Geschichten aus dem Leben Justus von Liebigs*; Kinderbuchverlag Berlin: Berlin, Germany, 1965.
3. Rocke, A.J. *Nationalizing Science: Adolphe Wurtz and the Battle for French Chemistry*; MIT Press: Cambridge, MA, USA, 2001.
4. Website of the Liebig Museum. Available online: https://www.liebig-museum.de or https://www.gdch.de/fileadmin/downloads/GDCh/historische_staetten/liebig.pdf (accessed on 19 April 2023).
5. Website of the Museum of Kazan School of Chemistry. Available online: http://www.ksu.museum.ru/chmku/eng/s2.htm (accessed on 19 April 2023).
6. Wikipedia on Chloralhydrat. Available online: https://de.wikipedia.org/wiki/Chloralhydrat#cite_note-14 (accessed on 19 April 2023).
7. Usselman, M.; Rocke, A.; Reinhart, C.; Foulser, K. Restaging Liebig: A Study in the Replication of Experiments. *Ann. Sci.* **2012**, *62*, 1–55. [CrossRef]
8. Everts, S. A Most Important Artifact. *C&E News*, 2015, Volume 93. Available online: https://cen.acs.org/articles/93/i35/Important-Artifact.html (accessed on 5 May 2023).

Disclaimer/Publisher's Note: The statements, opinions and data contained in all publications are solely those of the individual author(s) and contributor(s) and not of MDPI and/or the editor(s). MDPI and/or the editor(s) disclaim responsibility for any injury to people or property resulting from any ideas, methods, instructions or products referred to in the content.

Article

Coinage Metal Complexes Containing Perfluorinated Carboxylates

Robin Piani [1], Björn B. Beele [1], Jörg Rust [2], Christian W. Lehmann [2] and Fabian Mohr [1,*]

[1] Anorganische Chemie, Fakultät für Mathematik und Naturwissenschaften, Bergische Universität Wuppertal, Gaußstr. 20, 42119 Wuppertal, Germany

[2] Chemische Kristallographie und Elektronenmikroskopie, Max-Planck-Institut für Kohlenforschung, Kaiser-Wilhelm-Platz 1, 45470 Mülheim an der Ruhr, Germany

* Correspondence: fmohr@uni-wuppertal.de; Tel.: +49-202-439-3641

Abstract: A variety of coinage-metal complexes containing perfluorinated carboxylate ligands, together with their structures and thermal behavior, are reported. The silver(I) salts were accessible from the direct reaction of Ag_2O with the acids in toluene. Their gold(I) phosphine counterparts formed in high yields by transmetallation using the silver(I) salts. Some structurally unique, mixed-metal (Au,Ag) complexes formed upon combining solutions of the silver(I) salts with the gold(I) phosphine carboxylates. The reduction of dinuclear copper(II) compounds containing perfluorinated carboxylates with triphenylphosphine resulted in the formation of the corresponding copper(I) tris(phosphine) complexes. X-ray structures of representative complexes, together with IR- and TGA data, are reported.

Keywords: silver; gold; copper; perfluorinated carboxylates; molecular structures; TGA

1. Introduction

Silver(I) salts of perfluorocarboxylic acids have been known for decades and are typically prepared from the sodium salt of the respective acid and $AgNO_3$ in water or from an aqueous solution of the acid and a basic silver(I) compound such as Ag_2O or Ag_2CO_3 [1–4]. More recently, a water-free method using AgF in THF or hexane has also been reported [5]. The molecular structure of silver(I) perfluorobutyrate was determined by X-ray diffraction as early as 1956 [6]. Although the structure is of poor quality by modern standards, the atom connectivity could be unambiguously established. The salt exists in the solid state as a centrosymmetric dimer with a short (2.90 Å) Ag–Ag distance and O,O-bridging carboxylato ligands. These dinuclear eight-membered rings form a coordination polymer through additional Ag–O bonds between neighboring dimers (Figure 1, top left). The structure of silver(I) trifluoroacetate is analogous [7].

In contrast, the silver salt of pentafluorobenzoic acid (crystallized from water) has a slightly different structure. The carboxylic acid bridges two silver atoms in an alternating fashion, resulting in a chain-type structure (Figure 1, top right) [8]. The 2,4,6-trifluorobenzoato silver salt, also crystallized from water, forms the eight-membered ring dimer structure, but the coordination polymer is assembled through hydrogen bonding between bridging water molecules and not through Ag–O carboxylate bonding (Figure 1, bottom) [9]. In contrast to most silver(I) salts, silver perfluorocarboxylates are soluble in a variety of solvents, including water and organic solvents such as CH_2Cl_2, $CHCl_3$, alcohols, acetone, benzene, toluene and diethyl ether. Indeed, Swarts commented on the high solubility of silver(I) trifluoroacetate in both water and benzene in his pioneering study of metal trifluoroacetates from 1939 [10]. This report inspired us to attempt a synthesis of silver(I) perfluorocarboxylates in toluene to avoid the use of water as a solvent. With these silver salts in hand, we wished to prepare the corresponding gold(I) phosphine complexes and to examine the possibility of isolating bimetallic Ag/Au-complexes. To complete our investigation, we also examined the structures and reactions of some copper(II) and copper(I) complexes of perfluorinated carboxylic acids.

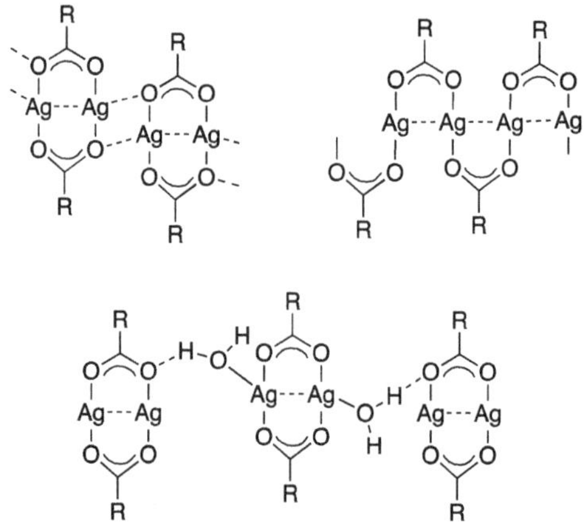

Figure 1. Schematic illustration of known coordination polymers of silver(I) carboxylates.

2. Materials and Methods

Unless specified otherwise, reactions were carried out under ambient conditions using HPLC-grade solvents with protection from light for reactions involving Ag and Au compounds. Chemicals and solvents were commercial products and were used as received. [AuCl(PPh$_3$)] [11], the gold(I) carboxylates [12], as well as the copper(II) carboxylates [13–15], were prepared by published methods. NMR spectra were measured on a Bruker Avance 400 instrument. Spectra were referenced externally to Me$_4$Si (^{1}H), 85% H$_3$PO$_4$ (^{31}P) and CFCl$_3$ (^{19}F). Copies of the NMR spectra may be found in the supplementary material. IR spectra were recorded on a Nicolet iS5 spectrometer equipped with an iD5 diamond ATR unit. TGA and DSC measurements were simultaneously carried out using a Netzsch STA 449 F5 Jupiter instrument. Experiments were conducted in 40 µL alumina crucibles, which were closed with alumina lids. Samples were heated from 25 to 1100 °C with a heating rate of 5 K/min in a nitrogen atmosphere, applying a constant nitrogen flow of 25 mL/min during the experiment.

2.1. Preparation of the Silver(I) Carboxylates

To a suspension of Ag$_2$O in toluene, the appropriate perfluorocarboxylic acid was added. Within a few minutes at ambient temperature, most of the silver oxide was dissolved. After about 10 min., anhydrous Na$_2$SO$_4$ (ca. 0.5 g) was added to absorb the water. The mixture was filtered through Celite and the filtrate was subsequently concentrated in vacuum to a small volume. The addition of pentane precipitated a colourless solid, which was isolated by filtration and was subsequently dried in air.

2.1.1. [Ag(O$_2$CCF$_3$)(η^2-MeC$_6$H$_5$)]$_n$ (**1**)

This was prepared as described above using Ag$_2$O (0.310 g, 1.34 mmol) and trifluoroacetic acid (0.20 mL, 2.59 mmol). A total of 0.349 g (0.654 mmol, 49%) product was isolated. IR (ATR): 1614 ν_a(OCO), 1449 ν_s(OCO) cm^{-1}. ^{19}F-NMR (CD$_3$OD): δ = −75.65 (s, CF$_3$). ^{1}H-NMR (CD$_3$OD): δ = 7.09–7.30 (m, 5 H, C$_6$H$_5$), 2.35 (s, 3 H, Me). Colourless, X-ray-quality crystals formed upon the slow evaporation of a toluene solution in the dark.

2.1.2. [Ag$_2$(O$_2$CCF$_2$CF$_3$)$_2$(η^2-MeC$_6$H$_5$)]$_n$ (**2**)

This was prepared as described above using Ag$_2$O (0.330 g, 1.42 mmol) and perfluoropropionic acid (0.29 mL, 2.79 mmol). A total of 0.503 g (0.793 mmol, 56%) product was isolated. IR (ATR): 1616 ν_a(OCO), 1416 ν_s(OCO) cm^{-1}. ^{19}F-NMR (CD$_2$Cl$_2$): δ = −83.36 (t, J = 1.7 Hz, CF$_3$), −118.43 (q, J = 1.7 Hz, CF$_2$). ^{1}H-NMR (CD$_2$Cl$_2$): δ = 7.30–7.36 (m, 2 H,

C_6H_5), 7.19–7.28 (m, 3 H, C_6H_5), 2.39 (s, 3 H, Me). X-ray-quality crystals formed by the slow evaporation of a toluene solution of the compound.

2.1.3. [Ag{$O_2C(CF_2)_2CF_3$}]$_n$ (3)

This was prepared as described above using Ag_2O (0.270 g, 1.17 mmol) and perfluorobutyric acid (0.3 mL, 2.31 mmol). A total of 0.563 g (0.877 mmol, 76%) product was isolated. IR (ATR): 1607 ν_a(OCO), 1417 ν_s(OCO) cm^{-1}. ^{19}F-NMR (CD_2Cl_2): δ = −80.45 (t, J = 8.9 Hz, CF_3), −115.06 (sext., J = 8.9 Hz, β-CF_2), −126.29 (m, α-CF_2).

2.1.4. [Ag($O_2CC_6F_5$)(MeC_6H_5)]$_n$ (4)

Pentafluorobenzoic acid (0.320 g, 1.51 mmol) was dissolved in toluene (10 mL) with slight heating, and then Ag_2O (0.175 g, 0.755 mmol) was added. The mixture was held at ca. 80 °C for 10 min. During this time, most of the solid was dissolved. Workup as described above afforded 0.392 g (0.537 mmol, 71%) colourless product. IR (ATR): 1654 ν_a(OCO), 1489 ν_s(OCO) cm^{-1}. ^{19}F-NMR (CD_3OD): δ = −144.04 (dd, J = 7.9, 14.5 Hz, o-F), −159.56 (t, J = 19.7 Hz, m-F), −165.18 (m, p-F). ^{1}H-NMR (CD_3OD): δ = 7.09–7.27 (m, 5 H, C_6H_5), 2.34 (s, 3 H, Me).

2.1.5. [Ag(O_3SCF_3)]$_n$ (5)

This was prepared as described above using Ag_2O (0.363 g, 1.57 mmol) and trifluoromethanesulfonic acid (0.30 mL, 3.39 mmol). A total of 0.573 g (2.23 mmol, 66%) product was isolated. IR (ATR): 1213 ν_a(SO_3), 1017 ν_s(SO_3) cm^{-1}. ^{19}F-NMR (CD_3OD): δ = −79.88 (s, CF_3).

2.2. Preparation of the Gold(I) Carboxylates

The complexes were prepared based on the procedure reported in the literature [12], with minor modifications. As an example, the preparation of [(Ph_3P)Au(O_2CCF_3)] is given in detail. The other mono- and dinuclear derivatives were prepared similarly.

2.2.1. [(Ph_3P)Au(O_2CCF_3)] (1a)

[Ag(O_2CCF_3)(η^2-MeC_6H_5)]$_n$ (0.045 g, 0.202 mmol) in CH_2Cl_2 (10 mL) was added to a solution of [AuCl(PPh_3)] (0.100 g, 0.202 mmol) in CH_2Cl_2 (10 mL). Immediately, a colourless solid (AgCl) precipitated. After stirring the mixture at room temperature for ca. 1 h, it was passed through Celite and concentrated under reduced pressure. The addition of pentane precipitated a colourless solid. This was isolated by filtration and was dried in air. A total of 0.098 g (0.171 mmol, 85%) product was isolated. IR (ATR): 1695 ν_a(OCO) cm^{-1}. ^{31}P{^{1}H}-NMR (CD_2Cl_2): δ = 27.36 (s). ^{19}F-NMR (CD_2Cl_2): δ = −74.32 (s, CF_3).

2.2.2. [(Ph_3P)Au($O_2CCF_2CF_3$)] (2a)

This was prepared as described above using [AuCl(PPh_3)] (0.100 g, 0.202 mmol) and [Ag($O_2CCF_2F_3$)(η^2-MeC_6H_5)]$_n$ (0.055 g, 0.202 mmol). A total of 0.078 g (0.125 mmol, 62%) colourless product was isolated. IR (ATR): 1710 ν_a(OCO) cm^{-1}. ^{31}P{^{1}H}-NMR (CD_2Cl_2): δ = 27.28 (s). ^{19}F-NMR (CD_2Cl_2): δ = −83.32 (t, J = 1.7 Hz, CF_3), −119.52 (m, CF_2).

2.2.3. [(Ph_3P)Au{$O_2C(CF_2)_2CF_3$}] (3a)

This was prepared as described above using [AuCl(PPh_3)] (0.100 g, 0.202 mmol) and [Ag{$O_2C(CF_2)_2CF_3$}]$_n$ (0.065 g, 0.202 mmol). A total of 0.088 g (0.131 mmol, 65%) colourless product was isolated. IR (ATR): 1699 ν_a(OCO), 1482 ν_s(OCO) cm^{-1}. ^{31}P{^{1}H}-NMR (CD_2Cl_2): δ = 27.24 (s). ^{19}F-NMR (CD_2Cl_2): δ = −81.08 (t, J = 8.9 Hz, CF_3), −116.91 (m, β-CF_2), −127.08 (m, α-CF_2).

2.2.4. [(Ph$_3$P)Au(O$_2$CC$_6$F$_5$)}] (**4a**)

This was prepared as described above using [AuCl(PPh$_3$)] (0.100 g, 0.202 mmol) and [Ag(O$_2$CC$_6$F$_5$)]$_n$ (0.064 g, 0.202 mmol). A total of 0.119 g (0.179 mmol, 89%) colourless product was isolated. IR (ATR): 1650 ν_a(OCO) cm^{-1}. ^{31}P{^{1}H}-NMR (CD$_2$Cl$_2$): δ = 27.45 (s). ^{19}F-NMR (CD$_2$Cl$_2$): δ = −141.34 (dd, *J* = 8.1, 23.1 Hz, *o*-F), −155.87 (m, *m*-F), −163.00 (m, *p*-F).

2.2.5. [(dppb)(AuO$_2$CCF$_3$)$_2$] (**5a**)

This was prepared as described above using [Au$_2$Cl$_2$(dppb)] (0.050 g, 0.055 mmol) and [Ag(O$_2$CCF$_3$)]$_n$ (0.024 g, 0.110 mmol). A total of 0.058 g (0.054 mmol, 98%) colourless product was isolated. IR (ATR): 1686 ν_a(OCO) cm^{-1}. ^{31}P{^{1}H}-NMR (CD$_2$Cl$_2$): δ = 18.97 (s). ^{19}F-NMR (CD$_2$Cl$_2$): δ = −74.07 (s, CF$_3$).

2.2.6. [(dppb)(AuO$_2$CC$_6$F$_5$)$_2$] (**6a**)

This was prepared as described above using [Au$_2$Cl$_2$(dppb)] (0.050 g, 0.055 mmol) and [Ag(O$_2$CC$_6$F$_5$)]$_n$ (0.035 g, 0.110 mmol). A total of 0.063 g (0.049 mmol, 91%) colourless product was isolated. IR (ATR): 1649 ν_a(OCO) cm^{-1}. ^{31}P{^{1}H}-NMR (CD$_2$Cl$_2$): δ = 19.45 (s). ^{19}F-NMR (CD$_2$Cl$_2$): δ = −141.02 (dd, *J* = 8.3, 23.1 Hz, *o*-F), −156.49 (t, *J* = 20.6 Hz, *m*-F), −163.55 (m, *p*-F).

2.2.7. [(Xantphos)(AuO$_2$CCF$_2$CF$_3$)$_2$] (**7a**)

This was prepared as described above using [Au$_2$Cl$_2$(Xantphos)] (0.050 g, 0.044 mmol) and [Ag(O$_2$CCF$_2$CF$_3$)(η^2-MeC$_6$H$_5$)]$_n$ (0.024 g, 0.088 mmol). A total of 0.044 g (0.034 mmol, 77%) colourless product was isolated. ^{31}P{^{1}H}-NMR (CD$_2$Cl$_2$): δ = 18.01 (s). ^{19}F-NMR (CD$_2$Cl$_2$): δ = −83.21 (m, CF$_3$), −119.52 (m, CF$_2$).

2.2.8. [(Xantphos){AuO$_2$C(CF$_2$)$_2$CF$_3$}$_2$] (**8a**)

This was prepared as described above using [Au$_2$Cl$_2$(Xantphos)] (0.050 g, 0.044 mmol) and [Ag(O$_2$CCF$_2$CF$_3$)(η^2-MeC$_6$H$_5$)]$_n$ (0.028 g, 0.088 mmol). A total of 0.060 g (0.043 mmol, 98%) colourless product was isolated. ^{31}P{^{1}H}-NMR (CD$_2$Cl$_2$): δ = 18.02 (s). ^{19}F-NMR (CD$_2$Cl$_2$): δ = −81.07 (t, *J* = 9.1 Hz, CF$_3$), −117.00 (m, β-CF$_2$), −126.98 (m, α-CF$_2$).

2.3. Preparation of the Bimetallic Silver(I)-Gold(I) Carboxylates

2.3.1. [(CF$_3$CO$_2$)$_2$AgAu(PPh$_3$)]$_2$ (**9a**)

Solid [Ag(O$_2$CCF$_3$)(η^2-MeC$_6$H$_5$)]$_n$ (0.016 g, 0.074 mmol) was added to a solution of [(Ph$_3$P)Au(O$_2$CCF$_3$)] (0.042 g, 0.074 mmol) in CH$_2$Cl$_2$ (10 mL). After stirring the mixture at room temperature for ca. 2 h, it was concentrated to small volume under reduced pressure. The addition of pentane precipitated a colourless solid. This was isolated by filtration and was dried in air. A total of 0.056 g (0.036 mmol, 49%) product was isolated. IR (ATR): 1669, 1639 ν_a(OCO) cm^{-1}. ^{31}P{^{1}H}-NMR (CD$_2$Cl$_2$): δ = 27.06 (s). ^{19}F-NMR (CD$_2$Cl$_2$): δ = −73.58 (s, CF$_3$). X-ray-quality crystals formed upon the slow evaporation of a solution of the compound in CH$_2$Cl$_2$ layered with hexanes.

2.3.2. [(CF$_3$CF$_2$CO$_2$)$_2$AgAu(PPh$_3$)]$_2$ (**10a**)

This was prepared as described above using [(Ph$_3$P)Au(O$_2$CCF$_2$CF$_3$)] (0.046 g, 0.074 mmol) and [Ag$_2$(O$_2$CCF$_2$CF$_3$)$_2$(η^2-MeC$_6$H$_5$)]$_n$ (0.047 g, 0.074 mmol). A total of 0.131 g (0.074 mmol, 99%) product was isolated. IR (ATR): 1679, 1644 ν_a(OCO) cm^{-1}. ^{31}P{^{1}H}-NMR (CD$_2$Cl$_2$): δ = 26.90 (s). ^{19}F-NMR (CD$_2$Cl$_2$): δ = −83.34 (t, *J* = 1.7 Hz, CF$_3$), −118.83 (m, CF$_2$).

2.3.3. [{CF$_3$(CF$_2$)$_2$CO$_2$}$_2$AgAu(PPh$_3$)]$_2$ (**11a**)

This was prepared as described above using [(Ph$_3$P)Au{O$_2$C(CF$_2$)$_2$CF$_3$}] (0.050 g, 0.074 mmol) and [Ag{O$_2$C(CF$_2$)$_2$CF$_3$}]$_n$ (0.024 g, 0.074 mmol). A total of 0.041 g (0.021 mmol, 55%) product was isolated. IR (ATR): 1657, 1648 ν_a(OCO) cm^{-1}. ^{31}P{^{1}H}-NMR (CD$_2$Cl$_2$): δ = 26.73 (s). ^{19}F-NMR (CD$_2$Cl$_2$): δ = −81.06 (t, *J* = 8.9 Hz, CF$_3$), −116.29 (m, β-CF$_2$),

−126.99 (m, α-CF$_2$). X-ray-quality crystals formed upon the slow evaporation of a solution of the compound in CH$_2$Cl$_2$ layered with hexanes.

2.3.4. [(CF$_3$CO$_2$)$_2$AgAu(dppb)]$_2$ (**12a**)

This was prepared as described above using [(dppb)(AuO$_2$CCF$_3$)$_2$] (0.038 g, 0.035 mmol) and [Ag(O$_2$CCF$_3$)]$_n$ (0.016 g, 0.070 mmol). A total of 0.043 g (0.029 mmol, 90%) colourless product was isolated. IR (ATR): 1641 ν$_a$(OCO) cm^{-1}. ^{1}H-NMR (CD$_2$Cl$_2$): δ = 7.60–7.43 (m, 22 H, dbbp), 7.29–7.23 (m, 2 H, dppb). ^{31}P{^{1}H}-NMR (CD$_2$Cl$_2$): δ = 18.74 (s). ^{19}F-NMR (CD$_2$Cl$_2$): δ = −73.55 (s, CF$_3$).

2.3.5. [(C$_6$F$_5$CO$_2$)$_2$AgAu(dppb)]$_2$ (**13a**)

This was prepared as described above using [(dppb)(AuO$_2$CC$_6$F$_5$)$_2$] (0.036 g, 0.029 mmol) and [Ag(O$_2$CC$_6$F$_5$)]$_n$ (0.019 g, 0.058 mmol). A total of 0.053 g (0.028 mmol, 97%) colourless product was isolated. IR (ATR): 1597 ν$_a$(OCO) cm^{-1}. ^{1}H-NMR (CD$_2$Cl$_2$): δ = 7.63–7.55 (m, 4 H, dppb), 7.38 (t, J = 7.3 Hz, 4 H, p-PPh$_2$), 7.18 (t, J = 7.2 Hz, 4 H, m-PPh$_2$), 7.12–7.09 (m, 8 H, o-PPh$_2$). ^{31}P{^{1}H}-NMR (CD$_2$Cl$_2$): δ = 21.46 (s). ^{19}F-NMR (CD$_2$Cl$_2$): δ = −114.55 (m, o-F), −151.57 (t, J = 21.2 Hz, m-F), −162.88 (m, p-F).

2.3.6. [{CF$_3$(CF$_2$)$_2$CO$_2$}$_2$AgAu(Xantphos)]$_2$ (**14a**)

This was prepared as described above using [(Xantphos){AuO$_2$C(CF$_2$)$_2$CF$_3$}$_2$] (0.059 g, 0.042 mmol) and [Ag{O$_2$C(CF$_2$)$_2$CF$_3$}]$_n$ (0.027 g, 0.084 mmol). A total of 0.079 g (0.039 mmol, 83%) colourless product was isolated. IR (ATR): 1649 ν$_a$(OCO) cm^{-1}. ^{1}H-NMR (CD$_2$Cl$_2$): δ = 7.73 (dd, J = 7.8, 1.3 Hz, 2 H, Xantphos), 7.45–7.51 (m, 4 H, p-Ph$_2$P), 7.20–7.34 (m, 16 H, m-Ph$_2$P, o-Ph$_2$P), 7.12 (dd, J = 7.8, 1.6 Hz, 2 H, Xantphos), 6.50 (ddd, J = 13.5, 7.8, 1.5 Hz, 2 H, Xantphos), 1.72 (s, 3 H, Me). ^{31}P{^{1}H}-NMR (CD$_2$Cl$_2$): δ = 17.84 (s). ^{19}F-NMR (CD$_2$Cl$_2$): δ = −81.05 (t, J = 8.9 Hz, CF$_3$), −116.43 (m, β-CF$_2$), −126.98 (m, α-CF$_2$).

2.4. Preparation of the Perfluorinated Copper Complexes

2.4.1. [Cu$_2$(O$_2$CCF$_3$)$_4$(dioxane)]$_n$ (**16**)

A 1,4-dioxane solution (10 mL) of copper(II) acetate monohydrate (1.00 g, 5.01 mmol) and trifluoroacetic acid (0.8 mL, 10.38 mmol) was heated to reflux for ca. 2 h. After this time, the solvent was removed in vacuum, affording a turquoise solid (1.51 g, 90%). IR (ATR): 1689 ν$_a$(OCO) cm^{-1}. X-ray-quality turquoise plates were obtained by recrystallisation from toluene. When basic copper carbonate was used under the same conditions, a blue solution was obtained, which deposited crystals upon standing. The material was identified as the trinuclear complex [Cu$_3$(O$_2$CCF$_3$)$_6$(H$_2$O)$_4$(dioxane)$_2$]·(dioxane) (**15**) by X-ray diffraction.

2.4.2. [Cu(O$_2$CCF$_3$)(PPh$_3$)$_3$] (**17**)

Ph$_3$P (0.337 g, 1.285 mmol) was added to a solution of [Cu$_2$(O$_2$CCF$_3$)$_4$(dioxane)]$_n$ (0.130 g, 0.195 mmol) in MeOH (10 mL). The mixture was heated to reflux for ca. 2 h, by which time the colour changed from blue to colourless. Upon standing the solution at −20 °C, colourless crystals were deposited after 24 h. These were isolated by filtration, washed with Et$_2$O and dried. A total of 0.111 g (0.158 mmol, 38%) colourless crystals were obtained. IR (ATR): 1674 ν$_a$(OCO), 1480 ν$_s$(OCO) cm^{-1}. ^{31}P{^{1}H}-NMR (CD$_2$Cl$_2$): δ = −1.80 (s). ^{19}F-NMR (CD$_2$Cl$_2$): δ = −75.03 (s, CF$_3$). X-ray-quality crystals were picked out of the MeOH solution before filtration and drying.

2.4.3. [Cu(O$_2$CCF$_2$CF$_3$)(PPh$_3$)$_3$]·MeOH (**18**)

This was prepared as described above using [Cu$_2$(O$_2$CCF$_2$CF$_3$)$_4$(H$_2$O)$_2$] (0.280 g, 0.343 mmol) and Ph$_3$P (0.660 g, 2.52 mmol). A total of 0.414 g (0.396 mmol, 58%) colourless crystals were obtained. IR (ATR): 3453 ν(OH), 1685 ν$_a$(OCO), 1433 ν$_s$(OCO) cm^{-1}. ^{31}P{^{1}H}-NMR (CD$_2$Cl$_2$): δ = −2.43 (s). ^{19}F-NMR (CD$_2$Cl$_2$): δ = −83.02 (t, J = 1.7 Hz, CF$_3$), −119.86 (q, J = 1.7 Hz, CF$_2$). ^{1}H-NMR (CD$_2$Cl$_2$): δ = 3.45 (s, 3 H, MeOH), 7.20–7.43 (m, 45 H, Ph$_3$P).

2.4.4. [Cu{O$_2$C(CF$_2$)$_2$CF$_3$}(PPh$_3$)$_3$] (19)

This was prepared as described above using [Cu$_2${O$_2$C(CF$_2$)$_2$CF$_3$}$_4$] (0.150 g, 0.153 mmol) and Ph$_3$P (0.281 g, 1.07 mmol). A total of 0.247 g (0.232 mmol, 76%) colourless crystals were obtained. IR (ATR): 1693 ν_a(OCO), 1433 ν_s(OCO) cm^{-1}. ^{31}P{^{1}H}-NMR (CD$_2$Cl$_2$): δ = −2.40 (s). ^{19}F-NMR (CD$_2$Cl$_2$): δ = −81.09 (t, J = 8.9 Hz, CF$_3$), −117.18 (sext., J = 9.0 Hz, β-CF$_2$), −126.94 (m, α-CF$_2$). X-ray-quality crystals were picked out of the MeOH solution before filtration and drying.

2.4.5. [Cu(O$_2$CC$_6$F$_5$)(PPh$_3$)$_3$] (20)

This was prepared as described above using [Cu$_2$(O$_2$CC$_6$F$_5$)$_4$] (0.190 g, 0.196 mmol) and Ph$_3$P (0.346 g, 1.319 mmol). A total of 0.247 g (0.233 mmol, 60%) product was isolated. IR (ATR): 1648 ν_a(OCO), 1336 ν_s(OCO) cm^{-1}. ^{31}P{^{1}H}-NMR (CD$_2$Cl$_2$): δ = −1.30 (s). ^{19}F-NMR (CD$_2$Cl$_2$): δ = −142.61 (q, J = 8.6, 15.0 Hz, o-F), −158.56 (t, J = 20.5 Hz, m-F), −163.80 (m, p-F). X-ray quality-crystals of the MeOH solvate were picked out of the MeOH solution before filtration and drying.

2.4.6. [Cu(O$_2$CCF$_2$CF$_3$)(PPh$_3$)$_2$] (21)

This was prepared as described above using [Cu$_2$(O2CCF$_2$CF$_3$)$_4$(H$_2$O)$_2$] (0.100 g, 0.128 mmol) and Ph$_3$P (0.168 g, 0.640 mmol). A total of 0.097 g (0.129 mmol, 50%) colourless crystals were obtained. IR (ATR): 1675 ν_a(OCO) cm^{-1}. ^{31}P{^{1}H}-NMR (CD$_2$Cl$_2$): δ = −1.85 (s). ^{19}F-NMR (CD$_2$Cl$_2$): δ = −83.11 (t, J = 1.6 Hz, CF$_3$), −119.89 (m, CF$_2$). X-ray-quality crystals of the MeOH solvate were picked out of the MeOH solution before filtration and drying.

2.5. X-ray Crystallography

Diffraction data were collected at 90 K using a Rigaku Oxford Diffraction Gemini E Ultra-diffractometer, or at 100 K using either a Bruker AXS Enraf-Nonius KappaCCD with 0.2 × 2 mm^2 focus rotating anode or a Bruker-AXS Kappa Mach3 APEX-II with Iµs microfocus radiation source using Mo Kα radiation (λ = 0.71073 nm). The dataset of compound **14a** was collected at 100 K at beamline P24 at PETRA III DESY, Hamburg, Germany, with an energy of 20 keV (λ = 0.619900 Å). Data integration, scaling and empirical absorption correction were carried out using the program packages XDS (synchrotron data) [16], CrysAlis Pro [17], DATCOL [18], SADABS [19] or APEX3 [20]. The structures were solved using SHELXT [21] and refined with SHELXL [22], operated through the Olex2 interface [23]. The fluorine atoms in one of the CF$_3$-groups in the structure of [Cu$_2$(O$_2$CCF$_3$)$_4$(dioxane)]$_n$ (**16**) were found to be disordered over two positions. This was successfully modelled with a 50:50 occupancy, and appropriate restraints (EADP) were applied. Similarly, in **14a** one CF$_3$-group displayed disordering, which was best modelled with a 75:25 occupancy and EDAP restraints. Important crystallographic and refinement details are collected in Table 1.

Table 1. Crystallographic and refinement details for compounds herein.

	1	2	2a	4a	6a	7a	9a	11a
CCDC code	2167914	2167908	2167925	2167926	2167923	2167921	2167924	2167915
Empirical formula	$C_{11}H_8F_5O_4Ag_2$	$C_{13}H_8F_{10}O_4Ag$	$C_{21}H_{15}F_5O_2PAu$	$C_{25}H_{15}F_5O_2PAu$	$C_{44}H_{24}F_{10}O_4P_2Au_2$	$C_{45}H_{32}F_{10}O_6P_2Au_2$	$C_{44}H_{30}F_{12}O_8P_2Ag_2Au_2$	$C_{52}H_{30}F_{28}O_8P_2Ag_2Au_2$
Formula weight	533.91	633.93	622.27	670.31	1262.50	1298.58	1586.29	1986.37
Crystal system	Triclinic	Monoclinic	Monoclinic	Monoclinic	Triclinic	Triclinic	Triclinic	Triclinic
Space group	P-1	$P2_1/c$	$P2_1/n$	$P2_1/n$	P-1	P-1	P-1	P-1
Unit cell dimensions	a = 9.1704(5) Å b = 9.6753(6) Å c = 10.1292(6) Å α = 61.546(6)° β = 64.823(5)° γ = 85.495(5)°	a = 12.1816(9) Å b = 13.4270(12) Å c = 10.6126(9) Å α = 90.0° β = 94.470(7)° γ = 90.0°	a = 11.8549(10) Å b = 11.7981(10) Å c = 14.7984(13) Å α = 90.0° β = 104.747(3)° γ = 90.0°	a = 13.2968(6) Å b = 6.6460(5) Å c = 25.3473(14) Å α = 90.0° β = 97.9436(6)° γ = 90.0°	a = 11.9016(13) Å b = 13.5895(7) Å c = 14.8220(7) Å α = 65.9066(6)° β = 83.007(5)° γ = 89.395(6)°	a = 10.243(3) Å b = 13.205(7) Å c = 17.993(7) Å α = 70.86(3)° β = 87.40(3)° γ = 71.03(4)°	a = 7.9232(5) Å b = 11.8065(9) Å c = 13.2352(12) Å α = 85.362(5)° β = 73.848(8)° γ = 73.859(7)°	a = 10.3296(5) Å b = 11.8286(6) Å c = 13.5824(7) Å α = 75.056(4)° β = 75.850(4)° γ = 70.362(5)°
Volume Å^3	706.34(8)	1730.5(2)	2001.6(3)	2218.5(2)	2170.1(3)	2169.3(17)	1142.33(16)	1487.29(14)
Z	2	4	4	4	2	2	1	1
ρ_{calc} g/cm^3	2.510	2.433	2.065	2.007	1.932	1.988	2.306	2.218
μ mm^{-1}	2.857	2.385	7.491	6.767	6.911	6.918	7.418	5.759
F(000)	508.0	1208.0	1184.0	1280.0	1196.0	1240.0	748.0	940.0
Crystal size	0.02 × 0.05 × 0.10	0.02 × 0.05 × 0.11	0.022 × 0.057 × 0.059	0.011 × 0.025 × 0.045	0.08 × 0.09 × 0.12	0.04 × 0.08 × 0.08	0.035 × 0.04 × 0.16	0.03 × 0.05 × 0.07
2θ range for data collection	4.97 to 64.904°	5.768 to 58.862°	3.946 to 67.612°	4.697 to 72.105°	5.31 to 74.062°	5.48 to 62.874°	5.554 to 72.200°	4.834 to 58.974°
Reflections collected	7127	9243	74,095	53,830	104,934	52,648	52,950	13,672
Independent reflections	7127	4023	8037	10,522	22,074	14,289	8942	6916
Data/restraints/parameters	7127/0/210	4023/0/263	8037/0/271	10,522/0/307	22,074/0/559	14,289/0/589	8942/0/316	6916/0/424
Goodness-of-fit on F^2	0.942	1.044	1.056	1.019	1.027	1.024	1.015	1.044
Final R indices [I > 2σ(I)]	R_1 = 0.0296 wR_2 = 0.0657	R_1 = 0.0226 wR_2 = 0.0431	R_1 = 0.0164 wR_2 = 0.0353	R_1 = 0.0304 wR_2 = 0.0481	R_1 = 0.0306 wR_2 = 0.0639	R_1 = 0.0699 wR_2 = 0.1600	R_1 = 0.0274 wR_2 = 0.0511	R_1 = 0.0300 wR_2 = 0.0448
Final R indices [all data]	R_1 = 0.0429 wR_2 = 0.0684	R_1 = 0.0286 wR_2 = 0.0451	R_1 = 0.0221 wR_2 = 0.0371	R_1 = 0.0589 wR_2 = 0.0534	R_1 = 0.0472 wR_2 = 0.0681	R_1 = 0.1181 wR_2 = 0.1861	R_1 = 0.0429 wR_2 = 0.0544	R_1 = 0.0435 wR_2 = 0.0491
Largest difference peak/hole e/Å^3	1.84/−1.06	0.60/−0.53	1.4/−0.6	1.0/−1.3	2.8/−2.7	3.8/−3.9	0.9/−1.9	1.00/−1.29
	14a	15	16	17	19	20	21	
CCDC code	2167922	2167916	2167909	2167913	2167912	2167910	2167911	
Empirical formula	$C_{55}H_{32}F_{28}O_8P_2Ag_2Au_2$	$C_{24}H_{22}F_{18}O_{22}Cu$	$C_{12}H_8F_{12}O_{10}Cu_2$	$C_{56}H_{45}F_5O_2P_3Cu$	$C_{58}H_{45}F_7O_2P_3Cu$	$C_{63}H_{53}F_5O_4P_3Cu$	$C_{39}H_{30}F_5O_2P_2Cu$	

Table 1. Cont.

Formula weight	2042.42	1205.11	667.26	963.37	1063.39	1125.50	751.11
Crystal system	Monoclinic	Monoclinic	Monoclinic	Triclinic	Triclinic	Triclinic	Triclinic
Space group	$P2_1/c$	$P2_1/n$	$P2_1/n$	P-1	P-1	P-1	P-1
Unit cell dimensions	a = 20.778(3) Å b = 15.689(3) Å c = 18.248(3) Å α = 90.0° β = 97.595(14)° γ = 90.0°	a = 12.7528(7) Å b = 10.3794(5) Å c = 18.5367(14) Å α = 90.0° β = 108.717(7)° γ = 90.0°	a = 8.7675(2) Å b = 14.5136(4) Å c = 16.7461(4) Å α = 90.0° β = 98.800(2)° γ = 90.0°	a = 12.9761(3) Å b = 17.9813(4) Å c = 21.516(06) Å α = 82.311(2)° β = 83.5472(2)° γ = 88.1828(19)°	a = 11.3935(4) Å b = 12.2780(4) Å c = 18.7548(6) Å α = 87.683(3)° β = 78.047(3)° γ = 72.226(3)°	a = 13.0559(3) Å b = 13.1731(4) Å c = 18.7364(4) Å α = 95.532(2)° β = 105.660(2)° γ = 114.835(3)°	a = 12.1576(8) Å b = 12.9136(10) Å c = 13.1932(10) Å α = 89.326(6)° β = 66.955(7)° γ = 63.060(7)°
Volume Å^3	5896.1(16)	2323.9(3)	2105.82(9)	4942.9(2)	2443.47(15)	2733.51(14)	1664.2(2)
Z	4	2	4	4	2	2	2
ρ_{calc} g/cm^3	2.299	1.722	2.105	1.295	1.445	1.367	1.499
μ mm^{-1}	5.816	1.504	2.174	0.591	0.616	0.553	0.816
F(000)	3872.0	1202.0	1304.0	1992.0	1092.0	1164.0	768.0
Crystal size	0.04 × 0.05 × 0.12	0.05 × 0.07 × 0.09	0.02 × 0.07 × 0.08	0.06 × 0.07 × 0.12	0.07 × 0.09 × 0.09	0.06 × 0.09 × 0.11	0.04 × 0.06 × 0.08
2θ range for data collection	5.2 to 52.744°	4.666 to 58.686°	4.922 to 59.016°	4.978 to 59.088°	4.81 to 58.838°	4.652 to 59.004°	5.458 to 58.654°
Reflections collected	80,334	13,733	11,030	52,923	23,872	27,958	15,883
Independent reflections	12,054	5405	4885	23,105	11,311	12,725	7621
Data/restraints/parameters	12,054/36/912	5405/0/298	4885/24/344	23,105/0/1171	11,311/0/640	12,725/0/689	7621/0/442
Goodness-of-fit on F^2	1.052	1.050	1.042	1.023	1.026	1.028	1.039
Final R indices [$I > 2\sigma(I)$]	R_1 = 0.0370 wR_2 = 0.0746	R_1 = 0.0921 wR_2 = 0.2197	R_1 = 0.0317 wR_2 = 0.0709	R_1 = 0.0470 wR_2 = 0.1073	R_1 = 0.0329 wR_2 = 0.0722	R_1 = 0.0387 wR_2 = 0.0956	R_1 = 0.0273 wR_2 = 0.0648
Final R indices [all data]	R_1 = 0.0569 wR_2 = 0.0831	R_1 = 0.1190 wR_2 = 0.2417	R_1 = 0.0420 wR_2 = 0.0757	R_1 = 0.0619 wR_2 = 0.1155	R_1 = 0.0417 wR_2 = 0.0768	R_1 = 0.0478 wR_2 = 0.1011	R_1 = 0.0321 wR_2 = 0.0671
Largest difference peak/hole e/Å^3	1.93/−1.66	3.43/−1.94	0.49/−0.69	2.79/−1.62	0.42/−0.47	2.02/−0.62	0.46/−0.33

3. Results and Discussion
3.1. Silver(I) Complexes with Perfluorinated Carboxylato Ligands

The addition of various perfluorocarboxylic acids (CF_3COOH, CF_3CF_2COOH, $CF_3(CF_2)_2COOH$ and C_6F_5COOH) to a toluene suspension of Ag_2O at room temperature resulted in dissolution of the silver oxide within minutes, accompanied by a slight turbidity due to the water produced by the reaction. In the case of pentafluorobenzoic acid, slight heating was required since the acid itself is not very soluble in toluene at room temperature. Work-up consisted of the addition of anhydrous Na_2SO_4 (although this is not strictly necessary), filtration and evaporation. With this simple procedure, high-purity anhydrous silver(I) salts of trifluoroacetic acid, perfluoropropionic acid, perfluorobutyric acid and pentafluorobenzoic acid (**1–4**) were obtained in good yields within a short time as colourless solids (Scheme 1). This procedure can also be extended to perfluorinated sulfonic acids, as exemplified by the synthesis of [$Ag(O_3SCF_3)$] from triflic acid and Ag_2O in toluene.

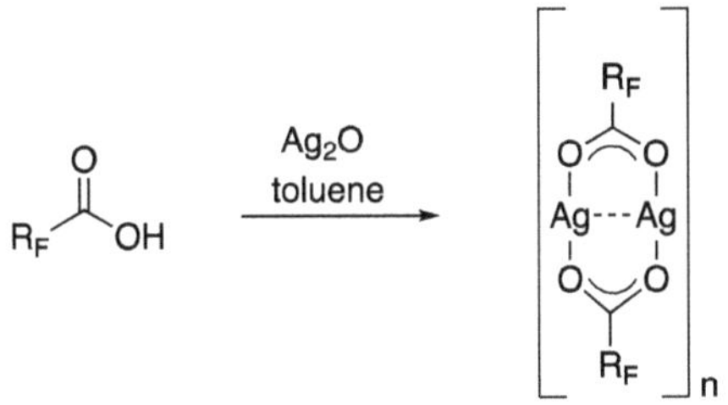

R_F = CF_3 (**1**), CF_2CF_3 (**2**), $(CF_2)_2CF_3$ (**3**), C_6F_5 (**4**)

Scheme 1. Synthesis of silver(I) perfluorocarboxylates.

The silver carboxylates **1–4** were characterised by ^{19}F-NMR spectroscopy and IR spectroscopy and, in the case of the trifluoroacetate and perfluoropropionate salts **1** and **2**, by X-ray diffraction (see below). The most characteristic features of the IR spectra of the complexes are the symmetric- and asymmetric carbonyl stretching frequencies, which were observed at around 1600 and 1400 cm^{-1}, respectively. These values are similar to those reported in the literature [24]. The proton NMR spectra of compounds (**1**), (**2**) and (**4**) show signals corresponding to toluene, which was confirmed to be present in the structures by X-ray diffraction and/or thermogravimetric analysis (see below). The perfluorobutyrate salt (**3**), however, was isolated in toluene-free form, based on NMR spectroscopy and TGA. The molecular structures of (**1**) and (**2**) were confirmed by X-ray diffraction experiments. The structure of the trifluoroacetate salt (**1**) consists of a pair of silver atoms that are O,O-bridged by two trifluoroacetato ligands, resulting in an Ag–Ag distance of 2.8951(4) Å. The eight-membered ring is considerably bent along the silver–silver axis, with an O–Ag–O angle of about 137°. A molecule of toluene is positioned such that it forms a η^2-bond to Ag1 and a η^1-bond to Ag2 (Figure 2 top) with two shorter Ag–C distances of 2.488(3) Å and 2.512(3) Å (η^2 bonding to Ag1) and one longer one of 2.712(3) Å (η1 bonding to Ag2). These values are typical for arene–silver interactions, which fall in the range of 2.16–2.92 Å [25]. The silver dimers are connected through additional Ag–O bonds, forming a coordination polymer (Figure 2 bottom). The toluene-free structure of silver(I) trifluoroacetate has been reported by two different groups [7,26]. In both cases, the overall motif (a coordination polymer of dinuclear eight-membered rings) and the Ag–Ag distances are very similar to that of our toluene adduct. The major difference is the much flatter eight-membered ring ($\angle$O–Ag–O = 158°) in the toluene-free structure.

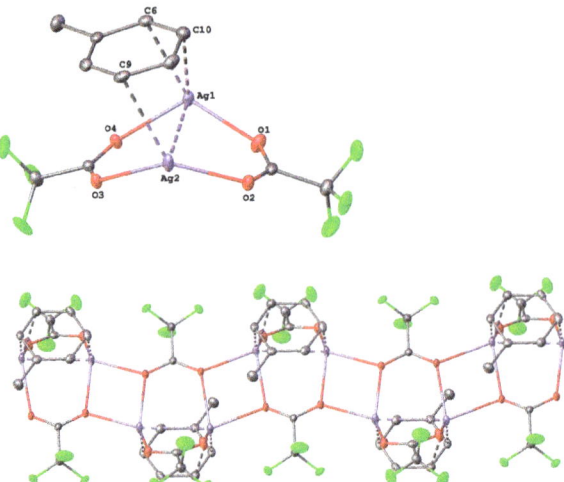

Figure 2. Asymmetric unit of [Ag$_2$(O$_2$CCF$_3$)$_2$(MeC$_6$H$_5$)]$_n$ (**1**) (**top**). Ellipsoids show 50% probability levels. Hydrogen atoms have been omitted for clarity. The image on the bottom shows the coordination polymer.

The perfluoropropionate analogue (**2**) (Figure 3) is structurally similar, except that the eight-membered ring is much flatter, with an O–Ag–O angle of 164°. In addition, the toluene shown here is only η^2-bound to Ag1, with silver–carbon bond distances of 2.593(2) Å and 2.735(2) Å; the shortest Ag2–C distance is greater than 3.5 Å. The Ag–Ag distance within the dinuclear complex is with 2.9333(3) Å, which is also slightly larger than that observed in (**1**). Such Ag···Ag interactions, known as argentophilicity, are often observed in solid-state structures of silver compounds; typically, the distances range from 2.9 to 3.1 Å [27].

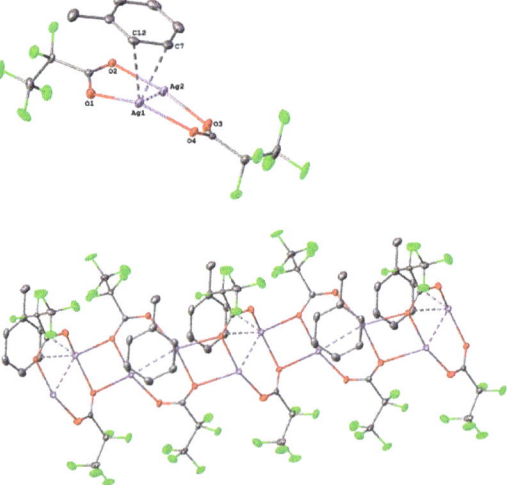

Figure 3. Asymmetric unit of [Ag$_2$(O$_2$CCF$_2$CF$_3$)$_2$(MeC$_6$H$_5$)]$_n$ (**2**) (**top**). Ellipsoids show 50% probability levels. Hydrogen atoms have been omitted for clarity. The image on the bottom shows the coordination polymer.

Thermogravimetric analysis of (**2**) (Figure 4 left) shows loss of the coordinated toluene occurring at around 100 °C, followed by a single decomposition step to metallic silver commencing at 328 °C. In the case of the toluene-free compounds (**3**) and (**4**), only a single decomposition step (onset at 295 °C and 224 °C respectively) can be observed (Figure 4 centre and right). These observations are consistent with those reported for other silver perfluorocarboxylates, which also undergo a decomposition to metallic Ag [24,28].

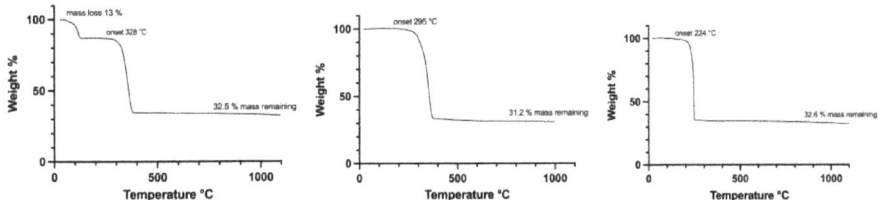

Figure 4. TGA traces of complexes **2** (**left**), **3** (**centre**) and (**4**) (**right**).

3.2. Gold(I) Complexes with Perfluorinated Carboxylato Ligands

Silver(I) carboxylates **1–4** were subsequently used to prepare a series of mono- and dinuclear phosphine-gold(I) perfluorocarboxylates by a metathesis reaction from the corresponding phosphine gold chlorides (Scheme 2).

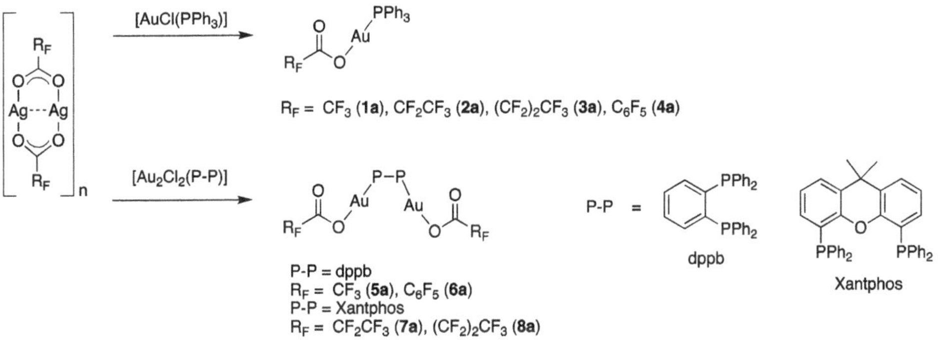

Scheme 2. Synthesis of gold(I) perfluorocarboxylates.

The gold(I) compounds were isolated as colourless solids in good yields and were characterised by various spectroscopic methods. Several of these gold(I) compounds (**1a–4a** and **5a**) are known compounds, but their molecular structures (see below) are described here for the first time. The IR spectra of the complexes are similar to previous reports [12,29,30]. Although not used in this work, it should be mentioned here that the very first gold(I) phosphine complex with a perfluorinated carboxylato ligand was the perfluorooctanoato complex [(Ph$_3$P)Au{O$_2$C(CF$_2$)$_6$CF$_3$}] reported by Beck in 1987 [31]. The molecular structures of the mononuclear complexes **2a** and **4a**, as well as the dinuclear species **6a** and **7a**, are depicted in Figures 5 and 6 respectively.

In each case, the gold atom is linearly coordinated by a phosphorus atom from the phosphine ligand and an oxygen atom of the monodentate carboxylate group. In the case of the bis(phosphine) complexes (**6a**) and (**7a**), the two gold atoms are forced into close proximity, resulting in Au···Au distances of 2.9195(4) Å and 2.8938(11) Å, respectively. These values are slightly shorter than those observed in the trifluoroacetate gold complexes of dppb and Xantphos [30,32]. In the mononuclear species (**2a**) and (**4a**), there are no intermolecular gold–gold contacts, which are frequently observed in solid-state structures of phosphine gold(I) complexes [33].

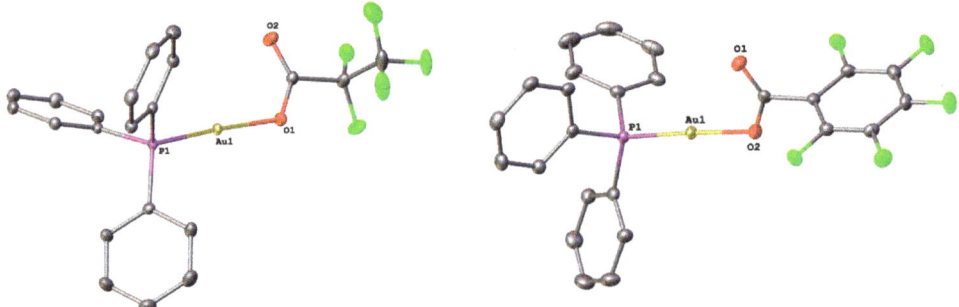

Figure 5. Molecular structures of [Au(O$_2$CCF$_2$CF$_3$)(PPh$_3$)] (**2a**) (**left**) and [Au(O$_2$CC$_6$F$_5$)(PPh$_3$)] (**4a**) (**right**). Ellipsoids show 50% probability levels. Hydrogen atoms have been omitted for clarity.

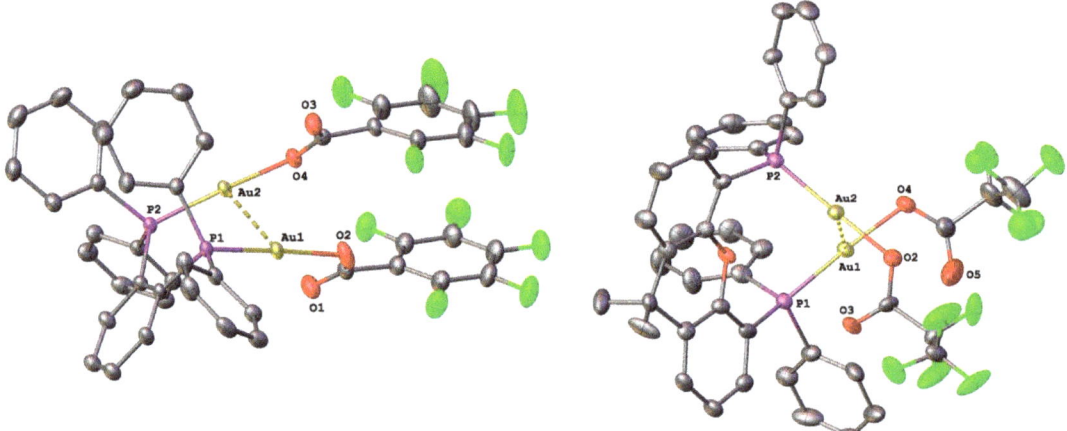

Figure 6. Molecular structures of [(dppb){Au(O$_2$CC$_6$F$_5$)}$_2$] (**6a**) (**left**) and [(Xantphos){Au(O$_2$CCF$_2$CF$_3$)}$_2$] (**7a**) (**right**). Ellipsoids show 50% probability levels. Hydrogen atoms have been omitted for clarity.

3.3. Bimetallic Silver/Gold Complexes with Perfluorinated Carboxylato Ligands

The oxygen atom of the carboxylato ligand not bound to gold in the complexes discussed above could potentially act as a donor atom towards an additional, different, metal centre. Together with potential metal–metal interactions, this could lead to the formation of multimetallic supramolecular assemblies. We therefore reacted the phosphine–gold(I) carboxylates with the silver(I) carboxylates in a 1:1 ratio (Scheme 3) and isolated colourless solids, whose ^{31}P NMR chemical shifts were different from those of the phosphine gold(I) carboxylates themselves. The observed signals in the ^{19}F NMR spectra of the products were very similar to those of the starting materials.

The IR spectra of the compounds featured two CO bands, at different wavenumbers than those of the individual precursors, suggesting the presence of carboxylate groups in different chemical environments. The Ph$_3$P complexes **9a** and **10a** resulted in single crystals that were suitable for X-ray diffraction, allowing us to unambiguously determine their molecular structures (Figure 7).

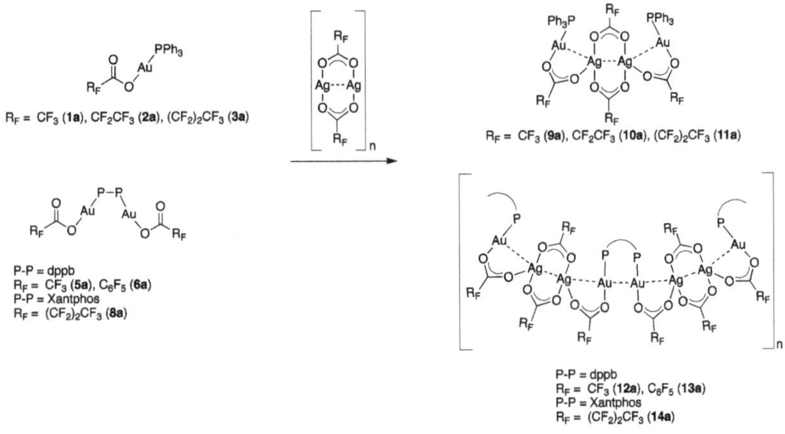

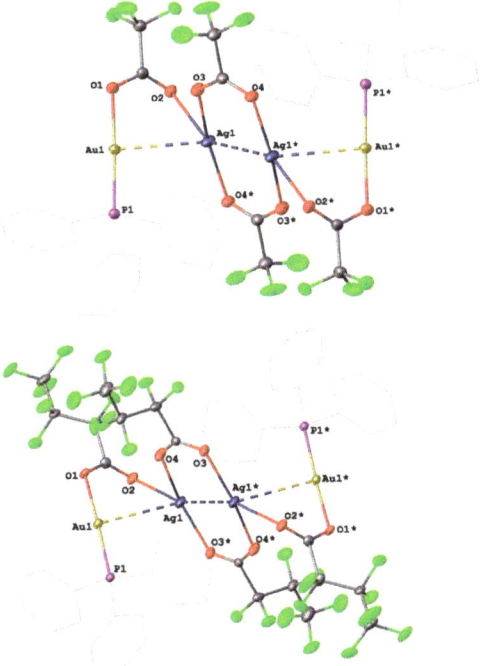

Scheme 3. Synthesis of bimetallic Ag/Au complexes.

Figure 7. Molecular structures of [(CF$_3$CO$_2$)$_2$AgAu(PPh$_3$)]$_2$ **9a** (**top**) and [{CF$_3$(CF$_2$)$_2$CO$_2$}$_2$AgAu(PPh$_3$)]$_2$ **10a** (**bottom**). Ellipsoids show 50% probability levels. Hydrogen atoms have been omitted for clarity and the phenyl rings of the Ph$_3$P ligands are drawn in wireframe style. Atoms generated by symmetry are labelled with an asterisk.

The structures were tetranuclear Au$_2$Ag$_2$ complexes containing two distinct bridging carboxylato ligands. One of them bridges two silver atoms, while the other bridges a gold- and a silver atom, resulting in Ag–Ag and Au–Ag distances of 2.9688(6) Å and 3.1820(4) Å, respectively. The values fall within the range of aurophilic or argentophilic interactions [27,33]. The gold atoms show, as expected for Au$^{(I)}$, a linear ($\angle$P–Au–O = 177°)

coordination, whilst the silver has three coordinates with a distorted T-shaped geometry. The structure can be interpreted as two phosphine gold carboxylates acting as *Au,O*-chelating ligands towards a silver atom located within a dinuclear, eight-membered-ring structure. As a result, the Ag–Ag distances are slightly increased compared to those of the corresponding silver salts discussed above. In addition, the Au–Ag–Ag–Au chain is bent, with two 117° angles at each silver centre. The bite angle (∠O–Ag–Au) of the chelating ligands is, thus, approximately 65°. Similar structures have been reported for the tetranuclear Au_2Ag_2 complexes $[(CF_3CF_2CO_2)_2AgAu\{(4\text{-}Me_2NC_6H_4)PPh_2\}]_2$ and $[\{CF_3(CF_2)_6CO_2\}_2AgAu(PPh_3)]_2$ [4,34]. The gold–chalcogen chelation of silver atoms has also been observed in the tetranuclear complex $[(CF_3CO_2)_2AgAu(C_6F_5)(PhSCH_2PPh_2)]_2$, in which the sulfur and gold atoms of a $PhSCH_2PPh_2AuC_6F_5$ unit coordinate with the silver [35]. Analogous reactions with the dinuclear phosphine-gold(I) carboxylates and the silver carboxylates also afforded colourless solids, with similar spectroscopic properties to the products discussed above. In the case of the Xantphos derivative (**14a**), X-ray-quality crystals were obtained (Figure 8).

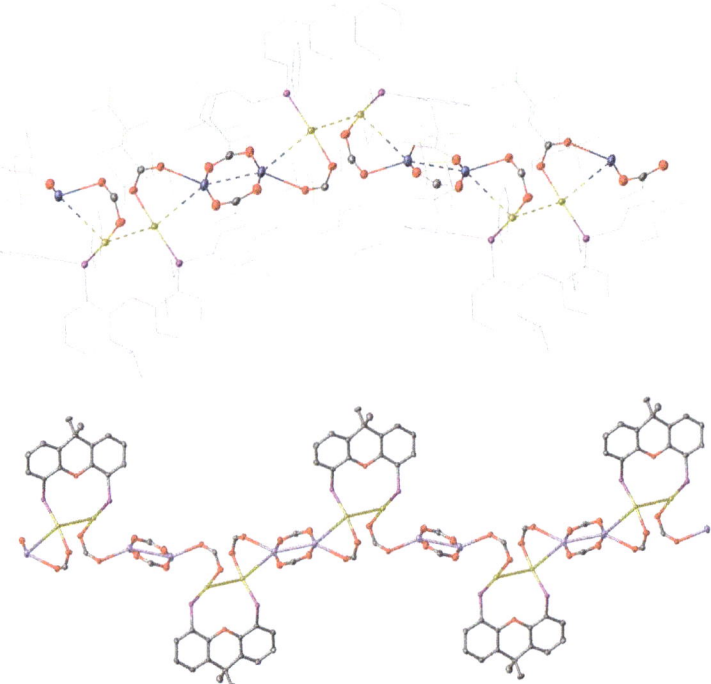

Figure 8. Molecular structure of $[\{CF_3(CF_2)_2CO_2\}_2AgAu(Xantphos)]_n$ (**14a**) (top). Hydrogen atoms have been omitted for clarity. The Xantphos ligands as well as the $CF_3(CF_2)_2$ chains are drawn in wireframe style. The bottom image shows the polymer chain with hydrogen atoms and the perfluorinated groups omitted.

The basic molecular structure is identical to that discussed above (*Au,O*-chelating a silver centre in an eight-membered ring dimer), except that this structure is polymeric due to the presence of the diphosphine ligand, which bridges two gold atoms (Figure 8 bottom). The Ag–Ag distance [2.9715(9) Å] shown here is very similar to that observed in the Ph_3P complex discussed above, whilst the Ag–Au distance [3.3159 Å] is considerably longer. The Au–Au separation in the Ag,Au-polymer is also slightly larger than in the gold(I)–carboxylato complex **7a**. The structurally related coordination polymer

[(CF$_3$CO$_2$)$_2$AgAu(binap)]$_n$ [binap = 2,2′-bis(diphenylphosphino)-1,1′-binaphthyl] was adventitiously isolated when solutions of [Au$_2$(O$_2$CCF$_3$)$_2$(binap)] were left to stand in the presence of excess [AgO$_2$CCF$_3$] [36]. In this compound, there are two unequal Ag–Au bond lengths of ca. 3.3 and 3.0 Å, the latter of which is similar to that observed in the structure of **14a**.

3.4. Copper Complexes with Perfluorinated Carboxylato Ligands

Copper(II) carboxylates generally exist as dinuclear paddlewheel complexes of the type [Cu$_2$(O$_2$CR)$_4$(L)$_2$], featuring *O,O*-bridging carboxylato ligands and short Cu–Cu distances [37]. In the case of the perfluorocarboxylates, there are a few structurally authenticated examples also showing this arrangement. These include the coordination polymers [Cu$_2$(O$_2$CC$_6$F$_5$)$_4$(dioxane)]$_n$ [38] and [Cu$_2$(O$_2$CCF$_3$)$_4$]$_n$ [39], as well as the dinuclear species [Cu$_2$(O$_2$CCF$_3$)$_4$(L)$_2$] (L = ^{i}PrOC$_2$H$_4$OH [40], quinoline [41], MeCN [42], Et$_2$O [43], Bz$_2$O [44], and ^{t}Bu$_2$S [45]), [Cu$_2${O$_2$C(CF$_2$)$_2$CF$_3$}$_4$(THF)$_2$] [40], [Cu$_2${O$_2$C(CF$_2$)$_7$CF$_3$}$_4$(acetone)$_2$] [46] and [Cu$_2$(O$_2$CC$_6$F$_5$)$_4$(HO$_2$CC$_6$F$_5$)$_2$] [47].

When reproducing the procedure by Krupoder to prepare [Cu$_2$(O$_2$CCF$_3$)$_4$(dioxane)] [15], we isolated two different products depending on the copper-precursor used: from the reaction of basic copper carbonate with trifluoroacetic acid in 1,4-dioxane, we isolated pale blue crystals upon the slow evaporation of the solution. These crystals turned out to be the unique trinuclear Cu$^{(II)}$ complex [Cu$_3$(O$_2$CCF$_3$)$_6$(H$_2$O)$_4$(dioxane)$_2$]·(dioxane) shown in Figure 9.

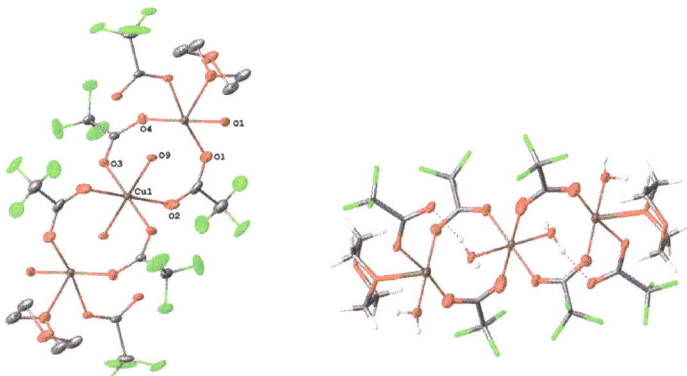

Figure 9. Molecular structure of [Cu$_3$(O$_2$CCF$_3$)$_6$(H$_2$O)$_4$(dioxane)$_2$]·(dioxane) (**15**). Ellipsoids show 30% probability levels. Hydrogen atoms, as well as the dioxane of solvation, have been omitted for clarity. Only one part of the disordered coordinated dioxane molecules is shown. The image on the right depicts the intramolecular hydrogen bonds.

The compound consists of two different copper(II) sites, one octahedral featuring two water molecules trans to each other and four equatorial oxygen atoms from the bridging trifluoroacetate groups. The other two copper sites are square pyramidal, with a dioxane-oxygen atom at the tip of the pyramid and four oxygen atoms from water, two bridging and one *O*-bound trifluoroacetate at the base. In addition, there are intramolecular H-bonds between the coordinated water molecules at the octahedral Cu-site and the carbonyl-oxygen atom of the trifluoroacetate ligands at the square pyramidal copper (Figure 9 right). When trifluoroacetic acid was reacted with copper(II) acetate monohydrate in 1,4-dioxane, however, a turquoise, water-free material (based on IR spectroscopy) was obtained after evaporation in vacuum. Recrystallization from toluene afforded fine plates, which were identified as the Cu$^{(II)}$ coordination polymer [Cu$_2$(O$_2$CCF$_3$)$_4$(dioxane)]$_n$ (**16**) by X-ray diffraction (Figure 10).

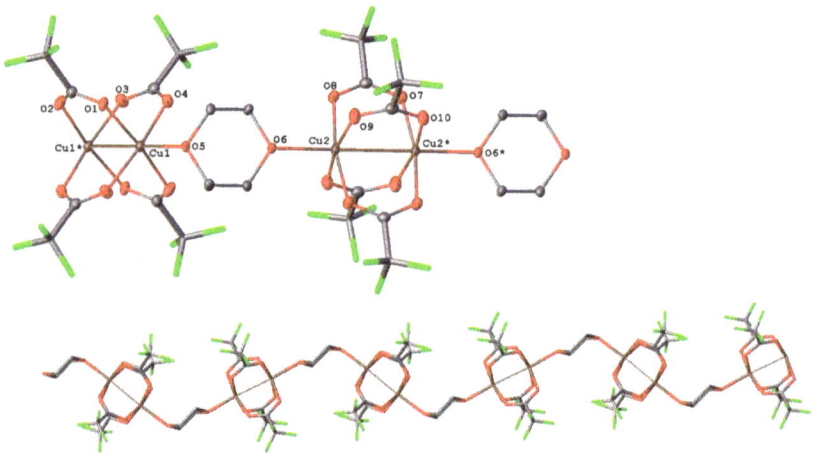

Figure 10. Molecular structure of the coordination polymer [Cu$_2$(O$_2$CCF$_3$)$_4$(dioxane)]$_n$ (**16**) (top). Ellipsoids show 50% probability levels. Hydrogen atoms have been omitted for clarity and only one part of the disordered CF$_3$ groups is shown. The CF$_3$ groups are drawn as capped sticks. Atoms generated by symmetry are labelled with an asterisk. The image on the bottom shows the zig-zag shape of the polymer chain.

The structure Is a typical paddlewheel structure with a Cu–Cu distance of 2.6381(5) Å and bridging 1,4-dioxane molecule. This structural motif is also found in the solid-state structures of several other copper(II) coordination polymers derived from ^{t}BuCOOH, ^{t}BuCH$_2$COOH [48], EtCOOH [49], 2-IC$_6$H$_4$COOH [50], PhCOOH [51] and C$_6$F$_5$COOH [38]. In complex **16**, the orientation of the chair-configured dioxane molecules leads to a zig-zag-shaped coordination polymer (Figure 10 bottom). In all the other published structures, the dioxane is oriented such that step-shaped polymers are formed.

It is known that copper(II) perfluorocarboxylates react with an excess of Ph$_3$P (2.5 or 3.5 equivalents per Cu), forming the colourless, diamagnetic Cu$^{(I)}$ phosphine complexes [Cu(O$_2$CR$_F$)(PPh$_3$)$_2$] and [Cu(O$_2$CR$_F$)(PPh$_3$)$_3$], respectively [52]. In most cases, recrystallisation of the tris-complex results in the bis-species, suggesting an equilibrium in solution. The phosphine acts as both the reducing agent and ligand. It was proposed that in both the bis- and tris-complexes, the copper is four-coordinate, with either monodentate or bidentate carboxylate ligands. Experimental data also suggested that when the pKa of the parent acid is less than 3.5, only tris(triphenylphosphine) complexes can be formed. This, however, was proven to be incorrect by Edwards and White, who successfully isolated bis(triphenylphosphine) complexes containing perfluoropropionato, perfluorobutyrato and pentafluorobenzoato ligands [13]. The tetrahedral coordination geometry, as well as the presence of a bidentate carboxylato ligand, was confirmed by an X-ray structure determination of the bis(triphenylphosphine) complex with trifluoroacetate, which was prepared by an electrochemical synthesis from copper metal, Ph$_3$P and trifluoroacetic acid in MeCN [53]. Given that the corresponding tris(triphenylphosphine) complexes are unknown, we attempted their preparation and characterization with modern spectroscopic methods. The reaction of the copper(II) carboxylates with 3.5 equivalents (per Cu) of Ph$_3$P in MeOH afforded colourless crystals upon cooling to a reaction mixture to $-20\,^\circ$C (Scheme 4).

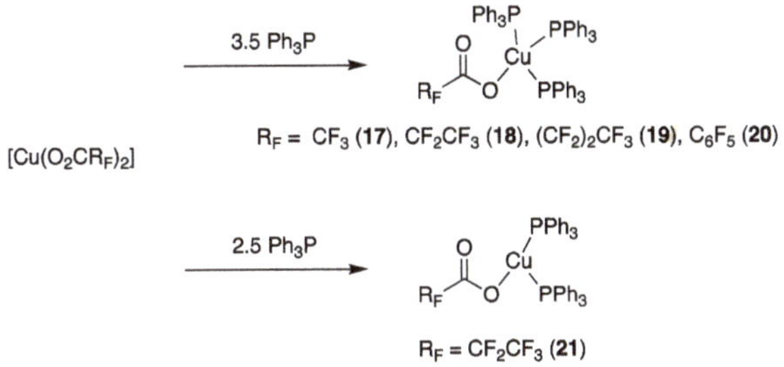

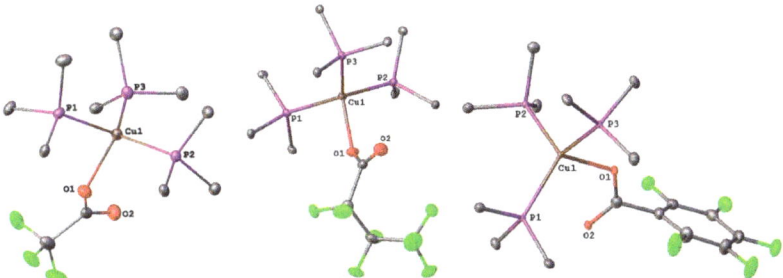

Scheme 4. Synthesis of copper(I) complexes with perfluorinated carboxylates.

The perfluorinated complexes are quite soluble in MeOH even at −20 °C, which explains the low yields obtained in some cases. If the reaction mixtures are evaporated to dryness, however, the products are contaminated with Ph$_3$PO, which can only be removed with great difficulty and loss of material. In the ^{31}P-NMR spectra of the compounds, sharp singlets in the range from −1.3 to −2.4 ppm are observed, which are typical for (triphenylphosphine)copper(I) complexes. The IR spectra display intense bands for the asymmetric CO stretching frequency between 1648 and 1685 cm^{-1}. These values are different to those observed in their corresponding bis(triphenylphosphine) counterparts [13]. The molecular structures of the trifluoroacetate- (**17**), perfluorobutyrate- (**19**) and pentafluorobenzoate (**20**) derivatives were confirmed by X-ray diffraction experiments (Figure 11).

Figure 11. Molecular structures of [Cu(O$_2$CCF$_3$)(PPh$_3$)$_3$] **17**·(left), [Cu{O$_2$C(CF$_2$)$_2$CF$_3$}(PPh$_3$)$_3$]· **19** (centre) and [Cu(O$_2$CC$_6$F$_5$)(PPh$_3$)$_3$]·**20** (right). Ellipsoids show 50% probability levels. Hydrogen atoms have been omitted for clarity. Only one of the two independent molecules is shown and only the *ipso*-carbon atoms of the Ph$_3$P ligands are depicted.

In these three compounds, the copper atom has four coordinates, with three phosphorus atoms from the Ph$_3$P ligands as well as an oxygen atom from the monodentate carboxylates in a trigonal pyramidal arrangement. The monodentate coordination of the carboxylate is confirmed by one short [2.1121(16), 2.1047(12) and 2.1015(14) Å] and one long (>3 Å) Cu–O distance. There are only three structurally characterized tris(triphenylphosphine) copper(I) carboxylates containing phthalic acid [54], benzene-1,2-dioxyacetic acid [55] and ethylphenylmalonic acid [56]. In these compounds, the Cu-coordination mode, as well as the bond distances, are very similar to those discussed above.

For comparison, the bis(triphenylphosphine) complex with perfluoropropionate was prepared by the same method (Scheme 4), except that less (2.5 equivalents per Cu) Ph$_3$P was used. In this case, a colourless solid was also isolated, with a slightly different chemical shift in its ^{31}P-NMR spectrum and a different CO stretching frequency when compared to the tris-complex. The observed IR spectrum agrees well with the data reported in the literature [13]. The molecular structure of **21** was studied by an X-ray diffraction experiment (Figure 12).

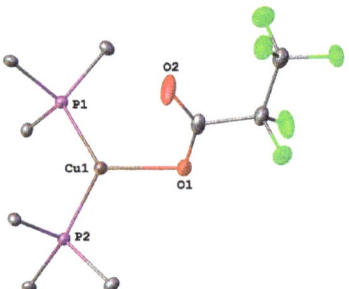

Figure 12. Molecular structure of [Cu{$O_2CCF_2CF_3$}(PPh_3)$_2$]·(**21**). Ellipsoids show 50% probability levels. Hydrogen atoms have been omitted for clarity. Only the *ipso*-carbon atoms of the Ph_3P ligands are shown.

The complex consists of a three-coordinate copper atom featuring two *P*-bound Ph_3P ligands as well as a monodentate, *O*-bound perfluoropropionato ligand in a nearly trigonal planar geometry. In the literature, such bis(triphenylphosphine) complexes were originally proposed to be four-coordinate copper(I) complexes with chelating carboxylato ligands [52]. The longer of the two Cu-O distances in the complex [ca. 2.7 Å] is shorter than that in the tris-species discussed above, but longer than that in [Cu(O_2CCF_3)(PPh_3)$_2$]·(2.5 Å) [53] and the non-fluorous complexes [Cu(O_2CCH_2COOH)(PPh_3)$_2$]·(2.5 Å) [57] and [Cu{$O_2CC_6H_3$(3,5-NO_2)$_2$}(PPh_3)$_2$]·(2.6 Å) [58]. The results thus clearly indicate that although the coordination number at copper differs in the bis- and tris-complexes, monodentate coordination of the carboxylato ligands is observed in both cases.

Thermal analysis of the bis(triphenylphosphine) perfluoropropionato copper(I) complex (Figure 13) shows a gradual decomposition commencing at 244 °C. The remaining mass of 19.1% is consistent with formation of Cu_2O and not elemental copper. A similar decomposition to Cu_2O (admixed with Cu and $Cu_2P_2O_7$) was observed in the related triphenylphosphite complexes of perfluorinated carboxylic acids [59].

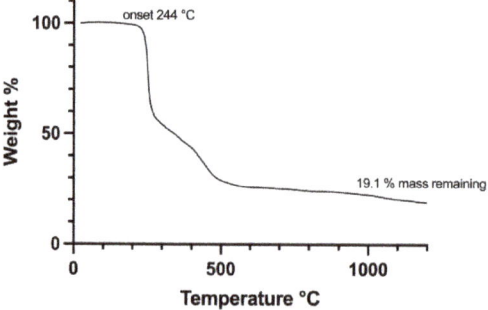

Figure 13. TGA curve of [Cu{$O_2CCF_2CF_3$}(PPh_3)$_2$] (**21**).

4. Conclusions

Silver(I) salts of perfluorinated carboxylic (and sulfonic acids) can be readily prepared from the acids and Ag_2O in toluene, avoiding the presence of water. A structural analysis of several compounds revealed that these silver(I) salts exist as coordination polymers assembled from dinuclear [$Ag_2(O_2CR_F)_2$] (R_F = perfluorinated group) building blocks. In some cases, toluene is either η^2- or η^3-bound to the silver. The silver(I) salts can be used to transmetallate the perfluorocarboxylate-group to a gold(I)-phosphine species. Several examples with both mono- and bis-phosphines were examined, and their solid-state structures were determined. The structures of the bis-phosphine complexes feature aurophilic interactions. The combination of silver(I) salts and gold(I) phosphine carboxylate leads to the formation of mixed-metal species containing both Ag and Au. Depending on the nature of the phosphines, either polymeric structures or discrete tetranuclear Ag_2Au_2 complexes are observed in the solid state. The reduction in dinuclear paddlewheel-type copper(II) perfluorocarboxylates with Ph_3P leads to the corresponding copper(I) complexes with either two or three phosphine ligands. In the molecular structures of these compounds, the coordination mode of the carboxylate ligands is monodentate, resulting in coordination numbers of three or four at the metal center.

Supplementary Materials: The following supporting information can be downloaded at: https://www.mdpi.com/article/10.3390/chemistry5020058/s1, Copies of ^{1}H-, ^{31}P- and ^{19}F-NMR spectra of the compounds.

Author Contributions: Conceptualization, F.M.; formal analysis, F.M., B.B.B.; C.W.L.; investigation, R.P., J.R., B.B.B.; writing—original draft preparation, R.P.; writing—review and editing, F.M.; supervision, F.M. All authors have read and agreed to the published version of the manuscript.

Funding: This research received no external funding.

Data Availability Statement: The data presented in this study are available from the authors.

Conflicts of Interest: The authors declare no conflict of interest. The funders had no role in the design of the study; in the collection, analyses, or interpretation of data; in the writing of the manuscript; or in the decision to publish the results.

References

1. Hauptschein, M.; Grosse, A.V. Perfluoroalkyl halides prepared from silver perfluoro-fatty acid salts. I. Perfluoroalkyl iodides. *J. Am. Chem. Soc.* **1951**, *73*, 2461–2463. [CrossRef]
2. Adams, S.K.; Edwards, D.A.; Richards, R. Silver(I) carboxylates. I. Mass spectra and low frequency infrared spectra. *Inorg. Chim. Acta* **1975**, *12*, 163–166.
3. Edwards, D.A.; Harker, R.M.; Mahon, M.F.; Molloy, K.C. Aerosol-assisted chemical vapour deposition (AACVD) of silver films from triorganophosphine adducts of silver carboxylates, including the structure of $[Ag(O_2CC_3F_7)(PPh_3)_2]$. *Inorg. Chim. Acta* **2002**, *328*, 134–146. [CrossRef]
4. Römbke, P.; Schier, A.; Schmidbaur, H.; Cronje, S.; Raubenheimer, H. Mono- and bimetallic gold(I) and silver(I) pentafluoropropionates and related compounds. *Inorg. Chim. Acta* **2004**, *357*, 235–242.
5. Torroba, J.; Aynsley, J.; Tuzimotoa, P.A.; Bruce, D.W. One-pot synthesis of anhydrous silver carboxylates from silver(I) fluoride. *RSC Adv.* **2012**, *2*, 12866–12869.
6. Blakeslee, A.E.; Hoard, J.L. The structure of silver perfluorobutyrate. *J. Am. Chem. Soc.* **1956**, *78*, 3029–3033. [CrossRef]
7. Griffin, R.G.; Ellett, J.D.; Mehring, M.; Bullitt, J.G.; Waugh, J.S. Single crystal study of the ^{19}F shielding tensors of a trifluoromethyl group. *J. Chem. Phys.* **1972**, *57*, 2147–2155.
8. Weigand, H.; Tyrra, W.; Naumann, D. The structure of silverpentafluorobenzoate monohydrate, $AgCO_2C_6F_5 \cdot H_2O$. *Z. Anorg. Allg. Chem.* **2008**, *634*, 2125–2126. [CrossRef]
9. Lamann, R.; Hülsen, M.; Dolg, M.; Ruschewitz, U. Syntheses, crystal structures and thermal behavior of five new complexes containing 2,4,6-trifluorobenzoate as ligand. *Z. Anorg. Allg. Chem.* **2012**, *638*, 1424–1431. [CrossRef]
10. Swarts, F. Sur quelques trifluoroacétates. *Bull. Soc. Chim. Belg.* **1939**, *48*, 176–191.
11. Bruce, M.I.; Nicholson, B.K.; Shawkataly, O.B. Synthesis of gold-containing mixed metal cluster compleces. *Inorg. Synth.* **1989**, *26*, 324.
12. Römbke, P.; Schier, A.; Schmidbaur, H. Gold(I) carboxylates and fluorocarboxylates. *Z. Naturforsch.* **2002**, *57b*, 605–609. [CrossRef]
13. Edwards, D.A.; White, J.W. Copper(II) pentafluorobenzoates: Preparations, magnetic and spectroscopic features, and reation with triphenylphosphine. *J. Inorg. Nucl. Chem.* **1978**, *40*, 1335–1339. [CrossRef]
14. Handa, M.; Ishitobi, Y.; Yakuwa, T.; Yoshioka, D.; Ishida, H.; Mikuriya, M.; Hiromitsu, I.; Tanaka, H.; Ikeue, T. A polymer complex $[Cu(O_2CC_6F_5)_2(pyz)]_n$ formed from copper(II) pentafluorobenzoate and pyrazine. *Bull. Chem. Soc. Jpn.* **2009**, *82*, 1277–1279. [CrossRef]
15. Krupoder, S.A.; Danilovich, V.S.; Miller, A.O.; Furin, G.G. Synthesis and properties of the volatile polyfluoroaryl and polyfluorocarboxy derivatives of metals and metalloids - Prospective materials for microelectronics. *Russ. J. Org. Chem.* **1994**, *30*, 1243–1251.
16. Kabsch, W. XDS. *Acta Cryst.* **2010**, *D66*, 125–132.
17. *CrysAlisPro 41.123a*, 40.53; Rigaku Oxford Diffraction Ltd.: Oxford, UK, 2022.
18. *DATCOL*, Bruker AXS: Billerica, MA, USA, 2006.
19. Sheldrick, G.M. *SADABS 2.03*; Universiät Göttingen: Göttingen, Germany, 2002.
20. *APEX3, Crystallography Software Suite*, Bruker AXS: Billerica, MA, USA, 2017.

21. Sheldrick, G.M. SHELXT - Integrated space-group and crystal-structure determination. *Acta Cryst.* **2015**, *A71*, 3–8. [CrossRef]
22. Sheldrick, G.M. SHELXL-97, Program for Crystal Structure Refinement. University of Göttingen: Göttingen, Germany, 1997.
23. Dolomanov, O.V.; Bourhis, L.J.; Gildea, R.J.; Howard, J.A.K.; Puschmann, H. OLEX2: A complete structure solution, refinement and analysis program. *J. Appl. Cryst.* **2009**, *42*, 339–341. [CrossRef]
24. Szlyk, E.; Lakomska, I.; Grodzicki, A. Thermal and spectroscopic studies of the Ag(I) salts with fluorinated carboxylic and sulfonic acid residues. *Thermochim. Acta* **1993**, *223*, 207–212. [CrossRef]
25. Taylor, I.F., Jr.; Hall, E.A.; Amma, E.L. Metal ion-aromatic complexes. VII. Crystal and molecular structure of bis(m-xylene)silver perchlorate. *J. Am. Chem. Soc.* **1969**, *91*, 5745–5749. [CrossRef]
26. Karpova, E.V.; Boltalin, A.I.; Korenev, Y.M.; Troyanov, S.I. Silver(I) mono- and trifluoroacetates: Thermal stability and crystal structure. *Russ. J. Coord. Chem.* **1999**, *25*, 65–68.
27. Schmidbaur, H.; Schier, A. Argentophilic interactions. *Angew. Chem., Int. Ed.* **2015**, *54*, 746–784.
28. Szczesny, R.; Szlyk, E. Thermal decomposition of some silver(I) carboxylates under nitrogen atmosphere. *J. Therm. Anal. Calorim.* **2013**, *111*, 1325–1330.
29. Szlyk, E.; Lakomska, I.; Grodzicki, A. Studies of Au(I) complexes with triphenylphosphine and perfluorinated carboxylates. *Polish J. Chem.* **1995**, *69*, 1103–1108.
30. Deák, A.; Tunyogi, T.; Tárkányi, G.; Király, P.; Pálinkás, G. Self-assembly of gold(i) with diphosphine and bitopic nitrogen donor linkers in the presence of trifluoroacetate anion: Formation of coordination polymerversus discrete macrocycle. *CrystEngComm* **2007**, *9*, 640–643.
31. Appel, M.; Schloter, K.; Heidrich, J.; Beck, W. Metallorganische Lewis-Säuren: XXVII. Metallorganische Verbindungen mit Anionen der Fluorsulfonsäure, sowie von perfluorierten Sulfon- und Carbonsäuren. *J. Organomet. Chem.* **1987**, *322*, 77–88. [CrossRef]
32. Tunyogi, T.; Deák, A. [m-4,5-Bis(diphenylphosphino)-9,9-di-methylxanthene]bis[(trifluoroacetato)-gold(I)] and its dichloromethane 0.58-solvate. *Acta Cryst.* **2010**, *C66*, m133–m136.
33. Schmidbaur, H.; Schier, A. A briefing on aurophilicity. *Chem. Soc. Rev.* **2008**, *37*, 1931–1951.
34. Bartlett, P.N.; Cheng, F.; Cook, D.A.; Hector, A.L.; Levason, W.; Reid, G.; Zhang, W. Synthesis and structure of [{$C_7F_{15}CO_2$}$_2$AgAu (PPh$_3$)]$_2$ and its use in electrodeposition of gold–silver alloys. *Inorg. Chim. Acta* **2010**, *363*, 1048–1051. [CrossRef]
35. Fernández, E.J.; López-de-Luzuriaga, J.M.; Monge, M.; Rodriguez, M.A.; Crespo, O.; Gimeno, M.C.; Laguna, A.; Jones, P.G. Heteropolynuclear complexes with the ligand Ph2PCH2SPh: Theoretical evidence for metallophilic Au-M interactions. *Chem. Eur. J.* **2000**, *6*, 636–644. [CrossRef]
36. Wheaton, C.A.; Jennings, M.C.; Puddephatt, R.J. Complexes of gold(I) with a chiral diphosphine ligand: A polymer with Au-Ag and Ag-Ag metallophilic bonds. *Z. Naturforsch.* **2009**, *64b*, 1469–1477. [CrossRef]
37. Catterick, J.; Thornton, P. Structures and physical properties of polynuclear carboxylates. *Adv. Inorg. Chem. Radiochem.* **1977**, *20*, 291–362.
38. Larionov, S.V.; Glinskaya, L.A.; Klevtsova, R.F.; Lvov, P.E.; Ikorskii, V.N. Preparation, crystal od molecular structure, and magnetic properties of an adduct of copper(II) pentafluorobenzoate with 1,4-dioxane. *Russ. J. Inorg. Chem.* **1991**, *36*, 1413–1415.
39. Karpova, E.V.; Boltalin, A.I.; Korenev, Y.M.; Troyanov, S.I. Synthesis and X-ray diffraction analysis of copper(II) fluoroacetates Cu(CF_3COO)$_2$ and Cu(CHF_2COO)$_2$ 0.5H$_2$O containing polymeric chains in their structures. *Russ. J. Coord. Chem.* **2000**, *26*, 361–366.
40. Zhang, J.; Hubert-Pfalzgraf, L.G.; Luneau, D. Synthesis, characterization and molecular structures of Cu(II) and Ba(II) fluorinated carboxylate complexes. *Polyhedron* **2005**, *24*, 1185–1195. [CrossRef]
41. Moreland, J.A.; Doedens, R.J. Synthesis, crystal structure, and magnetic properties of a dimeric quinoline adduct of copper(II) trifluoroacetate. *J. Am. Chem. Soc.* **1975**, *97*, 508–513. [CrossRef]
42. Karpova, E.V.; Boltalin, A.I.; Zakharov, M.A.; Sorokina, N.I.; Korenev, Y.M.; Troyanov, S.I. Synthesis and crystal structure of copper(II) trifluoroacetates, Cu$_2$(CF$_3$COO)$_4$ · 2 CH$_3$CN and Cu(CF$_3$COO)$_2$(H$_2$O)$_4$. *Z. Anorg. Allg. Chem.* **1998**, *624*, 741–744. [CrossRef]
43. Vives, G.; Mason, S.A.; Prince, P.D.; Junk, P.C.; Steed, J.W. Intramolecular versus intermolecular hydrogen bonding of coordinated acetate to organic acids: A neutron, X-ray, and database study. *Cryst. Growth Des.* **2003**, *3*, 699–704. [CrossRef]
44. Nefedov, S.E.; Kushan, E.V.; Yakovleva, M.A.; Chikhichin, D.G.; Kotseruba, V.A.; Levchenko, O.A.; Kamalov, G.L. Binuclear complexes with the "Chinese Lantern" geometry as intermediates in the liquid-phase oxidation of dibenzyl ether with atmospheric oxygen in the presence of copper(II) carboxylates. *Russ. J. Coord. Chem.* **2012**, *38*, 224–231. [CrossRef]
45. Gahlot, S.; Purohit, B.; Jeanneau, E.; Mishra, S. Coinage metal complexes with di-tertiary-butyl sulfide as precursors with ultra-low decomposition temperature. *Chem. Eur. J.* **2021**, *27*, 10826–10832. [CrossRef]
46. Motreff, A.; Correa da Costa, R.; Allouchi, H.; Duttine, M.; Mathonière, C.; Duboc, C.; Vincent, J.M. Dramatic solid-state humidity-induced modification of the magnetic coupling in a dimeric fluorous copper(II)−carboxylate complex. *Inorg. Chem.* **2009**, *48*, 5623–5625. [CrossRef] [PubMed]
47. Han, L.J.; Kong, Y.J.; Huang, M.M. Magnetic properties and crystal structures of two copper coordination compounds with pentafluorobenzoate ligand. *Inorg. Chim. Acta* **2021**, *514*, 120019.
48. Kani, Y.; Tsuchimoto, M.; Ohba, S.; Tokii, T. catena-Poly[[tetrakis(m-2,2-dimethyl-propionato-O:O′)dicopper(II)]-m-dioxane-O:O′] and catena-poly-[[tetrakis(m-3,3-dimethylbutyrato-O:O′)dicopper(II)]-m-dioxane-O,O′]. *Acta Cryst.* **2000**, *C56*, e80–e81.

49. Borel, M.M.; Leclaire, A. Etude des Propionates Métalliques. V. Détermination de la Structure du Bis-propionato Cuivre(II) 0,5 Dioxanne. *Acta Cryst.* **1976**, *B32*, 1275–1278. [CrossRef]
50. Smart, P.; Mínguez Espallargas, G.; Brammer, L. Competition between coordination network and halogen bond network formation: Towards halogen-bond functionalised network materials using copper-iodobenzoate units. *CrystEngComm* **2008**, *10*, 1335–1344. [CrossRef]
51. Boniak, L.; Borel, M.M.; Busnot, F.; Leclaire, A. Structure de [Cu(C$_6$H$_5$COO)$_2$(C$_4$H$_8$O$_2$)$_{0.5}$]$_2$ 2 C$_4$H$_8$O$_2$. Mise en évidence d'une conformation "twist" du dioxane. *Rev. Chim. Miner.* **1979**, *16*, 501–508.
52. Hammond, B.; Jardine, F.H.; Vohra, A.G. Carboxylatocopper(I) complexes. *J. Inorg. Nucl. Chem.* **1971**, *33*, 1017–1024. [CrossRef]
53. Hart, R.D.; Healy, P.C.; Hope, G.A.; Turner, D.W.; White, A.H. Electrochemical synthesis and structural characterization of bis(triphenylphosphine)copper(I) fluoroacetates. *J. Chem. Soc. Dalton Trans.* **1994**, 773–779. [CrossRef]
54. Vykoukal, V.; Halasta, V.; Babiak, M.; Bursik, J.; Pinkas, J. Morphology control in AgCu nanoalloy synthesis by molecular Cu(I) precursors. *Inorg. Chem.* **2019**, *58*, 15246–15254.
55. Devereux, M.; McCann, M.; Cronin, J.F.; Cardin, C.; Convery, M.; Quillet, V. Synthesis and X-ray crystal structure of the copper(I) dicarboxylic acid complex [Cu(η^1-bdoaH)(PPh$_3$)$_3$] (bdoaH$_2$ = benzene-1,2- dioxyacetic acid). *Polyhedron* **1994**, *13*, 2359–2366. [CrossRef]
56. Darensbourg, D.J.; Holtcamp, M.W.; Khandelwal, B.; Klausmeyer, K.K.; Reibenspies, J.H. A more intimate examination of the role of copper(I) in the decarboxylation of derivatives of malonic acid. Comparisons with zinc(II) analogs. *Inorg. Chem.* **1995**, *34*, 2389–2398. [CrossRef]
57. Darensbourg, D.J.; Holtcamp, M.W.; Khandelwal, B.; Reibenspies, J.H. Intramolecular and intermolecular hydrogen bonding in triphenylphosphine derivatives of copper(I) carboxylates, (Ph$_3$P)$_2$CuO$_2$C(CH$_2$)$_n$COOH. Role of copper(I) in the decarboxylation of malonic acid and its derivatives. *Inorg. Chem.* **1994**, *33*, 531–537. [CrossRef]
58. Mauro, A.E.; Porta, C.C.; De Simone, C.A.; Zukerman-Schpector, J.; Catellano, E.E. Synthesis and structural studies of (3,5-dinitrobenzoate)bis(triphenylphosphine)copper(I). *Polyhedron* **1993**, *12*, 1141–1143. [CrossRef]
59. Szlyk, E.; Szymanska, I. Studies of new volatile copper(I) complexes with triphenylphosphite and perfluorinated carboxylates. *Polyhedron* **1999**, *18*, 2941–2948. [CrossRef]

Disclaimer/Publisher's Note: The statements, opinions and data contained in all publications are solely those of the individual author(s) and contributor(s) and not of MDPI and/or the editor(s). MDPI and/or the editor(s) disclaim responsibility for any injury to people or property resulting from any ideas, methods, instructions or products referred to in the content.

Article

Phosphine Functionalized Cp^C Ligands and Their Metal Complexes

Florian Nährig, Yu Sun and Werner R. Thiel *

Fachbereich Chemie, RPTU Kaiserslautern-Landau, Erwin-Schrödinger-Straße 54, 67663 Kaiserslautern, Germany
* Correspondence: thiel@chemie.uni-kl.de; Tel.: +49-631-2052752

Abstract: Simple nucleophilic aliphatic substitution gives access to mono- and diphosphine ligands with a Cp^C group in the backbone. The monophosphine ligand coordinates to gold(I) via the phosphine site, to thallium(I) via the cyclopentadienyl site and to ruthenium(II) via a combination of both, resulting in an *ansa*-type structure. Coordination with the cyclopentadiene site is not possible for the diphosphine ligand. In this case, monodentate coordination to gold(I) and bidentate coordination to the [PdCl(μ^2-Cl)]$_2$, the [Rh(CO)(μ^2-Cl)]$_2$, and the Rh(CO)Cl fragment is observed, showing the variability in coordination modes possible for the long-chain diphosphine ligand. Ligands and complexes were characterized by means of NMR and IR spectroscopy, elemental analysis and X-ray structure analysis.

Keywords: cyclopentadienyl ligand; phosphine ligand; rhodium; palladium; gold; thallium

Citation: Nährig, F.; Sun, Y.; Thiel, W.R. Phosphine Functionalized Cp^C Ligands and Their Metal Complexes. *Chemistry* **2023**, *5*, 912–933. https://doi.org/10.3390/chemistry5020062

Academic Editor: Catherine Housecroft

Received: 15 March 2023
Revised: 4 April 2023
Accepted: 6 April 2023
Published: 18 April 2023

Copyright: © 2023 by the authors. Licensee MDPI, Basel, Switzerland. This article is an open access article distributed under the terms and conditions of the Creative Commons Attribution (CC BY) license (https:// creativecommons.org/licenses/by/ 4.0/).

1. Introduction

Both cyclopentadienyl derivatives and phosphines are among the most frequently used ligands in organometallic chemistry. It is therefore not surprising that Charrier and Mathey published the first example of a phosphine-functionalized cyclopentadiene as early as 1978. They obtained it by reacting sodium cyclopentadienide and chloromethyldiphenylphosphine [1]. The combination of a cyclopentadiene-derived nucleophile with an electrophile bearing a phosphine site remains the most important route to phosphine-functionalized cyclopentadienes. Alternatively, these compounds are accessible by nucleophilic addition of a phosphide to a fulvene or to a spiro [2.4]hepta-4,6-diene derivative [2–5]. Based on these strategies, a large number of such compounds have become accessible in recent decades.

When both the cyclopentadienyl and the phosphine site coordinate to one transition metal center, the resulting compounds exhibit increased stability compared to classical complexes with unfunctionalized cyclopentadienyl ligands, because the η^5→η^3→η^1 shift of the cyclopentadienyl ligand is suppressed by the *ansa*-bridging of the two donors. In addition, the phosphine donor remains in the immediate vicinity of the metal center after dissociation, which has a further stabilizing effect.

Among all phosphine functionalized cyclopentadienyl derivatives, ferrocenyl phosphines have probably attracted the most attention, in particular due to their use as mono- or bidentate ligands in transition metal catalyzed reactions [6–16].

Recently, we presented the C_2-symmetric and thus chiral cyclopentadiene derivative (Cp^C)$^{-1}$ (**1^{-1}**) as a ligand for early and late transition metal complexes [17–19]. The corresponding cyclopentadiene Cp^{C}H (**1**) is accessible in a few steps with high yield from dibenzosuberenone. In parallel to the transition metal chemistry of compound **1**, we searched for new derivatives functionalized at its cyclopentadiene core. By reacting the anion (Cp^C)$^{-1}$ (**1^{-1}**) with oxygen, the corresponding ketone is accessible in good yield. With this η^4-coordinating dienone, we were able to synthesize ruthenium and iron compounds whose structures and reactivities correspond to those of the Shvo and Knölker catalysts, respectively [20]. Here, we now report on novel phosphine-functionalized derivatives of **1** and their use as ligands in transition metal chemistry.

2. Materials and Methods

Chemicals were purchased from the following suppliers: ABCR GmbH, Fisher Scientific GmbH, Merck KGaA and Strem Chemicals GmbH. All commercially available starting materials were used without further purification. Compounds sensitive to air or moisture were handled and reacted under exclusion of oxygen and water in suitable Schlenk tubes. The solvents dichloromethane, diethyl ether, n-pentane and toluene were dried with a MBraun MB-SPS solvent drying system and degassed by passing nitrogen for 10 min. Acetonitrile and tetrahydrofuran were dried according to standard methods [21]. Glassware was heated three times with a heat gun under vacuum and refilled with dry nitrogen before use. When necessary, the purification of compounds was carried out by column chromatography on the CombiFlash Rf200 instrument from Teledyne Isco with pre-packed RediSept® columns. NMR spectra were recorded using Avance 400 and 600 devices from Bruker Corporation at a temperature of 293 K (20 °C). The resonances were only partly assignable (numbering according to Scheme 1). Air- or moisture-sensitive compounds were measured under an atmosphere of nitrogen using an NMR tube with a Teflon cap from VWR International GmbH. The anhydrous deuterated solvents CD_3CN, C_6D_6, $CDCl_3$ and CD_2Cl_2 were dried according to standard methods, re-condensed and stored under an atmosphere of nitrogen in Schlenk tubes. The evaluation of the NMR spectra was carried out with the software MestReNova 6.0.2-5475 © of Mestrelab Research. The infrared spectra were recorded on a Perkin-Elmer FT-ATR-IR 100 spectrometer with an ATR cell with diamond-coated zinc selenide windows. The IR spectra were processed using the Perkin-Elmer Spectrum 6.3.5 software from and the OriginLab Corporation software OriginPro 8G. Elemental analyses were measured out in the Analytical Laboratory of the Department of Chemistry at the TU Kaiserslautern. Compounds sensitive to air or moisture were filled in a glove box into tin capsules and sealed under an atmosphere of argon. The measurement of the elemental analyses was carried with a Vario Micro Cube analyser from Elementar-Analysetechnik. Cp^CH (1) was synthesized according to a published procedure [17]. (tht)AuCl, TlOEt, [(η^6-C_6H_6)$RuCl_2$]$_2$, $(CH_3CN)_2PdCl_2$ and $[(CO)_2Rh(\mu^2\text{-Cl})]_2$ were obtained commercially.

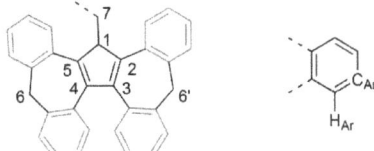

Scheme 1. Numbering for the assignment of NMR resonances (see below) of the Cp^C complexes.

$Cp^CHBzPPh_2$ (2): A total of 2.17 mL (3.47 mmol) of n-butyllithium (1.6 M solution in n-hexane) was slowly added at 0 °C to a solution of 1.24 g of Cp^CH (1) [17] (3.14 mmol) in 10 mL of toluene. The reaction mixture was stirred for 2 h at this temperature and then was slowly warmed to room temperature. During this time, the precipitation of a colorless solid was observed, which was dissolved by the addition of 1 mL of tetrahydrofurane. Then, 976 mg (3.14 mmol) of [2-(chloromethyl)phenyl](diphenyl)phosphane [22,23], dissolved in 10 mL of toluene, was slowly added at 0 °C. The reaction mixture was stirred for additional 2 h at this temperature. During this time, a color change from orange to dark red was observed and the reaction mixture was stirred for another 18 h at room temperature. A total of 15 mL of water was added, and the organic phase was separated, washed three times with 5 mL of water and dried over $MgSO_4$. The solvent was removed, and the crude red–brown product was dried under vacuum. Purification was carried out by column chromatography (MPLC) with a mixture of n-hexane and ethyl acetate (99:1). Compound 2 was obtained slightly contaminated as a light pink solid in the first fraction. As a minor product (16%), compound 3 can be isolated from the second fraction. Compound 2 was finally purified by precipitation from a saturated dichloromethane solution with *n*-pentane. Yield: 1.23 g (59%) of a colorless solid. Elemental analysis calcd. for $C_{50}H_{37}P$ (668.82 g/mol): C 89.79, H 5.58; found: C 89.87, H 5.70%. ^{1}H NMR (400 MHz, CDCl$_3$): δ 7.55 (d, $^3J_{HH}$ = 7.9 Hz, 1H, H$_{Ar}$), 7.34–7.28 (m, 2H, H$_{Ar}$), 7.25–7.05 (m, 15H, H$_{Ar}$), 6.98 (dt, $^3J_{HH}$ = 15.5, 7.2 Hz, 5H, H$_{Ar}$), 6.89–6.84 (m, 1H, H$_{Ar}$), 6.84–6.79 (m, 2H, H$_{Ar}$), 6.76 (t, $^3J_{HH}$ = 6.9 Hz, 2H, H$_{Ar}$), 6.60 (d, $^3J_{HH}$ = 7.1 Hz, 2H, H$_{Ar}$), 4.69 (t, $^3J_{HH}$ = 5.1 Hz, 1H, H1), 4.07 (dt, $^2J_{HH}$ = 16.7, $^3J_{HH}$ = 3.3 Hz, $^4J_{PH}$ = 3.3 Hz, 1H, H7), 3.80 (d, $^2J_{HH}$ = 12.7 Hz, 1H, H6 or H6′), 3.75–3.63 (m, 3H, H6 or H6′ and H7), 3.33 (d, $^3J_{HH}$ = 12.6 Hz, 1H, H6 or H6′). ^{13}C NMR (101 MHz, CDCl$_3$): δ 147.9, 144.9, 143.9, 143.1, 142.9, 141.6, 141.0, 139.8, 139.6, 136.8 (d, J_{PC} = 7.4 Hz), 136.7 (d, J_{PC} = 7.8 Hz), 136.2, 136.0, 134.2, 134.2, 134.1, 134.0, 133.9, 133.1, 132.9, 132.66, 132.1, 129.2, 128.9, 128.9, 128.8, 128.7, 128.6, 128.0, 128.0, 127.9, 127.4, 127.0, 126.6, 126.5, 126.3, 125.1, 124.8, 55.1 (C1), 42.2 (C6 or C6′), 41.8 (C6 or C6′), 34.6 (d, $^3J_{PC}$ = 24.1 Hz, C7). ^{31}P NMR (162 MHz, CDCl$_3$): δ −15.44 (s). IR (ATR, cm^{-1}): ṽ 3059 m, 2962 m, 2921 m, 2872 w, 1584 m, 1485 m, 1464 m, 1432 s, 1333 m, 1304 m, 1277 m, 1213 m, 1193 m, 1144 m, 1124 m, 1090 m, 1027 m, 940 m, 779 m, 765 s, 746 vs, 717 s, 697 vs, 674 s.

Cp^C(BzPPh$_2$)$_2$ (3): A total of 0.74 mL (1.18 mmol) of *n*-butyllithium (1.6 M in *n*-hexane) wasadded at 0 °C to a solution of 716 mg (1.07 mmol) of **2** in (10 mL) of toluene. The reaction mixture was stirred for 2 h at this temperature and afterwards slowly warmed to room temperature. During this time, the precipitation of a colorless solid was observed, which was dissolved by the addition of 1 mL of tetrahydrofurane. Then, 333 mg (1.07 mmol) of (diphenyl)phosphane [22,23] dissolved in 5 mL of toluene wasadded slowly at 0 °C and the reaction mixture was stirred for another 2 h at this temperature, whereby its color changed from pink to dark red. After warming to room temperature, the mixture was stirred for further 18 h. A total of 5 mL of water was added, the organic phase was separated, washed three times with 5 mL of water, and then dried over MgSO$_4$. The solvent was removed, and the light pink crude product was dried under vacuum. Finally, the crude product was purified by precipitation from a saturated dichloromethane solution with *n*-pentane. Colorless prismatic single crystals were obtained by slow diffusion of *n*-pentane into a saturated toluene solution. Yield: 840 mg (83%) of a colorless solid. Elemental analysis calcd. For $C_{69}H_{52}P_2$ (943.12 g/mol): C 87.87, H 5.56; found: C 87.86, H 5.9%. ^{1}H NMR (400 MHz, CDCl$_3$): δ 7.81 (d, $^3J_{HH}$ = 7.9 Hz, 2H, H$_{Ar}$), 7.42–7.30 (m, 18H, H$_{Ar}$), 7.25 (dd, $^3J_{HH}$ = 10.3, 5.5 Hz, 4H, H$_{Ar}$), 7.17 (t, 3 $^3J_{HH}$ = 7.4 Hz, 2H, H$_{Ar}$), 7.09 (dd, $^3J_{HH}$ = 10.6, 4.4 Hz, 4H, H$_{Ar}$), 7.02–6.94 (m, 6H, H$_{Ar}$), 6.90 (t, $^3J_{HH}$ = 7.6 Hz, 2H, H$_{Ar}$), 6.85 (t, $^3J_{HH}$ = 7.6 Hz, 2H, H$_{Ar}$), 6.71 (dd, $^3J_{HH}$ = 7.3, 3.8 Hz, 2H, H$_{Ar}$), 6.50 (d, $^3J_{HH}$ = 7.7 Hz, 2H, H$_{Ar}$), 4.33 (dd, $^2J_{HH}$ = 16.1, $^4J_{PH}$ = 8.0 Hz, 2H, H7), 3.84 (d, $^2J_{HH}$ = 16.1 Hz, 2H, H7), 3.39 (d, $^2J_{HH}$ = 12.5 Hz, 2H, H6), 2.48 (d, $^2J_{HH}$ = 12.4 Hz, 2H, H6). ^{13}C NMR (101 MHz, CDCl$_3$): δ 147.6, 145.0, 142.4, 141.7, 141.5, 141.1, 137.8, 137.7, 137.3, 137.2, 137.0, 134.4 (d, J_{PC} = 20.2 Hz), 133.9 (d, J_{PC} = 20.1 Hz), 133.3, 133.0, 132.3, 128.9, 128.7, 128.5 (d, J_{PC} = 19.4 Hz), 128.3, 128.1, 127.8, 127.5 (d, J_{PC} = 2.1 Hz), 126.8, 126.6, 126.5, 126.1, 124.7, 65.6 (t, $^4J_{PC}$ = 2.8 Hz, C1), 40.9 (br, C6), 40.1 (dt, $^3J_{PC}$ = 10.7, 5.7 Hz, C7). ^{31}P NMR (162 MHz, CDCl$_3$): δ −16.22 (s). IR (ATR, cm^{-1}): ṽ = 3053 m, 3023 w, 2999 w, 2927 w, 2905 w, 2867 w, 2843 w, 1586 m, 1480 m, 1461 m, 1434 s, 1264 w, 1157 w, 1120 w, 1092 m, 1065 m, 1026 m, 947 m, 904 m, 754 s, 738 vs, 694 vs, 680 m, 656 m.

[Cp^{C}HBzPPh$_2$(AuCl)] (4): The reaction was carried out in the absence of light. A total of 222 mg (0.33 mmol) of compound **2** and 136 mg (0.42 mmol) of (tetrahydrothiophene)gold(I) chloride was stirred for 18 h at room temperature in 15 mL of dichloromethane. A total of 30 mL of *n*-pentane was added to precipitate a colorless solid, which was filtered off, washed three times with 15 mL of diethyl ether and three times with 15 mL of *n*-pentane, and finally, dried under vacuum. Yield: 298 mg (78%) of a colorless solid. Prismatic single crystals containing one equivalent of toluene were obtained by slow diffusion of *n*-pentane into a saturated toluene solution. Elemental analysis calcd. for $C_{50}H_{37}AuClP$ (901.22 g/mol): C 66.64, H 4.14, found: C 66.53, H 4.10%. ^{1}H NMR (400 MHz, CDCl$_3$): δ 7.66–7.57 (m, 3H, H$_{Ar}$), 7.55–7.48 (m, 2 × $^3J_{HH}$ = 8.8 and 7.6 Hz, 1.9 Hz, 3H, H$_{Ar}$), 7.42–7.27 (m, 9H, H$_{Ar}$), 7.25–7.14 (m, 6H, H$_{Ar}$), 7.11–6.92 (m, 6H, H$_{Ar}$), 6.91–6.85 (m, 1H, H$_{Ar}$), 6.75–6.67 (m, 2H, H$_{Ar}$), 4.65 (t, $^3J_{HH}$ = 4.6 Hz, 1H, H1), 4.51 (dd, $^2J_{HH}$ = 18.3 Hz,

4.7 Hz, 1H, H7), 4.05 (dd, $^2J_{HH}$ = 18.7 Hz, 3.5 Hz, 1H, H7), 3.86 (d, $^2J_{HH}$ = 12.7 Hz, 1H, H6 or H6′), 3.79 (dd, $^2J_{HH}$ = 12.6 Hz, 7.2 Hz, 2H, H6 or H6′), 3.60 (d, $^2J_{HH}$ = 12.5 Hz, 1H, H6 or H6′). ^{13}C NMR (101 MHz, CDCl$_3$): δ 146.1, 144.2, 143.5, 143.3, 143.2, 142,0 141.6, 140.7, 139.7 (d, J_{PC} = 5.1 Hz), 135.2 (d, J_{PC} = 14.0 Hz), 134.2 (d, J_{PC} = 13.8 Hz), 133.4, 132.4 (d, J_{PC} = 2.5 Hz), 132.3, 132.2, 132.1 (d, J_{PC} = 1.9 Hz), 132.1, 131.8, 129.6 (d, J_{PC} = 12.0 Hz), 129.3, 129.2, 129.0, 129.0, 128.9, 128.4, 128.1, 128.0, 128.0, 127.9, 127.9, 127.7, 127.7, 127.5, 127.4, 127.0, 126.9, 126.9, 126.8, 126.7, 126.6 (d, J_{PC} = 9.8 Hz), 126.2, 125.0 (d, J_{PC} = 10.0 Hz), 53.8 (C1), 41.9 (C6 or C6′), 41.8 (C6 or C6′), 32.4 (d, J_{PC} = 15.8 Hz, C7). ^{31}P NMR (162 MHz, CDCl$_3$): δ 24.4. IR (ATR, cm^{-1}): ṽ 3054 w, 3014 w, 2980 m, 2886 w, 2831 w, 1589 w, 1480 m, 1436 s, 1351 w, 1250 w, 1185 w, 1158 w, 1121 w, 1099 m, 1039 w, 998 w, 945 w, 908 w, 771 s, 744 vs, 711 s, 691 vs.

[(η^5-Cp^{C}HBzPPh$_2$)Tl] (5): The reaction was carried out in the absence of light. A total of 1.33 g (1.99 mmol) of 2 was dissolved in 30 mL o dry benzene. A total of 874 mg (350 mmol) of thallium(I) ethanolate was added and the mixture was stirred for 30 min at room temperature. A bright yellow solid precipitated, which was filtered off, washed three times with 40 mL of benzene and dried under vacuum. Yield 1.44 g (83%) of a bright yellow microcrystalline solid, which was recrystallized by cooling a hot saturated solution in THF to room temperature. Elemental analysis calcd. for C$_{50}$H$_{36}$PTl (872.18 g/mol): C 68.85, H 4.16; found: C 68.82, H 4.26%. ^{1}H NMR (600 MHz, DMSO-d^6): δ 7.58–7.53 (m, 1H, H$_{Ar}$), 7.46–7.38 (m, 6H, H$_{Ar}$), 7.30 (d, $^3J_{HH}$ = 7.2 Hz, 3H, H$_{Ar}$), 7.26 (t, $^3J_{HH}$ = 7.1 Hz, 3H H$_{Ar}$), 7.17 (t, $^3J_{HH}$ = 7.2 Hz, 2H, H$_{Ar}$), 7.15–7.12 (m, 1H, H$_{Ar}$), 7.10–7.05 (m, 2H, H$_{Ar}$), 6.98–6.83 (m, 9H, H$_{Ar}$), 6.78–6.74 (m, 2H, H$_{Ar}$), 6.61 (dd, 2 × $^3J_{HH}$ = 6.9, 4.8 Hz, 1H, H$_{Ar}$), 5.05 (d, $^2J_{HH}$ = 18.4 Hz, 1H, H7), 4.39 (d, $^2J_{HH}$ = 18.4 Hz, 1H, H7), 4.14 (d, $^3J_{HH}$ = 11.6 Hz, 1H, H6 or H6′), 3.83 (br, 2H, H6 or H6′), 3.74 (d, $^3J_{HH}$ = 10.8 Hz, 1H, H6 or H6′). ^{13}C NMR (151 MHz, DMSO-d^6): δ 148.0, 147.8, 139.8, 139.36, 138.6, 136.4, 136.3, 136.2, 135.6, 134.4, 134.3, 133.6 (d, J_{PC} = 19.7 Hz), 133.4 (d, J_{PC} = 19.5 Hz), 131.8, 130.9, 130.2, 128.9, 128.8, 128.8, 128.7, 128.5, 128.2 (d, J_{PC} = 3.9 Hz), 128.0, 127.6, 127.1, 126.9, 126.3, 125.4, 125.1, 124.8, 124.5, 124.4, 124.4, 123.8, 123.8, 123.7, 123.5, 123.0, 122.8, 121.1, 118.7, 115.5 (d, J_{PC} = 3.7 Hz, C1), 43.1 (C6 or C6′), 41.3 (C6 or C6′), 30.1 (d, J_{PC} = 24.3 Hz, C7). ^{31}P NMR (243 MHz, DMSO-d^6): δ −15.6.

[(η^5-Cp^{C}BzPPh$_2$)Ru(NCMe)$_2$]PF$_6$ (6): A total of 579 mg (0.66 mmol) of 5, 150 mg (0.30 mmol) of [(η^6-C$_6$H$_6$)RuCl$_2$]$_2$ and 140 mg (0.76 mmol) of KPF$_6$ were dissolved in 20 mL of acetonitrile and was stirred for 18 h at room temperature, while the color of the mixture changed from yellow to orange. The precipitated solid (TlCl, KCl) was filtered off and the volume of the solution was reduced to about 5 mL by removing the solvent under vacuum. By the addition of 30 mL of diethyl ether, an orange colored solid precipitated, which was filtered off, washed three times with 15 mL of diethyl ether and three times with 15 mL of n-pentane, and finally, dried under vacuum. Yield: 530 mg (80%) of an orange colored solid. Elemental analysis calcd. for C$_{54}$H$_{42}$F$_6$N$_2$P$_2$Ru (1036.98 g/mol): C 65.12, H 4.25, N 2.81; found: C 64.89, H 4.48, N 2.74.^{1}H NMR (400 MHz, CD$_3$CN): δ 7.63–7.58 (m, 2H, H$_{Ar}$), 7.52–7.49 (m, 3H, H$_{Ar}$), 7.48–7.42 (m, 3H, H$_{Ar}$), 7.42–7.34 (m, 4H, H$_{Ar}$), 7.31 (d, $^3J_{HH}$ = 7.7 Hz, 1H, H$_{Ar}$), 7.27–7.21 (m, 3H, H$_{Ar}$), 7.19–7.13 (m, 3H, H$_{Ar}$), 7.11–6.99 (m, 6H, H$_{Ar}$), 6.96–6.91 (m, 1H, H$_{Ar}$), 6.90–6.82 (m, 2H, H$_{Ar}$), 6.63 (td, $^3J_{HH}$ = 7.7, $^4J_{HH}$ = 0.8 Hz, 1H, H$_{Ar}$), 6.17 (d, $^3J_{HH}$ = 7.7 Hz, 1H, H$_{Ar}$), 4.47 (d, $^2J_{HH}$ = 13.1 Hz, 1H, H6 or H6′), 4.38 (dd, $^2J_{HH}$ = 18.0, 4 J PH = 6.8 Hz, 1H, H7), 4.18 (d, $^2J_{HH}$ = 12.8 Hz, 1H, H6 or H6′), 4.12 (d, $^2J_{HH}$ = 17.9 Hz, 1H, H7), 3.78 (t, $^2J_{HH}$ = 13.2 Hz, 2H, H6 or H6′), 1.96 (s, 6H, CH$_3$CN). ^{13}C NMR (101 MHz, CD$_3$CN): δ = 147.0, 146.8, 144.5, 143.9, 143.7 (d, J_{PC} = 2.5 Hz), 142.1, 135.1 (d, J_{PC} = 3.2 Hz), 134.9, 134.8 (d, J_{PC} = 3.2 Hz), 134.4, 133.8 (d, J_{PC} = 10.1 Hz), 133.2, 132.9 (d, J_{PC} = 10.7 Hz), 132.6, 132.5 (d, J_{PC} = 17.0 Hz), 132.0 (d, J_{PC} = 1.8 Hz), 131.9 (d, J_{PC} = 1.8 Hz), 131.2, 131.1 (d, J_{PC} = 1.9 Hz), 130.7 (d, J_{PC} = 1.7 Hz), 130.5, 130.3, 130.2, 129.5, 129.4, 129.3, 129.2, 129.1, 128.9, 128.5 (d, J_{PC} = 8.7 Hz), 128.4, 128.3, 128.3, 127.1, 126.5, 126.5, 126.3 (d, J_{PC} = 9.0 Hz), 126.0, 105.6 (d, J_{PC} = 3.5 Hz, C1), 92.4 (d, J_{PC} = 9.6 Hz, C2-C5), 92.0 (C2-C5), 87.8 (C2-C5), 75.0 (C2-C5), 41.9 (C6 or C6′), 41.0 (C6 or C6′), 30.6 (d, J_{PC} = 5.3 Hz, C7). ^{31}P NMR (162 MHz, CD$_3$CN): δ 51.7 (s, PPh$_2$), −144.6 (hept., $^1J_{PF}$ = 706.3 Hz, PF$_6^-$). ^{19}F NMR (376 MHz, CD$_3$CN): δ −72.9 (d, $^1J_{PF}$ = 706.1 Hz, PF$_6^-$). ESI-MS (CD$_3$CN): m/z found (calcd.)

810.13 (810.19, $[C_{50}H_{36}PRu(CH_3CN)]^+$), 769.09 (769.16, $[C_{50}H_{36}PRu]^+$). IR (ATR, cm^{-1}): ṽ 3056 w, 3005 w, 2963 w, 1926 w, 2853 w, 2038 vw, 1599 w, 1483 m, 1435 s, 1371 w, 1261 w, 1162 w, 1093 m, 1039 w, 1028 w, 950 w, 919 w, 876 m, 834 vs, 754 s, 747 s, 738 s, 719 m, 696 s, 674 m.

[Cp^C(BzPPh$_2$(AuCl)$_2$)] (**7**): The reaction was carried analogous to the synthesis of the gold(I) complex **4** using 239 mg (0.25 mmol) of the diphosphine derivative **3** and 171 mg (0.53 mmol) of (tht)AuCl. After stirring the mixture for 18 h at room temperature in 15 mL of CH$_2$Cl$_2$, the colorless precipitate was filtered of, washed four times with 15 mL of diethyl ether and then dried under vacuum. Prismatic single crystals containing one equivalent of toluene were obtained by slow diffusion of *n*-pentane into a saturated toluene solution. Yield: 313 mg (83%) of a colorless solid. Due to the poor solubility of compound **7** in all organic solvents, it was impossible to obtain a ^{13}C NMR spectrum. Elemental analysis calcd. for C$_{69}$H$_{52}$Au$_2$Cl$_2$P$_2$ (1407.94 g/mol): C 58.86, H 3.72; found: C 58.62, H 3.83%. ^{1}H NMR (400 MHz, CDCl$_3$): δ 7.61–7.56 (m, $^3J_{HH}$ = 7.0 Hz, 2H, H$_{Ar}$), 7.52–7.45 (m, 8H, H$_{Ar}$), 7.40–7.27 (m, 20H, H$_{Ar}$), 7.22–7.17 (m, $^3J_{HH}$ = 9.6, 3.9 Hz, 4H, H$_{Ar}$), 7.08–7.04 (m, 4H, H$_{Ar}$), 6.91–6.83 (m, 4H, H$_{Ar}$), 6.60 (d, $^3J_{HH}$ = 7.6 Hz, 2H, H$_{Ar}$), 4.01 (d, $^2J_{HH}$ = 18.2 Hz, 2H, H7 or H7′), 3.65 (d, $^2J_{HH}$ = 17.1 Hz, 2H, H7 or H7′), 3.51 (d, $^2J_{HH}$ = 12.6 Hz, 2H, H6 or H6′), 2.89 (d, $^2J_{HH}$ = 12.5 Hz, 2H, H6 or H6′). ^{31}P NMR (162 MHz, CDCl$_3$): δ 24.9. IR (ATR, cm^{-1}): ṽ 3058 w, 3015 w, 2948 w, 2925 w, 2842 w, 1588 w, 1480 m, 1438 s, 1310 w, 1261 m, 1179 w, 1101 s, 1068 m, 1029 m, 999 m, 903 w, 847 w, 793 m, 757 vs, 743 vs, 689 m.

[Cp^C(BzPPh$_2$)$_2$(PdCl(μ^2-Cl))$_2$] (**8**): 183 mg (0.19 mmol) of ligand **3** and 101 mg (0.39 mmol) of (CH$_3$CN)$_2$Pd(Cl)$_2$ were dissolved in 6 mL of CH$_2$Cl$_2$. The mixture was stirred for 18 h at room temp. Then, 10 mL of *n*-pentane were added. An orange colored solid precipitated, which was filtered off, washed three times with 5 mL of diethyl ether and three times with 5 mL of *n*-pentane and dried under vacuum. Yield: 197 mg (78%) of a temperature-sensitive orange colored solid, which was stored at T < 10 °C. Single crystals were obtained by slow diffusion of *n*-pentane into a saturated solution of the compound in toluene. Elemental analysis calcd. for C$_{69}$H$_{52}$Cl$_4$P$_2$Pd$_2$ (1297.77 g/mol): C 63.86, H 4.04; found: C 63.56, H 3.93%. ^{1}H NMR (400 MHz, CDCl$_3$): δ 7.87 (dd, $^3J_{HH}$ = 12.5, 7.6 Hz, 4H, H$_{Ar}$), 7.83–7.73 (m, 4H, 2 × H$_{Ar}$ and 2×H7 or H7′), 7.60 (d, $^3J_{HH}$ = 8.2 Hz, 2H, H$_{Ar}$), 7.55–7.50 (m, 2H, H$_{Ar}$), 7.45 (d, $^3J_{HH}$ = 7.5 Hz, 2H, H$_{Ar}$), 7.42–7.37 (m, 6H, H$_{Ar}$), 7.33–7.29 (m, 2H, H$_{Ar}$), 7.25–7.21 (m, 2H, H$_{Ar}$), 7.16–7.07 (m, 4H, H$_{Ar}$), 7.05–7.00 (m, 2H, H$_{Ar}$), 6.98–6.88 (m, 9H, H$_{Ar}$), 6.85 (dd, $^3J_{HH}$ = 6.1, 0.7 Hz, 1H, H$_{Ar}$), 6.83–6.78 (m, 2H, H$_{Ar}$), 6.62 (dd, $^3J_{HH}$ = 12.5, 7.9 Hz, 4H, H$_{Ar}$), 4.80 (d, $^2J_{HH}$ = 17.7 Hz, 2H, H7 or H7′), 3.89 (d, $^2J_{HH}$ = 12.3 Hz, 2H, H6 or H6′), 3.82 (d, $^2J_{HH}$ = 12.5 Hz, 2H, H6 or H6′). ^{13}C NMR (101 MHz, CDCl$_3$): δ 147.9, 144.3, 141.6 (d, J_{PC} = 10.2 Hz), 141.0, 140.6, 137.8 (d, J_{PC} = 11.6 Hz), 133.0 (d, J_{PC} = 9.6 Hz), 132.3, 131.8, 130.4, 129.1, 128.8, 128.7, 128.6, 128.5, 128.4, 128.3, 128.0, 127.7, 127.4, 127.3, 127.2, 126.9 (d, J_{PC} = 21.3 Hz), 126.7, 126.6, 126.4, 126.3 (d, J_{PC} = 3.5 Hz), 125.1, 63.8 (C1), 48.2 (d, J_{PC} = 11.4 Hz, C7 or C7′), 41.88 (C6 or C6′). ^{31}P NMR (162 MHz, CDCl$_3$): δ 25.4 (s). IR (ATR, cm^{-1}): ṽ 3056 m, 2945 w, 2892 w, 2836 w, 1589 w, 1480 m, 1461 w, 1435 w, 1327 w, 1188 w, 1161 w, 1092 s, 999 w, 902 w, 850 w, 757 s, 741 s, 706 s, 690 s.

[Cp^C(BzPPh$_2$)$_2$)(Rh(μ^2-Cl(CO))$_2$] (**9**): A total of 107 mg (0.11 mmol) of ligand **3** and 45.3 mg (0.11 mmol) of [(CO)$_2$Rh(μ^2-Cl)]$_2$ was dissolved in 6 mL of dichloromethane. The mixture was stirred for 2 h at room temperature. Then, 10 mL of n-pentane was added. A bright yellow solid precipitated, which was filtered off, washed three times with 5 mL of *n*-pentane and dried under vacuum. Yield: 122 mg (85%) of a bright yellow microcrystalline solid, which was stored at temperatures below 10 °C to prevent decomposition. Yellow prismatic single crystals were obtained by slow diffusion of *n*-pentane into a saturated solution of the compound in toluene. Elemental analysis calcd. for C$_{71}$H$_{52}$Cl$_2$O$_2$P$_2$Rh$_2$ (1275.84 g/mol): C 66.84, H 4.11; found: C 67.10, H 4.42%. ^{1}H NMR (600 MHz, CDCl$_3$): δ 8.47 (d, $^3J_{HH}$ = 7.8 Hz, 1H, H$_{Ar}$), 7.80–7.70 (m, 2H, H$_{Ar}$), 7.65 (dd, $^3J_{HH}$ = 7.0, $^3J_{HH}$ = 5.6 Hz, 1H, H$_{Ar}$), 7.55–7.51 (m, $^3J_{HH}$ = 11.1, $^3J_{HH}$ = 4.1 Hz, 1H, H$_{Ar}$), 7.47–7.34 (m, 12H, H$_{Ar}$), 7.32 (t, $^3J_{HH}$ = 7.5 Hz, 1H, H$_{Ar}$), 7.31–7.25 (m, 6H, H$_{Ar}$), 7.24–7.21 (m, 2H, H$_{Ar}$), 7.18 (t, $^3J_{HH}$ = 7.7 Hz, 1H, H$_{Ar}$), 7.12–7.04 (m, 5H, 3 × H$_{Ar}$ and 2 × H7 or H7′), 7.02–6.95 (m, 5H, H$_{Ar}$), 6.90 (d, $^3J_{HH}$ = 7.6 Hz, 1H, H$_{Ar}$), 6.89–6.82 (m,

$^3J_{HH}$ = 8.3 Hz, 2H, H$_{Ar}$), 6.81 (d, $^3J_{HH}$ = 7.9 Hz, 1H, H$_{Ar}$), 6.71 (dd, $^3J_{HH}$ = 11.8, $^3J_{HH}$ = 7.7 Hz, 1H, H$_{Ar}$), 6.69–6.61 (m, 4H, H$_{Ar}$), 4.03 (d, $^2J_{HH}$ = 12.7 Hz, 1H, H6 or H6′), 3.87 (d, $^2J_{HH}$ = 12.8 Hz, 1H, H6 or H6′), 3.77–3.69 (m, 2H, H6 or H6′ and H7 or H7′), 3.64 (d, $^2J_{HH}$ = 12.7 Hz, 1H, H6 or H6′), 3.45 (d, $^2J_{HH}$ = 18.9 Hz, 1H, H7 + H7′). ^{13}C NMR (151 MHz, CDCl$_3$): δ 183.2 (dd, $^1J_{RhC}$ = 78.0, $^2J_{PC}$ = 20.6 Hz, CO), 182.3 (dd, $^1J_{RhC}$ = 78.8, $^2J_{PC}$ = 21.1 Hz, CO), 148.0 (d, J_{PC} = 38.4 Hz), 144.6 (d, J_{PC} = 32.7 Hz), 142.7 (d, J_{PC} = 10.0 Hz), 142.2 (d, J_{PC} = 9.6 Hz), 141.9, 141.1, 141.0, 140.7, 136.5 (b), 135.6 (d, J_{PC} = 11.6 Hz), 133.8 (d, J_{PC} = 10.3 Hz), 132.8, 132.6, 132.5, 132.5, 132.2, 132.1, 132.0, 132.0, 131.8, 131.5, 131.0 (d, J_{PC} = 1.5 Hz), 131.0, 130.9, 130.8 (d, J_{PC} = 1.8 Hz), 130.2, 130.0, 129.9, 129.4, 129.2, 129.1, 128.9, 128.88, 128.8, 128.8, 128.6, 128.5, 128.3, 128.2, 128.1, 128.0, 127.8 (d, J_{PC} = 4.3 Hz), 127.7 (d, J_{PC} = 4.2 Hz), 127.5, 127.5, 127.4, 126.9, 126.7, 126.7, 126.1, 126.0, 126.0, 125.7, 125.6, 125.3, 124.9, 63.5 (C1), 46.5 (d, J_{PC} = 15.5 Hz, C7 or C7′), 44.9 (d, J_{PC} = 13.8 Hz, C7 or C7′), 42.1 (C6 or C6′), 41.7 (C6 or C6′). ^{31}P NMR (162 MHz, CDCl$_3$): δ 39.8 (d, $^1J_{RhP}$ = 171.5 Hz), 35.9 (d, $^1J_{RhP}$ = 175.0 Hz). IR (ATR, cm^{-1}): ṽ 3055 w, 3017 w, 2948 w, 2895 w, 2863 w, 2839 w, 1991 vs, 1975 vs, 1590 w, 1482 m, 1462 w, 1435 s, 1327 w, 1310 w, 1187 m, 1160 m, 1119 m, 1093 s, 999 m, 902 m, 855 w, 764 m, 753 s, 737 vs, 720 m, 690 vs, 677 s, 662 m.

[Cp^C(BzPPh$_2$)$_2$(RhCl(CO))] (**10**): A total of 217 mg (0.23 mmol) of ligand **3** and 46.1 mg (0.12 mmol) of [(CO)$_2$Rh(μ^2-Cl)]$_2$ were dissolved in 10 mL of dichloromethane. The mixture was stirred for 18 h at 45 °C. Then, 10 mL of *n*-pentane was added. A shiny yellow solid precipitated, which was filtered off, was washed three times with 5 mL of diethyl ether and three times with 5 mL of *n*-pentane and was dried under vacuum. Yield: 157 mg (62%) of a shiny yellow solid. Yellow prismatic single crystals were obtained by slow diffusion of *n*-pentane into a saturated solution of the compound in toluene. Elemental analysis calcd. for C$_{70}$H$_{52}$ClOP$_2$Rh (1109.49 g/mol): C 75.78, H 4.72; found: C 75.48, H 4.68. ^{1}H NMR (400 MHz, CDCl$_3$): δ 8.11–8.05 (m, 2H, H$_{Ar}$), 7.87 (s, 1H, H$_{Ar}$), 7.58–7.52 (m, 2H, H$_{Ar}$), 7.48–7.44 (m, 4H, H$_{Ar}$), 7.41–7.28 (m, 9H, H$_{Ar}$), 7.24–7.18 (m, 3H, H$_{Ar}$), 7.17–7.05 (m, 5H, H$_{Ar}$), 7.04–6.86 (m, 7H, H$_{Ar}$), 6.85–6.71 (m, 5H, H$_{Ar}$), 6.65–6.59 (m, 2H, H$_{Ar}$), 6.57–6.48 (m, 4H, H$_{Ar}$), 6.23 (d, $^2J_{HH}$ = 16.8 Hz, 1H, H7 or H7′), 5.56 (d, $^2J_{HH}$ = 16.7 Hz, 1H, H7 or H7′), 4.94 (d, $^3J_{HH}$ = 13.5 Hz, 1H, H7 or H7′), 4.09 (q, $^3J_{HH}$ = 12.3 Hz, 2H, H6 or H6′), 3.55 (d, $^2J_{HH}$ = 12.4 Hz, 1H, H7 or H7′), 3.37 (d, $^2J_{HH}$ = 12.4 Hz, 2H, 1 × H6 or H6′ and 1 × H7 or H7′). ^{13}C NMR (101 MHz, CDCl$_3$): δ 190.0 (d, $^1J_{RhC}$ = 78.5 Hz), 149.3, 147.7, 143.7 (d, $^{xx}J_{PC}$ = 15.8 Hz), 143.6 (d, J_{PC} = 3.5 Hz), 142.3, 141.2, 140.8, 139.8, 138.0, 137.9, 137.6, 137.2, 136.8, 136.0 (d, J_{PC} = 11.9 Hz), 135.4, 135.1, 135.0 (d, J_{PC} = 1.8 Hz), 134.5, 133.9 (d, J_{PC} = 3.7 Hz), 133.7 (d, J_{PC} = 2.6 Hz), 133.3, 133.2, 132.9, 132.8 (d, J_{PC} = 2.0 Hz), 132.4 (d, J_{PC} = 1.6 Hz), 132.2, 132.1, 132.0, 131.9, 131.4 (d, J_{PC} = 3.8 Hz), 131.0, 130.9, 130.5, 130.2 (d, J_{PC} = 4.0 Hz), 129.8 (d, J_{PC} = 1.9 Hz), 129.2, 128.7, 128.6, 128.5 (d, J_{PC} = 2.5 Hz), 128.4, 128.1, 128.0, 128.0, 127.8, 127.6, 127.6, 127.2, 127.1, 126.9 (d, J_{PC} = 2.2 Hz), 126.5, 126.3, 126.2, 126.1, 125.7, 125.0, 124.8, 66.9 (C1), 49.3 (C7 or C7′), 46.8 (d, J_{PC} = 8.7 Hz, C7 or C7′), 42.9 (C6 or C6′), 41.8 (C6 or C6′). ^{31}P NMR (162 MHz, CDCl$_3$): δ 35.6 (dd, $^2J_{PP}$ = 340.2, $^1J_{RhP}$ = 127.5 Hz), 26.2 (dd, $^2J_{PP}$ = 339.7, $^1J_{RhP}$ = 123.2 Hz). ESI-MS (CD$_2$Cl$_2$): *m/z* found (calcd.) 1073.39 (1073.25, [C$_{70}$H$_{52}$OP$_2$Rh]$^+$). IR (ATR, cm^{-1}): ṽ 3056 w, 3010 w, 2947 w, 2837 w, 1961 vs, 1587 w, 1573 w, 1480 m, 1466 m, 1449 m, 1434 s, 1339 w, 1312 w, 1279 w, 1187 m, 1157 w, 1120 w, 1090 s, 1044 w, 1028 w, 999 w, 947 w, 908 w, 872 w, 841 w, 781 m, 748 vs, 690 vs, 673 m.

X-rax structure analysis: Crystal data and refinement parameters were collected and are presented in Tables 1 and 2. All structures were solved using direct method of SIR92 [24], completed by subsequent difference Fourier syntheses, and refined by full-matrix least-squares procedures [25]. Semi-empirical absorption correction from equivalents (Multiscan) was applied to ligand **3**, while analytical numeric absorption correction was carried out to all other complexes [26]. All non-hydrogen atoms were refined with anisotropic displacement parameters. All hydrogen atoms were placed in calculated positions and refined using a riding model. In the original solved structures of complexes **5**, **7**, **8** and **9**, severely disordered and/or partially occupied solvents were also found. The quality of the measured samples and the corresponding collected raw data were limited. Even with the help of lots of restraints, these disorders could not be treated satisfyingly. To obtain a better understanding of the main structures, the SQUEEZE process integrated in PLATON was used. Additionally, detailed information has been posted in the corresponding final CIF files. CCDC 2247305-2247311 contain the supplementary crystallographic data for this paper. These data can be obtained free of charge from The Cambridge Crystallographic Data Centre via www.ccdc.cam.ac.uk/data_request/cif (accessed on 4 April 2023).

Table 1. Crystallographic data, data collection and refinement.

	3	4	5
empirical formula	$C_{83}H_{68}P$	$C_{57}H_{45}AuClP$	$C_{50}H_{36}PTl$
formula weight	1127.31	993.31	872.13
crystal size [mm]	0.408 × 0.307 × 0.239	0.244 × 0.106 × 0.069	0.313 × 0.093 × 0.063
T [K]	293(2)	150(2)	150(2)
λ [Å]	1.54184	1.54184	1.54184
crystal system	triclinic	triclinic	monoclinic
space group	$P\bar{1}$	$P\bar{1}$	$P2_1/n$
a [Å]	14.1471(5)	10.3433(3)	18.7154(2)
b [Å]	14.5119(4)	15.2162(6)	12.2606(1)
c [Å]	17.7514(6)	15.5558(7)	20.4489(2)
α [°]	107.920(3)	106.743(4)	90
β [°]	109.661(3)	107.515(3)	97.058(1)
γ [°]	93.028(3)	96.040(3)	90
V [Å^3]	3215.3(2)	2185.84(16)	4656.69(8)
Z	2	2	4
ρ calcd. [g·cm^{-3}]	1.164	1.509	1.244
μ [mm^{-1}]	0.950	7.519	7.198
Θ-range [°]	3.251–62.736	3.608–62.729	3.417–62.775
refl. coll.	24,391	15,048	35,775
indep. refl.	10,246 [R_{int} = 0.0229]	6947 [R_{int} = 0.0481]	7447 [R_{int} = 0.0257]
data/restr./param.	10246/0/768	6947/0/542	7447/0/469
final R indices [I > 2σ(I)] [a]	0.0522, 0.1455	0.0367, 0.0946	0.0171, 0.0402
R indices (all data)	0.0573, 0.1524	0.0381, 0.0961	0.0179, 0.0407
GooF [b]	1.036	1.082	1.029
Δρmax/min (e·Å^{-3})	1.687/−0.283	1.444/−1.808	0.391/−0.266

[a] $R1 = \Sigma||F_o| - |F_c||/\Sigma|F_o|$, $\omega R2 = [\Sigma(F_o^2 - F_c^2)^2/\Sigma\omega F_o^2]^{1/2}$. [b] $GooF = [\Sigma\omega(F_o^2 - F_c^2)^2/(n-p)]^{1/2}$.

Table 2. Crystallographic data, data collection and refinement.

	7	8	9	10
empirical formula	$C_{69}H_{52}Au_2Cl_2P_2$	$C_{69}H_{52}Cl_4P_2Pd_2$	$C_{71}H_{52}Cl_2O_2P_2Rh_2$	$C_{70}H_{52}ClOP_2Rh$
formula weight	1407.88	1297.64	1275.78	1109.41
crystal size [mm]	0.216 × 0.095 × 0.079	0.199 × 0.101 × 0.044	0.272 × 0.151 × 0.137	0.233 × 0.121 × 0.059
T [K]	150(2)	150(2)	150(2)	150(2)
λ [Å]	1.54184	1.54184	1.54184	1.54184
crystal system	triclinic	monoclinic	triclinic	monoclinic
space group	$P\bar{1}$	$P2/n$	$P\bar{1}$	$P2_1/n$
a [Å]	13.1508(3)	21.0446(7)	13.2483(3)	13.3016(3)
b [Å]	13.5694(4)	13.2823(4)	14.1374(3)	21.0013(4)
c [Å]	22.2637(5)	23.4004(9)	20.4213(4)	20.0864(4)
α [°]	76.934(2)	90	89.220(2)	90
β [°]	87.525(2)	107.894(4)	73.556(2)	109.243(2)

Table 2. Cont.

	7	8	9	10
γ [°]	64.608(3)	90	75.287(2)	90
V [Å^3]	3489.67(17)	6224.5(4)	3541.10(14)	5297.7(2)
Z	2	4	2	4
ρ calcd. [g·cm^{-3}]	1.340	1.385	1.197	1.391
μ [mm^{-1}]	9.186	7.028	5.189	3.997
Θ-range [°]	3.705–62.698	3.327–62.748	3.238–62.724	3.140–62.754
refl. coll.	27,287	24,022	49,439	21,589
indep. refl.	11,102 [R$_{int}$ = 0.0308]	9919 [R$_{int}$ = 0.0550]	11,257 [R$_{int}$ = 0.0564]	8461 [R$_{int}$ = 0.0294]
data/restr./param.	11102/0/676	9919/0/695	11257/0/713	8461/0/676
final R indices [I > 2σ(I)] ^{a}	0.0296, 0.0721	0.0510, 0.1388	0.0310, 0.0790	0.0321, 0.0852
R indices (all data)	0.0349, 0.0739	0.0623, 0.1466	0.0335, 0.0807	0.0362, 0.0886
GooF ^{b}	1.037	1.082	1.023	1.032
$\Delta\rho$max/min (e·Å^{-3})	1.310/−1.262	1.253/−0.718	0.734/−0.840	0.713/−0.518

^{a} $R1 = \Sigma||F_o| - |F_c||/\Sigma|F_o|$, $\omega R2 = [\Sigma(F_o^2 - F_c^2)^2/\Sigma\omega F_o^2]^{1/2}$. ^{b} $GooF = [\Sigma\omega(F_o^2 - F_c^2)^2/(n-p)]^{1/2}$.

3. Results

3.1. Ligand Synthesis and Characterization

Phosphine functionalized derivatives of Cp^{C}H (**1**) were obtained by the deprotonation of the compound with *n*-butyl lithium and treatment of the intermediately generated LiCp^C with 2-[(chloromethyl)phenyl](diphenyl)phosphine (Scheme 2) [22,23] The process follows a protocol published by Li et al. in 2019 for the synthesis of a phosphine-functionalized tetraethylcyclopentadiene [27]. Astonishingly, there are almost no other reports on cyclopentadienyl compounds with a 2-diphenylphosphanylphenylmethyl unit. Knochel et al. reported the synthesis of chiral 1-(S$_{Fc}$)-diphenylphosphanyl-2-(o-diphenylphosphanylphenylmethyl)ferrocene via a multi-step process starting with the Friedel–Crafts acylation of ferrocene with 2-bromobenzoyl chloride [28].

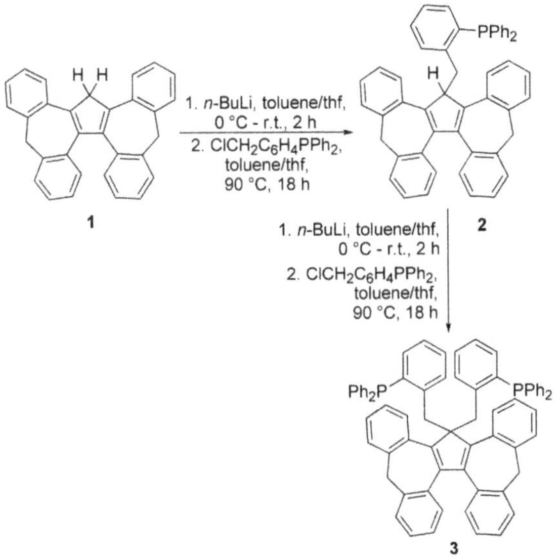

Scheme 2. Synthesis of ligands **2** and **3**.

Reacting the substrates in a 1:1 ratio with a slight excess of *n*-butyl lithium allowed us to isolate the mono-substituted derivative Cp^{C}HBzPPh$_2$ (**2**) in 59% yield after purification by column chromatography. As a minor product, the bi-functionalized compound

Cp^C(BzPPh$_2$)$_2$ (**3**) was obtained in 16% yield. This speaks for a protonation/deprotonation equilibrium between Cp^{C}H (**1**), LiCp^C, Cp^{C}HBzPPh$_2$ (**2**) and LiCp^{C}BzPPh$_2$ and a rather high reactivity of LiCp^{C}BzPPh$_2$ towards the electrophile. Accordingly, the treatment of phosphine **2** with another equivalent of *n*-butyl lithium and the electrophile gave the diphosphine **3** in 83% yield.

While the ^{31}P NMR spectra of **2** and **3** show rather similar chemical shifts of expected values [27,29], their ^{1}H and ^{13}C NMR spectra differ largely according to the symmetry of the compounds. In the ^{1}H NMR spectrum of C$_2$-symmetric **3**, there are four doublets in the aliphatic region, which are assigned to the diastereotopic protons of the two chemically inequivalent methylene groups of the molecule. In contrast, the ^{1}H and ^{13}C NMR spectra of C$_1$-symmetric **2** are more complicated. In the aliphatic region of the ^{1}H NMR spectrum, for example, there are four resonances for the two, now magnetically inequivalent methylene groups bridging the phenyl groups in the "wings" of the Cp backbone. Three further resonances (ABB′ spin system) are assigned to the proton at the Cp ring and the protons of the methylene group linking the Cp ring and the triphenyl phosphine moiety. These methylene protons can easily be identified by their coupling pattern, including a small long-range coupling to the phosphorous atom ($^4J_{PH}$ = 3.3 Hz) [29].

In the literature, there are only a few reports of 1,1-diphosphine-functionalized cyclopentadiene derivatives. In 1965, Keough and Grayson succeeded in adding fluorene to vinylphosphonium salts, resulting in dicationic phosphonium species [30]. Inagaki et al. published fluorene derivatives with two phosphinyl moieties directly bound to the sp^3 carbon atom of the fluorene backbone [31]. In addition, there are two patents dealing with such compounds [32,33].

By slow diffusion of *n*-pentane into a saturated toluene solution of compound **3**, single crystals suitable for an X-ray structure analysis could be obtained. Compound **3** crystallizes as colorless prisms of the triclinic space group P$\bar{1}$ with two equivalents of toluene in the unit cell. The two phosphine units are found arranged between the wings of the ligand, in an almost orthogonal position to the central cyclopentadiene ring. Figure 1 shows the structure of **3** and summarizes relevant structural parameters.

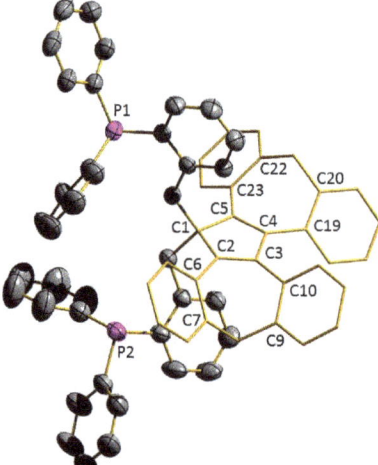

Figure 1. Molecular structure of compound **3** in the solid state. The two co-crystallizing molecules of toluene and all hydrogen atoms are omitted for clarity. The central Cp^C system is represented in stick-style for clarity, ellipsoids are drawn at the 50% level. Characteristic bond lengths [Å], angles [°] and dihedral angles [°]: C1-C2 1.525(2), C1-C5 1.532(2), C2-C3 1.358(2), C3-C4 1.476(2), C4-C5 1.359(3), P1···C1···P2 99.25(3), C3-C2-C6-C7 40.5(3), C2-C3-C10-C9 -48.2(3), C4-C5-C23-C22 36.8(3), C5-C4-C19-C20 -45.7(3).

3.2. Transition Metal Complexes

While ligand **3** is expected to act like a typical diphosphine, the coordination chemistry of mono-phosphine **2** allows more variations: it may either act as a typical monodentate phosphine, as an η^5-coordinating Cp-type ligand or may exhibit a combination of these two modes of metal–ligand interactions. To approve the first type of coordination, ligand **2** was treated with [(tht)AuCl] (tht = tetrahydrothiophene, Scheme 3). Gold(I) prefers linear coordination, which avoids involving the Cp ring of **2** in its coordination environment. The according gold(I) complex **4** was obtained in 78% yield as a colorless solid, which is stable under ambient conditions.

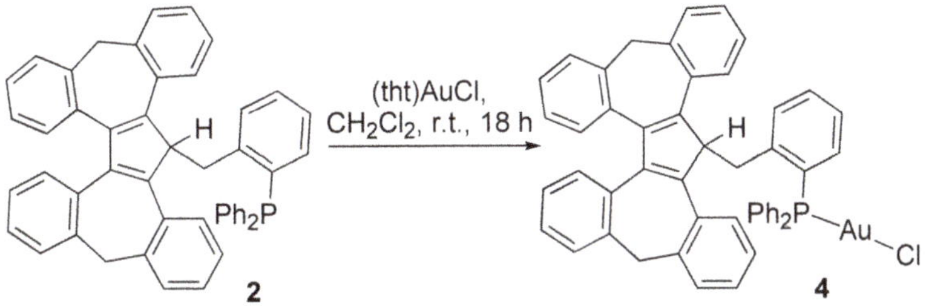

Scheme 3. Synthesis of the mononuclear gold(I) complex **4**.

While for typical ^{31}P NMR resonances of triarylposphine gold(I) chloride, complexes chemical shift of around 32 ppm are reported [34,35], the ^{31}P NMR resonance of compound **4** appears at 24.4 ppm. We assign this to some shielding effects of the aromatic environment of the Cp^C backbone. The coordination of the gold(I) site in addition leads to a better separation of the aliphatic ^{1}H NMR resonances compared to the metal-free ligand **2** (see the Supplementary Materials). Here, a triplet at 4.65 ppm ($^3J_{HH}$ = 4.6 Hz) is assigned to the proton located at the five-membered Cp ring.

By the slow diffusion of *n*-pentane into a saturated toluene solution of compound **4**, single crystals suitable for an X-ray structure analysis could be obtained. Compound **4** crystallizes as colorless prisms of the triclinic space group $P\bar{1}$ with one equivalent of toluene in the unit cell. Figure 2 shows the structure of **4** and summarizes relevant structural parameters.

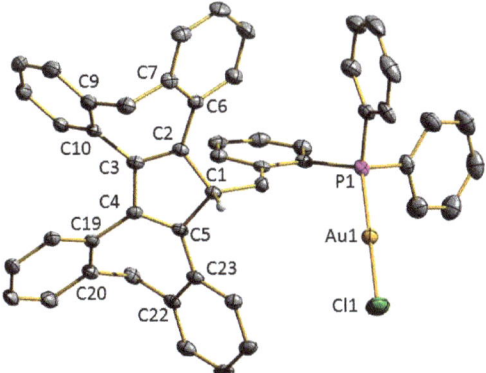

Figure 2. Molecular structure of compound **4** in the solid state. The co-crystallizing molecule of toluene and all hydrogen atoms except the hydrogen atom bound to C1 are omitted for clarity. Ellipsoids are drawn at the 50% level. Characteristic bond lengths [Å], angles [°] and dihedral angles [°]: Au1-P1 2.2427(10), Au1-Cl1 2.2865(10), C1-C2 1.516(5), C1-C5 1.515(5), C2-C3 1.356(6), C3-C4 1.486(5), C4-C5 1.369(6), P1-Au1-Cl1 179.42(4), C3-C2-C6-C7 42.61(7), C4-C5-C23-C22 36.71(7), C2-C3-C10-C9 −45.98(7), C5-C4-C19-C20 −42.35(7).

The crystal structure complex **4** shows the typical linear coordination geometry of gold(I) with an the P1-Au1-Cl1 angle of 179.42(4)°. The measured P1-Au1 (2.2427(10) Å) and Au1-Cl1 (2.2865(10) Å) distances are comparable to the data of (triphenylphosphine)gold(I) chloride and related gold(I) complexes [34,36]. In contrast to many other gold(I) complexes [37–39], there is no hint for a d^{10}–d^{10} interaction between the gold(I) sites of neighboring molecules. The phosphine substituent occupies an almost orthogonal position to the central five-membered ring of the Cp^C backbone, minimizing steric repulsions.

We recently published the synthesis of a series of late transition metal complexes wherein $TlCp^C$ is used as the transferring reagent for the $[Cp^C]^-$ ligand [19]. To achieve similar chemistry with a phosphine-functionalized Cp^C derivative, compound **2** was deprotonated in dry benzene with stoichiometric amounts of thallium(I) ethanolate in the absence of light (Scheme 4).

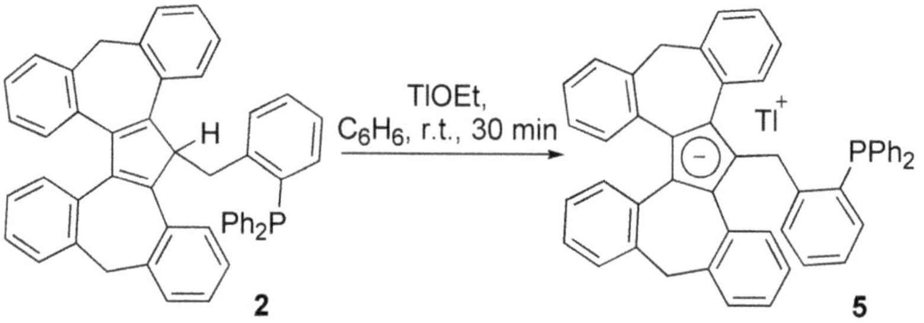

Scheme 4. Synthesis of the Cp^C thallium(I) complex **5**.

The thallium(I) compound **5** precipitated immediately after the two reactants were combined and was isolated by filtration with a yield of 83% as a pale yellow solid that can be handled in the air. The ^{1}H NMR spectrum of **5** differs clearly from the protonated ligand **2**. In addition to two doublets for the protons of the methylene group linking the Cp ring and the phosphine site ($^2J_{HH}$ = 18.4 Hz), there are four partially superimposed and slightly broadened resonances that are assigned to the four diastereotopic protons methylene groups in the "Cp^C-wings". This speaks for a C_1 symmetric structure on the NMR time scale and thus for a localization of the thalium(I) cation on one side of the Cp ring. The singlet at −15.56 ppm in the ^{31}P NMR spectrum is only very weakly shifted towards the lower field compared to the signal of the metal-free ligand **2**.

Crystallization from a saturated solution in the provided yellow, needle-like crystals with the monoclinic space group $P2_1/n$. Figure 3 shows the structure of **5** and summarizes relevant structural parameters.

There are numerous examples in the literature describing the coordination of thallium(I) to aromatic compounds [40–46], while only a few structurally characterized systems with thallium(I)–phosphorus coordination have been published. In 1974, Nakayama et al. published a gas-phase study on a series of thallium(I) complexes and stated that "No thallium(I) complexes with bismuth, arsenic nor phosphorus ligands has been isolated from a liquid phase so far." [47]. This changed during the past two decades, in particular due to the use of weakly coordinating anions [48,49]. The fact that thallium(I) phosphine complexes are hardly accessible as long as there is an alternative for the metal cation is confirmed by the solid-state structure of compound **5**. Compound **5** forms chains with thallium(I) centres bridging two $(Cp^C)^{-1}$ moieties, as shown in Figure 3, and there is no interaction with the phosphine site. In addition to the coordination to the Cp site, thallium(I) also coordinates to two of the six-membered rings, the phenylene ring bound to the phosphorus atom and one of the phenylene rings of the Cp^C moiety. Since the covalency of the bonds between the thallium(I) and the π-donating moieties is low, the Tl-C bond lengths differ largely as documented in the caption of Figure 3 exemplarily for the Cp unit. This allows the adoption of the positioning of the cation to the steric requirements of the solid-state structure. According to the electrostatic interaction between the Cp site and the thallium(I) cation, the Cp-Tl distance (2.6131(4) Å) is much shorter than the Ar-Tl distances (3.3139(4) and 3.2030(4) Å).

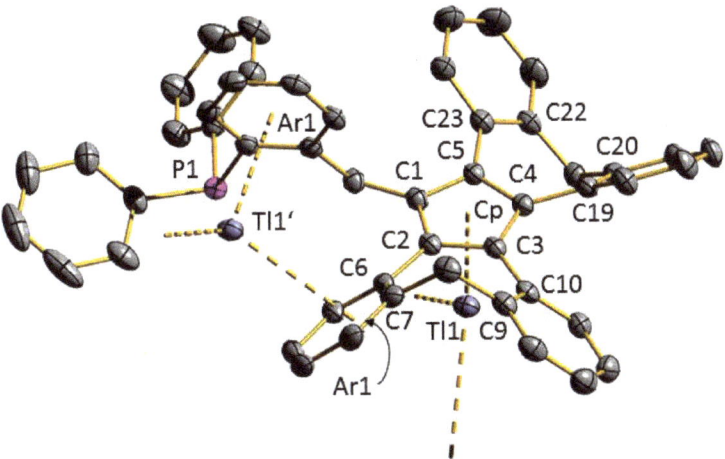

Figure 3. Molecular structure of compound **5** in the solid state. All hydrogen atoms are omitted for clarity, ellipsoids are drawn at the 50% level. Characteristic bond lengths [Å], angles [°] and dihedral angles [°]: C1-Tl1 2.8405(18), C2-Tl1 2.8137(18), C3-Tl1 2.8799(17), C4-Tl1 2.9341 (17), C5-Tl1 2.9342 (17), Cp-Tl1 2.6131(3), Ar1-Tl2 3.3139(4), Ar2-Tl2 3.2030(4), C1-C2 1.419(3), C1-C5 1.418(3), C2-C3 1.433(3), C3-C4 1.428(3), C4-C5 1.431(3), Cp-Tl1-Ar1 124.770(9), Cp-Tl1-Ar2 132.793(11), Ar1-Tl-Ar2 95.279(10), C3-C2-C6-C7 41.8(3), C4-C5-C23-C22 44.0(3), C2-C3-C10-C9 −37.3(2), C5-C4-C19-C20 −42.8(3); Cp denotes the centroid of the five-membered cyclopentadienyl ring. Ar1 and Ar2 denote the centroids of the two six-membered rings, which undergo coordination to the thallium(I) cation.

In 2003, Salzer et al. developed a protocol for the preparation of chiral cyclopentadienide ruthenium(II) complexes with an *ansa*-bridging phosphine group and two labile bound acetonitrile ligands) [50,51]. They started from cyclopentadiene derivatives bearing a chiral phosphine substituent and [RuCl(μ²-Cl)(η³:η³-C$_{10}$H$_{16}$)]$_2$, used Li$_2$CO$_3$ as the base and KPF$_6$ to provide a weakly coordinating counter anion. Due to our good experiences with TlCp^C as a [Cp^C]$^{-1}$ transfer reagent [19], we applied a combination of compound **5** and the ruthenium(II) precursor [(η⁶-C$_6$H$_6$)RuCl$_2$]$_2$ to generate an analogous Cp^C derivative (Scheme 5).

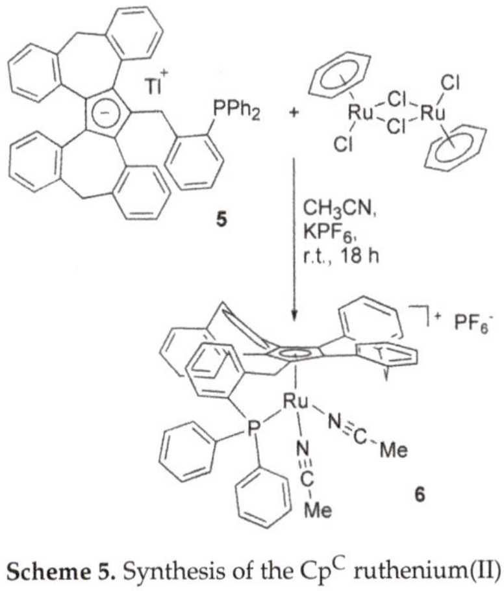

Scheme 5. Synthesis of the Cp^C ruthenium(II) complex **6**.

The reaction was carried out at room temperature in acetonitrile in the presence of a stoichiometric amount of KPF$_6$. After work-up, compound **6** was isolated in 80% yield as an orange-colored solid, which was characterized by means of NMR and IR spectroscopy, mass spectrometry and elemental analysis. Tiny crystals of insufficient quality were obtained by slow diffusion of diethylether into a solution of compound **6** in acetonitrile. The presence of the two acetonitrile ligands was confirmed by a singlet resonance at 1.96 ppm in the ^{1}H NMR spectrum and by an absorption band in the IR spectrum at 2038 cm^{-1}. The resonances of the four diastereotopic protons of the two methylene units in the "wings" of the Cp^C ligand are identified by their $^2J_{HH}$ coupling constants of about 13 Hz. For the two diastereotopic protons of the third methylene unit, a $^2J_{HH}$ coupling constant of 18 Hz is observed. One of these two protons shows an additional coupling ($^4J_{PH}$ = 6.8 Hz) to the phosphorus atom. The chemical shift of the ^{31}P resonance of 51.7 ppm unambiguously proves the coordination of the phosphine site to the ruthenium(II) center. A peak at m/z = 810 in the ESI mass spectrum corresponds to the mass of the fragment [M-CH$_3$CN]$^+$ a second peak at m/z = 769 is assigned to the fragment [M-2 × CH$_3$CN]$^+$.

As mentioned above, the diphosphine functionalized Cp^C derivative **3** can only act as a typical diphosphine donor. However, the two phosphine sites are linked by a chain of overall seven carbon atoms allowing either a chelating coordination of the diphosphine to one metal site or a bridging coordination to two metal sites. The simplest way of dealing with these alternatives is by coordination of two equivalents of gold(I) to compound **3**. The according gold(I) complex **7** was thus obtained in 83% yield in an analogous way as compound **4** using [(tht)AuCl] as the gold source (Scheme 6).

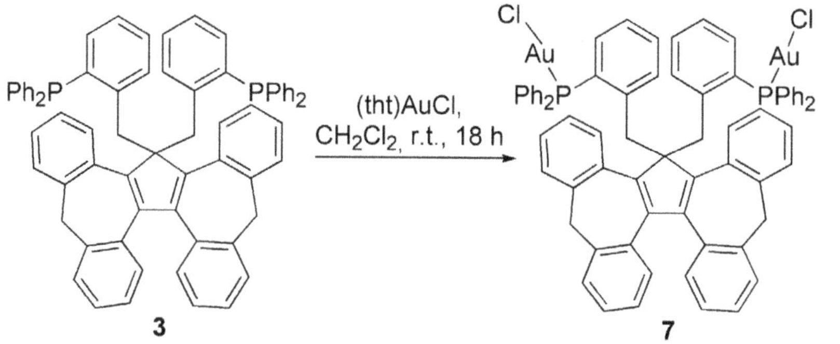

Scheme 6. Synthesis of the dinuclear gold(I) complex **7**.

In contrast to the mono nuclear gold(I) complex **4**, the dinuclear compound **7** is C$_2$ symmetric, which largely simplifies the ^{1}H NMR spectrum, leading to only four resonances for the diastereotopic protons of the two chemically inequivalent types of methylene groups with coupling constants of about 12.5 resp. 17.5 Hz. The ^{31}P NMR resonance at 24.9 ppm is close to the value obtained for compound **4** (24.4 ppm).

By the slow diffusion of n-pentane into a saturated toluene solution of compound **7**, single crystals suitable for an X-ray structure analysis could be obtained. Compound **7** crystallizes as colorless prisms of the triclinic space group P$\bar{1}$. Figure 4 shows the structure of **7** and summarizes relevant structural parameters.

Typical for gold(I) complexes, the two gold sites are coordinated almost linearly. To minimize steric interactions, in particular of the diphenylphosphinyl units, the two P-Au-Cl units point into opposite directions, which excludes an intramolecular Au···Au interaction. The bond parameters are in the same range as measured for complex **4**.

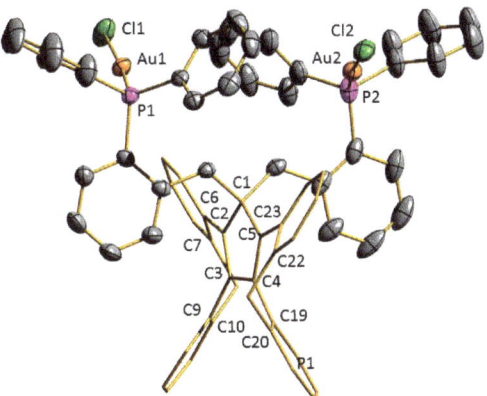

Figure 4. Molecular structure of compound **7** in the solid state. All hydrogen atoms are omitted for clarity. The central Cp^C system is represented in stick-style for clarity, ellipsoids are drawn at the 50% level Characteristic bond lengths [Å], angles [°] and dihedral angles [°]: Au1-Cl1 2.2878(13), Au2-Cl2 2.2807(13), Au1-P1 2.2352(11), Au2-P2 2.2295(13), C1-C2 1.532(6), C2-C3 1.358(6), C3-C4 1.467(7), C4-C5 1.367(6), C1-C5 1.530(6), P1-Au1-Cl1 171.50(4), P2-Au2-Cl2 173.53(5), C3-C2-C6-C7 33.7(7), C4-C5-C23-C22 −34.8(6), C2-C3-C10-C9 49.9(10), C5-C4-C19-C20 47.5(6).

In 2006 Gelman et al. obtained a mono nuclear chelate complex as well as a dinuclear, chlorido-bridged palladium(II) complex by the treatment of the diphosphine ligand 1,8-bis(diisopropylphosphino)triptycen with different amounts of [PdCl$_2$(CH$_3$CN)$_2$] [52,53]. In a series of analogue reactions, ligand **3** was reacted with different amounts of [PdCl$_2$(CH$_3$CN)$_2$]. In contrast to the results of Gelman et al., only the dinuclear palladium complex **8** could be isolated, independent from the ligand-to-metal ratio and from solvent and temperature. By applying a slight excess of the metal source, the best yield was obtained (78%, Scheme 7).

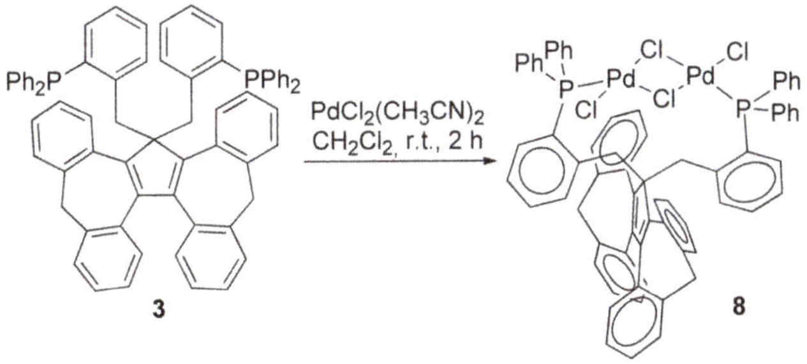

Scheme 7. Synthesis of the dinuclear, chlorido-bridged palladium(II) complex **8**.

^{1}H and ^{13}C NMR spectra of complex **8** clearly prove the presence of a C_2 symmetric compound. There are, e.g., only two doublets of doublets for the diastereotopic protons of the overall four methylene groups. While the methylene groups bridging the phenylene rings in the "wings" of the ligand give rise to two doublets at 3.82 und 3.89 ppm with a typical $^2J_{HH}$ coupling constant of about 12.4 Hz, the two methylene units linking the phosphine sites to the Cp^C core are observed at 4.80 and at about 7.78 ppm. The value of the latter chemical shift is rather unexpected. It was identified by means of a H,H-COSY NMR experiment and is explained by the fact that of two of these methylene proton direct towards the central Pd$_2$Cl$_4$ unit. According to the C_2 symmetry of compound **8**, there is only one resonance in the ^{31}P NMR spectrum, observed at 25.4 ppm, which is a typical value for this type of palladium(II) compounds [54].

By slow diffusion of *n*-pentane into a saturated toluene solution of complex **8**, single crystals suitable for an X-ray structure analysis could be obtained. Compound **8** crystallizes as colorless prisms of the monoclinic space group P2/$_n$ with two crystallographically independent molecules in the unit cell. Figure 5 shows the structure of one of these units of compound **8** and summarizes relevant structural parameters.

As expected on the basis of the NMR spectra of compound **8**, the phosphine donors coordinate to the central [(μ^2-Cl)Pd(Cl)]$_2$ unit in an *anti*-conformation that preserves the C_2-symmetry of the Cp^C moiety. The two palladium(II) sites occupy a distorted square-planar geometry. In comparison to structurally similar [(L)Pd(μ^2-Cl)(Cl)]$_2$ (L = σ-donor ligand) compounds, the Pd$\cdots$Pd distance (3.3098(6) Å) is by about 0.15–0.20 Å shorter, which we assign to the steric restrictions of the diphosphine ligand (see [55] and references cited therein). Due to the strongly σ-donating phosphine ligands the bond distances between the palladium(II) centers and the bridging chlorido ligands in *trans*-position to the phosphine donors are slightly longer than those in the *trans*-position to the apical chlorido ligands.

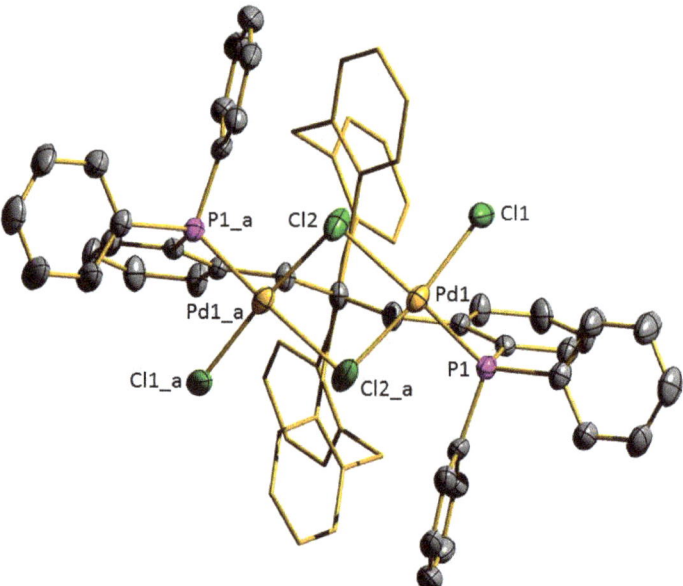

Figure 5. Molecular structure of compound **8** in the solid state. All hydrogen atoms are omitted for clarity. The central Cp^C system is represented in stick-style for clarity, ellipsoids are drawn at the 50% level. Characteristic bond lengths [Å], angles [°] and dihedral angles [°]: Pd$\cdots$Pd 3.3098(6), Pd1-Cl1 2.2880(16), Pd1-Cl2 2.3144(16), Pd1-P1 2.2499(14), C1-C2 1.533(6), C2-C3 1.360(7), C3-C4 1.475(7), Cl1-Pd1-Cl2 173.99(6), Cl1-Pd1-P1 89.96(5), Cl1-Pd1-Cl2_a 90.27(5), Cl2-Pd1-P1 95.71(5), Cl2-Pd1-Cl2_a 83.83(6), Cl2_a-Pd1-P1 171.23(5), C3-C2-C8-C9 35.5(5), C2-C3-C4-C13 47.1(6).

The fragment ClRh(CO) is well-known to prefer *trans*- over *cis*-coordination with chelating diphosphines as long as the linker between the phosphine sites allows this [56–63]. In ligand **3**, the two phosphine sites are linked by a chain of overall seven carbon atoms. Therefore, the formation of a *trans*-coordinated rhodium(I) complex seemed likely, when ligand **3** was reacted first with half an equivalent of [(μ^2-Cl)Rh(CO)$_2$]$_2$ at room temperature. However, spectroscopic characterization of the product unambiguously proved the generation of the dinuclear rhodium(I) complex **9** (Scheme 8). Reaction with one equivalent of the precursor [(μ^2-Cl)Rh(CO)$_2$]$_2$ provided complex **9** in excellent yields.

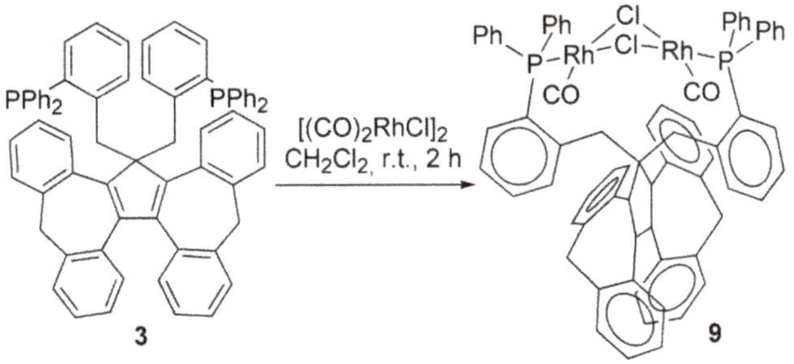

Scheme 8. Synthesis of the dinuclear, chlorido-bridged rhodium(I) carbonyl complex **9**.

The molecular structure of compound **9** resembles the structure of palladium(II) complex **8**. In contrast, in the case of **9** the phosphine donors are coordinated in a *syn*- and not in an *anti*-arrangement to the metal sites, which breaks the C_2-symmetry of the chelating ligand and leads to an overall C_1-symmetric structure.

This decrease in symmetry results in the typical pattern of resonances for the methylene protons as described above for other complexes with C_1-symmetry. As already observed for palladium complex **8**, the ^{1}H NMR resonances of two of the methylene protons are shifted largely towards lower field (7.00 ppm). The two phosphorus atoms are magnetically not equivalent and show couplings to the neighboring ^{103}Rh centers (δ = 35.9 ppm, $^1J_{RhP}$ = 171.5 Hz and δ = 39.8 ppm, $^1J_{RhP}$ = 175.0 Hz), which are close to values measured for other binuclear rhodium(I) complexes of similar structure [52,64,65]. In the ^{13}C NMR spectrum two doublets of doublets at δ = 183.2 (dd, $^1J_{RhC}$ = 78.0, $^2J_{PC}$ = 20.6 Hz) and δ = 182.3 (dd, $^1J_{RhC}$ = 78.8, $^2J_{PC}$ = 21.1 Hz) are assigned to the rhodium-bound carbonyl ligands. In the IR spectrum of compound **9**, there are two carbonyl absorptions at 1991 and 1975 cm^{-1} [52,64,65].

By the slow diffusion of *n*-pentane into a saturated toluene solution of complex **9**, single crystals suitable for an X-ray structure analysis could be obtained. Compound **9** crystallizes as yellow prisms of the triclinic space group $P\bar{1}$. Figure 6 shows the structure of **9** and summarizes relevant structural parameters.

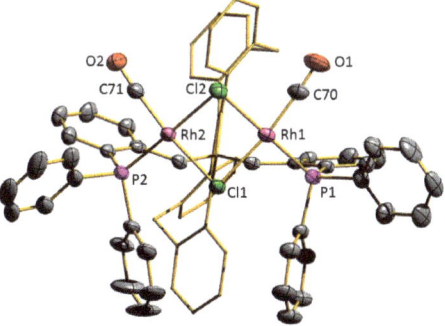

Figure 6. Molecular structure of compound **9** in the solid state. The central Cp^C system is represented in stick-style for clarity, ellipsoids are drawn at the 50% level. All hydrogen atoms are omitted for clarity. Characteristic bond lengths [Å], angles [°] and dihedral angles [°]: Rh⋯Rh 3.2797(5), Rh1-Cl1 2.3926(7), Rh1-Cl2 2.4127(7), Rh1-P1 2.2486(7), Rh1-C70 1.794(3), Rh2-Cl1 2.3880(7), Rh2-Cl2 2.4053(7), Rh2-P2 2.2431(7), Rh2-C71 1.805(3), Cl1-Rh1-Cl2 83.42(2), Cl1-Rh1-P1 96.09(2), Cl1-Rh1-C70 172.39(10), Cl2-Rh1-P1 176.67(3), Cl2-Rh1-C70 91.26(9), P1-Rh1-C70 89.53(9), Cl1-Rh2-Cl2 83.68(2), Cl1-Rh2-P2 97.62(3), Cl1-Rh2-C71 173.89(9), Cl2-Rh2-P2 177.68(3), Cl2-Rh2-C71 91.67(9), P2-Rh2-C71 86.90(9), Rh1-Cl1-Rh2 86.63(2), Rh1-Cl2-Rh2 85.80(2), C3-C2-C6-C7 38.2(3), C2-C3-C10-C9 46.6(4), C5-C4-C19-C20 45.7(4), C4-C5-C23-C22 -37.7(4).

Both rhodium(I) sites in compound **9** occupy a distorted square-planar geometry, which is typical for d^8-configured transition metal ions and has already been observed for the palladium(II) complex **8**. In contrast to **8**, however, the phosphines are coordinating the two rhodium(I) centers in a *syn*-configuration. This is in agreement with the NMR data of the complex **9** that correspond to a C_1-symmetric structure. The Rh-Cl distances in *trans*-position to the π-accepting carbonyl ligands are slightly shorter than those in *trans*-position to the phosphine donors. The large interplanar angle of 132.28° between the chlorido-bridged rhodium(I) sites, the P-P distance of 6.378 Å, which is somewhat shorter with respect to the metal-free ligand and the relative long Rh-Rh distance of 3.28 Å confirm calculations carried out by López-Valbuena et al. on structural features of diphosphane ligands with large bite-angles [64].

As mentioned above, the fragment ClRh(CO) tends to form *trans*-coordinated mononuclear complexes with chelating diphosphines possessing long linker units. Therefore, ligand **3** was reacted with half an equivalent of the precursor $[(\mu^2\text{-Cl})Rh(CO)_2]_2$. In 2006, Gelman et al. reported the preferential formation of a monometalic chelate complex using shorter reaction times [52,53]. In our case, however, the selective formation of the monorhodium(I) complex **10** was achieved by combining a prolonged reaction time and an elevated reaction temperature (Scheme 9).

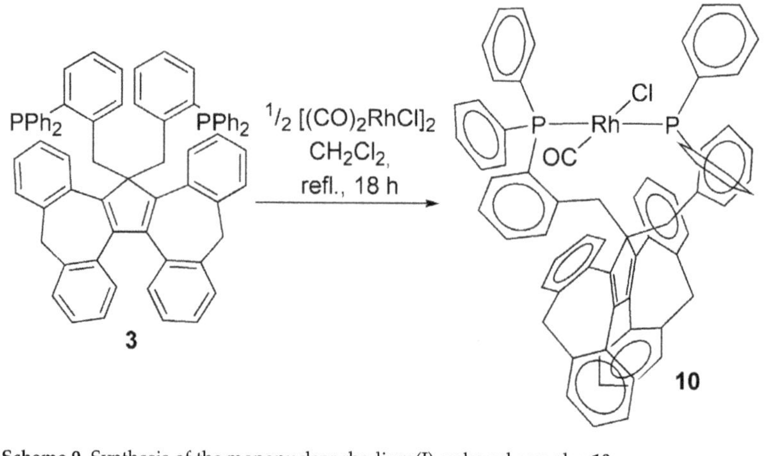

Scheme 9. Synthesis of the mononuclear rhodium(I) carbonyl complex **10**.

The C_2 symmetric ligand backbone in combination with a local C_S symmetry of the coordination environment at the rhodium(I) site led to an overall C_1 symmetry and thus, as expected, to eight doublets for the methylene protons in the ^{1}H NMR spectrum of compound **10**. Three of these resonances that are assigned according to their coupling constants to the methylene groups linking the Cp^C backbone with the phosphine sites are strongly shifted towards lower field. The X-ray structure (see below) of compound **10** shows that these three protons are oriented towards the central (*trans*-P)$_2$Rh(CO)Cl unit. The ^{31}P NMR spectrum of compound **10** reflects its C_1 symmetry: there are two doublets of doublets with a chemical shift of 26.2 und 35.6 ppm, each showing a large $^2J_{PP}$ coupling constant of 340 Hz and a smaller $^1J_{RhP}$ coupling constant (123.2 resp. 127.5 Hz). The large $^2J_{PP}$ coupling constant is typical for the *trans*-orientation of the two phosphine donors [66]. Compared to the chlorido-bridged complex **9**, the low field shift of the resonances of the two phosphorus atoms is not as pronounced. ESI mass spectrometry reveals a peak at $m/z = 1073.39$, which corresponds to the cationic fragment [M-Cl]$^+$ (calcd.: $m/z = 1073.25$). In the IR spectrum of compound **10**, there is an intense absorption $\tilde{\nu}$(CO) at 1961 cm^{-1}, which is typical for trans-coordinated, monomeric diphosphinerhodium carbonyl complexes [52,60–62,67,68].

By the slow diffusion of *n*-pentane into a saturated toluene solution of complex **10**, single crystals suitable for an X-ray structure analysis could be obtained. Compound **10** crystallizes as yellow prisms of the triclinic space group P2$_1$/n; Figure 7 shows the structure of **10** and summarizes relevant structural parameters.

As reported for other mononuclear diphosphine carbonyl(chlorido)rhodium(I) complexes with long-chained diphosphine ligands, the two phosphine donors are oriented in *trans*-position [53,61–63]. The central rhodium(I) site is coordinated in a distorted square-planar geometry. Due to the steric influence of the Cp^C moiety, the P1-Rh1-P2 angle is significantly smaller than 180° (167.27(3)°) as is the Cl1Rh1-C70 angle (169.08(9)°).

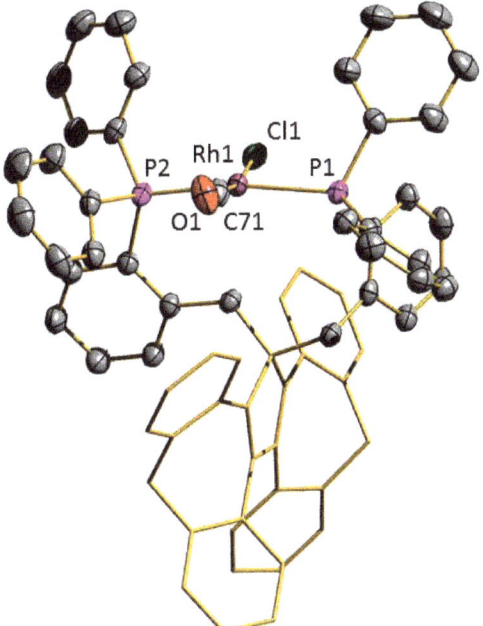

Figure 7. Molecular structure of compound **10** in the solid state. All hydrogen atoms are omitted for clarity. The central Cp^C system is represented in stick-style for clarity, ellipsoids are drawn at the 50% level. All hydrogen atoms are omitted for clarity. Characteristic bond lengths [Å], angles [°] and dihedral angles [°]: Rh1-Cl1 2.3953(7), Rh1-P1 2.3081(7), Rh1-P2 2.3356(7), Rh1-C70 1.813(3), Cl1-Rh1-P1 87.04(3), Cl1-Rh1-P2 87.15(2), Cl1-Rh1-C70 169.08(9), P1-Rh1-P2 167.27(3), P1-Rh1-C70 90.38(8), P2-Rh1-C70 97.37(8), C3-C2-C6-C7 35.9(4), C2-C3-C10-C9 47.3(3), C5-C4-C19-C20 45.3(3), C4-C5-C23-C22 39.5(3).

4. Discussion

The novel chiral cyclopentadienyl-type ligand Cp^{C}H (**1**) is available following a high yield protocol of overall five steps starting from cheap dibenzosuberenone. We have shown in this manuscript, that Cp^{C}H (**1**) has turned out to be an ideal precursor for the synthesis of Cp-functionalized phosphine ligands. Both the monophosphine ligand **2** and the diphosphine ligand **3** are obtained by high yield nucleophilic substitution reactions with 2-[(chloromethyl)phenyl](diphenyl)phosphine using deprotonated compound **1**. In particular, the simplicity of the synthesis of **3** providing a long-chained and chiral diphosphine is of interest, since syntheses of similar systems usually suffer from poor yields.

With compounds **2** and **3** overall six different types of coordination could be realized. The monophosphine **2** still has a cyclopentadiene backbone with one proton attached to the five-membered ring and is thus able to coordinate like a typical phosphine or a typical cyclopentadienide or by combining these two modes. Examples of all three modes of coordination could be obtained by selecting the appropriate (transition) metal precursor. In contrast to monophosphine **2**, diphosphine **3** cannot be deprotonated anymore at the cyclopentadiene site. It therefore reacts like a typical diphosphine. Bridging coordination to two gold(I) sites, as well as the *trans*-coordination to one rhodium(I) in a square-planar environment, are expected, in terms of the coordinative behavior of these two metal sites. In the latter case, a ten-membered ring is formed. However, we also found two examples for a coordination to two d^8 metal sites (palladium(II) and rhodium(I)) that are part of dinuclear chloride-bridged complexes. In these two cases, the ligand forms a twelve-membered ring including a $M(\mu^2\text{-Cl})_2M$ moiety. We assign this rather outstanding flexibility in coordination on one side to the long chain (seven carbon atoms) that connects the two phosphine centers. On the other side, there are four aromatic and one quaternary sp^3 hybridized carbon atoms in this chain, which reduce the number of possible conformations and thus stabilize the discussed modes of coordination.

In this manuscript, we have focused on aspects of the synthesis, spectroscopic data and structural elucidation of a series of complexes containing the new chiral phosphines **2** and **3**. The evaluation of, for example, the catalytical properties of selected compounds from this series is presently under investigation. Here, we concentrate on transformations that are known to require diphosphine ligands having a large bite angle and on transformations catalyzed by the *ansa*-bridged ruthenium complex **6**.

Supplementary Materials: The following supporting information can be downloaded at: https://www.mdpi.com/article/10.3390/chemistry5020062/s1. It contains figures of NMR and IR spectra of all compounds discussed in this manuscript.

Author Contributions: Investigation, F.N. and Y.S.; supervision and writing, W.R.T. All authors have read and agreed to the published version of the manuscript.

Funding: This research was funded by the Carl Zeiss Stiftung who donated a grant to F.N.

Data Availability Statement: Data is contained within the article and supplementary materials.

Conflicts of Interest: The authors declare no conflict of interest.

References

1. Charrier, C.; Mathey, F. La diphenyl-cyclopentadienylmethyl-phosphine et ses complexesferroceniques et cymantreniques. *Tetrahedron Lett.* **1978**, *19*, 2407–2410. [CrossRef]
2. Kettenbach, R.T.; Bonrath, W.; Butenschön, H. [ω-(Phosphanyl)alkyl]cyclopentadienyl Complexes. *Chem. Ber.* **1993**, *126*, 1657–1669. [CrossRef]
3. Butenschön, H. Cyclopentadienylmetal Complexes Bearing Pendant Phosphorus, Arsenic, and Sulfur Ligands. *Chem. Rev.* **2000**, *100*, 1527–1564. [CrossRef] [PubMed]
4. Bensley, D.M.; Mintz, E.A.; Sussangkarn, S.J. Synthesis of [C$_5$(CH$_3$)$_4$H]CH$_2$CH$_2$CH$_2$P(C$_6$H$_5$)$_2$: A novel heterodifunctional ligand possessing both a tetramethylcyclopentadiene and a remote diphenylphosphine functionality. *J. Org. Chem.* **1988**, *53*, 4417–4419. [CrossRef]
5. Bensley, D.M.; Mintz, E. 1,2,3,4,6-Pentamethylfulvene: A convenient precursor to substituted tetramethylcyclopentadienyl transition metal complexes. *J. Organomet. Chem.* **1988**, *353*, 93–102. [CrossRef]
6. Hayashi, T.; Kumada, M. Asymmetric synthesis catalyzed by chiral ferrocenylphosphine-transition metal complexes. 2. Nickel- and palladium-catalyzed asymmetric Grignard cross-coupling. *Acc. Chem. Res.* **1982**, *15*, 395–401. [CrossRef]
7. Fihri, A.; Meunier, P.; Hierso, J.-C. Performances of symmetrical achiral ferrocenylphosphine ligands in palladium-catalyzed cross-coupling reactions: A review of syntheses, catalytic applications and structural properties. *Coord. Chem. Rev.* **2007**, *251*, 2017–2055. [CrossRef]
8. Kataoka, N.; Shelby, Q.; Stambuli, J.P.; Hartwig, J.F. Air Stable, Sterically Hindered Ferrocenyl Dialkylphosphines for Palladium-Catalyzed C−C, C−N, and C−O Bond-Forming Cross-Couplings. *J. Org. Chem.* **2002**, *67*, 5553–5566. [CrossRef]
9. Ito, Y.; Sawamura, M.; Hayashi, T. Catalytic asymmetric aldol reaction: Reaction of aldehydes with isocyanoacetate catalyzed by a chiral ferrocenylphosphine-gold(I) complex. *J. Am. Chem. Soc.* **1986**, *108*, 6405–6406. [CrossRef]
10. Hayashi, T.; Mise, T.; Fukushima, M.; Kagotani, M.; Nagashima, N.; Hamada, Y.; Matsumoto, A.; Kawakami, S.; Konishi, M. Asymmetric Synthesis Catalyzed by Chiral Ferrocenylphosphine–Transition Metal Complexes. I. Preparation of Chiral Ferrocenylphosphines. *Bull. Chem. Soc. Jpn.* **1980**, *53*, 1138–1151. [CrossRef]
11. Hayashi, T.; Yamamoto, A.; Ito, Y.; Nishioka, E.; Miura, H.; Yanagi, K. Asymmetric Synthesis Catalyzed by Chiral Ferrocenylphosphine-Transition-Metal Complexes. 8. Palladium-Catalyzed Asymmetric Allylic Amination. *J. Am. Chem. Soc.* **1989**, *111*, 6301–6311. [CrossRef]

12. Hayashi, T.; Hayashizaki, K.; Kiyoi, T.; Ito, Y. Asymmetric synthesis catalyzed by chiral ferrocenylphosphine-transition-metal complexes. 6. Practical asymmetric synthesis of 1,1′-binaphthyls via asymmetric cross-coupling with a chiral [(alkoxyalkyl)ferrocenyl]monophosphine/nickel catalyst. *J. Am. Chem. Soc.* **1988**, *110*, 8153–8156. [CrossRef]
13. Chen, Y.; Yi, X.; Cheng, Y.; Huang, A.; Yang, Z.; Zhao, X.; Ling, F.; Zhong, W. Rh-Catalyzed Highly Enantioselective Hydrogenation of Functionalized Olefins with Chiral Ferrocenylphosphine-Spiro Phosphonamidite Ligands. *J. Org. Chem.* **2022**, *87*, 7864–7874. [CrossRef] [PubMed]
14. Barth, E.L.; Davis, R.M.; Mohadjer Beromi, M.; Walden, A.G.; Balcells, D.; Brudvig, G.W.; Dardir, A.H.; Hazari, N.; Lant, H.M.C.; Mercado, B.Q.; et al. Bis(dialkylphosphino)ferrocene-Ligated Nickel(II) Precatalysts for Suzuki–Miyaura Reactions of Aryl Carbonates. *Organometallics* **2019**, *38*, 3377–3387. [CrossRef]
15. Škoch, K.; Schulz, J.; Císařová, I.; Štěpnička, P. Pd(II) Complexes with Chelating Phosphinoferrocene Diaminocarbene Ligands: Synthesis, Characterization, and Catalytic Use in Pd-Catalyzed Borylation of Aryl Bromides. *Organometallics* **2019**, *38*, 3060–3073. [CrossRef]
16. Škoch, K.; Císařová, I.; Štěpnička, P. Synthesis of a Polar Phosphinoferrocene Amidosulfonate Ligand and Its Application in Pd-Catalyzed Cross-Coupling Reactions of Aromatic Boronic Acids and Acyl Chlorides in an Aqueous Medium. *Organometallics* **2016**, *35*, 3378–3387. [CrossRef]
17. Chung, J.-Y.; Schulz, C.; Bauer, H.; Sun, Y.; Sitzmann, H.; Auerbach, H.; Pierik, A.J.; Schünemann, V.; Neuba, A.; Thiel, W.R. Cyclopentadienide Ligand Cp^{C-} Possessing Intrinsic Helical Chirality and Its Ferrocene Analogues. *Organometallics* **2015**, *34*, 5374–5382. [CrossRef]
18. Chung, J.-Y.; Sun, Y.; Thiel, W.R. Titanium(IV) complexes bearing the $(Cp^C)^-$ ligand. *J. Organomet. Chem.* **2017**, *829*, 31–36. [CrossRef]
19. Nährig, F.; Gemmecker, G.; Chung, J.-Y.; Hütchen, P.; Lauk, S.; Klein, M.; Sun, Y.; Niedner-Schatteburg, G.; Sitzmann, H.; Thiel, W.R. Complexes of Platinum Group Elements Containing the Intrinsically Chiral Cyclopentadienide Ligand $(Cp^C)^{-1}$. *Organometallics* **2020**, *39*, 1934–1944. [CrossRef]
20. Nährig, F.; Nunheim, N.; Salih, K.S.M.; Chung, J.-Y.; Gond, D.; Sun, Y.; Becker, S.; Thiel, W.R. A Novel Cyclopentadienone and its Ruthenium and Iron Tricarbonyl Complexes. *Eur. J. Inorg. Chem.* **2021**, 4832–4841. [CrossRef]
21. Armarego, W.L.F.; Chai, C.L.L. *Purification of Laboratory Chemicals*, 6th ed.; Butterworth-Heinemann: Oxford, UK, 2009.
22. Herrmann, J.; Pregosin, P.S.; Salzmann, R.; Albinati, A. Palladium π-Allyl Chemistry of New P,S Bidentate Ligands. Selective but Variable Dynamics in the Isomerization of the η^3-C_3H_5 and η^3-PhCHCHCHPh π-Allyl Ligands. *Organometallics* **1995**, *14*, 3311–3318. [CrossRef]
23. Rauchfuss, T.B.; Patino, F.T.; Roundhill, D.M. Platinum Metal Complexes of Amine- and Ether-Substituted Phosphines. *Inorg. Chem.* **2002**, *14*, 652–656. [CrossRef]
24. Altomare, A.; Cascarano, G.; Giacovazzo, C.; Guagliardi, A.; Burla, M.C.; Polidori, G.; Camalli, M. SIR92—A program for automatic solution of crystal structures by direct methods. *J. Appl. Cryst.* **1994**, *27*, 435. [CrossRef]
25. Sheldrick, G.M. A short history of SHELX. *Acta Cryst.* **2008**, *A64*, 112–122. [CrossRef]
26. *Rigaku Oxford Diffraction*; Version 1.171.38.46; CrysAlisPro: Seattle, WA, USA, 2015.
27. Li, J.; Yin, J.; Wang, G.-X.; Yin, Z.-B.; Zhang, W.-X.; Xi, Z. Synthesis and reactivity of asymmetric Cr(i) dinitrogen complexes supported by cyclopentadienyl–phosphine ligands. *Chem. Commun.* **2019**, *55*, 9641–9644. [CrossRef] [PubMed]
28. Ireland, T.; Tappe, K.; Grossheimann, G.; Knochel, P. Synthesis of a New Class of Chiral 1,5-Diphosphanylferrocene Ligands and Their Use in Enantioselective Hydrogenation. *Chem. Eur. J.* **2002**, *8*, 843–852. [CrossRef]
29. Trampert, J.; Sun, Y.; Thiel, W.R. The reactivity of [{2-(diphenylphosphino)phenyl}methyl]-3-imidazol-2-ylidenes towards group VIII element precursors. *J. Organometal. Chem.* **2020**, *915*, 121222. [CrossRef]
30. Keough, P.T.; Grayson, M. Phosphonioethylation. Michael Addition to Vinylphosphonium Salts. *J. Org. Chem.* **1964**, *29*, 631–635. [CrossRef]
31. Matsusaka, Y.; Shitaya, S.; Nomura, K.; Inagaki, A. Synthesis of Mono-, Di-, and Trinuclear Rhodium Diphosphine Complexes Containing Light-Harvesting Fluorene Backbones. *Inorg. Chem.* **2017**, *56*, 1027–1030. [CrossRef]
32. Xie, L.; Zhao, S.; Zhang, M.; Wu, C.; Li, T.; Gao, M. Compound of Fluorene Class Containing Dual Hetero Atoms as well as Its Synthetic Method and Application. CN Patent 1480453A, 10 March 2004.
33. Liu, H.; Xie, L.; Jing, J.M. Solid Catalyst Composition for Olefinic Polymerization and Catalyst Thereof. CN Patent 1508159A, 30 June 2004.
34. Eger, T.R.; Munstein, I.; Steiner, A.; Sun, Y.; Niedner-Schatteburg, G.; Thiel, W.R. New cationic organometallic phosphane ligands and their coordination to gold(I). *J. Organomet. Chem.* **2016**, *810*, 51–56. [CrossRef]
35. Batchelor, L.K.; Păunescu, E.; Soudani, M.; Scopelliti, R.; Dyson, P.J. Influence of the Linker Length on the Cytotoxicity of Homobinuclear Ruthenium(II) and Gold(I) Complexes. *Inorg. Chem.* **2017**, *56*, 9617–9633. [CrossRef] [PubMed]
36. Borissova, A.O.; Korlyukov, A.A.; Antipin, M.Y.; Lyssenko, K.A. Estimation of Dissociation Energy in Donor–Acceptor Complex $AuCl\cdot PPh_3$ via Topological Analysis of the Experimental Electron Density Distribution Function. *J. Phys. Chem. A* **2008**, *112*, 11519–11522. [CrossRef] [PubMed]
37. Zi, W.; Dean Toste, F. Recent advances in enantioselective gold catalysis. *Chem. Soc. Rev.* **2016**, *45*, 4567–4589. [CrossRef]
38. Johnson, B.F.G. The chemistry of gold. *Gold Bull.* **1971**, *4*, 9–11. [CrossRef]

39. Scherbaum, F.; Grohmann, A.; Huber, B.; Krüger, C.; Schmidbaur, H. Use of the CH Acidity of 2,4,4-Trimethyl-4,5-dihydrooxazole to Synthesize Triauriomethanes and Novel Gold Clusters. *Angew. Chem. Int. Ed. Engl.* **1988**, *27*, 1544–1546. [CrossRef]
40. Janiak, C. (Organo)thallium (I) and (II) chemistry: Syntheses, structures, properties and applications of subvalent thallium complexes with alkyl, cyclopentadienyl, arene or hydrotris(pyrazolyl)borate ligands. *Coord. Chem. Rev.* **1997**, *163*, 107–216. [CrossRef]
41. Schumann, H.; Janiak, C.; Khani, H. Cyclopentadienylthallium(I) compounds with bulky cyclopentadienyl ligands. *J. Organomet. Chem.* **1987**, *330*, 347–355. [CrossRef]
42. Schumann, H.; Janiak, C.; Khan, M.A.; Zuckerman, J.J. Eine zweite ungewöhnliche Kristallmodifikation von pentabezylcyclopentadienylthallium(I), (PhCH$_2$)$_5$C$_5$Tl. *J. Organomet. Chem.* **1988**, *354*, 7–13. [CrossRef]
43. Frasson, E.; Menegus, F.; Panattoni, C. Chain Structure of the Cyclopentadienily of Monovalent Indium and Thallium. *Nature* **1963**, *199*, 1087–1089. [CrossRef]
44. Werner, H.; Otto, H.; Kraus, H.J. Die Kristallstruktur von TlC$_5$Me$_5$. *J. Organomet. Chem.* **1986**, *315*, C57–C60. [CrossRef]
45. Schmidbaur, H.; Bublak, W.; Riede, J.; Müller, G. [{1,3,5-(CH$_3$)$_3$H$_3$C$_6$}$_6$Tl$_4$]·[GaBr$_4$]$_4$—Synthese und Struktur eines gemischten Mono- und Bis(aren)thallium-Komplexes. *Angew. Chem.* **1985**, *97*, 402–403. [CrossRef]
46. Schmidbaur, H. Arenkomplexe von einwertigem Gallium, Indium und Thallium. *Angew. Chem.* **1985**, *97*, 893–904. [CrossRef]
47. Nakayama, H.; Nishijima, C.; Tachiyashiki, S. Ion-Molecule Reactions Betwee Thallium(I) and Various Ligands: Formation of 1:1 Complexes in Gas Phase. *Chem. Lett.* **1974**, *3*, 733–736. [CrossRef]
48. Betley, T.A.; Peters, J.C. The Strong-Field Tripodal Phosphine Donor, [PhB(CH$_2$PiPr$_2$)$_3$]$^-$, Provides Access to Electronically and Coordinatively Unsaturated Transition Metal Complexes. *Inorg. Chem.* **2003**, *42*, 5074–5084. [CrossRef]
49. Szlosek, R.; Ackermann, M.T.; Marquardt, C.; Seidl, M.; Timoshkin, A.Y.; Scheer, M. Coordination of Pnictogenylboranes Towards Tl(I) Salts and a Tl- Mediated P-P Coupling. *Chem. Eur. J.* **2023**, *29*, e202202911. [CrossRef]
50. Doppiu, A.; Englert, U.; Salzer, A. Cationic half-sandwich ruthenium(II) complexes with cyclopentadienyl–phosphine ligands. *Inorg. Chim. Acta* **2003**, *350*, 435–441. [CrossRef]
51. Doppiu, A.; Salzer, A. A New Route to Cationic Half-Sandwich Ruthenium(II) Complexes with Chiral Cyclopentadienylphosphane Ligands. *Eur. J. Inorg. Chem.* **2004**, 2244–2252. [CrossRef]
52. Azerraf, C.; Cohen, S.; Gelman, D. Roof-Shaped Halide-Bridged Bimetallic Complexes via Ring Expansion Reaction. *Inorg. Chem.* **2006**, *45*, 7010–7017. [CrossRef] [PubMed]
53. Azerraf, C.; Grossman, O.; Gelman, D. Rigid *trans*-spanning triptycene-based ligands: How flexible they can be? *J. Organomet. Chem.* **2007**, *692*, 761–767. [CrossRef]
54. Noskowska, M.; Śliwińska, E.; Duczmal, W. Simple fast preparation of neutral palladium (II) complexes with SnCl$^-$$_3$ and Cl$^-$ ligands. *Trans. Met. Chem.* **2003**, *28*, 756–759. [CrossRef]
55. Aullón, G.; Ujaque, G.; Lledós, A.; Alvarez, S.; Alemany, P. To Bend or Not To Bend: Dilemma of the Edge-Sharing Binuclear Square Planar Complexes of d^8 Transition Metal Ions. *Inorg. Chem.* **1998**, *37*, 804–813. [CrossRef]
56. Bunten, K.A.; Farrar, D.H.; Poë, A.J.; Lough, A. Stoichiometric and Catalytic Oxidation of BINAP by Dioxygen in a Rhodium(I) Complex. *Organometallics* **2002**, *21*, 3344–3350. [CrossRef]
57. Mondal, J.U.; Young, K.G.; Blake, D.M. Enthalpy changes in oxidative addition reactions of iodine with monomeric and dimeric rhodium(I) complexes. *J. Organomet. Chem.* **1982**, *240*, 447–451. [CrossRef]
58. Dyer, G.; Wharf, R.M.; Hill, W.E. ^{31}P NMR studies of *cis*-[RhCl(CO)(bis-phosphine)] complexes. *Inorg. Chim. Acta* **1987**, *133*, 137–140. [CrossRef]
59. Mann, B.E.; Masters, C.; Shaw, B.L. Nuclear magnetic resonance studies on metal complexes. Part VII. The ^{31}P n.m.r. spectra of some complexes of the type *trans*-RhCl-(CO)L$_2$ [L = tertiary phosphine or P(OMe)$_3$]. *J. Chem. Soc. A* **1971**, 1104–1106. [CrossRef]
60. Vallarino, L. Carbonyl complexes of rhodium. Part I. Complexes with triarylphosphines, triarylarsines, and triarylstibines. *J. Chem. Soc.* **1957**, 2287–2292. [CrossRef]
61. Marty, W.; Kapoor, P.N.; Bürgi, H.-B.; Fischer, E. Complexes of 3,3'-Oxybis[(diphenylphosphino)methylbenzene] with Ni(II), Pd(II), Pt(II), Rh(I), and Ag(I). How Important is Backbone Rigidity in the Formation of *trans*-Spanning Bidenatate Chelates? *Helv. Chim. Acta* **1987**, *70*, 158–170. [CrossRef]
62. Eberhard, M.R.; Heslop, K.M.; Orpen, A.G.; Pringle, P.G. Nine-Membered Trans Square-Planar Chelates Formed by a bisbi Analogue. *Organometallics* **2005**, *24*, 335–337. [CrossRef]
63. Reed, F.J.S.; Venanzi, L.M. Transition metal complexes with bidentate ligands spanning *trans*-positions. IV. Preparation and properties of some rhodium and iridium complexes of 2,11-bis(diphenylphosphinomethyl)benzo[c]phenanthrene. *Helv. Chim. Acta* **1977**, *60*, 2804–2814. [CrossRef]
64. López-Valbuena, J.M.; Escudero-Adan, E.C.; Benet-Buchholz, J.; Freixa, Z.; van Leeuwen, P.W.N.M. An approach to bimetallic catalysts by ligand design. *Dalton Trans.* **2010**, *39*, 8560–8574. [CrossRef]
65. Hierso, J.-C.; Lacassin, F.; Broussier, R.; Amardeil, R.; Meunier, P. Synthesis and characterisation of a new class of phosphine-phosphonite ferrocenediyl dinuclear rhodium complexes. *J. Organomet. Chem.* **2004**, *689*, 766–769. [CrossRef]
66. Meeuwissen, J.; Sandee, A.J.; de Bruin, B.; Siegler, M.A.; Spek, A.L.; Reek, J.N.H. Phosphinoureas: Cooperative Ligands in Rhodium-Catalyzed Hydroformylation? On the Possibility of a Ligand-Assisted Reductive Elimination of the Aldehyde. *Organometallics* **2010**, *29*, 2413–2421. [CrossRef]

67. Sanger, A.R.J. Reactions of di-μ-chloro-bis[cyclo-octa-1,5-dienerhodium(I)] with carbon mono-oxide and mono-, di-, or tri-tertiary phosphines, or 1,2-bis(diphenylarsino)ethane. *Chem. Soc. Dalton Trans.* **1977**, 120–129. [CrossRef]
68. Thurner, C.L.; Barz, M.; Spiegler, M.; Thiel, W.R. Ligands with cycloalkane backbones II. Chelate ligands from 2-(diphenylphosphinyl)cyclohexanol: Syntheses and transition metal complexes. *J. Organomet. Chem.* **1997**, *541*, 39–49. [CrossRef]

Disclaimer/Publisher's Note: The statements, opinions and data contained in all publications are solely those of the individual author(s) and contributor(s) and not of MDPI and/or the editor(s). MDPI and/or the editor(s) disclaim responsibility for any injury to people or property resulting from any ideas, methods, instructions or products referred to in the content.

Article

Stabilization of 2-Pyridyltellurium(II) Derivatives by Oxidorhenium(V) Complexes

Felipe Dornelles da Silva [1], Maximilian Roca Jungfer [2], Adelheid Hagenbach [2], Ernesto Schulz Lang [1,*] and Ulrich Abram [2,*]

[1] Department of Chemistry, Universidade Federal de Santa Maria, Avenida Roraima, n° 1000, Santa Maria 97105-900, Rio Grande do Sul, Brazil
[2] Institute of Chemistry and Biochemistry, Freie Universität Berlin, Fabeckstr. 34/36, 14195 Berlin, Germany
* Correspondence: eslang@ufsm.br (E.S.L.); ulrich.abram@fu-berlin.de (U.A.)

Abstract: Zwitterionic compounds such as pyridine-containing tellurenyl compounds are interesting building blocks for heterometallic assemblies. They can act as ambiphilic donor/acceptors as is shown by the products of reactions of the zwitterions HpyTeCl$_2$ or HCF$_3$pyTeCl$_2$ with the rhenium(V) complex [ReOCl$_3$(PPh$_3$)$_2$]. The products have a composition of [ReO$_2$Cl(pyTeCl)(PPh$_3$)$_2$] and [ReO$_2$Cl(CF$_3$pyTeCl)(PPh$_3$)$_2$] with central $\{O = Re = O \ldots Te(Cl)py\}^+$ units. The Re-O bonds in the products are elongated by approximately 0.1 Å compared with those to the terminal oxido ligands and establish Te . . . O contacts. Thus, the normally easily assigned concept of oxidation states established at the two metal ions becomes questionable (Re^V/Te^{II} vs. Re^{III}/Te^{IV}). A simple bond length consideration rather leads to a description with the coordination of a mesityltellurenyl(II) chloride unit to an oxido ligand of the Re(V) center, but the oxidation of the tellurium ion and the formation of a tellurinic acid chloride cannot be ruled out completely from an analysis of the solid-state structures. DFT calculations (QTAIM, NBO analysis) give clear support for the formation of a Re(V) dioxide complex donating into an organotellurium(II) chloride and the alternative description can at most be regarded as a less favored resonance structure.

Keywords: tellurium; rhenium; oxido bridge; DFT

1. Introduction

Organotellurium(II) compounds are valuable synthons in the organic chemistry of this element, but have also found increasing interest as components of coordination compounds [1–8]. They can act as Lewis-acidic metal centers or as donors similar to their lighter sulfur or selenium homologs. Particularly flexible are organotellurides, which contain additional donor positions in their scaffold allowing chelate formation together with the potential tellurium donor or an ambiphilic behavior with the Te(II) atom acting as Lewis acid. Such chelators are frequently established with phosphines [9–14], but also assemblies with amines, Schiff bases, or phenolates are known and are under discussion for potential applications as photoactive materials, catalysts and/or in material science [15–23]. Heavy chalcogens and halogens are known to establish non-covalent chalcogen-halogen, chalcogen-chalcogen, or halogen-halogen interactions, which frequently result in uncommon structural features and allow the modulation of the electronic situation in such compounds. A special situation is given, when a $\{TeX\}^+$ unit (X = halide) is bonded to a pyridine ring. The structures given in Figure 1 perfectly reflect the ambiphilic character of such pyridine-based tellurenyl halides. Already the solid-state structures of the unsubstituted [pyTeX] compounds (X = Cl, I) crystallize as dimers with the pyridine nitrogen donating to the tellurium(II) ion [24,25].

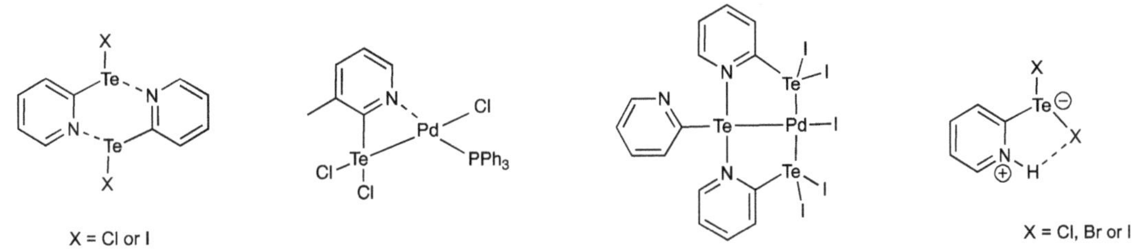

Figure 1. Pyridyltellurenyl halides and some of their complexes [23–27].

Contrastingly both nitrogen and tellurium act as donors in [{MepyTeCl$_2$}PdCl$_2$] [26]. The bonding situation in the second palladium compound of Figure 1 is more sophisticated, since all three of the tellurium atoms seem to donate to the transition metal, but one of them parallelly accepts electron density from the adjacent pyridine rings [27]. The zwitterionic acidification products of the pyridyltellurenyl chlorides [HpyTeX] have been shown to be versatile synthons for metal complexes [27], which stimulated us to perform experiments with the common oxidorhenium(V) precursor [ReOCl$_3$(PPh$_3$)$_2$]. With regard to our continuing interest in fluorinated ligand systems and the effects of fluorination on the coordination properties of such systems, we also incorporated CF$_3$-substituted pyridyltellurenyl halides in this study.

2. Materials and Methods

Unless otherwise stated, reagent-grade starting materials were purchased from commercial sources and either used as received or purified by standard procedures. Bis(2-pyridyl)ditellane, [HPyTeCl$_2$] and [ReOCl$_3$(PPh$_3$)$_2$] were synthesized according to published protocols [27,28]. The solvents were dried and deoxygenated according to standard procedures. NMR spectra were recorded at room temperature with JEOL 400 MHz ECS or ECZ multinuclear spectrometers. Chemical shifts are given relative to TeMe$_2$ (^{125}Te) and CFCl$_3$ (^{19}F). Elemental analyses were determined with a Heraeus Vario El III elemental analyzer. FTIR spectra were recorded on a Bruker Vertex spectrometer using attenuated total reflection (ATR). Confocal FT-Raman spectra were measured with a Bruker Senterra micro-Raman spectrometer using a 785 nm laser.

2.1. Syntheses

Bis(5-Trifluoromethyl-2-pyridyl)ditellane (**1**): Sodium borohydride (5 g, 132 mmol) was added to a mixture containing tellurium powder (2.55 g, 20 mmol) and sodium hydroxide (0.8 g, 20 mmol) in 200 mL of ethanol. The mixture was heated under reflux in a Schlenk flask under an argon atmosphere until the solution became colorless. Then, the system was cooled to room temperature, and 2-chloro-5-(trifluoromethyl)pyridine (7.261 g, 40 mmol) was added. The resulting mixture was heated on reflux for 6 h. After cooling to room temperature, the reaction mixture was extracted with chloroform (3 × 100 mL). The organic layers were collected, dried and the solvent was removed leaving a red, oily residue. Crystallization was performed from a CHCl$_3$/MeOH mixture. Orange red, crystalline solid. Yield: 87% (4.76 g) based on elemental tellurium. Elemental analysis: Calcd for C$_{12}$H$_6$F$_6$N$_2$Te$_2$: C 26.33, H 1.10, N 5.12%. Found: C 26.35, H 1.12, N 5.13%. ^{1}H NMR (CDCl$_3$, ppm): 8.65 (s, 2H), 8.09 (d, J = 8.2 Hz, 2H), 7.54 (d, J = 8.2, 2H). ^{13}C NMR (CDCl$_3$, ppm): 146.6 ppm (q, J = 4.1 Hz), 140.5 (q, J = 1.6 Hz), 133.4 (q, J = 3.4 Hz), 124.9 (q, J = 33.3 Hz), 123.2 (q, J = 272.5 Hz). ^{19}F NMR (CDCl$_3$, ppm): −62.5. ^{125}Te NMR (CDCl$_3$, ppm): 454.6. IR (cm^{-1}): 3046, 2902, 1583, 1553, 1315, 1061, 746, 487. Raman (cm^{-1}): 1332, 1061, 736, 487, 202.

[HCF$_3$pyTeCl$_2$] (**2**): A solution of **1** (109 mg, 0.2 mmol) in CHCl$_3$ (5 mL) was overlayered with hydrochloric acid (3 mL, 37%) and kept for four days. During this time, an orange-yellow solid precipitated. This solid was separated by filtration and dried in a vacuum. Orange-yellow crystals. Yield: 55 mg (40%). Elemental analysis: Calcd for $C_6H_4Cl_2F_3NTe$: C 20.85, H 1.17, N 4.05%. Found: C 20.80, H 1.10, N 4.08%. IR (cm^{-1}): 3422, 3086, 3072, 2997, 1560, 1535, 1242, 1025, 783, 492. Raman (cm^{-1}) 1518, 1231, 1058, 236, 284, 260.

[HCF$_3$pyTeBr$_2$] (3): This compound was prepared according to the procedure given for compound 2 using HBr (48%) instead of HCl. Orange-yellow crystals. Yield: 46%. Elemental analysis: Calcd for $C_6H_4Br_2F_3NTe$: C 16.59, H 0.93, N 3.22%. Found: C 16.51, H 0.98, N 3.19%. IR (cm^{-1}) 3405, 3088, 3070, 2983, 1587, 1529, 1235, 1022, 782, 490. Raman (cm^{-1}): 1233, 1026, 784, 268, 189, 167.

[ReO$_2$Cl(CF$_3$pyTeCl)(PPh$_3$)$_2$] (4): Solid [ReOCl$_3$(PPh$_3$)$_2$] (83 mg, 0.1 mmol) was added to a suspension of [HCF$_3$pyTeCl$_2$] (35 mg, 0.1mmol) in 6 mL of a DMF/EtOH mixture (1:1, v/v). The sparingly soluble solids slowly dissolved upon heating under reflux for approximately 15 min. The resulting, almost clear solution was filtered and red crystals were obtained after slow evaporation of the solvents. Red crystals. Yield: 53%. Elemental analysis (for a carefully dried sample to remove the co-crystallized solvent): Calcd for $C_{42}H_{33}Cl_2F_3NO_2P_2ReTe$: C 46.39, H 3.06, N 1.29%. Found: C 46.38, H 3.04, N 1.31%. IR (cm^{-1}): 2930, 2864, 1557, 1496, 1326, 1063, 998, 910, 751, 659, 446, 254. Raman (cm^{-1}): 1325, 1053, 1000, 656, 254.

[ReO$_2$Cl(pyTeCl)(PPh$_3$)$_2$] (5): The compound was prepared following the procedure given for complex 4 using [HpyTeCl$_2$] instead [HCF$_3$pyTeCl$_2$]. Red crystals. Yield: 66 mg (65%). Elemental analysis: Calcd for $C_{41}H_{34}Cl_2NO_2P_2ReTe$: C 48.31, H 3.36, N 1.37%. Found: C 48.38, H 3.36, N 1.40%. ^{125}Te NMR (CDCl$_3$, ppm): 1724.4. ^{31}P NMR (CDCl$_3$, ppm): −7.6 ppm. IR (cm^{-1}): 3080, 3056, 1558, 1544, 1051, 997, 913, 745, 669, 449, 259. Raman (cm^{-1}): 1050, 1000, 639, 249.

2.2. X-ray Crystallography

The intensities for the X-ray determinations were collected on an STOE IPDS II instrument with Mo Kα radiation or on Bruker Apex CCD diffractometers with Mo Kα or Ag Kα radiation. The space groups were determined by the detection of systematic absences. Absorption corrections were carried out by multiscan or integration methods [29,30]. Structure solution and refinement were performed with the SHELX program package using the OLEX2 platform [31–33]. Hydrogen atoms were derived from the final Fourier maps and refined, or placed at calculated positions and treated with the 'riding model' option of SHELXL. The representation of molecular structures was done using the program DIAMOND 4.2.2 [34].

2.3. Computational Details

DFT calculations were performed on the high-performance computing systems of the Freie Universität Berlin ZEDAT (Curta) using the program packages GAUSSIAN 09 and GAUSSIAN 16 [35,36]. The gas phase geometry optimization was performed using coordinates derived from the X-ray crystal structure using GAUSSVIEW [37]. The calculations were performed with the hybrid density functional B3LYP [38–40]. The double-ζ pseudopotential LANL2DZ basis set with the respective effective core potential (ECP) was applied to Re, while additional polarization functions (dp) were included for tellurium [41–43]. The 6-311++G** basis set was applied for all other atoms [44–48]. All basis sets as well as the ECPs were obtained from the basis set exchange (BSE) database [49]. Frequency calculations confirmed the optimized structures as minima. No negative frequencies were obtained for the given optimized geometries of all compounds. The NBO analysis was performed using the NBO6.0 functionality as implemented in GAUSSIAN. Further analysis of orbitals, charges, topology, etc., and their visualization was performed with the free multifunctional wavefunction analyzer Multiwfn [50,51]. Visualization of the mapped basins, which were calculated in Multiwfn, was done with GAUSSVIEW [37]. The visualization of the orbitals was done in Avogadro [52].

3. Results and Discussions

3.1. Bis(5-Trifluoromethyl-2-pyridyl)ditellane and the Zwitterions [HCF$_3$pyTeX$_2$] (X = Cl, Br)

The CF$_3$-substituted pyridylditellane **1** can readily be prepared following the general procedure for the non-substituted compound [27]. The treatment of 2-chloro-5-(trifluoromethyl)pyridine with two equivalents of elemental tellurium and an excess of NaBH$_4$ in boiling ethanol gives the ditelluride in excellent yields (Scheme 1). A crystalline product is obtained from a chloroform/ethanol mixture. The orange-red crystals are readily soluble in common organic solvents such as CHCl$_3$, acetonitrile, or THF. The purity of the ditelluride **1** can readily be checked by its ^{19}F and/or ^{125}Te NMR spectra. They give narrow signals at −62.5 ppm (^{19}F) and 454.6 ppm (^{125}Te). The ^{125}Te resonance appears close to the signal, which was previously obtained for the non-fluorinated ditelluride {pyTe}$_2$ (427.7 ppm) [27].

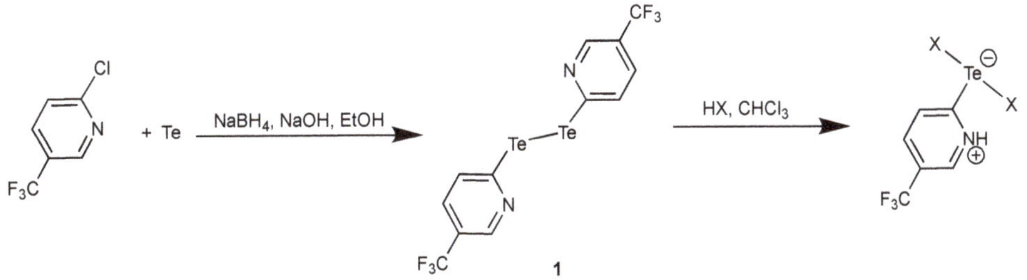

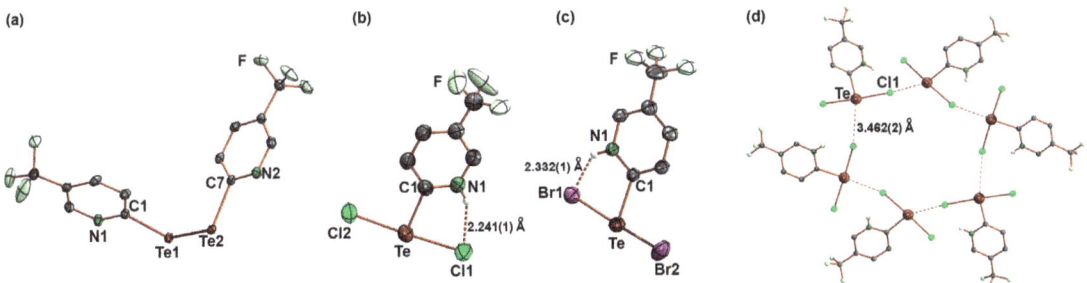

Scheme 1. Syntheses of {CF$_3$pyTe}$_2$ (**1**), [HCF$_3$pyTeCl$_2$] (**2**), and [HCF$_3$pyTeBr$_2$] (**3**).

Single crystals of **1** suitable for X-ray diffraction were obtained from the slow evaporation of a CHCl$_3$/EtOH mixture. An ellipsoid representation of the molecular structure of the CF$_3$-substituted ditellane is shown in Figure 2. The solid-state structure is unexceptional with a Te-Te bond length of 2.689(2) Å and C-Te-Te angles of 99.83(6) and 100.05(6)°. These values are close to those found for (2-pyTe)$_2$ [53]. More details about the crystallographic data are given in Supplementary Material.

Figure 2. Ellipsoid representations of the molecular structures of (**a**) {CF$_3$pyTe}$_2$ (**1**), (**b**) [HCF$_3$pyTeCl$_2$] (**2**) and (**c**) [HCF$_3$pyTeBr$_2$] (**3**), and (**d**) depiction of the assembly of compound **2** to hexameric units in the solid state due to Te . . . Cl long range interactions.

The treatment of a solution of compound **1** in CHCl$_3$ with HCl or HBr results in a cleavage of the Te-Te bond and the formation of zwitterionic compounds of the composition [HCF$_3$pyTeX$_2$] (X = Cl: **2**, X = Br: **3**). This synthetic route has been shown to be favorable for the synthesis of zwitterions, which precipitate directly from the reaction mixture. In this way, products of higher purity can be obtained than following the conventional route, where elemental halogens are used for the oxidation of tellurium [54–58]. Unfortunately, the low solubility of the [HCF$_3$pyTeX$_2$] zwitterions and their gradual decomposition in solution prevent from the recording of NMR spectra of sufficient quality. The presence of hydrogen bonds between the pyridine rings and the halides is supported by the detection of IR bands in the range around 2500 cm^{-1}. FT-Raman spectra of **2** and **3** allow the identification of some more characteristic bands as ν_{C-F} vibrations (1231 and 236 cm^{-1} for **2** and 1233 and 268 cm^{-1} for **3**). Bands at 284 and 260 cm^{-1} (compound **2**), and 189 and 169 cm^{-1} (compound **3**) can be assigned to the corresponding ν_{Te-X} stretches. The values agree with previous assignments on similar compounds [59–61].

Single crystals of **2** and **3** suitable for X-ray diffraction were obtained directly from the reaction mixtures. All our attempts to recrystallize the compounds did not result in crystals of better quality, since a gradual decomposition of the fluorinated products in solution was observed. Figure 2 contains ellipsoid plots of the two zwitterions. They show the expected T-shaped coordination environment of the tellurium atoms with C-Te-X angles between 87.1 and 90.7°. The Te-Cl bond lengths of 2.576(1) and 2.525(1) Å and the Te-Br bonds between 2.640(1) and 2.819(1) Å are in the usual range for tellurium(II) compounds. Hydrogen bonds are established between the pyridinium nitrogen atom and chlorine or bromine atoms. Summarizing, the structural features found in the molecular structures of [HCF$_3$pyTeCl$_2$] and [HCF$_3$pyTeBr$_2$] are similar to those of the non-substituted zwitterions [HpyTeX$_2$] [27,62].

The solid-state structures of compounds **2** and **3** are characterized by weak contacts between the tellurium and halogen atoms. Such non-covalent bonds are not unusual in the chemistry of heavy chalcogens and the nature of such chalcogen and/or halogen bonds is of permanent interest in different fields of chemical science [63–67]. Intermolecular tellurium-chlorine contacts of 3.462(2) Å produce hexameric assemblies in the rhombohedral structure of [HCF$_3$pyTeCl$_2$], while trimeric units with Te ... Br contacts of 3.561(1) Å are established in the monoclinic structure of compound **3**. A visualization of the latter contacts can be found in the Supplementary Material.

3.2. [ReO$_2$Cl(CF$_3$pyTeCl)(PPh$_3$)$_2$] (4) and [ReO$_2$Cl(pyTeCl)(PPh$_3$)$_2$] (5)

The zwitterionic tellurenyl compounds do not just show interesting bonding features in their solid-state structures but are also facile synthons in coordination chemistry. This has been shown with the synthesis of several copper and palladium complexes [25,26,68]. Since rhenium complexes with tellurium-containing ligands are still rare and are mainly restricted to telluroethers, tellurolates, and some ditellurides [69], we now performed a reaction of the common rhenium(V) precursor [ReOCl$_3$(PPh$_3$)$_2$] with [HCF$_3$pyTeCl$_2$] (Scheme 2). The sparingly soluble starting materials dissolve in a boiling mixture of DMF/EtOH within 15 min and red crystals deposit during slow evaporation of the solvents. They have a composition of [ReO$_2$Cl(CF$_3$pyTeCl)(PPh$_3$)$_2$] (**4**). The presence of a Re=O double bond is strongly indicated by an IR band at 910 cm^{-1}, which is the typical region for the *trans*-{ReO$_2$}$^+$ complexes, while the corresponding bands in mono-oxido complexes usually appear at higher wavenumbers [70].

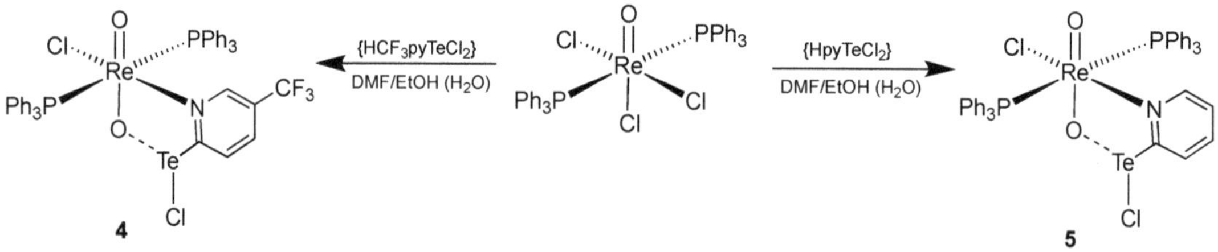

Scheme 2. Syntheses of [ReO$_2$Cl(CF$_3$pyTeCl)(PPh$_3$)$_2$] (**4**) and [ReO$_2$Cl(pyTeCl)(PPh$_3$)$_2$] (**5**).

Similar to the zwitterionic starting materials, the rhenium complex with the CF$_3$-substituted ligand is not stable in the solution. Its gradual decomposition allows the recording of ^{31}P and ^{19}F NMR spectra, but unfortunately not of ^{125}Te spectra of sufficient quality due to the long data acquisition times for this nucleus. This is particularly unfortunate since information about the electronic situation of the tellurium atom in the novel complex would be helpful to understand the bonding situation in the bimetallic compound. Luckily, the crystals, which were deposited from the reaction mixture, could be used for an X-ray diffraction study. The molecular structure of **4** is shown in Figure 3a and selected bond lengths and angles are summarized in Table 1. It becomes evident that the zwitterionic starting material deprotonates and coordinates with its pyridine ring in the equatorial coordination sphere of rhenium replacing a chlorido ligand. An additional and interesting interaction is established between the tellurium atom and an oxygen atom, which is bonded to rhenium. The bonding situation in the{Re-O2-Te(py)Cl}$^{3+}$ fragment is somewhat ambiguous, since it can be understood as a donation of electron density from an oxido ligand of the rhenium complex to the tellurium building block giving a "Re^V/Te^{II} situation" (Figure 3c), but also the formation of a tellurinic acid fragment, which donates with its oxygen atom to the sixth coordination position of rhenium, cannot be ruled out entirely. The latter case would produce a "Re^{III}/Te^{IV} situation".

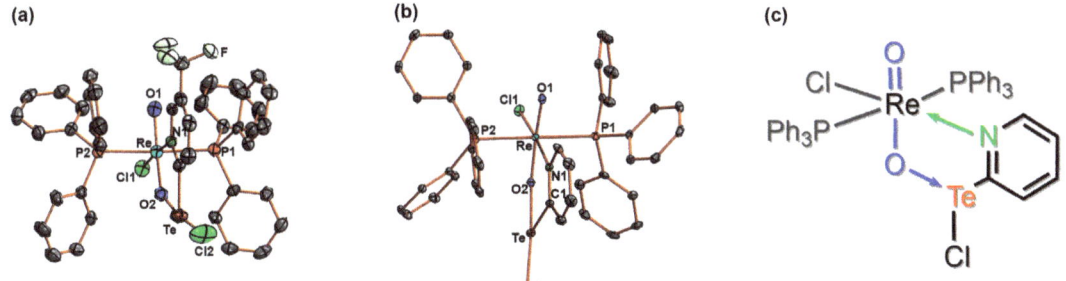

Figure 3. Ellipsoid representations of the molecular structures of (a) [ReO$_2$Cl(CF$_3$pyTeCl)(PPh$_3$)$_2$] (**4**), (b) [ReO$_2$Cl(pyTeCl)(PPh$_3$)$_2$] (**5**), and (c) a depiction of the bonding situation between the two metal centers derived from the experimental bond lengths and angles (Table 1).

Table 1. Selected bond lengths/Å and angles/° in ReO$_2$Cl(CF$_3$pyTeCl)(PPh$_3$)$_2$] (**4**) and (b) [ReO$_2$Cl(pyTeCl)(PPh$_3$)$_2$] (**5**).

	Re-O1	Re-O2	O2-Te	Te-Cl2	O1-Re-O2	Re-O2-Te	O2-Te-Cl2	O2-Te-C1
4	1.721(4)	1.822(3)	2.102(4)	2.578(2)	165.9(2)	135.2(2)	171.6(1)	81.0(2)
5	1.730(2)	1.824(2)	2.102(2)	2.5736(9)	168.42(9)	134.3(1)	169.99(6)	81.23(9)

In order to produce a second example for such compounds, which probably would be stable enough in solution to provide ^{125}Te NMR data for the evaluation of the bonding situation, we performed a reaction of [ReOCl$_3$(PPh$_3$)$_2$] with the unsubstituted zwitterion [HpyTeCl$_2$] (Scheme 2). The product, [ReO$_2$Cl(pyTeCl)(PPh$_3$)$_2$] (**5**), finally possesses the same basic structure as compound **4**. An ellipsoid representation of the molecular structure is shown in Figure 3b and selected bond lengths and angles are compared with the values in complex **4** in Table 1.

The arrangement of the two oxygen atoms in both complexes strongly suggests the presence of a *trans*-{ReO$_2$}$^+$ core, which is frequently found in rhenium(V) complexes with neutral co-ligands [70]. A slightly bent O-Re-O bond as well as the lengthening of the Re-O2 bonds compared with those to the terminal oxido ligands can be understood by the O-Te interactions, which are established from these atoms. Interactions between a lone-pair of an oxido ligand and Lewis acids are not without precedence. Typical examples are rhenium(V)-oxygen-boron bridges, which are readily formed e.g., to electron-deficient boranes [71–76], but also the formation of the {O=Re-O-Re=O}$^{4+}$ unit with a linear oxido bridge between two oxidorhenium(V) centers can be regarded in this sense [70]. More instructive is the bonding situation in a series of 6-diphenylphosphinoacenaphthyl-5-tellurenyl species, 6-Ph$_2$P-Ace-5-TeX (X = Cl, Mes), which have been studied by Beckmann and co-workers [77]. Oxidation of such compounds with H$_2$O$_2$ results in the formation of the corresponding phosphine oxides 6-Ph$_2$P(O)-Ace-5-TeX, in which P=O...Te interactions are established similar to those in the rhenium complexes of the present study. Interestingly, the nature of such interactions was found to be dependent on the residue X, in a way that only weak O...Te interactions a formed with X = Mes (O-Te distance: 2.837(2) Å), while 'dative bonds' were found for X = Cl (O-Te distance: 2.310(3) Å) [77]. The latter situation approximately describes what we found for the rhenium complexes **4** and **5**, where O-Te distances of 2.102 Å were detected.

[ReO$_2$Cl(pyTeCl)(PPh$_3$)$_2$] (**5**) is, fortunately, more stable in solution than its CF$_3$-substituted analog **4**. This allows the recording of ^{31}P and ^{125}Te NMR spectra (Figure 4). Particularly the ^{125}Te spectrum should be indicative for an evaluation of the bonding situation. At the first glance the measured chemical shift of 1724.4 ppm for **5** is surprising, since the value comes close to that observed for the monomeric tellurinic acid (ppy)Te^{IV}(O)OH (1469 ppm), where pph is (2-phenylazo)phenyl-C,N' [78]. But also with this point, the careful study of Beckmann et al. gives a plausible explanation. They also detected a strong dependence of the ^{125}Te chemical shift on the efficacy of the oxygen-tellurium orbital overlap [77]. They found values of 519.1 ppm for 6-Ph$_2$P(O)-Ace-5-TeMes (with only weak O... Te contacts), but 1622.5 ppm for 6-Ph$_2$P(O)-Ace-5-TeCl with the oxygen atom of the phosphine oxide donating to the Te(II) unit (*vide supra*) [77]. A recently published, detailed analysis of ^{125}Te NMR chemical shifts in organotellurium(II) compound confirmed the observed strong effects of substituents on the shielding of the tellurium nuclei in such compounds [79].

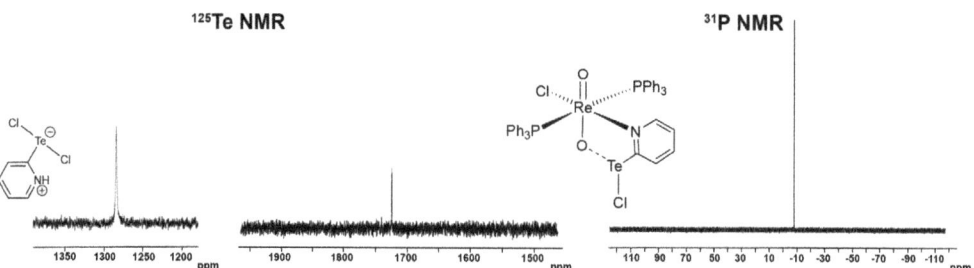

Figure 4. ^{125}Te and ^{31}P NMR spectra of {HpyTeCl$_2$} and [ReO$_2$Cl(pyTeCl)(PPh$_3$)$_2$] (**5**).

4. Computational Studies

The character of the O1-Re-O2-Te bonding is interesting for two reasons: (1) the oxidation states of rhenium and tellurium (although formal) should be fixed and (2) the bonding situation between these three atoms needs to be clarified. To answer these questions, we performed some DFT calculation on the B3LYP level. The gas-phase optimized geometry matches the experimentally observed geometry within 0.01 Å for the organic parts of the molecule, while the deviations are larger around the metals with an average deviation of 0.02 Å. The maximum deviation was found for the tellurium chlorine bond (0.058 Å). Such deviations around the metals are expected for a gas-phase calculation compared to the solid-state experimental data at this level of theory. The small deviations from the experimental data and the verification of the energetic minimum by a frequency calculation indicate that the obtained geometry is reasonable.

To understand about the oxidation state of the tellurium and the general charge distribution in the compound, we performed a QTAIM partitioning followed by integration of the electron density in the basins with a medium-sized grid and additional approximate refinement of the basin boundaries (giving the Bader-type charges) and a calculation of atomic dipole moment corrected Hirshfeld charges (ADCH) [5]. Several other local (i.e., calculated at the critical points in the topological analysis) and integral (i.e., calculated over the atomic basin in different space partitioning methods) topological descriptors of the electron density were employed in the past analyses of the bonding in molecules containing transition metals. Some of the most important local descriptors at the (3,−1) critical point, also referred to as bond critical point, are the electron density $\rho(r)$, the ellipticity $\varepsilon(r) = [\lambda_1(r)/\lambda_2(r)] - 1$ where $\lambda_1(r)$ and $\lambda_2(r)$ are the lowest and the second lowest eigenvalues of the Hessian matrix of $\rho(r)$, the ratio values between the perpendicular and the parallel curvatures–a covalency index $\eta(r) = |\lambda_1(r)|/\lambda_3(r)$ with $\lambda_1(r)$ and $\lambda_3(r)$ as the lowest and the highest eigenvalues of the Hessian matrix of $\rho(r)$, the Laplacian of the electron density ($\nabla^2\rho(r)$), the kinetic energy density ratio ($G(r)/\rho(r)$), and the total energy density ratio ($H(r)/\rho(r)$; $H(r) = G(r) + V(r)$) as well as $1/4\nabla^2\rho(r) = 2G(r) + V(r)$, where $V(r)$ is the potential energy density [5,80–82]. The delocalization index $\delta(A–B)$ is an integral property and indicates the number of electron pairs shared by the basins belonging to the atoms A and B [5,80–82]. Several of these descriptors, namely $\rho(r)$, $\varepsilon(r)$, $\eta(r)$, $\nabla^2\rho(r)$ and $H(r)$ were studied in the present system. Additionally, we calculated the Wiberg bond order matrix [5]. To rationalize these results, we also performed a natural bond orbital (NBO) analysis of the system and regarded the second-order perturbation analysis for metal-involving multiple bonding. Details about the results of these considerations (tabular material and visualizations) are given in the Supplementary Material.

First and foremost, the NBO analysis reveals only one lone-pair localized on rhenium and two lone-pairs localized on tellurium. Therefore, the overall Lewis structure depiction is consistent with a Re(V) center and a Te(II) center. The lone-pairs (LP) are shown in Figure 5. This finding is consistent with the respective Bader and Atomic Dipole Moment Corrected Hirshfeld atomic charges, which indicate a larger positive charge of rhenium compared to that of tellurium.

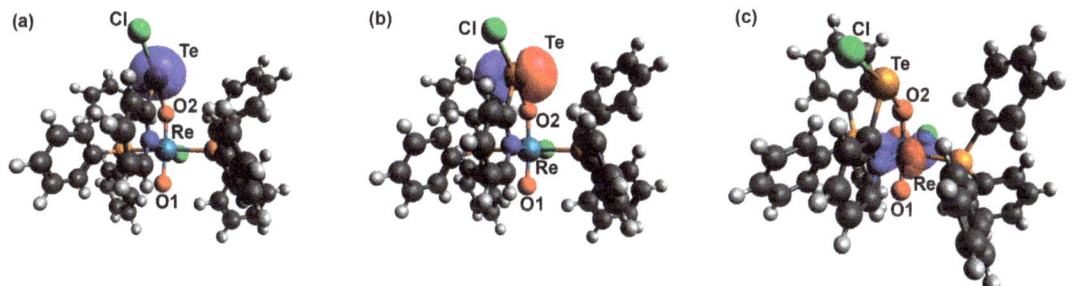

Figure 5. Mapping of (**a**) the s-type LP1 σ-orbital isosurface of Te, (**b**) the p-type π-orbital LP2 isosurface of Te and (**c**) the $d_{x^2-y^2}$-type δ-orbital LP1 orbital isosurface of Re at isosurface values of 0.06.

Regarding the bonding in the O1-Re-O2-Te fragment, the O1-Re bond represents a rather covalent double bond with significant triple bond character due to additional delocalization of LP1 and LP2 into the unoccupied LV1 orbital of rhenium. This observation is also expressed in the Wiberg bond order, which is bigger than 2 for this bond, and the valency of 3 for the oxygen atom O1. On the other hand, the Re-O2 bond can be described as a rather ionic single bond with double bond character due to a significant donation of LP3 of O1 into the Re-O1 π*-symmetry anti-bond, which essentially leads to the formation of a highly delocalized 3c4e bond. The O2-Te bond is again a rather ionic single bond due to stabilizing donation of the LP1, LP2 and LP4 lone pairs of O2 into the empty LV1 orbital on the tellurium atom. Albeit significant π- or double bond character is implied, the bond orders of O2 with Re and Te, respectively, are consistent with single bonds resulting in the overall valency of 2 for the oxygen atom O2. Due to the low mutual geometric accessibility (LP1 to LV1: $F = 0.095$) combined with the high energy difference of the respective orbitals (LP4 to LV1 $\Delta E = 0.58$ Hartree) of the tellurium atoms to form an additional π-bond with the oxygen, a tellurium-oxygen double bond cannot be evidenced despite some indicated delocalization. Conversely, the high delocalization energy from the geometrically accessible LP2 and LP4 orbitals of O2 into the LV1 of rhenium ($F = 0.176$ and $F = 0.186$) allows for the conclusion of a partial double or even triple bond between these two atoms, albeit with high energy differences between the donor and acceptor orbital. Overall, the bigger delocalization energy gain for the formal Re-O2 multiple-bonds compared to the Te-O2 multi-bond character, a Re(V)/Te(II) combination is evidenced albeit some resonance structures involving a Re(III)/Te(IV) structure appears valid. The donor and acceptor orbitals of the lone-pairs of the oxygen atom O2 are shown in the Supplementary Material.

A mapping of the electron localization function (ELF) in the Re-O-Te plane visualizes the degree of electron delocalization (Figure 6a). Critical points of the types (3,−3), (3,−1) and (3,1) are indicated as well as the paths connecting them. Details about their meaning and interpretation are given as Supplementary Information. For the present example, the Wiberg bond orders, the charges, and the NBO analysis of the system support the interpretation of (3,1) critical points, also referred to as ring critical points, as descriptors of an aromatically delocalized ring system involving Re and Te. On the contrary, the implied high ionicity of most of the participating electron pairs is consistent with a higher local concentration of the corresponding electrons around the donor atoms. This is also visualized by the ELF and the Laplacian maps in the Re-O-Te plane. Nevertheless, in the centers of the Re and Te involving ring systems, there is a region of considerable charge depletion around the corresponding (3,1) critical points, indicating the presence of ring-shaped, resonance-delocalized electron density around them. On the donor atoms, the directionality of the donor orbitals is clearly visible.

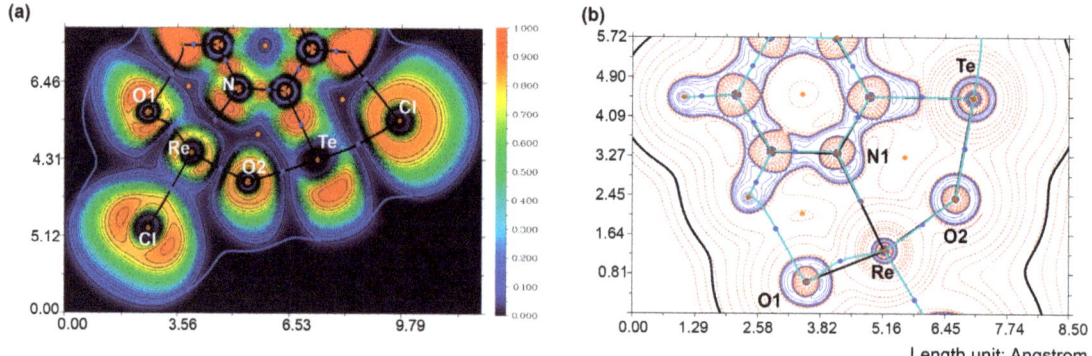

Figure 6. (a) ELF with critical points and bond paths: (3,−3) = brown, (3,−1) = blue, (3,1) = orange, bond path between (3,−3) and (3,−1) critical points = black. Contour lines of electron density isosurfaces and van der Waals radius (blue) are shown. (b) Laplacian map of the electron density (blue = negative; red = positive) with negative values corresponding to local electronic charge accumulation while positive values indicate regions of local electronic charge depletion in the O1-Re-O2-Te plane with topological descriptors and bond critical points: (3,−3) = brown, (3,−1) = blue, (3,1) = orange, bond path between (3,−3) and (3,−1) critical points = cyan. A contour line of the van der Waals radius is shown in black.

All bonds in the chelate ring feature some π-participation, given the bending of the corresponding orbitals towards each other. This is also presented by the ellipticities of ca. 0.2 for the bond critical points in the chelate ring, except for the Re-O2 bond, which has an ellipticity of only 0.04. This is especially surprising, given the large degree of delocalization of π-electrons according to the NBO analysis, and could be interpreted as a hint towards a Re(III)/Te(IV) resonance structure with less π-participation in the Re-O2 bond. Furthermore, an ellipticity of ca. 0.5 verifies the large π-character in the triple bond between Re and O1. Overall, the $\nabla^2\rho(r)>0$, the small $\eta(r)$, the small $\rho(r)$ and the small $\delta(A,B)$ for the Re-O2 and O2-Te bonds are consistent with a non-covalent closed-shell nature of these bonds [5,80–82]. In contrast, the Re-O1 bond, is more covalent than the bonds involving O2. The covalency, thus, increases in the order Te-O2 < Re-O2 < Re-O1.

Finally, we calculated the gradient vector field of the electron density and mapped it with electron density contour lines and topological features to learn about the directionality of the bonds under discussion (Figure 7). As a result, it can be stated that rhenium-ligand bonds are more directional than the tellurium-ligand bonds when looking at the ligand basins. However, in the metal basins, the rhenium shows a more directional and even distribution of electron density over the space with the ligand bonds located between the rhenium lone-pair. Conversely, tellurium is more polarized and less directional in this plane. The highly directional C-Te bond is an exception to the other rather dispersed tellurium-ligand bonds. Furthermore, it is evident, that the bonds involving O2 show very little directionality and are spread between the tellurium and rhenium rather evenly in this σ-bond plane.

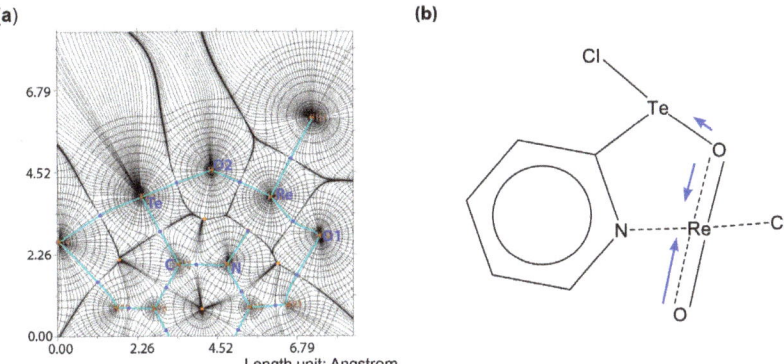

Figure 7. (a) Gradient vector field of the electron density in the O-Re-O-Te plane mapped on the electron density iso contour lines with topological descriptors. (3,3) = brown, (3,1) = blue, (3,−1) = orange, bond path between (3,3) and (3,1) critical points = cyan. (b) Approximated delocalized Lewis structure of the bonding situation in 5 (the lengths of the arrows represent the extent of the donation and the dotted lines indicate 3c4e hyperbonds.

Overall, the [ReO₂Cl(RpyTeCl)(PPh₃)₂] complexes are best described as rhenium(V) dioxido compounds with one of the oxido ligands donating into an organotellurium(II) chloride. However, a tellurinic(IV) acid chloride donating to a rhenium(III) oxido complex is a possible resonance structure. A definite Lewis structure is thus not completely representative of the bonding situation in this compound. A guess on a delocalized Lewis representation is given in Figure 7b, where the length of the arrows indicates the donor strength and dotted lines represent 3c4e hyperbonds. The concluding structural representation is well in accordance with the analysis of the bond lengths and angles determined experimentally by X-ray diffraction. It also fits with the diamagnetism of the complexes.

Supplementary Materials: The following supporting information can be downloaded at: https://www.mdpi.com/article/10.3390/chemistry5020063/s1, Figures S1–S6: Crystallographic data of (CF3pyTe)2 (**1**); (HCF3py)TeCl2 (**2**); (HCF3py)TeBr2 (**3**); [ReO2Cl(pyTeCl)(PPh3)2] (**5**) and [ReO2Cl(CF3pyTeCl)(PPh3)2] (**4**), Figures S7–S24: Spectroscopic Data, Figures S25–S35: Computational Chemistry, Table S1: Crystallographic data and data collection parameters; Table S2: Selected bond lengths (Å) and angles (°)

in (CF3pyTe)2 (**1**); Table S3: Selected bond lengths (Å) and angles (°) in (HCF3py)TeCl2 (**2**); Table S4: Selected bond lengths (Å) and angles (°) in (HCF3py)TeBr2 (**3**); Table S5: Selected bond lengths (Å) and angles (°) in [ReO2Cl(pyTeCl)(PPh3)2] (**5**); Table S6: Selected bond lengths (Å) and angles (°) in [ReO2Cl(CF3pyTeCl)(PPh3)2] (**4**); Table S7: Results of the charge analysis for selected atoms in [ReO2Cl(pyTeCl)(PPh3)2] (**5**). Wiberg bond order matrix for selected atoms, relevant bond orders are bold; Table S8; Lone-pair decomposition of Re and Te in [ReO2Cl(pyTeCl)(PPh3)2] (**5**); Table S9: Natural electron configuration of selected atoms in [ReO2Cl(pyTeCl)(PPh3)2] (**5**); Table S10: Selected parameters from the second order perturbation analysis of [ReO2Cl(pyTeCl)(PPh3)2] (**5**). Delocalization, which was interpreted as an ionic bond is bold; Table S11: Three-centered trans-bonds around Re (3c4e hyper-bonds) in [ReO2Cl(pyTeCl)(PPh3)2] (**5**); Table S12: Selected properties of the electron density at important bond critical points in [ReO2Cl(pyTeCl)(PPh3)2] (**5**). References [29,31–34,83] are cited in the supplementary materials.

Author Contributions: Conceptualization, U.A., E.S.L. and F.D.d.S.; methodology, F.D.d.S. and M.R.J.; validation, F.D.d.S., U.A. and M.R.J.; formal analysis, F.D.d.S., M.R.J., A.H. and U.A.; investigation, F.D.d.S., M.R.J. and U.A.; resources, U.A. and E.S.L.; data curation, F.D.d.S., A.H., M.R.J. and U.A.; writing—original draft preparation, U.A.; writing—review and editing, U.A., M.R.J., E.S.L. and F.D.d.S.; visualization and M.R.J.; supervision, U.A. and E.S.L.; project administration, F.D.d.S. and U.A.; funding acquisition, U.A. and E.S.L. All authors have read and agreed to the published version of the manuscript.

Funding: This work was supported by the Coordenadoria de Aperfeiçoamento de Pessoas de Nível Superior (CAPES/DAAD–Probral n. 88881.144118/2017 and CAPES/PRINT, n. 88881.310412/2018, Brazil), Conselho Nacional de Desenvolvimento Científico e Tecnológico (CNPq) and German Academic Exchange Service (DAAD, Germany). We gratefully acknowledge the High-Performance-Computing (HPC) Centre of the Zentraleinrichtung für Datenverarbeitung (ZEDAT) of the Freie Universität Berlin for computational time and support.

Institutional Review Board Statement: Not applicable.

Informed Consent Statement: Not applicable.

Data Availability Statement: Not applicable.

Acknowledgments: The publication of this article was funded by Freie Universität Berlin.

Conflicts of Interest: The authors declare no conflict of interest.

References

1. Gysling, H.J. The ligand chemistry of tellurium. *Coord. Chem. Rev.* **1982**, *42*, 133–244. [CrossRef]
2. Sudha, N.; Singh, H.B. Intramolecular coordination in tellurium chemistry. *Coord. Chem. Rev.* **1994**, *135–136*, 469–515. [CrossRef]
3. Chivers, T. Tellurium compounds of the main-group elements: Progress and prospects. *J. Chem. Soc. Dalton Trans.* **1996**, 1185–1194. [CrossRef]
4. Singh, A.K.; Sharma, S. Recent developments in the ligand chemistry of tellurium. *Coord. Chem.* **2000**, *209*, 49–98. [CrossRef]
5. Chivers, T.; Laitinen, R.S. Tellurium: A maverick among the chalcogens. *Chem. Soc. Rev.* **2015**, *44*, 1725–1739. [CrossRef]
6. Jones, J.S.; Gabbaï, F.P. Coordination and redox non-innocent behavior of hybrid ligands containing tellurium. *Chem. Lett.* **2016**, *45*, 376–384. [CrossRef]
7. Jain, V.K.; Chauhan, R.S. New vistas in the chemistry of platinum group metals with tellurium ligands. *Coord. Chem. Rev.* **2016**, *306*, 270–301. [CrossRef]
8. Arora, A.; Oswal, P.; Datta, A.; Kumar, A. Complexes of metals with organotellurium compounds and nanosized metal tellurides for catalysis, electrocatalysis and photocatalysis. *Coord. Chem. Rev.* **2022**, *459*, 214406. [CrossRef]
9. Gysling, H.J.; Luss, H.R. Synthesis and properties of the hybrid tellurium-phosphorus ligand phenyl o-(diphenylphosphino)phenyl telluride. X-ray structure of [Pt[PhTe(o-(PPh$_2$C$_6$H$_4$)]$_2$][Pt(SCN)$_4$]·2DMF. *Organometallics* **1984**, *3*, 596–598. [CrossRef]
10. Do, T.G.; Hupf, E.; Lork, E.; Mebs, S.; Beckmann, J. Bis(6-diphenylphosphinoacenaphth-5-yl)telluride as a ligand toward coinage metal chlorides. *Dalton Trans.* **2019**, *48*, 2635–2645. [CrossRef]
11. Nordheider, A.; Hupf, E.; Chalmers, B.A.; Knight, F.R.; Buhl, M.; Mebs, S.; Checinska, L.; Lork, E.; Camacho, P.S.; Ashbrook, E.S.; et al. Peri-substituted phosphorus–tellurium systems–An experimental and theoretical investigation of the P···Te through-space interaction. *Inorg. Chem.* **2015**, *54*, 2435–2446. [CrossRef] [PubMed]
12. Do, T.G.; Hupf, E.; Lork, E.; Beckmann, J. Bis(6-diphenylphosphinoacenaphth-5-yl)telluride as a ligand toward manganese and rhenium carbonyls. *Molecules* **2018**, *23*, 2805. [CrossRef] [PubMed]

13. Yang, H.; Lin, T.-P.; Gabbaï, F.P. Telluroether to telluroxide conversion in the coordination sphere of a metal: Oxidation-induced umpolung of a Te–Au bond. *Organometallics* **2014**, *33*, 4368–4373. [CrossRef]
14. Lin, T.-P.; Gabbaï, F.P. Two-electron redox chemistry at the dinuclear core of a TePt platform: Chlorine photoreductive elimination and isolation of a Te^{V}Pt^I complex. *J. Am. Chem. Soc.* **2012**, *134*, 12230–12238. [CrossRef]
15. Gupta, A.; Deka, R.; Srivastava, K.; Singh, H.B.; Butcher, R.J. Synthesis of Pd(II) complexes of unsymmetrical, hybrid selenoether and telluroether ligands: Isolation of tellura-palladacycles by fine tuning of intramolecular chalcogen bonding in hybrid telluroether ligands. *Polyhedron* **2019**, *172*, 95–103. [CrossRef]
16. Gupta, A.K.; Deka, R.; Singh, H.B.; Butcher, R.J. Reactivity of bis[{2,6-(dimethylamino)methyl}phenyl]telluride with Pd(II) and Hg(II): Isolation of the first Pd(II) complex of an organotellurenium cation as a ligand. *New J. Chem.* **2019**, *43*, 13225–13233. [CrossRef]
17. Ji, B.; Ding, K. Synthesis and crystallographic characterization of a palladium(II) complex with the ligand (4-ethoxyphenyl)[(2-amino-5-methyl)phenyl] telluride. *Inorg. Chem. Commun.* **1999**, *2*, 347–350. [CrossRef]
18. Panda, S.; Singh, H.B.; Butcher, R.J. Contrasting coordination behaviour of 22-membered chalcogenaaza (Se, Te) macrocylces towards Pd(II) and Pt(II): Isolation and structural characterization of the first metallamacrocyle with a C–Pt–Se linkage. *Chem. Commun.* **2004**, 322–323. [CrossRef]
19. Panda, S.; Zade, S.S.; Singh, H.B.; Butcher, R.J. The ligation properties of some reduced Schiff base selena/telluraaza macrocycles: Versatile structural trends. *Eur. J. Inorg. Chem.* **2006**, 172–184. [CrossRef]
20. Menon, S.C.; Panda, A.; Singh, H.B.; Patel, R.P.; Kulshreshtha, S.K.; Darby, W.L.; Butcher, R.J. Tellurium azamacrocycles: Synthesis, characterization and coordination studies. *J. Organomet. Chem.* **2004**, *689*, 1452–1463. [CrossRef]
21. Menon, S.C.; Panda, A.; Singh, H.B.; Butcher, R.J. Synthesis and single crystal X-ray structure of the first cationic Pd(II) complex of a tellurium-containing polyaza macrocycle: Contrasting reactions of Pd(II) and Pt(II) with a 22-membered macrocyclic Schiff base. *Chem. Commun.* **2000**, 143–144. [CrossRef]
22. Nakayama, Y.; Watanabe, K.; Ueyama, N.; Nakamura, A.; Harada, A.; Okuda, J. Titanium complexes having chelating diaryloxo ligands bridged by tellurium and their catalytic behavior in the polymerization of ethylene. *Organometallics* **2000**, *19*, 2498–2503. [CrossRef]
23. Takashima, Y.; Nakayama, Y.; Yasuda, H.; Nakamura, A.; Harada, A. Synthesis of *cis*-dichloride complexes of Group 6 transition metals bearing alkyne and chalcogen-bridged chelating bis(aryloxo) ligands as catalyst precursors for ring-opening metathesis polymerization. *J. Organomet. Chem.* **2002**, *654*, 74–82. [CrossRef]
24. Baranov, A.V.; Matsulevich, Z.V.; Fukin, G.F.; Baranov, E.V. Synthesis and structure of pyridine-2-tellurenyl chloride. *Russ. Chem. Bull.* **2010**, *59*, 581–583.
25. Chauhan, R.S.; Kedarnath, G.; Wadawale, A.; Slawin, A.M.Z.; Jain, V.K. Reactivity of 2-chalcogenopyridines with palladium–phosphine complexes: Isolation of different complexes. *Dalton Trans.* **2013**, *42*, 259–269. [CrossRef]
26. Cechin, C.N.; Razera, G.F.; Tirloni, B.; Piquini, P.C.; de Carvalho, L.M.; Abram, U.; Lang, E.S. Oxidation of crude palladium powder by a diiodine adduct of (2-PyTe)$_2$ to obtain the novel Pd^{II} complex [PdI(TePy-2)(I$_2$TePy-2)$_2$]. *Inorg. Chem. Commun.* **2020**, *118*, 107966. [CrossRef]
27. da Silva, F.D.; Simoes, C.A.D.P.; dos Santos, S.S.; Lang, E.S. Versatility of bis(2-pyridyl)ditellane. *ChemistrySelect* **2017**, *2*, 2708–2712. [CrossRef]
28. Parshall, G.W.; Shive, L.W.; Cotton, F.A. Phosphine complexes of rhenium. *Inorg. Synth.* **1977**, *17*, 110–112.
29. Coppens, P. *The Evaluation of Absorption and Extinction in Single-Crystal Structure Analysis*; Crystallographic Computing: Copenhagen, Denmark, 1979.
30. Sheldrick, G.M. *SADABS*; University of Göttingen: Göttingen, Germany, 1996.
31. Sheldrick, G.M. A short history of SHELX. *Acta Crystallogr.* **2008**, *64*, 112–122. [CrossRef]
32. Sheldrick, G.M. Crystal structure refinement with SHELXL. *Acta Crystallogr.* **2015**, *71*, 3–8.
33. Dolomanov, O.V.; Bourhis, L.J.; Gildea, R.J.; Howard, J.A.K.; Puschmann, H. OLEX2: A complete structure solution, refinement and analysis program. *J. Appl. Cryst.* **2009**, *42*, 339–341. [CrossRef]
34. Putz, H.; Brandenburg, K. *DIAMOND, Crystal and Molecular Structure Visualization Crystal Impact*; Version 4.6.5; Brandenburg GbR: Bonn, Germany, 2021.
35. Frisch, M.J.; Trucks, G.W.; Schlegel, H.B.; Scuseria, G.E.; Robb, M.A.; Cheeseman, J.R.; Scalmani, G.; Barone, V.; Petersson, G.A.; Nakatsuji, H.; et al. *Gaussian 16*; Revision B.01; Gaussian, Inc.: Wallingford, CT, USA, 2016.
36. Frisch, M.J.; Trucks, G.W.; Schlegel, H.B.; Scuseria, G.E.; Robb, M.A.; Cheeseman, J.R.; Scalmani, G.; Barone, V.; Petersson, G.A.; Nakatsuji, H.; et al. *Gaussian 09*; Revision A.02; Gaussian, Inc.: Wallingford, CT, USA, 2016.
37. Dennington, R.; Keith, T.A.; Millam, J.M. *GaussView*; Version 6; Semichem Inc.: Shawnee Mission, KS, USA, 2016.
38. Vosko, S.H.; Wilk, L.; Nusair, M. Accurate spin-dependent electron liquid correlation energies for local spin density calculations: A critical analysis. *Can. J. Phys.* **1980**, *58*, 1200–1211. [CrossRef]
39. Becke, A.D. Density-functional thermochemistry. III. The role of exact exchange. *J. Chem. Phys.* **1993**, *98*, 5648–5652. [CrossRef]
40. Lee, C.; Yang, W.; Parr, R.G. Development of the Colle-Salvetti correlation-energy formula into a functional of the electron density. *Phys. Rev.* **1988**, *37*, 785–789. [CrossRef] [PubMed]
41. Hay, P.J.; Wadt, W.R.; Willard, R.J. Ab initio effective core potentials for molecular calculations. Potentials for the transition metal atoms Sc to Hg. *Chem. Phys.* **1985**, *82*, 299–310. [CrossRef]

42. Wadt, W.R.; Hay, P.J.J. Ab initio effective core potentials for molecular calculations. Potentials for main group elements Na to Bi. *Chem. Phys.* **1985**, *82*, 284–298. [CrossRef]
43. Check, C.E.; Faust, T.O.; Bailey, J.M.; Wright, B.J.; Gilbert, T.M.; Sunderlin, L.S.J. Addition of Polarization and Diffuse Functions to the LANL2DZ Basis Set for P-Block Elements. *Phys. Chem. A* **2001**, *105*, 8111–8116. [CrossRef]
44. Clark, T.; Chandrasekhar, J.; Spitznagel, G.W.; Schleyer, P.V.R. Efficient diffuse function-augmented basis sets for anion calculations. III. The 3-21+G basis set for first-row elements, Li-F. *J. Comput. Chem.* **1983**, *4*, 294–301. [CrossRef]
45. Francl, M.M.; Pietro, W.J.; Hehre, W.J.; Binkley, J.S.; Gordon, M.S.; DeFrees, D.J.; Pople, J.A. Self-consistent molecular orbital methods. XXIII. A polarization-type basis set for second-row elements. *J. Chem. Phys.* **1982**, *77*, 3654–3665. [CrossRef]
46. Krishnan, R.; Binkley, J.S.; Seeger, R.; Pople, J.A. Self-consistent molecular orbital methods. XX. A basis set for correlated wave functions. *J. Chem. Phys.* **1980**, *72*, 650–654. [CrossRef]
47. McLean, A.D.; Chandler, G.S. Contracted Gaussian basis sets for molecular calculations. I. Second row atoms, Z = 11–18. *J. Chem. Phys.* **1980**, *72*, 5639–5648. [CrossRef]
48. Spitznagel, G.W.; Clark, T.; von Rague Schleyer, P.; Hehre, W.J. An evaluation of the performance of diffuse function-augmented basis sets for second row elements, Na-Cl. *J. Comput. Chem.* **1987**, *8*, 1109–1116. [CrossRef]
49. Pritchard, B.P.; Altarawy, D.; Didier, B.; Gibson, T.D.; Windus, T.L.J. New Basis Set Exchange: An Open, Up-to-Date Resource for the Molecular Sciences Community. *Chem. Inf. Model.* **2019**, *59*, 4814–4820. [CrossRef] [PubMed]
50. Lu, T.; Chen, F. Multiwfn: A multifunctional wavefunction analyzer. *J. Comput. Chem.* **2012**, *33*, 580–592. [CrossRef] [PubMed]
51. Lu, T.; Chen, F. Quantitative analysis of molecular surface based on improved Marching Tetrahedra algorithm. *J. Mol. Graph. Model.* **2012**, *38*, 314–323. [CrossRef]
52. Hanwell, M.D.; Curtis, D.E.; Lonie, D.C.; Vandermeersch, T.; Zurek, E.; Hutchison, G.R. Avogadro: An advanced semantic chemical editor, visualization, and analysis platform. *J. Cheminformatics* **2012**, *4*, 1–17. [CrossRef]
53. Bhasin, K.K.; Arora, V.; Klapötke, T.M.; Crawford, M.-J. One-Pot Synthesis of Pyridyltellurium Derivatives from a Reaction with Isopropylmagnesium Chloride and X-ray Crystal Structures of Various Pyridyl Ditellurides. *Eur. J. Inorg. Chem.* **2004**, 4781–4788. [CrossRef]
54. Hauge, S.; Vikane, O. Three-coordinated Divalent Tellurium Complexes: The Crystal Structures of Tetraphenylarsonium Diiodophenyltellurate(II) and Tetraphenylarsonium Bromoiodophenyltellurate(II). *Acta Chem. Scand.* **1983**, *A37*, 723–728. [CrossRef]
55. Du Mont, W.-W.; Meyer, H.-U.; Kubiniok, S.; Pohl, S.; Saak, W. Spaltung sperriger Diaryldltelluride mit Brom und Iod; Strukturbestimmung an Et$_4$N$^+$ 2,4,6-(i-C$_3$H$_7$)$_3$C$_6$H$_2$TeI$^-$$_2$. *Chem. Ber.* **1992**, *125*, 761–766. [CrossRef]
56. Faoro, E.; de Oliveira, G.M.; Schulz Lang, E. Synthesis and Structural Characterization of the novel T-shaped Organotellurium(II) Dihalides (PyH)[mesTeClBr] and (PyH)[mesTeX$_2$] (Py = pyridine; mes = mesityl; X = Cl, Br). *Z. Anorg. Allg. Chem.* **2006**, *632*, 2049–2052. [CrossRef]
57. Schulz Lang, E.; de Oliveira, G.M.; Casagrande, G.A. Synthesis of new T-shaped hypervalent complexes of tellurium showing Te–π-aryl interactions: X-ray characterization of [(mes)XTe(μ-X)Te(mes)(etu)] (X = Br, I) and [Ph(etu)Te(μ-I)Te(etu)Ph][PhTeI4] (mes = mesityl; etu = ethylenethiourea). *J. Organomet. Chem.* **2006**, *691*, 59–64. [CrossRef]
58. Faoro, E.; Oliveira, G.M.; Schulz Lang, E.; Pereira, C.B. Synthesis and structural features of new aryltellurenyl iodides. *J. Organomet. Chem.* **2010**, *695*, 1480–1486. [CrossRef]
59. Dance, H.S.; McWhinnie, W.R. Isotopic studies by vibrational spectroscopy of the tellurium–carbon bond in diaryltellurium dihalides. *J. Chem. Soc. Dalton Trans.* **1975**, 43–45. [CrossRef]
60. Lee, H.; Kim, I.-Y.; Han, S.-S.; Bae, B.-S.; Choi, M.K.; Yang, I.-S. Spectroscopic ellipsometry and Raman study of fluorinated nanocrystalline carbon thin films. *J. Appl. Phys.* **2001**, *90*, 813–818. [CrossRef]
61. Sandmann, D.J.; Li, L.; Tripathy, S.; Stark, J.C.; Acampora, L.A.; Foxman, B.M. Conformational polymorphism of di-2-naphthyl ditelluride. *Organometallics* **1994**, *13*, 348–353. [CrossRef]
62. Khrustalev, V.N.; Matsulevich, Z.V.; Lukiyanova, J.M.; Aysin, R.R.; Peregudov, A.S.; Leites, L.A.; Borisov, A.V. A Facile Route for Stabilizing Highly Reactive ArTeCl Species Through the Formation of T-Shaped Tellurenyl Chloride Adducts: Quasi-Planar Zwitterionic [HPy*]TeCl$_2$ and [HPm*]TeCl$_2$; Py* = 2-pyridyl, Pm* = 2-(4,6-dimethyl)pyrimidyl. *Eur. J. Inorg. Chem.* **2014**, 3582–3586. [CrossRef]
63. Beckmann, J.; Hesse, M.; Poleschner, H.; Seppelt, K. Formation of mixed-valent aryltellurenyl halides RX$_2$TeTeR. *Angew. Chem. Int. Ed.* **2007**, *46*, 8277–8280. [CrossRef] [PubMed]
64. Benz, S.; Poblador-Bahamonde, A.I.; Low-Ders, N.; Matile, S. Catalysis with pnictogen, chalcogen, and halogen bonds. *Angew. Chem. Int. Ed.* **2018**, *57*, 5408–5412. [CrossRef]
65. Vogel, L.; Wonner, P.; Huber, S.M. Chalcogen bonding: An overview. *Angew. Chem. Int Ed.* **2019**, *58*, 1880–1891. [CrossRef]
66. Bamberger, J.; Ostler, F.; Mancheno, O.G. Frontiers in Halogen and chalcoge-bond donor organocatalysis. *ChemCatChem* **2019**, *11*, 5198–5211. [CrossRef]
67. Ho, P.C.; Wang, J.Z.; Meloni, F.; Vargas-Baca, I. Chalcogen bonding in materials chemistry. *Coord. Chem. Rev.* **2020**, *422*, 213464. [CrossRef]
68. Da Silva, F.D.; Bortolotto, T.; Tirloni, B.; de Freitas Daudt, N.; Schulz Lang, E.; Cargnelutti, R. Bis(2-pyridyl)ditellane as a precursor to Co^{II}, Cu^{I} and Cu^{II} complex formation: Structural characterization and photocatalytic studies. *New J. Chem.* **2022**, *46*, 18165–18172. [CrossRef]

69. Noschang Cabral, B.; Fonseca, J.R.; Roca Jungfer, M.; Krebs, A.; Hagenbach, A.; Schulz Lang, E.; Abram, U. Oxidorhenium(V) and Rhenium(III) Complexes with Arylselenolato and -tellurolato Ligands. *Eur. J. Inorg. Chem.* **2022**, e202300023. [CrossRef]
70. Abram, U. Rhenium. In *Comprehensive Coordination Chemistry II*; McCleverty, J.A., Meyer, T.J., Eds.; Elsevier: Amsterdam, The Netherlands, 2003; Volume 5, pp. 271–403.
71. Smeltz, J.L.; Lilly, C.P.; Boyle, P.D.; Ison, E.A. The electronic nature of terminal oxo ligands in transition-metal complexes: Ambiphilic reactivity of oxorhenium species. *J. Am. Chem. Soc.* **2013**, *135*, 9433–9441. [CrossRef] [PubMed]
72. Lambic, N.S.; Sommer, R.D.; Ison, E.A. Transition-metal oxos as the Lewis basic component of frustrated Lewis pairs. *J. Am. Chem. Soc.* **2016**, *138*, 4832–4842. [CrossRef] [PubMed]
73. Miller, A.J.M.; Nabinger, J.A.; Bercaw, J.E. Homogeneous CO hydrogenation: Ligand effects on the Lewis acid-assisted reductive coupling of carbon monoxide. *Organometallics* **2010**, *29*, 4499–4516. [CrossRef]
74. Lambic, N.S.; Brown, C.A.; Sommer, R.D.; Ison, E.A. Dramatic increase in the rate of olefin insertion by coordination of Lewis acids to the oxo ligand in oxorhenium(V) hydrides. *Organometallics* **2017**, *36*, 2042–2051. [CrossRef]
75. Belanger, S.; Beauchamp, A.L. Oxo ligand reactivity in the [ReO$_2$L$_4$]+ complex of 1-methylimidazole. Preparation and crystal structures of salts containing the ReOL$_4$$^{3+}$ core and apical CH$_3$O$^-$, BF$_3$O$_2$$^-$, and (CH$_3$O)$_2PO_2$$^-$ groups. *Inorg. Chem.* **1997**, *36*, 3640–3647. [CrossRef]
76. Massaaki, A.; Tsuyoshi, M.; Hideki, S.; Akira, N.; Yoichi, S. Lewis acid trifluoroboron coordination to trans-dioxorhenium(V) moiety: Structural and spectroscopic characterization of trans-[Re^V(O)(OBF$_3$)(1-MeIm)$_4$](BF$_4$)(1-MeIm=1-methylimidazole). *Chem. Lett.* **1997**, *26*, 1073–1074.
77. Hupf, E.; Do, T.G.; Nordheider, A.; Wehrhahn, M.; Sanz Camacho, P.; Ashbrook, S.E.; Lork, E.; Slawin, A.M.Z.; Mebs, S.; Woollins, J.D.; et al. Selective oxidation and functionalization of 6-diphenylphosphinoacenaphthyl-5-tellurenyl species 6-Ph$_2$P-Ace-5-TeX (X = Mes, Cl, O$_3$SCF$_3$). Various types of P−E···Te(II,IV) bonding situations (E = O, S, Se). *Organometallics* **2017**, *36*, 1566–1579. [CrossRef]
78. Deka, R.; Sarkar, A.; Butcher, R.J.; Junk, P.C.; Turner, D.R.; Deacon, G.B.; Singh, H.B. Isolation of the novel example of a monomeric organotelllurinic acid. *Dalton Trans.* **2020**, *49*, 1173–1180. [CrossRef] [PubMed]
79. Pietrasiak, E.; Gordon, C.P.; Coperet, C.; Togni, A. Understanding ^{125}Te NMR chemical shifts in dissymmetric organo-telluride compounds from natural chemical shift analysis. *Phys. Chem. Chem. Phys.* **2020**, *22*, 2319–2326. [CrossRef] [PubMed]
80. Cabeza, J.A.; van der Maelen, J.F.; García-Granda, S. Topological Analysis of the Electron Density in the N-Heterocyclic Carbene Triruthenium Cluster [Ru$_3$(μ-H)$_2$(μ^3-MeImCH)(CO)$_9$] (Me2Im = 1,3-dimethylimidazol-2-ylidene). *Organometallics* **2009**, *28*, 3666–3672. [CrossRef]
81. Matito, E.; Solà, M. The role of electronic delocalization in transition metal complexes from the electron localization function and the quantum theory of atoms in molecules viewpoints. *Coord. Chem. Rev.* **2009**, *253*, 647–665. [CrossRef]
82. Poater, J.; Duran, M.; Solà, M.; Silvi, B. Theoretical Evaluation of Electron Delocalization in Aromatic Molecules by Means of Atoms in Molecules (AIM) and Electron Localization Function (ELF) Topological Approaches. *Coord. Chem. Rev.* **2005**, *105*, 3911–3947. [CrossRef]
83. Bruker. *APEX2, SAINT and SADABS*; Bruker AXS Inc.: Madison, WI, USA, 2009.

Disclaimer/Publisher's Note: The statements, opinions and data contained in all publications are solely those of the individual author(s) and contributor(s) and not of MDPI and/or the editor(s). MDPI and/or the editor(s) disclaim responsibility for any injury to people or property resulting from any ideas, methods, instructions or products referred to in the content.

Article

Synthesis of 5-Metalla-Spiro[4.5]Heterodecenes by [1,4]-Cycloaddition Reaction of Group 13 Diyls with 1,2-Diketones

Hanns M. Weinert [1], Christoph Wölper [1] and Stephan Schulz [1,2,*]

[1] Institute of Inorganic Chemistry, University of Duisburg-Essen, Universitätsstr. 5–7, 45141 Essen, Germany
[2] Center for Nanointegration Duisburg-Essen (CENIDE), University of Duisburg-Essen, Carl-Benz-Straße 199, 47057 Duisburg, Germany
* Correspondence: stephan.schulz@uni-due.de

Abstract: Monovalent group 13 diyls are versatile reagents in oxidative addition reactions. We report here [1,4]-cycloaddition reactions of β-diketiminate-substituted diyls LM (M = Al, Ga, In, Tl; L = HC[C(Me)NDipp]$_2$, Dipp = 2,6-^{i}Pr$_2$C$_6$H$_3$) with various 1,2-diketones to give 5-metalla-spiro[4.5]heterodecenes **1**, **4–6**, and **8–10**, respectively. In contrast, the reaction of LTl with acenaphthenequinone gave the [2,3]-cycloaddition product **7**, with Tl remaining in the +1 oxidation state. Compound **1** also reacted with a second equivalent of butanedione as well as with benzaldehyde in aldol-type addition reactions to the corresponding α,β-hydroxyketones **2** and **3**, while a reductive activation of a benzene ring was observed in the reaction of benzil with two equivalents of LAl to give the 1,4-aluminacyclohex-2,4-dien **12**. In addition, the reaction of L'BCl$_2$ (L' = HC[C(Me)NC$_6$F$_5$]$_2$) with one equivalent of benzil in the presence of KC$_8$ gave the corresponding 5-bora-spiro[4.5]heterodecene **13**, whereas the hydroboration reaction of butanedione with L'BH$_2$ (**14**), which was obtained from the reaction of L'BCl$_2$ with L-selectride, failed to give the saturated 5-bora-spiro[4.5]heterodecane.

Keywords: group 13 diyls; low-valent metal; cycloaddition; aldol addition

1. Introduction

Neutral monovalent, six-electron group 13 diyls LM (Al [1], Ga [2], In [3], Tl [4]; L = HC[C(Me)NDipp]$_2$; Dipp = 2,6-^{i}Pr$_2$C$_6$H$_3$) are group 13 analogues of singlet NHC-carbenes. In particular, alanediyl LAl, and gallanediyl LGa have received steadily increasing interest in recent years due to their interesting ambiphilic electronic nature [5–8], resulting from the presence of both a filled donor (HOMO) and an empty (p-type) acceptor orbital (LUMO). As a result, these neutral diyls often exhibit transition metal-like reactivity [9,10]. LAl is more reactive than LGa and has been found to readily undergo oxidative addition reactions with a wide variety of E-X σ-bonds, including thermodynamically very stable C-F bonds [11–13], whereas, LGa often reacted in a more selective way [14].

Cycloaddition reactions of monovalent group 13 diyls have been less explored. Reactions of a sterically demanding gallanediyl with (*p*-tolyl)NN(*p*-tolyl) gave the 1,2-diaza-3,4-dimetallacyclobutane (**I**) (Scheme 1) [15], while reactions of LAl with alkynes afforded cycloaluminapropenes (**II**) [16]. In addition, aluminum pinacolates were obtained from reactions of both LAl [16] and a diamidoalumanyl anion [17] with Ph$_2$CO (**III**) and Ph(CO)CH$_3$ (**IV**), respectively. Similarly, the digallane (dpp-Bian)Ga–Ga(dpp-Bian) (dpp-Bian = 1,2-bis[(2,6-diisopropylphenyl)imino]acenaphthene) reacted with benzaldehyde to the respective 1,2-diphenyl-1,2-ethaneoate adduct (**V**) [18], while its reaction with 3,6-di-tert-butyl-ortho-benzoquinone occurred with oxidation of the Ga (II) atoms and two dpp-Bian dianions to give the mononuclear catecholate **VI** [19], but this reaction most likely doesn't occur via cycloaddition. In contrast, to the best of our knowledge, the formation of [1,4]-cycloaddition products with diketones is limited to trapping experiments of an in situ formed monomeric in compound (**I**) with benzil derivatives (**VII**) [20].

Citation: Weinert, H.M.; Wölper, C.; Schulz, S. Synthesis of 5-Metalla-Spiro[4.5]Heterodecenes by [1,4]-Cycloaddition Reaction of Group 13 Diyls with 1,2-Diketones. *Chemistry* **2023**, *5*, 948–964. https://doi.org/10.3390/chemistry5020064

Academic Editors: Christoph Janiak, Sascha Rohn and Georg Manolikakes

Received: 27 March 2023
Revised: 14 April 2023
Accepted: 18 April 2023
Published: 20 April 2023

Copyright: © 2023 by the authors. Licensee MDPI, Basel, Switzerland. This article is an open access article distributed under the terms and conditions of the Creative Commons Attribution (CC BY) license (https://creativecommons.org/licenses/by/4.0/).

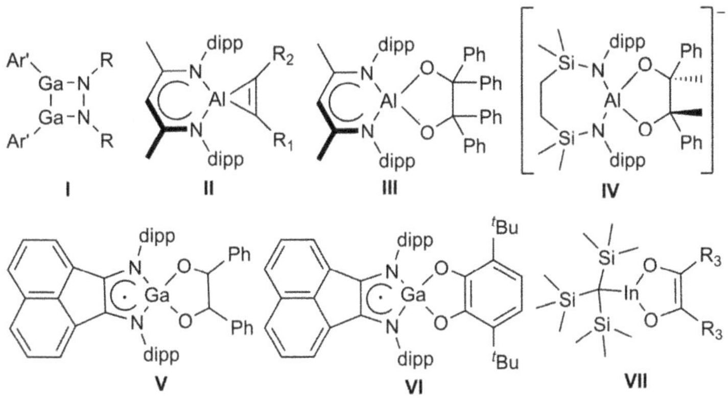

Scheme 1. Compounds formed by cycloaddition reactions of low-valent group 13 compounds. **II–VI**: dipp = -2,6-^{i}Pr$_2$C$_6$H$_3$; **I**: R = -4-CH$_3$-C$_6$H$_5$, Ar' = -C$_6$H$_3$-2,6-(C$_6$H$_3$-2,6-^{i}Pr$_2$)$_2$; **II**: R$_1$ = R$_2$ = -SiMe$_3$, R$_1$ = R$_2$ = -Ph, or R$_1$ = -SiMe$_3$ and R$_2$ = -Ph; **VII**: R$_3$ = -C$_6$H$_5$, 4-MeO-C$_6$H$_4$, or 4-Br-C$_6$H$_4$.

Our general interest in the reactivity of low valent group 13 diyls LM in σ-bond activation reactions [21–26] as well as of unsaturated main group element compounds in single electron transfer and cycloaddition reactions [27–30] let us now focus on reactions of group 13 diyls LM (M = Al, Ga, In, Tl) with 1,2-diketones. Both the group 13 elements and the substituents of the 1,2-diketones were found to influence the product formation. In addition to the expected [1,4]-cycloaddition reaction to 5-metalla-spiro[4.5]heterodecenes (**1, 4–6, 8–10**), we also observed a [2,3]-cycloaddition reaction of the β-diketiminate ligand as well as an activation reaction of a benzoyl group. Furthermore, 5-galla-spiro[4.5]heterodecene **1** was found to undergo aldol addition, and to commemorate Liebig's studies on "Radicals of Benzoic Acid" [31], we isolated compound **3** from the addition reaction of **1** with benzaldehyde.

2. Results and Discussion

2.1. Synthesis

The reaction of LGa with one equivalent of butanedione proceeded with [1,4]-cycloaddition and formation of the expected 5-metalla-spiro[4.5]heterodecene **1** (Scheme 2), while no defined product was isolated from analogous reactions of LAl and LIn. LAl was found to be too reactive, resulting in the formation of a rather complex product mixture, (Figure S50), while LIn is less reactive under these conditions, resulting in an incomplete conversion of LIn (Figure S51). In both cases, no defined compound could be isolated. In addition, the reaction of LTl with butanedione gave a colorless precipitate, which is presumed to be an insoluble thallium enolate, as well as LH according to in situ ^{1}H NMR spectroscopic studies (Figure S52). The reaction of **1** with a second equivalent of butanedione gave the aldol addition product **2**, which was also selectively formed in the reaction of LGa with two equivalents of butanedione. The formation of **1** is most likely kinetically favored due to a lower energy barrier for the cycloaddition reaction compared to the aldol addition, and hence the equimolar reaction mainly gave compound **1**. Moreover, **1** reacted with benzaldehyde to give the aldol addition product **3**. The formation of the aldol addition products as observed in the reactions with LGa is also expected to occur in the reactions with LAl and LIn, explaining the unselectivity of these reactions.

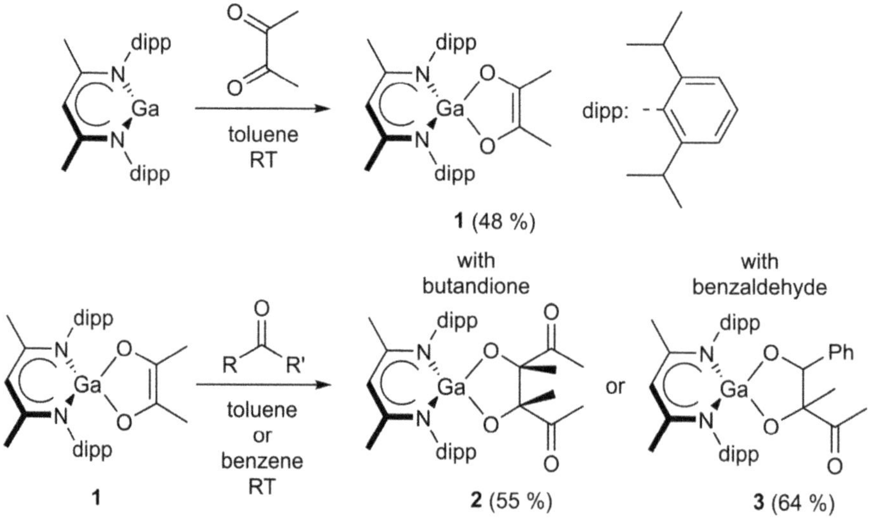

Scheme 2. The reaction of LGa with butanedione **1** and subsequent aldol addition reactions with butanedione (**2**) and benzaldehyde (**3**).

In order to kinetically stabilize the [1,4]-cycloaddition products and to prevent aldol addition side reactions, we increased the steric demand of the 1,2-diketone. LAl, LGa, and LIn reacted selectively with acenaphthenequinone to the corresponding [1,4]-cycloaddition products **4–6**, while LTl reacted to the [2,3]-cycloaddition product **7** with the unsaturated ligand backbone (Scheme 3). The isolation of compound **7** was hampered by its low solubility and its tendency to decompose in solution with the formation of LH.

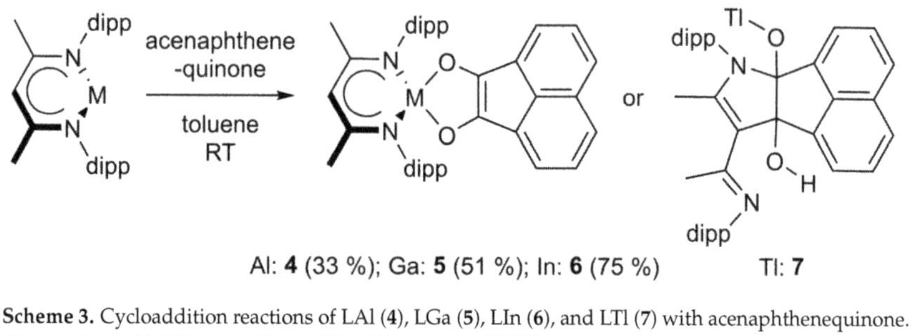

Scheme 3. Cycloaddition reactions of LAl (**4**), LGa (**5**), LIn (**6**), and LTl (**7**) with acenaphthenequinone. The reaction with Tl was performed in n-hexane at $-80\ °C$.

In addition, reactions of LAl and LGa with benzil afforded compounds **8** and **9**, respectively, while in situ ^{1}H NMR monitoring of the reaction of LIn and benzil revealed the formation of multiple products (Scheme 4). The [1,4]-cycloaddition product of the reaction with LIn **10** was finally selectively crystallized from a mixture of acetonitrile and benzene. In contrast, LTl did not react with benzil. This is in agreement with results from quantum chemical calculations, which showed that the energy level of the metal-centered electron lone pair decreases steadily with increasing atomic number, and finally falls below the ligand-centered orbitals in the case of the thallanediyl LTl [32,33].

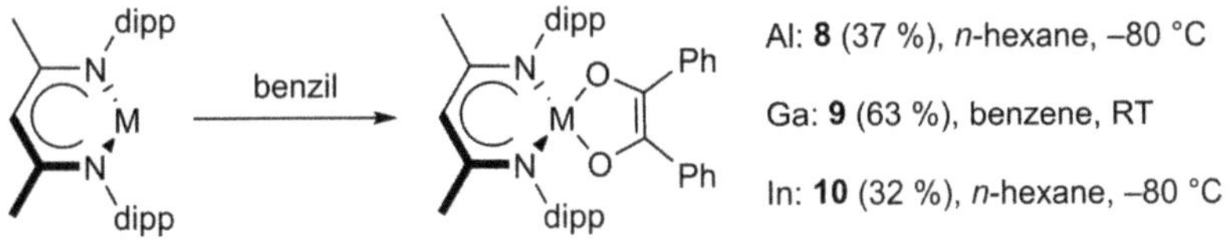
benzil
Al: 8 (37 %), n-hexane, –80 °C
Ga: 9 (63 %), benzene, RT
In: 10 (32 %), n-hexane, –80 °C

Scheme 4. The [1,4]-cycloaddition reactions of LAl (**8**), LGa (**9**), and LIn (**10**) with benzil.

Similar to the formation of compound **7**, the reaction of LAl with two equivalents of benzil resulted in an addition to the β-diketiminate ligand to give compound **11** (Scheme 5). In marked contrast, the slow addition of benzil to a concentrated solution of two equivalents of LAl proceeded with the activation of one of the two phenyl groups of the benzil molecule to finally give compound **12**. The reduction of a benzene ring demonstrates the high reducing power of the alanediyl LAl.

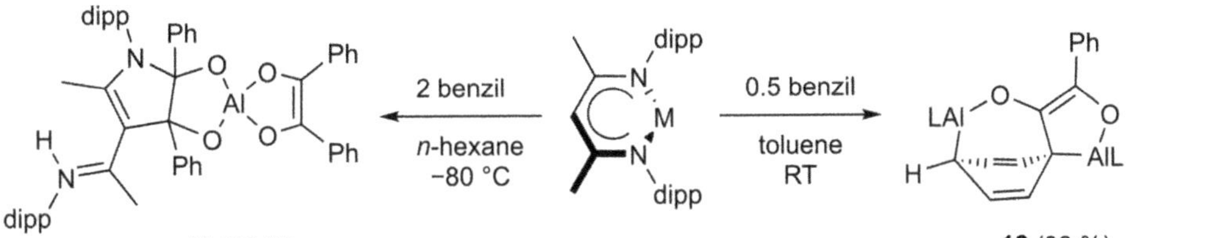
2 benzil
n-hexane
-80 °C
0.5 benzil
toluene
RT

11 (16 %)

12 (33 %)

Scheme 5. The reaction of LAl with two and one-half equivalents of benzil resulting in the formation of **11** and **12**, respectivly.

Surprisingly, the reaction of **8** with one equivalent of LAl does not yield **12**, demonstrating that **8** is not a reaction intermediate in the formation of compound **12**. As the E conformation of the benzil molecule is likely prevalent in solution, we suggest that compound **8′** might form in solution as the kinetically-controlled reaction intermediate of the two-electron reduction reaction of benzil with LAl (Scheme 6). **8′** can then either isomerize to compound **8** or reacts with LAl to compound **12** since it shows the correct conformation for the nucleophilic attack of a second alanediyl molecule.

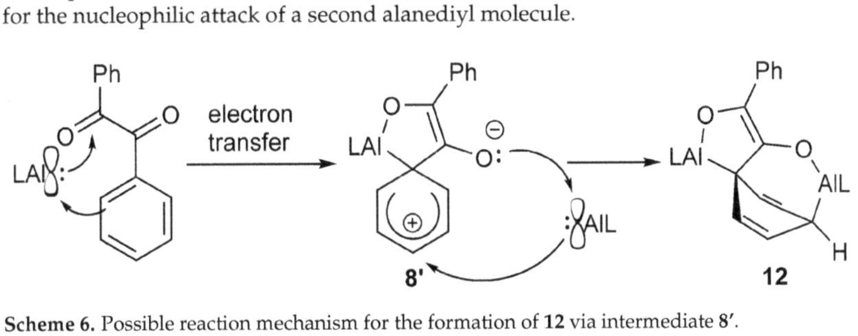
electron
transfer
8'
12

Scheme 6. Possible reaction mechanism for the formation of **12** via intermediate **8′**.

We also attempted to extend these studies to reactions of (in situ formed) boranediyl LB. Unfortunately, we failed to synthesize $LBCl_2$ by reaction of $LSiMe_3$ with BCl_3, whereas $L'BCl_2$ ($L' = HC[C(Me)NC_6F_5]_2$) was quantitatively formed by this approach [34]. The reaction of $L'BCl_2$ with KC_8 in the presence of benzil gave the corresponding 5-bora-spiro[4.5]heterodecene **13** (Scheme 7). On the other hand, the hydroboration reaction of butanedione with $L'BH_2$ (**14**), which was obtained from the reaction of $L'BCl_2$ with L-selectride, did not give the saturated 5-bora-spiro[4.5]heterodecane. No reaction took place at room temperature, while decomposition to yet unknown compounds was observed at 80 °C.

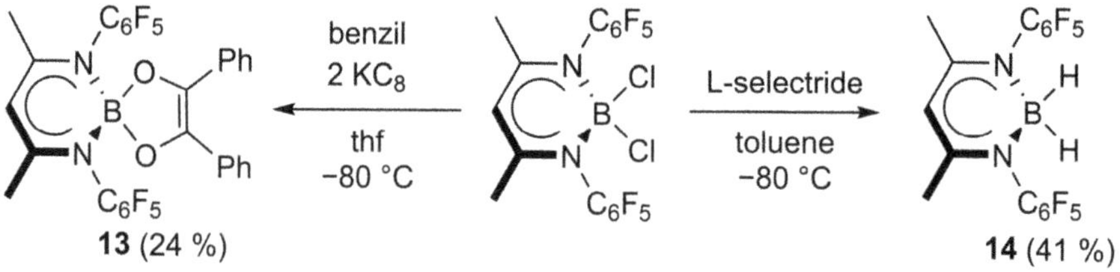

Scheme 7. The reaction of L'BCl$_2$ with benzil and KC$_8$ (**13**) as well as L-selectride (**14**).

Our studies demonstrate for the first time the ability of group 13 diyls to activate diketones by cycloaddition reaction to give unsaturated dialkolates of the respective group 13 elements. Furthermore, 5-galla-spiro[4.5]heterodecene **1** was found to be a valuable synthon for the synthesis of α,β-hydroxyketones via aldol addition reaction. The decreased reactivity and reduced power of the group 13 diyls with increasing atomic number (LAl > LGa > LIn > LTl) was also demonstrated: LAl was found to reduce one of the benzene rings of the benzil molecule to give the cyclohexadiene derivate **12**, while LTl did not react with benzil. Moreover, in product **7**, the Tl atom remains in the oxidation state +1.

2.2. Spectroscopic Characterization and Single Crystal X-ray Structures

The compounds are soluble in benzene and toluene except for compounds **6** and **7**, which were found to be soluble only in tetrahydrofuran and 2-methyltetrahydrofuran. All indium and thallium compounds were found to decompose slowly in solution with the formation of LH. The ^{1}H and ^{13}C NMR spectra of the 5-metalla-spiro[4.5]heterodecenes **1**, **4**, **5**, **8–10**, and **13** show comparable resonance patterns for the β-diketiminate ligand L at slightly shifted frequencies as well as the expected signals for the diketones, indicating that all compounds adopt the same symmetrical structures in solution. For compound **10**, this is true in thf, but in benzene, the resonances are significantly broadened, and no assignment to a defined product was possible (compare Figures S28 and S29).

The molecule symmetry in the aldol-addition products **2** and **3** is reduced, resulting in magnetically inequivalent protons of the β-diketiminate ligand L. In addition, the ^{1}H NMR spectrum of compound **2** shows broad resonances at room temperature. Temperature-dependent ^{1}H NMR spectra showed a reduced number of magnetically equivalent protons, demonstrating dynamic behavior for both **2** and **3** in solution (Figure S5).

IR spectra of **2** and **3** show bands for the CO stretching vibration of the ketones at 1706 and 1702 cm^{-1}, respectively. The activation of the benzene ring in compound **12** by the 1,4-addition of two Al centers is indicated by the shielding of the carbon atoms in the ^{13}C NMR spectrum (39.6 and 32.7 ppm) and the high field shift of the geminal CH proton (1.92 ppm) in the ^{1}H NMR spectrum.

The molecular structures of compounds **1**, **2**, and **4–13** in the solid state were determined by single crystal X-ray diffraction (sc-XRD). Suitable crystals were obtained by fractional crystallization from the reaction solutions except for compound **6**, which were obtained from a solution in 2-methyltetrahydrofuran layered with *n*-hexane. Compounds **1**, **4**, **5**, **6**, **8**, and **11** crystallize in monoclinic space groups, compounds **2** and **10** in orthorhombic space groups, and compounds **7**, **9**, **12**, and **13** in the triclinic space group $P\bar{1}$, respectively (see Figures 1–4 and S53–S57 and Table S1a–d).

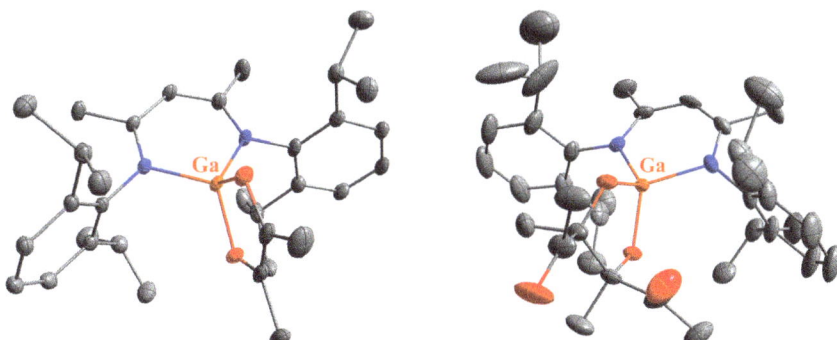

Figure 1. Molecular structures of **1** (**left**) and **2** (**right**) in their crystals. Displacement ellipsoids are drawn at the 50% probability level. H atoms are omitted, C atoms are displayed in grey, N atoms in blue, O atoms in red and Ga atoms in orange. Only the major compound of disorders are shown for clarity.

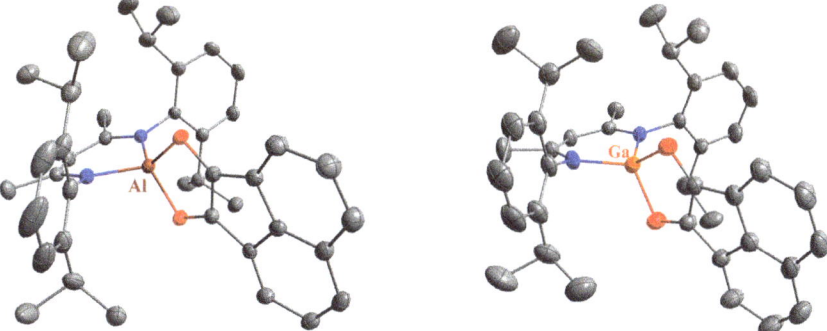

Figure 2. Molecular structures of **4** (**left**) and **5** (**right**) in their crystals. Displacement ellipsoids are drawn at the 50% probability level and H atoms are omitted for clarity. C atoms are displayed in grey, N atoms in blue, O atoms in red, Al atoms in dark red, and Ga atoms in orange.

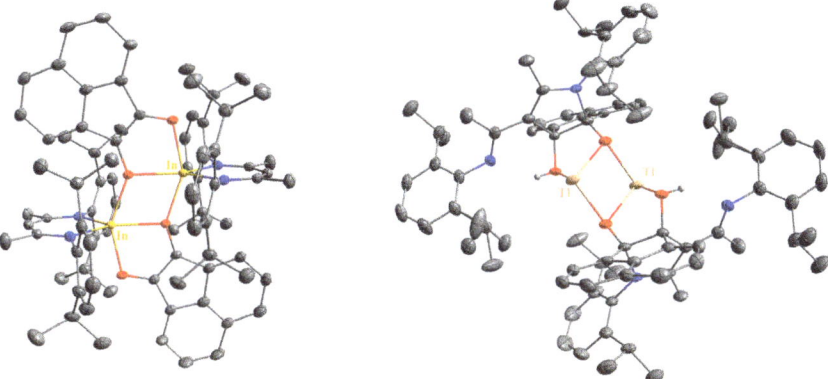

Figure 3. Molecular structures of **6** (**left**) and **7** (**right**) in their crystals. Displacement ellipsoids are drawn at the 50% probability level and H atoms (except NH and OH) are omitted for clarity. C atoms are displayed in grey, N atoms in blue, O atoms in red, In atoms in yellow, and Tl atoms in beige.

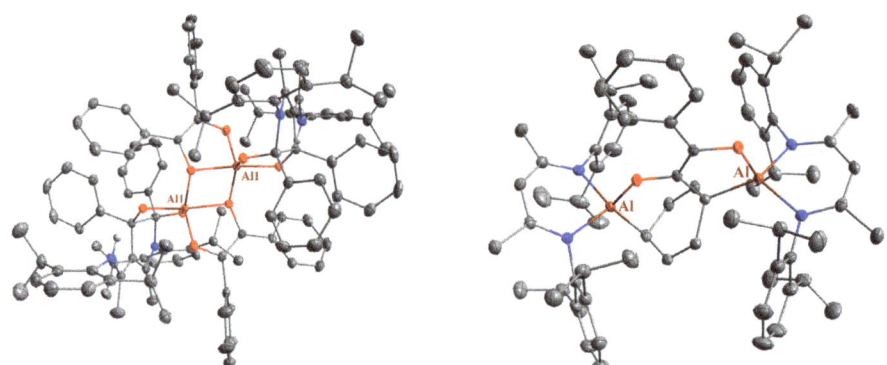

Figure 4. Molecular structures of **11** (**left**) and **12a** (**right**) in their crystals. Displacement ellipsoids are drawn at the 50% probability level and H atoms are omitted for clarity. C atoms are displayed in grey, N atoms in blue, O atoms in red, and Al atoms in dark red.

The O–Ga–O bond angles in the five-membered rings of compounds **1** and **2** are 7° more acute than the N–Ga–N angles, see Table 1. In addition, the 5-membered ring in compound **2** is saturated, and only the meso-isomer was found. The 5-metalla-spiro[4.5]heterodecenes (M = B (**13**), Al (**8**), Ga (**9**), In (**10**); see Figures S53–S57) show increasing M–O and M–N bond lengths with an increasing atomic number of the group 13 element, while the O–M–O (105.18(8)° **12**; 93.36(5)° **8**; 91.20(4)°, 91.73(4)° **9**; 82.29(4) **10**) and N–M–N (105.51(8) **13**; 98.21(6), 98.11(7) **8**; 98.99(5)°, 99.79(5)° **9**; 92.23(4) **10**) bond angles decrease. In the case of **10**, the In atom is further coordinated by an acetonitrile molecule, resulting in a penta-coordinated metal center. A dynamic behavior involving the acetonitrile molecule could explain the broad features in the ^{1}H NMR of **10** in benzene in contrast to the spectrum in the coordinating solvent thf (the chemical shift of MeCN in thf solution of **10** is found at the frequency expected for solvated MeCN at 1.94 ppm).

Table 1. Selected bond lengths [Å] and angles [°] of **1**, **2**, and **4**–**13**.

	O–M	O–M–O	N–M	N–M–N
1	1.8280(10), 1.8355(10)	92.89(5)	1.9122(12), 1.9176(12)	98.53(5)
2 [a]	1.827(3), 1.831(3); 1.836(3), 1.831(3)	91.88(13); 92,45(14)	1.920(3), 1,916(3); 1.919(3), 1.903(3)	98.84(15); 98.67(13)
4	1.7804(9), 1.7766(8)	97.18(4)	1.8582(10), 1.8597(10)	98.16(4)
5	1.8648(14), 1.8665(13)	94.80(5)	1.9055(14), 19150(13)	99.87(6)
6	2.0903(9), 2.2310(10)	79.12(4)	2.1537(11), 2.1580(11)	93.39(5)
7	2.7463(18) [b], 2.4536(15) [c]	61.51(5)	/	/
8 [a]	1.7502(14), 1.7544(14); 1.74958(14), 1.7489(14)	93.36(5); 92.92(5)	1.8648(15), 1.8664(14); 1.8641(14), 1.8620(14)	98.21(6); 98.11(7)
9 [a]	1.8336(10), 1.8315(10); 1.8293(10), 1.8364(10)	91.73(4); 91.20(4)	1.9056(12), 1.9025(11); 1.9050(11), 1.9055(11)	99.79(5); 98.99(5)
10	2.0684(10), 2.0474(10)	82.29(4)	2.1192(10), 2.1289(11)	92.23(4)
11	1.7830(16), 1.8087(16); 1.9530(16) [d], 1.7941(17)	88.15(7); 84.26(7)	/	/
12a	1.7393(8), 1.9885(11) [e]; 1.7496(8), 2.0136(11) [e]	108.90(4)[e]; 94.06(4) [e]	1.9095(9), 1.9115(9); 1.9125(9), 1.8974(9)	95.97(4); 95.31(4)
12b [a]	1.7332(11), 1.9851(15) [e]; 1.7507(11), 2.014(14) [e] 1.7355(11), 1.9830(15) [e]; 1.7492(11), 2.0155(15) [e]	108.17(6) [e], 93.88(5) [e]; 107.81(6) [e], 93.83(5) [e]	1.9127(13), 1.9022(13); 1.9120(13), 1.9070(13) 1.9148(13), 1.9060(13); 1.9059(13), 1.9151(13)	95.09(6); 94.83(5) 94.76(5); 95.11(5)
13	1.4573(13), 1.4587(13)	105.18(8)	1.5622(14), 1.5576(14)	105.51(8)

[a]: Two independent molecules in the unit cell; [b]: Tl–OH; [c]: 2.4280(15) Å to the second Tl atom; [d]: 1.8297(15) Å to the second Al atom; [e]: Al–C and O–Al–C instead of Al–O and O–Al–O.

The cycloaddition reaction products obtained from the reactions of acenaphthenequinone with LIn (**6**) and LTl (**7**) are dimeric in the solid state (Figure 3), in contrast to the monomeric 5-metalla-spiro[4.5]heterodecenes **4** and **5** (Figure 2), respectively. This structural difference agrees with the reduced solubility of **6** and **7** compared to **4** and **5** in hydrocarbons and the lower symmetry in solution as indicated by the double number of signals for the ^{i}Pr groups (Figure S19). As the NMR spectra of compound **10** in thf-d$_8$ solution retain a high symmetry (Figure S28) it is unlikely that compound **6** dissociates in thf solution to thf-coordinated monomeric species.

The heavier 5-inda-spiro[4.5]heterodecene **6** forms an O-bridged dimer in the solid state with penta-coordinated In centers (Figure 3), while the O-bridged thallium complex **7** has only three-coordinated Tl atoms. The O–M–O bond angles also decrease with the increasing atomic number of M (97.18(4)° **4**; 94.80(5)° **5**; 79.12(4)° **6**; 61.51(5)° **7**) as was observed for compounds **8**, **9**, **10**, and **13**. The two μ-bridging oxygen atoms in compounds **6** and **7** form a rhombus with the two metal centers. Again, the O–In–O angle (69.12(4)°) is more acute than the O–Tl–O angle (83.131(5)°), resulting in almost identical M–M distances (In–In: 3.6082(5), Tl–Tl: 3.6475(3) Å).

The structure of compound **11** represents an intermediate between the structures of compounds **6** and **7**. In **7**, one benzil unit bridges the N and the γ-C atoms as was observed in compound **7**. The second benzil unit forms a 5-membered alumina cycle with a bridging O atom, leading to a dimeric structure in the solid state analogous to that observed for compound **6**, also with a penta-coordinated metal center. However, the Al–Al distance in compound **7** is shorter (2.9442(13) Å) compared to that of compound **6**, and the O–Al–O bond angle (78.36(7)°) falls in between those observed for compounds **6** and **7**.

Finally, two different solvates of compound **12** were structurally characterized with either benzene (**12a**) or n-hexane (**12b**) in the unit cell. Surprisingly, a phenyl ring is activated by LAl in compound **12**, resulting in the formation of a 2-alumina-3-oxabicyclo[3.2.2]non-6,8-diene unit. The C–Al–O bond angle (about 108°, see Table 1) in the bicyclic molecule is more obtuse than the angle in the annulated five-membered ring (about 94°), while the Al–C bond length (1.98 Å) is shorter compared to the sum of the covalent bond radii (2.01 Å) [35].

3. Conclusions

In this comprehensive study, we demonstrated the general ability of group 13 diyls LM (M = (B), Al, Ga, In) to undergo [1,4]-cycloaddition reactions with diketones to give 5-metalla-spiro[4.5]heterodecenes **1**, **4**–**6**, and **8**–**10** and **13**, respectively. Galla-spiro[4.5]heterodecene **1** also undergoes aldol addition reactions to give α,β-hydroxyketones **2** and **3**. In contrast, thallanediyl LTl failed to undergo cycloaddition reactions at the metal center, but the reaction with acenaphthenequinone proceeded with [2,3]-cycloaddition of the ligand L. Moreover, the higher reactivity of LAl was demonstrated in the reaction of two equivalence LAl with benzil, leading to the activation of a phenyl ring in compound **12**.

4. Materials and Methods

All manipulations were performed using standard Schlenk and glovebox techniques under an argon atmosphere, dried by passage through preheated Cu_2O pellets and molecular sieve columns. Toluene, n-pentane, and n-hexane were dried using an MBraun solvent purification system (SPS). Benzene, deuterated benzene, thf, 2-methyltetrahydrofuran and deuterated thf were distilled from Na/K alloy and acetonitrile from CaH_2. Solvents were degassed and stored over activated molecular sieves. Starting reagents L'BCl$_2$ [34,36], LAl [1,37], LGa [2,38,39], LIn [3], and LTl [4] were prepared according to the (slightly modified) literature methods (LK was isolated and not prepared in situ). ^{1}H (300 MHz, 400 MHz, 600 MHz), ^{11}B{^{1}H} (128.5 MHz, 192.5 MHz), ^{13}C{^{1}H} (75.5 MHz, 100.7 MHz, 150.9 MHZ), and ^{19}F{^{1}H} (376.5 MHz) spectra were recorded with a Bruker Avance DPX-300, a Bruker Avance Neo 400 MHz or a Bruker Avance III HD 600 NMR spectrometer and are referenced to internal C_6D_6 (^{1}H: δ = 7.16, ^{13}C: δ = 128.06), and thf-d$_8$ (^{1}H: δ = 3.58;

^{13}C: δ = 67.21). Heteronuclear NMR measurements were performed protium decoupled unless otherwise noted. IR spectra were recorded in a glovebox using a BRUKER ALPHAT FT-IR spectrometer equipped with a single reflection ATR sampling module to ensure oxygen- and water-free conditions. Microanalysis was performed at the Elemental Analysis Laboratory of the University of Duisburg-Essen.

4.1. Synthesis of LGa($C_4H_6O_2$) (1)

A total of 35.3 mg of butanedione (410 µmol) was added to 200 mg of LGa (410 µmol) dissolved in 5 mL of toluene. The clear yellow solution turned orange and was stirred for 15 min. The solution was layered with 15 mL of n-hexane and stored overnight at room temperature. Small amounts of solids were separated by filtration and the filtrate was concentrated to about 0.5 mL. The product was obtained as an orange crystalline solid when stored at −30 °C. Yield: 112 mg (197 µmol, 48%).

Anal. Calcd. for $C_{33}H_{47}GaN_2O_2$: C, 69.12, H, 8.26; N, 4.88; Found: C, 69.3, H, 8.29; N, 4.86%. ATR-IR: υ 2960, 2866, 1660, 1587, 1555, 1529, 1462, 1440, 1382, 1316, 1256, 1214, 1176, 1125, 1101, 1022, 984, 932, 872, 803, 772, 761, 747, 668, 533, 533 cm^{-1}. ^{1}H NMR (400 MHz, C_6D_6, 25 °C): δ 7.04 (s, 6 H, C_6H_3-2,6^{i}Pr$_2$), 4.86 (s, 1 H, γ-CH), 3.32 (sept, $^3J_{HH}$ = 6.8 Hz, 4 H, CH(CH$_3$)$_2$), 1.74 (s, 6 H, Ga(OCCH$_3$)$_2$), 1.54 (d, $^3J_{HH}$ = 6.8 Hz, 12 H, CH(CH$_3$)$_2$), 1.54 (s, 6 H, ArNCCH$_3$), 1.12 (d, $^3J_{HH}$ = 6.9 Hz, 12 H CH(CH$_3$)$_2$). ^{13}C NMR (100.6 MHz, C_6D_6, 25 °C): δ 171.4 (ArNCCH$_3$), 144.2, 138.8, 128.1, 124.4 (ArC), 131.6 (Ga(OCCH$_3$)$_2$), 96.0 (γ-CH), 28.7 (CH(CH$_3$)$_2$), 24.8, 24.6 (CH(CH$_3$)$_2$), 23.3 (ArNCCH$_3$), 16.1 (Ga(OCCH$_3$)$_2$).

4.2. Synthesis of LGa($C_8H_{12}O_4$) (2)

8.8 mg of butanedione (103 µmol) was added to 20 mg of LGa (41 µmol) dissolved in 0.5 mL of C_6D_6. A colorless precipitate formed shortly after, which was redissolved by heating the suspension. Storage at room temperature gave the product a colorless crystalline solid. Yield: 15 mg (23 µmol, 55%).

Anal. Calcd. for $C_{37}H_{53}Ga_1N_2O_4$: C, 67.38, H, 8.10; N, 4.25; Found: C, 67.0, H, 8.88 N, 4.07%. ATR-IR: υ 2958, 2928, 2866, 1706, 1535, 1464, 1438, 1384, 1347, 1318, 1262, 1217, 1182, 1109, 1082, 1059, 1024, 995, 925, 875, 798, 761, 695, 668, 605, 542, 529 cm^{-1}. ^{1}H NMR (400 MHz, C_6D_6, 25 °C): δ 7.15-6.98 (m, 6 H, C_6H_3-2,6^{i}Pr$_2$), 4.86 (s, 1 H, γ-CH), 3.32 (sept, $^3J_{HH}$ = 6.8 Hz, 1 H, CH(CH$_3$)$_2$), 3.18 (m, 3 H, CH(CH$_3$)$_2$), 1.96 (s, 4.5 H, GaC$_8$H$_{12}$O$_4$), 1.80 (s, 1.5 H, GaC$_8$H$_{12}$O$_4$), 1.74 (s, 1.5 H, GaC$_8$H$_{12}$O$_4$), 1.53 (m, 18 H, CH(CH$_3$)$_2$, ArNCCH$_3$), 1.12 (d, $^3J_{HH}$ = 6.9 Hz, 3 H, CH(CH$_3$)$_2$), 1.04 (d, $^3J_{HH}$ = 6.9 Hz, 9 H, CH(CH$_3$)$_2$), 0.95 (s, 4.5 H, GaC$_8$H$_{12}$O$_{14}$). ^{1}H NMR (300 MHz, C_6D_6, 80 °C): δ 7.06 (s, 6 H, N-C_6H_5), 4.94 (s, 1 H, γ-CH), 3.31 (sept, $^3J_{HH}$ = 6.9 Hz, 4 H, CH(CH$_3$)$_2$), 1.86 (s(br), 6 H, GaC$_8$H$_{12}$O$_{14}$), 1.67 (s(br), 1.5 H, GaC$_8$H$_{12}$O$_{14}$), 1.63 (s,6 H, ArNCCH$_3$), 1.49 (d, $^3J_{HH}$ = 6.7 Hz, 12 H, CH(CH$_3$)$_2$), 1.14 (d, $^3J_{HH}$ = 6.9 Hz, 12 H, CH(CH$_3$)$_2$). ^{13}C NMR (100.6 MHz, C_6D_6, 25 °C): many resonances were not observed due to line broadening (cp. ^{1}H NMR spectrum) and low solubility in benzene.

4.3. Synthesis of LGa($C_{11}H_{12}O_3$) (3)

21.4 µL of benzaldehyde (22.2 mg, 209.3 µmol) was added to 100 mg of LGa($C_4H_6O_2$) (1) (174 µmol) dissolved in 1 mL of toluene. A rapid color change from orange to yellow was observed and a precipitate began to form. The suspension was stored in the freezer overnight, filtered, and washed with n-hexane, yielding 75 mg of LGa($C_{11}H_{12}O_3$). Yield: 70 mg (111 µmol, 64%).

Anal. Calcd. for $C_{40}H_{53}GaN_2O_3$: C, 70.69, H, 7.86; N, 4.12; Found: C, 70.6, H, 7.96; N, 3.95%. ATR-IR: υ 2960, 2927, 2868, 2814, 1702, 1537, 1438, 1394, 1367, 1359, 1320, 1268, 1182, 1142, 1096, 1051, 1028, 929, 877, 796, 757, 732, 699, 672, 604, 569, 492, 474 cm^{-1}. ^{1}H NMR (400 MHz, C_6D_6, 25 °C): δ 7.44, 7.23–7.00 (m, 11 H, C_6H_3-2,6^{i}Pr$_2$ and C_6H_5), 4.88 (s, 1 H, γ-CH), 4.60 (s, 1 H, GaOCH), 3.61 (sept, $^3J_{HH}$ = 6.7 Hz, 1 H, CH(CH$_3$)$_2$), 3.47 (sept, $^3J_{HH}$ = 6.8 Hz, 1 H, CH(CH$_3$)$_2$), 3.15 (sept, $^3J_{HH}$ = 6.8 Hz, 1 H, CH(CH$_3$)$_2$), 3.04 (sept, $^3J_{HH}$ = 6.8 Hz, 1 H, CH(CH$_3$)$_2$), 1.74 (s, 3 H, GaOC(CO)CH$_3$), 1.61 (s, 3 H, ArNCCH$_3$),

1.60 (s, 3 H, ArNCCH$_3$), 1.59 (d, $^3J_{HH}$ = 6.8 Hz, 3 H, CH(CH$_3$)$_2$), 1.51 (d, $^3J_{HH}$ = 6.8 Hz, 3 H, CH(CH$_3$)$_2$), 1.45 (d, $^3J_{HH}$ = 6.7 Hz, 3 H, CH(CH$_3$)$_2$), 1.26 (d, $^3J_{HH}$ = 6.8 Hz, 3 H, CH(CH$_3$)$_2$), 1.13 (d, $^3J_{HH}$ = 6.9 Hz, 3 H, CH(CH$_3$)$_2$), 1.10 (d, $^3J_{HH}$ = 6.9 Hz, 3 H, CH(CH$_3$)$_2$), 1.05 (d, $^3J_{HH}$ = 6.8 Hz, 3 H, CH(CH$_3$)$_2$), 0.92 (d, $^3J_{HH}$ = 6.8 Hz, 3 H, CH(CH$_3$)$_2$), 0.44 (s, 3 H, GaOCCH$_3$). ^{13}C NMR (100.6 MHz, C$_6$D$_6$, 25 °C): δ 171.4 (GaOC(CO)CH$_3$), 172.0 (ArNCCH$_3$), 172.0, 144.4, 144.4, 143.9, 143.9, 143.5, 139.7, 139.0, 127.6, 127.1, 126.6, 124.7, 124.7, 124.7, 124.6 (ArC), 95.8 (γ-CH), 84.7 (GaOCCH$_3$), 80.7 (GaOCH), 29.0, 28.9, 28.6, 28.5 (CH(CH$_3$)$_2$), 25.4, 25.0, 24.7, 24.7, 24.5, 24.2, 24.1, 23.7 (CH(CH$_3$)$_2$), 24.6 (GaOC(CO)CH$_3$), 23.4, 23.4 (ArNCCH$_3$), 17.9 (GaOCCH$_3$).

4.4. Synthesis of LAl(C$_{12}$H$_6$O$_2$) (4)

50 mg of LAl (112 μmol) and 20.5 mg of acenaphthenequinone (112 μmol) were dissolved in 5 mL of toluene and the resulting dark purple solution was stirred overnight at ambient temperature. All volatiles were removed, and the dark purple residue was washed with small amounts of n-hexane. Yield: 23 mg (37 μmol, 33%).

Anal. Calcd. for C$_{41}$H$_{47}$AlN$_2$O$_2$: C, 78.56, H, 7.56; N, 4.47; Found: C, 78.1, H, 7.37; N, 4.28%. ATR-IR: υ 2964, 2927, 2868, 1531, 1462, 1444, 1382, 1313, 1249, 1196, 1133, 1105, 1024, 917, 898, 803, 772, 759, 738, 703, 596, 482, 472, 445, 416 cm^{-1}. ^{1}H NMR (400 MHz, C$_6$D$_6$, 25 °C): δ 7.18 (d, $^3J_{HH}$ = 6.8 Hz, 2 H, AlC$_{12}$H$_6$O$_2$), 7.10 (d, $^3J_{HH}$ = 8.3 Hz, 2 H, AlC$_{12}$H$_6$O$_2$), 6.98 (dd, $^3J_{HH}$ = 6.7, 8.3 Hz, 2 H, AlC$_{12}$H$_6$O$_2$), 6.95–6.87 (m, 6 H, C$_6$H$_3$-2,6^{i}Pr$_2$), 5.02 (s, 1 H, γ-CH), 3.35 (sept, $^3J_{HH}$ = 6.8 Hz, 4 H, CH(CH$_3$)$_2$), 1.54 (s, 6 H, ArNCCH$_3$), 1.52 (d, $^3J_{HH}$ = 6.8 Hz, 12 H, CH(CH$_3$)$_2$), 1.10 (d, $^3J_{HH}$ = 6.9 Hz, 12 H CH(CH$_3$)$_2$). ^{13}C NMR (100.6 MHz, C$_6$D$_6$, 25 °C): δ 172.5 (ArNCCH$_3$), 144.4, 137.2, 128.4, 124.7 (ArC), 146.0, 133.5, 127.4, 127.1, 124.9, 123.4, 116.8 (AlC$_{12}$H$_6$O$_2$), 98.1 (γ-CH), 28.9 (CH(CH$_3$)$_2$), 24.9, 24.8 (CH(CH$_3$)$_2$), 23.0 (ArNCCH$_3$).

4.5. Synthesis of LGa(C$_{12}$H$_6$O$_2$) (5)

A total of 50 mg of LGa (103 μmol) and 18.7 mg of acenaphthenequinone (103 μmol) were dissolved in 5 mL of toluene and the resulting dark purple solution was stirred overnight at room temperature. Crystals suitable for sc-XRD were obtained from a highly concentrated solution after storage at −30 °C. However, the product can be more conveniently isolated as a purple powder by removing all volatiles and washing the residue with n-hexane. Yield: 35 mg (52 μmol, 51%).

Anal. Calcd. for C$_{41}$H$_{47}$GaN$_2$O$_2$: C, 73.55, H, 7.08; N, 4.18; Found: C, 73.3, H, 7.09; N, 4.49%. ATR-IR: υ 3061, 3034, 2964, 2923, 2866, 1526, 1462, 1442, 1382, 1313, 1256, 1192, 1180, 1131, 1106, 1055, 1024, 921, 900, 883, 800, 769, 759, 627, 596, 581, 526, 472, 441, 420 cm^{-1}. ^{1}H NMR (400 MHz, C$_6$D$_6$, 25 °C): δ 7.22 (d, $^3J_{HH}$ = 6.6 Hz, 2 H, GaC$_{12}$H$_6$O$_2$), 7.12 (d, $^3J_{HH}$ = 9.4 Hz, 2 H, GaC$_{12}$H$_6$O$_2$), 7.22 (dd, $^3J_{HH}$ = 6.6, 8.3 Hz, 2 H, GaC$_{12}$H$_6$O$_2$), 6.91 (s, 6 H, C$_6$H$_3$-2,6^{i}Pr$_2$), 4.89 (s, 1 H, γ-CH), 3.35 (sept, $^3J_{HH}$ = 6.7 Hz, 4 H, CH(CH$_3$)$_2$), 1.54 (s, 6 H, ArNCCH$_3$), 1.51 (d, $^3J_{HH}$ = 6.9 Hz, 12 H, CH(CH$_3$)$_2$), 1.11 (d, $^3J_{HH}$ = 6.9 Hz, 12 H CH(CH$_3$)$_2$). ^{13}C NMR (100.6 MHz, C$_6$D$_6$, 25 °C): δ 172.1 (ArNCCH$_3$), 144.0, 137.8, 128.5, 124.7 (ArC), 146.6, 134.1, 127.4, 127.1, 124.7, 123.2, 116.7 (GaC$_{12}$H$_6$O$_2$), 96.5 (γ-CH), 28.9 (CH(CH$_3$)$_2$), 24.8, 24.7 (CH(CH$_3$)$_2$), 23.2 (ArNCCH$_3$).

4.6. Synthesis of LIn(C$_{12}$H$_6$O$_2$) (6)

200 mg of LIn (376 μmol) and 68.4 mg of acenaphthenequinone (376 μmol) were dissolved in 5 mL of toluene and the resulting dark purple suspension was stirred overnight at ambient temperature. The product was obtained as a purple powder by filtration. Yield: 201 mg (281 μmol, 75%).

Anal. Calcd. for C$_{41}$H$_{47}$InN$_2$O$_2$: C, 68.91, H, 6.63; N, 3.92; Found: C, 68.7, H, 6.54; N, 3.66%. ATR-IR: υ 2967, 2951, 2925, 2866, 1547, 1522, 1458, 1438, 1409, 1382, 1367, 1328, 1316, 1266, 1172, 1130, 1101, 1079, 1055, 1030, 935, 902, 854, 798, 761, 730, 592, 536, 481, 428 cm^{-1}. ^{1}H NMR (400 MHz, thf-d$_8$, 25 °C): δ 7.52, 7.36, 7.24, 7.07 (m, 6 H, InC$_{12}$H$_6$O$_2$), 7.18, 7.07, 6.63 (m, 6 H, C$_6$H$_3$-2,6^{i}Pr$_2$), 5.40 (s, 1 H, γ-CH), 3.72 (sept, $^3J_{HH}$ = 6.7 Hz, 2 H, CH(CH$_3$)$_2$), 3.09

(sept, $^3J_{HH}$ = 6.8 Hz, 2 H, CH(CH$_3$)$_2$), 1.76 (s, 6 H, ArNCCH_3), 1.27 (d, $^3J_{HH}$ = 6.6 Hz, 6 H, CH(CH_3)$_2$), 1.19 (d, $^3J_{HH}$ = 6.7 Hz, 6 H CH(CH_3)$_2$), 0.20 (d, $^3J_{HH}$ = 6.7 Hz, 6 H, CH(CH_3)$_2$), −0.28 (d, $^3J_{HH}$ = 6.8 Hz, 6 H CH(CH_3)$_2$). ^{13}C NMR (100.6 MHz, thf-d$_8$, 25 °C): δ 172.7 (ArNCCH$_3$), 150.6, 145.2, 144.0, 143.5, 137.2, 135.7, 135.5, 127.6, 127.4, 127.1, 126.9, 126.1, 125.3, 125.2, 123.0, 121.3, 120.0, 116.9 (ArC + InC_{12}H$_6$O$_2$), 95.6 (γ-CH), 28.5, 28.0 (CH(CH$_3$)$_2$), 26.4, 25.5, 24.6, 22.2 (CH(CH$_3$)$_2$), 25.7 (ArNCCH$_3$).

4.7. Synthesis of LTl(C$_{14}$H$_{10}$O$_2$) (7)

A total of 100 mg of LTl (161 μmol) and 27.8 mg of acenaphthenequinone (153 μmol) were cooled to −80 °C and 2 mL of n-hexane was added. The resulting mixture was warmed to ambient temperature within 8 h, and the resulting off-white solid (60 mg) was separated by filtration. The ^{1}H NMR spectrum of this solid is essentially consistent with that obtained from isolated crystals from the reaction mixture, which were characterized by sc-XRD and are highly reproducible between different batches. All attempts to further purify compound **9** failed due to decomposition in the solution state, resulting in the formation of the protonated ligand LH.

^{1}H NMR (400 MHz, C$_6$D$_6$, 25 °C): δ 7.62–6.84 (m, 22 H, C$_6$H$_3$-2,6^{i}Pr$_2$ and TlC$_{12}$H$_6$O$_2$), 4.04, 3.20, 2.57, 1.98 (broad, 4 H, CH(CH$_3$)$_2$), 1.91, 1.85 (s, 6 H, ArNCCH_3), 1.24, 1.21, 1.20, 1.12, 1.11, 0.74, 0.49, −0.18 (d, 24 H, CH(CH_3)$_2$).

4.8. Synthesis of LAl(C$_{14}$H$_{10}$O$_2$) (8) and LAl(C$_{14}$H$_{10}$O$_2$)$_2$ (11)

LAl(C$_{14}$H$_{10}$O$_2$)$_2$ (**11**): 100 mg of LAl (225 μmol) and 94.6 mg of benzil (450 μmol) were cooled to −80 °C and 5 mL of *n*-hexane was slowly added and the mixture was warmed to ambient temperature overnight with stirring. The resulting solid was filtered off and dissolved in toluene. Storage at −30 °C afforded 30 mg of LAl(C$_{14}$H$_{10}$O$_2$)$_2$ as a crystalline yellow solid (suitable for sc-XRD). Yield: 40 mg (35 μmol, 16%).

Anal. Calcd. for C$_{57}$H$_{61}$AlN$_2$O$_4$: C, 79.14, H, 7.11; N, 3.24; Found: C, 78.7, H, 7.17; N, 3.23%. ATR-IR: υ 2964, 2928, 2868, 1578, 1529, 1462, 1442, 1371, 1334, 1316, 1253, 1214, 1131, 1057, 1023, 933, 839, 803, 776, 759, 747, 732, 709, 695, 664, 613, 577, 528, 480 cm^{-1}. ^{1}H NMR (600 MHz, C$_6$D$_6$, 25 °C): δ 11.76 (s, 1 H, ArNHCCH$_3$), δ 7.88 (d, $^3J_{HH}$ = 7.7 Hz, 2 H, C$_6$H$_5$), 7.82 (d, $^3J_{HH}$ = 7.7 Hz, 2 H, C$_6$H$_5$), 7.32 (d, $^3J_{HH}$ = 7.7 Hz, 1 H, C$_6$H$_5$), 7.27 (m, 3 H, C$_6$H$_5$), 7.22 (t, $^3J_{HH}$ = 7.6 Hz, 2 H, C$_6$H$_5$), 7.13-6.93 (m, 11 H, C$_6$H$_3$-2,6^{i}Pr$_2$ and C$_6$H$_5$), 6.88 (t, $^3J_{HH}$ = 7.3 Hz, 1 H, C$_6$H$_5$), 6.83 (t, $^3J_{HH}$ = 7.6 Hz, 1 H, C$_6$H$_5$), 6.80 (d, $^3J_{HH}$ = 7.7 Hz, 1 H, C$_6$H$_5$), 6.72 (t, $^3J_{HH}$ = 7.6 Hz, 2 H, C$_6$H$_5$), 6.65 (t, $^3J_{HH}$ = 7.4 Hz, 1 H, C$_6$H$_5$), 6.62 (t, $^3J_{HH}$ = 7.7 Hz, 1 H, C$_6$H$_5$), 6.07 (d, $^3J_{HH}$ = 7.9 Hz, 1 H, C$_6$H$_5$), 4.69 (sept, $^3J_{HH}$ = 6.7 Hz, 1 H, CH(CH$_3$)$_2$), 4.45 (sept, $^3J_{HH}$ = 6.7 Hz, 1 H, CH(CH$_3$)$_2$), 3.35 (sept, $^3J_{HH}$ = 6.5 Hz, 1 H, CH(CH$_3$)$_2$), 2.20 (sept, $^3J_{HH}$ = 6.8 Hz, 1 H, CH(CH$_3$)$_2$), 2.11 (s, 3 H, C$_6$H$_5$CH_3), 1.92 (d, $^3J_{HH}$ = 6.6 Hz, 3 H, CH(CH_3)$_2$), 1.78 (s, 36 H, ArNCCH_3), 1.69 (s, 3 H, ArNCCH_3), 1.47 (d, $^3J_{HH}$ = 6.8 Hz, 3 H, CH(CH_3)$_2$), 1.41 (d, $^3J_{HH}$ = 6.4 Hz, 3 H, CH(CH_3)$_2$), 1.22 (d, $^3J_{HH}$ = 6.5 Hz, 3 H, CH(CH_3)$_2$), 1.01 (d, $^3J_{HH}$ = 6.6 Hz, 3 H, CH(CH_3)$_2$), 0.92 (d, $^3J_{HH}$ = 6.9 Hz, 3 H, CH(CH_3)$_2$), 0.90 (d, $^3J_{HH}$ = 7.0 Hz, 3 H, CH(CH_3)$_2$), 0.45 (d, $^3J_{HH}$ = 6.7 Hz, 3 H, CH(CH_3)$_2$). Due to the low solubility of the sample, no meaningfull 2D NMR spectrum could be recorded. Signals in the aromatic region could not be assigned. ^{13}C NMR (150.9 MHz, C$_6$D$_6$, 25 °C): δ 168.8, 166.6 (ArNCCH$_3$), 137.9, 129.3, 128.6, 125.7, 21,4 (toluene), 150.7, 148.0, 146.6, 146.6, 145.9, 145.6, 143.0, 139.5, 137.0, 134.7, 133.4, 133.1, 131.4, 130.3, 130.2, 129.2, 129.0, 128.9, 127.9, 127.8, 126.9, 126.5, 126.5, 126.3, 126.1, 126.0, 125.4, 124.7, 124.3, 124.1, 124.0, 123.6, 115.1, 110.6 (C$_6$H$_3$-2,6^{i}Pr$_2$ and (COC_6H$_5$)$_2$), 89.0 (γ-CH), 29.1, 28.2, 28.1, 28.1 (CH(CH$_3$)$_2$), 27.4, 26.4, 25.6, 25.5, 24.2, 24.1, 23.4, 23.1, (CH(CH$_3$)$_2$), 19.1, 17.8 (ArNCCH$_3$).

LAl(C$_{14}$H$_{10}$O$_2$) (**8**): The mother liquor from the synthesis of LAl(C$_{14}$H$_{10}$O$_2$)$_2$ (**11**) was dried in vacuo and the resulting residue was washed with 20 mL of *n*-hexane, yielding 40 mg of LAl(C$_{14}$H$_{10}$O$_2$). The washing liquid was concentrated to 10 mL and stored at −30 °C to give a second fraction of LAl(C$_{14}$H$_{10}$O$_2$). Yield: 55 mg (84 µmol, 37%).

Anal. Calcd. for C$_{43}$H$_{51}$AlN$_2$O$_2$: C, 78.87, H, 7.85; N, 4.28; Found: C, 78.5, H, 7.79; N, 4.25%. ATR-IR: υ 3056, 3018, 2962, 2926, 2868, 1585, 1535, 1462, 1442, 1384, 1316, 1253, 1176, 1105, 1055, 1022, 925, 912, 900, 803, 790, 769, 759, 736, 695, 643, 546, 507, 447, 416 cm^{-1}. ^{1}H NMR (400 MHz, C$_6$D$_6$, 25 °C): δ 7.48 (d, $^3J_{HH}$ = 7.8 Hz, 4 H, C$_6$H$_5$), 7.02 (m, 6 H, C$_6$H$_3$-2,6^{i}Pr$_2$), 6.99 (m, 4 H, C$_6$H$_5$), 6.86 (tt, $^3J_{HH}$ = 7.2 Hz, $^4J_{HH}$ = 1.4 Hz, 2 H, C$_6$H$_5$), 5.03 (s, 1 H, γ-CH), 3.35 (sept, $^3J_{HH}$ = 6.7 Hz, 4 H, CH(CH$_3$)$_2$), 1.55 (s, 6 H, ArNCCH$_3$), 1.44 (d, $^3J_{HH}$ = 6.8 Hz, 12 H, CH(CH$_3$)$_2$), 1.08 (d, $^3J_{HH}$ = 6.7 Hz, 12 H, CH(CH$_3$)$_2$). ^{13}C NMR (100.6 MHz, C$_6$D$_6$, 25 °C): δ 172.5 (ArNCCH$_3$), 139.7 (COAl), 138.8, 127.9, 127.6, 125.6 (C$_6$H$_5$), 144.4, 138.0, 128.2, 124.6 (ArC), 98.2 (γ-CH), 28.9 (CH(CH$_3$)$_2$), 25.0, 24.7 (CH(CH$_3$)$_2$), 23.2 (ArNCCH$_3$).

4.9. Synthesis of LGa(C$_{14}$H$_{10}$O$_2$) (9)

A total of 100 mg of LGa (205 µmol) and 43.1 mg of benzil (205 µmol) were combined with 2 mL of benzene and the mixture was stirred overnight, during which the product precipitated as an orange crystalline solid. Yield: 90 mg (129 µmol, 63%).

Anal. Calcd. for C$_{43}$H$_{51}$GaN$_2$O$_2$: C, 74.03, H, 7.37; N, 4.02; Found: C, 73.6, H, 7.37; N, 4.15%. ATR-IR: υ 2962, 2927, 2865, 1585, 1533, 1491, 1464, 1442, 1384, 1318, 1260, 1180, 1105, 1065, 1051, 1022, 929, 912, 803, 765, 724, 699, 683, 627, 535, 481, 445 cm^{-1}. ^{1}H NMR (400 MHz, C$_6$D$_6$, 25 °C): δ 7.47 (d, $^3J_{HH}$ = 7.8 Hz, 4 H, C$_6$H$_5$), 7.01 (m, 10 H, C$_6$H$_5$, and C$_6$H$_3$-2,6^{i}Pr$_2$), 6.87 (tt, $^3J_{HH}$ = 7.1 Hz, $^4J_{HH}$ = 1.4 Hz, 2 H, C$_6$H$_5$), 4.90 (s, 1 H, γ-CH), 3.35 (sept, $^3J_{HH}$ = 6.8 Hz, 4 H, CH(CH$_3$)$_2$), 1.54 (s, 6 H, ArNCCH$_3$), 1.46 (d, $^3J_{HH}$ = 6.7 Hz, 12 H, CH(CH$_3$)$_2$), 1.09 (d, $^3J_{HH}$ = 6.9 Hz, 12 H, CH(CH$_3$)$_2$). ^{13}C NMR (100.6 MHz, C$_6$D$_6$, 25 °C): δ 172.1 (ArNCCH$_3$), 139.6 (COGa), 139.3, 128.6, 127.6, 125.4 (C$_6$H$_5$), 144.1, 138.5, 128.6, 124.5 (ArC), 96.4 (γ-CH), 28.9 (CH(CH$_3$)$_2$), 24.8, 24.6 (CH(CH$_3$)$_2$), 23.4 (ArNCCH$_3$).

4.10. Synthesis of LIn(C$_{14}$H$_{10}$O$_2$)·MeCN (10)

A total of 100 mg of LIn (188 µmol) and 39.4 mg of benzil (188 µmol) were balanced in a Schlenk flask and cooled to −80 °C. 5 mL of *n*-hexane was added to the solids, and the mixture was stirred overnight and allowed to warm to room temperature. All volatiles were removed under reduced pressure, the residue was dissolved in a mixture of 1 mL benzene and 10 mL acetonitrile and concentrated to about 1 mL. **10** was obtained as an orange crystalline solid (suitable for sc-XRD) after storage at 6 °C. Yield: 45 mg (61 µmol, 32%).

ATR-IR: υ 2962, 2924, 2868, 2296, 2268, 1592, 1519, 1435, 1384, 1356, 321, 1270, 1178, 1133, 1051, 1021, 924, 906, 863, 800, 757, 711, 693, 665, 609, 530, 439 cm^{-1}. ^{1}H NMR (400 MHz, thf-d$_8$, 25 °C): δ 7.24-7.12 (m, 6 H, C$_6$H$_3$-2,6^{i}Pr$_2$), 6.90 (d, $^3J_{HH}$ = 7.0 Hz, 4 H, C$_6$H$_5$), 6.74 (d, $^3J_{HH}$ = 7.4 Hz, 4 H, C$_6$H$_5$), 6.70 (t, $^3J_{HH}$ = 7.2 Hz, 2 H, C$_6$H$_5$), 5.19 (s, 1 H, γ-CH), 3.32 (sept, $^3J_{HH}$ = 6.9 Hz, 4 H, CH(CH$_3$)$_2$), 1.94 (s, 2.5 H, MeCN), 1.84 (s, 6 H, ArNCCH$_3$), 1.25 (d, $^3J_{HH}$ = 6.8 Hz, 12 H, CH(CH$_3$)$_2$), 1.23 (d, $^3J_{HH}$ = 6.9 Hz, 12 H, CH(CH$_3$)$_2$). ^{13}C NMR (100.6 MHz, thf-d$_8$, 25 °C): δ 172.5 (ArNCCH$_3$), 139.7 (COIn), 143.8, 143.3, 127.1, 124.4 (ArC), 142.9, 128.6, 126.9, 123.8 (C$_6$H$_5$), 96.9 (γ-CH), 28.8 (CH(CH$_3$)$_2$), 25.0, 24.5 (CH(CH$_3$)$_2$), 24.5 (ArNCCH$_3$), 117.3, 0.4 (MeCN). 100 mg LGa

4.11. Synthesis of (LAl)$_2$(C$_{14}$H$_{10}$O$_2$) (12)

A total of 23.6 mg of benzil (113 µmol) dissolved in 1 mL of toluene was added dropwise to a solution of 100 mg of LAl (225 µmol) in 1 mL of toluene. The mixture was stirred for 2 days, after which an orange precipitate formed. The suspension was stored overnight at −6 °C. 40 mg of **12** was obtained as an orange powder by filtration. The filtrate consists mainly of LAl and (LAl)(C$_{14}$H$_{10}$O$_2$) (**8**). Yield: 40 mg (36 µmol, 33%).

Anal. Calcd. for $C_{72}H_{92}Al_2N_4O_2$: C, 78.65, H, 8.43; N, 5.10; Found: C, 78.9, H, 8.51; N, 4.90%. ATR-IR: υ 2964, 2928, 2865, 1529, 1462, 1440, 1384, 1316, 1253, 1179, 1156, 1102, 1061, 1022, 972, 937, 899, 873, 800, 786, 774, 759, 734, 693, 618, 532, 480, 445, 422 cm^{-1}. ^{1}H NMR (400 MHz, C_6D_6, 25 °C): δ 8.53 (d, $^3J_{HH}$ = 7.9 Hz, 2 H, C_6H_5), 7.42 (dd, $^3J_{HH}$ = 7.0, 8.4 Hz, 2 H, C_6H_5), 7.24–7.11 (m, 8 H, C_6H_3-2,6^{i}Pr$_2$), 7.07 (t, $^3J_{HH}$ = 7.3 Hz, 1 H, C_6H_5), 7.00 (m, 4 H, C_6H_3-2,6^{i}Pr$_2$), 4.96 (s, 1 H, γ-CH), 4.88 (dd, $^3J_{HH}$ = 6.0, 8.4 Hz, 2 H, AlC(CHCH)$_2$CHAl), 4.84 (s, 1 H, γ-CH), 4.02 (d, $^3J_{HH}$ = 8.4 Hz, 2 H, AlC(CHCH)$_2$CHAl), 3.45 (sept, $^3J_{HH}$ = 6.7 Hz, 2 H, CH(CH$_3$)$_2$), 3.33–3.20 (m, 4 H, CH(CH$_3$)$_2$), 3.07 (sept, $^3J_{HH}$ = 6.7 Hz, 2 H, CH(CH$_3$)$_2$), 1.92 (t, $^3J_{HH}$ = 6.0, 1 H, AlC(CHCH)$_2$CHAl), 1.53 (s, 6 H, ArNCCH$_3$), 1.52 (s, 6 H, ArNCCH$_3$), 1.33 (d, $^3J_{HH}$ = 6.8 Hz, 6 H, CH(CH$_3$)$_2$), 1.30 (d, $^3J_{HH}$ = 6.7 Hz, 6 H, CH(CH$_3$)$_2$), 1.25 (d, $^3J_{HH}$ = 7.0 Hz, 6 H, CH(CH$_3$)$_2$), 1.16 (d, $^3J_{HH}$ = 6.8 Hz, 6 H, CH(CH$_3$)$_2$), 1.12 (d, $^3J_{HH}$ = 6.8 Hz, 6 H, CH(CH$_3$)$_2$), 0.96 (d, $^3J_{HH}$ = 6.7 Hz, 6 H, CH(CH$_3$)$_2$), 0.92 (d, $^3J_{HH}$ = 6.78 Hz, 6 H, CH(CH$_3$)$_2$), 0.82 (d, $^3J_{HH}$ = 6.7 Hz, 6 H, CH(CH$_3$)$_2$). ^{13}C NMR (100.6 MHz, C_6D_6, 25 °C): δ 170.8, 170.3 (ArNCCH$_3$), 148.5, 132.7 (COAl), 140.5, 126.9, 126.9, 123.0 (C_6H_5), 128.6, 123.3 (AlC(CHCH)$_2$CHAl), 146.2, 145.3, 143.3, 142.5, 141.4, 141.3, 127.4, 127.1, 125.3, 124.7, 123.8, 123.2 (ArC), 98.1, 97.4 (γ-CH), 39.6, 32.7 (AlC(CHCH)$_2$CHAl), 30.0, 28.9, 28.5, 28.2 (CH(CH$_3$)$_2$), 26.6, 26.5, 25.0, 24.9, 24.8, 24.6, 23.8, 23.6 (CH(CH$_3$)$_2$), 23.8 (ArNCCH$_3$). Peaks in italic were only observed in HSQC or HMBC respectively.

4.12. Synthesis of L'B($C_{14}H_{10}O_2$) (13)

A total of 56.9 mg of KC$_8$ (421 μmol) and 41.1 mg of benzil (198 μmol) were cooled to −80 °C, 5 mL of thf was added and the resulting mixture was warmed to room temperature upon stirring. The liquid phase was transferred through a filter cannula into a cooled (−80 °C) Schlenk flask containing 100 mg of L'BCl$_2$ (198 μmol). All volatiles were removed at room temperature in a vacuo, and the resulting residue was extracted with 5 mL of toluene. The extract was concentrated and stored at −30 °C, yielding 20 mg of an orange crystalline solid. The mother liquor was further concentrated to give a second fraction. Yield: 30 mg (46 μmol, 24%).

Anal. Calcd. for $C_{31}H_{17}BF_{10}N_2O_2$: C, 57.26, H, 2.64; N, 4.31; Found: C, 57.4, H, 2.63; N, 4.31%. ATR-IR: υ 1564, 1508, 1469, 1450, 1392, 1290, 1264, 1130, 1084, 1065, 1042, 1022, 987, 925, 871, 761, 724, 693, 663, 619, 597, 523, 457 cm^{-1}. ^{1}H NMR (400 MHz, C_6D_6, 25 °C): δ 7.62–7.57 (m, 4 H, C_6H_5), 6.98 (t, $^3J_{HH}$ = 7.8 Hz, 4 H, C_6H_5), 6.86 (t, $^3J_{HH}$ = 7.41 Hz, 2 H, C_6H_5), 4.69 (s, 1 H, γ-CH), 1.26 (s, 6 H, ArNCCH$_3$). ^{11}B NMR (128.4 MHz, C_6D_6, 25 °C): δ 7.0. ^{13}C NMR (100.6 MHz, C_6D_6, 25 °C): δ 166.2 (ArNCCH$_3$), 136.7 (COB), 133.4, 128.4, 127.1, 126.1 (C_6H_5), 97.6 (γ-CH), 20.6 (ArNCCH$_3$). ^{13}C {^{19}F} NMR (150.9 MHz, C_6D_6, 25 °C): δ 144.4, 141.0, 137.9, 116.4 (C_6F_5). ^{19}F NMR (376.5 MHz, C_6D_6, 25 °C): δ −143.5 (d, $^3J_{HH}$ = 17.6 Hz, 4 F, C_6F_5), −153.8 (t, $^3J_{HH}$ = 22.5 Hz, 2 F, C_6F_5), −162.3 (m, 4 F, C_6H_5).

4.13. Synthesis of L'BH$_2$ (14)

At −80 °C, 783 μL (783 μmol) of a 1 M (n-hexane) L-selectride solution was added to 200 mg of L'BCl$_2$ (391 μmol) dissolved in 10 mL of toluene. The reaction mixture was allowed to reach ambient temperature overnight. All volatiles were then removed in vacuo, and the resulting residue was extracted with 10 mL of n-pentane. The extract was concentrated and stored at −30 °C to give 70 mg of 14 as a yellow crystalline solid. Yield: 70 mg (158 μmol, 41%).

Anal. Calcd. for $C_{17}H_9BF_{10}N_2$: C, 46.19, H, 2.05; N, 6.34; Found: C, 46.6, H, 2.32; N, 6.16%. ATR-IR: υ 2964, 2428, 2370, 2291, 2192, 1582, 1549, 1504, 1467, 1438, 1392, 1344, 1260, 1206, 1069, 1051, 1024, 1009, 989, 939, 817, 788, 782, 726, 647, 561, 472 cm^{-1}. ^{1}H NMR (400 MHz, C_6D_6, 25 °C): δ 4.77 (s, 1 H, γ-CH), 3.77 (s (br), 1 H, BH$_2$), 3.46 (s (br), 1 H, BH$_2$), 1.24 (s, 6 H, ArNCCH$_3$). ^{11}B NMR (192.5 MHz, C_6D_6, 25 °C): δ 7.0 (B). ^{13}C NMR (100.6 MHz, C_6D_6, 25 °C): δ 167.6 (ArNCCH$_3$), 143.7 (m, $^2J_{HH}$ = 250 Hz, C_6F_5), 140.3 (m, $^2J_{HH}$ = 252 Hz, C_6F_5), 138.2 (m, $^2J_{HH}$ = 252 Hz, C_6F_5), 120.1 (td, J_{HH} = 17.8, 5.2 Hz, C_6F_5), 99.6 (γ-CH), 19.8 (ArNCCH$_3$). ^{19}F NMR (376.5 MHz, C_6D_6, 25 °C): δ −143.7 (d, $^3J_{HH}$ = 20.8 Hz, 4 F, C_6F_5), −156.2 (t, $^3J_{HH}$ = 22.3 Hz, 2 F, C_6F_5), −161.9 (td, J_{HH} = 23.3, 6.1 Hz, 4 F, C_6H_5).

4.14. Crystallographic Details

Crystals were mounted on nylon loops in inert oil. Data of were collected on a Bruker AXS D8 Kappa diffractometer (**2, 5, 7, 9, 10, 11, 12b**) with APEX2 detector (monochromated Mo$_{K\alpha}$ radiation, $\lambda = 0.71073$ Å) and on a Bruker AXS D8 Venture diffractometer (**1, 4, 6, 8, 12a, 13**) with Photon II detector (monochromated Cu$_{K\alpha}$ radiation, $\lambda = 1.54178$ Å, microfocus source) at 100(2) K. Structures were solved by Direct Methods (SHELXS-2013) [40] and refined anisotropically by full-matrix least-squares on F^2 (SHELXL-2014) [41–43]. Absorption corrections were performed semi-empirically from equivalent reflections based on multi-scans (Bruker AXS APEX2). Hydrogen atoms were refined using a riding model or rigid methyl groups.

1: An isopropyl group is disordered over two positions. C24 and C24′ were refined with common positions and displacement parameters (EXYZ, EADP). RIGU restraints were applied to the displacement parameters of the disordered atoms.

2: Two isopropyl groups are disordered over two positions. Their bond lengths and angles were restrained to be equal (SADI) and RIGU restraints were applied to their displacement parameters. Additional SIMU restraints were applied to the group in residue 2. In residue 1, the diol ligand is disordered over two positions. All corresponding bond lengths were restrained to be equal (SADI) and RIGU restraints were used to refine the displacement parameters. Atoms in close proximity (O4, O4′ and C33, C33′) were refined with common displacement parameters (EADP). One solvent molecule is disordered over two positions. All bond lengths and angles of the solvent molecules were restrained to be equal (SADI) and the molecule was restrained to be planar (FLAT). RIGU restraints were applied to the displacement parameters of the solvent atoms. The quantitative results of the disordered moieties should be scrutinized, and especially the data for the diol ligand may be unreliable. In addition, ice formed during the measurement and combined with a scattering of the mount/oil used as glue, clearly visible background scattering was found in the frames. This leads to some distorted intensities. The resolution of (090) could be related to the (103) reflection of water and was therefore omitted. Three reflections shaded by the beamstop were also ignored in the refinement. The remaining most disagreeable reflections have a resolution of >2 Å with F_{obs} higher than F_{calc}. **2** was refined as a 2-component inversion twin.

5: The crystal was a non-merohedral twin and the model was refined against de-twinned HKLF4 data. At low angles, some disagreeable reflections are found. These are either due to poor separation of overlaps in the integration or, more likely, to background scattering caused by the icing of the crystal. Three reflections shaded by the beam-stop were ignored in the refinement (OMIT).

6: The structure contains a 2-methyltetrahydrofuran molecule highly disordered over an inversion center. The final refinement was conducted with a solvent-free data set from a PLATON/SQUEEZE run [44]. The molecule was included in the sum formula for completeness.

7: An isopropyl group is disordered over two positions. Its bond lengths and angles were restrained to be equal (SADI) and RIGU restraints were applied to its displacement parameters. The structure contains two highly disordered n-hexane molecules. The final refinement was conducted with a solvent-free data set from a PLATON/SQUEEZE run [44]. The molecules were included in the sum formula for completeness.

9: A diisopropylphenyl group is disordered over two positions. RIGU restraints were applied to the displacement parameters of the corresponding atoms.

12a: A benzene molecule is disordered over two positions. RIGU restraints were applied to the displacement parameters of the atoms of the solvent molecules.

12b: An *n*-hexane molecule is disordered over a center of inversion. The bond lengths and bond angles of all solvent molecules were restrained to be equal and RIGU restraints were applied to the displacement parameters of their atoms. An additional SIMU restraint was used for the disordered molecule on the center of inversion. Its displacement parameters suggested further disorder that could not be resolved any further.

Supplementary Materials: The following supporting information can be downloaded at: https://www.mdpi.com/article/10.3390/chemistry5020064/s1, Heteronuclear NMR (^{1}H, ^{11}B, ^{13}C, ^{19}F) and IR data of all compounds (Figures S1–S48) as well as crystallographic details (Table S1a–d and Figures S53–S57).

Author Contributions: Conceptualization, H.M.W. and S.S.; investigation, H.M.W.; sc-XRD acquisition and refinement, C.W.; writing—original draft preparation, H.M.W.; writing—review and editing, S.S.; supervision, S.S. All authors have read and agreed to the published version of the manuscript.

Funding: This work was supported by the Deutsche Forschungsgemeinschaft DFG (INST 20876/282-1 FUGG) and the University of Duisburg-Essen.

Data Availability Statement: Spectroscopic data and crystallographic details are given in the electronic supplement. The structures of compounds **1**, **2**, and **4–13** in the solid state were determined by single-crystal X-ray diffraction and the crystallographic data have been deposited with the Cambridge Crystallographic Data Centre as supplementary publication nos. CCDC-2251780 (**1**), -2251781 (**2**), -2251782 (**4**), -2251783 (**5**), -2251784 (**6**), -2251785 (**7**), -2251786 (**8**), -2251787 (**9**), -2251788 (**10**), -2251789 (**11**), -2251790 (**12a**), -2251791 (**12b**), and -2251909 (**13**). Copies of the data can be obtained free of charge on application to CCDC, 12 Union Road, Cambridge, CB21EZ (fax: (+44) 1223/336033; e-mail: deposit@ccdc.cam-ak.uk).

Acknowledgments: We are thankful to A. Gehlhaar and Y. Schulte (University of Duisburg-Essen) for assistance with the sc-XRD data acquisition, to T. Schaller (University of Duisburg-Essen) for performing non-routine NMR experiments, and to J. Tewes (University of Duisburg-Essen) for providing L'BCl$_2$.

Conflicts of Interest: The authors declare no conflict of interest.

References

1. Cui, C.; Roesky, H.W.; Schmidt, H.-G.; Noltemeyer, M.; Hao, H.; Cimpoesu, F. Synthesis and Structure of a Monomeric Aluminum(I) Compound [{HC(CMeNAr)$_2$}Al] (Ar=2,6–*i*Pr$_2$C$_6$H$_3$): A Stable Aluminum Analogue of a Carbene. *Angew. Chem. Int. Ed.* **2000**, *39*, 4274–4276. [CrossRef]
2. Hardman, N.J.; Eichler, B.E.; Power, P.P. Synthesis and characterization of the monomer Ga{(NDippCMe)$_2$CH} (Dipp = C$_6$H$_3$Pr^i$_2$-2,6): A low valent gallium(I) carbene analogue. *Chem. Commun.* **2000**, 1991–1992. [CrossRef]
3. Hill, M.S.; Hitchcock, P.B. A mononuclear indium(I) carbene analogue. *Chem. Commun.* **2004**, 1818–1819. [CrossRef] [PubMed]
4. Hill, M.S.; Hitchcock, P.B.; Pongtavornpinyo, R. Neutral carbene analogues of the heaviest Group 13 elements: Consideration of electronic and steric effects on structure and stability. *Dalton Trans.* **2005**, 273–277. [CrossRef] [PubMed]
5. Liu, Y.; Li, J.; Ma, X.; Yang, Z.; Roesky, H.W. The chemistry of aluminum(I) with β-diketiminate ligands and pentamethylcyclopentadienyl-substituents: Synthesis, reactivity and applications. *Coord. Chem. Rev.* **2018**, *374*, 387–415. [CrossRef]
6. Weetman, C.; Xu, H.; Inoue, S. Recent Developments in Low-Valent Aluminum Chemistry. *EIBC* **2020**, 1–20. [CrossRef]
7. Zhong, M.; Sinhababu, S.; Roesky, H.W. The unique β-diketiminate ligand in aluminum(I) and gallium(I) chemistry. *Dalton Trans.* **2020**, *49*, 1351–1364. [CrossRef]
8. Hobson, K.; Carmalt, C.J.; Bakewell, C. Recent advances in low oxidation state aluminium chemistry. *Chem. Sci.* **2020**, *11*, 6942–6956. [CrossRef]
9. Power, P.P. Main-group elements as transition metals. *Nature* **2010**, *463*, 171–177. [CrossRef]
10. Weetman, C.; Inoue, S. The Road Travelled: After Main-Group Elements as Transition Metals. *ChemCatChem* **2018**, *10*, 4213–4228. [CrossRef]
11. Chu, T.; Boyko, Y.; Korobkov, I.; Nikonov, G.I. Transition Metal-like Oxidative Addition of C–F and C–O Bonds to an Aluminum(I) Center. *Organometallics* **2015**, *34*, 5363–5365. [CrossRef]
12. Bakewell, C.; White, A.J.P.; Crimmin, M.R. Reactions of Fluoroalkenes with an Aluminium(I) Complex. *Angew. Chem.* **2018**, *130*, 6748–6752. [CrossRef]
13. Crimmin, M.R.; Butler, M.J.; White, A.J.P. Oxidative addition of carbon-fluorine and carbon-oxygen bonds to Al(I). *Chem. Commun.* **2015**, *51*, 15994–15996. [CrossRef] [PubMed]

14. Chu, T.; Nikonov, G.I. Oxidative Addition and Reductive Elimination at Main-Group Element Centers. *Chem. Rev.* **2018**, *118*, 3608–3680. [CrossRef]
15. Wright, R.J.; Brynda, M.; Fettinger, J.C.; Betzer, A.R.; Power, P.P. Quasi-isomeric gallium amides and imides GaNR$_2$ and RGaNR (R = organic group): Reactions of the digallene, Ar′GaGaAr′ (Ar′ = C$_6$H$_3$-2,6-(C6H$_3$-2,6-Pr$^i_2$)$_2$) with unsaturated nitrogen compounds. *J. Am. Chem. Soc.* **2006**, *128*, 12498–12509. [CrossRef]
16. Cui, C.; Köpke, S.; Herbst-Irmer, R.; Roesky, H.W.; Noltemeyer, M.; Schmidt, H.G.; Wrackmeyer, B. Facile synthesis of cyclopropene analogues of aluminum and an aluminum pinacolate, and the reactivity of LAlη^2-C$_2$(SiMe$_3$)$_2$ toward unsaturated molecules (L = HC(CMe)(NAr)$_2$, Ar = 2,6-i-Pr$_2$C$_6$H$_3$). *J. Am. Chem. Soc.* **2001**, *123*, 9091–9098. [CrossRef]
17. Liu, H.-Y.; Hill, M.S.; Mahon, M.F. Diverse reactivity of an Al(I)-centred anion towards ketones. *Chem. Commun.* **2022**, *58*, 6938–6941. [CrossRef]
18. Sokolov, V.G.; Koptseva, T.S.; Moskalev, M.V.; Bazyakina, N.L.; Piskunov, A.V.; Cherkasov, A.V.; Fedushkin, I.L. Gallium Hydrides with a Radical-Anionic Ligand. *Inorg. Chem.* **2017**, *56*, 13401–13410. [CrossRef]
19. Fedushkin, I.L.; Skatova, A.A.; Dodonov, V.A.; Chudakova, V.A.; Bazyakina, N.L.; Piskunov, A.V.; Demeshko, S.V.; Fukin, G.K. Digallane with redox-active diimine ligand: Dualism of electron-transfer reactions. *Inorg. Chem.* **2014**, *53*, 5159–5170. [CrossRef]
20. Uhl, W.; Keimling, S.U.; Phol, S.; Saak, W.; Wartchow, R. Benzil Derivatives as Trapping Reagents for the Monomeric Alkylindium(i) Compound in–c(SiMe$_3$)$_3$. *Chem. Ber.* **1997**, *130*, 1269–1272. [CrossRef]
21. Ganesamoorthy, C.; Bläser, D.; Wölper, C.; Schulz, S. Temperature-Dependent Electron Shuffle in Molecular Group 13/15 Intermetallic Complexes. *Angew. Chem.* **2014**, *126*, 11771–11775. [CrossRef]
22. Ganesamoorthy, C.; Helling, C.; Wölper, C.; Frank, W.; Bill, E.; Cutsail, G.E., III; Schulz, S. From stable Sb- and Bi-centered radicals to a compound with a Ga=Sb double bond. *Nat. Commun.* **2018**, *9*, 87. [CrossRef] [PubMed]
23. Helling, C.; Wölper, C.; Schulz, S. Synthesis of a gallaarsene {HC[C(Me)N-2,6-i-Pr$_2$-C$_6$H$_3$]$_2$}GaAsCp* containing a Ga=As double bond. *J. Am. Chem. Soc.* **2018**, *140*, 5053–5056. [CrossRef]
24. Krüger, J.; Ganesamoorthy, C.; John, L.; Wölper, C.; Schulz, S. A General Pathway for the Synthesis of Gallastibenes containing Ga=Sb Double Bonds. *Chem. Eur. J.* **2018**, *24*, 9157–9164. [CrossRef]
25. Helling, C.; Cutsail, G.E.; Weinert, H.; Wölper, C.; Schulz, S. Ligand Effects on the Electronic Structure of Heteroleptic Antimony-Centered Radicals. *Angew. Chem.* **2020**, *132*, 7631–7638. [CrossRef]
26. Ganesamoorthy, C.; Schoening, J.; Wölper, C.; Song, L.; Schreiner, P.R.; Schulz, S. A silicon–carbonyl complex stable at room temperature. *Nat. Chem.* **2020**, *12*, 608–614. [CrossRef] [PubMed]
27. Weinert, H.M.; Wölper, C.; Haak, J.; Cutsail, G.E.; Schulz, S. Synthesis, structure and bonding nature of heavy dipnictene radical anions. *Chem. Sci.* **2021**, *12*, 14024–14032. [CrossRef]
28. Weinert, H.M.; Wölper, C.; Schulz, S. Synthesis of distibiranes and azadistibiranes by cycloaddition reactions of distibenes with diazomethanes and azides. *Chem. Sci.* **2022**, *13*, 3775–3786. [CrossRef]
29. Sharma, M.K.; Wölper, C.; Haberhauer, G.; Schulz, S. Reversible and Irreversible [2+2] Cycloaddition Reactions of Heteroallenes to a Gallaphosphene. *Angew. Chem.* **2021**, *133*, 21953–21957. [CrossRef]
30. Sharma, M.K.; Wölper, C.; Haberhauer, G.; Schulz, S. Multi-talented gallaphosphene for Ga–P–Ga heteroallyl cation generation, CO$_2$ storage and C(sp^3)–H bond activation. *Angew. Chem.* **2021**, *133*, 6859–6865. [CrossRef]
31. Wöhler, F.; von Liebig, J. Untersuchungen über das Radikal der Benzoesäure. *Ann. Pharm.* **1832**, *3*, 249–282. [CrossRef]
32. Chen, C.-H.; Tsai, M.-L.; Su, M.-D. Theoretical Study of the Reactivities of Neutral Six-Membered Carbene Analogues of the Group 13 Elements. *Organometallics* **2006**, *25*, 2766–2773. [CrossRef]
33. Hardman, N.J.; Phillips, A.D.; Power, P.P. Bonding and Reactivity of a β-Diketiminate, Gallium(I), Carbene Analogue. *ACS Symp. Ser.* **2002**, *822*, 2–15. [CrossRef]
34. Vidovic, D.; Findlater, M.; Cowley, A.H. A β-diketiminate-Supported Boron Dication. *J. Am. Chem. Soc.* **2007**, *129*, 8436–8437. [CrossRef]
35. Pyykkö, P.; Atsumi, M. Molecular single-bond covalent radii for elements 1-118. *Chem. Eur. J.* **2009**, *15*, 186–197. [CrossRef]
36. Panda, A.; Stender, M.; Wright, R.J.; Olmstead, M.M.; Klavins, P.; Power, P.P. Synthesis and characterization of three-coordinate and related beta-diketiminate derivatives of manganese, iron, and cobalt. *Inorg. Chem.* **2002**, *41*, 3909–3916. [CrossRef]
37. Qian, B.; Ward, D.L.; Smith, M.R. Synthesis, Structure, and Reactivity of β-Diketiminato Aluminum Complexes. *Organometallics* **1998**, *17*, 3070–3076. [CrossRef]
38. Stender, M.; Eichler, B.E.; Hardman, N.J.; Power, P.P.; Prust, J.; Noltemeyer, M.; Roesky, H.W. Synthesis and Characterization of HC{C(Me)N(C$_6$H$_3$-2,6-i-Pr$_2$)}$_2$MX$_2$ (M = Al, X = Cl, I; M = Ga, In, X = Me, Cl, I): Sterically Encumbered β-Diketiminate Group 13 Metal Derivatives. *Inorg. Chem.* **2001**, *40*, 2794–2799. [CrossRef]
39. Stender, M.; Wright, R.J.; Eichler, B.E.; Prust, J.; Olmstead, M.M.; Roesky, H.W.; Power, P.P. The synthesis and structure of lithium derivatives of the sterically encumbered β-diketiminate ligand [{(2,6-Pr$^i_2$H$_3$C$_6$)N(CH$_3$)C}$_2$CH]$^-$, and a modified synthesis of the aminoimine precursor. *Dalton Trans.* **2001**, 3465–3469. [CrossRef]
40. Sheldrick, G.M. Phase annealing in *SHELX*-90: Direct methods for larger structures. *Acta Crystallogr.* **1990**, *A46*, 467–473. [CrossRef]
41. Sheldrick, G.M. *SHELXL-2014, Program for the Refinement of Crystal Structures*; Univ. of Göttingen: Göttingen, Germany, 2014.
42. Sheldrick, G.M. A short history of SHELX. *Acta Crystallogr.* **2008**, *A64*, 112–122. [CrossRef] [PubMed]

43. Hübschle, C.B.; Sheldrick, G.M.; Dittrich, B. *ShelXle*: A Qt graphical user interface for *SHELXL*. *J. Appl. Cryst.* **2011**, *44*, 1281–1284. [CrossRef] [PubMed]
44. van der Sluis, P.; Spek, A.L. BYPASS: An effective method for the refinement of crystal structures containing disordered solvent regions. *Acta Cryst.* **1990**, *A46*, 194–201. [CrossRef]

Disclaimer/Publisher's Note: The statements, opinions and data contained in all publications are solely those of the individual author(s) and contributor(s) and not of MDPI and/or the editor(s). MDPI and/or the editor(s) disclaim responsibility for any injury to people or property resulting from any ideas, methods, instructions or products referred to in the content.

Article

Synthesis, Structure, and Spectroscopic Properties of Luminescent Coordination Polymers Based on the 2,5-Dimethoxyterephthalate Linker

Aimée E. L. Cammiade, Laura Straub, David van Gerven, Mathias S. Wickleder and Uwe Ruschewitz *

Department of Chemistry, University of Cologne, Greinstraße 6, D-50939 Cologne, Germany; aimee.cammiade@uni-koeln.de (A.E.L.C.)
* Correspondence: uwe.ruschewitz@uni-koeln.de

Abstract: We report on the synthesis and the crystal structure of the solvent-free coordination polymer $Co^{II}(2,5\text{-}DMT)$ (**1**) with 2,5-DMT ≡ 2,5-dimethoxyterephthalate which is isostructural to the already reported Mn^{II} and Zn^{II} congeners ($C2/c$, $Z = 4$). In contrast, for M = Mg^{II}, a MOF with DMF-filled pores is obtained, namely $Mg_2(2,5\text{-}DMT)_2(DMF)_2$ (**2**) ($P\bar{1}$, $Z = 2$). Attempts to remove these solvent molecules to record a gas sorption isotherm did not lead to meaningful results. In a comparative study, the thermal (DSC/TGA) and luminescence properties of all the four compounds were investigated. The compounds of the $M^{II}(2,5\text{-}DMT)$ composition show high thermal stability up to more than 300 °C, with the Zn^{II} compound having the lowest decomposition temperature. $M^{II}(2,5\text{-}DMT)$ with $M^{II} = Mn^{II}$, Zn^{II} and **2** show a bright luminescence upon blue light irradiation (λ = 405 nm), whereas Co^{II} in **1** quenches the emission. While Zn^{II} in $Zn^{II}(2,5\text{-}DMT)$ and Mg^{II} in **2** do not significantly influence the (blue) emission and excitation bands compared to the free 2,5-DMT ligand, Mn^{II} in $Mn^{II}(2,5\text{-}DMT)$ shows an additional metal-centred red emission.

Keywords: coordination polymers; fluorescence; metal–organic frameworks; methoxy substituents; terephthalates

Citation: Cammiade, A.E.L.; Straub, L.; van Gerven, D.; Wickleder, M.S.; Ruschewitz, U. Synthesis, Structure, and Spectroscopic Properties of Luminescent Coordination Polymers Based on the 2,5-Dimethoxyterephthalate Linker. *Chemistry* **2023**, *5*, 965–977. https://doi.org/10.3390/chemistry5020065

Academic Editor: Edwin Charles Constable

Received: 31 March 2023
Revised: 14 April 2023
Accepted: 19 April 2023
Published: 22 April 2023

Copyright: © 2023 by the authors. Licensee MDPI, Basel, Switzerland. This article is an open access article distributed under the terms and conditions of the Creative Commons Attribution (CC BY) license (https://creativecommons.org/licenses/by/4.0/).

1. Introduction

Since their discovery more than 20 years ago [1,2], the research on MOFs (metal–organic frameworks) at the border between coordination, solid state, and materials chemistry has continued to attract an ever-increasing amount of interest. The simple design starting from a large variety of different metal cations or metal–oxo clusters as nodes and an almost unlimited number of possible organic linker ligands has now led to more than 100,000 entries in the MOF subset [3] of the CSD database [4], although not all entries follow strictly the recommendations of the IUPAC terminology for MOFs [5]. Since their first introduction, many possible applications in the fields of gas storage [6] and gas separation/purification [7], catalysis [8], drug transport/delivery [9], sensing [10], or electronic applications such as proton [11] and lithium-ion conductivity [12] have been discussed for this class of materials.

With respect to sensing, fluorescent (or, more generally, luminescent) MOFs [13] are being focused upon as a changing wavelength or a diminishing signal upon uptake of an adsorbate is simple to detect. Luminescence can be either metal- or linker-based, including ligand-to-metal charge transfer (LMCT) and metal-to-ligand charge transfer (MLCT) processes [14]. For the former, lanthanide-based MOFs are most prominent, e.g., Eu^{3+}-based materials with a strong red or Tb^{3+}-based compounds with a strong green luminescence [15]. There is an almost uncountable number of publications in this field, but it is our impression that linker-based luminescence has rarely been investigated, although even simple and very frequently used conjugated linkers such as 1,3,5-benzenetricarboxylic acid (BTC) or 1,4-benzenedicarboxylic acid (BDC) show a weak emission at 440 nm [16] and

388 nm [17], respectively. For potential sensing applications, linker-based luminescence holds many promising perspectives due to a direct interaction between the linker and the adsorbate, whereas a luminescent metal cation is somewhat "hidden" in its coordination sphere, making the influence of a non-coordinating adsorbate on its emission properties apparently weaker.

As a very spectacular result, linker-based luminescence was applied to detect defects within crystalline MOFs with a high spatial resolution [18]. In this work, based on MOFs with the UiO-67 topology, the authors used bulky fluorescein isothiocyanate or rhodamine B isothiocyanate substituents for their approach. However, there are also linkers with much smaller substituents, which show a strong luminescence. Among them, 2,5-dimethoxyterephthalic acid (2,5-DMT) shows a strong blue emission at λ_{em} = 410 nm (λ_{ex} = 370 nm)[this work], which compares well with the published results on its dimethyl ester (λ_{em} = 402 nm with λ_{ex} = 320 nm) [19]. In the literature, several coordination polymers (CP) and MOFs have already been reported with the 2,5-DMT linker and the Mn^{2+} [20], Zn^{2+} [20–22], Co^{2+} [23], Cu^{2+} [24], Th^{4+} [25], and Eu^{3+} metal cations [26]. To our surprise, the luminescence properties of the resulting materials were not investigated in most of these reports. The only exception is the Eu^{3+}-based compound (doped with Tb^{3+}), where a mainly lanthanide-based emission was observed [26]. Mertens et al. reported solvent-free coordination polymers M^{II}(2,5-DMT) with M^{II} = Mn^{2+}, Zn^{2+} [20]. In the following, we add Co^{II}(2,5-DMT) to this series and compare the thermal stability as well as the luminescence behaviour of all the three compounds. Additionally, we present the first Mg^{2+}-based MOF of the Mg_2(2,5-DMT)$_2$(DMF)$_2$ composition, which is discussed in the context of the three aforementioned CPs.

2. Materials and Methods

2.1. Synthesis of the Linker

All reagents were purchased commercially and used without further purification.

General: The synthesis of 2,5-dimethoxy terephthalic acid mainly followed the protocol of a two-step synthesis provided in the literature [27], but with an increased reaction time of the first step to increase the overall yield.

Synthesis of diethyl 2,5-dimethoxybenzene-1,4-dicarboxylate: The compound was synthesised in dry glassware under an inert (argon) atmosphere: 1.337 g (5.26 mmol, 1.00 eq.) diethyl-2,5-dihydroxy-terephthalate and 2.326 g (16.8 mmol, 3.20 equiv.) potassium carbonate were dissolved in 16 mL dry acetone; 1.1 mL (2.508 g, 17.7 mmol, 3.36 eq.) CH_3I was added and the suspension was stirred for 48 h at 60 °C. The solvent was evaporated under reduced pressure and the resulting residue was dissolved in water and poured into a separation funnel. After the recombination of the aqueous solutions, they were re-extracted with ethyl acetate. The organic layers were collected and dried through the addition of $MgSO_4$. Filtration through a glass funnel and evaporation of the solvent under reduced pressure led to a colourless powder with a yield of 93% (1.38 g, 4.90 mmol).

^{1}H-NMR: 300 MHz, $CDCl_3$: ppm = 7.37 (s, 2H, H-3/6), 4.39 (q, J = 7.1 Hz, 4H, H-8/11), 3.89 (s, 6H, H-13/14), 1.40 (t, J = 7.1 Hz, 6H, H-9/12).

^{13}C-NMR: 75 MHz, $CDCl_3$: ppm = 165.7 (C-7/10), 152.5 (C-2/5), 124.5 (C-1/4), 115.5 (C-3/6), 61.5 (C-8/11), 57.0 (C-13/14), 14.4 (C-9/12).

2.2. Synthesis of Coordination Polymers/MOFs

Co^{II}(2,5-DMT) (1): 19.9 mg 2,5-dimethylterephthalic acid (0.087 mmol, 1.00 eq.) and 16.6 mg $Co(NO_3)_2 \cdot 6$ H_2O (0.090 mmol, 1.03 eq.) were dissolved in DMF (2.7 mL) in a glass vial (10 mL). The vial was sealed and heated at 100 °C for 48 h.

Mn^{II}(2,5-DMT): 20.3 mg 2,5-dimethylterephthalic acid (0.089 mmol, 1.00 eq.) and 15.8 mg $Mn(NO_3)_2 \cdot 4$ H_2O (0.88 mmol, 0.99 eq.) were dissolved in DMF (2.7 mL) in a glass vial (10 mL). The vial was sealed and heated at 100 °C for 48 h.

Zn^{II}(2,5-DMT): 19.4 mg 2,5-dimethylterephthalic acid (0.085 mmol, 1.00 eq.) and 16.7 mg $Zn(NO_3)_2 \cdot 6\ H_2O$ (0.088 mmol, 1.03 eq.) were dissolved in DMF (2.7 mL) in a glass vial (10 mL). The vial was sealed and heated at 100 °C for 48 h.

Mg_2(2,5-DMT)$_2$(DMF)$_2$ (**2**): 17.1 mg 2,5-dimethylterephthalic acid (0.075 mmol, 1.00 eq.) and 12.2 mg $Mg(NO_3)_2 \cdot 6\ H_2O$ (0.082 mmol, 1.09 eq.) were dissolved in DMF (2.7 mL) in a glass vial (10 mL). The vial was sealed and heated at 100 °C for 48 h.

2.3. Analytical Methods

PXRD patterns were recorded on a Rigaku MiniFlex 600-C diffractometer (Cu Kα radiation, Ni filter) (Tokyo, Japan) as flat samples. The typical recording times were 30 min. The obtained data were analysed and processed with the WinXPow programme package [28]. Additionally simulated patterns were generated from the Crystallographic Information Files of the respective substances with WinXPow [28]. The resulting data from the measurements were visualised with Gnuplot [29].

DSC/TG measurements were conducted on a Mettler Toledo (Gießen, Germany) DSC1 coupled with a TGA/DSC1 (Star System). The measured samples were placed and weighed in a corundum crucible under an argon stream (40 mL/min). Initially, the sample was heated to 30 °C and held at this temperature for 10 min. Subsequently, the sample was heated to 1000 °C with a heating rate of 10 °C/min. Finally, the data were evaluated with the STARe program package.

Luminescence measurements were carried out using a FLS980 spectrometer from Edinburgh Instruments (Livingston, UK) with a xenon lamp and a PMT detector. The measurements in the solution (2,5-DMT) were carried out in DMF using quartz glass cuvettes, the solid state measurements of 2,5-DMT and **2**—between two quartz glass plates. Mn^{II}(2,5-DMT) and Zn^{II}(2,5-DMT) were measured as KBr pellets. All the measurements were conducted at room temperature.

2.4. X-ray Single-Crystal Structure Analysis

The measurements were carried out on a Bruker D8 Venture diffractometer with either Ag Kα (λ = 0.56086 Å) or Mo Kα (λ = 0.71073 Å) radiation and a multi-layered mirror monochromator. For the reduction of the diffraction data by integration and absorption correction, the SAINT [30] and SADABS/TWINABS [31,32] programs from the APEX4 program package [33] were used. The determination of the space group and the starting model was carried out with the SHELXT program [34]. For further refinement, the SHELXL-18 [35] program was applied using the least squares method. All non-hydrogen atoms were refined with anisotropic displacement parameters. The hydrogen atoms were refined isotropically on the calculated positions using a riding model with their U_{iso} values constrained to 1.5 times the U_{eq} of their pivot atoms for terminal sp^3 carbon atoms and 1.2 times for all other carbon atoms. The data obtained from the SCXRD measurement of **2** revealed a non-merohedral crystal twin whose two domains were tilted by 4° with respect to each other. These two domains were integrated separately and subjected to twin absorption correction. Based on the reflections of the more dominant domain, an HKLF 4 file was generated, from which the structure was initially solved and refined. The final structure model was refined against the reflections of both domains (HKLF 5).

2.5. Further Software Programs

Visualisations of all crystal structures were made with the Diamond 4.6 program package [36]. Visualisation of the fluorescent data and DSC/TGA measurements was completed using the Origin 8.5 program package [37]. The ChemDraw Professional 15.0 program package [38] was used to visualise the organic molecules.

3. Results and Discussion

3.1. Synthesis and General Characterisation

Through a solvothermal reaction of (1:1) 2,5-dimethoxyterephthalic acid (2,5-DMT) and the respective metal nitrates ($M^{II}(NO_3)_2 \cdot x\,H_2O$ with M^{II} = Co, Mg, Zn, x = 6 and M^{II} = Mn, x = 4), coordination polymers M^{II}(2,5-DMT) with M^{II} = Co (**1**), Mn, Zn and the Mg_2(2,5-DMT)$_2$(DMF)$_2$ MOF (**2**) were obtained. The reactants were thoroughly mixed, dissolved in dimethylformamide (DMF), and heated for 48 h at 100 °C in a 10 mL glass vial. The resulting precipitates were filtered and dried. Powder X-ray diffractograms (PXRD, Figures S1–S4, Supporting Information) revealed that the M^{II}(2,5-DMT) coordination polymers with M^{II} = Mn^{II}, Zn^{II} ($C2/c$, Z = 4) [20] known from the literature and the new Co^{II} compound **1** crystallise isostructurally to each other. The PXRD patterns confirm that all the three compounds were obtained as samples with a high degree of purity (Table 1). The PXRD pattern of **2**, however, looks completely different to the other three, thus revealing that a material with a different structural arrangement was synthesised. After solving its crystal structure ($P1$, Z = 2; vide infra), it became evident that a MOF-type structure with DMF-filled channels was formed. Again, a very good correlation between the experimental PXRD pattern and the pattern simulated from the solved crystal structure confirms a high purity of this material. Single crystals of **1** and **2**, which exhibit a block-shaped habitus, were isolated from the precipitates mentioned above and measured on an X-ray single-crystal diffractometer. The crystals of **1** are transparent pinkish violet, whereas the crystals of **2** are—as expected—colourless.

Table 1. Calculated/observed values of the elemental analysis of M^{II}(2,5-DMT) with M^{II} = Co^{II} (**1**), Mn^{II}, Zn^{II} and Mg_2(2,5-DMT)$_2$(DMF)$_2$ (**2**) with an error tolerance of ±0.3%.

	Co^{II}(2,5-DMT) (**1**)	Mn^{II}(2,5-DMT)	Zn^{II}(2,5-DMT)	Mg_2(2,5-DMT)$_2$(DMF)$_2$ (**2**)
N	–/0.21	–/0.08	–/0.20	4.36/4.40
C	42.43/42.76	43.03/43.20	41.48/41.55	48.56/48.20
H	2.85/2.84	2.89/2.81	2.79/2.70	4.70/4.73
S	–/–	–/–	–/–	–/–

3.2. Crystal Structures

Complete crystallographic data can be found in the Supporting Information (Tables S1–S10). The three isostructural coordination polymers crystallise in the monoclinic space group $C2/c$ [20]; the novel Mg MOF crystallises in the triclinic space group $P1$. For comparison, selected crystallographic data of all the four compounds are given in Table 2.

The Co^{II} CP (**1**) crystallises isostructurally to the known Mn^{II} and Zn^{II} CPs, which have already been described in the literature [20]. Therefore, the crystal structure of **1** is only briefly discussed. Its asymmetric unit (ORTEP plot) is given as Figure S5 in the Supporting Information. In these compounds, the M^{II} cation is coordinated by four oxygen atoms stemming from the carboxylate groups of four different 2,5-DMT linkers (four shorter M^{II}–O distances given in Table 2). This leads to a distorted tetrahedral coordination, which is depicted for **1** in Figure 1. To quantify the distortion of the coordination spheres, we used the continuous shape measures approach (CShM) [40]; the respective values for a tetrahedral coordination (T-4) are given in Table 2. Values much larger than 1 indicate a severe distortion of the tetrahedral coordination. In Figure 1, two further oxygen atoms with significantly longer M^{II}–O bonds (cp. Table 2) are depicted stemming from the methoxy groups of two 2,5-DMT linkers. Thus, a distorted octahedral coordination can be assumed, but CShM values (OC-6) >> 1 again indicate a strong distortion. However, it is remarkable (cp. Table 2) that a decreasing distortion of the tetrahedral coordination from the Co compound to the Mn and the Zn compounds goes along with an increasing distortion of the octahedral coordination. Obviously, with these methoxy substituents in terephthalate-based linkers, pockets are formed to accommodate metal cations. The size of the metal cations seems to have a direct influence on the pocket's shape, more tetrahedral or more octahedral. This is also reflected in the Co^{II}–O distances of **1**, which show a large spread from 1.9834(7) Å to 2.3533(7) Å, while in the Co^{II} CPs with terephthalate ligands, a less distorted octahedral coordination with Co^{II}–O distances from 2.059(3) Å to 2.119(9) Å is found [41,42].

Table 2. Selected crystallographic data of M^{II}(2,5-DMT) with M^{II} = Co^{II} (**1**), Mn^{II}, Zn^{II} and Mg_2(2,5-DMT)$_2$(DMF)$_2$ (**2**).

	Co^{II}(2,5-DMT) (**1**)	Mn^{II}(2,5-DMT)	Zn^{II}(2,5-DMT)	Mg_2(2,5-DMT)$_2$(DMF)$_2$ (**2**)
Crystal system	monoclinic	monoclinic	monoclinic	triclinic
Space group, Z	C2/c, 4	C2/c, 4	C2/c, 4	P1, 2
a/Å	16.1305(5)	16.7686(6)	16.5936(6)	8.833(3)
b/Å	8.6024(3)	8.4646(3)	8.4438(3)	9.691(3)
c/Å	7.3426(2)	7.4464(3)	7.4838(3)	18.674(6)
α/°	90	90	90	98.274(7)
β/°	96.425(1)	99.093(1)	97.649(2)	93.305(10)
γ/°	90	90	90	107.308(9)
Volume/Å^3	1012.47(5)	1043.66(7)	1039.25(7)	1501.6(8)
Temp./K	100(2)	153(2)	153(2)	100(2)
Ionic radius, CN = 4 [39]	0.72 Å (Co^{2+}, hs)	0.80 Å (Mn^{2+}, hs)	0.74 Å (Zn^{2+})	0.71 Å (Mg^{2+})
M^{II}–O/Å	1.9834(7), 2× 2.0389(7), 2× 2.3532(7), 2×	2.0761(5), 2× 2.1391(6), 2× 2.5595(6), 2×	1.9547(13), 2× 2.0023(13), 2× 2.6223(14), 2×	Mg1: 1.961(3), 1.989(3), 2.007(3), 2.118(3), 2,198(3), 2.284(3) Mg2: 2.042(4), 2.048(4), 2.080(3), 2.082(3), 2.100(3), 2.107(3)
CShM values [40]	4.896 (T-4) 1.697(OC-6)	3.847 (T-4) 2.848 (OC-6)	2.428 (T-4) 3.280 (OC-6)	Mg1: 2.915 (OC-6) Mg2: 0.145 (OC-6)
Ref.	CCDC-2225418[this work]	CCDC-813469 [20]	CCDC-813470 [20]	CCDC-2225419[this work]

It should be noted that Mertens et al. chose a different description of their Mn^{II} and Zn^{II} CPs as they included two even more distant oxygen atoms, which led to CN = 8 and a distorted "super dodecahedron" [20]. None of these descriptions can be considered wrong or correct, they are more an expression of the flexibility of metal coordination in such compounds with 2,5-DMT ligands. $M^{II}O_n$ polyhedra are connected to chains running along [001], which are interconnected through 2,5-DMT linkers to form a 3D coordination network. The resulting topology shows some similarities with MOFs of the MIL series (sra topology). For a more detailed description of these crystal structures, see [20]. Neither Mertens et al. nor our group found an indication of permanent porosity in these materials (cp. Figure S6, Supporting Information, showing a space-filling representation of **1**).

Using the same reaction conditions that led to the formation of M^{II}(2,5-DMT) with M^{II} = Co^{II} (**1**), Mn^{II}, Zn^{II}, we obtained a completely different compound when Mg(NO$_3$)$_2$·6 H$_2$O was used as the starting material. It was shown that this structure was also formed when the cooling time was increased to 96 h and/or when lauric acid was added as a monocarboxylic additive to improve the crystallinity. The resulting crystal structure is shown in Figure 2. The asymmetric unit of Mg_2(2,5-DMT)$_2$(DMF)$_2$ (**2**) consists of two crystallographically distinguishable magnesium atoms (Mg1, Mg2), one complete and two half-linker anions as well as two coordinating DMF molecules.

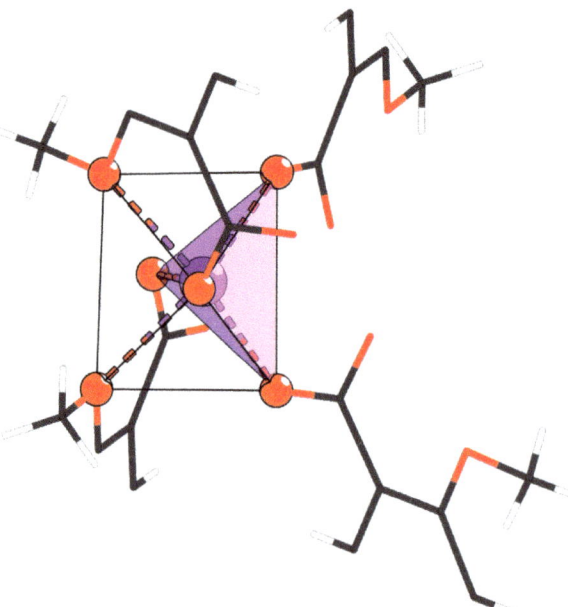

Figure 1. The coordination sphere of Co[II] in the crystal structure of Co[II](2,5-DMT), **1**; Co (purple sphere), O (red spheres), C (grey wireframes), and H (white wireframes). A tetrahedral Co[II]O$_4$ coordination is emphasized by purple shading, an extension to an octahedral Co[II]O$_6$ coordination—by thin dark grey lines.

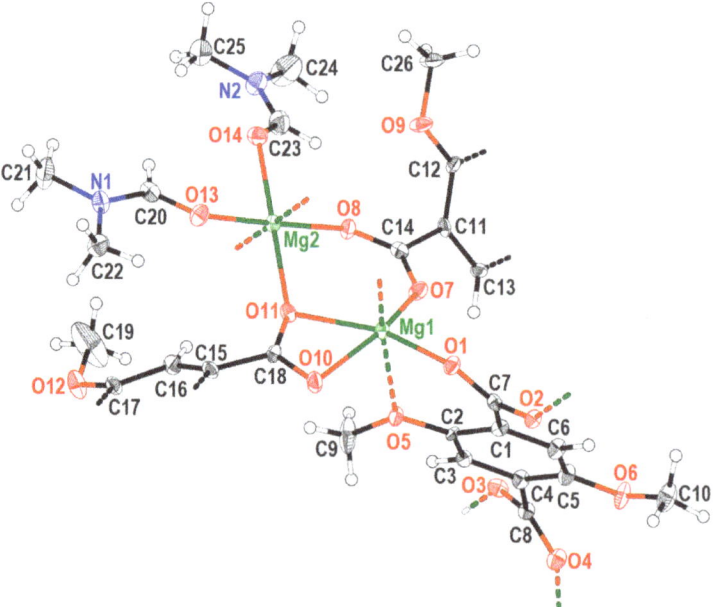

Figure 2. The asymmetric unit of Mg$_2$(2,5-DMT)$_2$(DMF)$_2$ (**2**) with atomic numbering; thermal ellipsoids are drawn with a 50% probability.

Both Mg atoms form distorted octahedral MgO$_6$ coordination spheres, which, however, show large differences. Mg1 is coordinated by five oxygen atoms which stem from the carboxylate groups of four different 2,5-DMT ligands (one carboxylate group coordinates in a bidentate chelating mode). The sixth oxygen atom belongs to a methoxy group of one of the four 2,5-DMT linkers. The resulting octahedron shows a severe distortion, as expressed by the large CShM value of 2.915 (Table 2). This confirms, as found for M^{II}(2,5-DMT) with M^{II} = Co^{II} (**1**), Mn^{II}, Zn^{II}, that, again, pockets are formed including the methoxy substituent, which, due to spatial restrictions, leads to distorted polyhedra if the metal cations do not fit perfectly into these pockets. In contrast, Mg2 is coordinated by four oxygen atoms of the carboxylate groups of four different 2,5-DMT anions and two oxygen atoms of two different DMF molecules. As there are no spatial restrictions, i.e., the oxygen atoms can be freely arranged, an almost undistorted octahedron is observed with a small CShM value (0.145). This is also reflected in the Mg–O distances: the Mg1–O distances range from 1.961(3) Å (O1) to 2.284(3) Å (O5), the Mg2–O distances—from 2.042(4) Å (O2) to 2.107(3) Å (O14). As expected, the longest bond is found between Mg1 and the oxygen atom O5 of the methoxy group. It is more than 0.1 Å longer than the typical Mg–O bond lengths found in the literature [43,44].

O11 bridges both MgO$_6$ octahedra, thus forming a corner-connected dimer. These dimeric units are interconnected through the 2,5-DMT ligands creating a three-dimensional network (Figure 3). This network forms channels into which the coordinating DMF molecules protrude. The pore sizes were calculated with the PLATON program package [45] taking the respective van der Waals radii into account, resulting in diameters from 3.10 Å to 4.64 Å. The channels (Figure 4) penetrate the whole framework of **2** in a wave-like fashion. It was assumed that these voids might be accessible to guest uptake after removal of the coordinating DMF molecules. This will be discussed in more detail below.

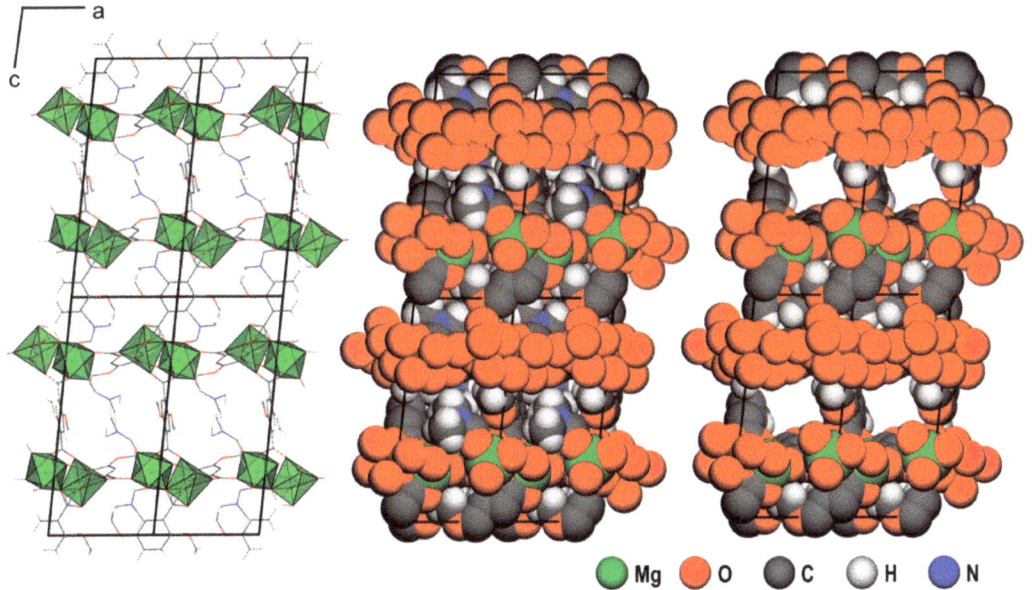

Figure 3. Section of the crystal structure of Mg$_2$(2,5-DMT)$_2$(DMF)$_2$ (**2**) in a view along [010]. (**Left**): representation of the linker anions as wireframes, the MgO$_6$ octahedra are emphasized in green; (**middle**): space-filling representation considering the van der Waals radii with the DMF molecules; (**right**): space-filling representation without the DMF molecules.

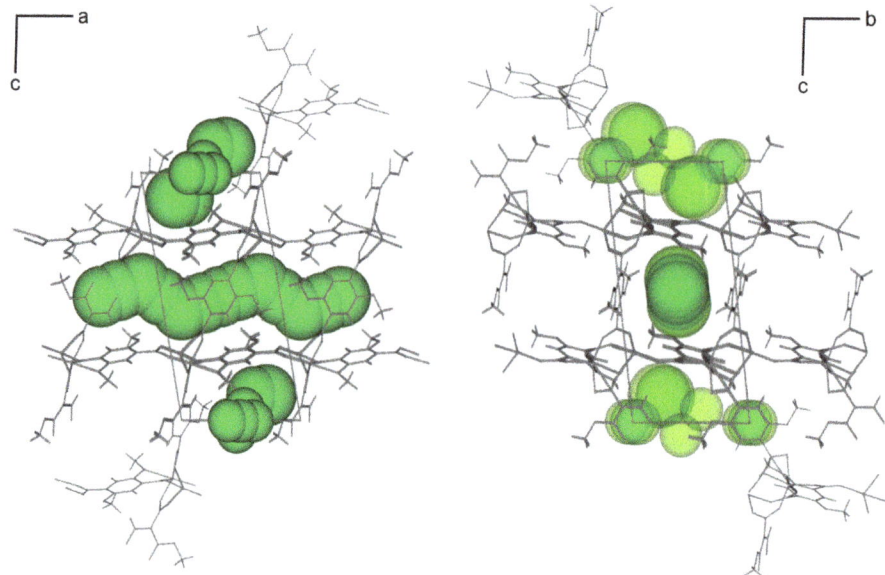

Figure 4. Visualisation of possible voids in Mg$_2$(2,5-DMT)$_2$(DMF)$_2$ (**2**) if the coordinating solvent molecules (DMF) can be removed; projections along [010] (**left**) and [100] (**right**).

It is remarkable that for Mg^{2+}, a completely different structure is observed when compared to Co^{2+}, Mn^{2+}, and Zn^{2+}. This might be explained by the smallest ionic radius of Mg^{2+} compared to the other three 3d metal cations Minor variations of the reaction conditions always lead to materials with the crystal structure of **2**. The major difference between the two different structure types is that there is no coordination of the solvent molecules in any of the three Co^{2+}, Mn^{2+}, and Zn^{2+} coordination polymers, whereas two DMF molecules coordinate to Mg$_2$ in the magnesium compound. This leads to small voids and wave-like channels with a potentially accessible porosity (vide infra).

3.3. Thermogravimetric Analyses

All the compounds M^{II}(2,5-DMT) with M^{II} = Co^{II} (**1**), Mn^{II}, Zn^{II} and **2** were investigated with regard to their thermal stability by means of coupled thermogravimetry (TG) and differential scanning calorimetry (DSC) using an inert Ar atmosphere and a 10 °C min^{-1} heating rate.

The TG curve of Mg$_2$(2,5-DMT)$_2$(DMF)$_2$ (**2**) shows four separated weight losses up to ~1000 °C (Figure 5, DSC curves are given as Figure S7, Supporting Information). The first mass loss of 4.64% was detected between 50 °C and 140 °C. It was followed by the second one between 140 °C and 250 °C. Calculations show that the sum of both fits almost perfectly to the release of the two coordinating DMF molecules (calc.: 22.7%, detected: 23.4%). The two following mass losses describe the decomposition of the framework starting above 300 °C.

For the three isostructural coordination polymers M^{II}(2,5-DMT) with M^{II} = Co^{II} (**1**), Mn^{II}, Zn^{II}, a very similar thermal stability was found (Figure 5). For Zn^{II}(2,5-DMT), the first mass loss with 30.1% occurred between 280 °C and 430 °C, which would fit a total decarboxylation of the linker molecule (calc.: 30.4%). Two further decreases were observed at higher temperatures between 430–550 °C and 550–800 °C, respectively. The first decrease could correspond to the release of a methanol molecule (obs.: 11.09%, calc.: 11.06%), which is formed from the methoxy group. Unfortunately, the amount of the residue after heating to 1000 °C was too small to record a PXRD pattern. In the case of Mn^{II}(2,5-DMT), a

TGA curve very similar to that of the Zn compound was observed (Figure 5). The first mass decreases were detected between 340–440 °C and 440–575 °C, i.e., at slightly higher temperatures compared to the Zn CP. The agreement between the calculated and observed values for a full decarboxylation and the cleavage of a methanol molecule is not as good as for the Zn compound (obs.: 43.02%, calc.: 38.4%). Co^{II}(2,5-DMT) (**1**) also shows thermal stability up to at least 300 °C. Here, the first mass loss occurred at ~430 °C, very similar to the Mn^{II} compound (obs.: 28.82%; calc.: 31.09% for a complete decarboxylation). Further decreases occurred between 420–510 °C and 510–630 °C, most likely due to the cleavage of the methoxy groups.

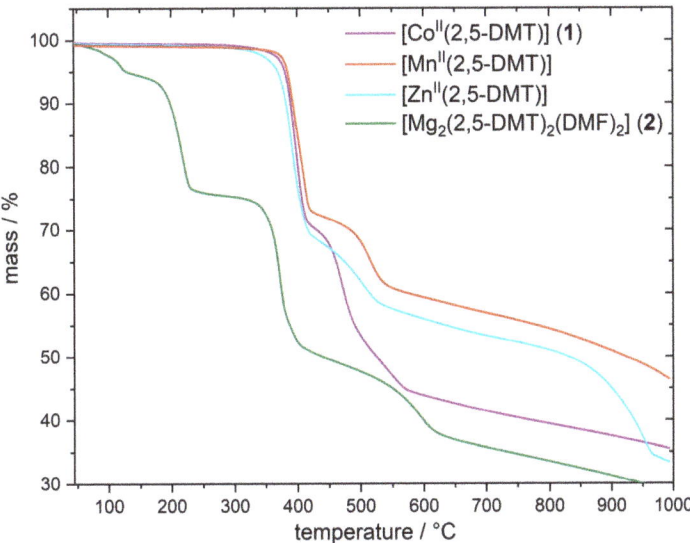

Figure 5. Comparison of the measured TG curves of Co^{II}(2,5-DMT) (violet curve, **1**), Mn^{II}(2,5-DMT) (red curve), Zn^{II}(2,5-DMT) (blue curve), and Mg_2(2,5-DMT)$_2$(DMF)$_2$ (green curve, **2**).

Comparison of all the M^{II}(2,5-DMT) compounds with M^{II} = Co^{II} (**1**), Mn^{II}, Zn^{II} shows that Mn^{II}(2,5-DMT) has the highest thermal stability, followed by Co^{II}(2,5-DMT) (**1**) with an only slightly decreased decomposition temperature, whereas Zn^{II}(2,5-DMT) shows a significantly lower thermal stability, by approx. 20 °C. As expected, the lowest thermal stability was found for Mg_2(2,5-DMT)$_2$(DMF)$_2$ (**2**), which is due to the release of the coordinating solvent molecules (DMF) upon heating. Remarkably, above 300 °C, the TG curve of **2** shows a very similar trend to the one found for the three solvent-free CPs starting at room temperature. It is therefore suggested that a coordination polymer Mg(2,5-DMT) isostructural to M^{II}(2,5-DMT) with M^{II} = Co^{II} (**1**), Mn^{II}, Zn^{II} (C2/c, Z = 4) might be formed after the release of the two DMF molecules. To confirm this assumption, a sample of **2** was heated at 280 °C in an argon stream for one hour. However, the PXRD pattern of the resulting material shows the same reflections as observed at room temperature, i.e., no structural change occurred under these conditions. Only the crystallinity of the material decreased significantly. Obviously, the framework of **2** collapses upon the release of the coordinating DMF guests. This also explains why we were unable to activate **2** and record a type I isotherm in the N_2 gas sorption measurements.

3.4. Luminescence Properties

Since coordination polymers M^{II}(2,5-DMT) with M^{II} = Co^{II} (**1**), Mn^{II}, Zn^{II} are isostructural compounds, the comparison of their emission and excitation bands could allow assumptions about the influence of different metal cations on the respective luminescence properties. The measured emission and excitation spectra are given in Figure 6 and compared with the respective spectra of the free 2,5-DMT ligand (grey reference). Table 3 summarises the optical properties of M^{II}(2,5-DMT) with M^{II} = Mn^{II}, Zn^{II}, **2**, and the free ligand. For Co^{II}(2,5-DMT) (**1**), no emission was observed as it was quenched by the Co^{2+} cations as known from the literature [46]. Pictures of the excited M^{II}(2,5-DMT) compounds with M^{II} = Co^{II} (**1**), Mn^{II}, Zn^{II} and **2** after blue light irradiation are shown in the Supporting Information (Figure S9).

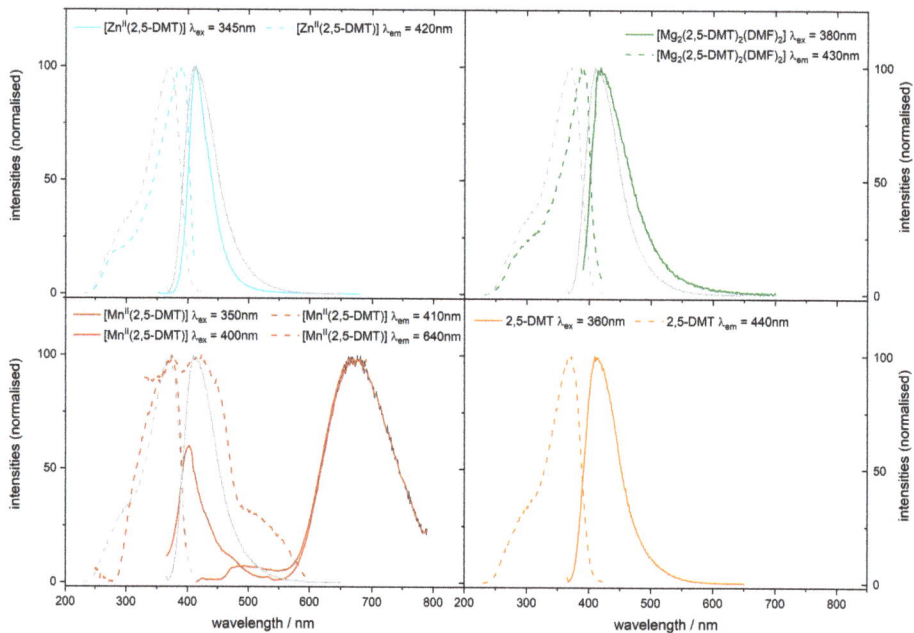

Figure 6. The emission (solid trace) and excitation spectra (dotted trace) of Zn^{II}(2,5-DMT) (**top left**, blue traces), Mg_2(2,5-DMT)$_2$(DMF)$_2$, **2** (**top right**, green traces), Mn^{II}(2,5-DMT) (**bottom left**, red traces), and 2,5-dimethoxyterephthalic acid (**bottom right**, orange traces) measured at 295 K. The emission and excitation spectra of the free 2,5-DMT linker are additionally plotted as a reference (grey traces) in each spectrum of the coordination polymers.

Table 3. Maximum excitation, emission and absorption wavelengths of M^{II}(2,5-DMT) with M^{II} = Mn^{II}, Zn^{II}, **2**, and the free 2,5-DMT ligand (s = shoulder).

	Max. Excitation/nm	Max. Emission/nm	Max. Absorption/nm
2,5-DMT	370	410	220, 250(s), 360
Mn^{II}(2,5-DMT)	370, 420	400, 660	200, 260(s), 370(s)
Zn^{II}(2,5-DMT)	390	410	205, 260(s), 390(s)
Mg_2(2,5-DMT)$_2$(DMF)$_2$, **2**	390	420	205, 250(s), 325

For Zn^{II}(2,5-DMT) and Mg_2(2,5-DMT)$_2$(DMF)$_2$ (**2**), the emission and excitation bands are similar to those observed for the free 2,5-DMT ligand. The excitation bands of both compounds are slightly red-shifted compared to the ligand's excitation band due to a weak interaction with the metal centre, while the emission band centres are almost identical for the coordination compounds and the ligand. This is consistent with the measured absorption spectra (Table 3), where the absorption bands of Zn^{II}(2,5-DMT) and Mg_2(2,5-DMT)$_2$(DMF)$_2$ (**2**) do not significantly differ from the absorption of the free ligand. Therefore, we attribute the emission bands in **2** and Zn^{II}(2,5-DMT) to a ligand-centred transition, while the metals do not participate in the radiative pathway in a significant way.

The ligand-centred emission band is also identified for the Mn^{II}(2,5-DMT) coordination polymer. In addition, a second emission band at 660 nm is observed after excitation at 410 nm, resulting in a red emission of the coordination polymer (cp. Figure S9). This emission band is attributed to a metal-centred transition of the Mn^{2+} ion as it has not been observed for the Zn^{2+} and Mg^{2+} coordination compounds and the free ligand. Through the red emission, it is possible to make an additional statement about the coordination, which cannot be unambiguously identified via X-ray single-crystal structure analysis as discussed above. While tetrahedrally coordinated Mn^{2+} emits green light, a red-light emission is observed for octahedrally coordinated Mn^{2+} ions caused by a transition from the excited $^4T_{1g}(^4G)$ state to the $^6A_{1g}(^6S)$ ground state [47,48]. Therefore, from these UV–vis spectra, one can conclude that the coordination sphere of Mn^{2+} in Mn^{II}(2,5-DMT) is best described as an MnO$_6$ octahedron.

4. Conclusions

We synthesized a new coordination polymer Co^{II}(2,5-DMT) (**1**) containing a fluorescent 2,5-DMT (2,5-dimethoxyterephthalate) linker; **1** is isostructural to the known Mn^{II} and Zn^{II} congeners. Attempts to synthesize an Mg^{2+} analogue led to the synthesis of Mg$_2$(2,5-DMT)$_2$(DMF)$_2$ (**2**) with coordinating DMF molecules and a MOF-type structure. Attempts to remove the DMF guests upon heating failed so that no permanent porosity could be proven. In a comparative study, thermal stability of all the four compounds was investigated. The solvent-free M^{II}(2,5-DMT) coordination polymers with M^{II} = Co^{II} (**1**), Mn^{II}, and Zn^{II} showed a very similar decomposition behaviour, with the Zn^{II} compound being a slightly less stable material. Nonetheless, all the three compounds decomposed clearly above 300 °C. M^{II}(2,5-DMT) with M^{II} = Mn^{II}, Zn^{II}, and **2** as well as the pristine linker 2,5-DMT exhibited a strong emission upon irradiation with blue/UV light, while in the Co^{2+}-containing material (**1**), the emission was quenched. Zn^{II}(2,5-DMT) and Mg$_2$(2,5-DMT)$_2$(DMF)$_2$ (**2**) showed a mainly ligand-based blue emission, while for Mn^{II}(2,5-DMT), an additional metal-based red emission was found. The latter points to an octahedral coordination of Mn^{2+}. This is remarkable from a structural point of view, as in solvent-free M^{II}(2,5-DMT) compounds, pockets around the M^{2+} cations are formed with an inner (distorted tetrahedral) and an outer coordination sphere. The inner sphere is solely formed by four oxygen atoms of carboxylate groups, whereas in the outer sphere, two oxygen atoms of the methoxy groups are also included, leading to a distorted octahedral coordination. Although the Mn–O$_{methoxy}$ distances are distinctively longer than the Mn–O$_{carboxylate}$ distances (by more than 0.4 Å), the UV/vis spectra of Mn^{II}(2,5-DMT) clearly indicate that there is still a significant Mn–O$_{methoxy}$ interaction.

We believe that the 2,5-DMT ligand is an attractive linker for the construction of luminescent coordination polymers and MOFs. Especially in MOFs with a permanent porosity, it might lead to interesting materials for sensing applications. The synthesis of such porous and luminescent MOFs is in the focus of our current research in this field.

Supplementary Materials: The following supporting information can be downloaded at: https://www.mdpi.com/article/10.3390/chemistry5020065/s1, Figure S1: PXRD pattern of [Co^{II}(2,5-DMT)] (**1**); Figure S2: PXRD pattern of [Mn^{II}(2,5-DMT)]; Figure S3: PXRD pattern of [Zn^{II}(2,5-DMT)]; Figure S4: PXRD pattern of [Mg$_2$(2,5-DMT)$_2$(DMF)$_2$] (**2**); Figure S5: Asymmetric unit of [Co^{II}(2,5-DMT)] (**1**); Figure S6: Space filling representation of [Co^{II}(2,5-DMT)] (**1**); Figure S7: DSC curves of [Mn^{II}(2,5-DMT)], [Zn^{II}(2,5-DMT)], [Co^{II}(2,5-DMT)] (**1**), and [Mg$_2$(2,5-DMT)$_2$(DMF)$_2$] (**2**); Figure S8: UV/vis measurements of 2,5-DMT and [Mn^{II}(2,5-DMT)], [Zn^{II}(2,5-DMT)], and [Mg$_2$(2,5-DMT)$_2$(DMF)$_2$] (**2**); Figure S9: Photographs of fluorescence of [Mn^{II}(2,5-DMT)], [Zn^{II}(2,5-DMT)], [Co^{II}(2,5-DMT)] (**1**), and [Mg$_2$(2,5-DMT)$_2$(DMF)$_2$] (**2**) upon blue light irradiation; Tables S1–S5: Crystallographic data of [Co^{II}(2,5-DMT)] (**1**); Tables S6–S10: Crystallographic data of [Mg$_2$(2,5-DMT)$_2$(DMF)$_2$] (**2**).

Author Contributions: A.E.L.C. synthesized all the compounds, conducted most of the measurements as well as analyses, and wrote parts of the original draft. L.S. helped A.E.L.C. with the luminescence measurements and wrote parts of that section. D.v.G. assisted A.E.L.C. in the single-crystal structure analysis. M.S.W. supervised L.S. and D.v.G. during their work. U.R. supervised the whole project, was in charge of administration and funding for the project, and wrote major parts of the original draft. All authors have read and agreed to the published version of the manuscript.

Funding: This research was funded by the Deutsche Forschungsgemeinschaft (DFG) under the grant number RU 546/12-1.

Data Availability Statement: Spectroscopic and DSC/TGA data as well as some crystallographic details are given in the electronic supplement. The crystal structures of compounds **1** and **2** were determined by single-crystal X-ray diffraction and the crystallographic data have been deposited with the Cambridge Crystallographic Data Centre as supplementary publication nos. CCDC-2225418 (**1**) and 2225419 (**2**). Copies of the data can be obtained free of charge on application to CCDC, 12 Union Road, Cambridge, CB21EZ (fax: (+44) 1223/336033; e-mail: deposit@ccdc.cam-ak.uk).

Acknowledgments: We acknowledge the help of C. Tobeck (DSC/TGA data acquisition) and R. Christoffels (gas sorption measurements).

Conflicts of Interest: The authors declare no conflict of interest.

References

1. Li, H.; Eddaoudi, M.; O'Keeffe, M.; Yaghi, O.M. Design and synthesis of an exceptionally stable and highly porous metal-organic framework. *Nature* **1999**, *402*, 276–279. [CrossRef]
2. Chui, S.S.Y.; Lo, S.M.F.; Charmant, J.P.H.; Orpen, A.G.; Williams, I.D. A Chemically Functionalizable Nanoporous Material [$Cu_3(TMA)_2(H_2O)_3$]$_n$. *Science* **1999**, *283*, 1148–1150. [CrossRef] [PubMed]
3. Moghadam, P.Z.; Li, A.; Wiggin, S.B.; Tao, A.; Maloney, A.G.P.; Wood, P.A.; Ward, S.C.; Fairen-Jimenez, D. Development of a Cambridge Structural Database Subset: A Collection of Metal–Organic Frameworks for Past, Present, and Future. *Chem. Mater.* **2017**, *29*, 2618–2625. [CrossRef]
4. Groom, C.R.; Bruno, I.J.; Lightfoot, M.P.; Ward, S.C. The Cambridge Structural Database. *Acta Crystallogr.* **2016**, *B72*, 171–179. [CrossRef] [PubMed]
5. Batten, S.R.; Champness, N.R.; Chen, X.-M.; Garcia-Martinez, J.; Kitagawa, S.; Öhrström, L.; O'Keeffe, M.; Paik Suh, M.; Reedijk, J. Terminology of metal–organic frameworks and coordination polymers (IUPAC Recommendations 2013). *Pure Appl. Chem.* **2013**, *85*, 1715–1724. [CrossRef]
6. Ma, S.; Zhou, H.-C. Gas storage in porous metal–organic frameworks for clean energy applications. *Chem. Commun.* **2010**, *46*, 44–53. [CrossRef]
7. Lin, R.-B.; Xiang, S.; Xing, H.; Zhou, W.; Chen, B. Exploration of porous metal–organic frameworks for gas separation and purification. *Coord. Chem. Rev.* **2019**, *378*, 87–103. [CrossRef]
8. Dhakshinamoorthy, A.; Asiri, A.M.; Garcia, H. Catalysis in Confined Spaces of Metal Organic Frameworks. *ChemCatChem* **2020**, *12*, 4732–4753. [CrossRef]
9. Yang, J.; Wang, H.; Liu, J.; Ding, M.; Xie, X.; Yang, X.; Peng, Y.; Zhou, S.; Ouyang, R.; Miao, Y. Recent advances in nanosized metal organic frameworks for drug delivery and tumor therapy. *RSC Adv.* **2021**, *11*, 3241–3263. [CrossRef]
10. Jin, J.; Xue, J.; Liu, Y.; Yang, G.; Wang, Y.-Y. Recent progresses in luminescent metal–organic frameworks (LMOFs) as sensors for the detection of anions and cations in aqueous solution. *Dalton Trans.* **2021**, *50*, 1950–1972. [CrossRef]
11. Ye, Y.; Gong, L.; Xiang, S.; Zhang, Z.; Chen, B. Metal-Organic Frameworks as a Versatile Platform for Proton Conductors. *Adv. Mater.* **2020**, *32*, 1907090. [CrossRef] [PubMed]
12. Sadakiyo, M.; Kitagawa, H. Ion-conductive metal–organic frameworks. *Dalton Trans.* **2021**, *50*, 5385–5397. [CrossRef] [PubMed]
13. Heine, J.; Müller-Buschbaum, K. Engineering metal-based luminescence in coordination polymers and metal–organic frameworks. *Chem. Soc. Rev.* **2013**, *42*, 9232–9242. [CrossRef] [PubMed]
14. Allendorf, M.D.; Bauer, C.A.; Bhakta, R.K.; Houk, R.J.T. Luminescent metal–organic frameworks. *Chem. Soc. Rev.* **2009**, *38*, 1330–1352. [CrossRef]
15. Sobieray, M.; Gode, J.; Seidel, C.; Poß, M.; Feldmann, C.; Ruschewitz, U. Bright luminescence in lanthanide coordination polymers with tetrafluoroterephthalate as a bridging ligand. *Dalton Trans.* **2015**, *44*, 6249–6259. [CrossRef]
16. Yang, H.; Zhang, H.-X.; Hou, D.-C.; Li, T.-H.; Zhang, J. Assembly between various molecular-building-blocks for network diversity of zinc–1,3,5-benzenetricarboxylate frameworks. *CrystEngComm* **2012**, *14*, 8684–8688. [CrossRef]
17. Fang, Q.; Zhu, G.; Xue, M.; Sun, J.; Sun, F.; Qiu, S. Structure, Luminescence, and Adsorption Properties of Two Chiral Microporous Metal–Organic Frameworks. *Inorg. Chem.* **2006**, *45*, 3582–3587. [CrossRef]
18. Schrimpf, W.; Jiang, J.; Ji, Z.; Hirschle, P.; Lamb, D.C.; Yaghi, O.M.; Wuttke, S. Chemical diversity in a metal–organic framework revealed by fluorescence lifetime imaging. *Nat. Commun.* **2018**, *9*, 1647. [CrossRef]

19. Shimizu, M.; Shigitani, R.; Kinoshita, T.; Sakaguchi, H. (Poly)terephthalates with Efficient Blue Emission in the Solid State. *Chem. Asian J.* **2019**, *14*, 1792–1800. [CrossRef]
20. Böhle, T.; Eissmann, F.; Weber, E.; Mertens, F.O.R.L. Poly[(μ_4-2,5-dimethoxybenzene-1,4-dicarboxylato)manganese(II)] and its zinc(II) analogue: Three-dimensional coordination polymers containing unusually coordinated metal centres. *Acta Crystallogr.* **2011**, *C67*, m5–m8. [CrossRef]
21. Henke, S.; Schneemann, A.; Kapoor, S.; Winter, R.; Fischer, R.A. Zinc-1,4-benzenedicarboxylate-bipyridine frameworks—Linker functionalization impacts network topology during solvothermal synthesis. *J. Mater. Chem.* **2012**, *22*, 909–918. [CrossRef]
22. Kim, D.; Ha, H.; Kim, Y.; Son, Y.; Choi, J.; Park, M.H.; Kim, Y.; Yoon, M.; Kim, H.; Kim, D.; et al. Experimental, Structural, and Computational Investigation of Mixed Metal–Organic Frameworks from Regioisomeric Ligands for Porosity Control. *Cryst. Growth Des.* **2020**, *20*, 5338–5345. [CrossRef]
23. Böhle, T.; Eißmann, F.; Weber, E.; Mertens, F.O.R.L. A Three-Dimensional Coordination Polymer Based on Co(II) and 2,5-Dimethoxyterephthalate Featuring MOF-69 Topology. *Struct. Chem. Commun.* **2011**, *2*, 91–94.
24. Guo, Z.; Reddy, M.V.; Goh, B.M.; San, A.K.P.; Bao, Q.; Loh, K.P. Electrochemical performance of graphene and copper oxide composites synthesized from a metal–organic framework (Cu-MOF). *RSC Adv.* **2013**, *3*, 19051–19056. [CrossRef]
25. Li, Z.-J.; Ju, Y.; Lu, H.; Wu, X.; Yu, X.; Li, Y.; Wu, X.; Zhang, Z.-H.; Lin, J.; Qian, Y.; et al. Boosting the Iodine Adsorption and Radioresistance of Th-UiO-66 MOFs via Aromatic Substitution. *Chem. Eur. J.* **2020**, *27*, 1286–1291. [CrossRef]
26. Cui, Y.; Xu, H.; Yue, Y.; Guo, Z.; Yu, J.; Chen, Z.; Gao, J.; Yang, Y.; Qian, G.; Chen, B. A Luminescent Mixed-Lanthanide Metal–Organic Framework Thermometer. *J. Am. Chem. Soc.* **2012**, *134*, 3979–3982. [CrossRef] [PubMed]
27. Ha, H.; Hahm, H.; Jwa, D.G.; Yoo, K.; Park, M.H.; Yoon, M.; Kim, Y.; Kim, M. Flexibility in metal–organic frameworks derived from positional and electronic effects of functional groups. *CrystEngComm* **2017**, *19*, 5361–5368. [CrossRef]
28. Win XPow, version 3.12 (12 February 2018); STOE & Cie GmbH: Darmstadt, Germany, 2018.
29. Gnuplot, version 4.6, Various Authors; 2014. Available online: http://www.gnuplot.info/ (accessed on 30 March 2023).
30. SAINT, version 8.40B; Bruker AXS Inc.: Madison, WI, USA, 2016.
31. Sheldrick, G.M. SADABS; University of Göttingen: Göttingen, Germany, 1996.
32. Krause, L.; Herbst-Irmer, R.; Sheldrick, G.M.; Stalke, D. Comparison of silver and molybdenum microfocus X-ray sources for single-crystal structure determination. *J. Appl. Crystallogr.* **2015**, *48*, 3–10. [CrossRef]
33. APEX4, version 2022.1-1; Bruker AXS Inc.: Madison, WI, USA, 2022.
34. Sheldrick, G.M. SHELXT—Integrated space-group and crystal-structure determination. *Acta Crystallogr. A* **2015**, *71*, 3–8. [CrossRef]
35. Sheldrick, G.M. A short history of SHELX. *Acta Crystallogr. A* **2008**, *64*, 112–122. [CrossRef]
36. Brandenburg, K. Diamond, version 4.6.8; Crystal Impact GbR: Bonn, Germany, 2022.
37. OriginPro, version 2022; OriginLab Corporation: Northhampton, MA, USA, 2022.
38. ChemDraw Professional 15.0 (RRID:SCR_016768). Available online: https://perkinelmerinformatics.com/products/research/chemdraw (accessed on 30 March 2023).
39. Shannon, R.D. Revised effective ionic radii and systematic studies of interatomic distances in halides and chalcogenides. *Acta Crystallogr.* **1976**, *A32*, 751–767. [CrossRef]
40. Casanova, D.; Cirera, J.; Llunell, M.; Alemany, P.; Avnir, D.; Alvarez, S. Minimal Distortion Pathways in Polyhedral Rearrangements. *J. Am. Chem. Soc.* **2004**, *126*, 1755–1763. [CrossRef] [PubMed]
41. Kurmoo, M.; Kumagai, H.; Green, M.A.; Lovett, B.W.; Blundell, S.J.; Ardavan, A.; Singleton, J. Two Modifications of Layered Cobaltous Terephthalate: Crystal Structures and Magnetic Properties. *J. Solid State Chem.* **2001**, *159*, 343–351. [CrossRef]
42. Liu, D.; Liu, Y.; Xu, G.; Li, G.; Yu, Y.; Wang, C. Two 3D Supramolecular Isomeric Mixed-Ligand CoII Frameworks—Guest-Induced Structural Variation, Magnetism, and Selective Gas Adsorption. *Eur. J. Inorg. Chem.* **2012**, *28*, 4413–4417. [CrossRef]
43. Yang, H.; Trieu, T.X.; Zhao, X.; Wang, Y.; Wang, Y.; Feng, P.; Bu, X. Lock-and-Key and Shape-Memory Effects in an Unconventional Synthetic Path to Magnesium Metal–Organic Frameworks. *Angew. Chem. Int. Ed.* **2019**, *58*, 11757–11762. [CrossRef]
44. Zhai, Q.-G.; Bu, X.; Mao, C.; Zhao, X.; Daemen, L.; Cheng, Y.; Ramirez-Cuesta, A.J.; Feng, P. An ultra-tunable platform for molecular engineering of high-performance crystalline porous materials. *Nat. Commun.* **2016**, *7*, 13645–13654. [CrossRef]
45. Spek, A.L. Single-crystal structure validation with the program PLATON. *J. Appl. Crystallogr.* **2003**, *36*, 7–13. [CrossRef]
46. Lustig, W.P.; Mukherjee, S.; Rudd, N.D.; Desai, A.V.; Li, J.; Ghosh, S.K. Metal–organic frameworks: Functional luminescent and photonic materials for sensing applications. *Chem. Soc. Rev.* **2017**, *46*, 3242–3285. [CrossRef] [PubMed]
47. Morad, V.; Cherniukh, I.; Pöttschacher, L.; Shynkarenko, Y.; Yakunin, S.; Kovalenko, M.V. Manganese(II) in Tetrahedral Halide Environment: Factors Governing Bright Green Luminescence. *Chem. Mater.* **2019**, *31*, 10161–10169. [CrossRef] [PubMed]
48. Artem'ev, A.V.; Davydova, M.P.; Berezin, A.S.; Brel, V.K.; Morgalyuk, V.P.; Bagryanskaya, I.Y.; Samsonenko, D.G. Luminescence of the Mn^{2+} ion in non-O$_h$ and T$_d$ coordination environments: The missing case of square pyramid. *Dalton Trans.* **2019**, *48*, 16448–16456. [CrossRef]

Disclaimer/Publisher's Note: The statements, opinions and data contained in all publications are solely those of the individual author(s) and contributor(s) and not of MDPI and/or the editor(s). MDPI and/or the editor(s) disclaim responsibility for any injury to people or property resulting from any ideas, methods, instructions or products referred to in the content.

Article

Divalent Europium, NIR and Variable Emission of Trivalent Tm, Ho, Pr, Er, Nd, and Ce in 3D Frameworks and 2D Networks of Ln–Pyridylpyrazolates

Heba Youssef [1,2], Jonathan Becker [1], Clemens Pietzonka [3], Ilya V. Taydakov [4,5], Florian Kraus [3] and Klaus Müller-Buschbaum [1,6,*]

[1] Institute of Inorganic and Analytical Chemistry, Justus-Liebig-University Giessen, Heinrich-Buff-Ring 17, 35392 Giessen, Germany; heba.youssef@anorg.chemie.uni-giessen.de (H.Y.)
[2] Department of Chemistry, Faculty of Science, Mansoura University, El Gomhouria, Mansoura Qism 2, Dakahlia Governorate, Mansoura 11432, Egypt
[3] Fachbereich Chemie, Philipps-University Marburg, Hans-Meerwein-Straße, 35032 Marburg, Germany
[4] Lebedev Physical Institute of the Russian Academy of Sciences, Leninskiy pr-t, 53, 119991 Moscow, Russia
[5] Basic Department of Chemistry of Innovative Materials and Technologies, G.V. Plekhanov Russian University of Economics, Stremyanny Per. 36, 117997 Moscow, Russia
[6] Center of Materials Research (LAMA), Justus-Liebig-University Giessen, Heinrich-Buff-Ring 16, 35392 Giessen, Germany
* Correspondence: klaus.mueller-buschbaum@anorg.chemie.uni-giessen.de

Abstract: The redox reactions of various lanthanide metals with 3-(4-pyridyl)pyrazole (4-PyPzH) or 3-(3-pyridyl)pyrazole (3-PyPzH) ligands yield the 2D network $^2_\infty$[Eu(4-PyPz)$_2$(Py)$_2$] containing divalent europium, the 3D frameworks $^3_\infty$[Ln(4-PyPz)$_3$] and $^3_\infty$[Ln(3-PyPz)$_3$] for trivalent cerium, praseodymium, neodymium, holmium, erbium, and thulium as well as $^3_\infty$[La(4-PyPz)$_3$], and the 2D networks $^2_\infty$[Ln(4-PyPz)$_3$(Py)] for trivalent cerium and thulium and $^2_\infty$[Ln$_2$(4-PyPz)$_6$]·Py for trivalent ytterbium and lutetium. The 18 lanthanide coordination polymers were synthesized under solvothermal conditions in pyridine (Py), partly acting as a co-ligand for some networks. The compounds exhibit a variety of luminescence properties, including metal-centered 4f–4f/5d–4f emission in the visible and near-infrared spectral range, metal-to-ligand energy transfer, and ligand-centered fluorescence and phosphorescence. The anionic ligands 3-PyPz$^-$ and 4-PyPz$^-$ serve as suitable antennas for lanthanide-based luminescence in the visible and near-infrared range through effective sensitization followed by emission through intra–4f transitions of the trivalent thulium, holmium, praseodymium, erbium, and neodymium. $^2_\infty$[Ce(4-PyPz)$_3$(Py)]$, $^3_\infty$[Ce(4-PyPz)$_3$], and $^3_\infty$[Ce(3-PyPz)$_3$] exhibit strong degrees of reduction in the 5d excited states that differ in intensity compared to the ligand-based emission, resulting in a distinct emission ranging from pink to orange. The direct current magnetic studies show magnetic isolation of the lanthanide centers in the crystal lattice of $^3_\infty$[Ln(3-PyPz)$_3$], Ln = Dy, Ho, and Er.

Keywords: divalent europium; cerium; NIR emitter; N-donor ligand

Citation: Youssef, H.; Becker, J.; Pietzonka, C.; Taydakov, I.V.; Kraus, F.; Müller-Buschbaum, K. Divalent Europium, NIR and Variable Emission of Trivalent Tm, Ho, Pr, Er, Nd, and Ce in 3D Frameworks and 2D Networks of Ln–Pyridylpyrazolates. *Chemistry* 2023, 5, 1006–1027. https://doi.org/10.3390/chemistry5020069

Academic Editors: Christoph Janiak, Sascha Rohn and Georg Manolikakes

Received: 27 March 2023
Revised: 20 April 2023
Accepted: 26 April 2023
Published: 28 April 2023

Copyright: © 2023 by the authors. Licensee MDPI, Basel, Switzerland. This article is an open access article distributed under the terms and conditions of the Creative Commons Attribution (CC BY) license (https://creativecommons.org/licenses/by/4.0/).

1. Introduction

Divalent europium, the mildest reducing agent of the redox-sensitive divalent lanthanide ions, has been successfully used in a wide variety of material applications such as medical imaging [1,2], photochemistry [3,4], lanthanide-activated phosphors [5,6], and sensing [7,8]. Trivalent lanthanides are known for their luminescence properties, with f–f based emission covering the spectrum from the ultraviolet (UV) to the near-infrared (NIR) spectral region, characteristic for each metal ion [9,10]. Several ions of typical NIR emitters also have possible transitions in the visible range [11], but these are usually too weak to be readily observed, especially for Tm^{3+} and Ho^{3+} [12–16]. NIR emitters have played an important role in many modern technologies such as organic light-emitting diodes (OLEDs) [17,18] and photovoltaics [19,20], which encouraged us to further study the photophysical properties of NIR emitters such as Tm, Ho, Nd, and Er [11,12,21,22].

In addition to the forbidden f–f transitions, 5d–4f transitions can also be detected among the trivalent and the divalent lanthanides such as Ce^{3+} and Eu^{2+}. The 5d–4f transitions have also been studied for decades on the luminescent mechanism and potential applications in various fields [23–25]. The emission occurs in the UV and/or in the blue spectral regions but can be shifted to a much longer wavelength depending on the environment of the Ln^{3+} ion [26,27]. Mostly, the 5d–4f transitions are absent due to thermal quenching by fast intersystem crossing from $4f^{n-1}5d^1$ to $4f^n$ configuration [4,28]. Pink-emitting cerium has rarely been detected in doped materials such as cerium-doped single-crystal aluminum nitride [29] and cerium–manganese-activated phosphor [30] and not for undoped systems. Furthermore, undoped red-emitting $^1_\infty[Ce(2\text{-PyPz})_3]$ and orange-emitting $[Ce(2\text{-PyPzH})_3Cl_3]$ (2-PyPzH = 3-(2-pyridyl)pyrazole) were just recently reported for orange emission [31,32]. These results inspired further investigations on the influence of changing the position of the nitrogen of the pyridyl ring in 3-(4-pyridyl)pyrazole (4-PyPzH) and 3-(3-pyridyl)pyrazole (3-PyPzH), which are presented in this work. The ligands 4-PyPzH and 3-PyPzH were used to synthesize homoleptic and highly luminescent trivalent lanthanide 3D coordination polymers with the formulas $^3_\infty[Ln(3\text{-PyPz})_3]$ and $^3_\infty[Ln(4\text{-PyPz})_3]$, Ln = Sm, Eu, Gd, Tb, Dy [33]. Neither 3-PyPzH nor 4-PyPzH as ligands have been explored for complexing divalent lanthanide ions. Following the reaction of europium metal with 4-PyPzH, a 2D network based on divalent europium was synthesized and presented in this work.

3-PyPzH was used to synthesize a wide variety of structures, from 3D and 2D networks to complexes of lanthanide trichlorides [14]. The weak ferromagnetic interaction for $^2_\infty[Ho_2(3\text{-PyPzH})_3Cl_6]\cdot 2MeCN$ encouraged us to study the magnetic properties of the presented Ln metal-based series.

2. Results and Discussion

Elemental lanthanides together with 3-(4-pyridyl)pyrazole (4-PyPzH) or 3-(3-pyridyl)pyrazole (3-PyPzH) in solvothermal synthesis-based reactions were used to obtain eighteen 3D frameworks and 2D networks (Scheme 1).

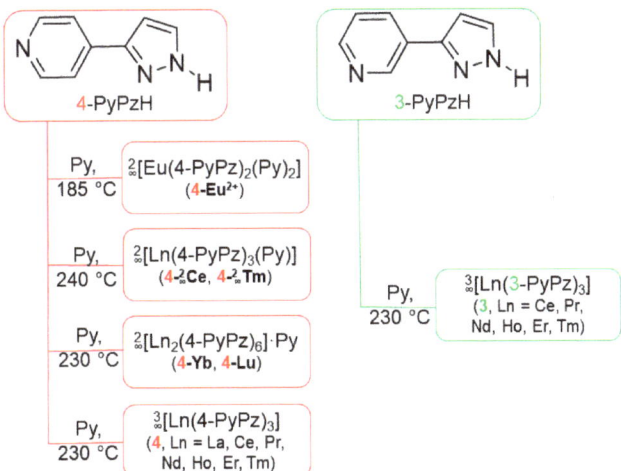

Scheme 1. Synthetic scheme for reactions of lanthanide metals with 4-PyPzH and 3-PyPzH to obtain 3D frameworks and 2D networks.

2.1. Structural Analysis

Structural diversity is observed along the series based on 4-PyPz$^-$, depending on the content of both 4-PyPz$^-$ and pyridine (Py) as linkers, all of which crystallize in the monoclinic crystal system and mostly with the space group $P2_1/n$. Exceptions are $^2_\infty$[Eu(4-PyPz)$_2$(Py)$_2$] (**4-Eu^{2+}**) and $^2_\infty$[Ln(4-PyPz)$_3$(Py)], Ln = Ce (**4-$^2_\infty$Ce**), Tm (**4-$^2_\infty$Tm**), which crystallize with the space groups $P2_1$ and Cc, respectively. The other 3-PyPz$^-$ based series $^3_\infty$[Ln(3-PyPz)$_3$], (**3**, Ln = Ce, Pr, Nd, Ho, Er, Tm) further crystallizes in the cubic crystal system with the space group $Pa\bar{3}$.

In $^2_\infty$[Eu(4-PyPz)$_2$(Py)$_2$] (**4-Eu^{2+}**), containing divalent europium, each Eu^{2+} ion coordinates to eight nitrogen atoms, six nitrogen atoms from four pyrazolate anions, and two nitrogen atoms from two pyridine molecules in a distorted pseudo-octahedral assembly (Figure 1), if the two nitrogen atoms of the pyrazolate anion are considered as one corner of the octahedron. The four pyrazolate anions act as bridges to the neighboring Eu^{2+} ions, forming a 2D coordination polymer. The Eu–N interatomic distances for divalent europium in $^2_\infty$[Eu(4-PyPz)$_2$(Py)$_2$] (**4-Eu^{2+}**) (254.9(4)–274.5(2) pm) are longer than those reported for the trivalent europium $^3_\infty$[Eu(4-PyPz)$_3$] (240.8–258.8 pm) [33], consistent with the difference in charge density and ionic radius [34]. Another comparison of the Eu–N of **4-Eu^{2+}** with the divalent europium complex [Eu(Ph$_2$pz)$_2$(Py)$_4$]·2Py (Ph$_2$pz = 3,5-diphenylpyrazolate, Eu–N = 253.8–274.1 pm) resulted in good agreement [35].

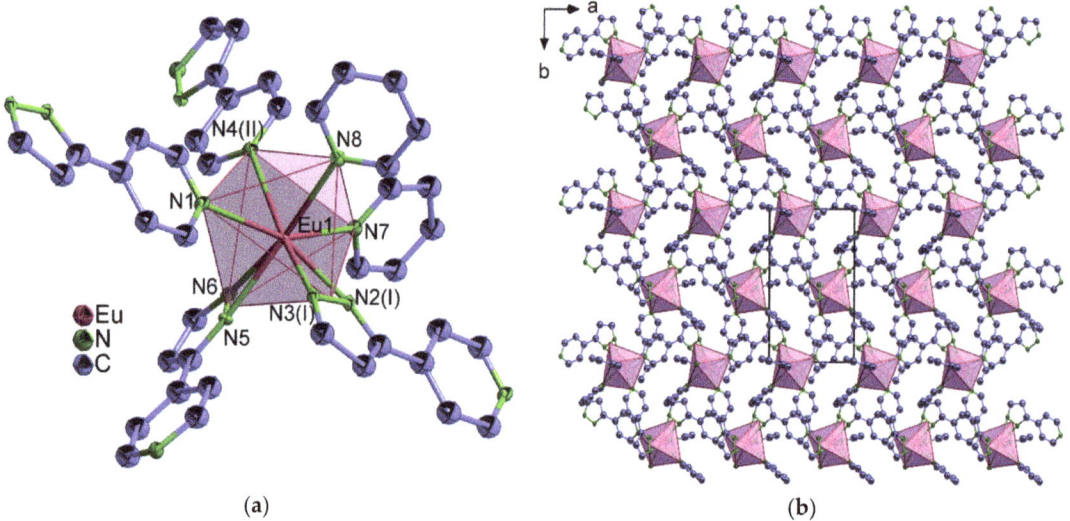

Figure 1. (a) Extended coordination sphere of the Eu^{2+} ion in $^2_\infty$[Eu(4-PyPz)$_2$(Py)$_2$] (**4-Eu^{2+}**); (b) crystal structure of **4-Eu^{2+}** with a view along [001]. The coordination polyhedra around Eu^{2+} are indicated in violet with thermal ellipsoids depicted at the 50% probability level. Symmetry operations: I x + 1, y, z II −x + 1, y + 1/2, −z. In all figures, the hydrogen atoms are omitted for clarity and the unit cell is depicted when required.

In $^2_\infty$[Ln(4-PyPz)$_3$(Py)], Ln = Ce^{3+} (**4-$^2_\infty$Ce**), Tm^{3+} (**4-$^2_\infty$Tm**), each Ln^{3+} ion coordinates to nine nitrogen atoms, eight nitrogen atoms from five pyrazolate anions, and a nitrogen atom from a pyridine molecule in a distorted pseudo-octahedral arrangement (Figure 2), if the two nitrogen atoms of the pyrazolate anion are viewed as one corner of the octahedron. The pyrazolate anions act as bridges to the neighboring Ln^{3+} ions, forming a 2D coordination polymer. Due to the lack of Tm^{3+}-nitrogen-based complexes and coordination polymers in the literature, only one example was comparable to **4-$^2_\infty$Tm**, [Tm(L^1)$_3$]$^{3+}$ (L^1 = 2,6-bis(5,6-dipropyl-1,2,4-triazin-3-yl)pyridine, Tm–N = 248.3–252.2 pm) [36], which is longer than **4-$^2_\infty$Tm** (Tm–N = 232.4–259.2 pm) due to the anionic character of the ligands, in the latter 4-PyPz$^-$.

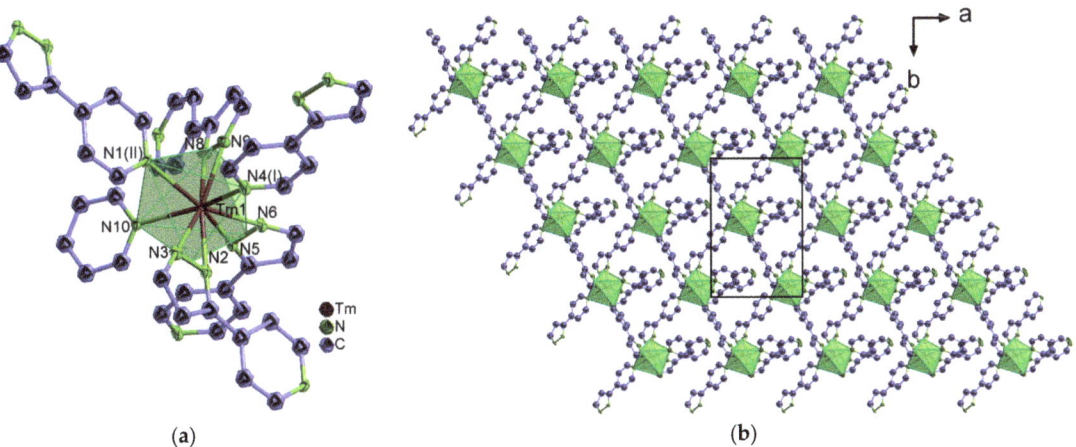

Figure 2. (a) Extended coordination sphere of Tm^{3+} in $^2_\infty[Tm(4\text{-PyPz})_3(Py)]$ ($4\text{-}^2_\infty Tm$); (b) crystal structure of $4\text{-}^2_\infty Tm$ with a view along [001]. The coordination polyhedra around Tm^{3+} are indicated in green with thermal ellipsoids depicted at the 50% probability level. Symmetry operations: I x − 1/2, y + 1/2, z II x + 1/2, y + 1/2, z.

The topology of the **4-Eu^{2+}** and **4-$^2_\infty$Tm** networks was determined according to the Reticular Chemistry Structure Resource (RCSR) and the Wells terminology [37,38] to result in an **sql** topology (Figure 3) with the Schläfli symbol $4^4 \cdot 6^2$ for both cases. This topology distinguishes from the rest of the series $^3_\infty[Ln(4\text{-PyPz})_3]$ (**4**, Ln = La, Ce, Pr, Nd, Ho, Er, Tm), which represent the **pcu** topology with the Schläfli symbol $4^{12} \cdot 6^3$ [33]. The **pcu** topology was also found for the isotypic series $^3_\infty[Ln(3\text{-PyPz})_3]$ (**3**, Ln = Ce, Pr, Nd, Ho, Er, Tm).

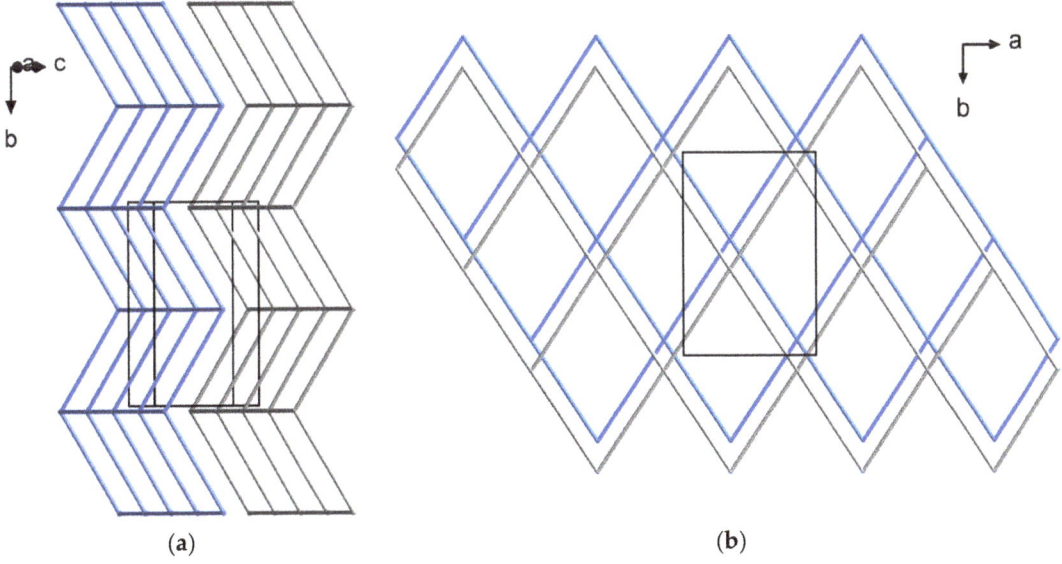

Figure 3. Topological representation of (a) $^2_\infty[Eu(4\text{-PyPz})_2(Py)_2]$ (**4-Eu^{2+}**); (b) $^2_\infty[Tm(4\text{-PyPz})_3(Py)]$ (**4-$^2_\infty$Tm**) as a uninodal 4-c net with **sql** topology.

The crystal structure of $^2_\infty[Ln_2(4\text{-PyPz})_6]Py$, Ln = Yb (**4-Yb**), Lu (**4-Lu**) contains two lanthanide sites. One site coordinates to nine nitrogen atoms from six pyrazolate anions in a distorted pseudo-octahedron, while the other coordinates to eight nitrogen atoms from five pyrazolate anions in a distorted trigonal bipyramid (Figure 4), if the two nitrogens of the pyrazolate anion are considered as a corner of the polyhedron. The two lanthanide sites are bridged through a pyrazolate anion, while each lanthanide site is simultaneously bridged through pyrazolate anions to adjacent identical sites to form a 2D layer extending along the bc plane. An anion coordinated to Ln1 does not act as a bridge through the nitrogen atom of its pyridine ring to another neighboring Ln ion (its position is pointed by arrows in Figure 4), making some ligands infinite and, thus, a correct topological analysis with the central atoms impossible.

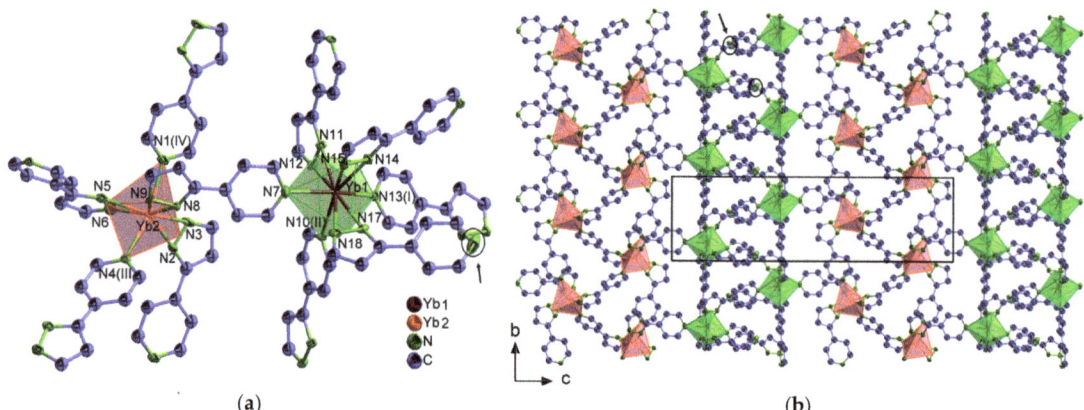

Figure 4. (a) Extended coordination sphere of Yb^{3+} in $^2_\infty[Yb_2(4\text{-PyPz})_6]$ (**4-Yb**); (b) crystal structure of **4-Yb** with a view along [100]. The coordination polyhedra around Yb1 and Yb2 are indicated in green and red, respectively. The arrows point to the uncoordinated nitrogen atom. The solvent molecules are omitted for clarity. Symmetry operations: I $-x + 1/2$, $y-1/2$, $-z + 1/2$ II x, y−1, z III $-x + 1/2, y - 1/2, - z + 3/2$ IV x, y + 1,z.

The different coordination numbers are also reflected by the Yb–N distances, which range from 227.7(6)–247.9(6) pm for CN = 8 (Yb2) to 234.1(6)–257.5(5) pm for CN = 9 (Yb1). Comparison of the Yb–N distances with [Yb(Ph$_2$pz)$_3$(Py)$_2$]·2(thf) (Ph$_2$pz = 3,5-diphenylpyrazolate, CN = 8, Yb–N = 225.7–244.3 pm) [39] and [YbL$_3$]·CH$_3$OH (HL = 2-(tetrazol-5-yl)-1,10-phenanthroline, CN = 9, Yb–N = 240.9–259.9) [40] shows good agreement for Yb–N. The Yb2–Npz (pz = pyrazolate nitrogen atom) is slightly shorter than the reported range, indicating the strength of the electrostatic interaction between the metal cation and the anionic pyrazolate ring.

The crystal structures of $^3_\infty[Ln(4\text{-PyPz})_3]$ (**4**, Ln = La, Ce, Pr, Nd, Ho, Er, Tm) and $^3_\infty[Ln(3\text{-PyPz})_3]$ (**3**, Ln = Ce, Pr, Nd, Ho, Er, Tm) are isotypic to the respective reported series of $^3_\infty[Ln(4\text{-PyPz})_3]$ and $^3_\infty[Ln(3\text{-PyPz})_3]$, Ln = Sm, Eu, Gd, Tb, Dy, respectively [33]. The extended coordination sphere of Ce^{3+} (CN = 9) in the two isotypic series $^3_\infty[Ln(4\text{-PyPz})_3]$ and $^3_\infty[Ln(3\text{-PyPz})_3]$ are shown in Figures S1 and S2. The volume of the unit cell and the average of Ln–N decrease with the increasing charge density along the two series $^3_\infty[Ln(4\text{-PyPz})_3]$ and $^3_\infty[Ln(3\text{-PyPz})_3]$ (Tables S10 and S11) as a direct consequence of the lanthanide contraction [41]. Tables with detailed crystallographic data and selected interatomic distances (pm) and angles (◦) of the studied compounds are given in the Supplementary Materials (Tables S1–S11).

The crystal structures were mostly determined by single-crystal X-ray diffraction (SCXRD), while the structures of $^2_\infty[Ce(4\text{-PyPz})_3(Py)]$ (**4-$^2_\infty$Ce**) and $^3_\infty[Tm(4\text{-PyPz})_3]$ (**4-Tm**) were characterized from microcrystalline products by powder X-ray diffraction (PXRD) and subsequent Pawley refinements (Figure 5a,b), confirming the isotypic character based on the SCXRD of $^2_\infty[Tm(4\text{-PyPz})_3(Py)]$ (**4-$^2_\infty$Tm**) and $^3_\infty[Er(4\text{-PyPz})_3]$ (**4-Er**), respectively.

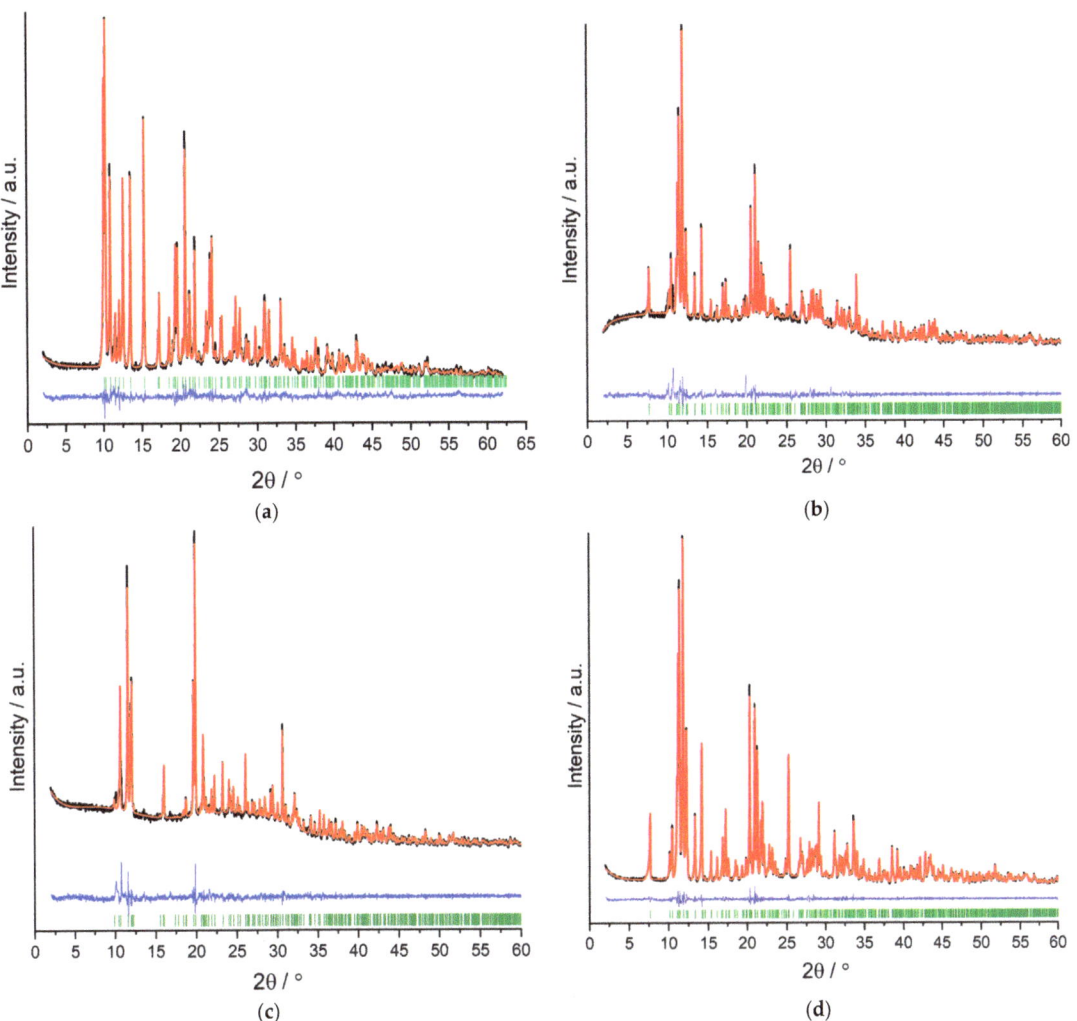

Figure 5. Pawley refinement results for PXRD of (a) $^2_\infty$[Ce(4-PyPz)$_3$(Py)] (**4-$^2_\infty$Ce**), (b) $^3_\infty$[Tm(4-PyPz)$_3$] (**4-Tm**), (c) $^2_\infty$[Eu(4-PyPz)$_2$(Py)$_2$] (**4-Eu^{2+}**), and (d) $^3_\infty$[Ce(4-PyPz)$_3$] (**4-Ce**), showing the experimental data (black) together with the Pawley fit (red), the corresponding difference plot (blue), as well as hkl position markers (green).

All bulk products were investigated by PXRD. For $^3_\infty$[Ln(3-PyPz)$_3$] (**3**, Ln = Ce, Pr, Nd, Ho, Er), the experimental diffraction patterns agree with the diffraction patterns simulated from the single-crystal data with no observation of additional reflections indicating the absence of crystalline byproducts (Figure S3). To account for the different measurement conditions of PXRD (298 K) and SCXRD (100 K), Pawley refinements for $^2_\infty$[Eu(4-PyPz)$_2$(Py)$_2$] (**4-Eu^{2+}**) (Figure 5c), $^3_\infty$[Ce(4-PyPz)$_3$] (**4-Ce**) (Figure 5d), and $^3_\infty$[Ln(4-PyPz)$_3$] (**4**, Ln = La, Pr, Nd, Ho, Er) (Figures S4–S8) were carried out, confirming the phase purity of the respective series of coordination polymers. The resulting difference plots show no significant deviations, and the refinement results (R_{wp}, GOF) are shown in Table S12. Other crystalline phases were found in the PXRD of $^2_\infty$[Tm(4-PyPz)$_3$(Py)] (**4-$^2_\infty$Tm**) and $^2_\infty$[Ln$_2$(4-PyPzH)$_6$]Py,

Ln = Yb (**4-Yb**), Lu (**4-Lu**) (Figure S9). Isolation of $^3_\infty$[Tm(3-PyPz)$_3$] (**3-Tm**) as single crystals was also possible.

2.2. Photophysical Properties

2.2.1. UV–VIS–NIR Absorption Spectra

Electronic absorption spectra were recorded in the solid state at room temperature (RT) for 4-PyPzH, $^2_\infty$[Eu(4-PyPz)$_2$(Py)$_2$] (**4-Eu^{2+}**), $^2_\infty$[Ce(4-PyPz)$_3$(Py)] (**4-$^2_\infty$Ce**), $^3_\infty$[Ln(4-PyPz)$_3$] (**4**, Ln = Ce, Pr, Nd, Ho, Er, Tm), and $^3_\infty$[Ln(3-PyPz)$_3$] (**3**, Ln = Ce, Pr, Nd, Ho, Er) (Figure 6). The absorption spectra of the free ligand 3-PyPzH was shown for the solid state in a range from about 200–270 and 570–305 nm corresponding to the intra-ligand transitions π–π* and/or n–π* [14]. The free ligand 4-PyPzH shows a broad band from 200 to 280 nm corresponding to the intra-ligand transitions. In the investigated coordination polymers, the intense wide absorption band corresponding to either ligand appears in the UV region. For $^2_\infty$[Eu(4-PyPz)$_2$(Py)$_2$] (**4-Eu^{2+}**), a broad absorption shoulder from 356–640 nm is associated with a metal-to-ligand charge transfer (MLCT) transition from the Eu^{2+} 4f orbitals to the π* orbitals of the coordinated ligands. For $^2_\infty$[Ce(4-PyPz)$_3$(Py)] (**4-$^2_\infty$Ce**), $^3_\infty$[Ce(4-PyPz)$_3$] (**4-Ce**), and $^3_\infty$[Ce(3-PyPz)$_3$] (**3-Ce**), the formation of shoulders at a higher wavelength from 320–470 nm is observed due to the transition from 4f to 5d. These absorption shoulders are compatible with the shoulders observed for the orange and the red emitters [Ce(2-PyPzH)$_3$Cl$_3$] and $^1_\infty$[Ce(2-PyPz)$_3$] [31]. Moreover, sharp and weak to medium bands can be assigned to the respective f–f transitions in both the VIS and NIR regions for $^3_\infty$[Ln(4-PyPz)$_3$], Ln = Pr (**4-Pr**), Nd (**4-Nd**), Ho (**4-Ho**), Er (**4-Er**), Tm (**4-Tm**) and $^3_\infty$[Ln(3-PyPz)$_3$], Ln = Pr (**3-Pr**), Nd (**3-Nd**), Ho (**3-Ho**), Er (**3-Er**), as assigned in Table 1 [11,42–45].

2.2.2. Emission and Excitation Spectra

The photoluminescence properties were recorded for all bulk products, $^2_\infty$[Eu(4-PyPz)$_2$(Py)$_2$] (**4-Eu^{2+}**), $^2_\infty$[Ce(4-PyPz)$_3$(Py)] (**4-$^2_\infty$Ce**), $^3_\infty$[Ln(4-PyPz)$_3$] (**4**, Ln = La, Ce, Pr, Nd, Ho, Er, Tm), and $^3_\infty$[Ln(3-PyPz)$_3$] (**3**, Ln = Ce, Pr, Nd, Ho, Er) in the solid state at RT and 77 K. The photoluminescence spectroscopy determinations for $^2_\infty$[Ce(4-PyPz)$_3$(Py)] (**4-$^2_\infty$Ce**), $^3_\infty$[Ce(4-PyPz)$_3$] (**4-Ce**), and $^3_\infty$[Ce(3-PyPz)$_3$] (**3-Ce**) (Figure 7) show interesting 5d–4f transitions with Ce^{3+}-centered light emission in the VIS range. Broad emission bands appear for **4-$^2_\infty$Ce**, **4-Ce**, and **3-Ce** from 520, 500, and 460 to 850 nm centered at 650, 650, and 641 nm, respectively, at RT, indicating large crystal field splitting and a large redshift for the emission wavelength reaches the red–orange visible region. The intensity of the ligand-based emission decreases from **4-$^2_\infty$Ce** through **4-Ce** to **3-Ce** which shifts the emission color from pink through orange pink to orange, the emission colors are represented in the CIE 1931 chromaticity diagram (Figure S24), and the color coordinates are listed in Table S13. In agreement with the absorption spectra, the excitation spectra show shoulders at higher wavelengths, which correlate with the lowest energy levels of the crystal field splitting bands of the 5d excited state of the Ce^{3+} ions.

The maximum excitation bands are at about 400 nm, corresponding to the respective coordinated pyrazolate anions. The lifetimes of **4-$^2_\infty$Ce** (1.08(2) ns), **4-Ce** (1.16(2) ns), and **3-Ce** (1.26(2) ns) are expected to be nanoseconds due to the parity allowed nature of the 5d–4f transition. These lifetimes are slightly shorter than the lifetimes of the reported red-emitting cerium $^1_\infty$[Ln(2-PyPz)$_3$] (2 ns, 2-PyPzH = 3-(2-pyridyl)pyrazole) [32] and orange-emitting cerium [Ce(2-PyPzH)$_3$Cl$_3$] (2.83 ns) [31].

The emission spectrum of $^2_\infty$[Eu(4-PyPz)$_2$(Py)$_2$] (**4-Eu^{2+}**) shows a ligand-based transition at 350 nm (Figure 7) along with some weak f–f transitions that can be assigned to a low content of trivalent Eu emission features. The ligand-based excitation band at around 335 nm for 4-PyPzH and 3-PyPzH [33] shows a hypochromic shift upon coordination to the investigated compounds to around 325 nm in **4-Pr**, **4-Nd**, **3-Nd**, **3-Ho**, **4-Er**, **3-Er**, and **4-Tm**. The blue shift for the ligand-based excitation band increases, reaching 318 nm for **3-Pr** and **4-Ho** and even below 300 nm for **4-Eu^{2+}**. For Pr^{3+}, Nd^{3+}, and Er^{3+}, additional direct f–f excitations from the ground states ^{3}H$_4$, ^{4}I$_{9/2}$, and ^{4}I$_{15/2}$, respectively, were also observed.

After coordination, the ligand-based emission band for both ligands at 405 nm shows a hypochromic shift to a value between 341 and 352 nm.

For $^3_\infty[\text{La}(4\text{-PyPz})_3]$ (**4-La**), an additional resolved broad band with λ_{onset} = 423 nm (~23,640 cm^{-1}), corresponding to the triplet state of the pyrazolate anion, was observed in the emission spectrum (Figure S27) at 77 K, which agrees well with the previously reported value [33].

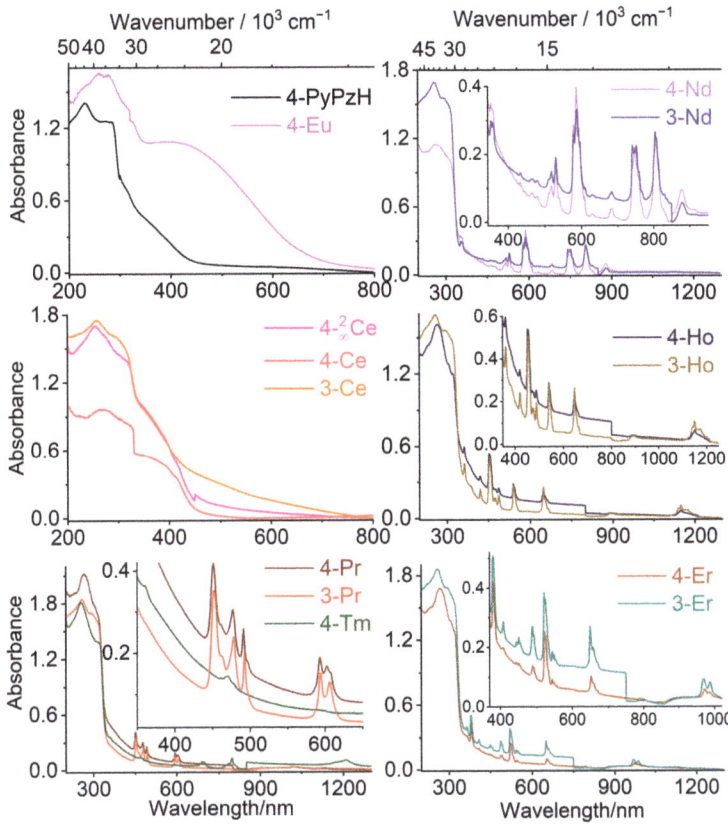

Figure 6. Solid-state absorption spectra of 4-PyPzH, $^2_\infty[\text{Eu}(4\text{-PyPz})_2(\text{Py})_2]$ (**4-Eu^{2+}**), $^2_\infty[\text{Ce}(4\text{-PyPz})_3(\text{Py})]$ (**4-$^2_\infty$Ce**), $^3_\infty[\text{Ln}(4\text{-PyPz})_3]$, (**4**, Ln = Ce, Pr, Nd, Ho, Er, Tm) and $^3_\infty[\text{Ln}(3\text{-PyPz})_3]$, (**3**, Ln = Ce, Pr, Nd, Ho, Er) in the solid state at room temperature.

Table 1. Absorption wavelengths of transitions of $^3_\infty[\text{Ln}(4\text{-PyPz})_3]$, (**4**, Ln = Pr, Nd, Ho, Er, Tm) and $^3_\infty[\text{Ln}(3\text{-PyPz})_3]$, (**3**, Ln = Pr, Nd, Ho, Er) in the solid state at room temperature.

ID	Intra–4f Absorption Transitions		λ_{max} (nm)
	Ground State	**Excited States**	
4-Pr	$^3H_4 \rightarrow$	$^3P_2, ^3P_1, ^3P_0, ^1D_2$	451, 477, 491, 593 nm
3-Pr	$^3H_4 \rightarrow$	$^3P_2, ^3P_1, ^3P_0, ^1D_2$	451, 477, 492, 593 nm
4-Nd	$^4I_{9/2} \rightarrow$	$^4D_{3/2}, ^2P_{1/2}, ^2K_{15/2}, ^2K_{13/2}, ^4G_{5/2}, ^4F_{9/2}, ^4F_{7/2}, ^4F_{5/2}, ^4F_{3/2}$	353, 432, 478, 529, 586, 681, 746, 803, 878 nm
3-Nd	$^4I_{9/2} \rightarrow$	$^2I_{11/2}, ^2P_{1/2}, ^2K_{15/2}, ^2K_{13/2}, ^4G_{5/2}, ^4F_{9/2}, ^4F_{7/2}, ^4F_{5/2}, ^4F_{3/2}$	353, 433, 479, 529, 588, 682, 740, 804, 878 nm
4-Ho	$^5I_8 \rightarrow$	$(^5G, ^3H)_5, (^5G, ^3G)_5, ^5G_6, ^5F_2, ^5F_3, ^5F4, ^5F_5, ^5I_5, ^5I_6$	362, 419, 452, 474, 486, 539, 646, 892, 1150 nm
3-Ho	$^5I_8 \rightarrow$	$(^5G, ^3H)_5, (^5G, ^3G)_5, ^5G_6, ^5F_2, ^5F_3, ^5F_4, ^5F_5, ^5I_5, ^5I_6$	362, 420, 455, 475, 486, 539, 646, 891, 1149 nm
4-Er	$^4I_{15/2} \rightarrow$	$^4G_{11/2}, (^2G, ^4F)_{9/2}, ^4F_{5/2}, ^4F_{7/2}, ^2H_{11/2}, ^4S_{8/2}, ^4F_{9/2}, ^4I_{11/2}$	379, 408, 451, 488, 521, 543, 652, 971 nm
3-Er	$^4I_{15/2} \rightarrow$	$^4G_{11/2}, (^2G, ^4F)_{9/2}, ^4F_{5/2}, ^4F_{7/2}, ^2H_{11/2}, ^4S_{8/2}, ^4F_{9/2}, ^4I_{11/2}$	379, 408, 450, 487, 520, 543, 649, 968 nm
4-Tm	$^3H_6 \rightarrow$	$^1D_2, ^1G_4, ^3F_3, ^3H_4, ^3H_5$	360, 470, 691, 796, 1209 nm

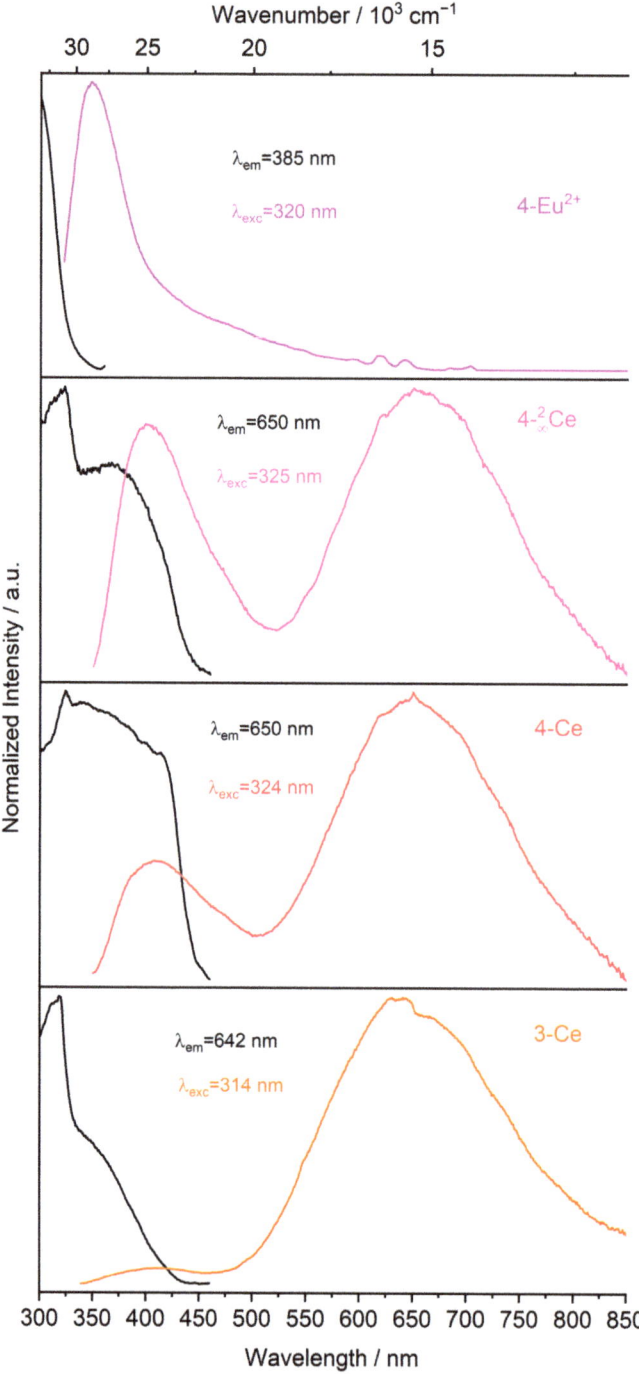

Figure 7. Normalized solid-state excitation (black) and emission (colored) spectra of $_\infty^2$[Eu(4-PyPz)$_2$(Py)$_2$] (**4-Eu^{2+}**), $_\infty^2$[Ce(4-PyPz)$_3$(Py)] (**4-$_\infty^2$Ce**), $_\infty^3$[Ce(4-PyPz)$_3$] (**4-Ce**), and $_\infty^3$[Ce(3-PyPz)$_3$] (**3-Ce**) at RT. Wavelengths for which the spectra were recorded are reported in the legends.

Although the excitation and emission spectra can provide a wealth of information, particularly about the coordination environment of the Ln^{3+} ions, it is uncommon to find the luminescence spectra for the Tm^{3+} and Ho^{3+}-based compounds. Nine-coordinated Tm^{3+} was only reported in three examples, and none of them investigated the photophysical properties, they mainly focused on the structural aspects [36,46,47]. In other cases, poor ligand-to-metal sensitization or back energy transfer occurs to allow only ligand-based luminescence, as in $[Tm_2(C_{15}H_{11}N_3)_2(C_7H_4BrO_2)_4(C_2O_4)]$, $(C_{15}H_{11}N_3 = 2,2':6'2''$-terpyridine and $C_7H_4BrO_2 = p$-bromobenzoic acid) [15], $Tm(bfa)_3phen$, (bfa = 4,4,4-trifluoro-1-phenyl-1,3-butanedione, phen = 1,10-phenanthroline) [12], and $Tm(ppa)_3 \cdot 2H_2O$, (ppa = 3-phenyl-2,4-pentanedionate) [13], where a significant ligand emission dominates the spectrum in addition to a single spectral band for the Tm^{3+}. Even nonefficient ligand sensitization with only a ligand emission band in the emission spectra was shown for $[(Tm-(TC)_3(H_2O)_2) \cdot (HPy \cdot TC)]_n$, (TC = 2-thiophenecarboxylate and HPy = pyridinium cation) [16]. In contrast, very good ligand-to-metal sensitization is observed for $^3_\infty[Tm(4-PyPz)_3]$ (**4-Tm**) (Figure 8). The transitions $^1G_4 \rightarrow ^3H_6, ^3F_4, ^3H_5$, and $^3H_4$ are readily observable at 480, 650, 787, and 1192 nm, respectively.

For $^3_\infty[Ho(4-PyPz)_3]$ (**4-Ho**) and $^3_\infty[Ho(3-PyPz)_3]$ (**3-Ho**), the $^5F_5 \rightarrow ^5I_8$ is observed at 648 nm in addition to the NIR transition $^5I_6 \rightarrow ^5I_8$ at 1155 nm. An additional NIR transition appears for **4-Ho** at 983, corresponding to the transition $^5F_5 \rightarrow ^5I_7$ and indicating more efficient ligand sensitization than in reported cases, such as $[(Ho-(TC)_3(H_2O)_2) \cdot (HPy \cdot TC)]_n$ with only ligand emission observable in the emission spectra [16]. The Ho^{3+}-based emission observed for both **3-Ho** and **4-Ho** is stronger than that of $^2_\infty[Ho_2(3-PyPzH)_3Cl_6] \cdot 2MeCN$ [14], which may be due to the absence of the vibrational energy of the chloride ligands.

For $^3_\infty[Pr(4-PyPz)_3]$ (**4-Pr**) and $^3_\infty[Pr(3-PyPz)_3]$ (**3-Pr**) (Figure 8), the highest intensity for the Pr^{3+}-based emission is found at 655 nm, corresponding to $^3P_0 \rightarrow ^3F_2$. NIR emission bands can also be observed at 738 and 1048 for **4-Pr** and at 736 and 1038 nm for **3-Pr**, corresponding to the transitions $^3P_0 \rightarrow ^3F_4$ and $^1D_2 \rightarrow ^3F_4$. Despite the ligand-based emission in **4-Pr** and **3-Pr** being more dominated than for the reported $^1_\infty[Pr(2-PyPz)_3]$ [32], the Pr^{3+}-based transitions are more characteristic than for other published cases, such as $^1_\infty[PrCl_3(ptpy)]$ and $[PrCl_3(ptpy)(py)]$, ptpy = 4'-phenyl-2,2':6',2''-terpyridine) [48].

For $^3_\infty[Er(4-PyPz)_3]$ (**4-Er**) and $^3_\infty[Er(3-PyPz)_3]$ (**3-Er**), the NIR transition $^4I_{13/2} \rightarrow ^4I_{15/2}$ is observed at about 1510 nm and the VIS transition at about 545 nm can also be observed for **4-Er** at RT and 77 K, while at 77 K for **3-Er**. Both **4-Er** and **3-Er** show Stark-level splitting in the emissive transitions in contrast, e.g., to the reported $^1_\infty[LnCl_3(bipy)(py)_2]py$, which shows no fine splitting [48].

For the NIR emitters $^3_\infty[Nd(4-PyPz)_3]$ (**4-Nd**) and $^3_\infty[Nd(3-PyPz)_3]$ (**3-Nd**), the transitions $^4F_{3/2} \rightarrow ^4I_{J/2}$, (J = 9, 11, 13) are observed at about 915, 1065, and 1345 nm [11].

Generally, for the VIS–NIR and the NIR emitters, the emissive energy levels of the Ln^{3+}-based transitions are populated by an antenna effect between the pyridylpyrazolate-based ligands and the lanthanide ions, which leads to ligand-to-metal energy transfer, as observed for 2-PyPzH [32] and 4,4'-bipyridine (bipy) [49].

The ligands' fluorescence emission band has lifetimes of 6.07 ns for 4-PyPzH and 3.4 ns for 3-PyPzH [33] which are shortened by coordination with different Ln^{3+} (τ = 0.93–1.04; Table S14). The lifetime increases with decreasing the temperature to 77 K, especially for **4-Nd**, which increases from 0.93(2) ns at RT to 3.04(9) ns at 77 K due to the decrease in thermal quenching. See the Supplementary Materials for half-page size absorption and photoluminescence spectra with designated 4f–4f transitions for the investigated compounds (Figures S10–S37).

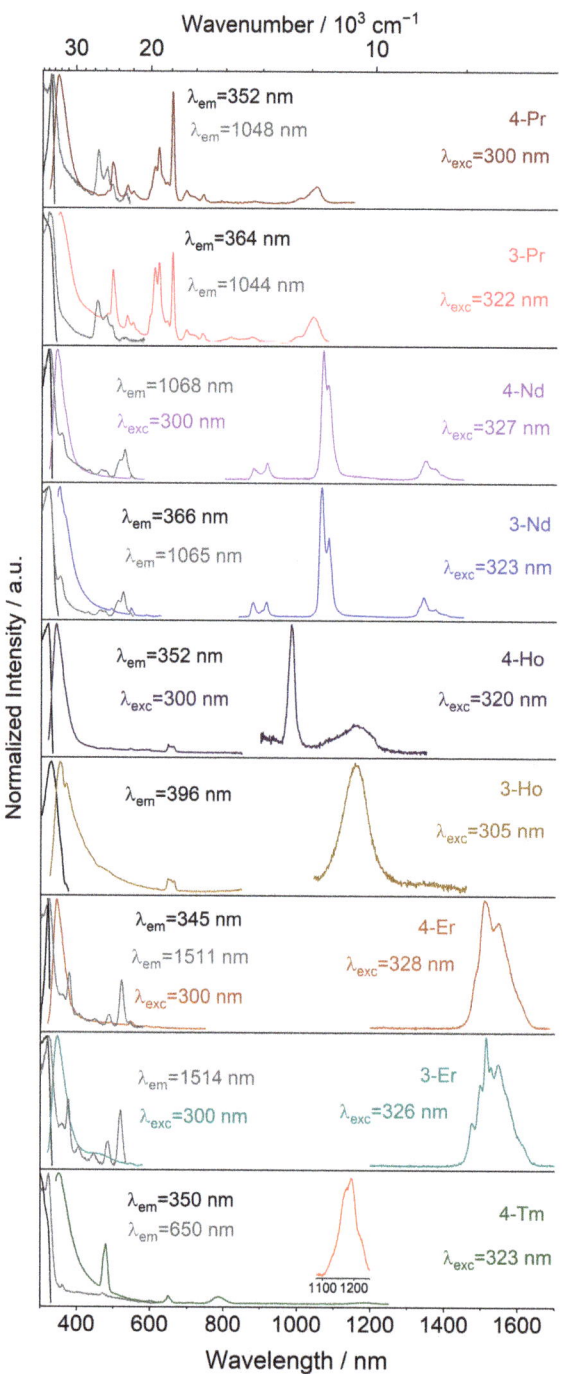

Figure 8. Normalized solid-state excitation (black) and emission spectra (colored) of $^3_\infty[\text{Ln}(4\text{-PyPz})_3]$, (**4**, Ln = Pr, Nd, Ho, Er, Tm) and $^3_\infty[\text{Ln}(3\text{-PyPz})_3]$, (**3**, Ln = Pr, Nd, Ho, Er) at RT. Wavelengths, for which the spectra were recorded, are reported in the legends.

2.2.3. Mechanism of Energy Transfer

To explain and understand the observed spectral results, a schematic diagram (Scheme 2) is shown, depicting the primary energy levels involved and the main energy transfer and relaxation pathways during the sensitization of lanthanide luminescence via the ligands. The ligands absorb energy and become excited from the singlet S_0 ground state to the singlet S_1 excited state by the absorption of visible light. The energy of the S_1 excited state is then transferred to the triplet-excited state (T) of the ligands through intersystem crossing (ISC). Competing processes include ligand fluorescence and nonradiative deactivation of the excited singlet state. Subsequently, the excitation energy is transferred to the excited 4f levels of the Ln^{3+} ions, resulting in the respective lanthanide ion emission to the respective 4f ground state [50]. According to Dexter's theory [51], the energy gap between the first excited energy level of the Ln^{3+} ions and the energy level of the triplet state of the respective ligand is important for an efficient energy transfer. If the energy gap is too large, the overlap between the ligand and the Ln^{3+} is reduced, and as a result, the energy transfer decreases sharply. On the other hand, if the energy gap is too small, energy transfer also occurs from the Ln^{3+} back to the resonance levels of the triplet states of the ligands, which also reduces the 4f-based emission.

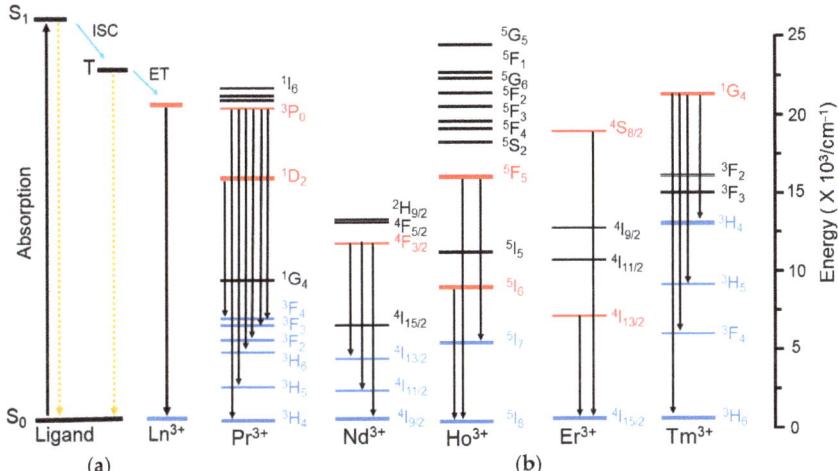

Scheme 2. (a) The Energy transfer mechanism; (b) energy level diagrams of Pr^{3+}, Nd^{3+}, Ho^{3+}, Er^{3+}, and Tm^{3+} ions.

In this study, the triplet-state energy levels of 4-PyPz$^-$ and 3-PyPz$^-$ were investigated by deduction from the phosphorescence spectra of the Gd-based coordination polymers and calculated to be 23,640 and 23,250 cm^{-1}, respectively [33]. This analogy is confirmed for 4-PyPz$^-$ through the spectra of **4-La**, as discussed before. The discussed 4f emission bands of the respective Ln-based CPs indicate that the triplet states of the ligands are suitable for a transfer of the absorbed light to the lanthanide ions via such an antenna effect. For instance, the energy difference (ΔE) between the ligand triplet state of 4-PyPz$^-$ (~23,640 cm^{-1}) and the energetic positions of Tm^{3+} ($^1G_4$ = ~21,300 cm^{-1}) results in an ΔE value in the optimal range. Pr^{3+} is slightly more complicated because it has two emission levels ($^3P_0$ and $^1D_2$). By considering the $^3P_0$ level as the main acceptor level with an energetic position of ~20,475 cm^{-1} [11], the ΔE values are calculated as 3165 and 2775 for **4-Pr** and **3-Pr**, respectively, both also being in the optimal range. For Ho^{3+}, the $^5F_4$, $^5S_2$, and $^5F_5$ levels are the main acceptor levels, and the emission from the $^5F_5$ and $^5I_6$ levels can partially be the result of a relaxation of the upper levels followed by transitions to the lower levels to give the characteristic NIR emission of the Ho^{3+} ion [52].

2.3. Magnetic Properties

Direct current (DC) magnetic susceptibility measurements were performed for $^3_\infty$[Ln(3-PyPz)$_3$], Ln = Ho (**3-Ho**), Er (**3-Er**) in a temperature range of 3 to 300 K and a magnetic field of 1T. As a link to the reported isotypic series of $^3_\infty$[Ln(3-PyPz)$_3$], Ln = Sm, Eu, Gd, Tb, Dy [33], the DC magnetic susceptibility measurements of $^3_\infty$[Dy(3-PyPz)$_3$] (**3-Dy**) were also performed.

The temperature dependence of the product of χT for all samples can be observed in Figure 9. At room temperature, the $\chi_M T$ (χ_M = molar magnetic susceptibility) values are 15.03, 14.05, and 10.70 for **3-Dy**, **3-Ho**, and **3-Er**, respectively. These experimental data are in satisfactory agreement with the theoretical values for the corresponding noninteracting Dy^{3+} ($^6H_{15/2}$, S = 5/2, L = 5, g = 4/3, χT = 14.17 cm^3 K mol^{-1}), Ho^{3+} ($^5I_8$, S = 5/2, L = 6, g = 5/4. χT = 14.07 cm^3 K mol^{-1}), and Er^{3+} ($^4I_{15/2}$, S = 3/2, L = 6, g = 6/5, χT = 11.48 cm^3 K mol^{-1}) [53].

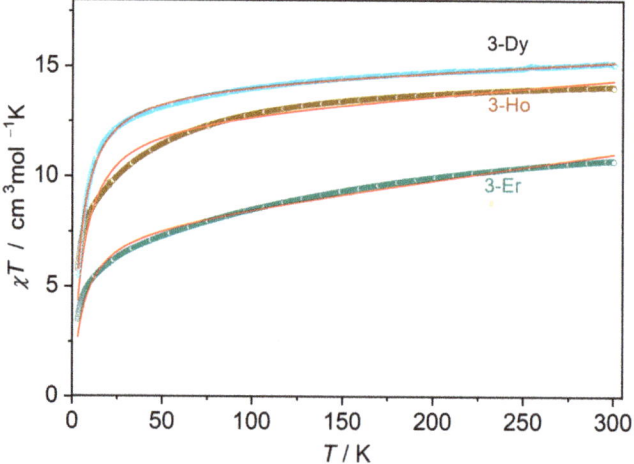

Figure 9. Variable temperature magnetic susceptibilities of $^3_\infty$[Ln(4-PyPz)$_3$], (**4**, Ln = Dy, Er), and $^3_\infty$[Ho(3-PyPz)$_3$] (**3-Ho**) in a temperature range from 3 to 300 K and a magnetic field of 1T.

For **3-Dy**, **3-Ho**, and **3-Er**, a monotonic slow decrease in the $\chi_M T$ product was observed upon cooling, which could be related to thermal depopulation within the m_J levels of the ground $^6H_{15/2}$, $^5I_8$, and $^4I_{15/2}$ multiplet, respectively. In addition, the $\chi_M T$ vs T plot did not display abrupt changes, which suggests a lack of magnetic interactions down to 45, 75, and 80 K, with $\chi_M T$ reaching 3.01, 5.79, and 3.49 cm^3 K mol^{-1}, respectively, because of the efficient magnetic isolation of lanthanide centers in the crystal lattice.

The data were fitted for **3-Dy**, **3-Ho**, and **3-Er** in the given temperature range with an effective magnetic moment μ_{eff} of 10.651(2), 10.09(1), and 7.855(9) µB and a Weiss constant θ of −4.41(1), −5.77(7), and −5.72(8) K, as well as a temperature-independent paramagnetic susceptibility χ_0 of 4.03(3) × 10^{-3}, 6.18(2) × 10^{-3}, and 11.52(8) × 10^{-3} cm^3 mol^{-1}. The small, negative Weiss constants θ are the results of spin–orbit coupling as well as the crystal field effect [54,55].

3. Materials and Methods

3.1. General Procedures

3-(4-pyridyl)pyrazole (4-PyPzH) and 3-(3-pyridyl)pyrazole (3-PyPzH) were synthesized as reported in the literature [56,57]. Lanthanide metals (holmium: 99.9%, Chempur, Karlsruhe, Germany; rest: >99.99%, Smart Elements, Vienna, Austria) were purchased and used as received. Pyridine (Py), dichloromethane (DCM), and cyclohexane (Cy) were purified by distillation and dried by standard procedures. All syntheses involving lanthanide elements were performed under argon or using vacuum lines, gloveboxes (MBraun Labmaster SP, Innovative Technology PureLab, Garching, Germany), Schlenk tubes, and Duran® glass ampoules (outer Ø 10 mm, wall thickness 1.5 mm). The solid reactants for the solvothermal reactions were mixed and sealed together with the solvent in an ampoule under reduced pressure ($p = 1.0 \times 10^{-3}$ mbar) after freezing the solvent with liquid nitrogen. Subsequently, the prepared ampoules were placed in heating furnaces based on Al_2O_3 tubes with Kanthal wire resistance heating and NiCr/Ni (Eurotherm 2416) temperature control elements, for which temperature programs and working steps according to the specific synthesis methods were used. After the solvents had been removed, the solid raw products were dried at RT in a dynamic vacuum ($p = 1.0 \times 10^{-3}$ mbar) before further steps. The bulk materials were characterized by PXRD and CHN analysis. The prepared 3D frameworks and 2D networks are air sensitive due to the known oxophilic behavior of the Ln-based CPs. It is expected that the CPs are insoluble in common organic solvents. We think the photostability tests are not significant in the possible applications of these synthesized CPs.

3.2. X-ray Crystallography

SCXRD determinations were performed on a Bruker AXS D8 Venture diffractometer (Karlsruhe, Germany) equipped with dual IµS microfocus sources, a collimating Quazar multilayer mirror, a Photon 100 detector, and an Oxford Cryosystems 700 low-temperature system (Mo–K$_\alpha$ radiation; λ = 71.073 pm). The structures were solved with direct methods and refined with the least squares method implemented in ShelX [58,59]. All nonhydrogen atoms were refined anisotropically. Hydrogen atoms were assigned to idealized geometric positions and included in structure factor calculations. Further, a ligand anion (4-PyPz$^-$) and the pyridine solvent in the asymmetric unit of **4-Yb** and **4-Lu** were found to be fully disordered and were refined with the help of restraints to achieve a proper structural model. For polymers **4-La, 4-Ce, 4-Pr, 4-Nd, 4-Ho**, and **4-Er**, the SQUEEZE [60] algorithm in PLATON [61–64] was used to include a bulk solvent model in the refinement. Two voids per unit cell were identified with SQUEEZE. The average volume was found to be 135×10^6 pm^3 for each void. The equivalent of 8 electrons for (**4-La, 4-Ce, 4-Pr**), 10 for (**4-Ho**), and 12 for (**4-Nd**) electrons per unit cell was also identified. ToposPro program package was used to determine the topology of the polymers [65]. Depictions of the crystal structures were created with Diamond [66]. The crystal structures have been deposited to the Cambridge Crystallographic Data Center (CCDC) as supplementary publication No. 2237763 (**4-Eu^{2+}**), 2237764 (**4-$^2_\infty$Tm**), 2237765 (**4-Yb**), 2237766 (**4-Lu**), 2237767 (**4-La**), 2237768 (**4-Ce**), 2237769 (**4-Pr**), 2237770 (**4-Nd**), 2237771 (**4-Ho**), 2237772 (for **4-Er**), 2237773 (**3-Ce**), 2237774 (**3-Pr**), 2237775 (**3-Nd**), 2237776 (**3-Ho**), 2237777 (for **3-Er**), and 2237778 (**3-Tm**). Crystallographic data and selected interatomic distances are listed in Tables S1–S11 for the investigated compounds.

PXRD analyses of the investigated compounds were carried out on a Stoe Stadi P diffractometer (Darmstadt, Germany) with a focusing Ge(111) monochromator and a Dectris Mythen 1K strip detector in Debye–Scherrer geometry. All powder samples were ground in a mortar and filled into Lindemann glass capillaries with 0.3 mm diameter under an inert gas atmosphere. All samples were measured in transmission geometry with Cu–K$_\alpha$ radiation (λ = 154.056 pm). Data collection was performed using the Stoe Powder Diffraction Software Package WinXPOW V3.0.2.1 and Pawley fits on the data were performed using TOPAS Academic V7 [67]. The data are presented in Figures 5 and S3–S9 in addition to Table S12.

3.3. Spectroscopical Investigations

3.3.1. Absorption Spectra

The UV–Vis–NIR absorption spectra were measured on solid-state products using a standard Agilent Cary 5000 UV–VIS–NIR spectrophotometer (Agilent Technologies,

Waldbronn, Germany) with a Praying Mantis accessory (Harrick Scientific Instruments, New York, NY, USA), which had been mounted and aligned for use with the DRP-ASC ambient chamber. The source, detector, and grating changeovers were at the standard position of 350, 800, and 800 nm, respectively, for all studied compounds except 4-$^2_\infty$Ce, 4-Nd, 4-Tm, and 3-Nd, the detector and grating changeovers were set to 850 nm, while 750 nm for 3-Er. For 4-Eu^{2+}, 4-$^2_\infty$Ce, 4-Ce, 4-Nd, 4-Tm, 3-Ce, and 3-Nd, the source was set to 320, 450, 330, 340, 335, 320, and 340 nm, respectively. The reference spectrum was collected on PTFE and the reference and samples were packed in the ambient chamber within the glovebox under inert conditions.

3.3.2. Photoluminescence Spectroscopy

The excitation and emission spectra were recorded for ground solid samples after filling them in quartz glass tubes under argon. The measurements were performed at room temperature as well as 77 K (latter using the liquid nitrogen-filled assembly FL-1013 of HORIBA) with a HORIBA Jobin Yvon Spex Fluorolog 3 spectrometer (Horiba-Jobin Yvon, Oberursel, Germany) equipped with a 450 W Xe short-arc lamp (USHIO INC., Tokyo, Japan), double-grated excitation, and emission monochromators, and a photomultiplier tube (R928P) using the FluoroEssence™ software V3.9. Excitation and emission spectra were corrected for the spectral response of the monochromators and the detector using spectral corrections provided by the constructor. In addition, a photodiode reference detector was used to correct the excitation spectra for the spectral distribution of the lamp intensity. An R5509-73 detector was used to collect the data in the NIR region. When required, the collection of the data was performed using an edge filter (Newport 20CGA-345, 395, 495 for the visible region and Reichmann Optics RG 830 long pass for the NIR region). Emission spectra with gating were recorded using a xenon flashlamp with a pulse repetition rate of 41 ms.

Photoluminescence overall decay process times were determined using the above-mentioned HORIBA Jobin Yvon Spex Fluorolog 3 spectrometer equipped with a dual lamp housing (FL-1040A), a UV xenon flashlamp (Exelitas FX-1102), and a TCSPC (time-correlated single-photon counting) upgrade, or picosecond pulsed laser diode. Emission decays were recorded using DataStation software V2.7. Exponential tail fitting was used for the calculation of resulting intensity decay using Decay Analysis Software 6. The quality of the fit was confirmed by χ^2 values being below 1.2.

3.4. PPMS Magnetic Measurements

Magnetic data were obtained with the application of the VSM option of a Quantum Design physical property measurement system (ppms). The data were corrected with respect to the contribution of the polypropylen sample holder as well as the diamagnetic contribution of the sample through utilization of both experimental data and Pascal constants (increment method). The total magnetic susceptibility is comprised of different parts: the diamagnetic contribution $\chi_{Diam.}$, the Curie paramagnetic contribution χ_{CW}, and a temperature-independent paramagnetic contribution χ_0.

$$\chi_{tot.} = \chi_{Diam.} + \chi_{Param.} = \chi_{Diam.} + \chi_{CW} + \chi_0$$

The Curie paramagnetic part is the ratio of the Curie constant C and the modified temperature $(T - \theta)$. θ is the Weiss temperature

$$\chi_{CW} = \frac{C}{T - \theta}$$

The Curie constant is given by the formula:

$$C = \mu_0 \frac{N_A \mu_B^2 n_{eff}^2}{3k_B}$$

with μ_0 = magnetic constant, N_A = Avogadro number, μ_B = Bohr magneton, n_{eff} = effective magnetic moment, k_B = Boltzmann constant.

The molar Curie paramagnetic contribution of the susceptibility is:

$$\chi_{molCW} = \frac{N_A \mu_B^2 n_{eff}^2}{3k_B(T-\theta)}$$

Or:

$$\chi_{molCW} = 0.1250 \frac{n_{eff}^2}{T-\Theta} \ / \ cm^3 mol^{-1}$$

To calculate the molar paramagnetic contribution of the susceptibility, we used the equation:

$$\chi_{mol} = 0.1250 \frac{n_{eff}^2}{T-\Theta} + \chi_0 \ / \ cm^3 mol^{-1}$$

For the analysis of the data the OriginPro software V2021b (Academic) was used.

3.5. Synthesis

3.5.1. Synthesis of $^2_\infty$[Eu(4-PyPz)$_2$(Py)$_2$] (4-Eu^{2+})

Freshly filed Eu metal (108.6 μmol) and an excess of 4-PyPzH (C$_8$H$_7$N$_3$, 220 μmol) were mixed with Py (0.6 mL) and sealed in an evacuated ampoule. The ampoule was heated to 185 °C in 1 h and maintained at this temperature for 72 h. The reaction mixture was then cooled to room temperature within 4 h. The excess ligand was washed away using a mixture of DCM and Cy. Suitable single crystals were selected for a SCXRD measurement. C$_{26}$H$_{22}$N$_8$Eu (598.47 g·mol^{-1}): C 51.59 (calcd. 52.18); H 3.04 (3.71); N 19.49 (18.72)%. Yield: 83%. FT-IR (ATR, Figure S39): $\tilde{\nu}$ = 3036 (w), 1065 (s), 1552 (w), 1522 (w), 1456 (w), 1439 (m), 1418 (w), 1404 (w), 1346 (w), 1293 (w), 1213 (m), 1187 (w), 1065 (w), 1044 (m), 999 (s), 963 (w), 951 (w), 925 (w), 871 (w), 830 (s), 761 (s), 690 (s), 659 (w), 614 (m), 532 (m), 451 (m) cm^{-1}.

3.5.2. Synthesis of $^2_\infty$[Ln(4-PyPz)$_3$(Py)] (4-$^2_\infty$Ce, 4-$^2_\infty$Tm)

The respective freshly filed Ln metal (81.4 μmol) and an excess of 4-PyPzH (C$_8$H$_7$N$_3$, 327.6 μmol) were mixed with Py and sealed in an evacuated ampoule. The ampoule was heated to 180 °C in 24 h then the temperature was raised to 240 °C within 48 h and maintained at this temperature for 72 h. The reaction mixture was then cooled to room temperature within 48 h. The excess ligand was washed away using a mixture of DCM and Cy. Suitable single crystals were selected for a SCXRD measurement. C$_{29}$H$_{23}$N$_{10}$Ce (651.67 g·mol^{-1}): C 52.63 (calcd. 53.45); H 2.67 (3.06); N 22.16 (21.49)%. Yield: 83%. FT-IR (ATR, Figure S40): $\tilde{\nu}$ = 3085 (w), 1698 (w), 1606 (s), 1523 (w), 1459 (w), 1440 (m), 1418 (w), 1347 (w), 1330 (w), 1295 (w), 1212 (s), 1099 (w), 1066 (w), 1047 (m), 1003 (s), 992 (m), 968 (w), 929 (m), 857 (w), 830 (m), 773 (s), 741 (w), 692 (s), 652 (m), 622 (w), 572 (m), 457 (s) cm^{-1}.

3.5.3. Synthesis of $^2_\infty$[Ln$_2$(4-PyPz)$_6$]Py (4-Yb, 4-Lu)

Freshly filed Yb (68.6 μmol) and an excess of 4-PyPzH (C$_8$H$_7$N$_3$, 275.6 μmol) were mixed with Py and sealed in an evacuated ampoule. The ampoule was heated to 180 °C in 24 h then the temperature was raised to 230 °C within 48 h and maintained at this temperature for 72 h. The reaction mixture was then cooled to room temperature within 48 h. Suitable single crystals were selected for a SCXRD measurement.

3.5.4. Synthesis of $^3_\infty$[Ln(4-PyPz)$_3$] (4, Ln = La, Ce, Pr, Nd, Ho, Er, Tm)

A mixture of the respective freshly filed Ln metal (91.1 μmol) and excess 4-PyPzH (C$_8$H$_7$N$_3$, 275.6 μmol), in 0.3 mL pyridine, was sealed in an evacuated ampoule. The temperature was raised to 230 °C in 48 h, held for 96 h, and then lowered to room temperature over a further 24 h. The excess ligand was washed away using a mixture of DCM and Cy. Colorless crystals were selected for SCXRD measurements.

$^3_\infty$[La(4-PyPz)$_3$] (**4-La**): C$_{24}$H$_{18}$N$_9$La (571.38 g·mol^{-1}): C 49.53 (calcd. 50.45); H 2.89 (3.18); N 21.50 (22.06)%. Yield: 80%. FT-IR (ATR, Figure S41): $\tilde{\upsilon}$ = 3096 (w), 1698 (w), 1607 (s), 1550 (w), 1526 (w), 1460 (w), 1447 (m), 1420 (w), 1348 (w), 1329 (w), 1215 (m), 1099 (w), 1066 (w), 1046 (m), 1005 (s), 968 (w), 927 (m), 831 (m), 763 (s), 739 (w), 694 (s), 652 (m), 527 (m), 459 (s) cm^{-1}.

$^3_\infty$[Ce(4-PyPz)$_3$] (**4-Ce**): C$_{24}$H$_{18}$N$_9$Ce (572.58 g·mol^{-1}): C 49.81 (calcd. 50.34); H 3.26 (3.17); N 21.73 (22.02)%. Yield: 85%. FT-IR (ATR, Figure S42): $\tilde{\upsilon}$ = 3101 (w), 1608 (s), 1526 (w), 1459 (w), 1447 (w), 1420 (w), 1348 (m), 1215 (m), 1066 (w), 1046 (m), 1006 (s), 969 (w), 927 (m), 831 (m), 763 (m), 739 (w), 695 (s), 652 (m), 527 (m), 460 (s) cm^{-1}.

$^3_\infty$[Pr(4-PyPz)$_3$] (**4-Pr**): C$_{24}$H$_{18}$N$_9$Pr (573.37 g·mol^{-1}): C 49.39 (calcd. 50.27); H 3.07 (3.16); N 21.16 (21.99)%. Yield: 84%. FT IR (ATR, Figure S43): $\tilde{\upsilon}$ = 3095 (w), 1607 (s), 1526 (w), 1460 (w), 1446 (w), 1420 (w), 1348 (w), 1328 (w), 1215 (m), 1099 (w), 1067 (w), 1046 (m), 1006 (s), 969 (w), 927 (m), 831 (m), 762 (s), 739 (w), 694 (s), 652 (m), 527 (m), 460 (s) cm^{-1}.

$^3_\infty$[Nd(4-PyPz)$_3$] (**4-Nd**): C$_{24}$H$_{18}$N$_9$Nd (576.70 g·mol^{-1}): C 49.33 (calcd. 49.98); H 2.73 (3.15); N 21.11 (21.86)%. Yield: 83%. FT-IR (ATR, Figure S44): $\tilde{\upsilon}$ = 3093 (w), 1609 (s), 1526 (w), 1459 (w), 1447 (m), 1420 (w), 1348 (w), 1215 (m), 1074 (w), 1047 (m), 1007 (s), 969 (w), 928 (m), 831 (m), 771 (s), 763 (s), 740 (w), 696 (s), 652 (w), 527 (w), 461 (s) cm^{-1}.

$^3_\infty$[Ho(4-PyPz)$_3$] (**4-Ho**): C$_{24}$H$_{18}$N$_9$Ho (597.39 g·mol^{-1}): C 47.44 (calcd. 48.25); H 2.75 (3.04); N 20.14 (21.10)%. Yield: 86%. FT-IR (ATR, Figure S45): $\tilde{\upsilon}$ = 3113 (w), 1697 (w), 1609 (s), 1549 (w), 1526 (w), 1460 (w), 1446 (w), 1419 (w), 1348 (w), 1328 (w), 1297 (w), 1214 (m), 1101 (w), 1075 (w), 1048 (m), 1007 (s), 970 (w), 929 (m), 845 (w), 831 (m), 770 (s), 760 (s), 739 (m), 696 (s), 652 (m), 528 (m), 461 (s) cm^{-1}.

$^3_\infty$[Er(4-PyPz)$_3$] (**4-Er**): C$_{24}$H$_{18}$N$_9$Er (599.72 g·mol^{-1}): C 48.85 (calcd. 48.07); H 3.34 (3.03); N 20.69 (21.02)%. Yield: 89%. FT-IR (ATR, Figure S46): $\tilde{\upsilon}$ = 3062 (w), 1696 (w), 1609 (s), 1549 (w), 1526 (w), 1460 (w), 1446 (w), 1419 (w), 1348 (w), 1328 (w), 1214 (m), 1101 (w), 1076 (m), 1048 (m), 1008 (s), 971 (w), 929 (m), 831 (m), 770 (s), 759 (s), 739 (m), 697 (s), 652 (m), 528 (m), 461 (s) cm^{-1}.

$^3_\infty$[Tm(4-PyPz)$_3$] (**4-Tm**): C$_{24}$H$_{18}$N$_9$Tm (601.39 g·mol^{-1}): C 48.90 (calcd. 47.93); H 3.11 (3.02); N 19.97 (20.96)%. Yield: 82%. FT-IR (ATR, Figure S47): $\tilde{\upsilon}$ = 3117 (w), 1697 (w), 1609 (s), 1550 (w), 1527 (m), 1460 (m), 1447 (m), 1419 (w), 1349 (m), 1329 (w), 1296 (w), 1214 (s), 1102 (w), 1077 (m), 1049 (m), 1008 (s), 971 (w), 930 (m), 845 (w), 830 (m), 770 (s), 759 (s), 740 (m), 696 (s), 652 (m), 528 (m), 461 (s) cm^{-1}.

3.5.5. Synthesis of $^3_\infty$[Ln(3-PyPz)$_3$] (3, Ln = Ce, Pr, Nd, Ho, Er)

A mixture of the respective freshly filed Ln metal (78.8 µmol) and an excess of 3-PyPzH (C$_8$H$_7$N$_3$, 240 µmol) together with 0.3 mL pyridine were sealed in an evacuated ampoule. The oven was heated to 180 °C in 24 h. Subsequently, the temperature was raised to 230 °C in 48 h. The temperature was held for 72 h and then lowered to room temperature over a further 48 h. Single crystals were selected for SCXRD measurements.

$^3_\infty$[Ce(3-PyPz)$_3$] (**3-Ce**): C$_{24}$H$_{18}$N$_9$Ce (572.58 g·mol^{-1}): C 50.50 (calcd. 50.34); H 3.11 (3.17); N 21.16 (22.02)%. Yield: 83%. FT-IR (ATR, Figure S48): $\tilde{\upsilon}$ = 3085 (w), 1596 (w), 1576 (m), 1509 (w), 1464 (m), 1453 (m), 1407 (m), 1359 (w), 1346 (w), 1250 (w), 1206 (m), 1186 (m), 1122 (w), 1099 (w), 1072 (m), 1039 (s), 963 (m), 928 (m), 859 (w), 818 (m), 779 (s), 716 (w), 702 (s), 656 (w), 635 (s), 510 (w), 461 (s) cm^{-1}.

$^3_\infty$[Pr(3-PyPz)$_3$] (**3-Pr**): C$_{24}$H$_{18}$N$_9$Pr (573.37 g·mol^{-1}): C 49.35 (calcd. 50.27); H 2.63 (3.16); N 22.89 (21.99)%. Yield: 81%. FT-IR (ATR, Figure S49): $\tilde{\upsilon}$ = 2897 (w), 1683 (m), 1596 (w), 1576 (w), 1508 (w), 1453 (m), 1407 (m), 1359 (w), 1345 (w), 1260 (w), 1249 (w), 1207 (m), 1186 (m), 1099 (w), 1074 (m), 1041 (s), 963 (m), 928 (m), 817 (w), 781 (s), 702 (s), 657 (w), 636 (s), 515 (w), 462 (s) cm^{-1}.

$^3_\infty$[Nd(3-PyPz)$_3$] (**3-Nd**): C$_{24}$H$_{18}$N$_9$Nd (576.70 g·mol^{-1}): C 48.28 (calcd. 48.98); H 2.84 (3.15); N 21.39 (21.86)%. Yield: 80%. FT-IR (ATR, Figure S50): $\tilde{\upsilon}$ = 3085 (w), 1576 (m), 1509 (w), 1465 (m), 1452 (m), 1408 (m), 1360 (m), 1347 (m), 1330 (w), 1249 (w), 1207 (m), 1186 (s),

1122 (w), 1099 (w), 1072 (m), 1041 (s), 963 (m), 928 (s), 818 (m), 779 (s), 717 (w), 702 (s), 656 (m), 636 (s), 509 (w), 463 (s) cm^{-1}.

$^3_\infty$[Ho(3-PyPz)$_3$] (**3-Ho**): C$_{24}$H$_{18}$N$_9$Ho (597.39 g·mol^{-1}): C 47.95 (calcd. 48.25); H 2.94 (3.04); N 20.16 (21.10)%. Yield: 86%. FT-IR (ATR, Figure S51): $\tilde{v}$ = 3086 (w), 1597 (w), 1577 (m), 1508 (w), 1465 (m), 1452 (m), 1408 (m), 1348 (m), 1248 (w), 1210 (m), 1187 (s), 1100 (w), 1075 (m), 1044 (s), 964 (m), 931 (m), 818 (m), 778 (s), 700 (s), 657 (m), 637 (s), 467 (s) cm^{-1}.

$^3_\infty$[Er(3-PyPz)$_3$] (**3-Er**): C$_{24}$H$_{18}$N$_9$Er (599.72 g·mol^{-1}): C 48.86 (calcd. 48.07); H 2.69 (3.03); N 20.77 (21.02)%. Yield: 87%. FT-IR (ATR, Figure S52): $\tilde{v}$ = 3086 (w), 1577 (w), 1508 (w), 1465 (w), 1453 (w), 1408 (w), 1348 (w), 1248 (w), 1210 (w), 1187 (m), 1100 (w), 1077 (m), 1045 (s), 964 (m), 931 (m), 818 (w), 780 (s), 700 (m), 657 (w), 637 (m), 513 (w), 468 (s) cm^{-1}.

3.5.6. Single Crystal of $^3_\infty$[Tm(3-PyPz)$_3$] (3-Tm):

Freshly filed Tm (68.7 μmol) and excess of 4-PyPzH (C$_8$H$_7$N$_3$, 210.5 μmol) were mixed with Py and sealed in an evacuated ampoule. The ampoule was heated to 180 °C in 1 h then the temperature was raised to 240 °C within 48 h and maintained at this temperature for 168 h. The reaction mixture was then cooled to room temperature within 48 h. Suitable single crystals were selected for a SCXRD measurement.

4. Conclusions

Divalent europium in the 2D network $^2_\infty$[Eu(4-PyPz)$_2$(Py)$_2$] and the trivalent lanthanide containing 3D frameworks $^3_\infty$[Ln(4-PyPz)$_3$] and $^3_\infty$[Ln(3-PyPz)$_3$], Ln = Ce^{3+}, Pr^{3+}, Nd^{3+}, Ho^{3+}, Er^{3+}, Tm^{3+}, $^3_\infty$[La(4-PyPz)$_3$], as well as the 2D networks $^2_\infty$[Ln(4-PyPz)$_3$(Py)], Ln = Ce^{3+}, Tm^{3+} and $^2_\infty$[Ln$_2$(4-PyPz)$_6$]Py, Ln = Yb^{3+}, Lu^{3+} were synthesized by redox reactions between the elemental lanthanide and the ligand 3-(4-pyridyl)pyrazole (4-PyPzH) or 3-(3-pyridyl)pyrazole (3-PyPzH). The 18 coordination polymers were synthesized in a solvothermal processes in pyridine, in which the latter can act as a co-ligand. Uncommon NIR emission for Tm^{3+} and Ho^{3+} was detected along with additional Pr^{3+}, Er^{3+}, and Nd^{3+} NIR emission benefited from a good ligand sensitizing effect. In addition, Ce^{3+}-based coordination polymers showed strong reductions in the 5d excited state, resulting in a distinctive pink to orange emission. Magnetic studies conducted with direct current (DC) showed magnetic isolation of the lanthanide centers in $^3_\infty$[Ln(3-PyPz)$_3$], Ln = Dy, Ho, Er. In summary, coordination polymers with pyridylpyrazolate ligands as N-donors can display a wide range of photoluminescent properties.

Supplementary Materials: The following supporting information can be downloaded at: https://www.mdpi.com/article/10.3390/chemistry5020069/s1, additional experimental details; Tables S1–S11: Crystallographic data and selected interatomic distances (pm) and angles (o) of $^2_\infty$[Eu(4-PyPz)$_2$(Py)$_2$] (**4-Eu^{2+}**), $^2_\infty$[Ln$_2$(4-PyPzH)$_6$]Py, Ln = Yb (**4-Yb**) and Lu (**4-Lu**), $^2_\infty$[Tm(4-PyPz)$_3$(Py)] (**4-$^2_\infty$Tm**), $^3_\infty$[Ln(4-PyPz)$_3$], Ln = La (**4-La**), Ce (**4-Ce**), Pr (**4-Pr**), Nd (**4-Nd**), Ho (**4-Ho**), Er (**4-Er**), and $^3_\infty$[Ln(3-PyPz)$_3$], Ln = Ce (**3-Ce**), Pr (**3-Pr**), Nd (**3-Nd**), Ho (**3-Ho**), Er (**3-Er**), Tm (**3-Tm**); Figure S1: Extended coordination sphere of Ce^{3+} in $^3_\infty$[Ce(4-PyPz)$_3$] (**4-Ce**) representing the series of isotypic framework compounds (**4**, Ln = La, Ce, Pr, Nd, Ho, Er, Tm). The coordination polyhedra around Ce^{3+} is indicated in green and the thermal ellipsoids describe a 50% probability level of the atoms. Symmetry operations: I x + 1/2, −y + 1/2, z + 1/2 II −x + 3/2, y − 1/2, −z + 3/2 III x + 1, y, z; Figure S2: Extended coordination sphere of Ce^{3+} in $^3_\infty$[Ce(3-PyPz)$_3$] (**3-Ce**) representing the series of isotypic framework compounds (**3**, Ln = Ce, Pr, Nd, Ho, Er, Tm). The coordination polyhedra around Ce^{3+} is indicated in green and the thermal ellipsoids describe a 50% probability level of the atoms. Symmetry operations: I −z + 1, x + 1/2, −y + 3/2 II y − 1/2, −z + 3/2, −x + 1 III x − 1/2, y, −z + 3/2 IV y − 1/2, z, −x + 3/2 V z − 1/2, x, −y + 3/2; Figure S3: Comparison of the experimental X-ray powder diffraction pattern of $^3_\infty$[Ln(3-PyPz)$_3$] (**3**, Ln = Pr, Nd, Ho, Er, Tm) at RT with the simulated pattern from the single-crystal X-ray data of $^3_\infty$[Er(3-PyPz)$_3$] (**3-Er**) at 100 K; Figures S4–S8: Pawley refinement results for PXRD of $^3_\infty$[Ln(4-PyPz)$_3$], Ln = La (**4-La**), Pr (**4-Pr**), Nd (**4-Nd**), Ho (**4-Ho**), and Er (**4-Er**) showing the experimental data (black) together with the Pawley fit (red), the corresponding difference plot (blue), as well as hkl position markers (green); Table S12: Pawley refinement results for $^2_\infty$[Eu(4-PyPz)$_2$(Py)$_2$] (**4-Eu^{2+}**), $^2_\infty$[Ce(4-PyPz)$_3$(Py)] (**4-$^2_\infty$Ce**), and $^3_\infty$[Ln(4-PyPz)$_3$] (**4**, Ln = La, Pr,

Nd, Ho, Er, Tm); Figure S9: Comparison of the experimental X-ray powder diffraction pattern of $^2_\infty$[Tm(4-PyPz)$_3$(Py)] (**4-$^2_\infty$Tm**), $^2_\infty$[Ln$_2$(4-PyPzH)$_6$]Py, Ln = Yb (**4-Yb**) and Lu (**4-Lu**) at RT with the respective simulated pattern from single-crystal X-ray data at 100 K; Figures S10–S23: Absorption spectra of 4-PyPzH, $^2_\infty$[Eu(4-PyPz)$_2$(Py)$_2$] (**4-Eu^{2+}**), $^2_\infty$[Ce(4-PyPz)$_3$(Py)] (**4-$^2_\infty$Ce**), $^3_\infty$[Ln(4-PyPz)$_3$], (**4**, Ln = Ce, Pr, Nd, Ho, Er, Tm) and $^3_\infty$[Ln(3-PyPz)$_3$], (**3**, Ln = Ce, Pr, Nd, Ho, Er) in the solid state at room temperature; Figure S24: Chromaticity coordinate diagram (CIE 1931) of the emission colors of $^2_\infty$[Ce(4-PyPz)$_3$(Py)] (**4-$^2_\infty$Ce**), $^3_\infty$[Ce(4-PyPz)$_3$] (**4-Ce**) and $^3_\infty$[Ce(3-PyPz)$_3$] (**3-Ce**); Table S13: Chromaticity coordinates (x,y) for $^2_\infty$[Ce(4-PyPz)$_3$(Py)] (**4-$^2_\infty$Ce**), $^3_\infty$[Ce(4-PyPz)$_3$] (**4-Ce**), and $^3_\infty$[Ce(3-PyPz)$_3$] (**3-Ce**); Figures S25–S38: Normalized excitation and emission spectra of $^2_\infty$[Eu(4-PyPz)$_2$(Py)$_2$] (**4-Eu^{2+}**), $^2_\infty$[Ce(4-PyPz)$_3$(Py)] (**4-$^2_\infty$Ce**), $^3_\infty$[Ln(4-PyPz)$_3$], (**4**, Ln = La, Ce, Pr, Nd, Ho, Er, Tm) and $^3_\infty$[Ln(3-PyPz)$_3$], (**3**, Ln = Ce, Pr, Nd, Ho, Er) at room temperature (top) and 77 K (bottom). Wavelengths at which the spectra were recorded are reported in the legends; Table S14: Photophysical data of $^2_\infty$[Eu(4-PyPz)$_2$(Py)$_2$] (**4-Eu^{2+}**), $^2_\infty$[Ce(4-PyPz)$_3$(Py)] (**4-$^2_\infty$Ce**), $^3_\infty$[Ln(4-PyPz)$_3$], (**4**, Ln = La, Ce, Pr, Nd, Ho, Er, Tm) and $^3_\infty$[Ln(3-PyPz)$_3$], (**3**, Ln = Ce, Pr, Nd, Ho, Er) in the solid state at room temperature and 77 K; Figures S39–S54: The infrared spectrum (ATR) of $^2_\infty$[Eu(4-PyPz)$_2$(Py)$_2$] (**4-Eu^{2+}**), $^2_\infty$[Ce(4-PyPz)$_3$(Py)] (**4-$^2_\infty$Ce**), $^3_\infty$[Ln(4-PyPz)$_3$], (**4**, Ln = La, Ce, Pr, Nd, Ho, Er, Tm), $^3_\infty$[Ln(3-PyPz)$_3$], (**3**, Ln = Ce, Pr, Nd, Ho, Er), 3-PyPzH, and 4-PyPzH.

Author Contributions: Conceptualization, K.M.-B. and H.Y.; methodology, H.Y.; software, H.Y. and J.B.; validation, H.Y.; formal analysis, H.Y. and C.P.; investigation, H.Y.; resources, I.V.T., F.K. and K.M.-B.; data curation, H.Y. and J.B.; writing—original draft preparation, H.Y.; writing—review and editing, H.Y., J.B., I.V.T., C.P., F.K. and K.M.-B.; visualization, H.Y.; supervision, K.M.-B.; project administration, K.M.-B. All authors have read and agreed to the published version of the manuscript.

Funding: This research was funded by the Deutsche Forschungsgemeinschaft DFG, grant No. MU-1562/7-2. H.Y. was awarded a PhD fellowship by the Egyptian Ministry of Higher Education (MoHE) and the German Academic Exchange Service (DAAD) within the German Egyptian Research Long-term Scholarship (GERLS) Program, 2017 (57311832), the funding agency is the German Academic Exchange Service Cairo. In the 14th round of applications, H.Y. was awarded a dissertation completion grant offered on the basis of JLU's Gender Equality Concept. The synthesis of the studied ligand was funded by the Russian Science Foundation (project No. 19–13–00272).

Data Availability Statement: CCDC 2237763 (**4-Eu^{2+}**), 2237764 (**4-$^2_\infty$Tm**), 2237765 (**4-Yb**), 2237766 (**4-Lu**), 2237767 (**4-La**), 2237768 (**4-Ce**), 2237769 (**4-Pr**), 2237770 (**4-Nd**), 2237771 (**4-Ho**), 2237772 (for **4-Er**), 2237773 (**3-Ce**), 2237774 (**3-Pr**), 2237775 (**3-Nd**), 2237776 (**3-Ho**), 2237777 (for **3-Er**), and 2237778 (**3-Tm**) contain the supplementary crystallographic data for this paper. These data can be obtained free of charge via http://www.ccdc.cam.ac.uk/conts/retrieving.html (or from the CCDC, 12 Union Road, Cambridge, CB2 1EZ, UK; Fax: +44 1223 336033; email: deposit@ccdc.cam.ac.uk).

Conflicts of Interest: The authors declare no conflict of interest.

References

1. Millward, J.M.; Ariza de Schellenberger, A.; Berndt, D.; Hanke-Vela, L.; Schellenberger, E.; Waiczies, S.; Taupitz, M.; Kobayashi, Y.; Wagner, S.; Infante-Duarte, C. Application of europium-doped very small iron oxide nanoparticles to visualize neuroinflammation with MRI and fluorescence microscopy. *Neuroscience* **2017**, *403*, 136–144. [CrossRef]
2. Lenora, C.U.; Carniato, F.; Shen, Y.; Latif, Z.; Haacke, E.M.; Martin, P.D.; Botta, M.; Allen, M.J. Structural features of europium(II)-containing cryptates that influence relaxivity. *Chem. Eur. J.* **2017**, *23*, 15404–15414. [CrossRef]
3. Acharjya, A.; Corbin, B.A.; Prasad, E.; Allen, M.J.; Maity, S. Solvation-controlled emission of divalent europium salts. *J. Photochem. Photobiol. A* **2022**, *429*, 113892–113899. [CrossRef]
4. Dorenbos, P. Anomalous luminescence of Eu^{2+} and Yb^{2+} in inorganic compounds. *J. Phys. Condens. Matter* **2003**, *15*, 2645–2665. [CrossRef]
5. Qin, X.; Liu, X.; Huang, W.; Bettinelli, M.; Liu, X. Lanthanide-activated phosphors based on 4f-5d optical transitions: Theoretical and experimental aspects. *Chem. Rev.* **2017**, *117*, 4488–4527. [CrossRef] [PubMed]
6. Zurawski, A.; Mai, M.; Baumann, D.; Feldmann, C.; Müller-Buschbaum, K. Homoleptic imidazolate frameworks $^3_\infty$[Sr$_{1-x}$Eu$_x$(Im)$_2$]—hybrid materials with efficient and tuneable luminescence. *Chem. Commun.* **2011**, *47*, 496–498. [CrossRef]
7. Kajdas, C.; Furey, M.J.; Ritter, A.L.; Molina, G.J. Triboemission as a basic part of the boundary friction regime. *Lubr. Sci.* **2002**, *14*, 223–254. [CrossRef]
8. Galimov, D.I.; Yakupova, S.M.; Vasilyuk, K.S.; Bulgakov, R.G. A novel gas assay for ultra-small amounts of molecular oxygen based on the chemiluminescence of divalent europium. *J. Photochem. Photobiol. A* **2021**, *418*, 113430–113437. [CrossRef]

9. Eliseeva, S.V.; Bünzli, J.-C.G. Lanthanide luminescence for functional materials and bio-sciences. *Chem. Soc. Rev.* **2010**, *39*, 189–227. [CrossRef]
10. Bünzli, J.C.G.; Piguet, C. Taking advantage of luminescent lanthanide ions. *Chem. Soc. Rev.* **2005**, *34*, 1048–1077. [CrossRef]
11. Carnall, W.T.; Fields, P.R.; Rajnak, K. Electronic energy levels in the trivalent lanthanide aquo Ions. I. Pr^{3+}, Nd^{3+}, Pm^{3+}, Sm^{3+}, Dy^{3+}, Ho^{3+}, Er^{3+}, and Tm^{3+}. *J. Chem. Phys.* **1968**, *49*, 4424–4442. [CrossRef]
12. Feng, J.; Zhang, H.-J.; Song, S.-Y.; Li, Z.-F.; Sun, L.-N.; Xing, Y.; Guo, X.-M. Syntheses, crystal structures, visible and near-IR luminescent properties of ternary lanthanide (Dy^{3+}, Tm^{3+}) complexes containing 4,4,4-trifluoro-1-phenyl-1,3-butanedione and 1,10-phenanthroline. *J. Lumin.* **2008**, *128*, 1957–1964. [CrossRef]
13. Serra, O.A.; Nassar, E.J.; Calefi, P.S.; Rosa, I.L.V. Luminescence of a new Tm^{3+} β-diketonate compound. *J. Alloys Compd.* **1998**, *275–277*, 838–840. [CrossRef]
14. Youssef, H.; Schäfer, T.; Becker, J.; Sedykh, A.E.; Basso, L.; Pietzonka, C.; Taydakov, I.V.; Kraus, F.; Müller-Buschbaum, K. 3D-Frameworks and 2D-networks of lanthanide coordination polymers with 3-pyridylpyrazole: Photophysical and magnetic properties. *Dalton Trans.* **2022**, *51*, 14673–14685. [CrossRef]
15. Ridenour, J.A.; Carter, K.P.; Butcher, R.J.; Cahill, C.L. RE-*p*-halobenzoic acid–terpyridine complexes, Part II: Structural diversity, supramolecular assembly, and luminescence properties in a series of *p*-bromobenzoic acid rare-earth hybrid materials. *CrystEngComm* **2017**, *19*, 1172–1189. [CrossRef]
16. Batrice, R.J.; Adcock, A.K.; Cantos, P.M.; Bertke, J.A.; Knope, K.E. Synthesis and characterization of an isomorphous lanthanide-thiophenemonocarboxylate series (Ln = La–Lu, except Pm) amenable to color tuning. *Cryst. Growth Des.* **2017**, *17*, 4603–4612. [CrossRef]
17. Kawamura, Y.; Wada, Y.; Hasegawa, Y.; Iwamuro, M.; Kitamura, T.; Yanagida, S. Observation of neodymium electroluminescence. *Appl. Phys. Lett.* **1999**, *74*, 3245–3247. [CrossRef]
18. Curry, R.; Gillin, W.P. 1.54 µm electroluminescence from erbium (III) tris(8-hydroxyquinoline) (ErQ)-based organic light-emitting diodes. *Appl. Phys. Lett.* **1999**, *75*, 1380–1382. [CrossRef]
19. Mehrdel, B.; Nikbakht, A.; Aziz, A.A.; Jameel, M.S.; Dheyab, M.A.; Khaniabadi, P.M. Upconversion lanthanide nanomaterials: Basics introduction, synthesis approaches, mechanism and application in photodetector and photovoltaic devices. *Nanotechnology* **2021**, *33*, 082001. [CrossRef] [PubMed]
20. Fischer, S.; Ivaturi, A.; Fröhlich, B.; Rüdiger, M.; Richter, A.; Krämer, K.W.; Richards, B.S.; Goldschmidt, J.C. Upconverter silicon solar cell devices for efficient utilization of sub-band-gap photons under concentrated solar radiation. *IEEE J. Photovolt.* **2013**, *4*, 183–189. [CrossRef]
21. Bünzli, J.-C.G.; Eliseeva, S.V. Basics of lanthanide photophysics. In *Springer Series on Fluorescence: Lanthanide Luminescence: Photophysical, Analytical and Biological Aspects*; Wolfbeis, O.S., Hof, M., Eds.; Springer: Berlin/Heidelberg, Germany, 2011; Volume 7, pp. 1–46.
22. Davies, G.M.; Aarons, R.J.; Motson, G.R.; Jeffery, J.C.; Adams, H.; Faulkner, S.; Ward, M.D. Structural and near-IR photophysical studies on ternary lanthanide complexes containing poly (pyrazolyl) borate and 1,3-diketonate ligands. *Dalton Trans.* **2004**, *4*, 1136–1144. [CrossRef]
23. Bao, S.; Liang, Y.; Wang, L.; Xu, L.; Wang, Y.; Liang, X.; Xiang, W. Superhigh-luminance Ce:YAG phosphor in glass and phosphor-in-glass film for laser lighting. *ACS Sustain. Chem. Eng.* **2022**, *10*, 8105–8114. [CrossRef]
24. Li, J.; Wang, L.; Zhao, Z.; Sun, B.; Zhan, G.; Liu, H.; Bian, Z.; Liu, Z. Highly efficient and air-stable Eu(II)-containing azacryptates ready for organic light-emitting diodes. *Nat. Commun.* **2020**, *11*, 5218–5225. [CrossRef] [PubMed]
25. Meyer, L.V.; Schönfeld, F.; Zurawski, A.; Mai, M.; Feldmann, C.; Müller-Buschbaum, K. A blue luminescent MOF as a rapid turn-off/turn-on detector for H_2O, O_2 and CH_2Cl_2, MeCN: $^3_\infty$[Ce(Im)$_3$ImH]·ImH. *Dalton Trans.* **2015**, *44*, 4070–4079. [CrossRef]
26. Matthes, P.R.; Müller-Buschbaum, K. Synthesis and characterization of the cerium(III) UV-emitting 2D-coordination polymer $^2_\infty$[Ce$_2$Cl$_6$(4,4′-bipyridine)$_4$]·py. *Z. Anorg. Allg. Chem.* **2014**, *640*, 2847–2851. [CrossRef]
27. Zhao, Z.; Wang, L.; Zhan, G.; Liu, Z.; Bian, Z.; Huang, C. Efficient rare earth cerium(III) complex with nanosecond d–f emission for blue organic light-emitting diodes. *Natl. Sci. Rev.* **2021**, *8*, nwaa193. [CrossRef] [PubMed]
28. Frey, S.T.; Horrocks Jr., W.D. Complexation, luminescence, and energy transfer of Ce^{3+} with a series of multidentate amino phosphonic acids in aqueous solution. *Inorg. Chem.* **1991**, *30*, 1073–1079. [CrossRef]
29. Wang, Q.; Wu, W.; Zhang, J.; Zhu, G.; Cong, R. Formation, photoluminescence and ferromagnetic characterization of Ce doped AlN hierarchical nanostructures. *J. Alloys Compd.* **2019**, *775*, 498–502. [CrossRef]
30. Vadan, M.; Popovici, E.J.; Ungur, L.; Vasilescu, M.; Macarovici, D. Synthesis of luminescent strontium-magnesium orthophosphate activated with cerium and manganese. In Proceedings of the SIOEL'99: Sixth Symposium on Optoelectronics, Bucharest, Romania, 22–24 September 1999; pp. 111–116.
31. Youssef, H.; Sedykh, A.E.; Becker, J.; Taydakov, I.V.; Müller-Buschbaum, K. 3-(2-pyridyl)pyrazole based luminescent 1D-coordination polymers and polymorphic complexes of various lanthanide chlorides including orange-emitting cerium(III). *Inorganics* **2022**, *10*, 254. [CrossRef]
32. Youssef, H.; Becker, J.; Sedykh, A.E.; Schäfer, T.; Taydakov, I.V.; Müller-Buschbaum, K. Red emitting cerium(III) and versatile luminescence chromaticity of 1D-coordination polymers and heterobimetallic Ln/AE pyridylpyrazolate complexes. *Z. Anorg. Allg. Chem.* **2022**, *648*, e202200295. [CrossRef]

33. Youssef, H.; Sedykh, A.E.; Becker, J.; Schäfer, T.; Taydakov, I.V.; Li, H.R.; Müller-Buschbaum, K. Variable luminescence and chromaticity of homoleptic frameworks of the lanthanides together with pyridylpyrazolates. *Chem. Eur. J.* **2021**, *27*, 16634–16641. [CrossRef]
34. Shannon, R.D. Revised effective ionic radii and systematic studies of interatomic distances in halides and chalcogenides. *Acta Crystallogr. Sect. A Cryst. Phys. Diffr. Theor. Gen. Crystallogr.* **1976**, *32*, 751–767. [CrossRef]
35. Guo, Z.; Blair, V.L.; Deacon, G.B.; Junk, P.C. Europium is different: Solvent and ligand effects on oxidation state outcomes and C-F activation in reactions between europium metal and pentafluorophenylsilver. *Chem. Eur. J.* **2022**, *28*, e202103865. [CrossRef] [PubMed]
36. Drew, M.G.B.; Guillaneux, D.; Hudson, M.J.; Iveson, P.B.; Russell, M.L.; Madic, C. Lanthanide(III) complexes of a highly efficient actinide(III) extracting agent—2,6-bis(5,6-dipropyl-1,2,4-triazin-3-yl)pyridine. *Inorg. Chem. Commun.* **2001**, *4*, 12–15. [CrossRef]
37. O'Keeffe, M.; Peskov, M.A.; Ramsden, S.J.; Yaghi, O.M. The reticular chemistry structure resource (RCSR) database of, and symbols for, crystal nets. *Acc. Chem. Res.* **2008**, *41*, 1782–1789. [CrossRef]
38. Wells, A.F. *Three-Dimensional Nets and Polyhedra*; Wiley-Interscience: New York, NY, USA, 1977.
39. Guo, Z.; Luu, J.; Blair, V.; Deacon, G.B.; Junk, P.C. Replacing mercury: Syntheses of lanthanoid pyrazolates from free lanthanoid metals, pentafluorophenylsilver, and pyrazoles, aided by a facile synthesis of polyfluoroarylsilver compounds. *Eur. J. Inorg. Chem.* **2019**, *2019*, 1018–1029. [CrossRef]
40. Jiménez, J.R.; Díaz-Ortega, I.F.; Ruiz, E.; Aravena, D.; Pope, S.J.A.; Colacio, E.; Herrera, J.M. Lanthanide tetrazolate complexes combining single-molecule magnet and luminescence properties: The effect of the replacement of tetrazolate N_3 by β-diketonate ligands on the anisotropy energy barrier. *Chem. Eur. J.* **2016**, *22*, 14548–14559. [CrossRef]
41. Hughes, I.D.; Däne, M.; Ernst, A.; Hergert, W.; Lüders, M.; Poulter, J.; Staunton, J.B.; Svane, A.; Szotek, Z.; Temmerman, W.M. Lanthanide contraction and magnetism in the heavy rare earth elements. *Nature* **2007**, *446*, 650–653. [CrossRef]
42. Carnall, W.T.; Fields, P.R.; Rajnak, K. Electronic energy levels of the trivalent lanthanide aquo ions. IV. Eu^{3+}. *J. Chem. Phys.* **1968**, *49*, 4450–4455. [CrossRef]
43. Carnall, W.T.; Fields, P.R.; Rajnak, K. Electronic energy levels of the trivalent lanthanide aquo ions. III. Tb^{3+}. *J. Chem. Phys.* **1968**, *49*, 4447–4449. [CrossRef]
44. Seidel, C.; Lorbeer, C.; Cybińska, J.; Mudring, A.-V.; Ruschewitz, U. Lanthanide coordination polymers with tetrafluoroterephthalate as a bridging ligand: Thermal and optical properties. *Inorg. Chem.* **2012**, *51*, 4679–4688. [CrossRef]
45. Huskowska, E.; Turowska-Tyrk, I.; Legendziewicz, J.; Riehl, J.P. The structure and spectroscopy of lanthanide(III) complexes with 2,2′-bipyridine-1,1′-dioxide in solution and in the solid state: Effects of ionic size and solvent on photophysics, ligand structure and coordination. *New J. Chem.* **2002**, *26*, 1461–1467. [CrossRef]
46. Fernández-Fernández, M.d.C.; Bastida, R.; Macías, A.; Pérez-Lourido, P.; Platas-Iglesias, C.; Valencia, L. Lanthanide(III) complexes with a tetrapyridine pendant-armed macrocyclic ligand: ^1H NMR structural determination in solution, X-ray diffraction, and density-functional theory calculations. *Inorg. Chem.* **2006**, *45*, 4484–4496. [CrossRef] [PubMed]
47. Bochkarev, M.N.; Khoroshenkov, G.V.; Schumann, H.; Dechert, S. A novel bis(imino)amine ligand as a result of acetonitrile coupling with the diiodides of Dy(II) and Tm(II). *J. Am. Chem. Soc.* **2003**, *125*, 2894–2895. [CrossRef]
48. Sedykh, A.E.; Kurth, D.G.; Müller-Buschbaum, K. Two series of lanthanide coordination polymers and complexes with 4′-phenylterpyridine and their luminescence properties. *Eur. J. Inorg. Chem.* **2019**, *2019*, 4564–4571. [CrossRef]
49. Matthes, P.R.; Eyley, J.; Klein, J.H.; Kuzmanoski, A.; Lambert, C.; Feldmann, C.; Müller-Buschbaum, K. Photoluminescent one-dimensional coordination polymers from suitable pyridine antenna and $LnCl_3$ for visible and near-IR emission. *Eur. J. Inorg. Chem.* **2015**, *2015*, 826–836. [CrossRef]
50. Bünzli, J.-C.G.; Comby, S.; Chauvin, A.-S.; Vandevyver, C.D.B. New Opportunities for Lanthanide Luminescence. *J. Rare Earths* **2007**, *25*, 257–274. [CrossRef]
51. Dexter, D.L. A theory of sensitized luminescence in solids. *J. Chem. Phys.* **1953**, *21*, 836–850. [CrossRef]
52. Dang, S.; Yu, J.; Wang, X.; Sun, L.; Deng, R.; Feng, J.; Fan, W.; Zhang, H. NIR-luminescence from ternary lanthanide [Ho^{III}, Pr^{III} and Tm^{III}] complexes with 1-(2-naphthyl)-4,4,4-trifluoro-1,3-butanedionate. *J. Lumin.* **2011**, *131*, 1857–1863. [CrossRef]
53. Kahn, O. *Molecular Magnetism*; Wiley-VCH: Weinheim, Germany, 1993.
54. Pham, Y.H.; Trush, V.A.; Carneiro Neto, A.N.; Korabik, M.; Sokolnicki, J.; Weselski, M.; Malta, O.L.; Amirkhanov, V.M.; Gawryszewska, P. Lanthanide complexes with N-phosphorylated carboxamide as UV converters with excellent emission quantum yield and single-ion magnet behavior. *J. Mater. Chem. C* **2020**, *8*, 9993–10009. [CrossRef]
55. Benelli, C.; Gatteschi, D. Magnetism of lanthanides in molecular materials with transition-metal ions and organic radicals. *Chem. Rev.* **2002**, *102*, 2369–2388. [CrossRef] [PubMed]
56. Del Giudice, M.R.; Mustazza, C.; Borioni, A.; Gatta, F.; Tayebati, K.; Amenta, F.; Tucci, P.; Pieretti, S. Synthesis of 1-methyl-5-(pyrazol-3-and-5-yl)- and 1,2,4-triazol-3- and 5-yl)-1,2,3,6-tetrahydropyridine derivatives and their evaluation as muscarinic receptor ligands. *Arch. Pharm. Pharm. Med. Chem.* **2003**, *336*, 143–154. [CrossRef] [PubMed]
57. Bauer, V.J.; Dalalian, H.P.; Fanshawe, W.J.; Safir, S.R.; Tocus, E.C.; Boshart, C.R. 4-[3(5)-Pyrazolyl]pyridinium salts. A new class of hypoglycemic agents. *J. Med. Chem.* **1968**, *11*, 981–984. [CrossRef] [PubMed]
58. Sheldrick, G.M. SHELXT—Integrated space-group and crystal-structure determination. *Acta Crystallogr. Sect. A Found. Crystallogr.* **2015**, *71*, 3–8. [CrossRef] [PubMed]
59. Sheldrick, G.M. Crystal structure refinement with SHELXL. *Acta Crystallogr. Sect. C Struct. Chem.* **2015**, *71*, 3–8. [CrossRef]

60. Spek, A.L. PLATON SQUEEZE: A tool for the calculation of the disordered solvent contribution to the calculated structure factors. *Acta Crystallogr. Sect. C Struct. Chem.* **2015**, *71*, 9–18. [CrossRef]
61. Spek, A.L. Structure validation in chemical crystallography. *Acta Crystallogr. Sect. D Biol. Crystallogr.* **2009**, *65*, 148–155. [CrossRef]
62. Spek, A.L. What makes a crystal structure report valid? *Inorg. Chim. Acta* **2018**, *470*, 232–237. [CrossRef]
63. Spek, A.L. CheckCIF validation ALERTS: What they mean and how to respond. *Acta Crystallogr. Sect. E Crystallogr. Commun.* **2020**, *76*, 1–11. [CrossRef]
64. Spek, A.L. Single-crystal structure validation with the program PLATON. *J. Appl. Crystallogr.* **2003**, *36*, 7–13. [CrossRef]
65. Blatov, V.A.; Shevchenko, A.P.; Proserpio, D.M. Applied topological analysis of crystal structures with the program package topospro. *Cryst. Growth Des.* **2014**, *14*, 3576–3586. [CrossRef]
66. Pennington, W.T. DIAMOND—Visual crystal structure information system. *J. Appl. Crystallogr.* **1999**, *32*, 1028–1029. [CrossRef]
67. Coelho, A.A. TOPAS and TOPAS-Academic: An optimization program integrating computer algebra and crystallographic objects written in C++. *J. Appl. Crystallogr.* **2018**, *51*, 210–218. [CrossRef]

Disclaimer/Publisher's Note: The statements, opinions and data contained in all publications are solely those of the individual author(s) and contributor(s) and not of MDPI and/or the editor(s). MDPI and/or the editor(s) disclaim responsibility for any injury to people or property resulting from any ideas, methods, instructions or products referred to in the content.

Article

Synthesis, Structures and Photophysical Properties of Tetra- and Hexanuclear Zinc Complexes Supported by Tridentate Schiff Base Ligands

Tobias Severin [1], Viktoriia Karabtsova [1], Martin Börner [1], Hendrik Weiske [2], Agnieszka Kuc [3] and Berthold Kersting [1],*

[1] Institut für Anorganische Chemie, Universität Leipzig, Johannisallee 29, 04103 Leipzig, Germany; tobias.severin@uni-leipzig.de (T.S.); karabtsovavictoria@gmail.com (V.K.); martin.boerner@uni-leipzig.de (M.B.)

[2] Willhelm-Ostwald-Institut für Physikalische und Theoretische Chemie, Universität Leipzig, Linnéstraße 2, 04103 Leipzig, Germany; hendrik.weiske@uni-leipzig.de

[3] Helmholtz-Zentrum Dresden-Rossendorf, Abteilung Ressourcenökologie, Forschungsstelle Leipzig, Permoserstr. 15, 04318 Leipzig, Germany; a.kuc@hzdr.de

* Correspondence: b.kersting@uni-leipzig.de; Fax: +49-(0)-341-97-36199

Abstract: The synthesis, structure and photophysical properties of two polynuclear zinc complexes, namely $[Zn_6L_2(\mu_3\text{-}OH)_2(OAc)_8]$ (**1**) and $[Zn_4L_4(\mu_2\text{-}OH)_2](ClO_4)_2$ (**2**), supported by tridentate Schiff base ligand 2,6-bis((N-benzyl)iminomethyl)-4-*tert*-butylphenol (**HL**) are presented. The synthesized compounds were investigated using ESI-MS, IR, NMR, UV-vis absorption spectroscopy, photoluminescence spectroscopy and single-crystal X-ray crystallography. The hexanuclear neutral complex **1** comprises six-, five- and four-coordinated Zn^{2+} ions coordinated by O and N atoms from the supporting ligand and OH- and acetate ligands. The Zn^{2+} ions in complex cation $[Zn_4L_4(\mu_2\text{-}OH)_2]^{2+}$ of **2** are all five-coordinated. The complexation of ligand **HL** by Zn^{2+} ions leads to a six-fold increase in the intensity and a large blue shift of the ligand-based $^1(\pi\text{-}\pi)^*$ emission. Other biologically relevant ions, i.e., Na^+, K^+, Mg^{2+}, Ca^{2+}, Mn^{2+}, Fe^{2+}, Co^{2+}, Ni^{2+} and Cu^{2+}, did not give rise to a fluorescence enhancement.

Keywords: salicylaldiminato ligands; zinc; coordination geometry; photophysical properties

Citation: Severin, T.; Karabtsova, V.; Börner, M.; Weiske, H.; Kuc, A.; Kersting, B. Synthesis, Structures and Photophysical Properties of Tetra- and Hexanuclear Zinc Complexes Supported by Tridentate Schiff Base Ligands. *Chemistry* **2023**, *5*, 1028–1045. https://doi.org/10.3390/chemistry5020070

Academic Editor: Roland C. Fischer

Received: 3 April 2023
Revised: 24 April 2023
Accepted: 28 April 2023
Published: 2 May 2023

Copyright: © 2023 by the authors. Licensee MDPI, Basel, Switzerland. This article is an open access article distributed under the terms and conditions of the Creative Commons Attribution (CC BY) license (https:// creativecommons.org/licenses/by/ 4.0/).

1. Introduction

Zinc as an essential trace metal for all living organisms is omnipresent in biogenic systems, such as in the cells of the human body. Meanwhile, it is well-known that Zn^{2+} greatly contributes to biological processes, for example, as part of the immune system, as an active site or co-factor in enzymes or as a regulator for proteins, and it can also provide a stabilizing function to their structure [1–10]. The Zn^{2+} ions can, however, also have toxic effects when toothier concentrations are too high, and it is presumably involved in the biochemical mechanisms of neuronal diseases [4,11]. Proper detection of free zinc, also in living cells, is highly desirable, but it still remains a challenge for common spectroscopic methods, with which free zinc is not obtainable. As a d^{10} metal, Zn^{2+} cannot be observed using UV-vis spectroscopy, but it is particularly suitable for fluorescence spectroscopy. For this technique, the CHEF effect (*chelation-enhanced fluorescence effect*) is exploited for a strong augmentation in the fluorescence intensity of the sensor molecule. Usually, the PET (*photoinduced electron transfer*) quenches the luminescence of some chemosensors, and it is turned off by the complexation of the closed-shell d^{10} metal ion [12–23]. This combines two important factors: the high sensitivity of fluorescence spectroscopy and the excellent selectivity for Zn^{2+}. A set of molecules, which provide these abilities, were already designed and explored: for example, 8-aminoquinoline-based sensor molecules such as 6-methoxy-(8-*p*-toluenesulfonamide)quinoline (TSQ), Zinquin (ZQ) and 2-(hydroxymethyl)-4-methyl-6-((quinolinyl-8-imino)methyl)phenol (HMQP) [24–30]. However, what is often left out from consideration is the influence of different anions that may disturb the mechanism of complexation with zinc and the sensitivity of the method.

Recently, we have reported the synthesis, structures and properties of some discrete four- and five-coordinated Zn^{2+} complexes supported by salicylaldiminato ligands [31–35]. As an extension, here, we report the synthesis of two zinc complexes derived from new Schiff base ligand **HL** (Scheme 1) bearing two imine functions and a phenolic oxygen as a N,O,N-donor set. The effect of the metal–ligand ratio variations on the coordination and luminescence properties is reported. The results are supported by accompanying DFT calculations of UV/vis spectra.

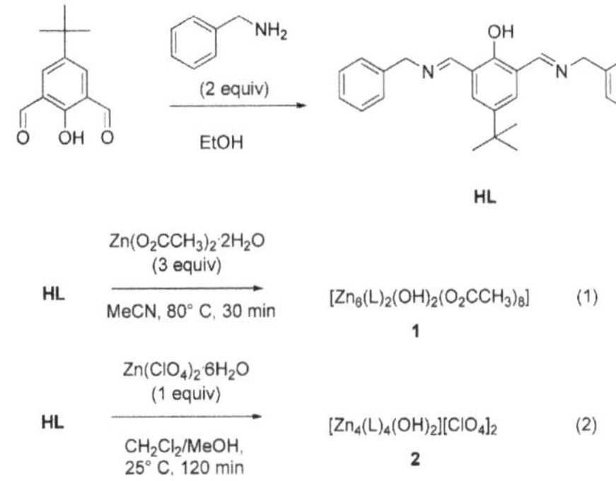

Scheme 1. Synthesis of ligand **HL** and zinc complexes **1** and **2**.

2. Materials and Methods

All reagents were purchased and used without any further purification. 4-*tert*-butyl-2,6-diformylphenol was purchased from Sigma Aldrich. Solvents were of HPLC grade and not additionally purified. UV/vis spectra were recorded with a V-670 spectrophotometer from JASCO and analyzed with SPECTRA MANAGER v2.05.03 software. HELLMA 110-QS quartz cells with 10 mm path length were used as cuvettes. Measurements were made within the range of 190–650 nm. Fluorescence spectra were recorded on a Perkin Elmer FL6500 spectrometer using a constant slit width. Data acquisition was performed using the FL WINLAB V3.00 program from PERKINELMER [36]. Measurements were performed using precision SUPRASIL type 111-10-40-QS fused silica cells (10 mm cell diameter, 3500 µL volume) from HELLMA ANALYTICS. The infrared spectra were recorded with a Bruker Vertex 80V FTIR spectrometer utilizing KBr pellets. The ORIGINPRO 8G program from ORIGINLAB CORPORATION was used for data analysis and a graphical display of the spectra [37]. ^{1}H and ^{13}C NMR spectra measurements were recorded on BRUKER model DPX-400 (^{1}H: 400 MHz; ^{13}C: 100 MHz) or VARIAN Gemini 300 instruments (^{1}H: 300 MHz; ^{13}C: 75 MHz). The solvent signal served as the internal standard. The MestReNova v11.0 program from MESTRELAB RESEARCH S. L. was utilized for data analyses and graphical displays [38]. ESI mass spectra were recorded using MICROTOF or IMPACT II mass spectrometers from BRUKER DALTONIK GMBH. Elemental analyses were measured using an ELEMENTAR VARIO EL instrument from ELEMENTAR ANALYSESYSTEME GMBH.

X-ray crystallography. Single-crystal X-ray diffraction experiments were carried out on a STOE STADIVARI, equipped with an X-ray micro-source (Cu-Kα, λ = 1.54186 Å) and a DECTRIS Pilatus 300K detector at 180(2) K. The diffraction frames were processed with the STOE X-RED software package [39]. The structures were solved by direct methods [40] and refined by full-matrix least-squares techniques on the basis of all data against F^2 using SHELXL-2018/3 [41]. PLATON was used to search for higher symmetry [42]. All non-hydrogen atoms were refined anisotropically. H atoms were placed in calculated positions and allowed to ride on their respective C atoms and treated isotropically using the 1.2- or 1.5-fold U_{iso} value of the parent C atoms. All non-hydrogen atoms were refined anisotropically. All calculations were performed using the Olex2 crystallographic platform [43]. ORTEP-3 and POV-RAY were used for the artwork of the structures [44]. CCDC 2248758-2248760 contains the supplementary crystallographic data for this paper.

Crystallographic data for 2,6-(BzN=CH)-4-*t*Bu-C$_6$H$_2$OH. C$_{26}$H$_{28}$N$_2$O, M_r = 384.50 g/mol, triclinic, space group P-1, a = 8.3627(2) Å, b = 10.9002(3) Å, c = 11.8113(4) Å, α = 100.761(2)°, β = 94.116(2)°, γ = 91.320(2)°, V = 1054.27(5) Å^3, Z = 2, ρ_{calcd} = 1.211 g/cm^3, T = 180(2) K, μ(Cu-K$_\alpha$), λ = 1.54186 Å, crystal size 0.326 × 0.307 × 0.256 mm^3, 19,262 reflections measured, 3908 unique, 3518 with $I > 2\sigma(I)$. Final R_1 = 0.0372 ($I > 2\sigma(I)$), wR_2 = 0.1057 (3908 refl.), 267 parameters and 0 restraints, min./max. residual electron density = $-0.181/0.254$ e$^-$/Å^3.

Crystallographic data for **1**·3MeCN·0.5H$_2$O. C$_{74}$H$_{90}$N$_7$O$_{20.5}$Zn$_6$, M_r = 1806.88 g/mol, triclinic, space group P-1, a = 9.9260(3) Å, b = 15.0044(5) Å, c = 28.7383(7) Å, α = 98.943(2)°, β = 92.514(2)°, γ = 104.406(2)°, V = 4079.9(2) Å^3, Z = 2, ρ_{calcd} = 1.471 g/cm^3, T = 180(2) K, μ(Cu-K$_\alpha$), λ = 1.54186 Å, crystal size 0.160 × 0.090 × 0.050 mm^3, 70,648 reflections measured, 14,641 unique, 7769 with $I > 2\sigma(I)$. Final R_1 = 0.0577 ($I > 2\sigma(I)$), wR_2 = 0.1521 (14641 refl.), 1140 parameters and 402 restraints, min./max. residual electron density = $-0.609/1.000$ e$^-$/Å^3. The *tert*-butyl groups are disordered over two positions (0.66/0.34 and 0.53/0.47) and restrained with SADI, RIGU, DELU, and SIMU. One benzyl group of each ligand is disordered over two positions (0.51/0.49 and 0.72/0.28), restrained with SADI, SIMU, DELU, and RIGU. AFIX 66 was used to fix both benzene rings. One methylene group was constrained with EADP. One MeCN is disordered over two positions (0.74/0.26), restrained with DFIX (1.137 and 2.593), SIMU, and RIGU. The nitrogen atom was constrained with EADP. The solvent water molecule is disordered over a special position and was modeled with PART-1. Crystals of **1**·3MeCN·0.5H$_2$O grown by recrystallization of **1**·4H$_2$O from MeCN solution quickly lost solvate molecules upon standing in air and became turbid.

Crystallographic data for **2**·3MeOH·0.5MeCN. C$_{108}$H$_{123.5}$Cl$_2$N$_{8.5}$O$_{17}$Zn$_4$, M_r = 2145.03 g/mol, monoclinic, space group P2$_1$/n, a = 17.9667(4) Å, b = 30.1926(5) Å, c = 19.4877(4) Å, α = 90°, β = 92.875(2)°, γ = 90°, V = 10,558.0(4) Å^3, Z = 4, ρ_{calcd} = 1.349 g/cm^3, T = 180(2) K, μ(Cu-K$_\alpha$), λ = 1.54186 Å, crystal size 0.186 × 0.154 × 0.121 mm^3, 101,721 reflections measured, 19,944 unique, 10,877 with $I > 2\sigma(I)$. Final R_1 = 0.0477 ($I > 2\sigma(I)$), wR_2 = 0.1328 (19,944 refl.), 1403 parameters and 492 restraints, min./max. residual electron density = $-0.572/0.767$ e$^-$/Å^3. One *tert*-butyl group is disordered over two positions (0.72/0.28), restrained with DFIX (1.54), SADI, SIMU, and RIGU. One benzyl group is disordered over two positions (0.72/0.28), restrained with SIMU and RIGU. AFIX 66 was used to fix the benzene rings, and the methylene group was constrained with EADP. The perchlorate ion is disordered over two positions (0.59/0.41) and restrained with SADI, SIMU, and RIGU. Crystals of **2**·3MeOH·0.5MeCN grown by the recrystallization of **2**·H$_2$O from the MeCN/MeOH solution quickly lost solvate molecules upon standing in air and became turbid.

Synthesis of 2,6-bis-((N-benzyl)iminomethyl)-4-*tert*-butylphenol (**HL**). Benzylamine (546 mg, 5.09 mmol, 2.10 eq.) was added to a solution of 5-*tert*-butyl-2-hydroxybenzene-1,3-dialdehyde (500 mg, 2.42 mmol, 1.00 eq.) in EtOH (20 mL). The resulting yellow solution was stirred for 2 h at ambient temperature to produce a yellow solid, which was isolated by filtration, washed with little cold EtOH and dried at 60 °C. The compound was recrystallized once from EtOH to produce a pure product: yellow X-ray quality crystals. Yield: 915 mg (98.2%). M.p. 122 °C. ^{1}H-NMR (400 MHz, CD$_2$Cl$_2$): δ = 1.32 (s, 9H, CH$_3$-^{t}Bu), 4.81 (s, 4H, CH$_2$), 7.24–7.39 (m, 12H, CH^{ar}), 8.70 (s, 2H, N=CH), 13.91 (s, 1H, OH^{ar}). ^{13}C{^{1}H}-NMR (100 MHz, DMSO, 90 °C): δ = 30.7 (CH$_3$), 33.3 (C(CH$_3$)$_3$), 62.4 (CH$_2$), 120.5 (*o*-C^{ar}), 126.5 (*p*-Bz-C^{ar}), 127.4 (*o*-Bz-C^{ar}), 127.9 (*m*-Bz-C^{ar}), 128.4 (*m*-C^{ar}), 138.7 (*ipso*-Bz-C^{ar}), 140.0 (mboxemphp-C^{ar}-^{t}Bu), 158.8 (*ipso*-C^{ar}-OH), 161.4 (HC=N). FT-IR (KBr): $\tilde{\nu}$/cm^{-1} = 3083/3059/3028 (w, νC-H^{ar}),

2954/2904/2837 (s-w, νC-H), 1633 (s, νC=N), 1598 (w, νC=C), 1472 (s, $\delta_{asym.}$ -CH$_3$, ^{t}Bu), 1439 (s, νC^{ar}-O), 1388/1372 (w, $\delta_{sym.}$ -CH$_3$, ^{t}Bu), 1287 (m), 1260 (m), 1225 (m), 1203 (w), 1118 (w), 1062 (m), 1027 (w), 942 (w), 882 (w), 844 (w), 823 (w), 750 (s), 718 (w), 699 (s). m/z (ESI+, MeCN): C$_{26}$H$_{29}$N$_2$O (384.23) [M + H$^+$]$^+$ calcd: 385.24; found 385.23. UV-vis (DCM/MeOH (3:2/v:v); 5.0 × 10^{-5} M): λ_{max}/nm (ε/M^{-1}cm^{-1}) = 248 (28,090), 348 (5480), 451 (5610); (MeCN; 2.5 × 10^{-5} M): λ_{max}/nm (ε/M^{-1}cm^{-1}) = 193 (56,650), 244 (43,210), 347 (7950), [444 (500)]. Elemental analysis for C$_{26}$H$_{28}$N$_2$O (384.51) calc. C 81.21, H 7.34, N 7.29%; found. C 80.92, H 7.36, N 7.23%.

Synthesis [Zn$_6$(L)$_2$(OH)$_2$(OAc)$_8$] (1). A solution of Zn(OAc)$_2$·2H$_2$O (190 mg, 866 mmol, 3.00 eq.) in 3 mL of MeCN was added to a solution of **HL** (111 mg, 289 mmol, 1.00 eq.) in MeCN (20 mL). The yellow solution intensified its color and was stirred for 30 min at 80 °C. The solution was left to stand for 2 weeks. The resulting solid was isolated by filtration and dried in air. Yield: 420 mg (86.9%). Elemental analysis for [Zn$_6$L$_2$(OH)$_2$(OAc)$_8$]·4H$_2$O (C$_{68}$H$_{88}$N$_4$O$_{24}$Zn$_6$) (1737.8) calc. C 47.00, H 5.10, N 3.22%; found. C 46.97, H 4.62, N 3.10%. This so isolated powder (1.4H$_2$O) was used for all spectroscopic measurements. ^{1}H-NMR (300 MHz, CD$_2$Cl$_2$): δ = 1.25/1.29 (s, 18H, CH$_3$-^{t}Bu), 1.73 (s, 24H, CH$_3$-OAc), 4.84/5.08 (s, 8H, CH$_2$), 7.11–7.42 (m, 20 H, Bz-CH^{ar}) 7.47–7.54, (m, 4H, m-CH^{ar}), 8.31/8.36 (s, 4H, N=CH) ppm. FT-IR (KBr pellet): $\tilde{\nu}$/cm^{-1} = 3029 (w, νC-H, ar), 2961 (w, νC-H), 1604/1580 (s, antisymm. νC-O, COO$^-$), 1415 (s, symm. νC-O, COO$^-$), 1344 (w), 1233 (w, νC^{ar}-O), 1066 (w), 1026 (w), 843 (w), 823 (w), 776 (w), 758 (w), 698 (w), 670 (w). m/z (ESI+, MeCN): For [Zn$_2$L$_2$OAc]$^+$ as C$_{54}$H$_{57}$N$_4$O$_4$Zn$_2^+$ (956.83) [M]$^+$ calcd: 957.29; found 957.28. UV-vis (DCM/MeOH (3:2 ratio); 2.5 × 10^{-5} M): λ_{max}/nm (ε/M^{-1}cm^{-1}) = [221 (16,160)], 260 (30,700), 387 (10,060).

Synthesis of [Zn$_4$L$_4$(OH)$_2$](ClO$_4$)$_2$ (2). A solution of Zn(ClO$_4$)$_2$·6H$_2$O (107 mg, 286 mmol, 1.10 eq.) in a few drops of acetonitrile was added to a solution of **HL** (100 mg, 260 mmol, 1.00 eq.) in 20 mL of DCM/MeOH (3:2). The yellow solution intensified its colour and was stirred for 2 h at 40 °C. The solution was left to stand for 2 weeks. The resulting solid was isolated by filtration and dried in air. Yield: 456 mg (86.4%). Elemental analysis for [Zn$_4$L$_4$(OH)$_2$](ClO$_4$)$_2$·H$_2$O (2046.50) calc. C 61.04, H 5.52, N 5.48%; found. C 61.15, H 5.81, N 5.61%. This isolated powder (2·H$_2$O) was used for all spectroscopic measurements. ^{1}H-NMR (300 MHz, CD$_2$Cl$_2$): δ = 1.39 (s, 36H, CH$_3$-^{t}Bu), 3.55/3.81/4.36 (m, 16H, CH$_2$), 6.38–7.46 (m, 48 H, CH^{ar}), 7.83/8.13 (s, 8H, N=CH) ppm. FTIR (KBr): $\tilde{\nu}$/cm^{-1} = 3063/3031 (w, νC-H, ar), 2958/2921 (m, νC-H), 1653/1631 (s, νC=N), 1542 (s, νC=C), 1454 (m, $\delta_{antisymm}$ -CH$_3$), 1398 (w), 1363 (w), 1233 (m, νC^{ar}-O), 1090 (s, νCl-O, perchlorate), 928 (w), 878 (w), 840 (w), 759 (m), 700 (m), 625 (s, δCl-O, perchlorate). m/z (ESI+, DCM): For [Zn$_2$L$_2$ClO$_4$]$^+$ as C$_{52}$H$_{54}$ClN$_4$O$_6$Zn$_2^+$ (997.24) [M]$^+$ calcd: 997.23; found 997.22. UV-vis (DCM/MeOH (3:2 ratio); 2.5 × 10^{-5} M): λ_{max}/nm (ε/M^{-1}cm^{-1}) = [228 (18,650)], 257 (27,900), 387 (8800).

Time-dependent density functional theory. All structures were optimized using PBE/TZ2P-D3(BJ) [45–47], and the 20 lowest singlet excitations were calculated using TD-DFT [48] at the CAMY-B3LYP [49–51] level of the theory, as implemented in AMS [52–54]. Implicit solvation model COSMO [55–57] was included in all calculations.

3. Results

3.1. Synthesis and Characterization of Compounds

The new di-imine ligand **HL** was prepared according to the route provided in Scheme 1. A mixture of 5-*tert*-butyl-2-hydroxybenzene-1,3-dialdehyde and two equivalents of benzylamine were allowed to react in a 1:2 molar ratio at room temperature for 2 h in EtOH in order to provide an intense yellow solution, from which the pure product could be isolated as a yellow microcrystalline solid in an almost quantitative yield.

Two new zinc(II) complexes, **1** and **2**, were prepared according to the reactions in Scheme 1. The reaction of the ligand **HL** with Zn(OAc)$_2$·2H$_2$O in a 3:1 ratio in acetonitrile at 80 °C for 30 min produced a pale yellow solution, from which a pale yellow, microcrystalline powder of composition [(Zn$_6$(L)$_2$(OH)$_2$(OAc)$_8$]·4H$_2$O (**1**) was reproducibly obtained in ca 87% yield. The treatment of **HL** with Zn(ClO$_4$)$_2$·6H$_2$O in a 1:1 ratio in a CH$_2$Cl$_2$/MeOH/MeCN mixed solvent system furnished a yellow solution, from which a pale yellow solid of composition [Zn$_4$L$_4$(OH)$_2$](ClO$_4$)$_2$·H$_2$O (**2**) precipitated in an 86% yield. Both complexes exhibit good solubility in acetonitrile, DMSO and dichloromethane, but they are only sparingly soluble in alcohols and various hydrocarbons and virtually insoluble in water. The synthesized compounds resulted in a satisfactory elemental analyses, and their formulation was confirmed by electrospray ionization mass spectrometry (ESI-MS), FT-IR-spectroscopy, UV-vis and NMR spectroscopy as well as single-crystal X-ray diffractometric analyses.

The infrared spectrum of the free ligand shows a prominent band at 1634 cm^{-1}, which can be attributed to the C=N stretching frequencies of the imine groups (Figure S1). The O-H stretching appears as a very broad band with a maximum at around 2600 cm^{-1}, indicating the presence of intramolecular hydrogen bonding interactions between the OH and imine groups. The IR spectrum of zinc complex **1** reveals two bands at 1580 and 1415 cm^{-1} attributed to the antisymmetric and symmetric stretching vibrations of the acetato ligand (Figure S2). The imine group gives rise to two weak features at 1650 and 1630 cm^{-1}. The IR spectrum of perchlorate salt **2** (Figure S2) reveals two strong absorption bands at 1090 and 625 cm^{-1}, characteristic values for the vibrations of a ClO$_4^-$ counter anion. Again, as in **1**, there are two bands at 1653 and 1631 cm^{-1} attributable to the stretching vibrations of a coordinated imine group.

3.2. Crystallographic Characterization

Single crystals of **HL** obtained from EtOH are triclinic and fall into space group $P\bar{1}$. The structure shown in Figure 1 reveals that both imino groups are (*E*)-configured and coplanar with the phenol ring. The phenolic OH group forms an intramolecular hydrogen bond (O1-H1-N1) with the imine group as observed in other salicylaldimines. Figure S5 displays the packing diagram of **HL**. As observed, the phenol rings of two adjacent molecules are engaged in *face-to-face* π–π stacking interactions, as manifested by an interplanar distance of 3.4286(11) Å. The DFT-optimized geometry of the **HL** ligand at the PBE/TZ2P-D3(BJ) level of theory is in agreement with the experimental data (Figure S16).

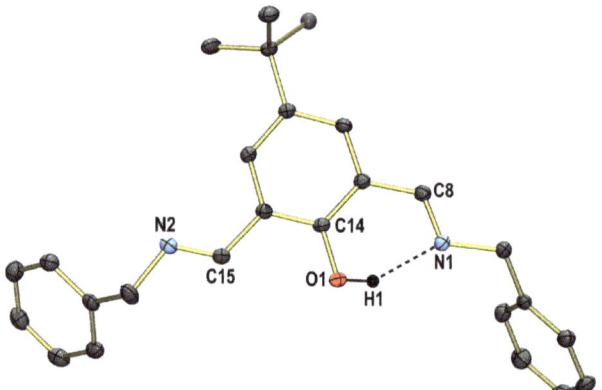

Figure 1. Molecular structure of the ligand **HL** in the crystal. Thermal ellipsoids are drawn at the 50% probability level. Hydrogen atoms are omitted for clarity, except one involved in intramolecular hydrogen bonding. Selected bond lengths (Å): N1-C8 1.2765(11), N2-C15 1.2677(16) and O1···N1 2.5905(12).

Crystals of [Zn$_6$L$_2$(OH)$_2$(OAc)$_8$]·3MeCN·0.5H$_2$O were grown by the slow evaporation of an acetonitrile solution. The crystal structure comprises a discrete, hexanuclear, mixed-ligand complex, [Zn$_6$L$_2$(OH)$_2$(OAc)$_8$] (**1**), and several co-crystallized MeCN and H$_2$O molecules. Figure 2 shows the structure of the neutral zinc complex. The zinc complex exhibits idealized C_2 symmetry (neglecting the four benzyl groups of the supporting ligand). Two nearly isostructural [Zn$_3$L(OH)(OAc)$_2$]$^{2+}$ moieties are connected by four $\mu_{1,3}$-briding acetate groups.

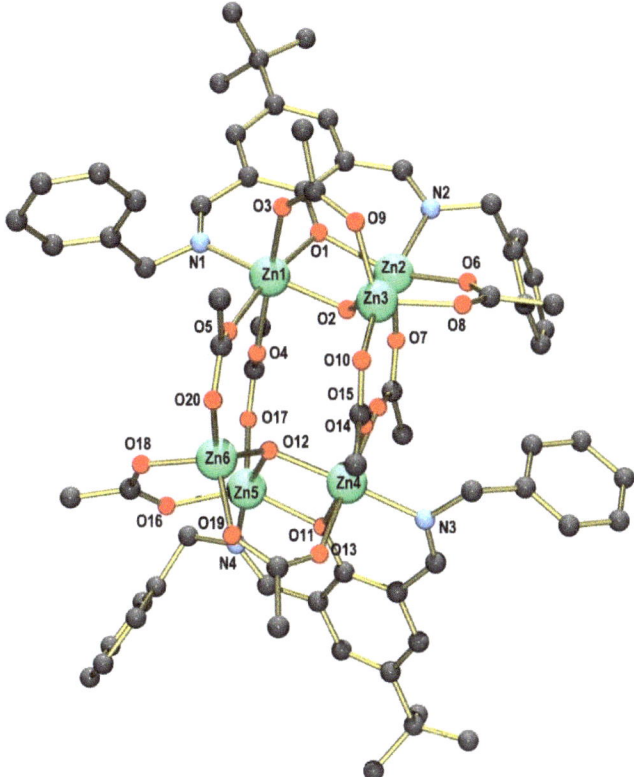

Figure 2. Ball and stick representation of the molecular structure of neutral complex **1** in crystals of [Zn$_6$L$_2$(OH)$_2$(OAc)$_8$]·3MeCN·0.5H$_2$O (**1**·3MeCN·0.5H$_2$O). Hydrogen atoms are omitted for clarity.

The three Zn^{2+} ions within the trinuclear subunits exhibit different coordination environments: Zn1 is six-coordinated by the imine N and phenolate O donors from the supporting ligand, three O atoms from three $\mu_{1,3}$-bridging acetato ligands and one triply bridging OH group in a distorted octahedral fashion. The Zn2 atom, on the other hand, is five-coordinated, being surrounded by imine N and phenolate O donors, two O atoms from two $\mu_{1,3}$-bridging acetates and the OH bridge. The coordination geometry of Zn2 is severely distorted and lies between the ideal trigonal bipyramidal or square pyramidal environments, as suggested by the τ value of 0.49 [58], or the SHAPE symmetry factors (Table S1) [59]. Finally, one Zn atom is coordinated in a distorted tetrahedral geometry by three O atoms from three $\mu_{1,3}$-bridging acetate groups and the triply bridging OH group. A selection of bond lengths and angles is given in Table 1. The bond lengths and angles show no unusual features and compare well with those in other six-, five- and four-coordinated Zn complexes. The corresponding bond lengths and angles within the two trinuclear subunits do not differ much. The angles around Zn1, for example, range from 80.33(16)°

to 98.8(2)° and from 78.95(16)° to 98.98(19)° for the corresponding Zn4 atom in the other [Zn$_3$L(OH)(OAc)$_2$]$^{2+}$ fragment. The structure of **1** is very similar to that of the hexanuclear complex [Zn$_6$L′$_2$(OH)$_2$(OAc)$_8$] supported by the ligand 2,6-*bis*((*N*-benzyl)iminomethyl)-4-methylphenol. This complex is also composed of trinuclear [Zn$_3$L(OH)(OAc)$_2$]$^{2+}$ moieties [11]. Other polynuclear zinc carboxylato complexes are known; basic zinc acetate, Zn$_4$O(CH$_3$COO)$_6$, is a prominent example [60].

Table 1. Selected bond lengths (Å) and angles (°) for **1**·3MeCN·0.5H$_2$O.

Zn1-O1 2.114(4)	N1-Zn1-O1 87.90(19)	N3-Zn4-O11 88.40(18)
Zn1-N1 2.064(5)	N1-Zn1-O3 97.27(17)	N3-Zn4-O13 90.8(2)
Zn1-O2 2.016(4)	N1-Zn1-O4 86.15(16)	N3-Zn4-O14 88.41(19)
Zn1-O3 2.103(4)	N1-Zn1-O5 98.8(2)	N3-Zn4-O15 98.98(19)
Zn1-O4 2.264(4)	O1-Zn1-O4 94.19(16)	O11-Zn4-O14 93.16(16)
Zn1-O5 2.101(4)	O2-Zn1-N1 163.88(19)	O12-Zn4-N3 166.20(19)
Zn2-O1 2.063(4)	O2-Zn1-O1 78.83(16)	O12-Zn4-O11 78.95(16)
Zn2-N2 2.091(5)	O2-Zn1-O3 92.02(16)	O12-Zn4-O13 94.78(18)
Zn2-O2 2.065(4)	O2-Zn1-O4 85.74(15)	O12-Zn4-O14 86.82(17)
Zn2-O6 2.060(4)	O2-Zn1-O5 93.55(17)	O12-Zn4-O15 93.07(17)
Zn2-O7 1.996(4)	O3-Zn1-O1 90.56(17)	O13-Zn4-O11 90.37(18)
Zn3-O2 1.939(4)	O3-Zn1-O4 174.26(17)	O13-Zn4-O14 176.35(18)
Zn3-O8 1.976(4)	O5-Zn1-O1 170.98(16)	O15-Zn4-O11 170.48(16)
Zn3-O9 1.957(4)	O5-Zn1-O3 94.56(17)	O15-Zn4-O13 95.50(18)
Zn3-O10 1.969(4)	O5-Zn1-O4 80.33	O15-Zn4-O14 81.13(16)
Zn4-O11 2.113(4)	O1-Zn2-N2 88.07(19)	N4-Zn5-O11 87.19(19)
Zn4-N3 2.034(5)	O1-Zn2-O2 78.89(16)	O12-Zn5-N4 164.02(19)
Zn4-O12 2.012(4)	O2-Zn2-N2 156.16(18)	O12-Zn5-O11 79.39(16)
Zn4-O13 2.107(4)	O6-Zn2-N2 89.1(2)	O12-Zn5-O16 95.29(19)
Zn4-O14 2.234(4)	O6-Zn2-O1 148.89(19)	O16-Zn5-N4 88.1(2)
Zn4-O15 2.096(4)	O6-Zn2-O2 91.90(18)	O16-Zn5-O11 135.03(19)
Zn5-O12 2.045(4)	O7-Zn2-N2 105.0(2)	O17-Zn5-N4 96.70(19)
Zn5-N4 2.059(5)	O7-Zn2-O1 109.06(18)	O17-Zn5-O11 115.40(19)
Zn5-O11 2.063(4)	O7-Zn2-O2 98.14(17)	O17-Zn5-O12 96.87(19)
Zn5-O16 2.046(5)	O7-Zn2-O6 101.6(2)	O17-Zn5-O16 109.6(2)
Zn5-O17 2.000(4)	O2-Zn3-O8 107.92(19)	O12-Zn6-O18 106.5(2)
Zn6-O12 1.934(4)	O2-Zn3-O9 108.63(17)	O12-Zn6-O19 108.21(18)
Zn6-O18 1.952(5)	O2-Zn3-O10 122.87(18)	O12-Zn6-O20 121.79(19)
Zn6-O19 1.966(4)	O9-Zn3-O8 108.25(19)	O18-Zn6-O19 106.9(2)
Zn6-O20 1.940(4)	O9-Zn3-O10 107.01(18)	O20-Zn6-O18 107.7(2)
	O10-Zn3-O8 101.30(19)	O20-Zn6-O19 104.93(19)

Single crystals of [Zn$_4$L$_4$(μ-OH)$_2$](ClO$_4$)$_2$·3MeOH·0.5MeCN (**2**·3MeOH·0.5MeCN) suitable for X-ray crystallographic analyses were grown by the slow evaporation of a mixed DCM/MeOH/MeCN (3:2:1) solution. The perchlorate salt crystallizes in the monoclinic space group $P2_1/n$ with four formula units per unit cell. Figure 3 displays the structure of the tetranuclear Zn complex. Selected bond lengths and angles are listed in Table 2. The crystal structure is composed of tetranuclear [Zn$_4$L$_4$(μ-OH)$_2$]$^{2+}$ complexes, perchlorate ions and co-crystallized MeOH and MeCN molecules. The two [Zn$_2$L$_2$]$^{2+}$ fragments are joined via two bridging hydroxido ligands to generate the [Zn$_4$L$_4$(μ-OH)$_2$]$^{2+}$ complex's dication. The Zn^{2+} ions are all five-coordinated, and they are surrounded by two imine N and two bridging phenolate O atoms from the supporting ligand and one bridging hydroxido ligand. The τ values range from 0.49 to 0.52, indicating that the coordination geometries of the Zn^{2+} ions lie in between the ideal trigonal bipyramidal or square pyramidal environments. This is also supported by the SHAPE symmetry factors given in Table S1. The Zn-N bond distances between 2.048(3) and 2.118(3) Å as well as the Zn-O distances between 1.962(2) and 2.154(2) Å are all within the usual ranges for such complexes.

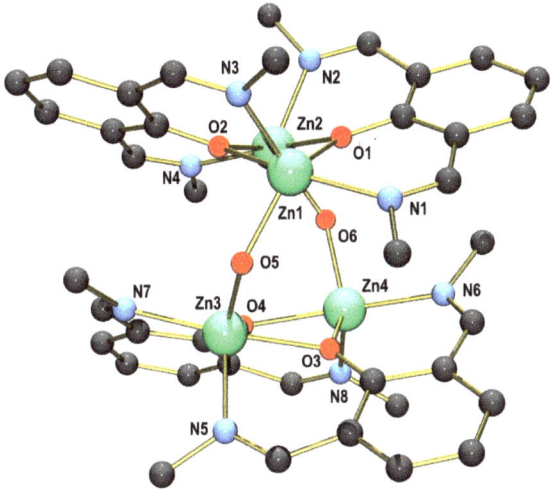

Figure 3. Molecular structure of the [Zn$_4$L$_4$(μ-OH)$_2$]$^{2+}$ complex cation in the crystals of (2·3MeOH·0.5MeCN). All hydrogen atoms, phenyl rings and t-Bu groups have been omitted for clarity.

Table 2. Selected bond lengths (Å) and angles (°) for 2·3MeOH·0.5MeCN.

Zn1-O1 2.018(3)	N1-Zn1-O2 159.01(11)	N5-Zn3-N7 118.88(11)
Zn1-O2 2.135(2)	N3-Zn1-N1 107.88(12)	N5-Zn3-O3 84.91(10)
Zn1-N1 2.117(3)	N3-Zn1-O2 85.17(12)	N7-Zn3-O3 157.33(11)
Zn1-N3 2.048(3)	O1-Zn1-N1 86.28(11)	O4-Zn3-N5 115.58(11)
Zn1-O5 1.956(2)	O1-Zn1-N3 119.00(12)	O4-Zn3-N7 86.56(11)
Zn2-O1 2.144(2)	O5-Zn1-N1 99.91(11)	O4-Zn3-O3 72.08(9)
Zn2-O2 2.012(3)	O5-Zn1-N1 99.91(11)	O5-Zn3-N5 108.29(12)
Zn2-N2 2.050(3)	O5-Zn1-N3 119.00(12)	O5-Zn3-N7 99.47(11)
Zn2-N4 2.118(3)	O5-Zn1-O1 129.36(10)	O5-Zn3-O3 88.85(10)
Zn2-O6 1.992(3)	O5-Zn1-O2 91.57(10)	O5-Zn3-O4 129.58(10)
Zn3-O3 2.154(2)	N2-Zn2-N4 110.11(11)	N6-Zn4-O4 160.17(11)
Zn3-O4 2.008(3)	N2-Zn2-O1 85.85(10)	N8-Zn4-N6 108.65(11)
Zn3-N5 2.053(3)	N4-Zn2-O1 158.57(11)	N8-Zn4-O4 84.43(11)
Zn3-N7 2.118(3)	O2-Zn2-N2 115.58(12)	O3-Zn4-N6 88.07(11)

Table 2. Cont.

Zn3-O5 1.962(2)	O2-Zn2-N4 87.08(12)	O3-Zn4-N8 116.03(11)
Zn4-O3 2.003(2)	O2-Zn2-O1 72.81(9)	O3-Zn4-O4 72.58(9)
Zn4-O4 2.135(2)	O6-Zn2-N2 110.14(12)	O6-Zn4-N6 98.18(11)
Zn4-N6 2.076(3)	O6-Zn2-N4 98.46(11)	O6-Zn4-N8 109.84(11)
Zn4-N8 2.055(3)	O6-Zn2-O1 88.76(10)	O6-Zn4-O3 128.75(10)
Zn4-O5 1.993(2)	O6-Zn2-O2 128.62(11)	O6-Zn4-O3 90.88(10)

3.3. Electronic Absorption and Emission Spectroscopy

The synthesized compounds were further characterized using UV-vis absorption and photoluminescence spectroscopy. The electronic absorption spectra for the free ligand **HL** and its deprotonated phenolate form, **L⁻**, were measured in a mixture of DCM/MeCN (3:2) at room temperature. The spectrum of the free ligand displays three main absorption maxima (Figure 4). The absorption maxima and corresponding extinction coefficients are listed in Table 3.

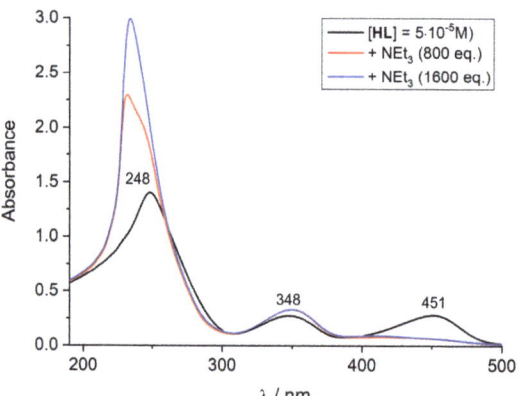

Figure 4. Measured UV-vis spectra of **HL** and its deprotonated form, **L⁻**. Experimental conditions: 5×10^{-5} M, CH$_2$Cl$_2$/MeOH (v:v/3:2), 298 K. The fully deprotonated ligand was obtained by adding an excess of NEt$_3$ to the solution of **HL**.

Table 3. UV-vis spectroscopic data for **HL** and the deprotonated ligand; luminescence spectroscopic data for **HL**, **L⁻** and **1**.

	λ_{max}/nm	ε_λ/[M^{-1}cm^{-1}]	Assignment
HL(a)	451	5600	π-π* (ArOH)
	348	5480	π-π*(ArOH)
	248	28,800	π-π*(ArOH+Ph)
L⁻(a,b)	349	6620	π-π* (ArO⁻)
	232	45,940	
L⁻(a,c)	349	6640	π-π* (ArO⁻)
	234	59,920	
	λ_{ex}/nm	λ_{em}/nm	Stokes shift/cm^{-1} (rel. intensity/a.u.)
HL(d)	447	497	2250 (5380)
L⁻(b,d)	346	563	11,139 (350)
1(d)	378	452	4331 (35,850)

(a) DCM/MeOH (3:2/v:v), 10^{-5} M sample concentration, (b) [NEt$_3$] = 8×10^{-3} M, (c) [NEt$_3$] = 16×10^{-3} M, (d) MeCN, 10^{-5} M sample concentration.

The broad band with an absorption maximum at 248 nm is attributed to the π-π* transitions localized on the phenol and phenyl rings of **HL**, according to time-dependent density functional theory (TDDFT) calculations (see details below). The lower-energy bands with absorption maxima at 348 nm and 451 nm, on the other hand, are assigned to electronic transitions within a more extended π system involving the phenolate ring and the two imine groups. The UV-vis spectrum of the deprotonated ligand differs significantly from that of the protonated ligand. It reveals two well-resolved absorption maxima at 234 and 349 nm. Instead of the band at 450 nm observed in **HL**, there is a tail on the 349 nm band that extends into the visible range but with no well-developed maximum. Thus, deprotonation has a strong effect on the spectroscopic properties of **HL**.

To study the complexation reactions of **HL** with Zn^{2+} ions in the solution, UV-vis spectrophotometric batch titrations were carried out. The method of YOE and JONES was used here, in which the ligand concentration is kept constant over the entire range and the amount of Zn^{2+} is raised stepwise until a stable complex species is formed [61–64]. Tetra-*n*-butylammonium hexafluorophosphate (10 mM) was added to the used solvent mixture (3:2—DCM/MeOH) to ensure a constant ionic strength during the complexation reactions.

The addition of aliquots of $Zn(OAc)_2$ (from 0.1 to 10 equiv.) leads to clear changes in the UV-vis spectrum of the ligand (Figure 5). Thus, the bands at 350 and 450 nm for **HL** vanish with increasing Zn^{2+} concentrations, and a new band with a maximum at 387 nm emerges (at a Zn/L ratio of 1.0 equiv.), providing clear support that Zn^{2+} binding to the bis(iminophenolate)ligand L^- occurs. Adding the further equivalents of $Zn(OAc)_2$ leads to a weak but significant hyperchromic effect manifested by the slight increase in the intensity of the 387 nm band. The YOE and JONES method was applied in order to determine the stoichiometric ratio of the complex. As observed from the inset of Figure 5, the plot of the absorbance value versus stoichiometric ratio $[Zn^{2+}]/[\mathbf{HL}]$ increases very steeply up to 1:1, and then there is a further, but much less pronounced, increase in intensity, suggesting that additional Zn^{2+} ions will not directly interact with the donor atoms of **HL**. The spectroscopic titration of **HL** with $Zn(ClO_4)_2 \cdot 6H_2O$ behaved in a very similar fashion (SI). Overall, these findings suggest that the ligand examined here has a strong propensity to form Zn^{2+} complexes with a 1:1 metal/ligand ratio. This ratio would be consistent with the 2:2 stoichiometries found in the solid state and implies that such 2:2 complexes also exist in the solution state. In the presence of additional co-ligands, this ratio may be increased, but it is accompanied by much less pronounced changes in the spectroscopic features of supporting ligand **HL**. These findings are also consistent with those made by fluorescence spectroscopy described below.

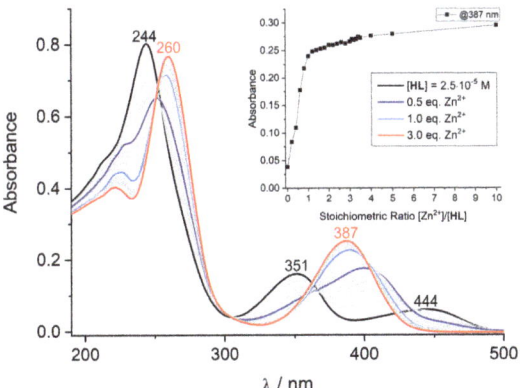

Figure 5. Spectrophotometric titration of **HL** with $Zn(OAc)_2 \cdot 2H_2O$ in CH_2Cl_2/MeOH (3:2/*v*:*v*) at a 10^{-5} M concentration and constant ionic strength (10^{-2} M N(*n*-Bu)$_4$PF$_6$, $T = 295$ K). The red curve corresponds to the final Zn^{2+}/**HL** stoichiometric ratio of 3.0. The inset shows the evolution of selected absorbance values versus the $[Zn^{2+}]/[\mathbf{HL}]$ stoichiometric ratio.

The complexation reactions were further examined using fluorescence spectroscopy. The emission spectra of compounds **HL**, **L⁻**, and of a 1:3 solution of **HL** and Zn(OAc)$_2$ are displayed in Figure 6 along with the corresponding excitation spectra. The emission spectra were measured in acetonitrile solution at the 10^{-5} M concentration. Table 3 lists the data. The free ligand fluorescence is blue-green, reflected in a single emission band with a maximum at 497 nm when excited at 447 nm. The excitation spectrum of **HL** monitored at 497 nm corresponds to the absorption spectrum and reveals an absorption band at 447 nm that is assigned to a transition into the $^1(\pi\text{-}\pi^*)$ state of the bis(iminomethyl)phenol system. The deprotonation of the ligand has a strong effect on the luminescence properties. This can be detected by the naked eye. Thus, in contrast to **HL**, **L⁻** exhibits a very weak, intraligand $^1(\pi\text{-}\pi)^*$ fluorescence emission band at 563 nm. The luminescence intensity decreases by a factor of 15, and the Stokes shift increases from 2250 cm^{-1} for **HL** to 11,139 cm^{-1} in **L⁻**. Remarkably, the complexation of the ligand by Zn^{2+} leads to a circa six-fold enhancement of the luminescence intensity, which is also accompanied by a hypsochromic shift in the emission band to 452 nm. This behavior has been observed for other Zn^{2+} complexes supported by *o*-hydroxy-aryl SCHIFF base ligands [65–68]. The enhancement of the emission intensity can be explained by the CHEF effect (*chelation-enhanced fluorescence effect*) due to the inhibition of photo-induced electron transfer processes [69,70].

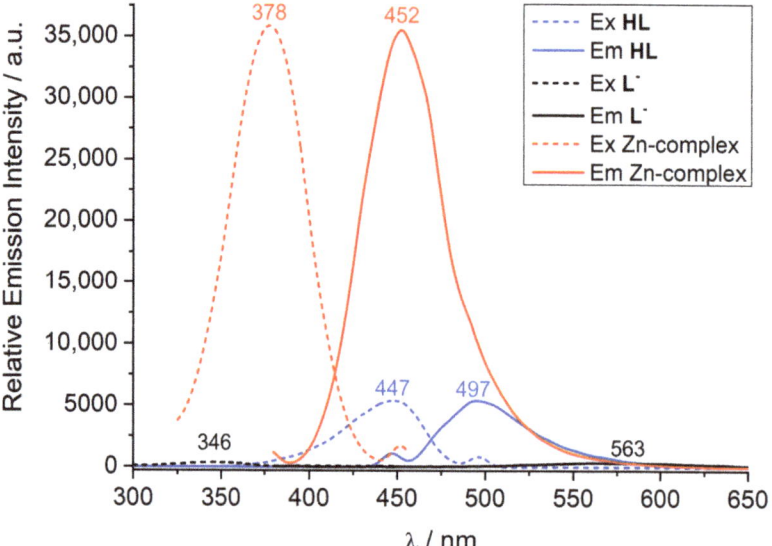

Figure 6. Excitation (dashed) and emission spectra (solid) of **HL**, **L⁻**, and a 1:3 mixture of HL/Zn(OAc)$_2$ in MeCN (10^{-5} M, 298 K).

Figure 7 displays a titration of **HL** with Zn(OAc)$_2$ monitored by fluorescence spectroscopy at 0.01 mM concentration. As observed, the intensity of the emission band at around 450 nm increases steadily with increasing Zn^{2+} concentrations. An inflection point occurs at a metal-to-ligand ratio (M/L) of 1:1. At higher M/L ratios, the emission intensity remains nearly constant. These observations are in good agreement with the conclusions provided by UV vis spectroscopy and provide strong support for the presence of Zn^{2+} complexes with 1:1 M/L ratio. The slight blue shift of the maximum of the emission band from 450 to 445 nm may be due to the change in the coordination geometry of the initial coordination geometry, such as an attachment of additional acetate or hydroxide ligands, although other factors cannot be ruled out.

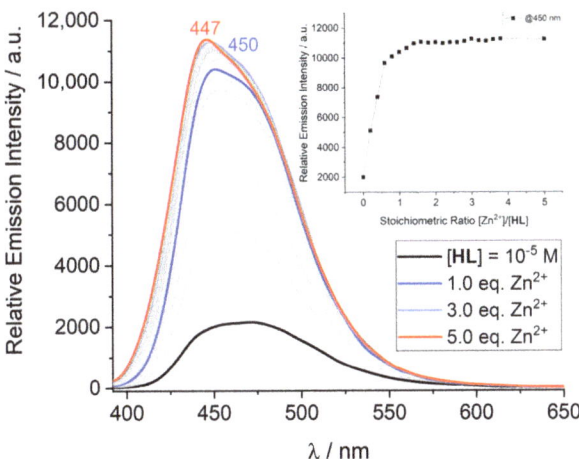

Figure 7. Fluorescence titration of **HL** with Zn(OAc)$_2$·2H$_2$O in CH$_2$Cl$_2$/MeOH (3:2/v:v) at a 10^{-5} M concentration and constant ionic strength (10^{-2} M N(n-Bu)$_4$PF$_6$, T = 295 K, λ_{exc} = 378 nm). The blue and red curves correspond to Zn^{2+}/**HL** stoichiometric ratios of 1.0 and 5.0, respectively. The inset shows the evolution of the fluorescence intensity of the emission band at 450 nm versus the [Zn^{2+}]/[**HL**] stoichiometric ratio.

The luminescence properties of the ligand **HL** were investigated in the presence of other metal ions in view of reports from the literature that o-hydroxy-aryl compounds allow for the sensitive optical detection of Zn^{2+} ions among other metal cations [11]. Thus, methanolic solutions containing the equimolar quantities of the ligand and a series of abundant and biologically relevant p- and d-block metals were prepared, and their emission properties were determined. Figure 8 shows that none of the investigated metal ions reveals a fluorescence enhancement. This is either attributed to the poor binding affinity of some metal ions to the SCHIFF-base ligands (in the case of Na$^+$, K$^+$, Mg^{2+} and Ca^{2+}) or the quenching of the excited singlet states by electron transfer reactions involving the d-states of open-shell transition metal ions (in case of Mn^{2+}, Fe^{2+}, Co^{2+}, Ni^{2+} and Cu^{2+}).

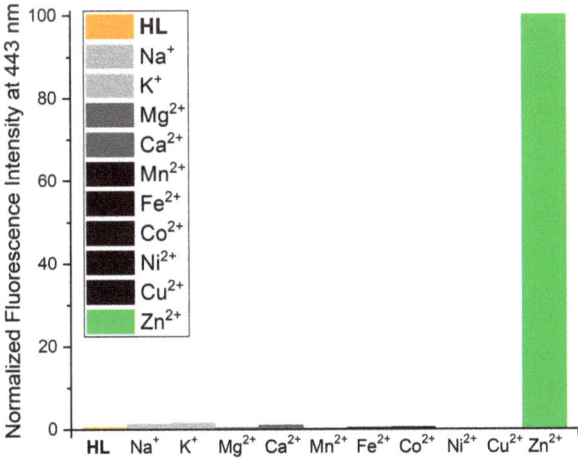

Figure 8. Normalized emission intensity at 443 nm of 10^{-5} M solutions containing equimolar quantities of **HL** and various metal ions (methanolic solution, concentration: ligand 1×10^{-5} M, metal salt: 3×10^{-5} M, T = 295 K); excitation wavelength 378 nm.

Several other fluorescent OFF–ON sensors for Zn^{2+} ions based on salicylidene aldimine binding sites have been reported [11,20,65]. Zn^{2+} binding enhances the rather weak intraligand $^1(\pi\text{-}\pi)$ fluorescence emission of the free ligands significantly due to the inhibition of photo-induced electron transfer processes [65]. Zn^{2+} ion binding also rigidifies the ligand architecture, thereby decreasing the probability of the vibrational deactivation of the excited singlet state [34,35].

The fluorescence intensity of HL as a function of the Zn^{2+} concentration was investigated in order to determine the LOD value (limit of detection value), which provides the lowest concentration that can be measured. Fluorescence intensity was found to be linear between 1×10^{-6} and 1×10^{-7} M Zn^{2+} and produced an LOD value of 0.24(1) ppm. This value is comparable to those of other Zn^{2+} fluorescence sensor systems featuring imino-phenolate units [11,34,35].

In order to support experimental findings, we have carried out TD-DFT calculations on ligand **HL**, differently (de-)protonated species and a $[Zn_2L_2]^{2+}$ complex to simulate the UV-vis absorption spectra (for details, see the Experimental Section). By considering various forms of **HL**, dynamic and solvent effects could be accounted for, allowing the full assignment of absorption bands in the experimental spectra (see Figure 9).

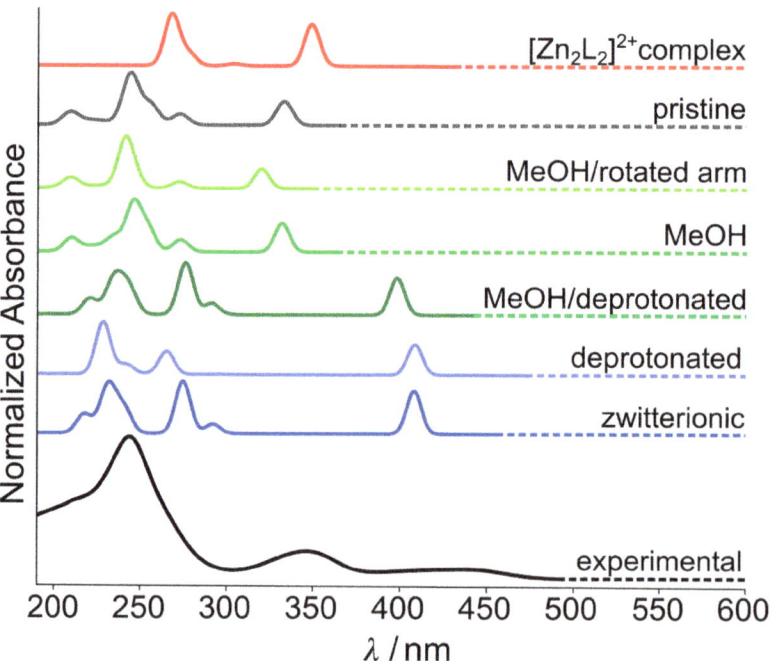

Figure 9. Simulated UV-vis spectra of different **HL** forms. Included are the zwitterionic form, explicitly coordinated methanol, deprotonated **HL**, the pristine ligand and the $[Zn_2L_2]^{2+}$ complex. The spectra were obtained using AMS code [53].

The band with the largest intensity was assigned to various $\pi\text{-}\pi^*$ transitions, while the band at around 350 nm was assigned to a $\pi\text{-}\pi^*$ transition with a contribution from the n-orbitals of the OH group. The lowest excitations at 450 nm were assigned to a $\pi\text{-}\pi^*$ transition with contributions from the deprotonated phenol group, and it was observed that the deprotonation resulted in a large redshift into the visible light range. As it is shown in Figure 10 the experimental results are in good agreement with the calculated ones.

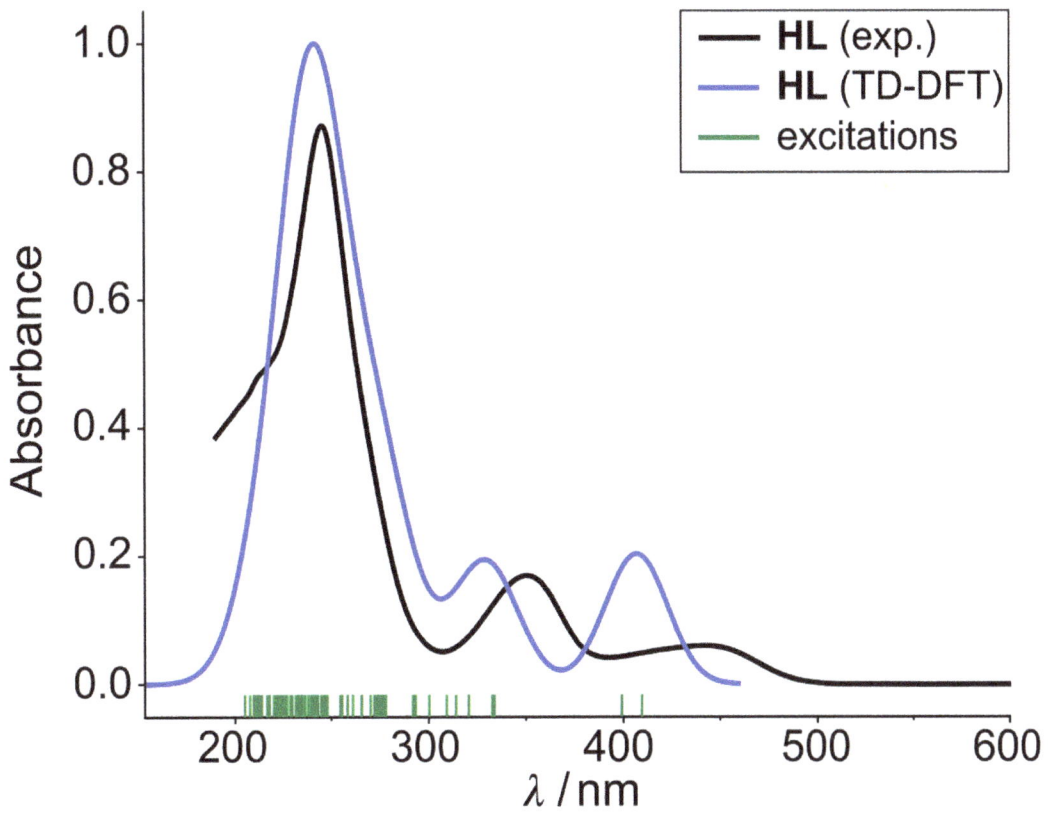

Figure 10. Summation of the calculated excitations for all species in Figure 9. The excitations are convolved with a 30 nm wide Gaussian to fit the experimental spectrum well.

4. Conclusions

The syntheses of the new tridentate Schiff-base ligand 2,6-bis((N-benzyl)iminomethyl)-4-*tert*-butylphenol (**HL**) and the corresponding polynuclear zinc complexes [$Zn_6L_2(\mu_3\text{-}OH)_2(OAc)_8$] (**1**) and [$Zn_4L_4(\mu_2\text{-}OH)_2$](ClO$_4$)$_2$ (**2**) are described. The exact composition of the two complexes was confirmed via X-ray crystallography, which revealed the influence of the coordinating acetate anion. The molecular structure of **1** contains Zn^{2+} ions, which are six-, five- and four-coordinated by the O and the two N atoms of the ligand and OH$^-$ and acetate ligands. In the discrete complex cation [$Zn_4L_4(\mu_2\text{-}OH)_2$]$^{2+}$ of **2** all Zn^{2+} ions, there are five-coordinated with τ values ranging from 0.49 to 0.52. The absorption properties of the ligand have been investigated with UV-vis spectroscopy. The lower energy band at 348 nm and the visible transition at 451 nm are designated as electronic transitions within the extended π system of the iminophenolate. Via the deprotonation of the ligand, the band at 451 nm disappears, whereby the spectroscopic properties of the ligand change significantly. The UV-vis titrations with the acetate and the perchlorate salts of Zn^{2+} exhibit the tendency of **HL** to form complexes with a stoichiometric ratio of 1:1. The results of luminescence spectroscopy studies support this observation. The complexation of Zn^{2+} leads to a further enhancement of the fluorescence intensity by factor six (CHEF effect), which was not detected for Na^+, K^+, Mg^{2+}, Ca^{2+}, Mn^{2+}, Fe^{2+}, Co^{2+}, Ni^{2+} or Cu^{2+}.

Supplementary Materials: The following supporting information can be downloaded at: https://www.mdpi.com/article/10.3390/chemistry5020070/s1. Figure S1: FTIR spectrum of **HL** (KBr pellet); Figure S2: FTIR spectra of [$Zn_6L_2(OH)_2(OAc)_8$] (**1**, KBr pellet) and [$Zn_4L_4(OH)_2$](ClO_4)$_2$ (**2**, KBr pellet); Figure S3. Overlay of the FTIR spectra of the ligand (**HL**) and the zinc complexes **1** and **2** (1700–1500 cm^{-1}); Figure S4. Overlay of the FTIR spectra of the zinc complexes **1** and **2** (1750–450 cm^{-1}); Figure S5: Packing diagram of **HL** showing intermolecular π-π stacking (offset face-to-face) interactions between adjacent ligand molecules; Figure S6: ^{13}C-NMR spectrum of **HL** recorded in dimethylsulfoxide-d^6 at 363 K; multiplicity of C-H coupling displayed; parts of the spectrum without signals cut out for better observation of the multiplicities; Figure S7: ^{1}H-NMR spectrum of [$Zn_6L_2(OH)_2(OAc)_8$] recorded in CD_3CN at ambient temperature; Figure S8: Spectrophotometric titration of **HL** with $Zn(ClO_4)_2 \cdot 6H_2O$ in CH_2Cl_2/MeOH (3:2/v:v) at a 10^{-5} M concentration and constant ionic strength (10^{-2} M N(n-Bu)$_4$PF$_6$, T = 295 K). The blue curve corresponds to a final Zn^{2+}/**HL** molar ratio of 1:1. The inset shows the evolution of selected absorbance values versus the [Zn^{2+}]/[**HL**] molar ratio; Figure S9. Excitation (blue) and emission spectrum (green) of HL in acetonitrile (c(HL) = 1×10^{-5} M, 298 K). MeCN; 1×10^{-5} M; λ_{Ex} = 447 nm; λ_{Em} = 497 nm; Figure S10: Excitation (violet) and emission spectrum (yellow) of L^- in acetonitrile in the presence of base (c(HL) = 1×10^{-5} M, c(NEt3) = 1×10^{-5} M, 298 K). MeCN; c(HL) = 1×10^{-5} M; c(NEt3) = 1×10^{-5} M; λ_{Ex} = 346 nm; λ_{Em} = 563 nm; Figure S11. Excitation (azure) and emission spectrum (turquoise) of [**Zn$_6$L$_2$(OH)$_2$(OAc)$_8$**] in acetonitrile (c(**complex**) = 1×10^{-5} M, 298 K). λ_{Ex} = 378 nm; λ_{Em} = 452 nm; Figure S12: Major contributions to the observed UV-Vis bands of **HL** with the single orbital transition contribution in parentheses. a) HOMO→LUMO: 334 nm (97%). b) HOMO-1→LUMO+1: 245 nm (67%); Figure S13: Major contributions to the observed UV-Vis bands of **HL$^-$** with the single orbital transition contribution in parentheses. a) HOMO→LUMO: 409 nm (98%). b) HOMO-2→LUMO+1: 229 nm (36%); Figure S14: Major contributions to the observed UV-Vis bands of [Zn_2L_2]$^{2+}$ with the single orbital transition contribution in parentheses. a) HOMO→LUMO: 348 nm (57%); Figure S15: PBE-D3(BJ) optimized structure of **HL** used for the simulation of UV-Vis spectra (pristine) and basis for the other used structures; Figure S16: Fluorescence spectra **HL** as a function of Zn^{2+} ion concentration; Table S1. Shape symmetry factors for complex **1**.

Author Contributions: Conceptualization, T.S. and V.K.; methodology, T.S.; validation, T.S., B.K. and A.K.; formal analysis, H.W. and A.K.; investigation, V.K. and T.S.; resources, B.K.; data curation, B.K.; writing—original draft preparation, T.S.; writing—review and editing, M.B. and B.K.; visualization, T.S., M.B. and B.K.; supervision, B.K.; project administration, B.K.; funding acquisition, B.K. All authors have read and agreed to the published version of the manuscript.

Funding: This research was funded by the German Science Foundation (Priority programme 2102, "Light Controlled Reactivity of Metal Complexes", project KE 585/9-1).

Data Availability Statement: Deposition Numbers 2248758-2248760 contain the supplementary crystallographic data for this paper. These data are provided free of charge by the joint Cambridge Crystallographic Data Centre and Fachinformationszentrum Karlsruhe Access Structures service. TD-DFT data will be available soon and will be announced.

Acknowledgments: We are thankful to H. Krautscheid for providing facilities for X-ray crystallographic measurements.

Conflicts of Interest: The authors declare no conflict of interest.

References

1. Follmer, C.; Real-Guerra, R.; Wasserman, G.E.; Olivera-Severo, D.; Carlini, C.R. Jackbean, soybean and Bacillus pasteurii ureases. *Eur. J. Biochem.* **2004**, *271*, 1357–1363. [CrossRef] [PubMed]
2. Sumner, J.B. The Isolation and Crystallization of the Enzyme Urease. *J. Biol. Chem.* **1926**, *69*, 435–441. [CrossRef]
3. Bertagnolli, H.; Kaim, W. The Dinuclear CuA Center in Cytochrome c Oxidase and N2O Reductase-A Metal–Metal Bond in Biology? *Angew. Chem. Int. Ed. Engl.* **1995**, *34*, 771–773. [CrossRef]
4. Lv, Y.; Cao, M.; Li, J.; Wang, J. A Sensitive Ratiometric Fluorescent Sensor for Zinc(II) with High Selectivity. *Sensors* **2013**, *13*, 3131–3141. [CrossRef] [PubMed]
5. Fernley, R.T. Non-cytoplasmic carbonic anhydrases. *Trends Biochem. Sci.* **1988**, *13*, 356–359. [CrossRef]
6. Hewett-Emmett, D.; Tashian, R.E. Functional diversity, conservation, and convergence in the evolution of the alpha-, beta-, and gam-ma-carbonic anhydrase gene families. *Mol. Phylogenetics Evol.* **1996**, *5*, 50–77. [CrossRef] [PubMed]
7. Thiel, G.; Lietz, M. Regulator neuronaler Gene: Zinkfingerprotein REST. *Biol. Unserer Zeit* **2004**, *34*, 96–101. [CrossRef]

8. Tashian, R.E. The carbonic anhydrases: Widening perspectives on their evolution, expression and function. *Bioessays* **1989**, *10*, 186–192. [CrossRef]
9. Liu, M.-J.; Bao, S.; Gálvez-Peralta, M.; Pyle, C.J.; Rudawsky, A.C.; Pavlovicz, R.E.; Killilea, D.W.; Li, C.D.; Nebert, W.; Wewers, M.D.; et al. ZIP8 Regulates Host Defense through Zinc-Mediated Inhibition of NF-κB. *Cell Rep.* **2013**, *3*, 386–400. [CrossRef]
10. Ternes, W. *Biochemie der Elemente—Anorganische Chemie Biologischer Prozesse, Springer Spektrum*; Springer: Berlin/Heidelberg, Germany, 2013.
11. Roy, P.; Dhara, K.; Manassero, M.; Ratha, J.; Banerjee, P. Selective Fluorescence Zinc Ion Sensing and Binding Behavior of 4-Methyl-2,6-bis(((phenylmethyl)imino)methyl)phenol: Biological Application. *Inorg. Chem.* **2007**, *46*, 6405–6412. [CrossRef]
12. Yu, G.; Liu, Y.; Song, Y.; Wu, X.; Zhu, D. A new blue light-emitting material. *Synth. Met.* **2001**, *117*, 211–214. [CrossRef]
13. Di Bella, S.; Fragala, I. Two-dimensional characteristics of the second-order nonlinear optical response in dipolar donor–acceptor coor-dination complexes. *New J. Chem.* **2002**, *26*, 285–290.
14. Leung, A.C.W.; Chong, J.H.; Patrick, B.O.; MacLachlan, M.J. Poly(salphenyleneethynylene)s: A New Class of Soluble, Conjugated, Metal-Containing Polymers. *Macromolecules* **2003**, *36*, 5051–5054. [CrossRef]
15. Wang, P.; Hong, Z.; Xie, Z.; Tong, S.; Wong, O.; Lee, C.-S.; Wong, N.; Hung, L.; Lee, S. A bis-salicylaldiminato Schiff base and its zinc complex as new highly fluorescent red dopants for high performance organic electroluminescence devices. *Chem. Commun.* **2003**, *14*, 1664–1665. [CrossRef]
16. Rigamonti, L.; Demartin, F.; Forni, A.; Righetto, S.; Pasini, A. Copper(II) Complexes of salen Analogues with Two Differently Sub-stituted (Push−Pull) Salicylaldehyde Moieties. A Study on the Modulation of Electronic Asymmetry and Nonlinear Optical Properties. *Inorg. Chem.* **2006**, *45*, 10976–10989. [CrossRef]
17. Zhang, G.; Yang, G.; Wang, S.; Chen, Q.; Ma, J.S. A Highly Fluorescent Anthracene-Containing Hybrid Material Exhibiting Tunable Blue–Green Emission Based on the Formation of an Unusual "T-Shaped" Excimer. *Chem. Eur. J.* **2007**, *13*, 3630–3635. [CrossRef]
18. Kotova, O.; Semenov, S.; Eliseeva, S.; Troyanov, S.; Lyssenko, K.; Kuzmina, N. New Helical Zinc Complexes with Schiff Base Derivatives of β-Diketonates or β-Keto Esters and Ethylenediamine. *Eur. J. Inorg. Chem.* **2009**, *2009*, 3467–3474. [CrossRef]
19. Hai, Y.; Chen, J.-J.; Zhao, P.; Lv, H.; Yu, Y.; Xu, P.; Zhang, J.-L. Luminescent zinc salen complexes as single and two-photon fluo-rescence subcellular imaging probes. *Chem. Commun.* **2011**, *47*, 2435–2437. [CrossRef]
20. Dumur, F.; Contal, E.; Wantz, G.; Gigmes, D. Photoluminescence of Zinc Complexes: Easily Tunable Optical Properties by Variation of the Bridge Between the Imido Groups of Schiff Base Ligands. *Eur. J. Inorg. Chem.* **2014**, *2014*, 4186–4198. [CrossRef]
21. Su, Q.; Wu, Q.-L.; Li, G.-H.; Liu, X.-M.; Mu, Y. Bis-salicylaldiminato zinc complexes: Syntheses, characterization and luminescent properties. *Polyhedron* **2007**, *26*, 5053–5060. [CrossRef]
22. Cheng, J.; Ma, X.; Zhang, Y.; Liu, J.; Zhou, X.; Xiang, H. Optical Chemosensors Based on Transmetalation of Salen-Based Schiff Base Complexes. *Inorg. Chem.* **2014**, *53*, 3210–3219. [CrossRef]
23. Malthus, S.J.; Cameron, S.A.; Brooker, S. Improved Access to 1,8-Diformyl-carbazoles Leads to Metal-Free Carbazole-Based [2 + 2] Schiff Base Macrocycles with Strong Turn-On Fluorescence Sensing of Zinc(II) Ions. *Inorg. Chem.* **2018**, *57*, 2480–2488. [CrossRef] [PubMed]
24. Zhou, X.; Yu, B.; Guo, Y.; Tang, X.; Zhang, H.; Liu, W. Both Visual and Fluorescent Sensor for Zn^{2+} Based on Quinoline Platform. *Inorg. Chem.* **2010**, *49*, 4002–4007. [CrossRef]
25. Frederickson, C.J.; Kasarskis, E.J.; Ringo, D.; Frederickson, R.E. A quinoline fluorescence method for visualizing and assaying the histochemically reactive zinc (bouton zinc) in the brain. *J. Neurosci. Methods* **1987**, *20*, 91–103. [CrossRef] [PubMed]
26. Nasir, M.S.; Fahrni, C.J.; Suhy, D.A.; Kolodsick, K.J.; Singer, C.P.; O'Halloran, T.V. The chemical cell biology of zinc: Structure and intracellular fluorescence of a zinc-quinolinesulfonamide complex. *JBIC J. Biol. Inorg. Chem.* **1999**, *4*, 775–783. [CrossRef] [PubMed]
27. Coyle, P.; Zalewski, P.D.; Philcox, J.C.; Forbes, I.J.; Ward, A.D.; Lincoln, S.F.; Mahadevan, I.; Rofe, A.M. Measurement of zinc in hepatocytes by using a fluorescent probe, zinquin: Relationship to metallothionein and intracellular zinc. *Biochem. J.* **1994**, *303*, 781–786. [CrossRef]
28. Zalewski, P.D.; Forbes, I.J.; Betts, W.H. Correlation of apoptosis with change in intracellular labile Zn(II) using zinquin [(2-methyl-8-p-toluenesulphonamido-6-quinolyloxy)acetic acid], a new specific fluorescent probe for Zn(II). *Biochem. J.* **1993**, *296*, 403–408. [CrossRef] [PubMed]
29. Mahadevan, I.B.; Kimber, M.C.; Lincoln, S.F.; Tiekink, E.R.T.; Ward, A.D.; Betts, W.H.; Forbes, I.J.; Zalewski, P.D. The Synthesis of Zinquin Ester and Zinquin Acid, Zinc(II)-Specific Fluorescing Agents for Use in the Study of Biological Zinc(II). *Aust. J. Chem.* **1996**, *49*, 561–568. [CrossRef]
30. Toroptsev, I.V.; Eshchenko, V.A. Histochemical detection of zinc using fluorescent 8-(arensulfonilamino)-quinolines. *Tsitologiia* **1970**, *12*, 1481–1484.
31. Ullmann, S.; Börner, M.; Kahnt, A.; Abel, B.; Kersting, B. Green-Emissive Zn^{2+} Complex Supported by a Macrocyclic Schiff-Base/Calix[4]arene-Ligand: Crystallographic and Spectroscopic Characterization. *Eur. J. Inorg. Chem.* **2021**, *36*, 3691–3698. [CrossRef]
32. Laube, C.; Taut, J.A.; Kretzschmar, J.; Zahn, S.; Knolle, W.; Ullman, S.; Kahnt, A.; Kersting, B.; Abel, B. Light controlled oxidation by supramolecular Zn(ii) Schiff-base complexes. *Inorg. Chem. Front.* **2020**, *7*, 4333–4346. [CrossRef]

33. Klose, J.; Severin, T.; Hahn, P.; Jeremies, A.; Bergmann, J.; Fuhrmann, D.; Griebel, J.; Abel, B.; Kersting, B. Coordination chemistry and photoswitching of dinuclear macrocyclic cadmium-, nickel-, and zinc complexes containing azobenzene carboxylato co-ligands. *Beilstein J. Org. Chem.* **2019**, *15*, 840–851. [CrossRef]
34. Ullmann, S.; Schnorr, R.; Laube, C.; Abel, B.; Kersting, B. Photoluminescence properties of tetrahedral zinc(ii) complexes supported by calix[4]arene-based salicyldiminato ligands. *Dalton Trans.* **2018**, *47*, 5801–5811. [CrossRef]
35. Ullmann, S.; Schnorr, R.; Handke, M.; Laube, C.; Abel, B.; Matysik, J.; Findeisen, M.; Rüger, R.; Heine, T.; Kersting, B. Zn2+-Ion Sensing by Fluorescent Schiff Base Calix[4]arene Macrocycles. *Chem. Eur. J.* **2017**, *23*, 3824–3827. [CrossRef] [PubMed]
36. *Spectrum FL*; PerkinElmer: Waltham, MA, USA, 2019.
37. *OriginPro 8G*; OriginLab Corporation: Northampton, MA, USA, 2009.
38. *MestReNova 14.1.0*; Mestrelab Research S.L.: Santiago de Compostela, Spain, 2019.
39. *X-AREA and X-RED 32*; V1.35; STOE & Cie GmbH: Darmstadt, Germany, 2006.
40. Sheldrick, G.M. Phase annealing in SHELX-90: Direct methods for larger structures. *Acta Crystallogr. Sect. A Found. Crystallogr.* **1990**, *46*, 467–473. [CrossRef]
41. Sheldrick, G.M. Crystal structure refinement with SHELXL. *Acta Crystallogr. Sect. C Struct. Chem.* **2015**, *71*, 3–8. [CrossRef] [PubMed]
42. Spek, A.L. *PLATON—A Multipurpose Crystallographic Tool*; Utrecht University: Utrecht, The Netherlands, 2000.
43. Dolomanov, O.V.; Bourhis, L.J.; Gildea, R.J.; Howard, J.A.K.; Puschmann, H. OLEX2: A complete structure solution, refinement and analysis program. *J. Appl. Crystallogr.* **2009**, *42*, 339–341. [CrossRef]
44. Farrugia, L.J. ORTEP-3 for Windows—A version of ORTEP-III with a Graphical User Interface (GUI) by J. Farrugia. *J. Appl. Cryst.* **1997**, *30*, 565. [CrossRef]
45. Van Lenthe, E.; Baerends, E.J. Optimized Slater-type basis sets for the elements 1-118. *J. Comput. Chem.* **2003**, *24*, 1142–1156. [CrossRef]
46. Perdew, J.P.; Burke, K.; Ernzerhof, M. Generalized Gradient Approximation Made Simple. *Phys. Rev. Lett.* **1996**, *18*, 3865–3868. [CrossRef]
47. Grimme, S.; Ehrlich, S.; Goerigk, L. Effect of the damping function in dispersion corrected density functional theory. *J. Comput. Chem.* **2011**, *32*, 1456–1465. [CrossRef] [PubMed]
48. van Gisbergen, S.; Snijders, J.; Baerends, E. Implementation of time-dependent density functional response equations. *Comput. Phys. Commun.* **1999**, *118*, 119–138. [CrossRef]
49. Seth, M.; Ziegler, T. Range-Separated Exchange Functionals with Slater-Type Functions. *J. Chem. Theory Comput.* **2012**, *8*, 901–907. [CrossRef] [PubMed]
50. Marques, M.A.; Oliveira, M.J.; Burnus, T. Libxc: A library of exchange and correlation functionals for density functional theory. *Comput. Phys. Commun.* **2012**, *183*, 2272–2281. [CrossRef]
51. Lehtola, S.; Steigemann, C.; Oliveira, M.J.; Marques, M.A. Recent developments in libxc—A comprehensive library of functionals for density functional theory. *Softwarex* **2018**, *7*, 1–5. [CrossRef]
52. Rüger, R.; Franchini, M.; Trnka, T.; Yakovlev, A.; van Lenthe, E.; Philipsen, P.; van Vuren, T.; Klumpers, B.; Soin, T. *AMS 2021.104 SCM, Theoretical Chemistry*; Vrije Universiteit: Amsterdam, The Netherlands, 2021.
53. Baerends, E.J.; Ziegler, T.; Atkins, A.J.; Autschbach, J.; Baseggio, O.; Bashford, D.; Bérces, A.; Bickelhaupt, F.M.; Bo, C.; Boerrigter, P.M.; et al. *ADF, 2021.104, SCM, Theoretical Chemistry*; Vrije Universiteit: Amsterdam, The Netherlands, 2021.
54. te Velde, G.; Bickelhaupt, F.M.; Baerends, E.J.; Fonseca Guerra, C.; van Gisbergen, S.J.A.; Snijders, J.G.; Ziegler, T. Chemistry with ADF. *J. Comput. Chem.* **2001**, *22*, 931–967. [CrossRef]
55. Louwen, J.N.; Pye, C.C.; van Lenthe, E.; Austin, N.D.; McGarrity, E.S.; Xiong, R.; Sandler, S.I.; Burnett, R.I. *AMS 2021.104 COS-MO-RS, SCM, Theoretical Chemistry*; Vrije Universiteit: Amsterdam, The Netherlands, 2021.
56. Pye, C.C.; Ziegler, T.; van Lenthe, E.; Louwen, J.N. An implementation of the conductor-like screening model of solvation within the Amsterdam density functional package—Part II. COSMO for real solvents. *Can. J. Chem.* **2009**, *87*, 790–797. [CrossRef]
57. Pye, C.C.; Ziegler, T. An implementation of the conductor-like screening model of solvation within the Amsterdam density functional package. *Theor. Chem. Acc.* **1999**, *101*, 396–408. [CrossRef]
58. Addison, A.W.; Rao, T.N.; Reedijk, J.; Van Rijn, J.; Verschoor, G.C. Synthesis, structure, and spectroscopic properties of copper(II) compounds containing nitrogen–sulphur donor ligands; the crystal and molecular structure of aq-ua[1,7-bis(N-methylbenzimidazol-2′-yl)-2,6-dithiaheptane]copper(II) perchlorate. *J. Chem. Soc. Dalton Trans.* **1984**, *7*, 1349–1356. [CrossRef]
59. Llunell, M.; Casanova, D.; Cirera, J.; Alemany, P.; Alvarez, S. *SHAPE*; University of Barcelona: Barcelona, Spain, 2013.
60. Hirozo, K.; Yoshihiko, S. The Crystal Structure of Zinc Oxyacetate, $Zn_4O(CH_3COO)_6$. *Bull. Chem. Soc. Jap.* **1954**, *27*, 112–114.
61. Yoe, J.H.; Jones, A.L. Colorimetric Determination of Iron with Disodium-1,2-dihydroxybenzene-3,5-disulfonate. *Ind. Eng. Chem. Anal. Ed.* **1944**, *16*, 111–115. [CrossRef]
62. Yoe, J.H.; Harvey, A.E. Colorimetric Determination of Iron with 4-Hydroxybiphenyl-3-carboxylic Acid. *J. Am. Chem. Soc.* **1948**, *70*, 648–654. [CrossRef]
63. Meyer, A.S.; Ayres, G.H. The Mole Ratio Method for Spectrophotometric Determination of Complexes in Solution. *J. Am. Chem. Soc.* **1957**, *79*, 49–53. [CrossRef]
64. Meyer, A.S.; Ayres, G.H. The Interaction of Platinum(II) and Tin(II) Chlorides. *J. Am. Chem. Soc.* **1955**, *77*, 2671–2675. [CrossRef]

65. Mitra, A.; Hinge, V.K.; Mittal, A.; Bhakta, S.; Guionneau, P.; Rao, C.P. A Zinc-Sensing Glucose-Based Naphthyl Imino Conjugate as a Detecting Agent for Inorganic and Organic Phosphates, Including DNA. *Chem. Eur. J.* **2011**, *17*, 8044–8047. [CrossRef]
66. Pathak, R.K.; Dessingou, J.; Rao, C.P. Multiple Sensor Array of Mn2+, Fe2+, Co2+, Ni2+, Cu2+, and Zn2+ Complexes of a Triazole Linked Imino-Phenol Based Calix[4]arene Conjugate for the Selective Recognition of Asp, Glu, Cys, and His. *Anal. Chem.* **2012**, *84*, 8294–8300. [CrossRef]
67. Mum-midivarapu, V.V.S.; Tabbasum, K.; Chinta, J.P.; Rao, C.P. 1,3-Di-amidoquinoline conjugate of calix[4]arene (L) as a ratiometric and colorimetric sensor for Zn2+: Spectroscopy, microscopy and computational studies. *Dalton Trans.* **2012**, *41*, 1671–1674. [CrossRef]
68. Mum-midivarapu, V.V.S.; Bandaru, S.; Yarramala, D.S.; Samanta, K.D.; Mhatre, S.; Rao, C.P. Binding and Ratiometric Dual Ion Recognition of Zn2+ and Cu2+ by 1,3,5-Tris-amidoquinoline Conjugate of Calix[6]arene by Spectroscopy and Its Supramolecular Features by Microscopy. *Anal. Chem.* **2015**, *87*, 4988–4995. [CrossRef]
69. Nugent, J.W.; Lee, H.; Lee, H.-S.; Reibenspies, J.H.; Hancock, R.D. Mechanism of chelation enhanced fluorescence in complexes of cadmium(ii), and a possible new type of anion sensor. *Chem. Commun.* **2013**, *49*, 9749–9751. [CrossRef]
70. Joseph, R.; Chinta, J.P.; Rao, C.P. Lower Rim 1,3-Diderivative of Calix[4]arene-Appended Salicylidene Imine (H2L): Experimental and Computational Studies of the Selective Recognition of H2L toward Zn2+ and Sensing Phosphate and Amino Acid by [ZnL]. *J. Org. Chem.* **2010**, *75*, 3387–3395. [CrossRef]

Disclaimer/Publisher's Note: The statements, opinions and data contained in all publications are solely those of the individual author(s) and contributor(s) and not of MDPI and/or the editor(s). MDPI and/or the editor(s) disclaim responsibility for any injury to people or property resulting from any ideas, methods, instructions or products referred to in the content.

Article

Surface Functionalization of Calcium Phosphate Nanoparticles via Click Chemistry: Covalent Attachment of Proteins and Ultrasmall Gold Nanoparticles

Kathrin Kostka and Matthias Epple *

Inorganic Chemistry and Center for Nanointegration Duisburg-Essen (CENIDE), University of Duisburg-Essen, Universitaetsstr. 5-7, 45141 Essen, Germany; kathrin.kostka@uni-due.de
* Correspondence: matthias.epple@uni-due.de

Abstract: Calcium phosphate nanoparticles (60 nm) were stabilized with either polyethyleneimine (PEI; polycationic electrolyte) or carboxymethylcellulose (CMC; polyanionic electrolyte). Next, a silica shell was added and terminated with either azide or alkyne groups via siloxane coupling chemistry. The particles were covalently functionalized by copper-catalyzed azide-alkyne cycloaddition (CuAAC; click chemistry) with proteins or gold nanoparticles that carried the complementary group, i.e., either alkyne or azide. The model proteins hemoglobin and bovine serum albumin (BSA) were attached as well as ultrasmall gold nanoparticles (2 nm). The number of protein molecules and gold nanoparticles attached to each calcium phosphate nanoparticle was quantitatively determined by extensive fluorescent labelling and UV–Vis spectroscopy on positively (PEI) or negatively (CMC) charged calcium phosphate nanoparticles, respectively. Depending on the cargo and the nanoparticle charge, this number was in the range of several hundreds to thousands. The functionalized calcium phosphate particles were well dispersible in water as shown by dynamic light scattering and internally amorphous as shown by X-ray powder diffraction. They were easily taken up by HeLa cells and not cytotoxic. This demonstrates that the covalent surface functionalization of calcium phosphate nanoparticles is a versatile method to create transporters with firmly attached cargo molecules into cells.

Keywords: calcium phosphate; nanoparticles; surface functionalization; proteins; drug delivery; click chemistry; gold

Citation: Kostka, K.; Epple, M. Surface Functionalization of Calcium Phosphate Nanoparticles via Click Chemistry: Covalent Attachment of Proteins and Ultrasmall Gold Nanoparticles. *Chemistry* **2023**, *5*, 1060–1076. https://doi.org/10.3390/chemistry5020072

Academic Editors: Sofia Lima, Christoph Janiak, Sascha Rohn and Georg Manolikakes

Received: 31 March 2023
Revised: 29 April 2023
Accepted: 5 May 2023
Published: 7 May 2023

Copyright: © 2023 by the authors. Licensee MDPI, Basel, Switzerland. This article is an open access article distributed under the terms and conditions of the Creative Commons Attribution (CC BY) license (https://creativecommons.org/licenses/by/4.0/).

1. Introduction

Nanoparticles are widely used in many areas, as not only their size but also their shape and surface functionalization can be varied. Due to their small size, they are easily taken up by cells [1,2]. Therefore, multifunctional nanoparticles find many interesting applications as drug carriers in modern biomedicine, also with an appropriate surface functionalization to target cells or tissues [3–7]. As adsorption leads only to a weak bond between a nanoparticle and its cargo, a covalent attachment is preferable as it creates a more stable attachment [8,9]. Calcium phosphate nanoparticles are especially promising drug carriers in biomedicine [10–13] due to their high similarity to human hard tissues (bones and teeth). These are both rich in hydroxyapatite, $Ca_5(PO_4)_3OH$, the most common calcium phosphate mineral [14,15]. To ensure their colloidal stability, also in salt- and protein-containing biological media, calcium phosphate nanoparticles are usually surface-coated with polymers or polyelectrolytes. After subsequent coating with a silica shell [16–19], they can be equipped with thiol, azide or alkyne groups via siloxane chemistry. This permits a covalent attachment of molecules to their surface.

Here, we report a fundamental study where two model proteins (hemoglobin and bovine serum albumin; BSA) functionalized with alkyne groups were attached to the azide-terminated surface of calcium phosphate nanoparticles. Proteins represent important cargo molecules which often cannot penetrate the cell membrane on their own [20]. Thus, we chose hemoglobin and BSA to serve as models for other, more functional proteins. These particles can also serve as models for nanoparticles with a protein corona, although were not designed for that purpose. As abundant plasma protein, albumin plays a prominent role in protein coronas which determine the pathway of nanoparticles in the body [21,22].

In addition, we demonstrate how azide-terminated ultrasmall gold nanoparticles can be covalently attached to alkyne-terminated calcium phosphate nanoparticles. The surface conjugation was performed by copper-catalyzed azide-alkyne cycloaddition (CuAAC; click chemistry) [23–27]. By extensive fluorescent labelling of all components, it was possible to confirm the integrity and to elucidate the chemical composition of these nanocarriers. The charge of the nanoparticles was adjusted to either positive (by polyethyleneimine; PEI) or negative (by carboxymethylcellulose; CMC). Furthermore, their cytocompatibility and their ability to cross the cell membrane were demonstrated in HeLa cell culture.

2. Materials and Methods

2.1. Reagents

The following reagents were used: albumin bovine fraction V (BSA, Serva, Heidelberg, Germany, M_W = 66.5 kDa); AlexaFluor®-647-succinimidyl ester (Invitrogen, Life Technologies, Carlsbad, CA, USA); AlexaFluor®-488-azide (Jena Bioscience, Jena, Germany, 95%); AlexaFluor®-647-alkyne (Jena Bioscience); AlexaFluor®-568-phalloidin (Invitrogen); aminoguanine (Alfa Aesar, Haverhill, MA, USA); aqueous ammonia solution (Carl Roth, Karlsruhe, Germany, 7.8%); (3-azidopropyl)triethoxysilane (SelectLab Chemicals, Münster, Germany, 97%); calcium lactate (Merck, Darmstadt, Germany); carboxymethylcellulose (CMC, Sigma-Aldrich, St. Louis, MO, USA, branched, M_W = 70 kDa); copper sulfate ($CuSO_4$, Sigma-Aldrich, p.a.); diammonium hydrogen phosphate (Merck, p.a.); 3-(4,5-dimethylthiazol-2-yl)-2,5-diphenyltetrazolium bromide (MTT, Sigma-Aldrich); D-(+)-trehalose dihydrate (Sigma-Aldrich, p.a.); Dulbecco's modified Eagle's medium (DMEM, Gibco, ThermoFisher Scientific, Waltham, MA, USA); Dulbecco's phosphate buffered saline (DPBS, Gibco); ethanol (VWR, Darmstadt, Germany, p.a.); 3.7% formaldehyde solution (Merck, p.a.); glutathione (GSH, Sigma-Aldrich, 98%); hemoglobin (human, Sigma-Aldrich); methanol (VWR, p.a.); Hoechst-3342 dye (ThermoFisher Scientific); 3-methylsiloxy-1-propyne (AOB-Chem, Los Angeles, CA, USA, 95%); polyethyleneimine (PEI, Sigma-Aldrich, branched, M_W = 25 kDa); potassium carbonate (K_2CO_3, Sigma-Aldrich, p.a.); propargyl-PEG3-aminooxy linker (Broadpharm, San Diego, CA, USA); sodium ascorbate (Sigma-Aldrich, 99%); sodium borohydride ($NaBH_4$, Sigma-Aldrich, 96%); sodium hydroxide (NaOH, VWR, p.a.); tetraethylorthosilicate (TEOS, Sigma-Aldrich, 98%); tris(3-hydroxypropyl-triazolylmethyl)amine (THPTA, Sigma-Aldrich, 95%).

For all syntheses and purifications, ultrapure water (Purelab ultra instrument from ELGA, Celle, Germany) was used. All syntheses and analyses were carried out in water unless otherwise noted.

2.2. Instruments

All calcium phosphate nanoparticles described in the following syntheses were isolated and purified by centrifugation with a 5430/5430 R Centrifuge (Eppendorf AG, Hamburg, Germany) at room temperature for 30 min at 14,200 rpm in 5 mL tubes. The gold nanoparticles were purified for 40 min at 4000 rpm in centrifugal spin filters (Amicon® Ultra, 3 kDa MWCO, Merck). The particles were washed with water and redispersed with a sonotrode (UP50H, Sonotrode N7, amplitude 70%, pulse duration 0.12 s, Hielscher Ultrasonics GmbH, Teltow, Germany). To determine the calcium concentration in the dispersions, atomic absorption spectroscopy (AAS) with an M-Series AA spectrometer (ThermoElectron Corporation, Waltham, MA, USA) was performed. For this, 0.5 mL of the dispersion was mixed with 0.5 mL HCl (0.1 M) and filled up to 5 mL with water. To determine the gold concentration, 20 µL of the gold-containing sample was reacted with 980 µL aqua regia and filled up to 5 mL with water for the AAS measurements.

After characterization and for storage, the particles were lyophilized in a D-(+)-trehalose solution (20 mg mL^{-1}; cryoprotectant) with a Christ Alpha 2-4 LSC instrument (Martin Christ GmbH, Osterode am Harz, Germany). For subsequent investigations, the nanoparticles were redispersed in water.

Dynamic light scattering (DLS) for particle size determination (number distribution) and zeta potential measurement was performed with 760 µL of the nanoparticle dispersion in a dip cell kit (Malvern, UK) in a Zetasizer Nano ZS instrument (λ = 633 nm, Malvern Nano ZS ZEN 3600 (Malvern Panalytical; Malvern, UK). For scanning electron microscopy (SEM), 5 µL of the dispersion was placed on a sample holder, dried in air, sputtered with gold/palladium, and examined with an Apreo S LoVac microscope (ThermoFisher Scientific). Energy-dispersive X-ray spectroscopy (EDX; detector type: SUTW-Sapphire) was used together with the SEM. The samples were analyzed by EDX without prior sputtering. The average particle diameter was calculated from visual inspection of at least 50 nanoparticles in SEM images with the software ImageJ [28].

X-ray powder diffraction (XRD) was performed with Cu Kα radiation (λ = 1.54 Å) on silicon single crystal sample holders cut along (911) with a Bruker D8 instrument (Bruker, Billerica, MA, USA). A range of 5–90 °2Θ, a step size of 0.01 °2Θ, a counting time of 0.6 s, and a nickel filter were used.

UV–Vis spectroscopy in a 400 µL Suprasil® quartz cuvette was carried out with a Varian Cary 300 Bio spectrophotometer (Agilent Technologies, Santa Clara, CA, USA). For fluorescence spectroscopy, the particles were dispersed in water and measured in a quartz cuvette with an Agilent Technologies Cary Eclipse spectrophotometer. Differential centrifugal sedimentation (DCS) with a DC24000 UHR instrument (CPS Instruments, Prairieville, LA, USA) was used to determine the size of the water-dispersed ultrasmall gold nanoparticles. Prior to the measurement, a sucrose gradient was created, and a calibration was performed with a PVC calibration dispersion standard (diameter 483 nm; CPS Instruments).

2.3. Synthesis of CaP/PEI/SiO$_2$-N$_3$ and CaP/CMC/SiO$_2$-N$_3$ Nanoparticles

The synthesis of PEI- or CMC-stabilized calcium phosphate (CaP) nanoparticles was carried out as follows, following a modified procedure first reported by Rojas et al. [18]. Extending these earlier results where only cationic nanoparticles with a primary PEI shell were prepared, we have extended the synthetic procedure to prepare anionic nanoparticles with CMC as primary shell. First, a calcium lactate solution (18 mmol L^{-1}, 5 mL min^{-1}, pH 10), a diammonium hydrogen phosphate solution (10.8 mmol L^{-1}, 5 mL min^{-1}, pH 10) and either a PEI solution (2 g L^{-1}, 7 mL min^{-1}) or a CMC solution (2 g L^{-1}, 7 mL min^{-1}) were pumped for 30 s into a beaker containing 10 mL water at room temperature with two peristaltic pumps. After stirring for 20 min, the silica shell was applied. 12 mL of the CaP-PEI or CaP-CMC nanoparticle dispersion, 60 µL tetraethylorthosilicate (TEOS), 120 µL ammonia solution (7.8%) and 48 mL ethanol were rapidly mixed under stirring. The dispersion was stirred for 16 h. The particles were purified by centrifugation and redispersed in 12 mL of water to obtain CaP/PEI/SiO$_2$ and CaP/CMC/SiO$_2$ nanoparticles.

For the attachment of azide groups to the particle surface, 12 mL of the redispersed CaP/PEI/SiO$_2$ or CaP/CMC/SiO$_2$ particles were mixed with 48 mL ethanol, 58.8 µL ammonia solution (7.8%), and 181 µL (3-azidopropyl)triethoxysilane and stirred for 6 h. The resulting CaP/PEI/SiO$_2$-N$_3$ and CaP/CMC/SiO$_2$-N$_3$ nanoparticles were purified by centrifugation.

2.4. Synthesis of CaP/PEI/SiO$_2$-Alkyne and CaP/CMC/SiO$_2$-Alkyne Nanoparticles

For the attachment of alkyne groups on the particle surface, the same procedure as in 2.3 was followed, except for the use of 3-trimethylsiloxy-1-propyne instead of (3-azidopropyl)triethoxysilane.

2.5. Synthesis of Fluorescent and Alkyne-Conjugated BSA and Hemoglobin

First, BSA and hemoglobin (Hem) (1 mg mL^{-1}) were conjugated with AF647-succinimidyl ester according to the manufacturer's instructions. Centrifugal spin filters were used for three times purification (Amicon® Ultra, 3 kDa MWCO) at 4000 rpm for 40 min. The labelling allowed to quantify Hem-AF647 and BSA-AF647 in subsequent synthesis steps by UV spectroscopy after recording calibration curves as the protein concentration was known (1 mg mL^{-1}).

Next, the fluorescent proteins were conjugated with alkyne groups. To attach alkyne groups to BSA-AF647 and Hem-AF647, the fluorescently labelled proteins (1 mg mL^{-1}) were incubated with 125 µM propargyl-PEG3-aminooxy linker for 12 h at 37 °C under stirring. Subsequently, BSA-AF647-alkyne and Hem-AF647-alkyne were purified twice with centrifugal spin filters (3 kDa MWCO) at 4000 rpm for 40 min to remove excess linker and redispersed in water.

2.6. Conjugation of BSA-AF647-Alkyne and Hem-AF647-Alkyne to CaP/PEI/SiO$_2$-N$_3$ and CaP/CMC/SiO$_2$-N$_3$ Nanoparticles

BSA-AF647-alkyne and Hem-AF647-alkyne were coupled to CaP/PEI/SiO$_2$-N$_3$ and CaP/CMC/SiO$_2$-N$_3$ nanoparticles by CuAAC, respectively. For this, 0.5 mL of CaP/PEI/SiO$_2$-N$_3$ or CaP/CMC/SiO$_2$-N$_3$ nanoparticle dispersion were used: 95 mg mL^{-1} CaP/PEI/SiO$_2$-N$_3$ and 93 mg mL^{-1} CaP/CMC/SiO$_2$-N$_3$ for the conjugation with BSA-AF647-alkyne; 58 mg mL^{-1} CaP/PEI/SiO$_2$-N$_3$ and 56 mg mL^{-1} CaP/CMC/SiO$_2$-N$_3$ for the conjugation with Hem-AF647-alkyne, respectively. Each dispersion was mixed with 0.35 µL NaOH (0.1 M), 38.5 µL aminoguanidine (1 mg mL^{-1}), 36.5 µL sodium ascorbate (100 mM), 3.65 µL of a solution containing CuSO$_4$ (40 µM) and THPTA (200 µM), and 25 µL of BSA-AF647-alkyne (1 mg mL^{-1}) or 25 µL of Hem-AF647-alkyne (1 mg mL^{-1}). Each dispersion was stirred for 1 h at room temperature under light exclusion. The particles were purified by spin filtration and redispersed in 0.5 mL water by ultrasonication for 4 s.

2.7. Conjugation of AF488-N$_3$ to CaP/PEI/SiO$_2$-Alkyne and CaP/CMC/SiO$_2$-Alkyne Nanoparticles

For the click reaction of AF488-N$_3$ to CaP/PEI/SiO$_2$-alkyne and CaP/CMC/SiO$_2$-alkyne nanoparticles, 5 mL of the nanoparticle dispersion was mixed with 167 µL AF488-N$_3$, 385 µL aminoguanidine (1 mg mL^{-1}), 3.5 µL NaOH (0.1 M), 36.5 µL of a solution containing CuSO$_4$ (40 µM) and THPTA (200 µM), and 365 µL sodium ascorbate (100 mM) and stirred for 1 h at RT under light exclusion. The amount of AF488-N$_3$ was set to leave some alkyne groups on the calcium phosphate nanoparticles unconjugated. The particles were purified by spin filtration and redispersed in 5 mL water by ultrasonication for 4 s.

2.8. Synthesis of Fluorescent Ultrasmall Gold Nanoparticles (2 nm; Au-AF647-N$_3$)

Ultrasmall gold nanoparticles carrying AF647 were prepared as described earlier by clicking AF647-alkyne to the surface of azide-functionalized glutathione-coated gold nanoparticles [29]. In order to load the surface of the Au-GSH-N$_3$ nanoparticles with fluorescent AF647-alkyne molecules, but to leave some azide groups free for subsequent coupling, only 1/3 of AF647-alkyne of the synthesis protocol reported in [29] was used. The particles had a diameter of the gold core of 2 nm and were characterized by UV spectroscopy, fluorescence spectroscopy, atomic absorption spectroscopy, NMR spectroscopy, and differential centrifugal sedimentation, giving the same results as in the standard synthesis reported in [29].

2.9. Conjugation of Ultrasmall Au-AF647-N$_3$ Nanoparticles to CaP/PEI/SiO$_2$-AF488-Alkyne or CaP/CMC/SiO$_2$-AF488-Alkyne Nanoparticles

50 µL of Au-AF647-N$_3$ nanoparticle dispersion (1 mg mL^{-1}) was used for conjugation to CaP/PEI/SiO$_2$-AF488-alkyne and CaP/CMC/SiO$_2$-AF488-alkyne nanoparticles, respectively. CuAAC was carried out under the same conditions as during the clicking of the alkyne-terminated proteins to the azide-terminated calcium phosphate nanoparticles

described above, leading to CaP/PEI/SiO$_2$-AF488-Au-AF647 and CaP/CMC/SiO$_2$-AF488-Au-AF647 nanoparticles. The particles were purified by spin filtration and redispersed in 0.5 mL water by ultrasonication for 4 s.

2.10. Cell Culture Studies

HeLa cells were cultured at 37 °C in 5% CO$_2$ atmosphere in cell culture flasks and then seeded onto the well plates for microscopy. Dulbecco's modified Eagle's medium (DMEM) with 10% fetal bovine serum, 100 U mL^{-1} penicillin, and 100 U mL^{-1} streptomycin served as cell culture medium.

An MTT assay was performed by spectrophotometric analysis. HeLa cells were transferred into a 24-well plate with 5×10^4 cells/well in 500 µL DMEM and cultivated for 24 h. 30 µL (1 mg mL^{-1}) of the different nanoparticle dispersions was added, respectively, for incubation for 24 h. Subsequently, the cells were washed three times with 500 µL DPBS, treated with 300 µL 3-(4,5-dimethylthiazol-2-yl)-2,5-diphenyltetrazolium bromide (MTT, 1 mg mL^{-1}) and incubated for 1 h at 37 °C. Then, the MTT solution was removed and replaced with 300 µL DMSO for 30 min. Finally, 100 µL of the DMSO solution was put into a 96-well plate for spectrophotometric analysis at 570 nm with a multiscan FC instrument (Thermo Fisher Scientific). The relative cell viability was normalized to the control group (mock), i.e., untreated cells.

For nanoparticle uptake, HeLa cells (5×10^4 cells/well) were cultured in an 8-well plate in 250 µL of DMEM for 24 h prior to incubation with the nanoparticles. Subsequently, 30 µL of the nanoparticle dispersion (1 mg mL^{-1}) diluted with 220 µL of DMEM was added. The incubation time was 24 h. Then, the cells were washed twice with DPBS to remove dead cells and free nanoparticles. For cell fixation, 150 µL of a 3.7% formaldehyde solution was added for 15 min at room temperature. The fixed cells were incubated with AF568-phalloidin (30 µL in 2 mL DPBS; 230 µL/well) for 20 min for actin staining. For nuclear staining, a Hoechst-3342 solution (4 µL in 2 mL DPBS; 230 µL/well) was applied for 15 min. The cells were washed twice with DPBS after each fixation and staining step. Images were taken with a fluorescence microscope (Keyence Biorevo BZ-9000, Neu-Isenburg, Germany) with a TRITC filter (Ex. 540/25 nm, DM 565 nm, BA 605/55 nm) and a FITC filter (Ex. 470/40 nm, DM 495 nm, BA 535/50 nm). A confocal laser scanning microscope (Leica TCS SP8X FALCON, Wetzlar, Germany) with laser wavelengths of 405 nm (nuclear staining; Hoechst-3342), 488 nm (AF488), 568 nm (AlexaFluor568 phalloidin), 647 nm (AF647), and a pulsed laser WLL (470–670 nm) was used for higher resolution, together with an HC PL APO UVIS CS2 63X/1.2 water immersion lens.

3. Results and Discussion

The calcium phosphate nanoparticles were prepared with either positive charge (by PEI) or negative charge (by CMC) to elucidate the effect of the particle charge on the biomolecule attachment and the subsequent particle effect on cells. A shell of silica to permit a covalent surface functionalization by azide or alkyne groups was added. Copper-catalyzed azide-alkyne cycloaddition (CuAAC; click chemistry) was used for subsequent covalent attachment of molecules and gold nanoparticles [30]. For this, BSA-AF647 and Hem-AF647 proteins were previously conjugated with alkyne groups, and ultrasmall gold nanoparticles were functionalized with azide groups. In the following, we denote the particles in abbreviated form, indicating the particle charge and omitting the omnipresent silica shell. Thus, the particle type CaP/CMC/SiO$_2$-AF488 is abbreviated as CaP^{neg}-AF488, the particle type CaP/PEI/SiO$_2$-BSA is abbreviated as CaP^{pos}-BSA, etc. Figure 1 shows all steps of the particle synthesis.

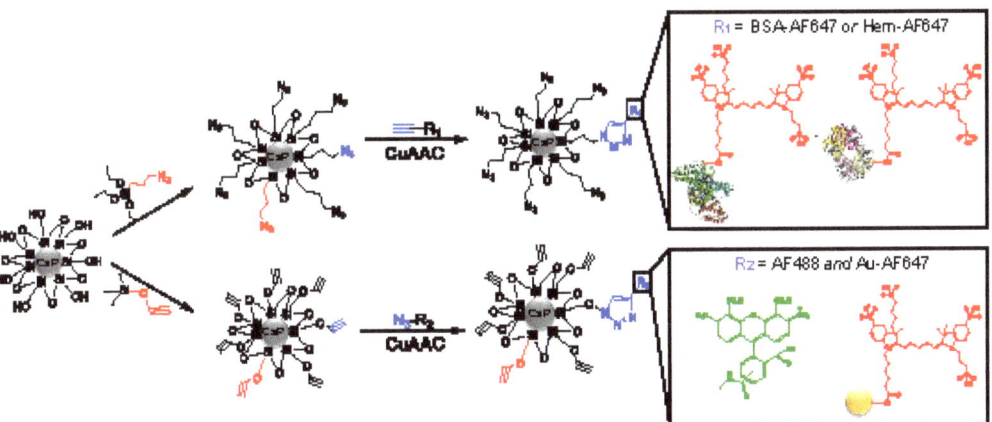

Figure 1. Steps of the surface functionalization of calcium phosphate nanoparticles (CaP) with Hem-AF647-, BSA-AF647-, and AF647-labelled ultrasmall gold nanoparticles (Au-AF647).

The attachment of fluorescently labelled bovine serum albumin (BSA-AF647) led to CaP^{pos}-BSA-AF647 and CaP^{neg}-BSA-AF647 nanoparticles, and the attachment of fluorescently labelled hemoglobin (Hem-AF647) led to CaP^{pos}-Hem-AF647 and CaP^{neg}-Hem-AF647 nanoparticles. Each BSA molecule carried 3.6 AF647 molecules, and each hemoglobin molecule carried 3.0 AF647 molecules. These labelling degrees were later used to compute the number of protein molecules on each calcium phosphate nanoparticle.

We also attached AF647-carrying ultrasmall gold nanoparticles (Au-AF647) together with the dye AF488, leading to CaP^{pos}-AF488-Au-AF647 and CaP^{neg}-AF488-Au-AF647 nanoparticles. As fluorescent control particles without cargo, CaP^{pos}-AF488 and CaP^{neg}-AF488 nanoparticles were prepared. All particles were thoroughly purified and characterized. Figure 2 schematically shows the synthetic pathways to all prepared nanoparticles.

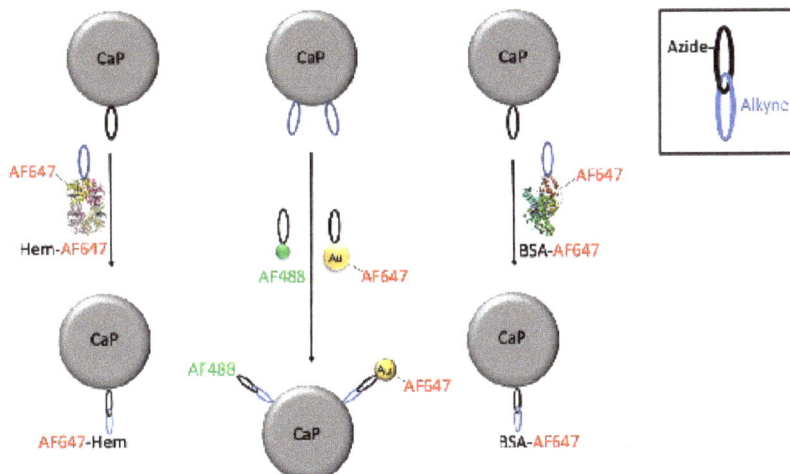

Figure 2. Summary of the surface conjugation of Hem-AF647-, BSA-AF647-, and AF647-carrying ultrasmall gold nanoparticles to calcium phosphate nanoparticles. All conjugated moieties were fluorescently labelled for detection and quantification.

The composition of the particles was quantitatively determined by UV–Vis spectroscopy (Figure 3). The recording of calibration curves and the determination of the calcium concentration in the dispersion by AAS permitted to compute the number of molecules attached to each nanoparticle (see below).

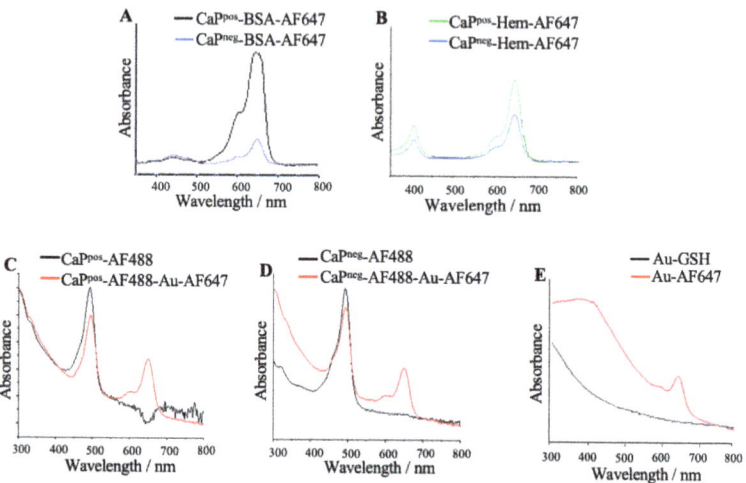

Figure 3. UV–Vis spectra of the functionalized calcium phosphate nanoparticles (**A–D**) and of ultrasmall gold nanoparticles before (Au-GSH; GSH = glutathione) and after (Au-AF647) the attachment of AF647 (**E**). All fluorescent labels were clearly detectable in the UV spectra and later used to quantify the individual species. Note the absorption of the protein hemoglobin itself at approximately 400 nm in (**B**).

All particles showed the strong fluorescence of the attached molecules (Figure 4). The green color depicts the presence of AF488 and the red color the presence of AF647. In general, the particle charge does not play a significant role for the optical properties and also does not affect the conjugation.

The solid core diameter of the particles was determined by scanning electron microscopy (SEM). All calcium phosphate nanoparticles had an approximately spherical (or globular) shape (Figure 5). The particle size varied between 39 nm and 90 nm, depending on the synthesis, but did not depend on the particle charge (see also Table 1). The calcium phosphate nanoparticles were agglomerated to some degree in the SEM due to the drying process. For the particle size distribution, we used the particle diameter that a spherical particle would have (by visual inspection). The gold nanoparticles were too small (2 nm) to be detectable by SEM.

Dynamic light scattering (DLS) was performed to determine the hydrodynamic diameter and the water dispersibility of the nanoparticles. The increasing hydrodynamic diameter confirmed the successful attachment of the fluorescent molecules and of gold nanoparticles (Figure 6). The hydrodynamic diameter was between 129 nm and 340 nm, i.e., always larger than the solid core diameter as determined by SEM, indicating a moderate agglomeration in dispersion. In general, the gold nanoparticles were too small for DLS [29]; therefore, they were characterized by differential centrifugal sedimentation (DCS).

To determine the elemental composition of the functionalized nanoparticles, energy-dispersive X-ray spectroscopy (EDX) was performed. All expected elements were found (Figure 7). An EDX mapping of similar silica-coated calcium phosphate nanoparticles was reported in [16], showing a homogeneous distribution of calcium, phosphate, and silicon.

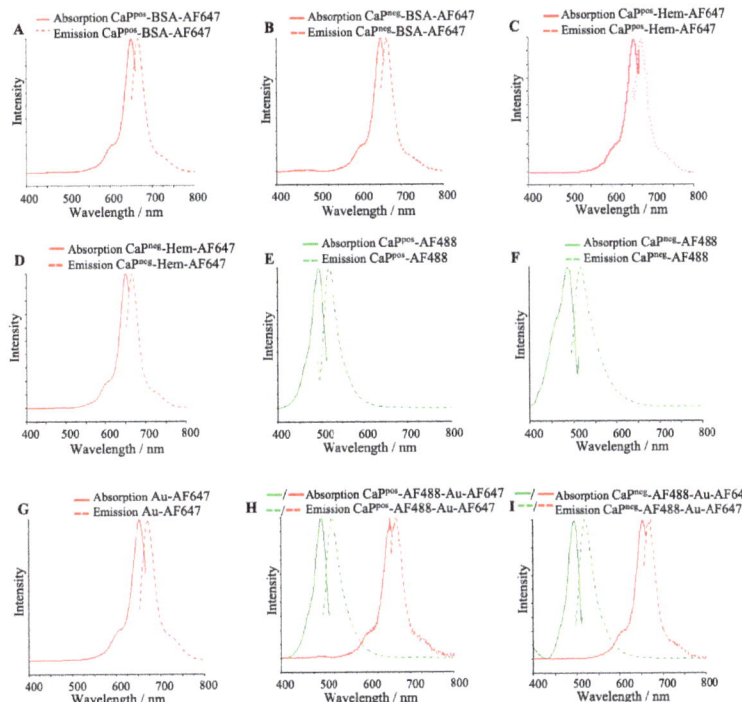

Figure 4. Absorption and emission spectra of functionalized calcium phosphate nanoparticles (**A–F**) and of ultrasmall gold nanoparticles before (**G**) and after attachment to calcium phosphate (**H,I**). In all cases, the expected fluorescence signals were present, giving a green (AF488) or a red (AF647) fluorescence.

X-ray powder diffraction was carried out to determine the internal crystallinity of the nanoparticles (Figure 8). The nanoparticles were almost X-ray amorphous (extreme broadening of the peaks) and did not show the characteristic peaks of the most prominent calcium phosphate phase, i.e., hydroxyapatite. According to our experience and the literature, it is likely that they consist of X-ray amorphous calcium phosphate with a stoichiometry close to hydroxyapatite [31–33].

From the calcium concentration in a given dispersion by AAS and the solid core diameter by SEM, the particle concentration can be computed if uniform spherical particles are assumed. Here, the solid core diameters of the nanoparticles with the same stabilization (PEI or CMC) were averaged/normalized for both syntheses and used for the calculations. The concentration of calcium in the samples was converted into the concentration of stoichiometric hydroxyapatite, $Ca_5(PO_4)_3OH$, and then into a volume by the density of hydroxyapatite (3140 kg m^{-3}). The combination of the calcium concentration (giving the total particle volume) and the average particle diameter (giving the volume of one particle) gave the concentration of calcium phosphate nanoparticles (see [18] for details of the computation). All particles had the expected zeta potential, i.e., the cationic polyelectrolyte PEI and the anionic polyelectrolyte CMC gave the particles their characteristic charge. The attachment of the cargo did not reverse the particle charge, i.e., cationic nanoparticles remained positively charged and anionic nanoparticles remained negatively charged. This led to well-dispersed particles as indicated by the hydrodynamic diameter as determined by DLS, despite some agglomeration as the comparison with the solid core diameter by SEM shows. Table 1 summarizes the characterization data of all functionalized calcium phosphate nanoparticles.

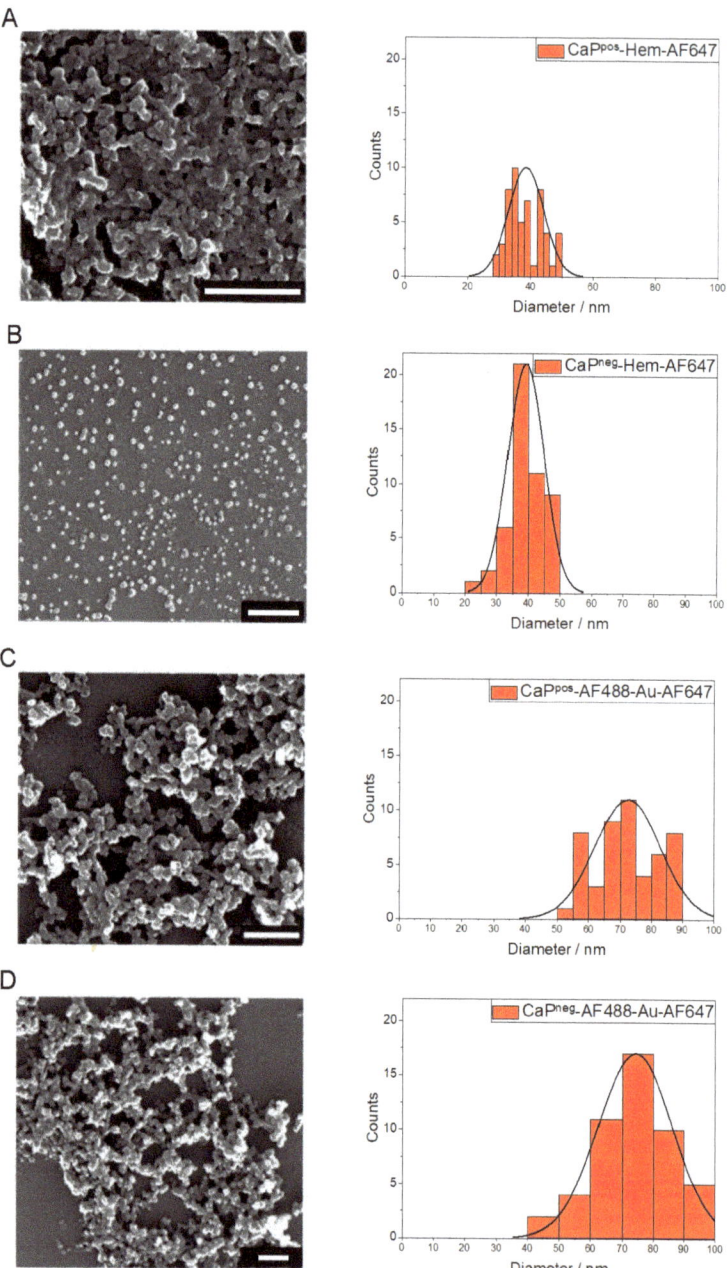

Figure 5. Representative SEM images of surface-functionalized calcium phosphate nanoparticles. Scale bars 500 nm. (**A**) CaP^{pos}-Hem-AF647, (**B**) CaP^{neg}-Hem-AF647, (**C**) CaP^{pos}-AF488-Au-AF647 and (**D**) CaP^{neg}-AF488-Au-AF647 nanoparticles. The corresponding particle size distributions indicate the polydispersity of the particles, based on the assumption of spherical particles.

Table 1. Properties of PEI- and CMC-stabilized calcium phosphate nanoparticles with different conjugated molecules.

	CaP^{pos}-BSA-AF647	CaP^{neg}-BSA-AF647	CaP^{pos}-Hem-AF647	CaP^{neg}-Hem-AF647	CaP^{pos}-AF488	CaP^{neg}-AF488	CaP^{pos}-AF488-Au-AF647	CaP^{neg}-AF488-Au-AF647
Solid core diameter by SEM/nm	64 ± 6	49 ± 5	39 ± 6	39 ± 6	72 ± 11	74 ± 12	72 ± 11	74 ± 12
$w(Ca^{2+})$ in dispersion by AAS/kg m^{-3}	0.092	0.090	0.056	0.053	0.096	0.092	0.046	0.040
$w(HAP)$ in dispersion/kg m^{-3} (computed)	0.231	0.226	0.140	0.133	0.241	0.231	0.115	0.100
N(nanoparticles) in dispersion/m^{-3} (computed)	5.35×10^{17}	1.17×10^{18}	1.44×10^{18}	1.36×10^{18}	3.92×10^{17}	3.46×10^{17}	1.88×10^{17}	1.51×10^{17}
Hydrodynamic particle diameter by DLS/nm	181 ± 9	149 ± 10	126 ± 9	206 ± 16	200 ± 8	291 ± 15	301 ± 7	340 ± 21
Polydispersity index (PDI) by DLS	0.32 ± 0.02	0.33 ± 0.01	0.29 ± 0.07	0.35 ± 0.02	0.22 ± 0.14	0.31 ± 0.02	0.29 ± 0.06	0.24 ± 0.12
Zeta potential by DLS/mV	+17 ± 2	−16 ± 1	+20 ± 4	−24 ± 6	+20 ± 1	−17 ± 1	+14 ± 1	−12 ± 6

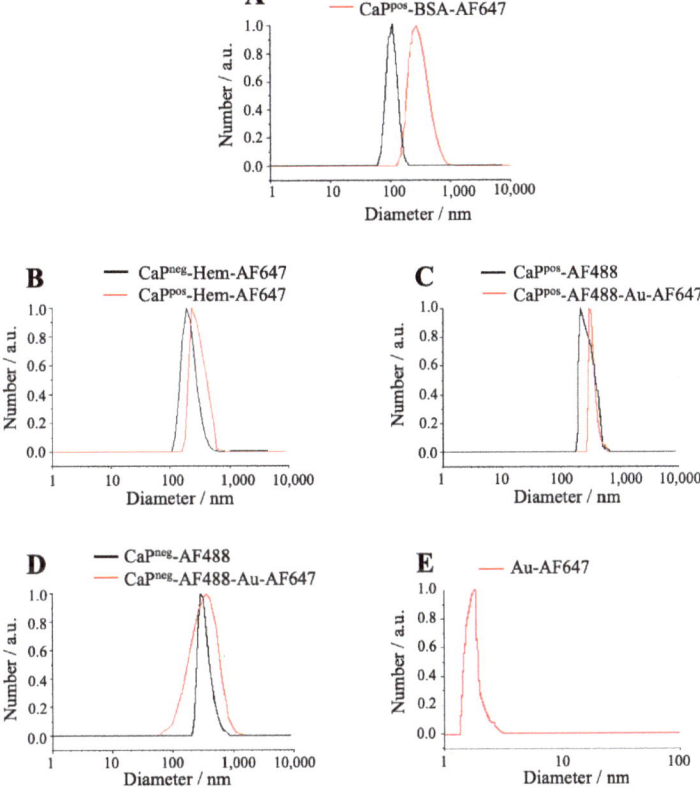

Figure 6. Particle size distribution (by number) of calcium phosphate nanoparticles by dynamic light scattering (DLS) (**A**–**D**), and differential centrifugal sedimentation (DCS) of Au-AF647 nanoparticles (**E**).

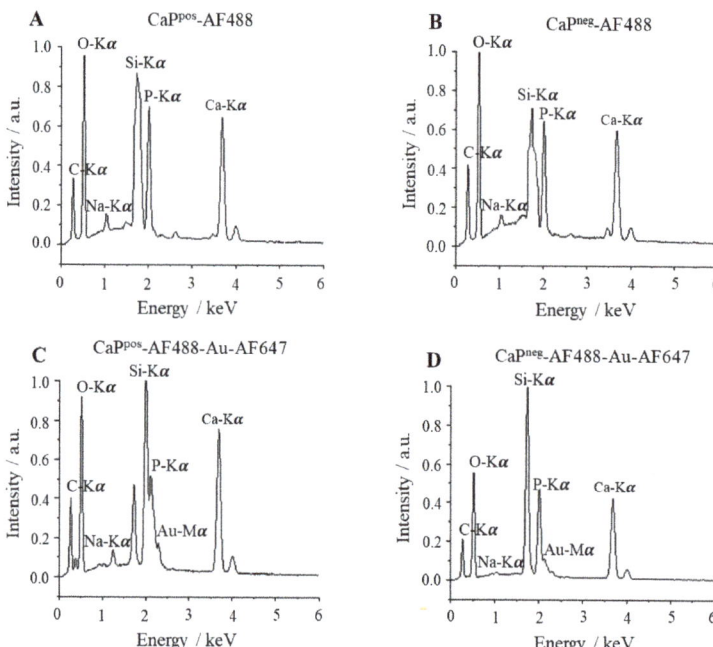

Figure 7. Representative energy-dispersive X-ray spectra (EDX) of calcium phosphate nanoparticles carrying AF488 (**A,B**) and AF488 together with Au-AF647 nanoparticles (**C,D**).

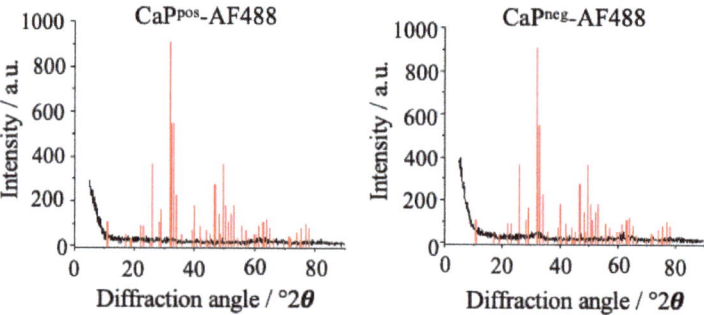

Figure 8. Representative X-ray powder diffractograms of functionalized calcium phosphate nanoparticles. The peaks of crystalline hydroxyapatite (JCPDS No. 00-009-0432) are shown for comparison as red bars.

The composition of the functionalized nanoparticles comprises the number of attached cargo molecules on each nanoparticle. This was possible by recording UV–Vis spectra of all dispersions and quantifying the absorption of each fluorophore with Lambert-Beer's law. Previously, calibration curves were recorded for each cargo compound. The molar masses of AF647-alkyne (910 Da), BSA-AF647-alkyne (67.4 kDa), Hem-AF647-alkyne (65.9 kDa), AF488-alkyne (774 Da), AF647-azide (940 Da) and AF488-azide (659 Da), were used to convert mass concentrations into molar concentrations. The molar ratio of cargo molecules to calcium phosphate nanoparticles then gave the desired loading number.

The number of conjugated gold nanoparticles was determined in the same way. First, the concentration of gold nanoparticles in the dispersion was computed with the average gold core diameter of 2 nm that was measured earlier [29]. Each gold nanoparticle contains approximately 250 gold atoms if assumed as spherical and weighs approximately $250 \times M(\text{Au})/N_A = 250 \times 197 \text{ g mol}^{-1}/6.023 \times 10^{23} \text{ mol}^{-1} = 8.2 \times 10^{-20}$ g. Again, the concentration of AF647 in the dispersion was determined by quantitative UV–Vis spectroscopy. The molar ratio of AF647 molecules and gold nanoparticles gave the number of AF647 on each gold nanoparticle with 2.3 (Table 2). The number of gold nanoparticles attached to each calcium phosphate nanoparticle then followed from the concentrations of gold and calcium in the dispersion.

Table 2. Gold concentration in dispersion and the number of AF647 molecules attached to each gold nanoparticle (Au-NP; 2 nm), measured for the functionalized calcium phosphate nanoparticles CaP^{pos}-AF488-Au-AF647 and CaP^{neg}-AF488-Au-AF647.

	w(Au) in Dispersion (AAS)/ kg m^{-3}	c(Au) in Dispersion (AAS)/ mol m^{-3}	N(Au-NP) in Dispersion/m^{-3}	c(Au-NP) in Dispersion/ mol m^{-3}	c(AF647) in Dispersion/ mol m^{-3}
CaP^{pos}-AF488-Au-AF647	4.00×10^{-2}	2.03×10^{-1}	4.95×10^{20}	8.22×10^{-4}	1.89×10^{-3}
CaP^{neg}-AF488-Au-AF647	5.75×10^{-3}	2.92×10^{-2}	7.11×10^{19}	1.18×10^{-4}	2.71×10^{-4}

Table 3 shows the numbers of protein molecules and gold nanoparticles attached to each calcium phosphate nanoparticle. The number of attached protein molecules differed considerably between BSA and Hem. The lower number of attached BSA molecules on negatively charged calcium phosphate nanoparticles may be explained by its negative charge at the moderately alkaline pH during clicking: The isoelectric point of BSA is 4.5 to 5 [34], and that of hemoglobin is 7.1 to 7.3 [35], i.e., hemoglobin is less negatively charged than BSA at neutral pH.

Table 3. Number of cargo proteins and nanoparticles on the functionalized calcium phosphate nanoparticles as computed by calcium phosphate particle concentrations (Table 1) and conjugated molecule concentrations (by UV spectroscopy).

	CaP^{pos}-BSA-AF647	CaP^{neg}-BSA-AF647	CaP^{pos}-Hem-AF647	CaP^{neg}-Hem-AF647	CaP^{pos}-AF488	CaP^{neg}-AF488	CaP^{pos}-AF488-Au-AF647	CaP^{neg}-AF488-Au-AF647
Protein molecules on each CaP nanoparticle	385	9	127	99	-	-	-	-
Au nanoparticles on each CaP nanoparticle	-	-	-	-	-	-	2717	452
AF647 molecules on each CaP nanoparticle	1385	34	380	297	-	-	6250	1040
AF488 molecules on each CaP nanoparticle	-	-	-	-	9320	7920	1310	1210
Molar ratio Ca:Au by AAS	-	-	-	-	-	-	85:15	97:3
Molar ratio Ca:Au by EDX	-	-	-	-	-	-	93:7	95:5

For ultrasmall gold nanoparticles, the ratio of calcium to gold as determined by AAS and EDX was similar. Due to the presence of glutathione on the surface of the gold nanoparticles, they are expected to carry a negative charge as well at moderately alkaline pH. This explains why fewer gold nanoparticles are conjugated to negatively charged calcium phosphate nanoparticles. In contrast, the number of AF488 molecules clicked to calcium phosphate did not depend on the particle charge.

Of course, the numbers given in Tables 2 and 3 must be considered as being associated with a considerable uncertainty. This arises from a number of assumptions such as monodisperse and spherical nanoparticles. Furthermore, the applied analytical methods (UV–Vis, AAS) all carry some instrumental uncertainty. It is not possible to quantitatively determine all these uncertainties, but it is realistic to assume that the cargo numbers given in Tables 2 and 3 carry an uncertainty of at least ±30%. Nevertheless, they give at least an impression on the conjugated cargo and allow to identify trends in the conjugation efficiency.

To test the efficiency of the calcium phosphate nanoparticles to transport their cargo across the cell membrane into cells, uptake studies with HeLa cells were performed (Figure 9). All types of nanoparticles were well taken up by the cells, with a slight preference for positively charged nanoparticles. The cargo (Hem, BSA, Au-NPs) was well transported into the cells as indicated by the corresponding fluorescence signals. It has been shown earlier that many dissolved proteins alone cannot cross the cell membrane [36–39] but ultrasmall gold nanoparticles are well able to do that [40]. The co-localization of AF488 and AF647 indicates that the gold nanoparticles remained attached to the calcium phosphate nanoparticles during the uptake, i.e., that they are not separated and taken up by cells separately.

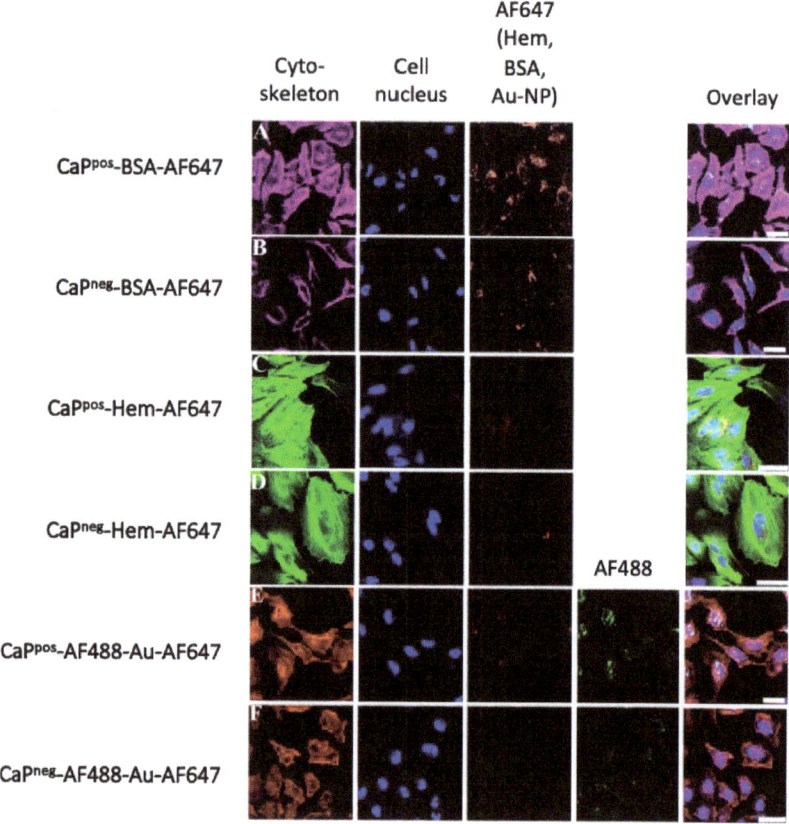

Figure 9. CLSM images demonstrate the uptake of functionalized calcium phosphate nanoparticles by HeLa cells. All particle types easily entered the cells. Actin staining with AF568-phalloidin (magenta; red) or AF488-phalloidin (green), nuclear staining with HOECHST33342 (blue). (**A**) CaP^{pos}-BSA-AF647, (**B**) CaP^{neg}-BSA-AF647, (**C**) CaP^{pos}-Hem-AF647, (**D**) CaP^{neg}-Hem-AF647, (**E**) CaP^{pos}-AF488-Au-AF647 and (**F**) CaP^{neg}-AF488-Au-AF647. Scale bars 20 µm.

Calcium phosphate nanoparticles enter eukaryotic cells via endocytosis [41] and end up in endolysosomes [36–38]. It was suggested that they will dissolve there under the acidic conditions in an endolysosome to release their cargo. It is very likely that the endolysosome bursts due to increased osmotic pressure after the dissolution of calcium phosphate [42], thus the cargo will eventually reach the cytosol.

Finally, it was shown by an MTT assay that the functionalized calcium phosphate nanoparticles were not cytotoxic (Figure 10).

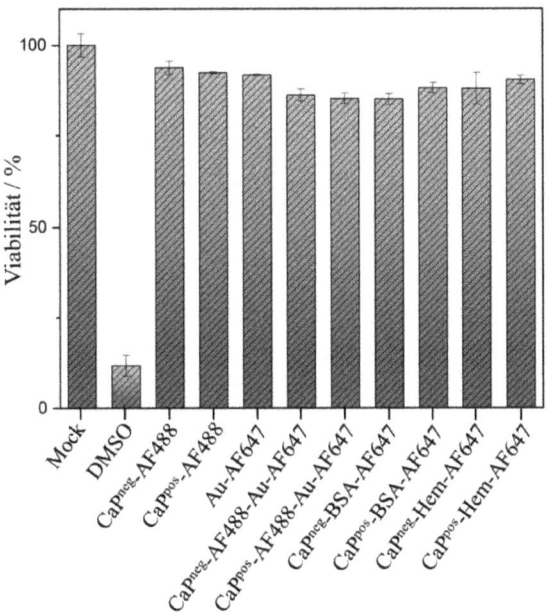

Figure 10. Viability of HeLa cells after 24 h incubation. Control groups: Mock = positive control, i.e., cells without treatment with nanoparticles. DMSO = negative control, where DMSO was added to the cells in DMEM instead of nanoparticles. The data show the average of three independent assays for each sample.

4. Conclusions

The conjugation of proteins to the surface of calcium phosphate nanoparticles via copper-catalyzed azide-alkyne cycloaddition (CuAAC, click chemistry) leads to a stable attachment. This was shown for the model proteins hemoglobin (Hem) and bovine serum albumin (BSA) and can be extended to other proteins, e.g., to antibodies or immunostimulatory proteins. To confirm the binding and to quantify the number of attached protein molecules, they were fluorescently labelled before the conjugation. Of course, this process can be skipped if functional particles shall be prepared, e.g., for therapeutic purpose. The prerequisite for CuAAC conjugation is the presence of an azide or an alkyne group on the protein. As calcium phosphate nanoparticles can be prepared with the complementary group, either one is possible. This procedure can be extended to coat large calcium phosphate nanoparticles (70 nm) with small gold nanoparticles (2 nm). There is a covalent bond between the two types of nanoparticles, again established via click chemistry. These super-particles are stable and not separated under cell culture conditions. This opens the pathway for the synthesis of multifunctional calcium phosphate particles that carry a cargo inside and functional gold nanoparticles, e.g., carrying biomolecules such as siRNA or peptides, on the outside.

The charge of the calcium phosphate nanoparticle and the cargo compounds under the moderately alkaline conditions during clicking influences the conjugation efficiency. If nanoparticle and cargo have opposite charges, the clicking is less efficient and leads to fewer clicked cargo units. Thus, a synthesis must be fine-tuned for every protein and potential cargo molecule or nanoparticle.

Finally, the fact that calcium phosphate nanoparticles were first conjugated with AF488 molecules and then with ultrasmall gold nanoparticles that attached to the remaining azide groups shows that it is possible to attach more than one species to the calcium phosphate surface. Thus, calcium phosphate nanoparticles that carry a range of species on their surface (e.g., a protein, a dye, and a gold nanoparticles) are easily conceivable. Nevertheless, their full characterization will involve careful purification, component labelling, and elemental analysis to ensure a reproducible synthesis with all cargo molecules in a defined ratio and concentration.

Author Contributions: Conceptualization, M.E.; methodology, M.E.; validation, K.K. and M.E.; formal analysis, K.K. and M.E.; investigation, K.K.; resources, M.E.; data curation, K.K. and M.E.; writing—original draft preparation, K.K.; writing—review and editing, M.E.; visualization, K.K.; supervision, M.E.; project administration, M.E. All authors have read and agreed to the published version of the manuscript.

Funding: This research received no external funding.

Data Availability Statement: Data are available from the authors upon request.

Acknowledgments: The authors thank Ursula Giebel for scanning electron microscopy and Dietrich Tönnes for X-ray powder diffraction. In addition, the authors thank the Imaging Center Campus Essen (ICCE) for access to confocal laser scanning microscopy on the Leica TCS SP8X Falcon CLSM, which was funded by the DFG (Großgeräteprogramm nach Art. 91b GG, INST 20876/294-1 FUGG to Shirley Knauer).

Conflicts of Interest: The authors declare no conflict of interest.

References

1. Carrillo-Carrion, C.; Bocanegra, A.I.; Arnaiz, B.; Feliu, N.; Zhu, D.C.; Parak, W.J. Triple-labeling of polymer-coated quantum dots and adsorbed proteins for tracing their fate in cell cultures. *ACS Nano* **2019**, *13*, 4631–4639. [CrossRef]
2. Behzadi, S.; Serpooshan, V.; Tao, W.; Hamaly, M.A.; Alkawareek, M.Y.; Dreaden, E.C.; Brown, D.; Alkilany, A.M.; Farokhzad, O.C.; Mahmoudi, M. Cellular uptake of nanoparticles: Journey inside the cell. *Chem. Soc. Rev.* **2017**, *46*, 4218–4244. [CrossRef] [PubMed]
3. Jing, X.; Zhang, Y.; Li, M.; Zuo, X.; Fan, C.; Zheng, J. Surface engineering of colloidal nanoparticles. *Mater. Horiz.* **2023**, *10*, 1185–1209. [CrossRef]
4. Lin, Z.; Aryal, S.; Cheng, Y.H.; Gesquiere, A.J. Integration of in vitro and in vivo models to predict cellular and tissue dosimetry of nanomaterials using physiologically based pharmacokinetic modeling. *ACS Nano* **2022**, *16*, 19722–19754. [CrossRef] [PubMed]
5. Jiang, Y.; Fan, M.; Yang, Z.; Liu, X.; Xu, Z.; Liu, S.; Feng, G.; Tang, S.; Li, Z.; Zhang, Y.; et al. Recent advances in nanotechnology approaches for non-viral gene therapy. *Biomater. Sci.* **2022**, *10*, 6862–6892. [CrossRef]
6. Chung, S.; Lee, C.M.; Zhang, M. Advances in nanoparticle-based mRNA delivery for liver cancer and liver-associated infectious diseases. *Nanoscale Horiz.* **2022**, *8*, 10–28. [CrossRef] [PubMed]
7. Fleming, A.; Cursi, L.; Behan, J.A.; Yan, Y.; Xie, Z.; Adumeau, L.; Dawson, K.A. Designing functional bionanoconstructs for effective in vivo targeting. *Bioconjug. Chem.* **2022**, *33*, 429–443. [CrossRef]
8. Patel, S.; Kim, J.; Herrera, M.; Mukherjee, A.; Kabanov, A.V.; Sahay, G. Brief update on endocytosis of nanomedicines. *Adv. Drug Deliv. Rev.* **2019**, *144*, 90–111. [CrossRef]
9. Pelaz, B.; Alexiou, C.; Alvarez-Puebla, R.A.; Alves, F.; Andrews, A.M.; Ashraf, S.; Balogh, L.P.; Ballerini, L.; Bestetti, A.; Brendel, C.; et al. Diverse applications of nanomedicine. *ACS Nano* **2017**, *11*, 2313–2381. [CrossRef]
10. Tsikourkitoudi, V.; Karlsson, J.; Merkl, P.; Loh, E.; Henriques-Normark, B.; Sotiriou, G.A. Flame-made calcium phosphate nanoparticles with high drug loading for delivery of biologics. *Molecules* **2020**, *25*, 1747. [CrossRef]
11. Uskoković, V.; Tang, S.; Nikolić, M.G.; Marković, S.; Wu, V.M. Calcium phosphate nanoparticles as intrinsic inorganic antimicrobials: In search of the key particle property. *Biointerphases* **2019**, *14*, 031001. [CrossRef]
12. Degli Esposti, L.; Carella, F.; Adamiano, A.; Tampieri, A.; Iafisco, M. Calcium phosphate-based nanosystems for advanced targeted nanomedicine. *Drug Dev. Ind. Pharm.* **2018**, *44*, 1223–1238. [CrossRef]
13. Sokolova, V.; Epple, M. Biological and medical applications of calcium phosphate nanoparticles. *Chem. Eur. J.* **2021**, *27*, 7471–7488. [CrossRef] [PubMed]
14. Dorozhkin, S.V.; Epple, M. Biological and medical significance of calcium phosphates. *Angew. Chem. Int. Ed.* **2002**, *41*, 3130–3146. [CrossRef]
15. Dorozhkin, S.V. Calcium orthophosphates in nature, biology and medicine. *Materials* **2009**, *2*, 399–498. [CrossRef]

16. Kozlova, D.; Chernousova, S.; Knuschke, T.; Buer, J.; Westendorf, A.M.; Epple, M. Cell targeting by antibody-functionalized calcium phosphate nanoparticles. *J. Mater. Chem.* **2012**, *22*, 396–404. [CrossRef]
17. Rojas-Sanchez, L.; Loza, K.; Epple, M. Synthesis and intracellular tracing surface-functionalized calcium phosphate nanoparticles by super-resolution microscopy (STORM). *Materialia* **2020**, *12*, 100773. [CrossRef]
18. Rojas-Sanchez, L.; Sokolova, V.; Riebe, S.; Voskuhl, J.; Epple, M. Covalent surface functionalization of calcium phosphate nanoparticles with fluorescent dyes by copper-catalysed and by strain-promoted azide-alkyne click chemistry. *ChemNanoMat* **2019**, *5*, 436–446. [CrossRef]
19. Damm, D.; Kostka, K.; Weingärtner, C.; Wagner, J.T.; Rojas-Sanchez, L.; Gensberger-Reigl, S.; Sokolova, V.; Überla, K.; Epple, M.; Temchura, V. Covalent coupling of HIV-1 glycoprotein trimers to biodegradable calcium phosphate nanoparticles via genetically encoded aldehyde-tags. *Acta Biomater.* **2022**, *140*, 586–600. [CrossRef]
20. Kopp, M.; Kollenda, S.; Epple, M. Nanoparticle–protein interactions: Therapeutic approaches and supramolecular chemistry. *Acc. Chem. Res.* **2017**, *50*, 1383–1390. [CrossRef]
21. Wang, Y.F.; Zhou, Y.; Sun, J.; Wang, X.; Jia, Y.; Ge, K.; Yan, Y.; Dawson, K.A.; Guo, S.; Zhang, J.; et al. The Yin and Yang of the protein corona on the delivery journey of nanoparticles. *Nano Res.* **2023**, *16*, 715–734. [CrossRef] [PubMed]
22. Stauber, R.H.; Westmeier, D.; Wandrey, M.; Becker, S.; Docter, D.; Ding, G.B.; Thines, E.; Knauer, S.K.; Siemer, S. Mechanisms of nanotoxicity—Biomolecule coronas protect pathological fungi against nanoparticle-based eradication. *Nanotoxicology* **2020**, *14*, 1157–1174. [CrossRef] [PubMed]
23. McKay, C.S.; Finn, M.G. Click chemistry in complex mixtures: Bioorthogonal bioconjugation. *Chem. Biol.* **2014**, *21*, 1075–1101. [CrossRef]
24. Kolb, H.C.; Finn, M.G.; Sharpless, K.B. Click chemistry: Diverse chemical function from a few good reactions. *Angew. Chem. Int. Ed.* **2001**, *40*, 2004–2021. [CrossRef]
25. Devaraj, N.K. The future of bioorthogonal chemistry. *ACS Cent. Sci.* **2018**, *4*, 952–959. [CrossRef]
26. Qie, Y.; Yuan, H.; von Roemeling, C.A.; Chen, Y.; Liu, X.; Shih, K.D.; Knight, J.A.; Tun, H.W.; Wharen, R.E.; Jiang, W.; et al. Surface modification of nanoparticles enables selective evasion of phagocytic clearance by distinct macrophage phenotypes. *Sci. Rep.* **2016**, *6*, 26269. [CrossRef] [PubMed]
27. Yi, G.; Son, J.; Yoo, J.; Park, C.; Koo, H. Application of click chemistry in nanoparticle modification and its targeted delivery. *Biomater. Res.* **2018**, *22*, 13. [CrossRef]
28. Rueden, C.T.; Schindelin, J.; Hiner, M.C.; DeZonia, B.E.; Walter, A.E.; Arena, E.T.; Eliceiri, K.W. ImageJ2: ImageJ for the next generation of scientific image data. *BMC Bioinform.* **2017**, *18*, 529. [CrossRef]
29. Klein, K.; Loza, K.; Heggen, M.; Epple, M. An efficient method for covalent surface functionalization of ultrasmall metallic nanoparticles by surface azidation, followed by copper-catalyzed azide-alkyne cycloaddition. *ChemNanoMat* **2021**, *7*, 1330–1339. [CrossRef]
30. Meldal, M.; Tornoe, C.W. Cu-catalyzed azide−alkyne cycloaddition. *Chem. Rev.* **2008**, *108*, 2952–3015. [CrossRef]
31. Dorozhkin, S.V. Synthetic amorphous calcium phosphates (ACPs): Preparation, structure, properties, and biomedical applications. *Biomater. Sci.* **2021**, *9*, 7748–7798. [CrossRef] [PubMed]
32. Dorozhkin, S.V. Nanosized and nanocrystalline calcium orthophosphates. *Acta Biomater.* **2010**, *6*, 715–734. [CrossRef] [PubMed]
33. Urch, H.; Vallet-Regi, M.; Ruiz, L.; Gonzalez-Calbet, J.M.; Epple, M. Calcium phosphate nanoparticles with adjustable dispersability and crystallinity. *J. Mater. Chem.* **2009**, *19*, 2166–2171. [CrossRef]
34. Phan, H.T.; Bartelt-Hunt, S.; Rodenhausen, K.B.; Schubert, M.; Bartz, J.C. Investigation of bovine serum albumin (BSA) attachment onto self-assembled monolayers (SAMs) using combinatorial quartz crystal microbalance with dissipation (QCM-D) and spectroscopic ellipsometry (SE). *PLoS ONE* **2015**, *10*, e0141282. [CrossRef]
35. Luner, S.J.; Kolin, A. A new approach to isoelectric focusing and fractionation of proteins in a pH gradient. *Proc. Natl. Acad. Sci. USA* **1970**, *66*, 898–903. [CrossRef]
36. Kollenda, S.; Kopp, M.; Wens, J.; Koch, J.; Schulze, N.; Papadopoulos, C.; Pöhler, R.; Meyer, H.; Epple, M. A pH-sensitive fluorescent protein sensor to follow the pathway of calcium phosphate nanoparticles into cells. *Acta Biomater.* **2020**, *111*, 406–417. [CrossRef]
37. Rotan, O.; Severin, K.N.; Pöpsel, S.; Peetsch, A.; Merdanovic, M.; Ehrmann, M.; Epple, M. Uptake of the proteins HTRA1 and HTRA2 by cells mediated by calcium phosphate nanoparticles. *Beilstein J. Nanotechnol.* **2017**, *8*, 381–393. [CrossRef]
38. Kopp, M.; Rotan, O.; Papadopoulos, C.; Schulze, N.; Meyer, H.; Epple, M. Delivery of the autofluorescent protein R-phycoerythrin by calcium phosphate nanoparticles into four different eukaryotic cell lines (HeLa, HEK293T, MG-63, MC3T3): Highly efficient, but leading to endolysosomal proteolysis in HeLa and MC3T3 cells. *PLoS ONE* **2017**, *12*, e0178260. [CrossRef]
39. Sokolova, V.; Rotan, O.; Klesing, J.; Nalbant, P.; Buer, J.; Knuschke, T.; Westendorf, A.M.; Epple, M. Calcium phosphate nanoparticles as versatile carrier for small and large molecules across cell membranes. *J. Nanopart. Res.* **2012**, *14*, 910. [CrossRef]
40. Zarschler, K.; Rocks, L.; Licciardello, N.; Boselli, L.; Polo, E.; Garcia, K.P.; De Cola, L.; Stephan, H.; Dawson, K.A. Ultrasmall inorganic nanoparticles: State-of-the-art and perspectives for biomedical applications. *Nanomedicine* **2016**, *12*, 1663–1701. [CrossRef]

41. Sokolova, V.; Kozlova, D.; Knuschke, T.; Buer, J.; Westendorf, A.M.; Epple, M. Mechanism of the uptake of cationic and anionic calcium phosphate nanoparticles by cells. *Acta Biomater.* **2013**, *9*, 7527–7535. [CrossRef] [PubMed]
42. Neuhaus, B.; Tosun, B.; Rotan, O.; Frede, A.; Westendorf, A.M.; Epple, M. Nanoparticles as transfection agents: A comprehensive study with ten different cell lines. *RSC Adv.* **2016**, *6*, 18102–18112. [CrossRef]

Disclaimer/Publisher's Note: The statements, opinions and data contained in all publications are solely those of the individual author(s) and contributor(s) and not of MDPI and/or the editor(s). MDPI and/or the editor(s) disclaim responsibility for any injury to people or property resulting from any ideas, methods, instructions or products referred to in the content.

Article

Activated Carbon from Sugarcane Bagasse: A Low-Cost Approach towards Cr(VI) Removal from Wastewater

Rana Ahmed [1], Inga Block [1], Fabian Otte [1], Christina Günter [2], Alysson Duarte-Rodrigues [3], Peter Hesemann [3], Amitabh Banerji [1] and Andreas Taubert [1,*]

1 Institute of Chemistry, University of Potsdam, D-14476 Potsdam, Germany
2 Institute of Geosciences, University of Potsdam, D-14476 Potsdam, Germany
3 ICGM, Université de Montpellier-CNRS-ENSCM, 34095 Montpellier, France
* Correspondence: ataubert@uni-potsdam.de; Tel.: +49-(0)331-977-5773

Abstract: The potential of pretreated sugarcane bagasse (SCB) as a low-cost and renewable source to yield activated carbon (AC) for chromate CrO_4^{2-} removal from an aqueous solution has been investigated. Raw sugarcane bagasse was pretreated with H_2SO_4, H_3PO_4, HCl, HNO_3, KOH, NaOH, or $ZnCl_2$ before carbonization at 700 °C. Only pretreatments with H_2SO_4 and KOH yield clean AC powders, while the other powders still contain non-carbonaceous components. The point of zero charge for ACs obtained from SCB pretreated with H_2SO_4 and KOH is 7.71 and 2.62, respectively. Batch equilibrium studies show that the most effective conditions for chromate removal are a low pH (i.e., below 3) where >96% of the chromate is removed from the aqueous solution.

Keywords: wastewater; water treatment; Cr(VI); heavy metals; adsorption; sugarcane bagasse; activated carbon; low-cost

1. Introduction

Chromium is a naturally occurring element and the 21st most abundant element in the earth's crust [1]. Although chromium in its metallic form was first described in 1797 [2], its technical applications were limited until the middle of the 20th century. This is due to the high toxicity of chromium compounds, which was discovered early on. However, with the advent of the modern industrialized society in the mid to late 1800s, there was a rapidly increasing need for heat- and corrosion-resistant alloys. Moreover, the textile industry, another rapidly growing sector, was actively looking into advanced chemical processes for tanning and dyeing [3–6]. In their well-known book "Handwörterbuch der reinen und angewandten Chemie" Liebig, Poggendorf and Wöhler stated about the element in 1842 [7]:

"Von dem metallischen Chrom wird noch keine Anwendung gemacht; um so wichtiger aber sind durch ihre technischen Anwendungen mehrere seiner Verbindungen geworden. Auf den Organismus wirken sie als Gifte, indessen hat man sie noch nicht als Arzneimittel anzuwenden versucht."

"Metallic chromium is not yet used, but several of its compounds have become all the more important through their technical applications. They act as poisons on the organism, but no attempt has yet been made to use them as medicines."

Though numerous applications have been identified and developed for metallic chromium and its salts since the days of Justus von Liebig, chromium compounds obviously remain harmful. Despite this, chromium compounds can be found in the environment all over the planet but foremost in developing countries, where high chromium levels dramatically endanger entire eco systems [1,6,8]. This is a development that could probably not have been foreseen by the scientists in Liebig's time but needs to be faced and resolved nowadays.

In terms of toxicity, one of the main challenges of Cr is its chemical versatility. Chromium can adopt oxidation states between –II and +VI, but Cr(III) and Cr(VI) are most prominent in natural environments [6]. Cr(III) is an essential nutrient for the human body and necessary for controlling blood glucose levels and insulin action in tissues [9]. Cr(VI) is a carcinogen, highly mobile in aqueous environments, highly soluble in water, and 500 times more toxic than Cr(III) [5,10]. Because of its high water solubility and high mobility in the environment, Cr(VI) can enter the terrestrial food chain and finally end up in higher animals and humans. Once taken up, Cr(VI) causes liver and kidney damage, asthma, and skin ulcerations, along with immunotoxic, genotoxic, and neurotoxic effects, among others [3].

Plants also take up Cr(VI). Consequently, Cr(VI) can be found in plant roots, shoots, stems, leaves, and seeds [11,12]. Cr(VI) inhibits germination and affects root, stem, and leaf growth [6]. Moreover, the quality of flowers, crop yields, photosynthesis, respiration, and symbiotic nitrogen fixation are severely disturbed by Cr(VI) [11]. Therefore, the reduction of Cr(VI) levels in water bodies is a significant and large-scale problem needing reliable treatments and solutions [6,13].

Electroplating, leather tanning, dye and pigment, steel and alloy, automobile, ammunition, paint, and textile manufacturing industries are the most common sources of Cr(VI) [3–6]. Many of these industries release significant amounts of Cr(VI) into surface water bodies, which leads to critical Cr(VI) levels around these manufacturing sites [3,10].

Some of the most common methods to control Cr(VI) levels in water are ion exchange, precipitation, flocculation, reverse osmosis, electrocoagulation, electrodialysis, membrane filtration, solvent extraction, and adsorption [3,5,10,14,15]. Each of these methods has advantages and drawbacks. For instance, ion exchange, electrocoagulation, reverse osmosis, electrodialysis, and membrane filtration require a high amount of energy and regular maintenance [3,6,10]. Industries in developing countries are typically not able to manage the high costs of those treatments. Additionally, some of these treatments produce tremendous amounts of sludge as a byproduct. Deposited sludge waste then triggers secondary pollution and requires further management [6,10]. As a result, and because a large number of these manufacturing sites are located in developing countries [13,16,17], there is an ever-growing need for effective yet (very) low-cost approaches towards Cr(VI) removal.

Activated carbons (ACs) are very effective adsorbents [13,17,18], but the very high demand overall and the significant cost of ACs limit their use, especially in developing countries. Hence, low-cost adsorbents and effective yet cheap activation of suitable raw materials have become a substantial focus of science and technology [17].

Generally, the effective activation of carbonaceous raw materials involves chemical or physical modifications that significantly improve the quality of the adsorbents. The enhanced formation of micro- and mesopores, generation of large surface areas, and the addition of functional groups to the adsorbent surface are key parameters for improved performance of (low-cost) ACs [19]. Acids such as H_3PO_4 [20,21], H_2SO_4 [22–24], HCl [20], or HNO_3 [19,25], strong bases such as KOH [21,24] or NaOH [20,21], and salts such as $ZnCl_2$ [21,26], $CaCl_2$ [27], or $FeCl_3$ [21] are popular activators for AC production. Depending on the treatment, different types of materials are obtained. For instance, NaOH or HCl pretreatments reduce the silica and ash content, while H_3PO_4 pretreatments yield P-containing functional groups and highly microporous materials [20]. H_2SO_4 produces acidic surface oxygen species and increases the specific surface area [28], while HNO_3 pretreatment yields cellulose nitrate groups and high pore volume [19]. $ZnCl_2$ [26] and KOH [29] yield smaller pore sizes and higher surface areas in the ACs. Interestingly, the reasons why these different treatments produce different pore sizes are still largely unclear.

Nowadays, numerous raw materials have been identified as suitable sources for ACs for water treatment, including agricultural waste [4,30]. Popular raw materials include pomegranate peel, orange peel, banana peel, corn cobs, rice husks, sugarcane bagasse (SCB), sawdust, plant leave waste, tea leaves, cottonseed, coconut shell, and rice straw [13,16].

Further studies suggest that adsorbents prepared from raw or carbonized SCB [31–35], coconut shell [36], or rice husks [30,37] effectively remove metal ions and dyes.

The current study focuses on SCB as a raw material for ACs for Cr(VI) removal from water. SCB is one of the most common and cheapest agricultural wastes in the world and is a major byproduct of the sugar industry [33]. Annually, approximately 54 million tons of dry SCB are produced worldwide. To avoid further management costs, the majority of the SCB is burned directly on the sugarcane fields, causing significant air pollution [33].

However, SCB is an interesting raw material because of its high lignin and cellulose content, which is rich in carbonyl, hydroxyl, ether, phenol, amine, and sulfhydryl groups [33,38]. Being agricultural waste, raw SCB also contains impurities that often require thorough pretreatment and modification to boost the adsorption capacities of the resulting ACs [38].

The functional groups present on the surface of SCB-based ACs act as active adsorption sites that bind different heavy metal ions to the surface [35,38]. Therefore, utilizing SCB waste as a source of ACs not only combats the global wastewater crisis but also reduces the volume of agricultural waste. Hence, considering the increasing demand for a low-cost, sustainable, and easy wastewater treatment process, SCB is an attractive starting material [33].

A final reason for selecting SCB as the starting material is the fact that there are regions, such as Bangladesh, that offer a large amount of sugarcane fields and a large textile industry in close proximity. As a result, SCB as highly abundant agricultural waste and Cr(VI) contamination, which poses severe environmental and health problems, occurs in exactly the same geographic region. The current study, therefore, builds on the fact that local agricultural waste could be used to improve local water quality.

2. Materials and Methods

Chemicals and apparatus. Sulfuric acid (>95% H_2SO_4), zinc chloride dihydrate ($ZnCl_2 \ast 2\,H_2O$), ortho-phosphoric acid (85% H_3PO_4), nitric acid (65% HNO_3), hydrochloric acid (37% HCl), potassium hydroxide pellets (≥90% KOH), sodium hydroxide (≥98% NaOH), potassium dichromate (≥99% $K_2Cr_2O_7$), sodium chloride (≥99.99% NaCl), 1,5-diphenylcarbazide, and acetone were of analytical grade quality and supplied by Merck (Darmstadt, Germany). All chemicals were used without further purification.

Collection and purification of SCB. Raw SCB was collected from local sugarcane-juice vendors of Mymensingh, Bangladesh, using a white polyethylene bag for transport. The raw SCB was cut into small pieces with scissors and soaked in tap water to remove dirt and sugar. Afterwards, the SCB was washed thoroughly with distilled water until the residual water was dirt-free (i.e., clear). Next, the washed SCB was dried in a preheated oven at 110 °C for 24 h. After drying, the SCB was ground with a Philips HR3655/00 Blender until the particle size was below 0.50 mm. The particle size below 0.50 mm was ensured by passing the SCB powder through a 0.50 mm sieve (VEB Metallweberei, Neustadt/Orla). The ground SCB powders were then stored in an airtight glass bottle until further use.

Pre-carbonization treatment of SCB. A total of 25 g of the clean and dry SCB powder was mixed with 250 mL of 3M H_2SO_4 in a three-necked round bottom flask with a condenser in an oil bath. The oil bath was maintained at 80 °C, and the mixture was stirred at 200 rpm for 24 h. Then the mixture was filtered using vacuum filtration to separate the pretreated SCB. The SCB was then washed with hot distilled water until the pH was 6–7. After that, the powder was dried overnight at 105 °C, and the dry, treated SCB was kept in an airtight bottle until further use. The same process was used for treatment with 3M HCl, 30% (v/v) H_3PO_4, 30% (v/v) HNO_3, 30% (w/v) $ZnCl_2$, 1M KOH, and 1M NaOH.

Preparation of activated carbon. All SCB samples were loaded in a custom-made pyrolysis oven described previously [39], and the oven was then heated to 700 °C at 5 °C/min under Argon. The powders were kept for 1 h at 700 °C, then the oven was switched off and left overnight to cool to room temperature. After cooling, all carbonized

SCB (cSCB) powders were stored in airtight bottles until further use. Table 1 summarizes all materials studied in this work.

Table 1. Overview of cSCB powders studied in this article.

Material	Pretreatment
SCB	None (raw SCB after washing and drying, no further pretreatment)
SB1	3M H_2SO_4
SB2	30% H_3PO_4
SB3	30% $ZnCl_2$
SB4	30% HNO_3
SB5	3M HCl
SB6	1M NaOH
SB7	1M KOH

Characterization and analysis. *Infrared spectroscopy* was conducted at room temperature on a Nicolet iS5 (Thermo Scientific, Waltham MA, USA) with an iD7 attenuated total reflection (ATR) unit, a resolution of 1 cm^{-1}, and 32 scans per measurement from 400–4000 cm^{-1}.

Scanning electron microscopy (SEM) was conducted on a JEOL JSM-6510 (JEOL, Freising, Germany) SEM operated at 5 kV. Prior to imaging, all samples were sputter-coated with Au/Pd for 75 s and 18 mA using an SC7620 mini sputter coater (Quorum Technologies, Lewes, UK).

X-ray powder diffraction (XRD) data were collected on a PANalytical Empyrean powder X-ray diffractometer in a Bragg–Brentano geometry. It is equipped with a PIXcel1D detector using Cu K_α radiation (λ = 1.5419 Å) operating at 40 kV and 40 mA. θ/θ scans were run in a 2θ range of 4–70° with a step size of 0.0131° and a sample rotation time of 1 s. The diffractometer is configured with a programmable divergence and anti-scatter slit and a large Ni-beta filter. The detector was set to continuous mode with an active length of 3.0061°.

Surface area and pore sizes were determined via nitrogen sorption at 77 K using a Micromeritics Tristar (Micromeritics Instrument Corp., Norcross, GA, USA). Prior to all measurements, the materials were degassed to about 2 Pa at 353 K for 10 h. The specific surface area (SSA) was calculated via the Brunauer–Emmett–Teller (BET) approach. Average pore sizes were estimated from the adsorption branch of the isotherm using the Barrett–Joyner–Halenda (BJH) method. The pore volume was determined at $P/P_0 > 0.99$.

Determination of the point of zero charge, pH_{pzc}, was performed according to published protocols [40,41]. First, 50 mL of a 0.01M NaCl solution was transferred to a 100 mL Erlenmeyer flask, and the initial pH (pH_i) was adjusted to 2–12 using (0.1 to 1M) H_2SO_4 and NaOH. The pH was recorded with a Digital pH meter GPH014 (PCE Instruments UK Ltd., Hamble-le-Rice, UK, refillable electrode). Then 0.15 g of cSCB was added to the solution, and the mixture was agitated on a magnetic stirrer for 24 h at 200 rpm. Thereafter the solution was vacuum–filtered with Whatman® Grade 1 filter paper (diameter 11 cm). The final pH (pH_f) of the solution was measured, and $\Delta pH = pH_f - pH_i$ was plotted vs. pH_i. The point where ΔpH and pH_i are zero is pH_{pzc} for the corresponding cSCB adsorbent.

Batch adsorption experiments. All batch adsorption studies were carried out in 50 mL Erlenmeyer flasks containing 25 mL of an aqueous $K_2Cr_2O_7$ solution with a concentration of 1 g/L. After the addition of the cSCB adsorbent, the mixtures were placed on a magnetic stirrer and stirred at 200 rpm at 30 °C for 24 h. The suspension was filtered, and the residual concentration of Cr(VI) was determined with a Vernier® UV-VIS Spectrophotometer (Vernier Software & Tech SE, Vernier Science Education, Beaverton, OR, USA) via Method 7196A as published by the US EPA [42]. Calibration was performed with a dilution series of aqueous $K_2Cr_2O_7$ with concentrations between 10 and 100 mg/L ($R^2 > 0.998$). The impact of solution pH on adsorption was evaluated from adsorption experiments with 20 mg/L $K_2Cr_2O_7$ solutions with starting pH from 1 to 11.

The fraction of Cr(VI) removed was calculated using Equation (1)

$$\%R = \frac{(C_0 - C_f)100}{C_0} \quad (1)$$

where %R is the percentage of Cr(VI) removed, while C_0 and C_f are the initial and final Cr(VI) concentrations (mg/L) of the solutions before and after treatment.

The influence of the initial Cr(VI) concentration and the influence of the adsorbent dose were studied using analogous approaches by either varying the $K_2Cr_2O_7$ concentration or the adsorbent dose while keeping all other conditions constant. The adsorption kinetics were studied by variation of the contact time of the adsorbent and $K_2Cr_2O_7$ solution. All experiments were repeated three times, and all spectrophotometric measurements were duplicates yielding six measurements per experimental condition. Values reported are mean and standard deviations obtained from these six values.

3. Results

Figure 1 shows X-ray diffraction (XRD) patterns obtained from the raw SCB and the chemically activated SCBs after pyrolysis. The XRD pattern of raw SCB shows three broad halos at 16.34 (110), 22.05 (002), and 34.68° (004) 2θ. These signals can be assigned to the presence of cellulose I [43,44].

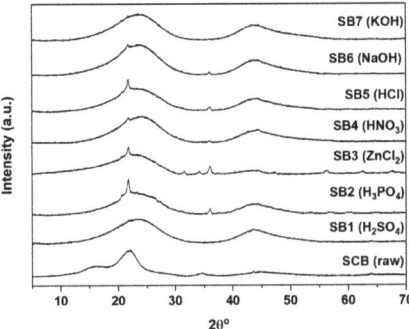

Figure 1. X-ray diffraction patterns obtained from SCB adsorbents after pyrolysis.

All pretreated and carbonized SCBs (cSCBs) only show two broad halos (not three as in the raw SCB) at around 24 and around 44° 2θ that can be assigned to graphitized carbon [20,45,46]. Moreover, sharp reflections at 21.79 (011) and 36.01° (112) 2θ in the patterns obtained from SB2, SB4, SB5, and SB6 can be assigned to cristobalite (ICDD 98-003-4933). Additional sharp reflections observed in the XRD patterns obtained from SB3 at 2θ = 31.67 (010), 36.02 (011), 56.199 (110), 62.52, (020), and 67.57° (112) 2θ can be assigned to wurtzite ZnO (ICDD 98-018-2355).

Only SB1 and SB7 (see Table 1 for assignments) show no additional reflections, indicating that there are no further crystalline (inorganic) compounds present in the materials. The halos at 2θ = 23.61 (002) and 43.82° (011) can be assigned to graphite 2H (ICDD 98-007-6767). Based on the XRD data, therefore, SB1 and SB7 are the purest carbons produced here (i.e., materials without further components visible in the XRD data). As a result, these two materials were chosen for an in-depth investigation along with the raw carbonized SCB for comparison.

Figure 2 shows representative IR spectra obtained from raw SCB, SB1, and SB7. Spectra of raw SCB show a broad peak centered at 3614 cm^{-1}, which can be assigned to the O-H stretching vibration of intramolecular hydrogen bonds in carbohydrates and lignin [47]. Bands centered at 3029 cm^{-1} stem from C-H bond vibrations [48]. A small band at 1770 cm^{-1} can be assigned to C=O bond vibrations of aldehydes, ketones, or carboxyl groups [48,49]. Furthermore, bands at 1612 and 1525 cm^{-1} are attributed to the aromatic rings present in lignin and to adsorbed water, while a band at 1281 cm^{-1} can be assigned to C-O stretching vibrations in lignin [44,47,49].

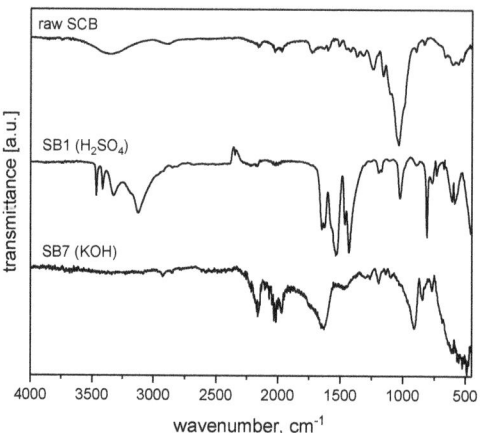

Figure 2. IR spectra obtained from raw SCB, SB1, and SB7 after pyrolysis. IR spectra of all materials can be found in the Supporting Information. While the spectra vary from material to material, all IR bands can be assigned to C-C, C=C, C=O, and C-H vibrations stemming from the organic and carbonized components of the materials. SB6 and SB7 exhibit rather noisy spectra, but also here, all bands and shoulders can be assigned to vibrations in the activated carbonaceous material.

IR spectra of SB1 show bands at 3323 and 1526 cm^{-1}, which correspond to the N-H stretching vibration of amines and amides along with nitro groups [50]. A band at 3118 cm^{-1} originates from C-H stretching vibrations, while a band at 1428 cm^{-1} can be attributed to C-H deformation vibrations [47,51]. Bands at 2390 and 1636 cm^{-1} indicate the presence of aromatic C=C bonds and C=O bonds in carboxyl groups [13,52]. Furthermore, the presence of aromatic moieties and amines in SB1 is further corroborated by bands at 1192 and 1024 cm^{-1} [34,50,51,53].

IR spectra of SB7 show a broad signal centered at 3375 cm^{-1}, which indicates the presence of O-H bonds, including hydrogen bonding [54]. A band at 1635 cm^{-1} again indicates the presence of C=O bonds, likely from ketones, aldehydes, or carboxylates [26]. Further bands between 1200 and 1100 cm^{-1} can be assigned to C-O vibrations in ether, alcohol, phenol, acid, or ester moieties [54], while a broad and poorly resolved group of bands at around 618 cm^{-1} is from C-H and C-C stretching vibrations [52].

Figure 3 shows representative scanning electron microscopy (SEM) data obtained from raw SCB, SB1, and SB7. SEM images of SCB show a large variety of particle sizes with numerous different morphologies. The surface of the particles is rather dense without much substructure.

SB1 is quite different from raw SCB. SB1 exhibits a very broad size distribution along with a fraction of smaller particles with sharper edges than those observed in SCB. Moreover, all particles also exhibit some surface roughness, and there are also large particles with micrometer-sized pores. All particles show cracks and appear to have a larger fraction of open surface than raw SCB. The opening and surface roughening can be assigned to the pre-activation with H_2SO_4 combined with thermal treatment that removes lignin and inorganic components, which leaves pores and cavities on the adsorbent surface (and likely also inside the powders), consistent with previous studies [13,22,52].

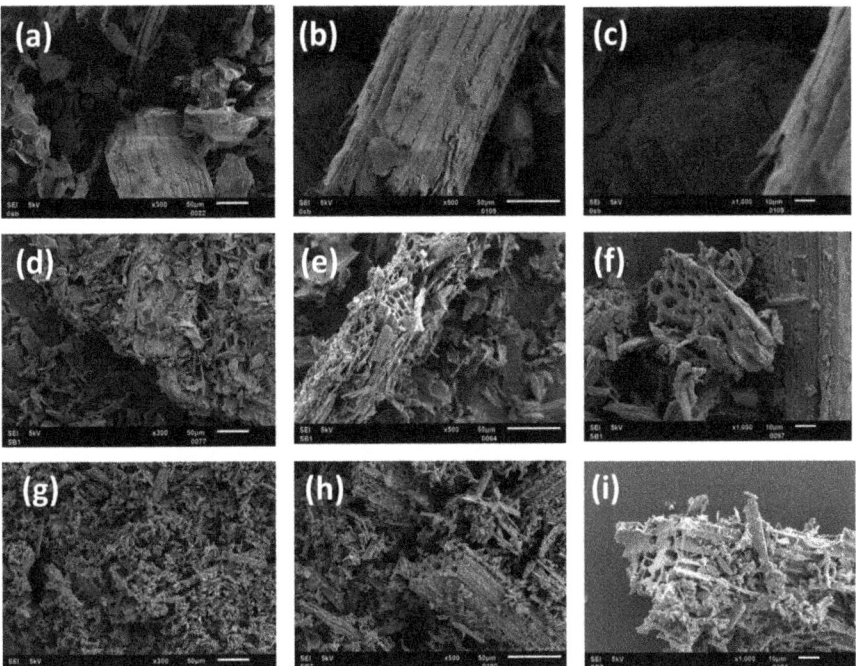

Figure 3. SEM images of SCB (**a**–**c**), SB1 (**d**–**f**), and SB7 (**g**–**i**) at different magnifications.

Likewise, SB7 is quite different from SCB. Again, smaller particles are visible. They appear finer and somewhat smaller than in SB1, but overall, the material exhibits a very broad particle size distribution. Additionally, SB7 exhibits sharp edges of the particles, and cracks and pores are visible in all particles.

Figure 4 shows complementary nitrogen sorption data for the three materials, SCB, SB1, and SB7. All datasets are very similar and only show a very limited N_2 uptake and a correspondingly small surface area of all materials. SCB is essentially non-porous (note that nitrogen sorption does not detect macropores, which are visible in the SEM images). The surface area determined for SB1 is 6–8 m^2/g, and SB7 has a surface area of around 30 m^2/g. Overall, these data indicate that the materials are essentially macroporous powders with negligible mesopore and micropore fractions.

The rather low surface areas observed for these materials are comparable with many other examples from the literature [14,16,18,19]. Many ACs made from agricultural waste show surface areas on the order of 10 to ca. 100 m^2/g. This likely stems from the fact that all these ACs are essentially based on carbohydrates and that the original materials have micrometer-sized features in the plants. Apparently, most treatments are not able to open up significant amounts of micro- or mesopores, but rather the treatments seem to modify the surfaces chemically, which in turn then alters the adsorption behavior.

Numerous studies have shown that pH is a crucial parameter when it comes to adsorption [4,55,56]. Figure 5a illustrates the adsorption efficiency of SB1 and SB7 vs. the initial solution pH. SB1 removes ca. 96% of the chromate within 24 h at pH 1–3. This is confirmed by visual inspection: solutions treated with SB1 at pH 1–3 show essentially no remaining color, Figure 5b, indicating an almost complete Cr(VI) removal. At pH 4 and higher, the fraction of Cr(VI) that is removed decreases and reaches ca. 20% at pH 9 and higher.

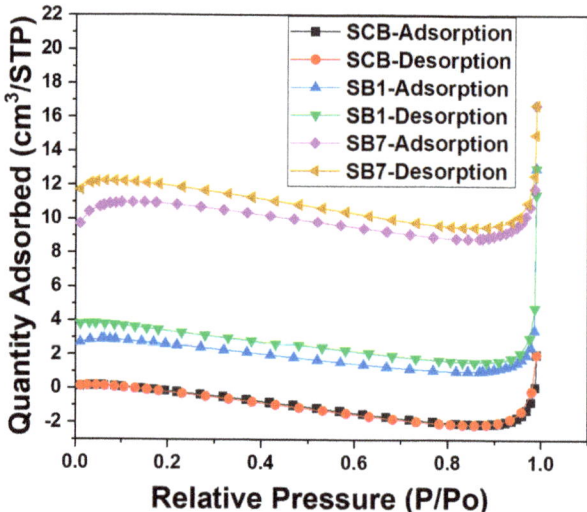

Figure 4. Nitrogen sorption isotherms obtained from SCB, SB1, and SB7. Adsorption and desorption branches are identified separately. The somewhat unusual shape of the isotherms with a decrease at intermediate pressures can be assigned to the very low surface areas of all materials studied here.

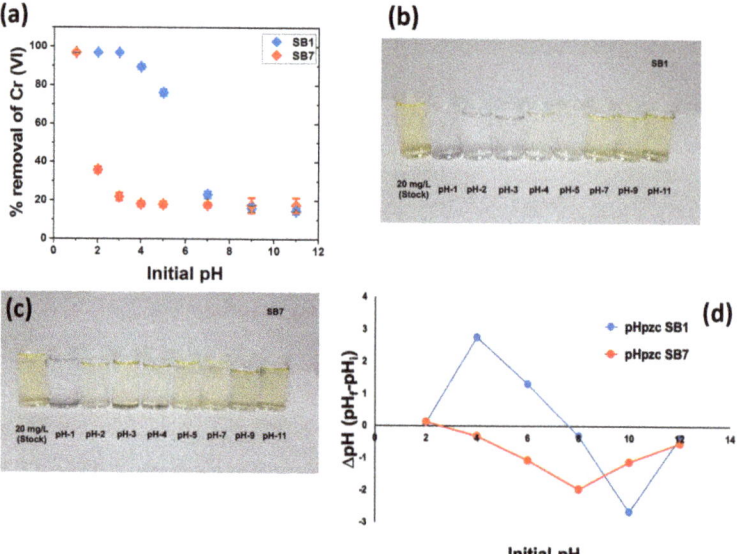

Figure 5. (a) Cr(VI) removal vs. initial pH for SB1 and SB7 (C_0 = 20 mg/L, m = 0.1 g, V = 25 mL, t = 24 h, T = 30 °C, rpm = 200). (b) Solution color vs. pH for treatment with SB1. (c) Solution color vs. pH for treatment with SB7. (d) Determination of pH_{pzc} for SB1 and SB7 (C_0 = 0.01 M NaCl, m = 0.15 g, V = 50 mL, t = 48 h, T = 25 °C, rpm = 200).

In contrast, SB7 only shows a high removal rate of ca. 96% Cr(VI) at pH 1. Already at pH 3, the removal rate drops to below 40% and reaches ca. 20% at pH 4. Again, this is corroborated by visual inspection: solutions treated at pH 2 and higher are still yellow, indicating incomplete Cr(VI) removal, Figure 5c.

To better quantify and understand this observation, the point of zero charge (pH_{pzc}) was determined, Figure 5d. The pH_{pzc} is 7.71 for SB1, showing that the surface is positively charged below pH ca. 7.7. In contrast, the pH_{pzc} of SB7 is 2.62, indicating that SB 7 is only positively charged below ca. pH 2.6, which is much lower than what is observed for SB1.

Figure 6 shows the effect of the initial Cr(VI) concentration with the SB1 adsorbent at an initial pH of 3. After 24 h, the highest Cr(VI) removal rate is observed at low concentrations. Generally, the removal rate decreases with increasing initial Cr(VI) concentration. This is consistent with previous data [4,13,50,56].

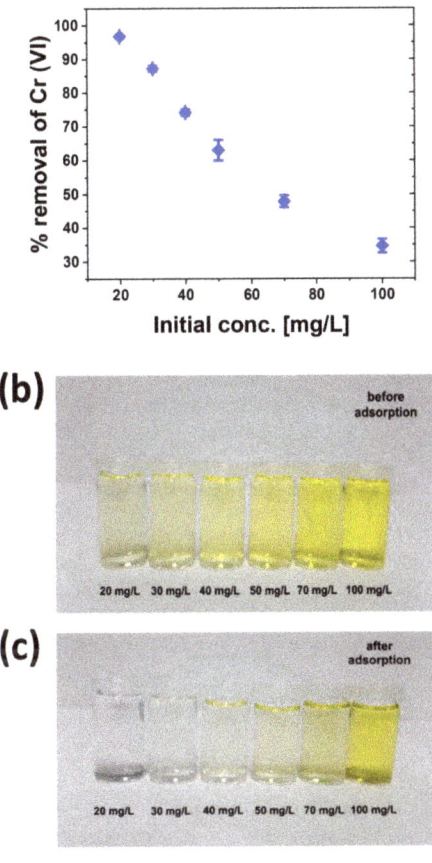

Figure 6. (a) Effect of initial SB1 concentration on Cr(VI) removal (pH = 3, m_{SB1} = 0.1 g, V = 25 mL, t = 24 h, T = 30 °C, rpm = 200). (b) Color of the Cr(VI) solutions before treatment with SB1. (c) Color of the Cr(VI) solutions after treatment with SB1.

Figure 7 shows the effects of the adsorbent dose on Cr(VI) removal. Generally, Cr(VI) removal increases as the adsorbent dose increases. At a dose of 0.1 g/25 mL and higher, no further color changes (or changes in the absorption data) are observed. The solution is essentially colorless, and >96% of the Cr(VI) is removed from the solution. Therefore, 0.1 g/25 mL was subsequently used for further batch experiments.

Figure 8 shows the effect of contact time on Cr(VI) removal. The data show a monotonous increase in Cr(VI) removal up to ca. 8 h. After that, the Cr(VI) removal is much slower and finally levels off at ca. 96% at around 24 h.

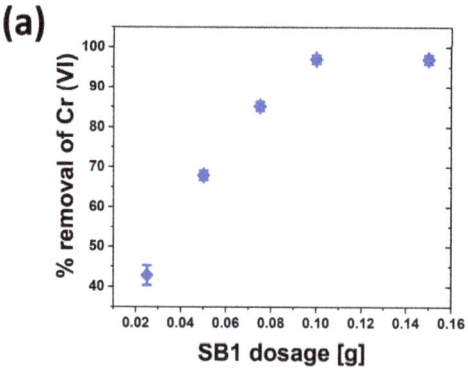
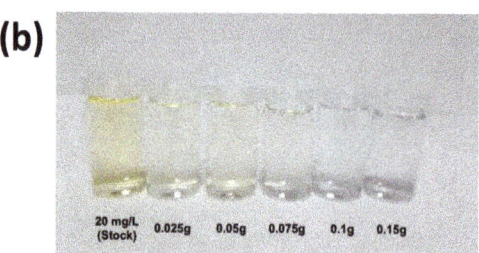

Figure 7. (a) SB1 dose vs. Cr(VI) removal (C_0(Cr) = 20 mg/L, pH = 3, V = 25 mL, t = 24 h, T = 30 °C, rpm = 200). (b) Photograph of the solutions after treatment with increasing adsorbent dose.

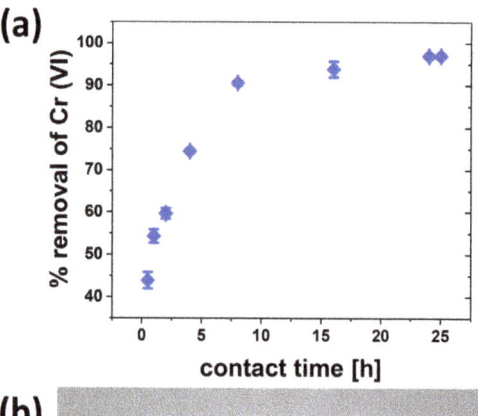
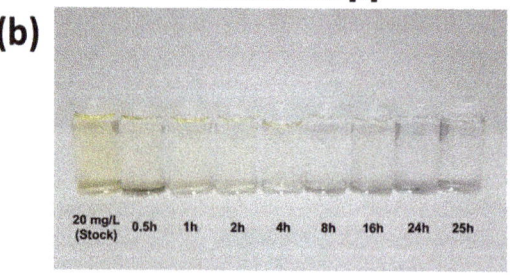

Figure 8. (a) Contact time vs. Cr(VI) removal (C_0(Cr) = 20 mg/L, pH = 3, m = 0.1 g, V = 25 mL, t = 24 h, T = 30 °C, rpm = 200). (b) Photograph of solutions after treatment.

4. Discussion

As stated in the introduction, low-cost and low-tech approaches towards Cr(VI) removal from aqueous solutions are highly important. The current study proposes a new approach based on activated carbons from sugarcane bagasse, a high-volume agricultural waste. In particular, we have identified two promising materials (SB1 and SB7) that are clean carbon materials that effectively remove dichromate from an aqueous solution. Both materials show a strong pH dependence on Cr(VI) removal: while SB1 removes up to 96% of Cr(VI) up to pH 3, SB7 is only effective at pH 1, Figure 5a.

This behavior can be assigned to the different points of zero charge in the two materials, Figure 5d. Considering the pH_{pzc} of 2.62 that is observed for SB7, these data indicate that already at pH 2, the charge density and overall charge of the SB7 surface is too low to produce a strong interaction between the adsorbent surface and the chromate or dichromate ions in solution. In contrast, the much higher pH_{pzc} of ca. 7.2 of SB7 suggests that here a much larger window of positive charge exists, and thus, there is a much wider pH window where the electrostatic interaction between the positively charged adsorbent surface and the dissolved anions is effective. Such an observation is consistent with previous studies [4,55–58]. Similarly, low pH_{pzc} values are also consistent with the literature; these effects have been assigned to surface adsorption of OH$^-$ from the pretreatment, with hydroxides then being released and acting as a buffer and lowering the pH_{pzc} [59].

Once the pH is high enough to supersede the pH_{pzc}, the surface of the adsorbent is negatively charged. This should result in electrostatic repulsion between adsorbent and anions in solution, which drives the adsorption capacities down at higher pH. Likely, the remaining low adsorption of below 20% is then due to effects such as hydrogen bonding or direct interaction with individual surface groups such as amines.

These data clearly show that acid treatment is more attractive in the current case because this treatment keeps the pH_{pzc} higher and thus provides a material that is effective in Cr(VI) removal from pH 1–3, similar to a previous study showing that net negative charges are advantageous for Cr(VI) removal [23]. Evaluation of the surface areas, Figure 4, also shows that the dominating effect is indeed the pH_{pzc} and not the surface areas, as they are very low in all cases investigated here.

As an interesting and technologically relevant observation, it must be noted that the pH of Cr(VI) containing industrial effluent occurs mostly at around 3 [4,60,61] or even below 3 [62]. As a result, especially SB1 is a prime candidate for direct use without any further modifications for Cr(VI) removal from industrial wastewaters. The one remaining challenge is the fact that the Cr(VI) concentrations in real wastewaters are often much higher than even the highest concentration studied here, Figure 6. Moreover, the adsorbent doses and contact times may need to be adjusted to account for a large-scale, real-life system and may therefore be different from the contact times studied here, Figures 7 and 8. There is thus a further need to improve the materials and the overall process to be able to directly use the SB1 material for Cr(VI) removal, but in spite of this, the ease of production, the very good performance under near-realistic pH conditions, and the low cost of the raw materials make the materials and the process an attractive candidate for further development.

5. Conclusions

Sugarcane bagasse (SCB) is a suitable raw material for the fabrication of activated carbons for Cr(VI) removal from synthetic wastewater at conditions that are reminiscent of real wastewaters from tanning and other industries. Pretreatments with KOH and H_2SO_4 remove impurities from the raw materials. After carbonization of these pretreated raw materials, effective adsorbents for Cr(VI) removal at low pH can be obtained. Acid activation yields materials that can be used between pH 1 and 3 with good to excellent removal rates; materials obtained from KOH-treated SCB only remove Cr(VI) at a very low pH of 1. Overall, pretreated SCB is a cheap and abundant carbon source that can effectively be converted to AC to remove Cr(VI) from wastewater.

Supplementary Materials: The following supporting information can be downloaded at: https://www.mdpi.com/article/10.3390/chemistry5020077/s1. IR spectra of SB1 to SB7.

Author Contributions: Conceptualization, R.A., A.B. and A.T.; methodology, all authors; software, R.A. and A.D.-R.; validation, all authors (R.A., I.B., F.O., C.G., A.D.-R., P.H., A.B. and A.T.); formal analysis, all authors except F.O.; investigation, all authors; resources, A.B., P.H. and A.T.; data curation, all authors; writing—original draft preparation, R.A, A.B. and A.T.; writing—review and editing, all authors; visualization, R.A., A.D.-R., P.H. and C.G.; supervision, A.B. and A.T.; project administration, A.B. and A.T.; funding acquisition, A.B., P.H. and A.T. All authors have read and agreed to the published version of the manuscript.

Funding: This research was funded by the University of Potsdam (grant #53170000, #53140000).

Data Availability Statement: The data presented in this study are available in the current article.

Conflicts of Interest: The authors declare no conflict of interest. The funders had no role in the design of the study; in the collection, analyses, or interpretation of data; in the writing of the manuscript; or in the decision to publish the results.

References

1. Ertani, A.; Mietto, A.; Borin, M.; Nardi, S. Chromium in Agricultural Soils and Crops: A Review. *Water. Air. Soil Pollut.* **2017**, *228*, 190. [CrossRef]
2. Ukhurebor, K.E.; Aigbe, U.O.; Onyancha, R.B.; Nwankwo, W.; Osibote, O.A.; Paumo, H.K.; Ama, O.M.; Adetunji, C.O.; Siloko, I.U. Effect of hexavalent chromium on the environment and removal techniques: A review. *J. Environ. Manag.* **2021**, *280*, 111809. [CrossRef] [PubMed]
3. Mitra, S.; Sarkar, A.; Sen, S. Removal of chromium from industrial effluents using nanotechnology: A review. *Nanotechnol. Environ. Eng.* **2017**, *2*, 11. [CrossRef]
4. Itankar, N.; Patil, Y. Management of hexavalent chromium from industrial waste using low-cost waste biomass. *Procd. Soc. Behav.* **2014**, *133*, 219–224. [CrossRef]
5. Joshi, K.M.; Shrivastava, V.S. Photocatalytic degradation of Chromium (VI) from wastewater using nanomaterials like TiO_2, ZnO, and CdS. *Appl. Nanosci.* **2011**, *1*, 147–155. [CrossRef]
6. Tumolo, M.; Ancona, V.; De Paola, D.; Losacco, D.; Campanale, C.; Massarelli, C.; Uricchio, V.F. Chromium pollution in European water, sources, health risk, and remediation strategies: An overview. *Int. J. Environ. Res. Public Health* **2020**, *17*, 5438. [CrossRef]
7. Liebig, J.; Poggendorff, J.C.; Wöhler, F. Chrom. In *Handwörterbuch der Reinen und Angewandten Chemie*, 1st ed.; Vieweg und Sohn: Braunschweig, France, 1842; Volume 2, pp. 264–284.
8. Wang, Y.; Su, H.; Gu, Y.; Song, X.; Zhao, J. Carcinogenicity of chromium and chemoprevention: A brief update. *Oncotargets Ther.* **2017**, *10*, 4065–4079. [CrossRef]
9. Laschinsky, N.; Kottwitz, K.; Freund, B.; Dresow, B.; Fischer, R.; Nielsen, P. Bioavailability of chromium(III)-supplements in rats and humans. *Biometals* **2012**, *25*, 1051–1060. [CrossRef]
10. Nur-E-Alam, M.; Mia, M.A.S.; Ahmad, F.; Rahman, M.M. An overview of chromium removal techniques from tannery effluent. *Appl. Water Sci.* **2020**, *10*, 205. [CrossRef]
11. Stambulska, U.Y.; Bayliak, M.M.; Lushchak, V.I. Chromium(VI) toxicity in legume plants: Modulation effects of rhizobial symbiosis. *Biomed Res. Int.* **2018**, *2018*, 8031213. [CrossRef]
12. Oliveira, H. Chromium as an Environmental Pollutant: Insights on Induced Plant Toxicity. *J. Bot.* **2012**, *2012*, 375843. [CrossRef]
13. Wang, Q.; Zhou, C.; Kuang, Y.; Jiang, Z.; Yang, M. Removal of hexavalent chromium in aquatic solutions by pomelo peel. *Water Sci. Eng.* **2020**, *13*, 65–73. [CrossRef]
14. Ahmed, R.; Moisy, P.; Banerji, A.; Hesemann, P.; Taubert, A. Monitoring and Management of Anions in Polluted Aqua Systems: Case Studies on Nitrate, Chromate, Pertechnetate and Diclofenac. In *Progress and Prospects in the Management of Oxoanion Polluted Aqua Systems*; Oladoja, N., Unuabonah, E.I., Eds.; Springer International Publishing: Cham, Switzerland, 2021; pp. 293–348.
15. Lofu, A.; Mastrorilli, P.; Dell'Anna, M.M.; Mali, M.; Sisto, R.; Vignola, R. Iron(II) modified natural zeolites for hexavalent chromium removal from contaminated water. *Arch. Environ. Prot.* **2016**, *42*, 35–40. [CrossRef]
16. Kumar, R.; Arya, D.K.; Singh, N.; Vats, H.K. Removal of Cr (VI) Using Low Cost Activated Carbon Developed by Agricultural Waste. *IOSR J. Appl. Chem.* **2017**, *10*, 76–79. [CrossRef]
17. Ai, T.; Jiang, X.; Liu, Q. Chromium removal from industrial wastewater using Phyllostachys pubescens biomass loaded Cu-S nanospheres. *Open Chem.* **2018**, *16*, 842–852. [CrossRef]
18. Saleem, J.; Bin Shahid, U.; Hijab, M.; Mackey, H.; McKay, G. Production and applications of activated carbons as adsorbents from olive stones. *Biomass Convers. Biorefin.* **2019**, *9*, 775–802. [CrossRef]
19. Ademiluyi, F.T.; David-West, E.O. Effect of Chemical Activation on the Adsorption of Heavy Metals Using Activated Carbons from Waste Materials. *ISRN Chem. Eng.* **2012**, *2012*, 674209. [CrossRef]
20. Luo, Y.; Li, D.; Chen, Y.; Sun, X.; Cao, Q.; Liu, X. The performance of phosphoric acid in the preparation of activated carbon-containing phosphorus species from rice husk residue. *J. Mater. Sci.* **2019**, *54*, 5008–5021. [CrossRef]

21. Bedia, J.; Peñas-Garzón, M.; Gómez-Avilés, A.; Rodriguez, J.J.; Belver, C. Review on Activated Carbons by Chemical Activation with FeCl3. *J. Carbon Res.* **2020**, *6*, 21. [CrossRef]
22. Low, L.W.; Teng, T.T.; Ahmad, A.; Morad, N.; Wong, Y.S. A novel pretreatment method of lignocellulosic material as adsorbent and kinetic study of dye waste adsorption. *Water Air Soil Pollut.* **2011**, *218*, 293–306. [CrossRef]
23. Elangovan, R.; Philip, L.; Chandraraj, K. Biosorption of chromium species by aquatic weeds: Kinetics and mechanism studies. *J. Hazard. Mater.* **2008**, *152*, 100–112. [CrossRef] [PubMed]
24. Wu, F.C.; Wu, P.H.; Tseng, R.L.; Juang, R.S. Preparation of novel activated carbons from H2SO4-Pretreated corncob hulls with KOH activation for quick adsorption of dye and 4-chlorophenol. *J. Environ. Manag.* **2011**, *92*, 708–713. [CrossRef] [PubMed]
25. Elkady, M.; Shokry, H.; Hamad, H. New activated carbon from mine coal for adsorption of dye in simulated water or multiple heavy metals in real wastewater. *Materials* **2020**, *13*, 2498. [CrossRef] [PubMed]
26. Luo, X.; Cai, Y.; Liu, L.; Zeng, J. Cr(VI) adsorption performance and mechanism of an effective activated carbon prepared from bagasse with a one-step pyrolysis and ZnCl2 activation method. *Cellulose* **2019**, *26*, 4921–4934. [CrossRef]
27. Kadam, A.A.; Lade, H.S.; Patil, S.M.; Govindwar, S.P. Low cost CaCl2 pretreatment of sugarcane bagasse for enhancement of textile dyes adsorption and subsequent biodegradation of adsorbed dyes under solid state fermentation. *Bioresour. Technol.* **2013**, *132*, 276–284. [CrossRef]
28. Jiang, Z.; Liu, Y.; Sun, X.; Tian, F.; Sun, F.; Liang, C.; You, W.; Han, C.; Li, C. Activated carbons chemically modified by concentrated H2SO4 for the adsorption of the pollutants from wastewater and the dibenzothiophene from fuel oils. *Langmuir* **2003**, *19*, 731–736. [CrossRef]
29. Hunsom, M.; Autthanit, C. Preparation of sludge-derived KOH-activated carbon for crude glycerol purification. *J. Mater. Cycles Waste Manag.* **2017**, *19*, 213–225. [CrossRef]
30. Bansal, M.; Garg, U.; Singh, D.; Garg, V.K. Removal of Cr(VI) from aqueous solutions using pre-consumer processing agricultural waste: A case study of rice husk. *J. Hazard. Mater.* **2009**, *162*, 312–320. [CrossRef]
31. Ullah, I.; Nadeem, R.; Iqbal, M.; Manzoor, Q. Biosorption of chromium onto native and immobilized sugarcane bagasse waste biomass. *Ecol. Eng.* **2013**, *60*, 99–107. [CrossRef]
32. Harripersadth, C.; Musonge, P.; Makarfi Isa, Y.; Morales, M.G.; Sayago, A. The application of eggshells and sugarcane bagasse as potential biomaterials in the removal of heavy metals from aqueous solutions. *S. Afr. J. Chem. Eng.* **2020**, *34*, 142–150. [CrossRef]
33. Siqueira, T.C.A.; da Silva, I.Z.; Rubio, A.J.; Bergamasco, R.; Gasparotto, F.; de Souza Paccola, E.A.; Yamaguchi, N.U. Sugarcane bagasse as an efficient biosorbent for methylene blue removal: Kinetics, isotherms and thermodynamics. *Int. J. Environ. Res. Public Health* **2020**, *17*, 526. [CrossRef]
34. Giusto, L.A.R.; Pissetti, F.L.; Castro, T.S.; Magalhães, F. Preparation of Activated Carbon from Sugarcane Bagasse Soot and Methylene Blue Adsorption. *Water Air Soil Pollut.* **2017**, *228*, 249. [CrossRef]
35. Shah, G.M.; Nasir, M.; Imran, M.; Bakhat, H.F.; Rabbani, F.; Sajjad, M.; Farooq, A.B.U.; Ahmad, S.; Song, L. Biosorption potential of natural, pyrolysed and acid-assisted pyrolysed sugarcane bagasse for the removal of lead from contaminated water. *PeerJ* **2018**, *6*, e5672. [CrossRef] [PubMed]
36. Abushawish, A.; Almanassra, I.W.; Backer, S.N.; Jaber, L.; Khalil, A.K.A.; Abdelkareem, M.A.; Sayed, E.T.; Alawadhi, H.; Shanableh, A.; Atieh, M.A. High-efficiency removal of hexavalent chromium from contaminated water using nitrogen-doped activated carbon: Kinetics and isotherm study. *Mater. Chem. Phys.* **2022**, *291*, 126758. [CrossRef]
37. Elham, A.; Hossein, T.; Mahnoosh, H. Removal of Zn(II) and Pb (II) ions Using Rice Husk in Food Industrial Wastewater. *J. Appl. Sci. Environ. Manag.* **2010**, *14*, 159–162. [CrossRef]
38. Sarker, T.S.; Azam, S.M.G.G.; El-Gawad, A.M.A.; Gaglione, S.A.; Bonanomi, G. Sugarcane bagasse: A potential low-cost biosorbent for the removal of hazardous materials. *Clean Technol. Environ.* **2017**, *19*, 2343–2362. [CrossRef]
39. Block, I.; Guenter, C.; Duarte-Rodrigues, A.; Paasch, S.; Hesemann, P.; Taubert, A. Carbon adsorbents from spent coffee for removal of methylene blue and methyl orange from water. *Materials* **2021**, *14*, 3996. [CrossRef]
40. Zeydouni, G.; Rodriguez Couto, S.; Nourmoradi, H.; Basiri, H.; Amoatey, P.; Esmaeili, S.; Saeidi, S.; Keishams, F.; Mohammadi, M.J.; Khaniabadi, Y.O. H2SO4-modified Aloe vera leaf shells for the removal of P-chlorophenol and methylene blue from aqueous environment. *Toxin Rev.* **2020**, *39*, 57–67. [CrossRef]
41. Khaniabadi, Y.O.; Heydari, R.; Nourmoradi, H.; Basiri, H.; Basiri, H. Low-cost sorbent for the removal of aniline and methyl orange from liquid-phase: Aloe Vera leaves wastes. *J. Taiwan Inst. Chem. Eng.* **2016**, *68*, 90–98. [CrossRef]
42. US EPA. SW-846 Test Method 7196A: Chromium, Hexavalent (Colorimetric). 1992. Available online: https://www.epa.gov/sites/production/files/2015-12/documents/7196a.pdf. (accessed on 23 March 2021).
43. Evans, S.K.; Wesley, O.N.; Nathan, O.; Moloto, M.J. Chemically purified cellulose and its nanocrystals from sugarcane baggase: Isolation and characterization. *Heliyon* **2019**, *5*, e02635. [CrossRef]
44. Kumar, A.; Singh Negi, Y.; Choudhary, V.; Kant Bhardwaj, N. Characterization of Cellulose Nanocrystals Produced by Acid-Hydrolysis from Sugarcane Bagasse as Agro-Waste. *J. Mater. Phys. Chem.* **2014**, *2*, 1–8. [CrossRef]
45. Meng, J.; Li, S.; Niu, J. Crystallite Structure Characteristics and Its Influence on Methane Adsorption for Different Rank Coals. *ACS Omega* **2019**, *4*, 20762–20772. [CrossRef]
46. Kim, B.-H.; Wazir, A.H.; Yang, K.S.; Bang, Y.H.; Kim, S.R.; Info, A. Molecular structure effects of the pitches on preparation of activated carbon fibers from electrospinning Review Articles. *Carbon Lett.* **2011**, *12*, 70–80. [CrossRef]

47. Athira, G.; Bahurudeen, A.; Appari, S. Thermochemical Conversion of Sugarcane Bagasse: Composition, Reaction Kinetics, and Characterisation of By-Products. *Sugar Tech* **2020**, *23*, 433–452. [CrossRef]
48. Corrales, R.C.; Mendes, F.M.; Perrone, C.C.; Sant'Anna, C.; de Souza, W.; Abud, Y.; Bon, E.P.; Ferreira-Leitão, V. Structural evaluation of sugar cane bagasse steam pretreated in the presence of CO_2 and SO_2. *Biotechnol. Biofuels* **2012**, *5*, 36–43. [CrossRef] [PubMed]
49. Mohtashami, S.-A.; Asasian Kolur, N.; Kaghazchi, T.; Asadi-kesheh, R.; Soleimani, M. Optimization of sugarcane bagasse activation to achieve adsorbent with high affinity towards phenol. *Turkish J. Chem.* **2018**, *42*, 1720–1735. [CrossRef]
50. Rai, M.K.; Shahi, G.; Meena, V.; Chakraborty, S.; Singh, R.S.; Rai, B.N. Removal of hexavalent chromium Cr (VI) using activated carbon prepared from mango kernel activated with H_3PO_4. *Resour.-Effic. Technol.* **2016**, *2*, S63–S70. [CrossRef]
51. Savou, V.; Grause, G.; Kumagai, S.; Saito, Y.; Kameda, T.; Yoshioka, T. Pyrolysis of sugarcane bagasse pretreated with sulfuric acid. *J. Energy Inst.* **2019**, *92*, 1149–1157. [CrossRef]
52. Labied, R.; Benturki, O.; Eddine Hamitouche, Y.; and Donnot, A. Adsorption of hexavalent chromium by activated carbon obtained from a waste lignocellulosic material (*Ziziphus jujuba* cores): Kinetic, equilibrium, and thermodynamic study. *Adsorpt. Sci. Technol.* **2018**, *36*, 1066–1099. [CrossRef]
53. Cabassi, F.; Casu, B.; Perlin, A.S. Infrared absorption and raman scattering of sulfate groups of heparin and related glycosaminoglycans in aqueous solution. *Carbohyd. Res.* **1978**, *63*, 1–11. [CrossRef]
54. Bedin, K.C.; Martins, A.C.; Cazetta, A.L.; Pezoti, O.; Almeida, V.C. KOH-activated carbon prepared from sucrose spherical carbon: Adsorption equilibrium, kinetic and thermodynamic studies for Methylene Blue removal. *Chem. Eng. J.* **2016**, *286*, 476–484. [CrossRef]
55. Suksabye, P.; Thiravetyan, P.; Nakbanpote, W.; Chayabutra, S. Chromium removal from electroplating wastewater by coir pith. *J. Hazard. Mater.* **2007**, *141*, 637–644. [CrossRef]
56. Vo, A.T.; Nguyen, V.P.; Ouakouak, A.; Nieva, A.; Doma, B.T., Jr.; Nguyen Tran, H.; Chao, H.-P. Efficient Removal of Cr(VI) from Water by Biochar and Activated Carbon Prepared through Hydrothermal Carbonization and Pyrolysis: Adsorption-Coupled Reduction Mechanism. *Water* **2019**, *11*, 1164. [CrossRef]
57. Singh, V.K.; Tiwari, P.N. Removal and Recovery of Chromium(VI) from Industrial Waste Water. *J. Chem. Technol. Biotechnol.* **1997**, *69*, 376–382. [CrossRef]
58. Álvarez, P.; Blanco, C.; Granda, M. The adsorption of chromium (VI) from industrial wastewater by acid and base-activated lignocellulosic residues. *J. Hazard. Mater.* **2007**, *144*, 400–405. [CrossRef] [PubMed]
59. Cardoso, B.; Mestre, A.S.; Carvalho, A.P.; Pires, J. Activated carbon derived from cork powder waste by KOH activation: Preparation, characterization, and VOCs adsorption. *Ind. Eng. Chem. Res.* **2008**, *47*, 5841–5846. [CrossRef]
60. Chowdhury, M.; Mostafa, M.G.; Biswas, T.K.; Mandal, A.; Saha, A.K. Characterization of the Effluents from Leather Processing Industries. *Environ. Process.* **2015**, *2*, 173–187. [CrossRef]
61. Genawi, N.M.; Ibrahim, M.H.; El-Naas, M.H.; Alshaik, A.E. Chromium removal from tannery wastewater by electrocoagulation: Optimization and sludge characterization. *Water* **2020**, *12*, 1374. [CrossRef]
62. Bilgiç, A.; Çimen, A. Removal of chromium(vi) from polluted wastewater by chemical modification of silica gel with 4-acetyl-3-hydroxyaniline. *RSC Adv.* **2019**, *9*, 37403–37414. [CrossRef]

Disclaimer/Publisher's Note: The statements, opinions and data contained in all publications are solely those of the individual author(s) and contributor(s) and not of MDPI and/or the editor(s). MDPI and/or the editor(s) disclaim responsibility for any injury to people or property resulting from any ideas, methods, instructions or products referred to in the content.

Article

Platform Chemicals from Ethylene Glycol and Isobutene: Thermodynamics "Pays" for Biomass Valorisation and Acquires "Cashback"

Sergey P. Verevkin [1,2,*] and Aleksandra A. Zhabina [3]

1. Competence Centre CALOR, Department Life, Light & Matter, Faculty of Interdisciplinary Research, University of Rostock, 18059 Rostock, Germany
2. Department of Physical Chemistry, Kazan Federal University, 420008 Kazan, Russia
3. Chemical Department, Samara State Technical University, 443100 Samara, Russia; aazhab@gmail.com
* Correspondence: sergey.verevkin@uni-rostock.de

Abstract: Ethylene glycol (EG) produced from biomass is a promising candidate for several new applications. In this paper, EG derivatives such as mono- and di-tert-butyl ethers are considered. However, accurate thermodynamic data are essential to optimise the technology of the direct tert-butyl ether EG synthesis reaction or reverse process isobutene release. The aim of this work is to measure the vapour pressures and combustion energies for these ethers and determine the vaporisation enthalpies and enthalpies of formation from these measurements. Methods based on the First and Second Law of Thermodynamics were combined to discover the reliable thermodynamics of ether synthesis reactions. The thermochemical data for ethylene glycol tert-butyl ethers were validated using structure–property correlations and quantum chemical calculations. The literature results of the equilibrium study of alkylation of EG with isobutene were evaluated and the thermodynamic functions of ethylene glycol tert-butyl ethers were derived. The energetics of alkylation determined according to the "First Law" and the "Second Law" methods agree very well. Some interesting aspects related to the entropy of ethylene glycol tert-butyl ethers were also revealed and discussed.

Keywords: ethers; combustion calorimetry; enthalpy of formation; transpiration method; vapour pressure; enthalpy of vaporisation; enthalpy of reaction; quantum chemical calculations

Citation: Verevkin, S.P.; Zhabina, A.A. Platform Chemicals from Ethylene Glycol and Isobutene: Thermodynamics "Pays" for Biomass Valorisation and Acquires "Cashback". *Chemistry* 2023, 5, 1171–1189. https://doi.org/10.3390/chemistry5020079

Academic Editors: Catherine Housecroft and Edwin Charles Constable

Received: 28 March 2023
Revised: 1 May 2023
Accepted: 5 May 2023
Published: 9 May 2023

Copyright: © 2023 by the authors. Licensee MDPI, Basel, Switzerland. This article is an open access article distributed under the terms and conditions of the Creative Commons Attribution (CC BY) license (https://creativecommons.org/licenses/by/4.0/).

1. Introduction

Producing energy, fuels, and chemicals from renewable biomass is crucial to preventing global warming by reducing the atmospheric CO_2 emissions caused by fossil fuel consumption. Cellulose, the most abundant biomass source, is currently seen as a promising alternative to fossil fuels because it is not edible and does not negatively affect the food market. The most attractive and sustainable route for the use of cellulose could be the direct conversion into platform chemicals [1]. The catalytic conversion of cellulose into ethylene glycol (EG), propylene glycol, xylitol, sorbitol, mannitol, etc. offers an alternative route to these important chemicals, which are used in the food industry as functional additives, as intermediates in the pharmaceutical industry and as monomers in the plastics industry [2]. An effective route for the one-pot production of EG from renewable cellulose using various non-expensive catalysts has attracted great interest from both academia and industry [2,3]. Compared to the petroleum-dependent multi-step process, the biomass route offers the outstanding advantages of a one-pot process and a renewable feedstock.

EG is used in the production of polyester fibres and resins (e.g., polyethylene terephthalate) and is widely applied as a component of antifreeze or coolant systems in cars and in de-icing fluids for aircraft [4]. Currently, EG is manufactured from petroleum-derived ethylene in several steps by cracking, epoxidation, and hydration [5]. The fact that EG can be derived not only from fossil resources but also from biomass [6] makes it a promising candidate for various new applications, e.g., for the synthesis of platform chemicals such as coating solvents and oxygenate additives for gasoline blends. For example, the reaction of isobutene with EG in the presence of an acid catalyst gives a mixture of mono- and di-tert-butyl ethers (see Figure 1).

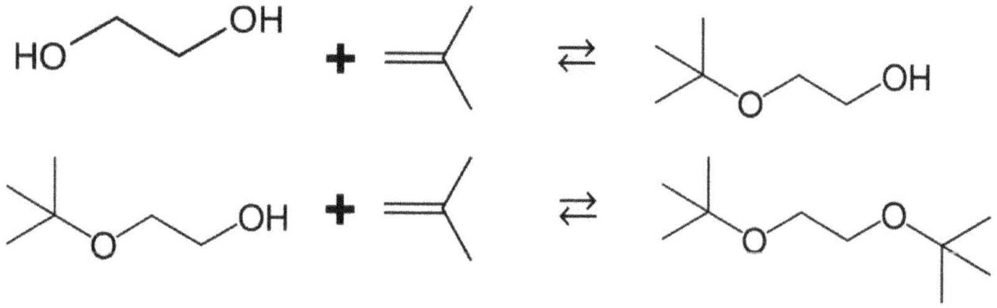

Figure 1. Alkylation of ethylene glycol with isobutene: synthesis of ethylene glycol mono-tert-butyl ether (EGM) in the first line and the synthesis of ethylene glycol di-tert-butyl ether (EGD) in the second line.

The latter are good solvents for paints, inks, and coatings with low toxicity [7]. The tert-butyl ethers of diols are also potential oxygenate additives for motor gasoline, exhibiting high anti-knock properties [8]. Isobutene is usually produced by the steam cracking of high-boiling petroleum fractions and must be separated from the C4 mixture in an additional purification step by using either sulphuric acid extraction or molecular sieves [9]. An interesting alternative to these conventional processes is the use of EG as a very good capture agent for isobutylene from the C4 cracking fraction [10]. The tert-butyl ethers formed can then be used directly as solvents or oxygenate additives or can be readily cracked to yield very pure isobutylene, which is needed for the production of a variety of chemicals and polymers [11].

The kinetics and catalysis of the tert-butylation of EG have been intensively studied since the 1970s [10–13]. The influence of various homogeneous and heterogeneous acid catalysts and the optimisation of reaction conditions were the focus of these studies. However, the thermodynamic aspects were not of interest, although the alkylation and dealkylation of EG are reversible processes and the equilibrium of these reactions takes place under thermodynamic control. This means that the equilibrium of these reactions can be shifted towards a high yield of the desired product according to Le Chatelier's principle. Therefore, accurate thermodynamic data (e.g., reaction enthalpies, reaction entropies, heat capacities, etc.) are indispensable to optimise the technology of the direct tert-butyl ether synthesis reaction or isobutene release in the reverse process. For the distillation of tert-butyl ethers from the reaction mixture as well as for their purification, reliable data on vapour pressures and evaporation enthalpies of the pure compounds are also needed. It has turned out that no thermodynamic data for tert-butyl ethers are available in the literature.

The aim of this work is, therefore, to measure the vapour pressures and combustion energies of tert-butyl ethers (see Figure 1) and determine the vaporisation enthalpies and enthalpies of formation from these measurements.

The focus of this study is on the thermochemical properties that are responsible for the energetics of chemical reactions, as well as for the liquid–gas phase change enthalpy. The common textbook equations relate these thermochemical properties:

$$\Delta_f H_m^o(g) = \Delta_f H_m^o(\text{liq}) + \Delta_l^g H_m^o \qquad (1)$$

where $\Delta_f H_m^o(g)$ is the gas phase standard molar enthalpy of formation, $\Delta_f H_m^o(\text{liq})$ is the liquid state standard molar enthalpy of formation, and $\Delta_l^g H_m^o$ is the standard molar enthalpy of vaporisation. In thermochemistry, it is common to adjust all of the enthalpies involved in Equation (1) to an arbitrary but common reference temperature. In this work, we have chosen $T = 298.15$ K as the reference temperature. The energetics of the synthesis reactions of tert-butyl ethers according to Equations (2) and (3):

$$\text{Ethylene glycole} + \text{isobutylene} = \text{EGM} \qquad (2)$$

$$\text{EGM} + \text{isobutylene} = \text{EGD} \qquad (3)$$

is essential for chemical-engineering calculations.

There are two ways to derive the energetics of chemical reactions: according to the First Law of Thermodynamics or according to the Second Law of Thermodynamics [14]. If the reaction enthalpies are derived from the calorimetrically measured enthalpies of formation, they are considered to be calculated from the First Law of Thermodynamics. For thermochemistry, Hess proposed to estimate the enthalpy of reaction as the difference between the enthalpies of the formation of the reactants and products, which is also called Hess's law. If the reaction enthalpies are derived from the temperature dependence of equilibrium constants, they are considered to be calculated from the Second Law of Thermodynamics. However, both methods complement each other. The results for a given reaction obtained by the "First Law" and "Second Law" methods can serve as a valuable test of the thermodynamic consistency and reliability of the experimental results.

In this work, we combined the methods of the "First Law" and the "Second Law" to obtain the reliable thermodynamics of the synthesis reactions of ethylene glycol tert-butyl ethers. The enthalpy of the formation of mono-tert-butyl ether was measured with high-precision combustion calorimetry and the vaporisation enthalpy was measured with transpiration and static methods. A sample of ethylene glycol di-tert-butyl ether with sufficient purity for thermochemical measurements was not available, so the thermochemical data for this ether were derived from the structure–property correlations and quantum chemical calculations. The results of the equilibrium study of Reactions (1) and (2) from the literature [15] were evaluated and the thermodynamic functions of these reactions were derived according to the "Second Law" method. The energetics of Reactions (1) and (2) determined according to the "First Law" and the "Second Law" methods agree very well. Finally, some interesting aspects related to the entropy of ethylene glycol tert-butyl ethers were revealed and discussed.

2. Experimental Methods

2.1. Materials

The sample of ethylene glycol mono-tert-butyl ether (EGM) with a purity >0.99 mass fraction was of commercial origin (TCI, # E0354). It was further purified by fractional distillation under reduced pressure. Purity was determined by capillary gas chromatography using a flame ionisation detector (FID). No impurities (greater than mass fraction 0.0004) could be detected in the sample used for the thermochemical experiments. Residual water in the sample was measured using a Mettler Toledo DL38 Karl Fischer titrator with HYDRANAL™ solvent. The mass fraction of water (865 ppm) was used for the corrections of the sample masses in the combustion calorimetry experiments.

2.2. Experimental and Theoretical Thermochemical Methods

The vapour pressures over the liquid sample of EGM were measured at different temperatures using the transpiration method [16]. About 0.5 g of the sample was used to cover the small glass beads to provide sufficient contact surface with the transporting gas and to eliminate hydraulic resistance. These covered glass spheres were loaded into the U-shaped saturator. A stream of nitrogen at a well-defined flow rate was passed through the saturator at a constant temperature (± 0.1 K), and the transported material was collected in a cold trap. The amount of substance condensed in the cold trap was determined by gas chromatography and used to derive the vapour pressure according to the ideal gas law. The details of this method can be found elsewhere [16–18]. A brief description and the necessary details can be found in ESI.

Independently, the vapour pressures above the liquid sample of EGM were measured at different temperatures using the static method [19]. The stainless-steel cylindrical cell with the sample was kept at a constant temperature. The sample cell was connected to high-temperature capacitance manometers, which can measure vapour pressures from 0.1 to 10^5 Pa. The details of this method can be found elsewhere [19,20]. A brief description and the necessary details can be also found in ESI.

The experimental vapour pressure–temperature dependencies measured by either the transpiration or static method were used to derive the enthalpies of vaporisation enthalpies, $\Delta_l^g H_m^o$, and inserted into Equation (1).

The energy of the combustion of EGM was measured with an isoperibol bomb calorimeter. The liquid sample was transferred (in the glove box) with a syringe into the polyethylene bulb (Fa. NeoLab, Heidelberg, Germany). The neck of the bulb was compressed with special tweezers and sealed by heating the neck near a glowing wire. The bulb with the liquid sample was placed in a crucible and burnt in oxygen at a pressure of 3.04 MPa, according to a procedure described in detail previously [21,22]. The combustion gases were analysed for carbon monoxide (Dräger tubes) and unburnt carbon, neither of which could be detected. To detect traces of CO in the combustion exhaust gases, they were passed through a Dräger tube (glass vial containing a chemical reagent that reacts with CO, with detection limits of 5 to 150 ppm). The energy equivalent of the calorimeter ε_{calor} was determined with a standard reference sample of benzoic acid (sample SRM 39j, N.I.S.T.). The auxiliary quantities used for the data acquisition of combustion experiments are compiled in Table S1. For the conversion of the energy of the actual bomb process into that of the isothermal process, and the reduction to standard state, the conventional procedure was applied [23]. The necessary details can be found in ESI. The standard molar enthalpy of formation of the liquid state, $\Delta_f H_m^o$(liq), was derived from their energies of combustion and inserted into Equation (1).

Quantum chemical (QC) calculations were carried out using the Gaussian 09 series software [24]. The energies of the most stable conformers were calculated using the G4 method [25]. The H_{298} values were finally converted to the theoretical $\Delta_f H_m^o$(g, 298.15 K)$_{theor}$ values and discussed. Calculations were performed under the assumption of "rigid rotator"–"harmonic oscillator" and the general procedure was described elsewhere [26].

3. Results and Discussion
3.1. Absolute Vapour Pressures

The systematic vapour pressure measurements on EGM were carried out for the first time. The primary experimental vapour pressures, p, of EGM at various temperatures measured by the transpiration method are given in Table 1.

Table 1. The absolute vapour pressures, p, and the standard vaporisation enthalpies, $\Delta_l^g H_m^o$, and vaporisation entropies, $\Delta_l^g S_m^o$, as determined by the transpiration method.

T/ K [a]	m/ mg [b]	V(N$_2$) [c]/ dm^3	T$_a$/ K [d]	Flow/ dm$^3 \cdot$h^{-1}	p/ Pa [e]	u(p)/ Pa [f]	$\Delta_l^g H_m^o$/ kJ·mol^{-1}	$\Delta_l^g S_m^o$/ J·K^{-1}·mol^{-1}
ethylene glycol mono-tert-butyl ether (EGM): $\Delta_l^g H_m^o$(298.15 K) = (53.0 ± 0.6) kJ·mol^{-1}								
$\Delta_l^g S_m^o$ (298.15 K) = (129.5 ± 1.1) J·mol^{-1}·K^{-1}								
$\ln(p/p_{ref}) = \frac{(306.9 \pm 2.1)}{R} - \frac{(77402 \pm 648)}{RT} - \frac{81.7}{R} \ln \frac{T}{298.15}$; $p_{ref} = 1$Pa								
288.3	10.60	1.540	292.0	2.05	143.7	3.6	53.9	132.4
291.3	10.41	1.195	292.8	2.05	181.7	4.6	53.6	131.6
293.2	10.76	1.084	294.8	1.05	207.9	5.2	53.5	131.0
296.2	9.88	0.816	293.2	2.04	251.5	6.3	53.2	129.9

Table 1. Cont.

$T/$ K [a]	$m/$ mg [b]	$V(N_2)$ [c]/ dm^3	$T_a/$ K [d]	Flow/ $dm^3 \cdot h^{-1}$	$p/$ Pa [e]	$u(p)/$ Pa [f]	$\Delta_l^g H_m^o/$ $kJ \cdot mol^{-1}$	$\Delta_l^g S_m^o/$ $J \cdot K^{-1} \cdot mol^{-1}$
298.1	8.22	0.582	293.6	1.06	293.4	7.4	53.1	129.5
301.0	9.98	0.578	293.0	2.04	357.1	9.0	52.8	128.6
303.0	9.29	0.458	293.2	1.06	419.2	10.5	52.7	128.3
306.0	9.74	0.400	292.2	1.04	501.0	12.5	52.4	127.3
308.0	10.09	0.353	293.4	1.06	589.5	14.8	52.2	127.0
311.0	10.24	0.305	294.0	1.05	690.9	17.3	52.0	125.9
313.0	11.61	0.301	294.4	1.06	794.5	19.9	51.8	125.4
317.9	14.14	0.268	295.6	1.07	1086.5	27.2	51.4	124.2
322.8	19.30	0.261	293.4	1.08	1503.3	37.6	51.0	123.2
327.8	25.70	0.260	293.8	1.08	2006.5	50.2	50.6	122.0

[a] Saturation temperature measured with the standard uncertainty ($u(T) = 0.1$ K). [b] Mass of transferred sample condensed at $T = 273$ K. [c] Volume of nitrogen ($u(V) = 0.005$ dm^3) used to transfer m ($u(m) = 0.0001$ g) of the sample. Uncertainties are given as standard uncertainties. [d] T_a is the temperature of the soap bubble meter used for the measurement of the gas flow. [e] Vapour pressure at temperature T, calculated from the m and the residual vapour pressure at the condensation temperature calculated by an iteration procedure. [f] Standard uncertainties were calculated with $u(p_i/\text{Pa}) = 0.025 + 0.025(p_i/\text{Pa})$ and are valid for pressures from 5 to 3000 Pa. The standard uncertainties for T, V, p, and m are standard uncertainties with 0.683 confidence levels. The uncertainty of the vaporisation enthalpy $U(\Delta_l^g H_m^o)$ is the expanded uncertainty (0.95 level of confidence) calculated according to the procedure described elsewhere [17,18]. Uncertainties include uncertainties from the experimental conditions and the fitting equation and vapour pressures as well as uncertainties from the adjustment of vaporisation enthalpies to the reference temperature $T = 298.15$ K.

The primary experimental vapour pressures, p, of EGM at various temperatures measured by the static method are given in Table 2.

The vapour pressures measured for EGM using the transpiration and static methods are compared in Figure S1 and are hardly distinguishable from each other. No vapour pressure measurements for mono and di-tert-butyl ethers can be found in the literature. Therefore, any additional information is valuable. Thus, we searched for individual experimental boiling temperatures at different pressures, which are sometimes given in the literature as the results of the distillation of reaction mixtures after synthesis. Since these data do not come from specific physico–chemical investigations, the temperatures are usually given in the range of a few degrees and the pressures are measured with non-calibrated manometers. However, in our previous work, we showed that reasonable trends can generally be derived even from such raw data [27]. The experimentally determined boiling temperatures for EGM at different pressures (see Figure S1) confirm this conclusion qualitatively (and quantitatively for $\Delta_l^g H_m^o$ (298.15 K) values as shown in Section 3.2). The individual experimental boiling temperatures at different pressures taken from the databases [28,29] are compiled in Table S2 and are used for the evaluation. The vapour pressure data sets given in Tables 1, 2 and S2 were approximated with the following equation [16]:

$$R \times \ln(p_i/p_{\text{ref}}) = a + \frac{b}{T} + \Delta_l^g C_{p,m}^o \times \ln\left(\frac{T}{T_0}\right) \quad (4)$$

In this equation, $R = 8.31,446$ $J \cdot K^{-1} \cdot mol^{-1}$; $p_{\text{ref}} = 1$Pa; and a and b are adjustable parameters and $\Delta_l^g C_{p,m}^o$ is the difference between the molar heat capacity of the liquid and gas phases; $T_0 = 298.15$ K was adopted in this work. The values of $\Delta_l^g C_{p,m}^o$ were estimated for the compound of interest using an empirical correlation [30] based on heat capacities $C_{p,m}^o$(liq). The latter values were of experimental origin or were assessed using a group-contribution method [31] (see Table S3). The parameters a and b fitted to the experimental vapour pressures were used to derive the thermodynamic functions of vaporisation, as shown in the next section.

Table 2. The absolute vapour pressures, p, and the standard vaporisation enthalpies, $\Delta_l^g H_m^o$, and vaporisation entropies, $\Delta_l^g S_m^o$, as determined by the static method.

T/K	p/Pa	u(p)/Pa [a]	$\Delta_l^g H_m^o$/kJ·mol^{-1}	$\Delta_l^g S_m^o$/J·K^{-1}·mol^{-1}
	ethylene glycol mono-tert-butyl ether (EGM): $\Delta_l^g H_m^o$(298.15 K) = (53.1 ± 0.2) kJ·mol^{-1}
	$\Delta_l^g S_m^o$(298.15 K) = (130.2 ± 0.2) J·mol^{-1}·K^{-1}
	$\ln(p/p_{ref}) = \frac{(307.6 \pm 0.2)}{R} - \frac{(77513 \pm 62)}{RT} - \frac{81.7}{R} \ln \frac{T}{298.15}$; p_{ref} = 1Pa
284.01	105.0	0.6	54.3	134.2
284.03	105.1	0.6	54.3	134.2
286.39	127.0	0.7	54.1	133.5
288.70	152.1	0.8	53.9	132.8
288.74	152.8	0.8	53.9	132.8
291.10	183.1	1.0	53.7	132.2
293.51	219.7	1.1	53.5	131.5
293.54	219.2	1.1	53.5	131.5
295.90	261.9	1.4	53.3	130.8
298.26	311.6	1.6	53.1	130.2
298.30	310.7	1.6	53.1	130.1
298.33	312.0	1.6	53.1	130.1
300.66	368.2	1.9	52.9	129.5
300.70	370.1	1.9	52.9	129.5
303.10	435.4	2.2	52.7	128.8
303.13	436.7	2.2	52.7	128.8
305.46	513.9	2.6	52.6	128.2
307.93	606.2	3.1	52.4	127.6
307.95	608.2	3.1	52.4	127.6
310.26	707.5	3.6	52.2	127.0
312.74	831.2	4.2	52.0	126.3
312.76	830.7	4.2	52.0	126.3
315.14	967.1	4.9	51.8	125.7
317.51	1120.2	5.7	51.6	125.1

[a] The uncertainties of the experimental vapor pressures were calculated according to the following equation: $u(p/\text{Pa}) = 0.05 + 0.005(p/\text{Pa})$ and valid for $p > 12$ Pa. The uncertainties of the vaporisation enthalpies are expressed as the expanded uncertainty (0.95 level of confidence, k = 2). They include uncertainties from the fitting equation and uncertainties from the temperature adjustment to T = 298.15 K. The uncertainties in the temperature adjustment of vaporisation enthalpies to the reference temperature T = 298.15 K are estimated to account for 20% of the total adjustment.

3.2. Thermodynamics of Vaporisation

The standard molar enthalpies of the vaporisation of EGM and EGD at temperature T were derived from the temperature dependence of the vapour pressures using the following equation:

$$\Delta_l^g H_m^o(T) = -b + \Delta_l^g C_{p,m}^o \times T \tag{5}$$

The standard molar vaporisation entropies at temperature T were also derived from the temperature dependences of the vapour pressures using Equation (6):

$$\Delta_l^g S_m^o(T) = \Delta_l^g H_m^o/T + R \times \ln(p/p^o) \tag{6}$$

The primary data on the vapour pressures, coefficients a and b (see Equation (4)), $\Delta_l^g H_m^o(T)$, and $\Delta_l^g S_m^o(T)$ values are compiled in Tables 1, 2 and S2. The vaporisation enthalpies derived indirectly from vapour pressure measurements are usually referenced to T_{av}, which is defined as the average temperature of the range under study. For comparison, $\Delta_l^g H_m^o(T_{av})$ values are commonly adjusted to T = 298.15 K. The temperature adjustment was performed according to Kirchhoff's law:

$$\Delta_l^g H_m^o(298.15 \text{ K}) = \Delta_l^g H_m^o(T_{av}) + \Delta_l^g C_{p,m}^o (298.15 \text{ K} - T_{av}) \tag{7}$$

A summary of the results for $\Delta_l^g H_m^o$ (298.15 K) is given in Table 3, column 5.

Table 3. Compilation of the enthalpies of vaporisation, $\Delta_l^g H_m^o$, derived for ethylene glycol tert-butyl ethers.

	M [a]	T-Range	$\Delta_l^g H_m^o$ (T_{av})	$\Delta_l^g H_m^o$ (298.15 K) [b]	Ref.
		K	kJ·mol^{-1}	kJ·mol^{-1}	
ethylene glycol mono-tert-butyl ether (EGM) 7580-85-0	T	288.3–327.8	52.4 ± 0.6	53.0 ± 0.7	Table 1
	S	284.0–317.5	53.0 ± 0.1	53.1 ± 0.2	Table 2
	BP	323.0–427.0	44.9 ± 0.9	50.8 ± 2.2	Table S2
	T_b			52.9 ± 0.5	this work
	SP			53.1 ± 0.5	this work
				53.1 ± 0.2 [c]	
ethylene glycol di-tert-butyl ether (EGD)	BP	335.0–444.0	44.3 ± 0.2	53.5 ± 1.8	Table S2
	T_b			51.7 ± 2.0	this work
	SP			52.3 ± 0.5	this work
				52.3 ± 0.2 [c]	

[a] Methods: T = transpiration method; S = static method; T_b = derived from correlation with the normal boiling points (see text); BP = derived from the individual boiling temperatures at different pressures (see text); SP = derived from empirical structure–property correlations (see text). [b] Vapour pressures available in the literature were treated using Equations (4) and (5) with the help of the heat capacity differences from Table S3 to evaluate the enthalpy of vaporisation at 298.15 K in the same way as our own results in Table 1. The uncertainty of the vaporisation enthalpy u($\Delta_l^g H_m^o$) is the expanded uncertainty (at 0.95 level of confidence, k = 2) calculated according to a procedure described elsewhere [9,10]. It includes uncertainties from the transpiration experimental conditions, uncertainties of vapour pressure, uncertainties from the fitting equation, and uncertainties from temperature adjustment to T = 298.15 K. [c] Weighted mean value (the uncertainties of the vaporisation enthalpies were used as a weighting factor). Values in bold are recommended for further thermochemical calculations.

The combined uncertainties of the vaporisation enthalpies $\Delta_l^g H_m^o$(298.15 K) from the transpiration method include the uncertainties of the experimental transpiration conditions, uncertainties of the vapour pressures, and uncertainties due to the temperature adjustment to the reference temperature T = 298.15 K, as developed elsewhere [17,18]. The combined uncertainties of the vaporisation enthalpies $\Delta_l^g H_m^o$(298.15 K) from the static method include uncertainties from the fitting equation and uncertainties from the temperature adjustment to T = 298.15 K. Uncertainties in the temperature adjustment of vaporisation enthalpies to the reference temperature T = 298.15 K are estimated to account for 20% of the total adjustment. The combined uncertainties of the vaporisation enthalpies $\Delta_l^g H_m^o$(298.15 K) obtained from the vapour pressure data collected in the literature (see Table S2) were calculated in the same way as for the static method.

As shown in Table 3 for EGM, our results for $\Delta_l^g H_m^o$(298.15 K) derived from transpiration and static methods are practically indistinguishable. The $\Delta_l^g H_m^o$(298.15 K) value estimated from the individual boiling temperatures found in the literature at different pressures also agrees with our results and is within the experimental uncertainties. No comparative data for $\Delta_l^g H_m^o$(298.15 K) of EGD were found in the literature.

3.3. Validation of the Vaporisation Enthalpies Using Structure–Property Correlations

The absence of data on vapour pressures and vaporisation thermodynamics for tert-butyl ethers has prompted an extended validation of the enthalpies of vaporisation using structure–property correlations as follows.

3.3.1. Correlation with the Normal Boiling Temperatures T_b

A correlation of the enthalpies of vaporisation of organic molecules with their normal boiling temperatures successfully serves to mutually validate these thermal data [32]. The chemical family of alkyl ethers is thermally stable even at elevated temperatures near the boiling points, therefore, numerous reliable normal boiling temperatures for many ethers have been found in the literature (see Tables 4 and 5).

Table 4. Correlation of the vaporisation enthalpies $\Delta_l^g H_m^o$ (298.15 K) of ethylene glycol mono-alkyl ethers with their T_b (normal boiling temperatures).

R-CH$_2$CH$_2$OH	T_b/ [a] K	$\Delta_l^g H_m^o$(298.15 K)$_{exp}$/ kJ·mol^{-1}	$\Delta_l^g H_m^o$(298.15 K)$_{calc}$/ [b] kJ·mol^{-1}	Δ/ [c] kJ·mol^{-1}
CH$_3$-CH$_2$CH$_2$OH	397.3	45.2	45.5	−0.3
Et-CH$_2$CH$_2$OH	408.1	48.2	48.1	0.1
Pr-CH$_2$CH$_2$OH	422.9	52.1	51.7	0.4
Bu-CH$_2$CH$_2$OH	444.2	56.6	56.9	−0.3
tBu-CH$_2$CH$_2$OH	428.0		52.9	

[a] Normal boiling temperatures and experimental vaporisation enthalpies are from [33]. The uncertainties of T_b are ± 0.5 K and the uncertainties of $\Delta_l^g H_m^o$ (298.15 K) are ±0.2 kJ·mol^{-1} (expressed as two times the standard deviation). [b] Calculated using Equation (8). [c] Difference between the experimental and calculated values.

Table 5. Correlation of the vaporisation enthalpies $\Delta_l^g H_m^o$ (298.15 K) of di-alkyl ethers and ethylene glycol di-alkyl ethers with their T_b (normal boiling temperatures).

R-O-R	T_b/ [a] K	$\Delta_l^g H_m^o$(298.15 K)$_{exp}$/ kJ·mol^{-1}	$\Delta_l^g H_m^o$(298.15 K)$_{calc}$/ [b] kJ·mol^{-1}	Δ/ [c] kJ·mol^{-1}
Et-O-Et	307.6	27.4	26.4	1.0
Pr-O-Pr	363.1	35.8	36.7	−0.9
Bu-O-Bu	413.5	45.0	46.0	−1.0
tBu-O-tBu	379.9	37.7	39.8	−2.1
RO-CH$_2$CH$_2$-OR				
CH$_3$O-CH$_2$CH$_2$-OCH$_3$	358.0	36.5	35.8	0.7
EtO-CH$_2$CH$_2$-OEt	392.5	43.3	42.1	1.2
PrO-CH$_2$CH$_2$-OPr	436.4	50.6	50.2	0.4
BuO-CH$_2$CH$_2$-OBu	479.0	58.8	58.1	0.7
tBuO-CH$_2$CH$_2$-OtBu	444.0	-	51.7	

[a] Normal boiling temperatures and experimental vaporisation enthalpies are from [33]. The uncertainties of T_b are ±0.5 K and the uncertainties of $\Delta_l^g H_m^o$ (298.15 K) are ± 0.2 kJ·mol^{-1} (expressed as two times the standard deviation). [b] Calculated using Equation (9). [c] Difference between the experimental and calculated values.

Numerous reliable experimental data on $\Delta_l^g H_m^o$(298.15 K) are also available in the literature (see Tables 4 and 5) and are ready for correlation with normal boiling temperatures.

For ethylene glycol mono-alkyl ethers R-CH$_2$CH$_2$OH, the following linear correlation was derived:

$$\Delta_l^g H_m^o(298.15\ K)/(kJ\cdot mol^{-1}) = -50.9 + 0.2426 \times T_b \text{ with } (R^2 = 0.9956) \quad (8)$$

The uncertainty of the enthalpies of vaporisation calculated with Equation (8) is estimated to be 0.5 kJ·mol^{-1} (see Table 4).

For ethylene glycol di-alkyl ethers RO-CH$_2$CH$_2$-OR, the following linear correlation was derived:

$$\Delta_l^g H_m^o(298.15\ K)/(kJ\cdot mol^{-1}) = -30.4 + 0.1848 \times T_b \text{ with } (R^2 = 0.9855) \quad (9)$$

The uncertainty of the enthalpies of vaporisation calculated with Equation (9) is estimated to be 2.0 kJ·mol^{-1} (see Table 5). The very high R^2 correlation coefficients of Equations (8) and (9) are evidence of the consistency of the data sets on alkyl ethers included in the correlations. The "empirical" results derived from Equations (8) and (9) are listed in Table 3 and labelled T_b. These results are valuable in supporting the enthalpy of vaporisation of EGM derived from other methods, especially for EGD for which no data are available.

3.3.2. Correlation with the Enthalpies of Vaporisation of the Parent Structures

Structure–property correlations are a valuable tool to establish the consistency of experimental data in the series of parent homologues. It is evident that the ethylene glycol mono-tert-butyl ether studied in this work belongs to a general family of ethylene glycol mono-alkyl ethers, R-CH$_2$CH$_2$OH, which are structurally parent to the di-alkyl ethers, R-O-R. In this work, we correlated the experimental $\Delta_l^g H_m^o$(298.15 K) values for the homologue series of R-CH$_2$CH$_2$OH with the experimental vaporisation enthalpies for the homologue series of di-alkyl ethers, R-O-R. The compilation of experimental data involved in this correlation is shown in Table 6.

Table 6. Correlation of the vaporisation enthalpies, $\Delta_l^g H_m^o$ (298.15 K), of ethylene glycol mono-alkyl ethers and di-alkyl ethers (in kJ·mol^{-1}) [a].

R-O-R	$\Delta_l^g H_m^o$ (298.15 K)$_{exp}$	R-CH$_2$CH$_2$OH	$\Delta_l^g H_m^o$ (298.15 K)$_{exp}$	$\Delta_l^g H_m^o$ (298.15 K)$_{calc}$ [b]	Δ [c]
CH$_3$-O-CH$_3$	19.3	CH$_3$-CH$_2$CH$_2$OH	45.2	44.9	0.3
Et-O-Et	27.4	Et-CH$_2$CH$_2$OH	48.2	48.5	−0.3
Pr-O-Pr	35.8	Pr-CH$_2$CH$_2$OH	52.1	52.3	−0.2
Bu-O-Bu	45.0	Bu-CH$_2$CH$_2$OH	56.6	56.4	0.2
tBu-O-tBu	37.7	tBu-CH$_2$CH$_2$OH		**53.1 ± 0.5** [d]	

[a] The experimental vaporisation enthalpies of both series are taken from [33]. The uncertainties of $\Delta_l^g H_m^o$(298.15 K) are ±0.2 kJ·mol^{-1} (expressed as two times the standard deviation). [b] Calculated according to Equation (10). [c] Difference between columns 4 and 5 in this table. [d] The uncertainty is estimated to be 0.5 kJ·mol^{-1} (expressed as two times the standard deviation). The value given in bold was recommended for thermochemical calculations.

A very good linear correlation was found for these structurally related series:

$$\Delta_l^g H_m^o (R-CH_2CH_2OH, 298.15\ K)/kJ\cdot mol^{-1} = 36.3 + 0.4463 \times \Delta_l^g H_m^o(R-O-R, 298.15\ K)\ \text{with}\ (R^2 = 0.9963) \quad (10)$$

The very high R^2 correlation coefficient can be taken as evidence for the general consistency of the evaluated vaporisation enthalpies of R-CH$_2$CH$_2$OH with the well-established set of data for d-alkyl ethers. Consequently, the enthalpy of vaporisation of ethylene glycol mono-tert-butyl ether (EGM) was derived using Equation (10) as an independent and complementary result (see Table 6).

A similar correlation was applied to EGD, as follows. The ethylene glycol di-tert-butyl ether (EGD) belongs to a general family of ethylene glycol di-alkyl ethers, RO-CH$_2$CH$_2$-OR, which are also structurally related to the series of di-alkyl ethers, R-O-R. Therefore, we correlated the experimental $\Delta_l^g H_m^o$(298.15 K) values for the homologue series of RO-CH$_2$CH$_2$-OR with the experimental vaporisation enthalpies for the homologue series of di-alkyl ethers, R-O-R. The compilation of experimental data involved in this correlation is shown in Table 7.

Table 7. Correlation of the vaporisation enthalpies, $\Delta_l^g H_m^o$ (298.15 K), of ethylene glycol di-alkyl ethers and di-alkyl ethers (in kJ·mol^{-1}) [a].

R-O-R	$\Delta_l^g H_m^o$(298.15 K)$_{exp}$	RO-CH$_2$CH$_2$-OR	$\Delta_l^g H_m^o$(298.15 K)$_{exp}$	$\Delta_l^g H_m^o$ (298.15 K)$_{calc}$ [b]	Δ [c]
CH$_3$-O-CH$_3$	19.3	CH$_3$O-CH$_2$CH$_2$-OCH$_3$	36.5	36.4	0.1
Et-O-Et	27.4	EtO-CH$_2$CH$_2$-OEt	43.3	43.4	−0.1
Pr-O-Pr	35.8	PrO-CH$_2$CH$_2$-OPr	50.6	50.7	−0.1
Bu-O-Bu	45.0	BuO-CH$_2$CH$_2$-OBu	58.8	58.7	0.1
tBu-O-tBu	37.7	tBuO-CH$_2$CH$_2$-OtBu		**52.3 ± 0.5** [d]	

[a] The experimental vaporisation enthalpies of both series are taken from [33]. The uncertainties of $\Delta_l^g H_m^o$(298.15 K) are ±0.2 kJ·mol^{-1} (expressed as two times the standard deviation). [b] Calculated according to Equation (10). [c] Difference between columns 4 and 5 in this table. [d] The uncertainty is estimated to be 0.5 kJ·mol^{-1} (expressed as two times the standard deviation). The value given in bold was recommended for thermochemical calculations.

In this case, too, an almost perfect linear correlation (with $R^2 = 0.9998$) was derived for these structural parent series:

$$\Delta_{l}^{g}H_{m}^{o}(RO-CH_{2}CH_{2}-OR, 298.15\ K)/kJ\cdot mol^{-1} = 19.6 + 0.8682 \times \Delta_{l}^{g}H_{m}^{o}(R-O-R, 298.15\ K) \text{ with } \left(R^{2}=0.9998\right) \quad (11)$$

The enthalpy of vaporisation of ethylene glycol di-tert-butyl ether (EGD) was, therefore, estimated according to Equation (11) and complements the results obtained in this work with other methods. The "empirical" results obtained from Equations (10) and (11) are given in Table 3 and marked SP (structure–property). These results help to ascertain the enthalpy of vaporisation of EGM and EGD determined by other methods.

As a conclusion from the validation of the vaporisation enthalpies performed in Section 3.3, completely different structure–property correlations were applied to check the data consistency for EGM and EGD (see Table 3). It was found that the experimental and empirical results agree very well for both ethers. To provide more confidence, a weighted average value was calculated for each compound (bold values in Table 3), and these values were recommended for further thermochemical calculations according to Equation (3) to derive the gas phase formation enthalpies using the $\Delta_{f}H_{m}^{o}(liq)$ values measured by combustion calorimetry, as shown in the following section.

3.4. Standard Molar Enthalpies of Formation in the Liquid Phase

The standard specific energy of combustion $\Delta_{c}u^{o}(liq)$ of liquid ethylene glycol mono-tert-butyl ether was determined from six experiments. The results of the combustion experiments are shown in Table 8.

Table 8. The results of the combustion experiments at $T = 298.15$ K ($p^{o} = 0.1$ MPa) for ethylene glycol mono-tert-butyl ether (EGM) [a].

Experiment	Exp. 1	Exp. 2	Exp. 3	Exp. 4	Exp. 5	Exp. 6
m (substance)/g	0.325060	0.281828	0.241385	0.325939	0.272169	0.230654
m' (cotton)/g	0.000962	0.001073	0.001054	0.000959	0.000957	0.001093
m'' (polyethylene)/g	0.311926	0.301513	0.389646	0.399094	0.400959	0.424149
$\Delta T_{c}/K$ [b]	1.69089	1.56374	1.75149	1.96599	1.85448	1.83677
$(\varepsilon_{calor})\cdot(-\Delta T_{c})/J$	−25,020.3	−23,138.8	−25,917	−29,091	−27,441	−27,179
$(\varepsilon_{cont})\cdot(-\Delta T_{c})/J$	−27.69	−25.42	−28.98	−33.14	−30.92	−30.54
$\Delta U_{decomp}\ HNO_{3}/J$	44.2	42.41	46.59	52.56	50.77	50.17
$\Delta U_{corr}/J$	7.63	6.97	7.93	9.13	8.47	8.35
$-m'\cdot\Delta_{c}u'/J$	16.3	18.18	17.86	16.25	16.22	18.52
$-m''\cdot\Delta_{c}u''/J$	14,460.05	13,977.33	18,062.94	18,500.92	18,587.38	19,662.4
$\Delta_{c}u^{o}(liq)/(J\cdot g^{-1})$	−32,362.8	−32,357.8	−32,357.9	−32,353.6	−324,366.2	−32,386.4
$\Delta_{c}u^{o}(liq)/(J\cdot g^{-1})$ [a]			32,364.1 ± 4.8 [c]			
$\Delta_{c}H_{m}^{o}(liq)/(kJ\cdot mol^{-1})$ [b]			−3833.3 ± 1.5 [d]			
$\Delta_{f}H_{m}^{o}(liq)/(kJ\cdot mol^{-1})$ [b]			−528.6 ± 1.7 [d]			

[a] Results are referenced to $T = 298.15$ K ($p^{o} = 0.1$ MPa). The definition of the symbols are assigned according to [23] as follows: m (substance), m' (cotton) and m'' (polyethylene) are, respectively, the mass of compound burnt, the mass of fuse (cotton), and the mass of auxiliary polyethylene used in each experiment; masses were corrected for buoyancy; V(bomb) = 0.33 dm³ is the internal volume of the calorimetric bomb; p^{i}(gas) = 3.04 MPa is the initial oxygen pressure in the bomb; $m^{i}(H_{2}O)$ = 1.00 g is the mass of water added to the bomb for dissolution of combustion gases; ε_{calor} = (14,797.1 ± 1.0) J·K^{-1}; $\Delta T_{c} = T^{f} - T^{i} - \Delta T_{corr}$ is the corrected temperature rise from initial temperature T^{i} to the final temperature T^{f}, with the correction ΔT_{corr} for heat exchange during the experiment; ε_{cont} is the energy equivalents of the bomb contents in their initial ε^{i}_{cont} and final states ε^{f}_{cont}, the contribution for the bomb content is calculated with $(\varepsilon_{cont})\cdot(-\Delta T_{c}) = (\varepsilon^{i}_{cont})\cdot(T^{i} - 298.15) + (\varepsilon^{f}_{cont})\cdot(298.15 - T^{f} + \Delta T_{corr}.)$; $\Delta U_{decomp}\ HNO_{3}$ is the energy correction for the nitric acid formation; and ΔU_{corr} is the correction to standard states. Auxiliary data are given in Table S1. [b] The heat exchange correction, ΔT_{corr}, was automatically applied to final temperatures so that the corrected temperature rise was directly calculated as $T^{f} - T^{i}$. [c] The uncertainty of combustion energy is expressed as the standard deviation of the mean. [d] The uncertainties are expressed as the twice standard deviation of the mean.

These $\Delta_c u^o$(liq) values were used to calculate the experimental standard molar enthalpy of combustion, $\Delta_c H_m^o$, (see Table 8) of ethylene glycol mono-tert-butyl ether, which refers to the reaction:

$$C_6H_{14}O_2 + 9.5\ O_2 = 6\ CO_2 + 7\ H_2O \tag{12}$$

The standard molar enthalpy of formation, $\Delta_f H_m^o$(liq), (see Table 8) was calculated based on the $\Delta_c H_m^o$ values of the reaction according to Equation (12) using Hess's Law and the molar enthalpies of the formation of H_2O(liq) and CO_2(g) assigned by CODATA [34]. The total uncertainties of the $\Delta_c H_m^o$ and $\Delta_f H_m^o$ values were calculated according to the guidelines presented by Hubbard et al. [23] and Olofsson [35]. The uncertainty of combustion energy, $\Delta_c u^o$(liq), was expressed as the standard deviation of the mean. According to the thermochemical practice, the uncertainties assigned to the $\Delta_f H_m^o$(liq) values are twice the overall standard deviations and include the uncertainties of the calibration, the combustion energies of the auxiliary materials, and the uncertainties of the enthalpies of formation of the reaction products H_2O and CO_2. A summary of the thermochemical data for the ethylene tert-butyl ethers is given in Table 9.

Table 9. Thermochemical data at T = 298.15 K for the ethylene tert-butyl ethers (p^o = 0.1 MPa, in kJ·mol^{-1}) [a].

Compound	$\Delta_f H_m^o$(liq)$_{exp}$	$\Delta_l^g H_m^o$ [b]	$\Delta_f H_m^o$(g)$_{exp}$	$\Delta_f H_m^o$(g)$_{theor}$ [c]
EGM	−528.6 ± 1.7	53.1 ± 0.2	−475.5 ± 1.7	−477.2 ± 3.5
EGD	(−605.8 ± 3.6) [d]	52.3 ± 0.6	-	−553.5 ± 3.5

[a] The uncertainties are given as the twice standard deviation. [b] From Table 3. [c] Theoretical value calculated with the G4 method according to the atomisation procedure. [d] Calculated as the difference between columns 5 and 3 in this table to give an estimate of the missing experimental value.

The combustion experiments with ethylene glycol mono-tert-butyl ether were carried out for the first time.

3.5. Standard Molar Enthalpies of Formation in the Gas Phase

The *experimental* standard molar enthalpy of formation, $\Delta_f H_m^o$(liq, 298.15 K), of ethylene glycol mono-tert-butyl ether given in Table 9 was used together with the *experimental* vaporisation enthalpy, $\Delta_l^g H_m^o$(298.15 K), recommended in Table 3, to derive the *experimental* standard molar enthalpy of formation in the gas phase, $\Delta_f H_m^o$(g, 298.15 K), (see Table 9, column 4) according to Equation (3). This experimental result can now be used for comparison with the theoretical value calculated by the high-level quantum-chemical (QC) composite method G4 [25]. Nowadays, QC methods have acquired the status of a valuable tool for the mutual consistency of thermochemical results. Therefore, the agreement or disagreement between the *theoretical* and *experimental* $\Delta_f H_m^o$(g, 298.15 K) values could help with the attestation of the quality of the experimental and computational procedures.

In a number of our recent studies [36], we have shown that the G4 method is capable of providing reliable gas phase enthalpies of formation $\Delta_f H_m^o$(g, 298.15 K) using the atomisation procedure [26]. Therefore, in the current study, the H_{298} enthalpies of the most stable conformers of EGM and EGD were calculated using the G4 method and converted to the theoretical values using the atomisation procedure (see Table 9, column 5). The *theoretical* value $\Delta_f H_m^o$(g, 298.15 K)$_{theor}$ = −477.2 ± 3.5 kJ·mol^{-1} for EGM agrees very well with the experimental result $\Delta_f H_m^o$(g, 298.15 K)$_{exp}$ = −475.5 ± 1.7 kJ·mol^{-1} and this creates confidence in the calculation method. Consequently, we used the *theoretical* value $\Delta_f H_m^o$(g, 298.15 K)$_{theor}$ = −553.5 ± 3.5 kJ·mol^{-1} for EGD and the experimental result for $\Delta_l^g H_m^o$(298.15 K) = 52.3 ± 0.6 kJ·mol^{-1} to calculate the missing liquid phase enthalpy of formation $\Delta_f H_m^o$(liq, 298.15 K) = −605.8 ± 3.6 kJ·mol^{-1} for EGD (see Table 9, column 2).

This completes the development of the data set of thermochemical values (see Table 9) required for the interpretation of the energetics of chemical reactions according to Equations (1) and (2), which represent the synthesis of platform chemicals from ethylene glycol and isobutene. These energetics are derived and discussed in the next section.

3.6. Energetics of Ethylene Glycol Alkylation Reactions from the "First Law" Method

It is well known that ether synthesis reactions from alcohols and olefins are highly exothermic [36], and accurate knowledge of the energetics of these reactions is essential for the safety of industrial processes. The liquid phase reaction enthalpies, $\Delta_r H_m^o$(298.15 K), of the EGM and EGD synthesis from EG and isobutene were calculated using Hess's Law according to the following equations:

$$\Delta_r H_m^o (\text{liq}) = \Delta_f H_m^o (\text{liq})_{(\text{EGM})} - \Delta_f H_m^o (\text{liq})_{(\text{isobutene})} - \Delta_f H_m^o (\text{liq})_{(\text{EG})} = -(36.3 \pm 2.9) \text{ kJ·mol}^{-1} \quad (13)$$

$$\Delta_r H_m^o (\text{liq}) = \Delta_f H_m^o (\text{liq})_{(\text{EGD})} - \Delta_f H_m^o (\text{liq})_{(\text{isobutene})} - \Delta_f H_m^o (\text{liq})_{(\text{EGM})} = -(39.6 \pm 4.5) \text{ kJ·mol}^{-1} \quad (14)$$

The experimental data on $\Delta_f H_m^o$(liq), required for the calculations according to Equations (13) and (14), are obtained in this work (see Table 9) or taken from the literature (see Table S4). The uncertainties of these reaction enthalpies include the uncertainties of all reactants. The enthalpies of Reactions (13) and (14) are quite similar in their experimental uncertainties. However, the reaction enthalpy of Reaction (14) is slightly more negative compared to Reaction (13), which is probably because the ethylene glycol di-tert-butyl ether is more branched and strained than the mono-tert-butyl substituted precursor. Consequently, more energy is required to form this compound. The energetics of Reactions (13) and (14) are moderate, but large enough to cause a possible reactor temperature runaway if the mixing of the reactants is disturbed. Therefore, the reliable thermodynamics of Reactions (13) and (14) evaluated in this work are useful for determining the appropriate temperature management of chemical reactors.

Another perspective for using the thermodynamic data of EGM and EGD is related to an interesting technological idea of improving the conversion of iso-olefins in the production of methyl tert-butyl ether (MTBE) or methyl tert-amyl ether (TAME). It has been found that the overall conversion of an iso-olefin can be significantly increased if the iso-olefin is first reacted with ethylene glycol at 40–70 °C and 5–7 atm over sulfonated resin catalysts and then the resulting ethylene glycol tert-alkyl ethers are reacted with methanol at an elevated temperature and 1.5–2 atm over the same catalyst [11]. The chemical reactions for the second step are shown in Figure 2.

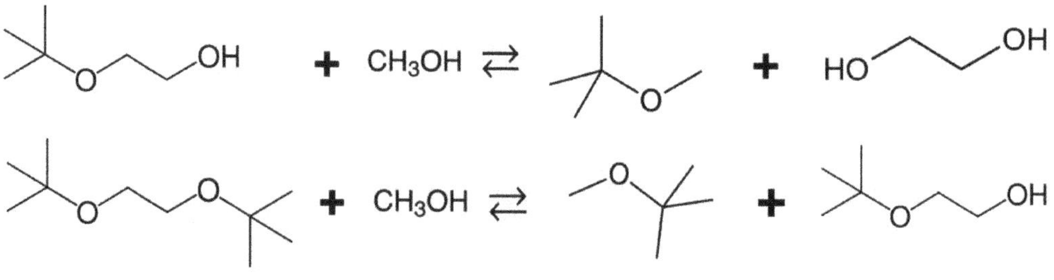

Figure 2. The liquid phase reactions of the MTBE synthesis from EGM or EGD and methanol.

The liquid phase reactions of the MTBE synthesis from EGM or EGD and methanol are given by Equations (15) and (16):

$$\text{EGM + Methanol = MTBE + Ethylene glycol} \quad (15)$$

$$\text{EGD + Methanol = MTBE + EGM} \quad (16)$$

The liquid phase reaction enthalpies, $\Delta_r H_m^\circ$(298.15 K), of the MTBE synthesis from EGM or EGD and methanol were calculated according to the following equations:

$$\Delta_r H_m^\circ(\text{liq}) = \Delta_f H_m^\circ(\text{liq})_{(\text{EG})} + \Delta_f H_m^\circ(\text{liq})_{(\text{MTBE})} - \Delta_f H_m^\circ(\text{liq})_{(\text{EGM})} - \Delta_f H_m^\circ(\text{liq})_{(\text{MeOH})} = -(0.7 \pm 3.0)\ \text{kJ·mol}^{-1} \quad (17)$$

$$\Delta_r H_m^\circ(\text{liq}) = \Delta_f H_m^\circ(\text{liq})_{(\text{EGM})} + \Delta_f H_m^\circ(\text{liq})_{(\text{MTBE})} - \Delta_f H_m^\circ(\text{liq})_{(\text{EGD})} - \Delta_f H_m^\circ(\text{liq})_{(\text{MeOH})} = -(2.6 \pm 4.5)\ \text{kJ·mol}^{-1} \quad (18)$$

The experimental data on $\Delta_f H_m^\circ$(liq) required for the calculations according to Equations (17) and (18) are obtained in this work (see Table 9) or taken from the literature (see Table S4). The uncertainties of these reaction enthalpies include the uncertainties of all reactants. It turned out that, according to our calculations, both Reactions (15) and (16) are practically thermoneutral and, therefore, technologically less demanding.

3.7. Thermodynamic Functions of Ethylene Glycol Alkylation from the "Second Law" Method

The results of the chemical equilibrium study of Reactions (1) and (2) in the presence of an acidic catalyst in the liquid phase were reported by Chang et al. [15]. The equilibrium concentrations of isobutene, EG, EGM, and EGD were measured by gas chromatography, and the concentration equilibrium constants for Reactions (1) and (2) were derived between 318 K and 393 K. The activity coefficients of the reactants were calculated with the commonly used UNIFAC method [37]. The concentration equilibrium constants were multiplied by the activity coefficients and the thermodynamic equilibrium constants K_a (1) and K_a (2) were estimated. The temperature dependences of K_a (1) and K_a (2) are shown in Figure 3.

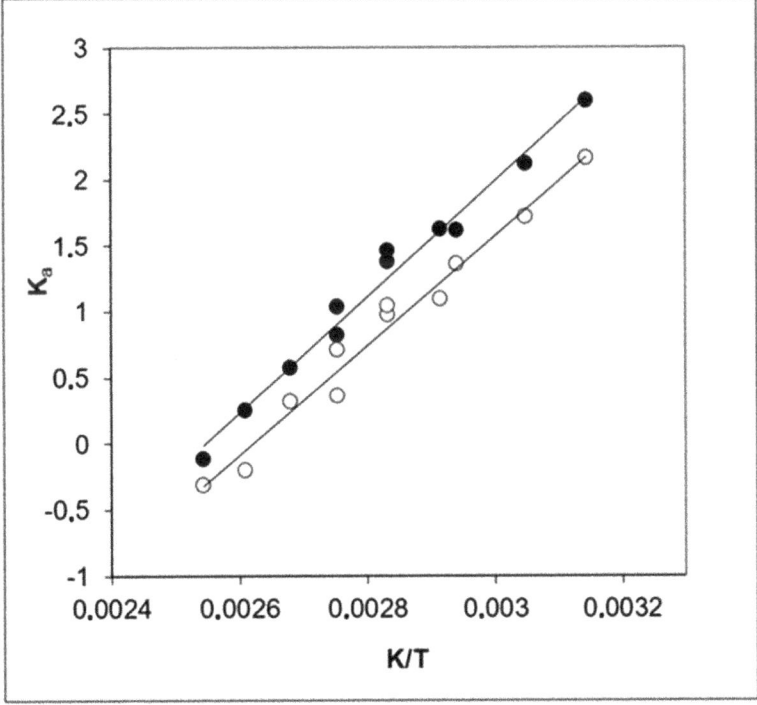

Figure 3. Temperature dependence of the thermodynamic equilibrium constants of the tert-butylation of ethylene glycol: ○—K_a (1) for the reaction according to Equation (1) and ●—K_a (2) for the reaction according to Equation (2). Experimental data are from Chang et al. [15]. Numerical data are compiled in Table S5.

The thermodynamic functions, the reaction enthalpies, $\Delta_r H_m^o$(liq), and reaction entropies, $\Delta_r S_m^o$(liq), were derived from the temperature dependences of K_a (1) and K_a (2). The results are given in Table 10.

Table 10. Thermodynamic functions $\Delta_r H_m^o$ and $\Delta_r S_m^o$ of Reactions (1) and (2) in the liquid phase and the temperature dependences $\ln K_a = a + b \cdot (T/K)^{-1}$ (with correlation coefficient R^2).

Reaction	<T-Range> [a] K	a	b	R^2	$\Delta_r H_m^o$ [b] kJ·mol^{-1}	$\Delta_r S_m^o$ [b] J·mol^{-1} K^{-1}
Equation (1)	318–393	−10.9	4160	0.9743	−34.6 ± 3.7	−91 ± 10
Equation (2)	318–393	−11.2	4395	0.9821	−36.5 ± 3.2	−93.1 ± 9.1

[a] The temperature range of the equilibrium study. [b] The values of the enthalpies $\Delta_r H_m^o$ and entropies $\Delta_r S_m^o$ of Reactions (1) and (2) were derived for the average temperature of the range given in column 2. It was assumed that the enthalpies of the reaction hardly change on passing from the average temperature of the experimental range to T = 298.15 K [36].

It turned out that the thermodynamic functions $\Delta_r H_m^o$ and $\Delta_r S_m^o$ of the synthesis of ethylene glycol mono-tert-butyl ether (Reaction 1) and the synthesis of ethylene glycol di-tert-butyl ether (Reaction 2) are hardly distinguishable within their experimental uncertainties. This finding is rather unexpected, therefore, it is now essential to compare the "First Law" reaction enthalpies, $\Delta_r H_m^o$(liq, 298.15 K), for the EGM and EGD synthesis from EG and isobutene (obtained according to Equations (15) and (16)), with the "Second Law" results derived from the equilibrium study of Reactions (1) and (2).

To our satisfaction, the energetics of Reactions (1) and (2) determined according to the "First Law" and the "Second Law" methods agree well and are within their experimental uncertainties. Even the slightly larger enthalpy of reaction for Reaction (1) observed according to the "First Law" is supported by a similar trend shown by the "Second Law" (see Table 10). Such good agreement provides confidence in the reliability of all of the thermodynamic data sets evaluated in this work and used for comparison according to the "First Law" and the "Second Law".

3.8. Entropies of Ethylene Glycol Tert-Butyl Ethers

Within the framework of this study, we were highly motivated by the practical significance of the thermodynamic functions of ethylene glycol tert-butyl ethers to optimise the technology of their production. One of the main goals was to show that, nowadays, only a reasonable combination of experimental and computational thermodynamics can reduce the costs of developing new technologies for biomass valorisation. The main focus of our thermochemical study was on the energetics of the reactions relevant to biomass valorisation. As shown in Table 10, the alkylation reactions are considerably exothermic so adequate thermal management is required to avoid the runaway of the chemical reactor. The reaction entropies $\Delta_r S_m^o$(liq) of Reactions (1) and (2) derived in Table 10 should be considered as by-products in this context. Fortunately, however, these new results open up an unexpected opportunity to gain deeper insights into the structural features of ethylene glycol tert-butyl ethers, which are investigated in this work.

Indeed, according to Hess's law, which is applied to Reactions (1) and (2), the following equations are responsible for their entropy changes:

$$\Delta_r S_m^o(\text{liq, reaction 1}) = S_m^o(\text{liq})_{(\text{EGM})} - S_m^o(\text{liq})_{(\text{isobutene})} - S_m^o(\text{liq})_{(\text{EG})} \quad (19)$$

$$\Delta_r S_m^o(\text{liq, reaction 2}) = S_m^o(\text{liq})_{(\text{EGD})} - S_m^o(\text{liq})_{(\text{isobutene})} - S_m^o(\text{liq})_{(\text{EGM})} \quad (20)$$

where the S_m^o(liq) values are the standard molar entropies of the corresponding reactants. Admittedly, the S_m^o(liq) values for the important platform chemicals isobutene and ethylene glycol are available in the literature [38–40]. This opens up possibilities for estimating the unknown values S_m^o(liq)$_{(\text{EGM})}$ and S_m^o(liq)$_{(\text{EGD})}$ according to Equations (19) and (20) using the experimental reaction entropies $\Delta_r S_m^o(\text{liq})$ given in Table 10. The entropies of isobutene and EG required for these calculations, as well as the results for EGM and EGD, are summarised in Table 11.

Table 11. Experimental and estimated entropies of Reactions (1) and (2) participants (at 298.15 K in $J \cdot K^{-1} \cdot mol^{-1}$).

	Isobutene	EG	EGM	EGD
σ	18	2	243	13122
$S_m^o(\text{liq})$: exp	194.0 [40]	166.9 [38]	269.9	370.8
add [a]	223.0	158.7	291.0	375.6
add + (σ-corr)	198.9	153.0	243.4	296.8
$S_m^o(\text{g})$: exp	293.6 [39]	323.6 [39]	400.1 [b]	498.0 [b]
add [a]	319.0	329.8	477.8	566.3
add + (σ-corr)	295.0	324.1	432.1	487.5
G4			443.8	571.3
G4 + (σ-corr)			398.1	492.2

[a] Calculated as the sum of corresponding additive contributions. [b] Calculated as the sum of $S_m^o(\text{liq})$ from this table and entropy of vaporisation $\Delta_l^g S_m^o(298.15 \text{ K})$ derived from the vapour pressure temperature dependences. Values given in bold were used for calculations in this work.

Considering that the values for $S_m^o(\text{liq})_{(\text{EGM})}$ and $S_m^o(\text{liq})_{(\text{EGD})}$ are new, it makes sense to use another complementary method to assess at least a possible level of these values. The correct estimation of the entropies of organic molecules is a challenging task. Various group contribution methods have been developed for this purpose, with Benson's type [41] of increments being the most popular in the thermodynamic community. The required contributions have been analysed and revised by Domalski and Hearing [42], and we used the numerical values from this compilation for the additive calculations of the entropies of ethylene glycol tert-butyl ethers. When calculating the entropy of a molecule, it is important to consider, together with the sum of the contributions, the global symmetry number σ, which is the number of superimposable configurations that include the outer symmetry and the inner free rotors; this is described in detail by Benson [41] and helpful examples are given in [43]. The results of the additive liquid phase and gas phase entropy calculations, including the σ values, are summarised in Table 11. However, these results are rather disappointing, because, for simple molecules such as isobutene with only two inner free rotors and σ = 18 and ethylene glycol with σ = 2, the agreement with the experimental values is still acceptable for both the liquid and the gas phase. For the more complicated molecules EGM and EGD, with a considerable number of free rotors and very large symmetry numbers, the additivity did not lead to entropies comparable to the experiment (see Table 11) in either the liquid or the gas phase. This significant disagreement raised the legitimate question of whether the experimental entropies were correct. To answer this question, we calculated the gas phase entropies of EGM and EGD using the quantum chemical G4 method. The results are shown in Table 11. At first glance, it was obvious that the entropies calculated by G4 did not match the experiment either (see Table 11), but subtracting the symmetry correction ($R \times \ln \sigma$) from the direct G4 result brought the calculated and experimental values into agreement (see Table 11):

$S_m^o(\text{g, EGM})_{\text{exp}} = 400.1 \text{ J} \cdot \text{K}^{-1} \cdot \text{mol}^{-1}$ is indistinguishable from $S_m^o(\text{g, EGM})_{\text{G4}} = 398.1 \text{ J} \cdot \text{K}^{-1} \cdot \text{mol}^{-1}$

$S_m^o(\text{g, EGD})_{\text{exp}} = 498.0 \text{ J} \cdot \text{K}^{-1} \cdot \text{mol}^{-1}$ is close to $S_m^o(\text{g, EGM})_{\text{G4}} = 492.0 \text{ J} \cdot \text{K}^{-1} \cdot \text{mol}^{-1}$

This very good agreement supports the reliability of the experimental liquid and gas phase entropies of EGM and EGD. At the same time, the difficulties in reconciling the experimental, additive, and quantum chemical entropies highlighted in this section provide important insights into the limitations of empirical methods and the positive experiences with quantum chemical methods. So, in this paper, the focus was not on the theory, rather, we *paid* attention to the practical thermodynamic aspects of biomass valorisation. Unexpectedly, however, we were confronted with theory and obtained "*cashback*" with theoretical aspects related to the entropy calculations of flexible molecules with a large number of free rotators.

3.9. Standard Molar Thermodynamic Functions of Ethylene Glycol Tert-Butyl Ethers

Due to the developments in "green chemistry", EG-tert-butyl ether has been considered as a large-scale platform chemical produced from cellulose [1]. However, thermodynamic modelling and optimisation of EG-tert-butyl ether synthesis require knowledge of the fundamental thermodynamic functions of these compounds. This work has contributed by evaluating the energetic and entropic properties of EG tert-butyl ethers. Thus, we compiled the available experimental values and calculated thermodynamic formation functions for EG tert-butyl ethers, which are listed in Table 12.

Table 12. Standard molar thermodynamic properties of ethylene glycol tert-butyl ethers at T = 298.15 K.

Compound	State	$\Delta_f H_m^o$ [a]	$\Delta_f S_m^o$	$\Delta_f G_m^o$	S_m^o [b]	$C_{p,m}^o$ [c]
		kJ·mol^{-1}	J·K^{-1}·mol^{-1}	kJ·mol^{-1}	J·K^{-1}·mol^{-1}	
EGM	liquid	−528.6 ± 1.7	−883.2	−265.4	269.9	273.5
	gas	−475.5 ± 1.7	−753.0	−250.8	400.1	191.8
EGD	liquid	−605.8 ± 3.6	−1327.4	−210.0	370.8	357.2
	gas	−553.5 ± 3.5	−1200.2	−195.7	498.0	253.7

[a] From Table 9. [b] From Table 11. [c] From Table S3.

The entropies of formation, $\Delta_f S_m^o$, were calculated based on Reactions (21) and (22):

$$6\,C_{graphite} + 7\,H_2(g) + O_2(g) = C_6H_{14}O_2 \text{ (g or l)} \tag{21}$$

$$10\,C_{graphite} + 11\,H_2(g) + O_2(g) = C_{10}H_{22}O_2 \text{ (g or l)} \tag{22}$$

using the following entropies of formation for $C_{graphite}$ (5.74 ± 0.13) J·K^{-1}·mol^{-1}, H$_2$(g) (130.52 ± 0.02) J·K^{-1}·mol^{-1}, and O$_2$(g) (205.04 ± 0.03) J·K^{-1}·mol^{-1} recommended in [44]. The Gibbs function of formation, $\Delta_f G_m^o$, was estimated according to Equation (23) from the values of $\Delta_f H_m^o$ and $\Delta_f S_m^o$ given Table 12:

$$\Delta_f G_m^o = \Delta_f H_m^o - T \times \Delta_f S_m^o \tag{23}$$

The standard molar thermodynamic functions in the liquid and the gas phase collected in Table 12 can be used for the optimisation of EG tert-butyl ethers to be processed into further valuable biobased fuel additivities and platform chemicals.

This article was written for the Justus von Leibig's 150 years theme issue; to build a bridge to the main topic, it is worth remembering that the "Liebig cooler" was not invented by Liebig, as is assumed, but became *popular* through Liebig. We have neither invented experimental thermochemical tools nor developed sophisticated quantum chemical methods. Nevertheless, through our work, we try to make "*popular*" the idea that the "judicious" combination of experimental, empirical, and quantum chemical methods" is the only modern way to obtain reliable practical and theoretical thermochemical results.

Supplementary Materials: The following supporting information can be downloaded at: https://www.mdpi.com/article/10.3390/chemistry5020079/s1. Table S1: Auxiliary quantities: formula, density, massic heat capacity, and expansion coefficients of the materials used in the present study [45,46]; Table S2: Compilation of data on molar heat capacities and heat capacity differences at T = 298.15 K [47]; Details on adjustment of vaporisation/sublimation enthalpies to the reference temperature T = 298.18 K; Details on transpiration method; Details on static methods; Table S3: The vapour pressures and standard vaporisation enthalpies and entropies obtained by the approximation of boiling points at different pressures available in the literature; Table S4: Thermochemical data for reference compounds [48,49]; Table S5: Experimental thermodynamic constants reported by Chang et al. [15]; Figure S1: Comparison of vapour pressures measured in this work over the liquid sample of ethylene glycol mono-tert-butyl ether.

Author Contributions: Conceptualization, methodology, validation, writing—original draft preparation, S.P.V. Investigation, validation, and writing—review and editing, A.A.Z. All authors have read and agreed to the published version of the manuscript.

Funding: S.P.V. acknowledges financial support from DFG, grant VE 265/12−1 "Glycolysis: thermodynamics and pathway predictions". A.A.Z. is grateful to Deutsche Akademische Austausch Dienst (DAAD) for a scholarship. This paper has been supported by the Kazan Federal University Strategic Academic Leadership Program ("PRIORITY-2030").

Data Availability Statement: All data used in this work are available from the text and supplementary materials file.

Conflicts of Interest: The authors declare no conflict of interest.

References

1. Fukuoka, A.; Dhepe, P.L. Catalytic conversion of cellulose into sugar alcohols. *Angew. Chem. Int. Ed. Engl.* **2006**, *45*, 5161–5163. [CrossRef] [PubMed]
2. Wang, A.; Zhang, T. One-pot conversion of cellulose to ethylene glycol with multifunctional tungsten-based catalysts. *Acc. Chem. Res.* **2013**, *46*, 1377–1386. [CrossRef] [PubMed]
3. Ji, N.; Zhang, T.; Zheng, M.; Wang, A.; Wang, H.; Wang, X.; Chen, J.G. Direct Catalytic Conversion of Cellulose into Ethylene Glycol Using Nickel-Promoted Tungsten Carbide Catalysts. *Angew. Chem. Int. Ed. Engl.* **2008**, *47*, 8510–8513. [CrossRef] [PubMed]
4. Yue, H.; Zhao, Y.; Ma, X.; Gong, J. Ethylene glycol: Properties, synthesis, and applications. *Chem. Soc. Rev.* **2012**, *41*, 4218–4244. [CrossRef] [PubMed]
5. Gao, Y.; Neal, L.; Ding, D.; Wu, W.; Baroi, C.; Gaffney, A.M.; Li, F. Recent Advances in Intensified Ethylene Production—A Review. *ACS Catal.* **2019**, *9*, 8592–8621. [CrossRef]
6. Kandasamy, S.; Samudrala, S.P.; Bhattacharya, S. The route towards sustainable production of ethylene glycol from a renewable resource, biodiesel waste: A review. *Catal. Sci. Technol.* **2019**, *9*, 567–577. [CrossRef]
7. Smith, R.L. Review of Glycol Ether and Glycol Ether Ester Solvents Used in the Coating Industry. *Environ. Health Perspect.* **1984**, *57*, 1–4. [CrossRef] [PubMed]
8. Samoilov, V.O.; Stolonogova, T.I.; Ramazanov, D.N.; Tyurina, E.V.; Lavrent'ev, V.A.; Porukova, Y.I.; Chernysheva, E.A.; Kapustin, V.M. tert-Butyl Ethers of Renewable Diols as Oxygenated Additives for Motor Gasoline. Part I: Glycerol and Propylene Glycol. Ethers. *Pet. Chem.* **2023**, 1–9. [CrossRef]
9. Viswanadham, N.; Saxena, S.K. Etherification of glycerol for improved production of oxygenates. *Fuel* **2013**, *103*, 980–986. [CrossRef]
10. Macho, V.; Kavala, M.; Kolieskova, Z. tert-Butyl ethers of polyhydric alcohols. *Neftekhimiya* **1979**, *19*, 821–827.
11. Jayadeokar, S.S.; Sharma, M.M. Ion exchange resin catalysed etherification of ethylene and propylene glycols with isobutylene. *React. Polym.* **1993**, *20*, 57–67. [CrossRef]
12. Klepáčová, K.; Mravec, D.; Kaszonyi, A.; Bajusl, M. Etherification of glycerol and ethylene glycol by isobutylene. *Appl. Catal. A Gen.* **2007**, *328*, 1–13. [CrossRef]
13. Liu, H.; Zhang, Z.; Tang, J.; Fei, Z.; Liu, Q.; Chen, X.; Cui, M.; Qiao, X. Quest for pore size effect on the catalytic property of defect-engineered MOF-808-SO4 in the addition reaction of isobutylene with ethylene glycol. *J. Solid State Chem.* **2019**, *269*, 9–15. [CrossRef]
14. Verevkin, S.P. *Gibbs Energy and Helmholtz Energy: Liquids, Solutions and Vapours*; Wilhelm, E., Letcher, T.M., Eds.; Royal Society of Chemistry: Cambridge, UK, 2021; ISBN 978-1-83916-201-5.
15. Chang, T.-K.; Hwang, J.-H.; Huang, H.-Y.; Su, W.-B.; Hong, C.-T. The thermodynamics properties for the liquid phase synthesis of ethylene glycol tert-butyl ether and ethylene glycol di-tert-butyl ether. *Shiyou Jikan* **2012**, *48*, 21–37.
16. Verevkin, S.P.; Emel'yanenko, V.N. Transpiration method: Vapor pressures and enthalpies of vaporization of some low-boiling esters. *Fluid Phase Equilibria* **2008**, *266*, 64–75. [CrossRef]

17. Verevkin, S.P.; Sazonova, A.Y.; Emel´yanenko, V.N.; Zaitsau, D.H.; Varfolomeev, M.A.; Solomonov, B.N.; Zherikova, K.V. Thermochemistry of halogen-substituted methylbenzenes. *J. Chem. Eng. Data* **2015**, *60*, 89–103. [CrossRef]
18. Emel'yanenko, V.N.; Verevkin, S.P. Benchmark thermodynamic properties of 1,3-propanediol: Comprehensive experimental and theoretical study. *J. Chem. Thermodyn.* **2015**, *85*, 111–119. [CrossRef]
19. Zaitsau, D.H.; Verevkin, S.P.; Sazonova, A.Y. Vapor pressures and vaporization enthalpies of 5-nonanone, linalool and 6-methyl-5-hepten-2-one. Data evaluation. *Fluid Phase Equilibria* **2015**, *386*, 140–148. [CrossRef]
20. Zaitsau, D.H.; Verevkin, S.P. Vapor Pressure of Pure Methyl Oleate—The Main Componene of Biodiesel. *Russ. J. Gen. Chem.* **2021**, *91*, 2061–2068. [CrossRef]
21. Verevkin, S.P.; Schick, C. Substituent effects on the benzene ring. Determination of the intramolecular interactions of substituents in tert-alkyl-substituted catechols from thermochemical measurements. *J. Chem. Eng. Data* **2000**, *45*, 946–952. [CrossRef]
22. Emel´yanenko, V.N.; Verevkin, S.P.; Heintz, A. The gaseous enthalpy of formation of the ionic liquid 1-butyl-3-methylimidazolium dicyanamide from combustion calorimetry, vapor pressure measurements, and ab initio calculations. *J. Am. Chem. Soc.* **2007**, *129*, 3930–3937. [CrossRef] [PubMed]
23. Hubbard, W.N.; Scott, D.W.; Waddington, G.; Rossini, F.D. Standard states and corrections for combustions in a bomb at constant volume. In *Experimental Thermochemistry: Measurements of Heats of Reactions*; Rossini, F.D., Ed.; Interscience: New York, NY, USA, 1956; Volume 1, pp. 75–128.
24. Frisch, M.J.; Trucks, G.W.; Schlegel, H.B.; Frisch, M.J.; Trucks, G.W.; Schlegel, H.B.; Scuseria, G.E.; Robb, M.A.; Cheeseman, J.R.; Scalmani, G.; et al. *Gaussian 09, Revision A.02*; Gaussian, Inc.: Wallingford, CT, USA, 2009.
25. Curtiss, L.A.; Redfern, P.C.; Raghavachari, K. Gaussian-4 theory. *J. Chem. Phys.* **2007**, *126*, 84108–84112. [CrossRef]
26. Verevkin, S.P.; Emel´yanenko, V.N.; Notario, R.; Roux, M.V.; Chickos, J.S.; Liebman, J.F. Rediscovering the wheel. Thermochemical analysis of energetics of the aromatic diazines. *J. Phys. Chem. Lett.* **2012**, *3*, 3454–3459. [CrossRef]
27. Samarov, A.A.; Verevkin, S.P. Hydrogen storage technologies: Methyl-substituted biphenyls as an auspicious alternative to conventional liquid organic hydrogen carriers (LOHC). *J. Chem. Thermodyn.* **2022**, *165*, 106648. [CrossRef]
28. SciFinder—Chemical Abstracts Service. Available online: https://scifinder.cas.org/ (accessed on 28 March 2023).
29. Reaxys. Available online: https://www.reaxys.com/ (accessed on 28 March 2023).
30. Chickos, J.S.; Acree, W.E. Enthalpies of sublimation of organic and organometallic compounds. 1910–2001. *J. Phys. Chem. Ref. Data* **2002**, *31*, 537–698. [CrossRef]
31. Chickos, J.S.; Hosseini, S.; Hesse, D.G.; Liebman, J.F. Heat capacity corrections to a standard state: A comparison of new and some literature methods for organic liquids and solids. *Struct. Chem.* **1993**, *4*, 271–278. [CrossRef]
32. Benson, S.W. Some Observations on the Structures of Liquid Alcohols and Their Heats of Vaporization. *J. Am. Chem. Soc.* **1996**, *118*, 10645–10649. [CrossRef]
33. Majer, V.; Svoboda, V. *Enthalpies of Vaporization of Organic Compounds: A Critical Review and Data Compilation*; Blackwell Scientific Publications: Oxford, UK, 1985; pp. 1–300.
34. Cox, J.D.; Wagman, D.D.; Medvedev, V.A. *CODATA Key Values for Thermodynamics*; Hemisphere Pub. Corp.: New York, NY, USA, 1989.
35. Olofsson, G. Assignment of uncertainties, Chapter 6. In *Experimental Chemical Thermodynamics. Combustion Calorimetry*; Sunner, S., Mansson, M., Eds.; Pergamon Press: Oxford, UK, 1976; Volume 1, pp. 137–161.
36. Verevkin, S.P.; Konnova, M.E.; Zherikova, K.V.; Pimerzin, A.A. Thermodynamics of glycerol and diols valorisation via reactive systems of acetals synthesis. *Fluid Phase Equilibria* **2020**, *510*, 112503. [CrossRef]
37. Fredenslund, A.; Jones, R.L.; Prausnitz, J.M. Group-contribution estimation of activity coefficients in nonideal liquid mixtures. *AIChE J.* **1975**, *21*, 1086–1099. [CrossRef]
38. Parks, G.S.; Kelley, K.K.; Huffman, H.M. Thermal data on organic compounds. V. A revision of the entropies and free energies of nineteen organic compounds. *J. Am. Chem. Soc.* **1929**, *51*, 1969–1973. [CrossRef]
39. Stull, D.R.; Westrum, E.F., Jr.; Sinke, G.C. *The Chemical Thermodynamics of Organic Compounds*; John Wiley and Sons, Inc.: New York, NY, USA, 1969.
40. Todd, S.S.; Parks, G.S. Thermal data on organic compounds. XV. Some heat capacity, entropy and free energy data for the isomeric butenes. *J. Am. Chem. Soc.* **1936**, *58*, 134–137. [CrossRef]
41. Benson, S.W. *Thermochemical Kinetics*, 2nd ed.; Wiley-Interscience: New York, NY, USA, 1976; pp. 47–50.
42. Domalski, E.S.; Hearing, E.D. Heat Capacities and Entropies of Organic Compounds in the Condensed Phase. Volume III. *J. Phys. Chem. Ref. Data* **1996**, *25*, 1. [CrossRef]
43. Domalski, E.S.; Hearing, E.D. Estimation of Thermodynamic Properties of Organic Compounds in the Gas, Liquid, and Solid Phases at 298.15 K. *J. Phys. Chem. Ref. Data* **1988**, *17*, 1637. [CrossRef]
44. Chase, M.W., Jr. NIST-JANAF Themochemical Tables, Fourth Edition. *J. Phys. Chem. Ref. Data* **1998**, *9*, 1–1951.
45. Evans, T.W.; Edlund, K.R. Tertiary alkyl ethers, preparation and properties. *Ind. Eng. Chem.* **1936**, *28*, 1186–1188. [CrossRef]
46. Verevkin, S.P.; Heintz, A. Determination of Vaporization Enthalpies of the Branched Esters from Correlation Gas Chromatography and Transpiration Methods. *J. Chem. Eng. Data* **1999**, *44*, 1240–1244. [CrossRef]
47. Atake, T.; Kawaji, H.; Tojo, T.; Kawasaki, K.; Ootsuka, Y.; Katou, M.; Koga, Y. Heat Capacities of Isomeric 2-Butoxyethanols from 13 to 300 K: Fusion and Glass Transition. *Bull. Chem. Soc. Jpn.* **2000**, *73*, 1987–1991. [CrossRef]
48. Pedley, J.P.; Naylor, R.D.; Kirby, S.P. *Thermochemical Data of Organic Compounds*, 2nd ed.; Chapman and Hall: London, UK, 1986.

49. Verevkin, S.P.; Emel'yanenko, V.N.; Nell, G. 1,2-Propanediol. Comprehensive experimental and theoretical study. *J. Chem. Thermodyn.* **2009**, *41*, 1125–1131. [CrossRef]

Disclaimer/Publisher's Note: The statements, opinions and data contained in all publications are solely those of the individual author(s) and contributor(s) and not of MDPI and/or the editor(s). MDPI and/or the editor(s) disclaim responsibility for any injury to people or property resulting from any ideas, methods, instructions or products referred to in the content.

Article

A Pathway for Aldol Additions Catalyzed by L-Hydroxyproline-Peptides via a β-Hydroxyketone Hemiaminal Intermediate

Lo'ay Ahmed Al-Momani [1,*], Heinrich Lang [2] and Steffen Lüdeke [3,4,*]

1. Department of Chemistry, Faculty of Science, The Hashemite University, Zarqa 13133, Jordan
2. Research Center for Materials, Architectures and Integration of Nanomembranes (MAIN), Research Group Organometallic Chemistry, Technische Universität Chemnitz, Rosenbergstraße 6, 09126 Chemnitz, Germany
3. Institute of Pharmaceutical Sciences, Albert-Ludwigs-Universität Freiburg, Albertstraße 25, 79104 Freiburg, Germany
4. Institute of Pharmaceutical and Biomedical Sciences (IPBS), Johannes Gutenberg-Universität Mainz, Staudingerweg 5, 55128 Mainz, Germany
* Correspondence: loay.al-momani@hu.edu.jo (L.A.M.); sluedeke@uni-mainz.de (S.L.)

Citation: Al-Momani, L.A.; Lang, H.; Lüdeke, S. A Pathway for Aldol Additions Catalyzed by L-Hydroxyproline-Peptides via a β-Hydroxyketone Hemiaminal Intermediate. *Chemistry* 2023, 5, 1203–1219. https://doi.org/10.3390/chemistry5020081

Academic Editor: Angelo Frongia

Received: 2 April 2023
Revised: 2 May 2023
Accepted: 8 May 2023
Published: 10 May 2023

Copyright: © 2023 by the authors. Licensee MDPI, Basel, Switzerland. This article is an open access article distributed under the terms and conditions of the Creative Commons Attribution (CC BY) license (https://creativecommons.org/licenses/by/4.0/).

Abstract: While the use of L-proline-derived peptides has been proven similarly successful with respect to enantioselectivity, the physico-chemical and conformational properties of these organocatalysts are not fully compatible with transition state and intermediate structures previously suggested for L-proline catalysis. L-Proline or L-4-hydroxyproline catalysis is assumed to involve proton transfers mediated by the carboxylic acid group, whereas a similar mechanism is unlikely for peptides, which lack a proton donor. Herein, we prepared an array of hydroxyproline-based dipeptides through amide coupling of Boc-protected *cis*- or *trans*-4-L-hydroxyproline (*cis*- or *trans*-4-Hyp) to benzylated glycine (Gly-OBn) and L-valine (L-Val-OBn) and used these dipeptides as catalysts for a model aldol reaction. Despite the lack of a proton donor in the catalytic site, we observed good stereoselectivities for the *R*-configured aldol product both with dipeptides formed from *cis*- or *trans*-4-Hyp at moderate conversions after 24 h. To explain this conundrum, we modeled reaction cycles for aldol additions in the presence of *cis*-4-Hyp, *trans*-4-Hyp, and *cis*- and *trans*-configured 4-Hyp-peptides as catalysts by calculation of free energies of conformers of intermediates and transition states at the density functional theory level (B3LYP/6-31G(d), DMSO PCM as solvent model). While a catalytic cycle as previously suggested with L-proline is also plausible for *cis*- or *trans*-4-Hyp, with the peptides, the energy barrier of the first reaction step would be too high to allow conversions at room temperature. Calculations on modeled transition states suggest an alternative pathway that would explain the experimental results: here, the catalytic cycle is entered by the acetone self-adduct 4-hydroxy-4-methylpentan-2-one, which forms spontaneously to a small extent in the presence of a base, leading to considerably reduced calculated free energy levels of transition states of reaction steps that are considered rate-determining.

Keywords: proline catalysis; stereoselective synthesis; catalytic cycle; quantum chemical calculations

1. Introduction

Similar to the catalytic site of an enzyme, the geometric arrangement of nitrogen and oxygen atoms in the amino acid L-Proline (**1**) has the potential to catalyze asymmetric reactions—a discovery that led to a concept awarded with the Nobel Prize in Chemistry in 2021. In the last forty years, **1** has been used as a catalyst, for example, in the asymmetric Robinson annulation [1,2], the aldol condensation [3,4], and in the stereoselective Mannich reaction [5,6]. **1** also catalyzes an unconventional reaction via aldol condensation of α-hydroxyketones and aldehydes to give α,β-dihydroxyketones [7,8]. Several aspects have been discussed, such as the small size of **1**, its rigidity, inexpensiveness, and ready availability [9].

The reaction path of aldol reactions catalyzed by **1** had previously been under debate with respect to the intermediates and transition states determining the overall reaction rates and the stereoselectivity, respectively. The Houk–List pathway suggests an enamine intermediate being formed from **1** and the carbonyl substrate (Scheme 1) [10,11]. In stereoselective C,C-bond formation, proline's carboxylic acid group helps positioning a second carbonyl substrate in close proximity to the enamine moiety. While this is the rate-limiting step of aldol reactions catalyzed by **1** [12], the enamine, which serves as the activated electron donor, is of particular importance in the understanding of catalysis by **1** and derivatives thereof. Its formation involves different intermediates and transition states including the formation of a hemiaminal after addition of the first carbonyl substrate and the tautomerization of a zwitterionic iminium intermediate to the desired enamine. The latter transition competes with the formation of a bicyclic intermediate [13]. A direct involvement of this intermediate in stereoselective C–C-coupling has been proposed [14], but would be incompatible with observed enantio- and diastereoselectivities; therefore, a bicyclic intermediate is considered a merely parasitic by-product that competes with enamine formation [15].

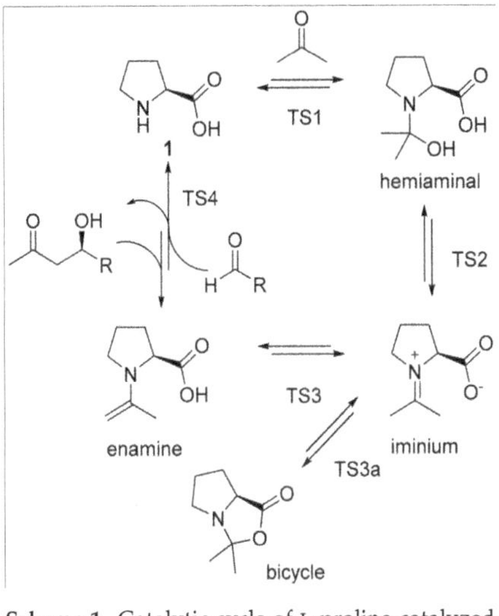

Scheme 1. Catalytic cycle of L-proline-catalyzed aldol reaction of an aldehyde and acetone with intermediates and transition states (TS1 to TS4). TS3a is the transition state for the formation of a bicyclic parasitic by-product from the iminium intermediate.

Drawbacks of L-proline catalysis are its only moderate stereoselectivity and the necessity for comparably large amounts of catalyst, as the catalytic activity is low. The use of N-terminal L-proline peptides as organocatalysts seems a promising alternative [16,17]. It has been shown that some N-terminal L-proline tripeptides considerably increased activity and stereoselectivity compared to catalysis by **1**; however, both activity and stereoselectivity strongly depended on the choice of the L- or D-amino acid residue in addition to L-proline [18]. Effects on catalytic activity were explained by an optimal relative disposition of the secondary amine in the terminal L-proline and the carboxylic acid group of an L- or D-aspartic acid amide in position three. The stereoselectivity depends on the conformation of the tripeptide [18]; in the case of peptides of the Pro-Pro-Xaa-NH$_2$-type, stereoselectivity clearly correlates with the *trans/cis* ratio of the Pro-Pro amide bond [19]. The notion of optimal distance and stereochemical arrangement in tripeptides is supported by the finding that neither activity nor stereoselectivity can be improved in tetrapeptides [20]. In the case of dipeptide catalysts, the reaction path of an aldol reaction may be more similar to the Houk–List pathway suggested for L-proline catalysis. The presence of a carboxylic acid group seems to be beneficial for catalytic activity as the conversion in reactions catalyzed by Pro-NH$_2$ or Pro-Xaa-NH$_2$, where Xaa can be any amino acid, is considerably lower [11,21,22]. Nevertheless, it could be shown that non-acidic N-terminal L-proline di-, tri-, tetra-, penta-, and hexapeptides whose C-terminal carboxylic acid group was masked as a methyl ester were still able to catalyze aldol reactions with moderate to high yield and moderate to high enantioselectivities [23,24].

Another important aspect is the increased conformational flexibility of peptide catalysts in comparison to **1** which might either stabilize or destabilize the geometry of a transition state or an intermediate throughout the catalytic cycle. However, if a defined chiral environment is still provided, it might even enhance selectivity [25,26]. Last but not least, both the chiral environment in the stereoselectivity-determining transition state and the rate-determining transition state are affected by the puckering of the proline ring. Proline adopts basically two preferred puckerings, C_γ-*exo* and C_γ-*endo* [27–30], which also applies to catalytic peptides with an N-terminal proline residue [26].

In *cis*-4-hydroxyproline (*cis*-4-Hyp, *cis*-**2**, Figure 1), C_γ-*endo* is the most preferred puckering that would also allow the formation of an intramolecular hydrogen bond between the *cis*-4-hydroxy group and the carboxylic acid group. Similarly, **1** has a slight preference for C_γ-*endo*. In contrast, in *trans*-4-hydroxyproline (*trans*-4-Hyp, *trans*-**2**, Figure 1), the 4-hydroxy group imposes a C_γ-*exo* pucker. This is in accordance with the gauche empirical rule, an effect that also contributes to the outstanding stability of the collagen triple helix [31,32]. Intuitively, the ring puckering should also affect the stereoselectivity of reactions catalyzed by hydroxyprolines and related peptides. Surprisingly, in different reactions tested, the stereochemical outcome of reactions (aldol, Mannich and Michael asymmetric reactions) catalyzed by *cis*-**2** or *trans*-**2** was similar to **1** catalysis or even lower [33]. These observations prompted us to study the reaction channels leading to either an *R*- or an *S*-configured product in Hyp-catalysis or catalysis by Hyp-derived dipeptides in further detail, in particular, with respect to the role of the conformer distribution in organocatalysts, transition states, and intermediates.

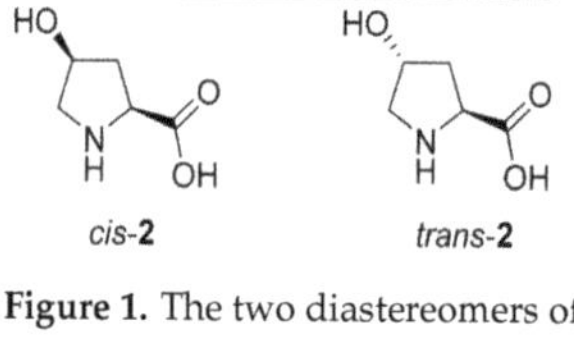

Figure 1. The two diastereomers of L-4-hydroxyproline (**2**).

We synthesized four Hyp-derived C-terminally protected dipeptides (*cis*-**6**, *trans*-**6**, *cis*-**7**, *trans*-**7**, Scheme 2) and performed the aldol addition of acetone to *p*-nitrobenzaldehyde (**8**) catalyzed by **6** or **7** as model reactions. To investigate whether catalysis of aldol additions with Hyp and Hyp-derived peptides follow a similar path as previously suggested for L-proline, we calculated the free energy differences of transition states and intermediates in Hyp and Hyp-dipeptide catalysis at the density functional theory level.

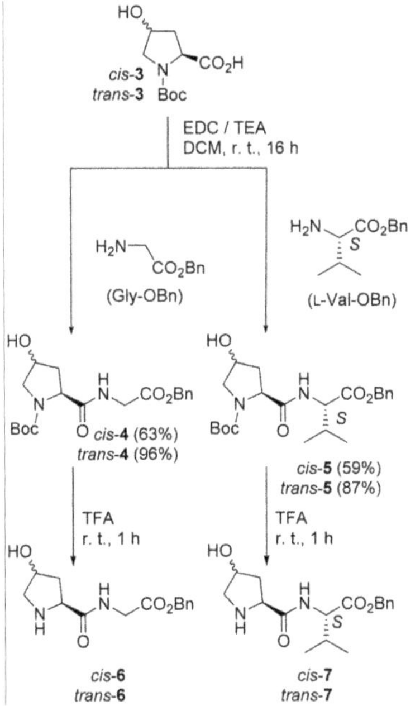

Scheme 2. Synthesis of dipeptide catalysts. A protection of the hydroxy groups of cis-3 or trans-3 was not necessary. Removal of the Boc group from dipeptides in the presence of anisole as a scavenger in dichloromethane was achieved in quantitative yields.

2. Materials and Methods

2.1. General Experimental Conditions

All reagents were of analytical grade. Solvents were dried by standard methods if necessary. TLC was carried out on aluminum sheets pre-coated with silica gel 60 F_{254} (Merck, Darmstadt, Germany). Detection was accomplished by UV light (λ = 254 nm). Preparative column chromatography was carried out on silica gel 60 (Merck, 40–63 μm). ^{1}H NMR (500.3 MHz) spectra were recorded with a Avance III 500 spectrometer (Bruker, Ettlingen, Germany). CDCl$_3$ (δ = 7.26 ppm), HDO (δ = 4.81 ppm), and DMSO (δ = 2.50 ppm) were used as internal standards. ^{13}C{^{1}H} NMR (125.8 MHz) spectra were calibrated with CDCl$_3$ (δ = 77.00 ppm) and DMSO-d$_6$ (δ = 39.43 ppm) as internal standard. IR spectra were recorded with a Nicolet IR200 FTIR-spectrometer (Thermo Fisher Scientific, Waltham, MA, USA). HR-MS spectra were measured with a micrOTOF QII mass spectrometer (Bruker Daltonics, Bremen, Germany). The enantiomeric excess (ee) values of the aldol reaction products were determined by chiral phase HPLC on an HP 1100 chromatography system (Agilent Technologies, Santa Clara, CA, USA)) equipped with a Chiralpak AS-H column (Daicel, Osaka, Japan) using an n-hexane/isopropanol mixture (70:30) as eluent (UV 254 nm, flow rate 0.7 mL min^{-1}, 25 °C).

2.2. Dipeptide Coupling

To a solution of 1.0 eq. of enantiomerically pure amino acid cis-3 or trans-3 (200 mg, 0.87 mmol) in dichloromethane (10 mL), 1.15 eq. of HOBt (50 mg, 1.0 mmol) were added, and the reaction mixture was cooled to 0 °C, followed by addition of 1.15 eq. of the N-terminally unprotected amino acids Gly-OBn (202 mg, 1.0 mmol) or L-Val-OBn (243 mg, 1.0 mmol). Afterwards, 3.0 eq. (263 mg, 2.61 mmol, 361 mL) of triethylamine (TEA) were added in a single portion and the reaction solution was allowed to stir for 10 min. at 0 °C. Then, 1.15 eq. (192 mg, 1.0 mmol) of 1-ethyl-3-(3-dimethylaminopropyl)carbodiimide (EDC) in dichloromethane (15 mL) were dropped gradually to the solution at 0 °C and stirred at room temperature overnight. The reaction mixture was extracted with 1 M HCl, sat. NaHCO$_3$ and finally with sat. NaCl. The collected dichloromethane layer was dried over anhyd. Na$_2$SO$_4$. The crude product was purified using column chromatography (SiO$_2$). All products **4** and **5** were characterized by FT-IR and ^{1}H, ^{13}C{^{1}H} NMR spectroscopy, and ESI-TOF mass-spectrometry (Figures S2–S17).

Dipeptides *cis*-**4** and *trans*-**4** were isolated as colorless viscous oils in a yield of 207 mg (63%) or 315 mg (96%).

cis-**4**: ^{1}H NMR (CDCl$_3$) δ = 7.62 (bs, 1H), 7.36–7.18 (m, 5H), 7.05 (bs, 1H), 5.08 (s, 2H), 4.85 (d, *J* = 8.5, 1H), 4.46–4.11 (m, 2H), 4.08–3.81 (m, 2H), 3.64–3.23 (m, 2H), 2.35–2.00 (m, 2H), 1.36 (s, 9H). ^{13}C NMR (CDCl$_3$, several rotamers) δ 174.09, 171.13, 169.52, 169.18, 155.45, 154.22, 135.20, 128.61, 128.47, 128.33, 80.72, 70.78, 69.81, 67.12, 60.36, 59.99, 59.34, 56.83, 56.00, 41.62, 38.62, 36.39, 28.33, 21.01, 14.18. IR (neat, cm^{-1}): 3307 (OH), 1742, 1667 (CO), 1189, 1131 (C-O). MS (C$_{19}$H$_{26}$N$_2$O$_6$): calcd. 379.1869 ([M+H]$^+$), exp. 379.1843 ([M+H]$^+$).

trans-**4**: ^{1}H NMR (CDCl$_3$) δ = 7.40 (bs, 1H), 7.29–7.19 (m, 5H), 6.86 (bs, 1H), 5.06 (s, 2H), 4.32 (bs, 2H), 4.08–3.84 (m, 2H), 3.82–3.25 (m, 3H), 2.36–1.90 (m, 2H), 1.33 (s, 9H). ^{13}C NMR (CDCl$_3$, several rotamers) δ = 172.34, 171.64, 170.29, 168.54, 154.78, 153.86, 134.21, 127.60, 127.48, 127.32, 79.79, 68.61, 68.15, 66.09, 59.43, 58.76, 57.68, 53.97, 40.18, 38.49, 36.47, 27.29, 20.02, 13.17. IR (neat, cm^{-1}): 3303 (OH), 1748, 1667 (C=O), 1159, 1127 (C-O). MS (C$_{19}$H$_{26}$N$_2$O$_6$): calcd. 379.1869 ([M+H]$^+$), exp. 379.1865 ([M+H]$^+$).

Dipeptides *cis*-**5** and *trans*-**5** were isolated as colorless viscous oils in a yield of 213 mg, (59%) or 320 mg (87%).

cis-**5**: ^{1}H NMR (CDCl$_3$) δ = 7.46 (d, *J* = 8.4, 1H), 7.32–7.20 (m, 5H), 6.77 (bs, 1H), 5.19–4.97 (m, 3H), 4.53–4.43 (m, 1H), 4.39 (d, *J* = 8.8, 1H), 4.25 (t, *J* = 20.2, 1H), 3.59–3.28 (m, 2H), 2.31–1.97 (m, 3H), 1.38 (s, 9H), 0.82 (m, 6H). ^{13}C NMR (CDCl$_3$, several rotamers) δ 172.18, 170.16, 154.52, 134.39, 127.56, 127.39, 127.33, 79.57, 69.72, 68.80, 65.93, 59.18, 58.52, 56.69, 56.40, 55.99, 34.54, 30.20, 27.32, 17.88, 16.30. IR (neat, cm^{-1}): 3283 (OH), 1737, 1701, 1664 (C=O), 1155, 1112 (C-O). MS (C$_{22}$H$_{32}$N$_2$O$_6$): calcd. 421.2339 ([M+H]$^+$), exp. 421.2338 ([M+H]$^+$).

trans-**5**: ^{1}H NMR (CDCl$_3$) δ = 7.47 (bs, 1H), 7.33–7.11 (m, 5H), 6.63 (bs, 1H), 5.07 (q, *J* = 12.2, 2H), 4.55-4.20 (m, 3H), 3.92–3.15 (m, 3H), 2.45–1.77 (m, 3H), 1.35 (s, 9H), 0.81 (d, *J* = 6.8, 3H), 0.78 (d, *J* = 6.9, 3H). ^{13}C NMR (CDCl$_3$, several rotamers) δ = 171.98, 170.95, 170.50, 154.92, 153.93, 134.68, 134.43, 127.55, 127.34, 79.62, 68.68, 68.10, 65.89, 65.58, 59.42, 58.94, 57.54, 56.42, 56.06, 53.98, 53.37, 52.48, 38.70, 35.50, 31.09, 30.15, 27.29, 25.88, 20.01, 18.18, 18.00, 16.84, 16.51, 16.14, 13.17. IR (neat, cm^{-1}): 3307 (OH), 1739, 1667 (C=O), 1158. MS (C$_{22}$H$_{32}$N$_2$O$_6$): calcd. 421.2339 ([M+H]$^+$), exp. 421.2333 ([M+H]$^+$).

2.3. Preparation of N-Deprotected Dipeptides

To a solution of 0.25 mmol of Boc-protected dipeptides *cis*-**4** (95 mg), *trans*-**4** (95 mg), *cis*-**5** (105 mg), or *trans*-**5** (105 mg) in 1 mL dichloromethane, 0.5 mL of anisole and then 0.5 mL trifluoroacetic acid (TFA) were added. The reaction mixture was stirred at room temperature for 1 h. All volatiles were removed at reduced pressure. Dipeptides **6** and **7** were obtained quantitatively as TFA salts without further purification and were characterized by ^{1}H NMR, ^{13}C{^{1}H} NMR, and FT-IR spectroscopy as well as MS spectrometry (Figures S18–S29).

cis-**6**: ^{1}H NMR (acetone-*d$_6$*) δ = 7.45–7.30 (m, 5H), 5.45 (dd, *J* = 25.9, 7.9, 1H), 5.21 (d, *J* = 4.2, 1H), 5.19–5.15 (m, 2H), 4.88–4.51 (m, 2H), 4.40–3.98 (m, 3H), 3.72–3.35 (m, 2H), 2.88–2.09 (m, 2H). ^{13}C NMR (CDCl$_3$) δ = 174.04, 169.26, 135.26, 128.73, 128.47, 80.89, 70.91, 67.29, 59.45, 56.97, 41.75, 36.28, 29.77, 28.43. IR (neat, cm^{-1}): 3088 (OH and NH), 1742, 1668 (C=O), 1189, 1133 (C-O). MS (C$_{14}$H$_{18}$N$_2$O$_4$): calcd. 279.1345 ([M+H]$^+$), exp. 279.1339 ([M+H]$^+$).

trans-**6**: ^{1}H NMR (CDCl$_3$) δ = 10.14 (bs, 1H), 8.46 (bs, 1H), 7.86 (bs, 1H), 7.38–7.22 (m, 5H), 5.53 (bs, 2H), 5.17–4.98 (m, 2H), 4.81 (bs, 1H), 4.52 (bs, 1H), 4.07 (bs, 2H), 3.63–3.11 (m, 2H), 2.52 (bs, 1H), 1.95 (bs, 1H). ^{13}C NMR (CDCl$_3$) δ 169.66, 169.45, 135.12, 128.76, 128.70, 128.39, 127.23, 70.70, 67.56, 58.89, 54.45, 53.56, 41.66, 38.86. IR (neat, cm^{-1}): 3088 (OH and

NH), 1745, 1667 (C=O), 1178, 1134 (C-O). MS ($C_{14}H_{18}N_2O_4$): calcd. 279.1345 ([M+H]$^+$), exp. 279.1339 ([M+H]$^+$).

cis-7: ^{1}H NMR (CDCl$_3$) δ = 8.03 (d, J = 9.2, 1H), 7.41–7.27 (m, 5H), 5.25–5.04 (m, 2H), 4.52 (dd, J = 9.3, 4.8, 1H), 4.41–4.26 (m, 1H), 3.86–3.71 (m, 1H), 3.12 (dd, J = 11.1, 4.6, 1H), 2.99 (dd, J = 11.1, 1.0, 1H), 2.62–2.36 (m, 2H), 2.33–2.11 (m, 2H), 2.03 (ddd, J = 26.9, 13.4, 10.9, 1H), 0.96–0.81 (m, 6H). ^{13}C NMR (CDCl$_3$) δ = 175.25, 171.98, 135.54, 128.65, 128.46, 128.39, 72.07, 66.97, 59.62, 56.86, 55.25, 39.82, 31.48, 19.12, 17.64. IR (neat, cm^{-1}): 3462, 3343, 3269 (OH and NH), 1741, 1643 (C=O), 1193, 1146 (C-O). MS ($C_{17}H_{24}N_2O_4$): calcd. 321.1814 ([M+H]$^+$), exp. 321.1809 ([M+H]$^+$).

trans-7: ^{1}H NMR (CDCl$_3$) δ = 8.24 (d, J = 9.2, 1H), 7.40–7.28 (m, 5H), 5.19 (d, J = 12.2, 1H) 5.12 (d, J = 12.2, 1H), 4.49 (dd, J = 9.3, 4.9, 1H), 4.38 (bs, 1H), 4.00 (t, J = 8.4, 1H), 3.03 (d, J = 12.3, 2H), 2.75 (m, 4H), 2.28 (dd, J = 13.8, 8.7, 1H), 2.22–2.11 (m, 1H), 1.92–1.80 (m, 1H), 0.86 (dd, J = 21.9, 6.9, 6H). ^{13}C NMR (CDCl$_3$) δ = 175.28, 171.86, 135.53, 128.67, 128.49, 128.42, 73.10, 67.03, 59.84, 56.64, 55.52, 40.32, 31.33, 19.23, 17.64. IR (neat, cm^{-1}): 3321 (OH and NH), 1737, 1652 (C=O), 1192, 1147 (C-O). MS ($C_{17}H_{24}N_2O_4$): calcd. 321.1814 ([M+H]$^+$), exp. 321.1809 ([M+H]$^+$).

2.4. Aldol Reactions

To **8** (151 mg, 1.0 mmol) dissolved in 2 mL of an acetone/DMSO mixture (ratio 1:5, v/v) or in pure acetone, 20 mol-% of the solid catalyst **2**, **6**, or **7** was added after stirring overnight at room temperature. In the case of **6** and **7**, TEA was added in equimolar amounts to release the free base from the TFA salt. The reaction mixture was diluted with ethyl acetate (5 mL) and washed twice with water (2 mL). The organic layer was dried over anhydrous Na$_2$SO$_4$. The conversion number was determined by ^{1}H NMR spectroscopy. The ee with respect to (R)-**9** was determined by chiral HPLC (see General Experimental Conditions). The absolute configuration of the major product was determined as R in accordance with the elution order in previously published data on chiral separations of **9**.

Aldol product **9**: ^{1}H NMR (CDCl$_3$): δ = 8.21 (d, J = 8.8 Hz, 2H, H-arom), 7.54 (d, J = 8.6 Hz. 2H, H-arom), 5.26 (dt, J = 3.7 Hz, 7.6 Hz, 1H, CH-O), 3.58 (d, J = 3.3 Hz, 1H, CO-C*H*H), 2.85 (dd, J = 3.4 Hz, 6.1 Hz, 1H, CO-CH*H*), 2.22 (s, 3H, CO-CH$_3$), 1.59 (bs, 1H, OH). ^{13}C NMR (CDCl$_3$): δ = 208.47 (C=O), 149.94 (C-arom), 147.26 (C-arom), 126.37 (2 × C-arom), 123.72 (2 × C-arom), 68.86 (C-O), 51.46 (CO-CH$_2$), 30.68 (CO-CH$_3$).

HPLC retention time (t_R): (R)-**9**: 12.2 min; (S)-**9**: 15.4 min (Figures S30–S39).

2.5. Quantum Chemical Calculations

Conformers of free catalysts and the intermediates of the aldol reaction cycle with cis-**2**, trans-**2**, cis-**6**, and trans-**6** were modeled using the conformer search algorithm in SPARTAN (Wavefunction, Inc., Irvine, CA, USA).). Geometries of transition states formed from cis-**2** and trans-**2** were modeled based on transition state geometries previously published for **1** [34,35]. All geometries were optimized at the B3LYP/6-31G(d) level with a polarized continuum model (PCM) for DMSO in GAUSSIAN 16 [36], whereby the level of theory was chosen as a compromise between accuracy and computational cost for the number of conformers that had to be optimized for this study. Geometries of transition states formed from cis-**6** and trans-**6** were modeled using the transition states from cis-**2** and trans-**2** as templates replacing the carboxylic acid –OH by Gly-OBn. Conformers were generated using a constrained conformer search with a frozen core structure, in which dihedrals were only freely rotatable for the 4-hydroxy group (provided it was not involved in the transition state) and the Gly-OBn group. Geometry optimizations at the density functional theory level were performed in a two-step procedure, first constrained (with frozen core structure) and thereafter unconstrained. Transition geometries without any precedence for L-proline such as the stereoselective step (TS4) and the alternative path for enamine formation presented in this study were modeled using the STQN method for locating transition structures [37]. TS4 geometries were modeled as transition states leading to an R-configured product (TS4$_{pro\,R}$) and to an S-configured product (TS4$_{pro\,S}$), respectively, where for TS4$_{\text{pro R}}$ and TS4$_{\text{pro S}}$ each two core structures were modeled: one with an *s-anti* orientation of the enamine, and one with an *s-syn* orientation. Additional conformers with respect to 4-OH and Gly-OBn were modeled as given above. Duplicate intermediate and transition structures were identified based on their individual interatomic distance distribution pattern as described elsewhere [38], and removed using a script in MATLAB (MathWorks, Inc., Natick, MA, USA). From the transition structures, only those were kept that exhibited one imaginary frequency after frequency calculations (same level as optimizations). From the transition states with a single imaginary frequency, only those were kept, where the frequency was clearly associated with a vibration along a bond to be formed or broken in the course of the transition. For the calculation of standard free energies (G) from statistical thermodynamics in GAUSSIAN, all frequencies were uniformly scaled by 0.97. Average values for G according to the Boltzmann weights of each contributing conformer and energy barriers were calculated in MATLAB.

Theoretical enantiomeric ratios (er) R:S were calculated from the ratios of the theoretical rate constants k_R and k_S for the R- or an S-selective reaction steps in each modeled reaction cycle [39]. These ratios were obtained from the difference of the energy barriers $\Delta G_{\text{pro R}}$ or $\Delta G_{\text{pro S}}$, i.e., the difference between free energies (Boltzmann-weighted average over different conformers) calculated for TS4$_{\text{pro R}}$ or TS4$_{\text{pro S}}$, respectively, and the free energy of the enamine intermediate:

$$\text{er} = \frac{k_R}{k_S} = e^{-\frac{\Delta G_{\text{pro R}} - \Delta G_{\text{pro S}}}{RT}}$$

For all theoretical er values, we assumed a reaction temperature of 298.15 K.

3. Results

As a model reaction for asymmetric aldol additions, we used the addition of acetone to **8** to give the chiral β-hydroxyketone **9** where acetone acted both as a solvent and as a reactant. The use of a solvent mixture with DMSO (ratio DMSO/acetone 5:1, *v/v*) ensured that all reactants and catalyst were fully dissolved, as it has been shown that the catalytic cycle only involves soluble proline complexes or soluble proline adducts [40]. All reactions were carried out in the presence of 20 mol-% of the catalyst as evidenced from earlier studies as the optimal catalyst concentration [33]. After 24 h of stirring at room temperature, the overall conversion was determined by ^{1}H NMR spectroscopy and the ee by chiral phase HPLC (see Experimental Section and Figures S30–S39). **1**, *trans*-4-Hyp (*trans*-**2**), and *cis*-4-Hyp (*cis*-**2**) were used as references to evaluate the efficiency of **6** and **7** as catalysts. The results are summarized in Table 1.

Like well-known aldol additions in the presence of **1**, catalysis with *cis*-**2** and *trans*-**2** results in virtually full conversion (>98%) after 24 h with moderate enantioselectivities in favor of the R-configured product (48% ee with *cis*-**2** and 40% with *trans*-**2**).

The highly similar stereochemical outcome of both conversions suggests only a little influence from the 4-hydroxy group on enantioselectivity confirming previous observations [33]. Theoretical models from previous studies on proline catalysis have shown that the stereoselectivity is determined by the relative arrangement of the double-bonded methylene group of the enamine intermediate (Scheme 1) and the carbonyl group of the substrate. Here, a coordination of the carbonyl oxygen by the carboxylic acid group of the enamine intermediate in TS4 clearly favors an R-configured product (TS4$_{\text{pro R}}$) as the energy barrier via TS4$_{\text{pro R}}$ is lower than via a transition state TS4$_{\text{pro S}}$ with opposite arrangement.

To evaluate the plausibility of a similar mechanism with *cis*-**2** and *trans*-**2**, in analogy to previously published models for catalysis with **1**, we modeled different conformers of the enamine intermediates and of transition states TS4$_{\text{pro R}}$ and TS4$_{\text{pro S}}$ toward product **9** with both *cis*-**2** and *trans*-**2** (lowest-energy conformers of TS4$_{\text{pro R}}$ are depicted in Figure 2). We calculated their respective standard free energies (G°) at the B3LYP/6-31G(d)-level with an implicit solvent model for DMSO and determined theoretical stereoselectivities for both catalysts from the $\Delta G°$ values calculated with both catalysts (Table 1). The full consideration of conformational space would require modeling numerous reaction channels for individual substrate, transition state, and product conformers and the calculation of rotational barriers to model the interconversion between different conformational species. Hence, even for systems with limited conformational space, any attempt of calculating a full reaction profile would be hardly manageable. Therefore, instead of considering free energy levels of individual conformers, we calculated free energy profiles based on Boltzmann-weighted average values of $G°$. This approach may be a source of error, if a reaction step going through a short-lived intermediate conformer would proceed considerably faster to the next intermediate than its accommodation in a lower-lying intermediate conformer, i.e., the energy barrier by the subsequent transition state is lower than the involved rotational barriers. In all other cases, focusing on the low-lying species within a limited set of conformers appears as a good approximation, at least to obtain a qualitative picture of the energy barriers in the catalytic cycle.

Table 1. Aldol reaction of acetone and 8 in DMSO/acetone solvent mixture of ratio 5:1 (v/v).

1: R = H; Z = OH
2: R = OH; Z = OH
6: R = OH; Z = NH-Gly-OBn
7: R = OH; Z = NH-L-Val-OBn

		Conversion (%) [a]	ee (%) [b]	er (exp.) [b]	er (calc.) [c]
1	1	>98	64	82:18	n. d.
2	cis-2	>98	48	74:26	91:9
3	trans-2	>98	40	70:30	93:7
4	cis-6	<5	98	99:1	>99:1
5	trans-6	40	98	99:1	>99:1
6	cis-7	36	94	97:3	n. d.
7	trans-7	42	98	99:1	n. d.

In the case of 6 and 7, triethylamine (TEA) was added in equimolar amounts to release the non-protonated catalysts from the TFA salt. [a] Determined by ^{1}H NMR. [b] Determined by chiral HPLC. [c] Calculated from forward rates of the stereoselective step obtained from DFT calculations.

(A)

(B)

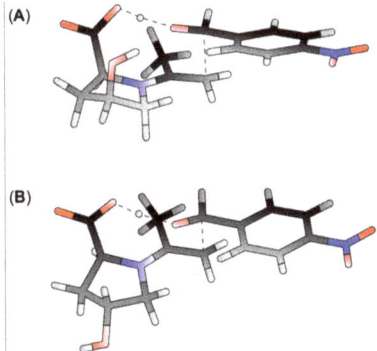

Figure 2. Transition states of the stereoselective step in the synthesis of 9 leading to an (R)-configured product (TS4$_{pro\ R}$) formed from cis-2 (**A**) and trans-2 (**B**).

As a first benchmark for the catalytic cycles with *cis*-**2** and *trans*-**2**, we calculated $\Delta G°$ between enamine intermediate/substrate and TS4$_{\text{pro }R}$ or TS4$_{\text{pro }S}$, respectively. The predicted er values according to the calculated $\Delta G°$ values are somewhat higher (91:9 with *cis*-**2** and 93:7 with *trans*-**2**) than what had been observed experimentally. Still, given the low level of theory used for our calculation and the above-mentioned focus on low-lying conformers, this is in the range of the stereochemical outcome expected for proline-catalyzed aldol reactions. Taken together, transition state geometries analogous to what had been suggested previously for aldol reactions catalyzed by **1** seem to apply as well for aldol additions catalyzed by *cis*- or *trans*-**2**. Depending on reaction conditions, however, the ee of such reactions may be moderate, as the difference between energy barriers going through TS4$_{\text{pro }R}$ and TS4$_{\text{pro }S}$ is small. Interestingly, but in accordance with the experiment, the 4-hydroxy-group has little impact on the transition states shown in Figure 2, yet may have implications on ring puckering, which leads to different geometries of the lowest-energy conformers of TS4$_{\text{pro }R}$ with *cis*-**2** (Figure 2A) and *trans*-**2** (Figure 2B).

With both *cis*-**2** and *trans*-**2**, we observed virtually full conversion (>98%) after 24 h. While C,C-bond formation is rate-limiting [12], the rate of this second-order step also depends on the concentration of the enamine intermediate, whose formation itself had previously been discussed as rate-limiting for proline-catalyzed aldol reactions [41]. To obtain a full picture of formation of enamine in the aldol reaction catalyzed by *cis*-**2** in comparison to *trans*-**2**, we also modeled geometries of the preceding intermediates and transition states (TS1, TS2, and TS3), from formation of the hemiaminal from catalyst and acetone through the iminium intermediate to the enamine (Figure 3). To provide a rough estimate of the error due to our treatment of the conformer problem, we also plotted respective $\Delta G°$ with highest- and lowest-energy conformers as 'whiskers' attached to each substrate, intermediate, transition, or product state. As given in Scheme 1, we also included a parasitic side path to a putatively formed bicyclic byproduct from the iminium intermediate via TS3a in our models of reaction cycles with *cis*-**2** and *trans*-**2**. The calculated energy barriers are reasonable in the boundaries of the accuracy of the model and are like what had been obtained for proline-catalyzed reactions previously. Differences in the energy profile of the *cis*-**2** and *trans*-**2** reaction cycle are small, except for the energy of the iminium intermediate being slightly lower in the *cis*-**2** cycle. However, as the energy associated with hemiaminal formation, whose rate is additionally limited by concentration of reactants and the corresponding frequency factor (not considered in Figure 3), is the highest, minor differences in subsequent reaction steps should not be rate-determining. Therefore, it is not surprising that we observe very similar conversion with both catalysts after 24 h.

Reactions with catalysts *cis*-**6** and *trans*-**6** show with 5 and 40% low conversion after 24 h yet are highly enantioselective (98% ee). Reactions catalyzed by *cis*-**7** and *trans*-**7** confirm the moderate conversion (36% and 42%) and high enantioselectivities (94% and 98% ee, respectively).

In analogy to *cis*-**2** and *trans*-**2**, we also modeled conformers of the enamine intermediates and the transition states TS4$_{\text{pro }R}$ and TS4$_{\text{pro }S}$ with *cis*-**6** and *trans*-**6**. Due to the flexible benzylglycinate moiety, in comparison to catalysis with **1** or **2**, conformational space is considerably larger. To take this into account, we modeled 47 conformers of the *cis*-enamine and 99 of the *trans*-species and altogether 12 transition structures for the *cis* species (4 for TS4$_{\text{pro }R}$ and 8 for TS4$_{\text{pro }S}$) and 24 for the *trans*-species (9 for TS4$_{\text{pro }R}$ and 15 for TS4$_{\text{pro }S}$). Different from **1** and **2** catalysis, with **6**, conformers may exist that are energetically favorable, but incompatible with the transition state, for example, due to steric interference of the flexible benzylglycinate terminus with the catalytic site. Nevertheless, the theoretical rates for formation of (*R*)-**9** vs. (*S*)-**9** calculated based on Boltzmann-averaged $\Delta G°$ show a clear preference for the *R*-product (er > 99:1), which agrees with the high enantioselectivity observed experimentally. While the energy barriers associated with TS4$_{\text{pro }R}$ and TS4$_{\text{pro }S}$ highly vary according to the respective transition geometry conformers, the aldol reaction seems to occur predominantly via low-lying conformers of TS4$_{\text{pro }R}$.

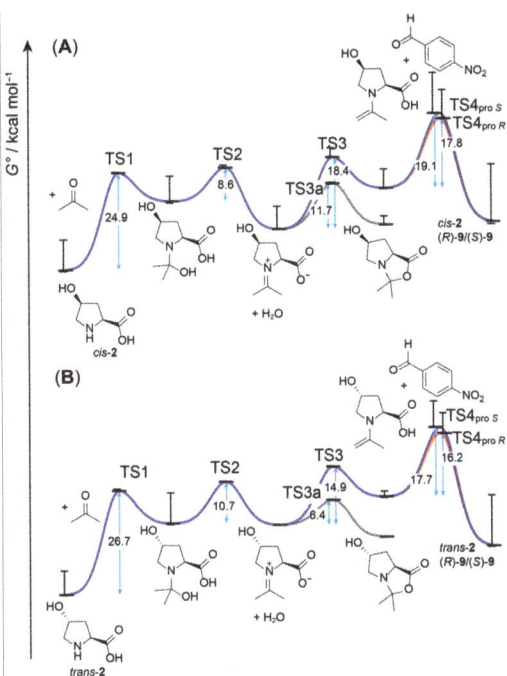

Figure 3. Standard transition free energies (ΔG°, represented by blue arrows) of the aldol reaction cycle catalyzed by cis-4-Hyp (cis-**2**) (**A**) and trans-4-Hyp (trans-**2**) (**B**) calculated at the B3LYP/6-31G(d) level with a solvent model for DMSO. The G° levels indicated by black horizontal bars are Boltzmann-averaged over the number of conformers given in Tables S1 and S2. The whiskers (vertical lines with bars) indicate G° values obtained with the lowest- and highest-energy conformers of each intermediate or transition state.

Since the proline catalysis-derived transition states of the stereoselective step of cis-**6** and trans-**6** catalysis successfully predict the actually observed enantioselectivities, in analogy to catalysis with **2**, we also modeled the preceding steps of aldol additions catalyzed by cis-**6** and trans-**6** (Figure 4). In comparison to **1** or **2**, the involved energy barriers are considerably higher, in particular, the one that corresponds to hemiaminal formation. The high barriers are due to the fact that proton transfers that are mediated by the carboxylic acid group in **1** and **2** must be undertaken by an amide in cis-**6** and trans-**6**. Given the high pK associated with amide deprotonation (approximately 25 in DMSO) [42], intermediary formation of an amidate in analogy to a carboxylate, as with **1** or **2**, is very unlikely. Such proton transfers would be involved in TS1, formation of the hemiaminal (Figure 5A), and in TS2, elimination of water from the hemiaminal to form an iminium zwitterion. Our attempts to model TS1 involving proton transfer from the amide failed (Figure S1); however, we were able to obtain an alternative transition geometry as previously suggested by Rankin et al. for proline [35], where the hydroxylate formed upon nucleophilic attack of the nitrogen to the carbonyl takes up the proton directly from the amine instead of the amide (Figure 5B). Still, given the high energy barrier of more than 50 kcal mol^{-1} that must be overcome for the formation of a hemiaminal from the peptide and acetone (Figure 4), a catalytic cycle as in Scheme 1 appears highly unlikely, as such a reaction would not take place at room temperature. Therefore, the whole catalytic cycle, starting with hemiaminal formation to formation of the crucial enamine intermediate, must be different in aldol additions catalyzed by peptides, in particular, those that do not feature a residue that could act as a proton donor.

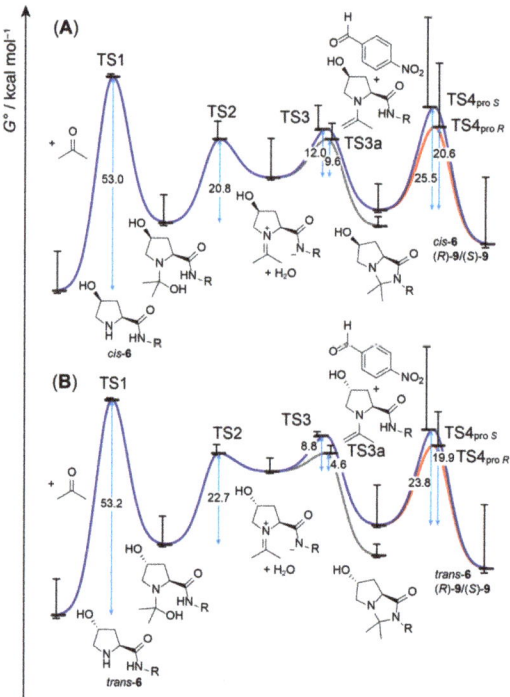

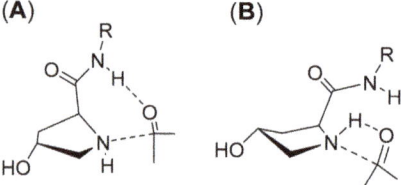

Figure 4. $\Delta G°$ profile of the aldol reaction cycle catalyzed by *cis*-**6** (**A**) and *trans*-**6** (**B**) calculated at the B3LYP/6-31G(d)/PCM (DMSO) level. The $\Delta G°$ values (blue arrows) are the differences between $G°$ levels, Boltzmann-averaged over the number of conformers given in Tables S3 and S4 (black horizontal bars). The whiskers indicate $G°$ values obtained with the lowest- and highest-energy conformers of each intermediate or transition state.

Figure 5. Models for transition structures towards hemiaminal formation (TS1) with a 4-Hyp peptide catalyst (here: peptide of *trans*-**2**). (**A**): hemiaminal formation involves proton transfer from the amide (in analogy to proton transfer from the carboxylic acid groups as with free amino acid catalysts). (**B**): direct proton transfer from the secondary amine as modeled for **1** in reference [35].

If proton transfer is not mediated by the peptide catalyst, it must occur from somewhere else. Considering the solvent DMSO as a poor proton donor, possible proton donor candidates present in the reaction solution are triethylammonium (from deprotonation of TFA salt present after Boc deprotection) or water molecules (released and taken up during reaction cycle). However, for statistical reasons, involvement of either of these very weak acids in TS1 is not very likely, as the collision frequency of three molecules in such a single third-order reaction step is expected to be very low. Furthermore, it has been shown that the presence of water has an inhibitory effect on proline-catalyzed aldol additions [43]. The last remaining candidate for a proton donor in TS1 would be the substrate, acetone, itself. A clear acetone dependence of conversion in aldol additions has been demonstrated for formation of **9** catalyzed by L-prolineamide, a catalyst lacking a proton-donating group such as **6** and **7**: while the yield was <10% at 20 vol% acetone [44], an increase to 80% had been observed in neat acetone [45]. As a transition state where one acetone molecule protonates the other in a third-order reaction step appears highly unlikely, we abandoned the idea of considering acetone itself as a proton donor. Nevertheless, an acetone content-dependent conversion would still be observed, if, instead of acetone itself, some kind of 'auxiliary substrate' with proton donor properties was fed into the catalytic cycle, whose concentration depends on acetone content. A promising candidate for such an auxiliary is the acetone self-adduct **10**, which is formed from two acetone molecules halfway to trimeric acetone **11** (Scheme 3). In the presence of a base (here **6**), **10** should be present to a certain extent (equilibrium constant of $K = 0.04$) [46,47].

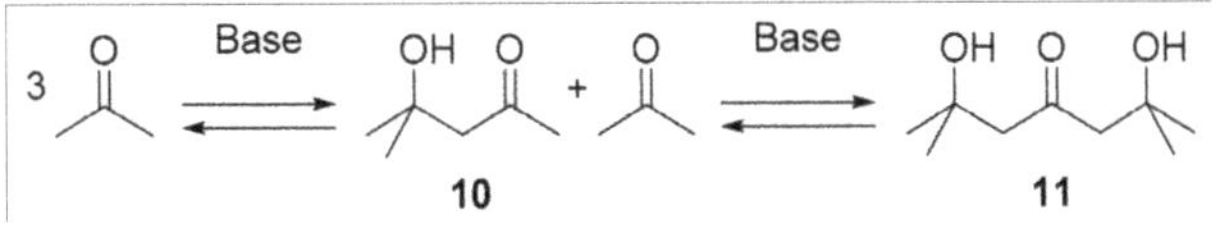

Scheme 3. Aldol self-reaction of acetone.

Assuming that with a peptide catalyst, a hemiaminal is not formed from catalyst and acetone, but from acetone and **10**, proton transfer would be conveniently mediated by the β-hydroxy group of **10**. Due to reduced symmetry of **10** in comparison to acetone, such a transition state could be modeled with four different configurations, of which two transition state configurations lead to an *R*-configured hemiaminal, and two to an *S*-configured hemiaminal (Figure 6). Each of the four configurations allow a comparably 'relaxed' transition geometry comprising a six-membered ring formed from the atoms that are involved in C,C-bond formation and proton transfer.

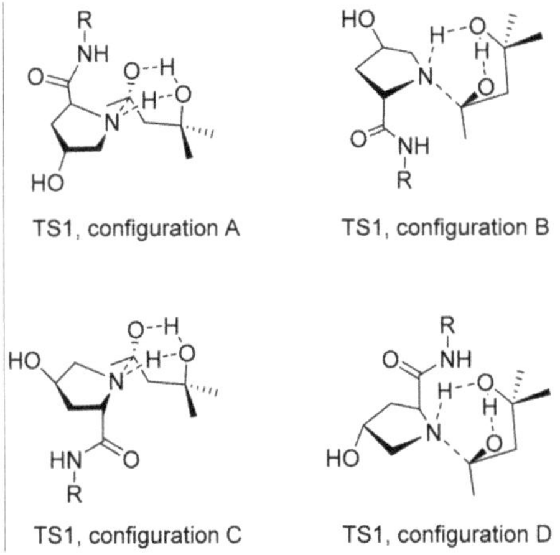

Figure 6. Four possible configurations (**A**, **B**, **C**, **D**) of TS1 (structure of transition state towards formation of a hemiaminal) formed from 4-Hyp-peptide catalyst (here shown for the *trans*-species) and **10**. Bonds that are formed or broken form a cyclic transition involving six atoms (dashed lines).

Similarly, starting from either an *R*- or an *S*-configured hemiaminal intermediate, formation of the enamine could as well occur via a cyclic transition state (Scheme 4). This way, the enamine is formed without the necessity of proton transfer via amide deprotonation, which would also make the formation of a parasitic bicyclic by-product less likely.

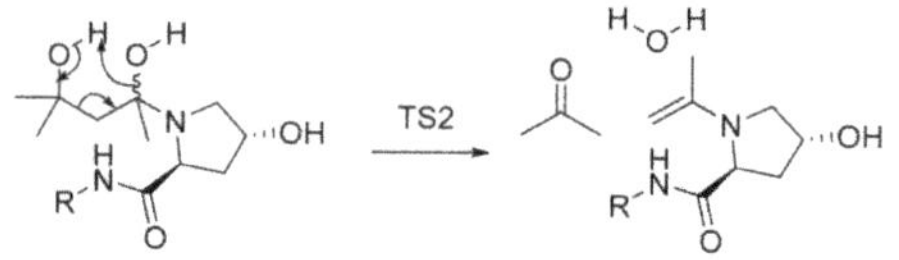

Scheme 4. Formation of enamine intermediate.

We calculated the transition states and intermediates from Figure 6 and Scheme 4 with 4-hydroxy-*N*-methyl-L-prolinamide (**12**) as a model peptide (Figure 7). Indeed, with 33.2 kcal/mol for TS1 (configuration A) with *cis*- and 36.6 kcal/mol for TS1 (configuration A) with *trans*-**12**, the energy barriers associated with hemiaminal formation are much lower than those found with transition geometries as in Figure 5B. Therefore, the mechanism involving **10** as a mediator does present a plausible alternative, albeit coming at the cost of a low proportion of active intermediates, as only small amounts of **10** are available. This is reflected by the considerably lower conversion observed with peptides instead of free amino acids (Table 1). With the concentration of **10** being a crucial determinant of rate and conversion, any parameter with an influence on the proportion of **10** may have an impact on conversion, such as temperature, catalyst solubility, acid/base conditions in the reaction vessel (avoiding the term 'pH' in the context of a DMSO solution), and the amount of excessive acetone. As particularly the latter can be easily modified, we also performed conversions in the presence of **6** and **7** in pure acetone (Table 2). As expected, the overall conversion slightly increases. Furthermore, differences in conversion between *cis*- and *trans*-configured catalysts vanish, which suggests that the very first step, the one that depends on acetone concentration, is rate-determining.

Table 2. Aldol reaction of acetone and **8** in pure acetone.

		Conversion (%) [a]	ee (%) [b]	er [b]
1	cis-**6**	45	98	99:1
2	trans-**6**	44	98	99:1
3	cis-**7**	85	6	47:53
4	trans-**7**	73	6	47:53

[a] Determined by ^{1}H NMR. [b] Determined by chiral HPLC.

For the aldol additions catalyzed by *cis*-**6** and *trans*-**6**, the high stereoselectivity reported above is fully retained in pure acetone. In contrast, it is virtually lost with both diastereomers of **7**, the catalyst with a sterically demanding L-Val sidechain. Furthermore, despite high substrate conversion, considerable formation of by-products was observed both in the presence of *cis*-**7** and *trans*-**7** (Figures S38 and S39). Possibly, in pure acetone, where higher amounts of **10** or even trimer **11** are present, alternative pathways may exist involving acetone self-adduct intermediates even at the stereoselective C,C-coupling step. If this was the case, different behavior between peptide catalysts with a more- and less sterically demanding environment around the catalytic site would be plausible.

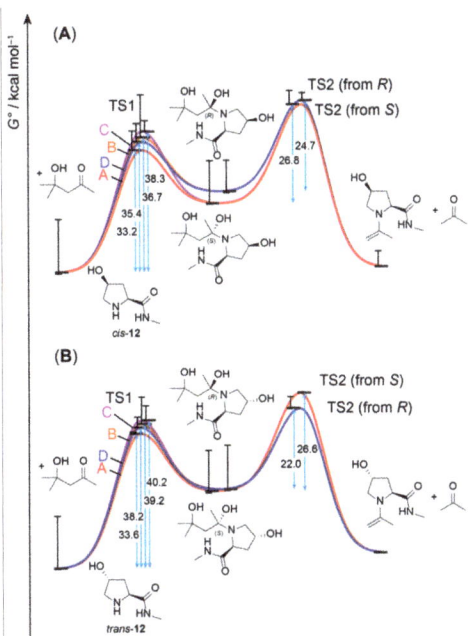

Figure 7. Energy barriers (ΔG°, blue arrows) in modeled catalytic pathways towards the enamine intermediates with *cis*-**12** (**A**) and *trans*-**12** (**B**) calculated at the B3LYP/6-31G(d)/PCM (DMSO) level. The different colors indicate different reaction channels through the four different configurations from Figure 6, two of which are going through a hemiaminal intermediate with *R*- and two with an *S*-configured hydroxy group. The whiskers represent G° values obtained with the lowest- and highest-energy conformers of each intermediate or transition state. In comparison to Figure 4, the energy barrier associated with formation of a hemiaminal from peptide catalyst and **10** is much lower than for the formation of a hemiaminal from acetone.

4. Discussion

Proline and hydroxyprolines catalyze stereoselective aldol additions through a highly stereospecific transition geometry consisting of a catalyst-enamine intermediate and the substrate. The enamine intermediate is formed via several reaction steps involving proton transfer to and from the carboxylic acid group of the catalyzing amino acid. An analogous catalytic cycle with peptide catalysts such as **6** and **7**, which lack a proton donor, is not very plausible, as the energy barrier for formation of a hemiaminal intermediate from peptide catalyst and the electrophilic aldol component (here: acetone) is too high to be overcome at room temperature. Still, aldol additions occur with excellent stereoselectivities, albeit lower conversion in model reactions after 24 h. Our DFT calculations suggest a modified pathway that is entered by an acetone self-adduct instead of acetone in peptide-catalyzed reactions. Here, the hydroxy group of the self-adduct acts as a 'proton-wire', thereby easing the transfer of the proton from the catalyst nitrogen to the substrate oxygen atom. Increasing the proportion of acetone provides more acetone self-adduct, thereby increasing conversion. However, this may come at the cost of decreased stereoselectivity, which suggests that the true catalytic cycle is even more complex than the one depicted by our calculations, which demands deeper computational and experimental kinetic analyses in the future. Still, the consideration of self-adduct intermediates provides a hitherto-neglected aspect in peptide-catalyzed aldol additions and might help in the design of synthetic pathways.

Supplementary Materials: The following supporting information can be downloaded at: https://www.mdpi.com/article/10.3390/chemistry5020081/s1: Supplementary computational data (Figure S1), analytical data (Figures S2–S29), chiral phase HPLC data (Figures S30–S39), and model geometries from DFT calculations (Tables S1–S6).

Author Contributions: Conceptualization, L.A.M. and S.L.; methodology, L.A.M. and S.L.; software, S.L.; formal analysis, L.A.M. and S.L.; investigation, L.A.M. and S.L.; resources, H.L.; writing—original draft preparation, S.L.; writing—review and editing, L.A.M., H.L. and S.L.; visualization, S.L. All authors have read and agreed to the published version of the manuscript.

Funding: Experiments were carried out during a research visit financially supported by the Deutsche Forschungsgemeinschaft (DFG) and the Deutsche Akademische Austauschdienst (DAAD). For computations, the authors acknowledge support by the state of Baden-Württemberg through bwHPC and the DFG through grant no INST 40/575-1 FUGG (JUSTUS 2 cluster).

Data Availability Statement: Data is contained within the article or Supplementary Materials.

Acknowledgments: L.A.M. would like to thank Tafila Technical University (TTU) for the traveling support. Many thanks go to the NMR and MS unit at TU Chemnitz, to Nadine Killius and Michael Müller at the University of Freiburg for help with chiral HPLC, and to Wolfgang Hüttel for carefully reading the manuscript.

Conflicts of Interest: The authors declare no conflict of interest.

References

1. Eder, U.; Sauer, G.; Wiechert, R. New Type of Asymmetric Cyclization to Optically Active Steroid CD Partial Structures. *Angew. Chem. Int. Ed.* **1971**, *10*, 496–497. [CrossRef]
2. Hajos, Z.G.; Parrish, D.R. Asymmetric Synthesis of Bicyclic Intermediates of Natural Product Chemistry. *J. Org. Chem.* **1974**, *39*, 1615–1621. [CrossRef]
3. Ahrendt, K.A.; Borths, C.J.; MacMillan, D.W.C. New Strategies for Organic Catalysis: The First Highly Enantioselective Organocatalytic Diels−Alder Reaction. *J. Am. Chem. Soc.* **2000**, *122*, 4243–4244. [CrossRef]
4. List, B. Proline-catalyzed asymmetric reactions. *Tetrahedron* **2002**, *58*, 5573–5590. [CrossRef]
5. Cordóva, A.; Notz, W.; Zhong, G.F.; Betancort, J.M.; Barbas, C.F., III. A Highly Enantioselective Amino Acid-Catalyzed Route to Functionalized α-Amino Acids. *J. Am. Chem. Soc.* **2002**, *124*, 1842–1843. [CrossRef]
6. Notz, W.; Tanaka, F.; Barbas, C.F., III. Enamine-Based Organocatalysis with Proline and Diamines: The Development of Direct Catalytic Asymmetric Aldol, Mannich, Michael, and Diels−Alder Reactions. *Acc. Chem. Res.* **2004**, *37*, 580–591. [CrossRef]
7. Mangion, I.K.; Northrup, A.B.; MacMillan, D.W.C. The Importance of Iminium Geometry Control in Enamine Catalysis: Identification of a New Catalyst Architecture for Aldehyde–Aldehyde Couplings. *Angew. Chem. Int. Ed.* **2004**, *43*, 6722–6724. [CrossRef]
8. Northrup, A.B.; MacMillan, D.W.C. The First Direct and Enantioselective Cross-Aldol Reaction of Aldehydes. *J. Am. Chem. Soc.* **2002**, *124*, 6798–6799. [CrossRef]
9. Jarvo, E.R.; Miller, S.J. Amino acids and peptides as asymmetric organocatalysts. *Tetrahedron* **2002**, *58*, 2481–2495. [CrossRef]
10. Hoang, L.; Bahmanyar, S.; Houk, K.N.; List, B. Kinetic and Stereochemical Evidence for the Involvement of Only One Proline Molecule in the Transition States of Proline-Catalyzed Intra- and Intermolecular Aldol Reactions. *J. Am. Chem. Soc.* **2003**, *125*, 16–17. [CrossRef]
11. List, B.; Lerner, R.A.; Barbas, C.F., III. Proline-Catalyzed Direct Asymmetric Aldol Reactions. *J. Am. Chem. Soc.* **2000**, *122*, 2395–2396. [CrossRef]
12. Zotova, N.; Broadbelt, L.J.; Armstrong, A.; Blackmond, D.G. Kinetic and mechanistic studies of proline-mediated direct intermolecular aldol reactions. *Bioorganic Med. Chem. Lett.* **2009**, *19*, 3934–3937. [CrossRef]
13. Schmid, M.B.; Zeitler, K.; Gschwind, R.M. The Elusive Enamine Intermediate in Proline-Catalyzed Aldol Reactions: NMR Detection, Formation Pathway, and Stabilization Trends. *Angew. Chem. Int. Ed.* **2010**, *49*, 4997–5003. [CrossRef]
14. Seebach, D.; Beck, A.K.; Badine, D.M.; Limbach, M.; Eschenmoser, A.; Treasurywala, A.M.; Hobi, R.; Prikoszovich, W.; Linder, B. Are Oxazolidinones Really Unproductive, Parasitic Species in Proline Catalysis?—Thoughts and Experiments Pointing to an Alternative View. *Helv. Chim. Acta* **2007**, *90*, 425–471. [CrossRef]
15. Sharma, A.K.; Sunoj, R.B. Enamine versus Oxazolidinone: What Controls Stereoselectivity in Proline-Catalyzed Asymmetric Aldol Reactions? *Angew. Chem. Int. Ed.* **2010**, *49*, 6373–6377. [CrossRef]
16. Martin, H.J.; List, B. Mining Sequence Space for Asymmetric Aminocatalysis: N-Terminal Prolyl-Peptides Efficiently Catalyze Enantioselective Aldol and Michael Reactions. *Synlett* **2003**, *2003*, 1901–1902. [CrossRef]
17. Shi, L.-X.; Sun, Q.; Ge, Z.-M.; Zhu, Y.-Q.; Cheng, T.-M.; Li, R.-T. Dipeptide-Catalyzed Direct Asymmetric Aldol Reaction. *Synlett* **2004**, *2004*, 2215–2217. [CrossRef]

18. Krattiger, P.; Kovasy, R.; Revell, J.D.; Ivan, S.; Wennemers, H. Increased Structural Complexity Leads to Higher Activity—Peptides as Efficient and Versatile Catalysts for Asymmetric Aldol Reactions. *Org. Lett.* **2005**, *7*, 1101–1103. [CrossRef]
19. Schnitzer, T.; Wennemers, H. Influence of the *Trans/Cis* Conformer Ratio on the Stereoselectivity of Peptidic Catalysts. *J. Am. Chem. Soc.* **2017**, *139*, 15356–15362. [CrossRef]
20. Schnitzer, T.; Wiesner, M.; Krattiger, P.; Revell, J.D.; Wennemers, H. Is more better? A comparison of tri- and tetrapeptidic catalysts. *Org. Biomol. Chem.* **2017**, *15*, 5877–5881. [CrossRef]
21. Kofoed, J.; Nielsen, J.; Reymond, J.L. Discovery of New Peptide-Based Catalysts for the Direct Asymmetric Aldol Reaction. *Bioorganic Med. Chem. Lett.* **2003**, *13*, 2445–2447. [CrossRef] [PubMed]
22. Metrano, A.J.; Chinn, A.J.; Shugrue, C.R.; Stone, E.A.; Kim, B.; Miller, S.J. Asymmetric Catalysis Mediated by Synthetic Peptides, Version 2.0: Expansion of Scope and Mechanisms. *Chem. Rev.* **2020**, *120*, 11479–11615. [CrossRef] [PubMed]
23. Tang, Z.; Yang, Z.-H.; Cun, L.-F.; Gong, L.-Z.; Mi, A.-Q.; Jiang, Y.-Z. Small Peptides Catalyze Highly Enantioselective Direct Aldol Reactions of Aldehydes with Hydroxyacetone: Unprecedented Regiocontrol in Aqueous Media. *Org. Lett.* **2004**, *6*, 2285–2287. [CrossRef] [PubMed]
24. Al-Momani, L.A.; Lataifeh, A. Novel O-ferrocenoyl hydroxyproline conjugates: Synthesis, characterization and catalytic properties. *Inorg. Chim. Acta* **2013**, *394*, 176–183. [CrossRef]
25. Crawford, J.M.; Sigman, M.S. Conformational Dynamics in Asymmetric Catalysis: Is Catalyst Flexibility a Design Element? *Synthesis* **2019**, *51*, 1021–1036. [CrossRef]
26. Rigling, C.; Kisunzu, J.K.; Duschmalé, J.; Häussinger, D.; Wiesner, M.; Ebert, M.O.; Wennemers, H. Conformational Properties of a Peptidic Catalyst: Insights from NMR Spectroscopic Studies. *J. Am. Chem. Soc.* **2018**, *140*, 10829–10838. [CrossRef]
27. Haasnoot, C.A.G.; De Leeuw, F.A.A.M.; De Leeuw, H.P.M.; Altona, C. Relationship Between Proton-Proton NMR Coupling Constants and Substituent Electronegativities. III. Conformational Analysis of Proline Rings in Solution Using a Generalized Karplus Equation. *Biopolymers* **1981**, *20*, 1211–1245. [CrossRef]
28. Kapitán, J.; Baumruk, V.; Kopecký, V.; Pohl, R.; Bouř, P. Proline Zwitterion Dynamics in Solution, Glass, and Crystalline State. *J. Am. Chem. Soc.* **2006**, *128*, 13451–13462. [CrossRef]
29. Lüdeke, S.; Pfeifer, M.; Fischer, P. Quantum-Cascade Laser-Based Vibrational Circular Dichroism. *J. Am. Chem. Soc.* **2011**, *133*, 5704–5707. [CrossRef]
30. Rüther, A.; Pfeifer, M.; Lórenz-Fonfría, V.A.; Lüdeke, S. pH Titration Monitored by Quantum Cascade Laser-Based Vibrational Circular Dichroism. *J. Phys. Chem. B* **2014**, *118*, 3941–3949. [CrossRef]
31. Shoulders, M.D.; Satyshur, K.A.; Forest, K.T.; Raines, R.T. Stereoelectronic and steric effects in side chains preorganize a protein main chain. *Proc. Natl. Acad. Sci. U.S.A.* **2010**, *107*, 559–564. [CrossRef]
32. Holmgren, S.K.; Taylor, K.M.; Bretscher, L.E.; Raines, R.T. Code for collagen's stability deciphered. *Nature* **1998**, *392*, 666–667. [CrossRef]
33. Al-Momani, L.A. Hydroxy-L-prolines as asymmetric catalysts for aldol, Michael addition and Mannich reactions. *ARKIVOC* **2012**, *6*, 101–111. [CrossRef]
34. Ashley, M.A.; Hirschi, J.S.; Izzo, J.A.; Vetticatt, M.J. Isotope Effects Reveal the Mechanism of Enamine Formation in L-Proline-Catalyzed α-Amination of Aldehydes. *J. Am. Chem. Soc.* **2016**, *138*, 1756–1759. [CrossRef]
35. Rankin, K.N.; Gauld, J.W.; Boyd, R.J. Density Functional Study of the Proline-Catalyzed Direct Aldol Reaction. *J. Phys. Chem. A* **2002**, *106*, 5155–5159. [CrossRef]
36. Frisch, M.J.; Trucks, G.W.; Schlegel, H.B.; Scuseria, G.E.; Robb, M.A.; Cheeseman, J.R.; Scalmani, G.; Barone, V.; Petersson, G.A.; Nakatsuji, H.; et al. *GAUSSIAN 16*; Revision C.01; Gaussian Inc.: Wallingford, CT, USA, 2019.
37. Peng, C.Y.; Ayala, P.Y.; Schlegel, H.B.; Frisch, M.J. Using Redundant Internal Coordinates to Optimize Equilibrium Geometries and Transition States. *J. Comput. Chem.* **1996**, *17*, 49–56. [CrossRef]
38. Temerinac, M.; Reisert, M.; Burkhardt, H. Invariant features for searching in protein fold databases. *Int. J. Comput. Math.* **2007**, *84*, 635–651. [CrossRef]
39. Gawley, R.E. Do the Terms "% ee" and "% de" Make Sense as Expressions of Stereoisomer Composition or Stereoselectivity? *J. Org. Chem.* **2006**, *71*, 2411–2416. [CrossRef]
40. Iwamura, H.; Wells, D.H.; Mathew, S.P.; Klussmann, M.; Armstrong, A.; Blackmond, D.G. Probing the Active Catalyst in Product-Accelerated Proline-Mediated Reactions. *J. Am. Chem. Soc.* **2004**, *126*, 16312–16313. [CrossRef]
41. List, B. Enamine catalysis is a powerful strategy for the catalytic generation and use of carbanion equivalents. *Acc. Chem. Res.* **2004**, *37*, 548–557. [CrossRef]
42. Bordwell, F.G. Equilibrium Acidities in Dimethyl Sulfoxide Solution. *Acc. Chem. Res.* **1988**, *21*, 456–463. [CrossRef]
43. Zotova, N.; Franzke, A.; Armstrong, A.; Blackmond, D.G. Clarification of the role of water in proline-mediated aldol reactions. *J. Am. Chem. Soc.* **2007**, *129*, 15100–15101. [CrossRef] [PubMed]
44. Sakthivel, K.; Notz, W.; Bui, T.; Barbas, C.F., III. Sakthivel_Amino Acid Catalyzed Direct Asymmetric Aldol Reactions—A Bioorganic Approach to Catalytic Asymmetric Carbon−Carbon Bond-Forming Reactions. *J. Am. Chem. Soc.* **2001**, *123*, 5260–5267. [CrossRef] [PubMed]
45. Tang, Z.; Jiang, F.; Cui, X.; Gong, L.Z.; Mi, A.Q.; Jiang, Y.Z.; Wu, Y.D. Enantioselective direct aldol reactions catalyzed by L-prolinamide derivatives. *Proc. Natl. Acad. Sci. USA* **2004**, *101*, 5755–5760. [CrossRef]

46. Guthrie, J.P. Equilibrium constants for a series of simple aldol condensations, and linear free energy relations with other carbonyl addition reactions. *Can. J. Chem.* **1978**, *56*, 962–973. [CrossRef]
47. Koelichen, K. Die chemische Dynamik der Acetonkondensation. *Z. Phys. Chem.* **1900**, *33U*, 129–177. [CrossRef]

Disclaimer/Publisher's Note: The statements, opinions and data contained in all publications are solely those of the individual author(s) and contributor(s) and not of MDPI and/or the editor(s). MDPI and/or the editor(s) disclaim responsibility for any injury to people or property resulting from any ideas, methods, instructions or products referred to in the content.

Article

Selective Fluorimetric Detection of Pyrimidine Nucleotides in Neutral Aqueous Solution with a Styrylpyridine-Based Cyclophane

Julika Schlosser, Julian F. M. Hebborn, Daria V. Berdnikova and Heiko Ihmels *

Department of Chemistry-Biology, Research Center of Micro- and Nanochemistry and (Bio)Technology (Cμ), University of Siegen, Adolf-Reichwein-Str. 2, 57068 Siegen, Germany; julika.schlosser@uni-siegen.de (J.S.)
* Correspondence: ihmels@chemie.uni-siegen.de; Tel.: +49-271-740-3440

Abstract: A styrylpyridine-containing cyclophane with diethylenetriamine linkers is presented as a host system whose association with representative nucleotides was examined with photometric and fluorimetric titrations. The spectrometric titrations revealed the formation of 1:1 complexes with log K_b values in the range of 2.3–3.2 for pyrimidine nucleotides TMP (thymidine monophosphate), TTP (thymidine triphosphate) and CMP (cytidine monophosphate) and 3.8–5.0 for purine nucleotides AMP (adenosine monophosphate), ATP (adenosine triphosphate), and dGMP (deoxyguanosine monophosphate). Notably, in a neutral buffer solution, the fluorimetric response to the complex formation depends on the type of nucleotide. Hence, quenching of the already weak fluorescence was observed with the purine bases, whereas the association of the cyclophane with pyrimidine bases TMP, TTP, and CMP resulted in a significant fluorescence light-up effect. Thus, it was demonstrated that the styrylpyridine unit is a useful and complementary fluorophore for the development of selective nucleotide-targeting fluorescent probes based on alkylamine-linked cyclophanes.

Keywords: fluorescent dyes; nucleotide recognition; heterocycles; host–guest chemistry

Citation: Schlosser, J.; Hebborn, J.F.M.; Berdnikova, D.V.; Ihmels, H. Selective Fluorimetric Detection of Pyrimidine Nucleotides in Neutral Aqueous Solution with a Styrylpyridine-Based Cyclophane. *Chemistry* **2023**, *5*, 1220–1232. https://doi.org/10.3390/chemistry5020082

Academic Editors: Christoph Janiak, Sascha Rohn, Georg Manolikakes and Robert B. P. Elmes

Received: 30 March 2023
Revised: 3 May 2023
Accepted: 8 May 2023
Published: 11 May 2023

Copyright: © 2023 by the authors. Licensee MDPI, Basel, Switzerland. This article is an open access article distributed under the terms and conditions of the Creative Commons Attribution (CC BY) license (https://creativecommons.org/licenses/by/4.0/).

1. Introduction

Nucleotides play a crucial role in several biological processes, for example as essential building blocks in DNA replication and RNA synthesis [1,2]. Furthermore, they are essential in cell signaling, metabolism, and enzyme reactions as cofactors for NAD$^+$ and FAD and as energy carriers in the form of triphosphate nucleotides [3,4]. Therefore, the detection and monitoring of nucleotides are important tasks to contribute to the assessment and understanding of biochemical processes in living organisms [5–9]. Along these lines, photometric and electrochemical analysis, as well as ^{1}H NMR spectroscopic analysis, are routinely used methods for nucleotide detection; however, elaborate protocols, relatively expensive equipment, and limited sensitivity are drawbacks of these methods [10,11]. For this purpose, fluorescence spectroscopy is a useful and easily accessible analytical tool because it enables the efficient and sensitive detection of biologically relevant analytes with suitable fluorescent probes (chemosensors), which change their emission properties upon analyte binding [12–20]. Along these lines, fluorescent probes that can detect nucleotides by means of emission quenching or emission enhancement (light up) have been reported [21–24]. However, selective chemosensors for particular nucleotides are still needed, so the development of such fluorescent probes still represents a rewarding and challenging research field in chemistry [25–27].

The most abundant nucleotide is adenosine triphosphate (ATP), which plays an important role in the energy transport in living organisms [28,29] and as a main biochemical component in cancer cells, where it can either enhance or suppress tumor growth, depending on the concentration [30]. Consequently, several different methods and approaches for the efficient and selective detection of ATP have been reported [31–38]. On the contrary, the selective analysis and sensing of other nucleotides has been scarcely reported so far.

For example, the selective photometric detection of thymidine triphosphate (TTP) relative to other mono-, di-and triphosphate nucleotides has been realized with gold nanoparticles and a p-xylylbis(Hg^{2+}-cyclen) complex [39]. Likewise, cytidine triphosphate (CTP) has been shown to induce selective luminescence quenching of a terbium(III)-organic framework [40], and a polyhydroxy-substituted Schiff base receptor has been reported to be a selective fluorescent chemosensor for CTP and ATP [41]. More recently, a bisnaphthalimide receptor with a pyridine spacer has been introduced as a selective fluorescent probe for CTP [42]. Moreover, anthracene derivatives with two appended imidazolium groups have been reported whose emission is efficiently quenched by GTP [43].

In this context, fluorescent cyclophanes have been established as useful scaffolds for the development of host systems for inorganic and organic anions and may be applied in chemosensing, bioimaging, and drug delivery [44–52]. In particular, several phenyl- [53,54], naphthalene- [55], anthracene- [56–58], and acridine-based [59] cyclophanes, as well as metallocyclophanes [60], have been reported as hosts that strongly bind to nucleotides [25,61]. In seminal work, bisnaphthalenophanes with six amino-functionalities in the linking chains have been introduced as ideal host molecules for the efficient recognition of organic and inorganic phosphates [55]. Likewise, it has been shown that similar anthracene- [62] and pyrene-based [63–65] cyclophanes have the ability to discriminate between different nucleotide triphosphates by the selective complex formation and that these interactions may be used for fluorimetric detection of nucleotides [44,61]. Besides the recognition of single nucleotides, cyclophanes are also apt to bind preferentially to nucleobases in mismatched and abasic site-containing DNA [66–69].

Although some cyclophanes are already available for fluorimetric nucleotide detection, there is still room for further development. Specifically, variations of the aromatic unit appear promising because this part of the host molecule provides an essential binding site for π stacking with the nucleic base. Surprisingly, most employed aromatic subunits are fused polycyclic fragments with limited conformational flexibility, such as naphthalene or anthracene, whereas more flexible scaffolds with resembling π surface, such as stilbenes or styryl-substituted hetarenes, have not been employed for this purpose, so far. Along these lines, we proposed that the known 2-styrylpyridine unit may serve as a useful, complementary aromatic component in nucleotide-binding cyclophanes because it provides a flexible aromatic surface, which may enable a more variable π stacking, along with a decent dipole, which may increase the binding affinity by dipole-dipole interactions with the nucleic base. Herein, we report on the synthesis of a bis-styrylpyridine-based cyclophane, and demonstrate that it may be used for fluorimetric detection and differentiation of nucleotides at physiological pH.

2. Results
2.1. Synthesis

The known dibromostyrylpyridine derivative **1** [70] was formylated by lithium-halogen exchange with *n*-BuLi and subsequent reaction with DMF to the corresponding styrylpyridine bis-carbaldehyde **2** in 63% yield (Scheme 1, see Supplementary Materials). Condensation of the latter with diethylenetriamine and subsequent reduction of the tetraimine intermediate **3** with $NaBH_4$ gave the macrocyclic polyamine **4** in a yield of 23%. The known derivative **1** was synthesized by a varied procedure and identified by comparison with the literature data [71], and the new compounds **2** and **4** were identified and fully characterized by NMR spectroscopy (^{1}H, ^{13}C, COSY, HSQC, and HMBC), elemental analyses, and mass spectrometry (Figures S2–S7). In all cases, the *E*-configuration of the alkene units in compounds **1, 2**, and **4** were indicated by characteristic coupling constants of the alkene protons ($^3J_{H-H}$ = 16 Hz).

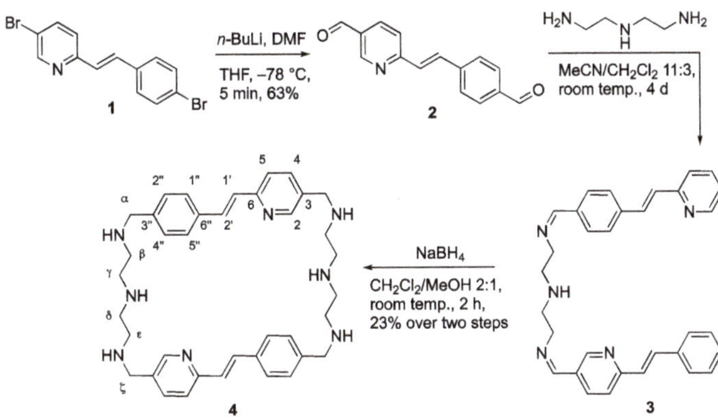

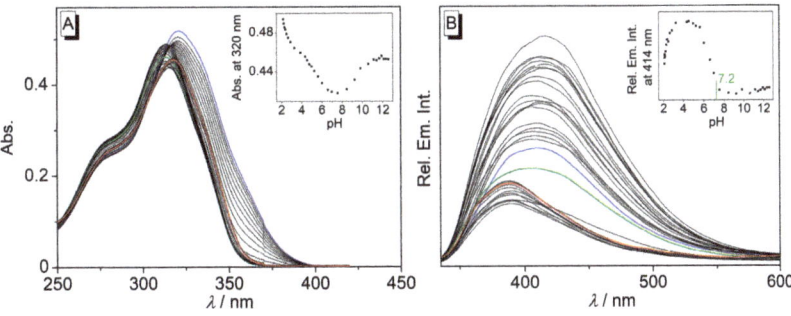

Scheme 1. Synthesis of cyclophane **4**.

2.2. Solvent and pH-Dependent Absorption and Emission Properties

In the MeOH solution, cyclophane **4** exhibited an absorption maximum at λ_{abs} = 314 nm and a fluorescence maximum at λ_{fl} = 379 nm with low emission quantum yield (<0.01) (see Supplementary Materials).

The pH dependence of the absorption properties of cyclophane **4** was determined by spectrometric acid-base titrations in Britton-Robinson buffer (Figure 1). At neutral pH, the absorption maximum was at λ_{abs} = 314 nm. The absorbance increased both at lower (pH < 5) and higher pH (>8) values, with the highest absorbance at pH 2. The absorption maximum also shifted with varying pH, from λ_{abs} = 321 nm at pH 2 to λ_{abs} = 314 nm at pH 7 and to λ_{abs} = 318 nm at pH 12. Furthermore, a slight shoulder at λ_{abs} = 364 nm was observed at pH 2, which steadily disappeared with increasing pH.

Figure 1. Photometric (**A**) and fluorimetric (**B**) acid–base titrations of cyclophane **4**. Blue line: beginning of titration (pH = 2.0); green line: absorbance and emission at pH 7.2; red line: end of titration (pH = 12.6). Insets: plot of the absorbance at 320 nm (**A**) and fluorescence at 414 nm (**B**) versus pH. In all cases: c = 10 µM, in Britton-Robinson buffer with 5% DMSO, λ_{ex} = 313 nm.

The data from the photometric titration were used to determine the pK_a values of 5.2 and 9.4. Another pK_a value was estimated to be in the range of 2–3, as has been usually observed for resembling cyclophanes with the same diethylenetriamine linker [62]; however, no adequate fit was obtained for this region, so a more accurate value was not available.

The emission spectrum of cyclophane **4** revealed a broad maximum at λ_{fl} = 410 nm at pH 2. With increasing pH to 4, the emission intensity firstly increased by a factor of ca. 2 and reached the highest intensity with a bathochromic shift of $\Delta\lambda_{fl}$ = 5 nm. With the further addition of base (pH > 4), the fluorescence was strongly quenched by about 70% with a hypsochromic shift of the emission maximum of $\Delta\lambda = 27$ nm at pH 8.5. At pH > 9, the emission intensity remained low, with a slight increase in the emission after pH > 10. Most notably, at neutral pH, the emission of the styrylpyridine is already sufficiently quenched so this compound may be used as a fluorescence light-up probe for target nucleotides at a physiological pH range, that is, under conditions usually found in real biological samples.

2.3. Nucleotide-Binding Properties of 4

The association of the macrocyclic polyamine **4** with selected nucleotides was investigated by photometric and fluorimetric titrations with adenosine monophosphate (AMP), ATP, deoxyguanosine monophosphate (dGMP), thymidine monophosphate (TMP), TTP, and CTP in cacodylate buffer solution at pH 7.2, that is, conditions at which the emission is already very low (Figures 2 and 3). Upon addition of AMP, ATP, and dGMP to **4**, the absorbance ($\lambda_{max} = 314$ nm) decreased with the formation of a red-shifted absorption band ($\Delta\lambda = 5$ nm) and isosbestic points at $\lambda = 323$ nm, 320 nm, and 320 nm, respectively (Figure 2A). In the presence of these nucleotides, the already weak fluorescence of the cyclophane **4** was further quenched with different efficiencies, that is, with I/I_0 of 0.46 (AMP), 0.59 (ATP), and 0.04 (dGMP) at saturation (Figures 2B and 4A). Moreover, the fluorescence maximum of styrylpyridine **4** was blue-shifted with $\Delta\lambda = 34$ nm on the addition of AMP and ATP, whereas no shift of the fluorescence maximum was observed with dGMP.

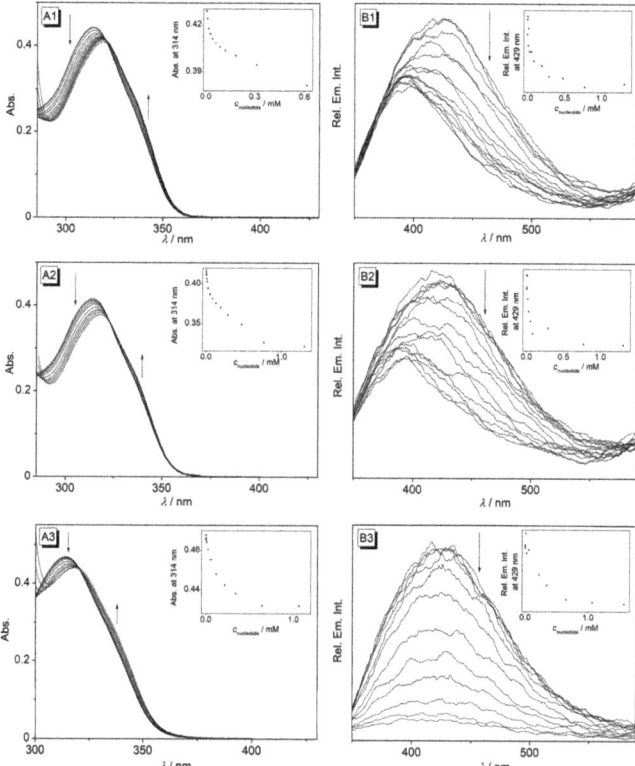

Figure 2. Photometric (**A**) and fluorimetric (**B**) titrations of cyclophane **4** ($c = 10$ μM) with ATP (**1**), AMP (**2**), and dGMP (**3**) in cacodylate buffer (pH 7.2); $T = 20$ °C; $\lambda_{ex} = 314$ nm. The arrows indicate the changes in absorption and emission upon the addition of nucleotides. Inset: plot of the absorption at 314 nm versus nucleotide concentration.

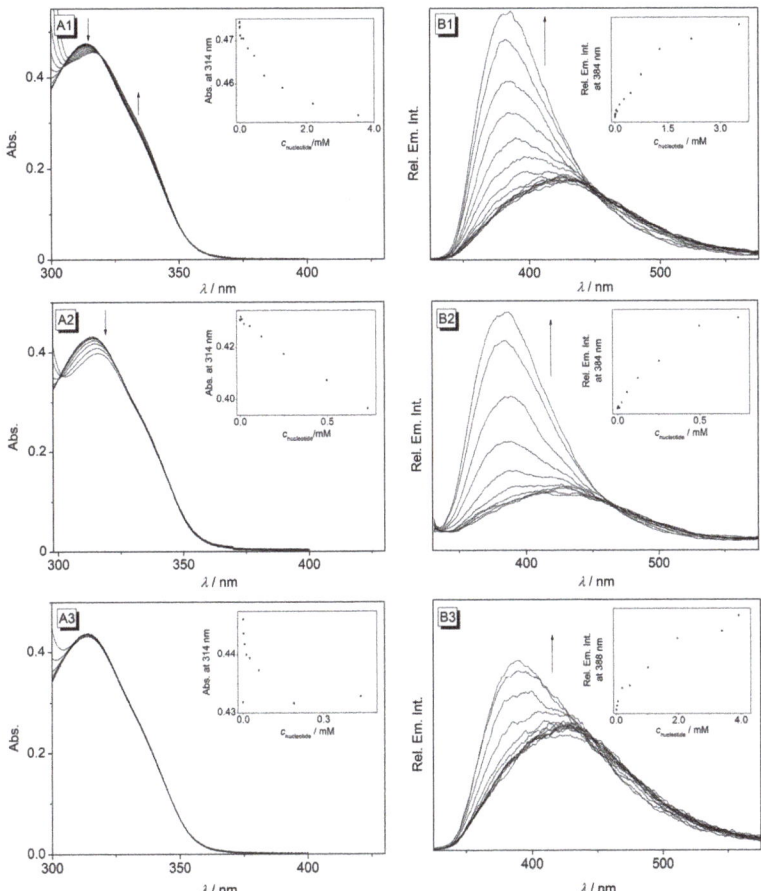

Figure 3. Photometric (**A**) and fluorimetric (**B**) titrations of cyclophane **4** ($c = 10$ μM) with TMP (**1**), TTP (**2**), and CMP (**3**) in cacodylate buffer (pH 7.2); $T = 20\,°C$; $\lambda_{ex} = 314$ nm. The arrows indicate the changes in absorption and emission upon the addition of nucleotides. Inset: plot of the absorption at 314 nm versus nucleotide concentration.

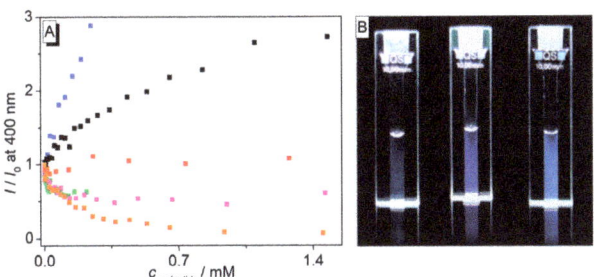

Figure 4. (**A**) Changes of relative emission intensity, I/I_0, of cyclophane **4** ($c = 10$ μM) upon addition of nucleotides, blue: TTP (detector limit reached at $c_{nucleotide} = 0.24$ mM), black: TMP, red: CMP, green: ATP (saturation reached at $c_{nucleotide} = 0.22$ mM), magenta: AMP, orange: dGMP; in cacodylate buffer (pH 7.2); $T = 20\,°C$; $\lambda_{ex} = 314$ nm. (**B**) Pictures of the emission color of **4** in cacodylate buffer (left) and in the presence of CMP (middle) and TMP (right).

The binding constants were determined from the experimental binding isotherms of the photometric titrations. Thus, the experimental data were reasonably fitted to a 1:1 binding stoichiometry of nucleotide and **4** with logK_b values of 4.1, 5.0, and 3.8 for AMP, ATP, and dGMP, respectively (Table 1). These values are in the same range of logK_b values for two resembling pyrene-based diethylenetriamine-cyclophanes with logK_b values of 3.00 and 4.15 with AMP, 5.48 and 5.55 with ATP and 3.51 and 4.50 with dGMP [63] and slightly higher than those observed with the resembling anthracene-based cyclophane with a logK_b value of 3.38 with ATP [62]. In comparison with mono- and triphosphate nucleotides, higher binding constants were also obtained with ATP as compared with AMP [63].

Table 1. Absorption and emission properties of cyclophane **4** and its complexes with nucleotides, and the corresponding binding constants, logK_b.

	λ_{abs}/nm [a]	$\Delta\lambda_{abs}$/nm	λ_{fl}/nm [a]	$\Delta\lambda_{fl}$/nm	logK_b [b]	I/I_0
4	314 (4.67) [c]	–	429 (<0.01) [d]	–	–	–
4/TMP	317	3	384	−45	2.8 ± 0.1	2.72
4/TTP	317	3	384	−45	3.2 ± 0.1	2.43
4/CMP	314	–	388	−41	2.3 ± 0.1	1.23
4/AMP	319	5	395	−34	4.1 ± 0.1	0.48
4/ATP	319	5	395	−34	5.0 ± 0.1	0.54
4/dGMP	319	5	429	–	3.8 ± 0.1	0.10

[a] In cacodylate buffer, pH 7.2; $T = 20\ °C$. [b] Determined from the analysis of the fluorimetric titration data with Specfit/32™ with adequate fits for complexes with **4**:nucleotide ratio 1:1. K in M^{-1}. [c] Molar extinction coefficient ε, given as lg ε, ε in cm^{-1} M^{-1}. [d] Fluorescence quantum yield, relative to naphthalene ($\phi_{fl} = 0.23$ in cyclohexane, ref. [72]), $\lambda_{ex} = 280$ nm, estimated error: ± 10% of the given value.

Titrations of the cyclophane **4** with TMP and TTP decreased the absorbance with red shifts of $\Delta\lambda = 3$ nm (Figure 3A). However, in contrast to titrations with the other nucleotides (see above), the addition of TMP and TTP resulted in a significant increase and blue shift ($\Delta\lambda = 45$ nm) of the fluorescence band (Figure 3B). The fluorescence light-up effect is more pronounced with TMP ($I/I_0 = 2.72$) than with TTP ($I/I_0 = 2.43$), respectively (Figure 4A). Upon the addition of CMP to **4**, the absorption band remained essentially unchanged. At the same time, a fluorescence light-up effect was observed upon the addition of CMP along with a blue shift of the fluorescence maximum of $\Delta\lambda = 41$ nm; however, the increase of the fluorescence intensity ($I/I_0 = 1.23$) was less pronounced than the one with TMP and TTP (Figure 4). Notably, the increased emission intensity of compound **4** upon complex formation with the pyrimidine nucleotides can be seen with the naked eye (Figure 4B). From the fluorimetric titration data, the limit of detection (LOD) of **4** was estimated to be 0.09 μM, 0.02 μM, and 0.04 μM for TMP, TTP, and CMP, respectively (Table S1).

The binding isotherms were determined from the fluorimetric titration data as logK_b = 2.8, 3.2, and 2.3 for 1:1 complexes with TMP, TTP, and CMP, respectively (Table 1, Figure S1). For comparison, the reported logK_b values of resembling pyrene- and anthracene-based cyclophanes are 4.77 and 5.16 [63], and 3.60 [62] for complexes with TTP, that is, somewhat higher than the values for cyclophane **4**. Furthermore, the binding constants for cyclophane **4** are higher for the complexes with purine nucleotides than for the pyrimidine nucleotides, which is in accordance with a literature-known pyrene-based cyclophane [63].

3. Discussion

The pK_a values of 2–3, 5.2, and 9.4 for cyclophane **4** are assigned to the eight available protonation sites, namely the amine and pyridine functionalities. Specifically, the pK_a values of the secondary amines fall in the range of the ones of similar, known amino-containing macrocyclic structures [63]. Accordingly, the pK_a values of the two central amino groups are estimated to be in the range of 2–3, and the pK_a value of 9.4 is assigned to the four lateral amino groups. In addition, the pK_a value of 5.2 relates to the two pyridine units, which is in accordance with the known pK_a value of 5.0 for 2-styrylpyridine [73].

Overall, the acid-base titrations revealed the expected protolytic equilibrium resulting from the protonation of the amino functionalities and the pyridine unit in an acidic medium (Scheme 2). In particular, as has been shown for resembling fluorophore-containing polyamine-linked cyclophanes [62,74], the emission of the styrylpyridine is efficiently quenched by a photoinduced electron transfer (PET) reaction of the electron-donating amine functionalities with the excited fluorophore, whereas upon protonation this deactivation pathway is suppressed and the emission intensity increases significantly [75]. Apparently, the pyridine unit does not interfere with this general process; however, under acidic conditions, the formation of the corresponding pyridinium may be responsible for the shifts of the emission maximum at lower pH values [76].

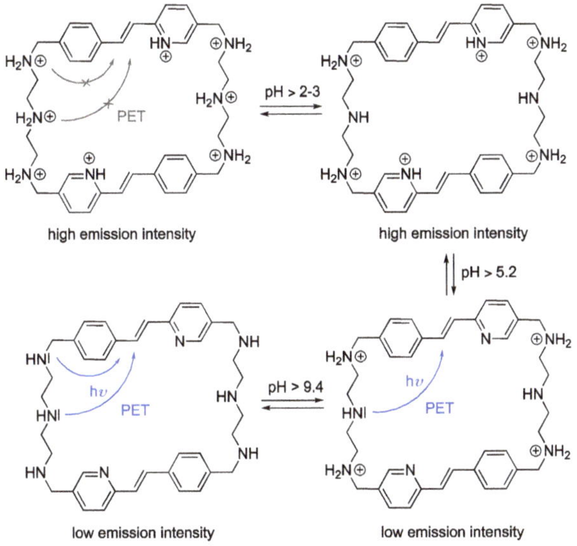

Scheme 2. Protolytic equilibrium of cyclophane 4.

As compared with resembling anthracene- and pyrene-based cyclophanes, which show a fluorescence light-up effect upon complexation of TTP, CTP, and ATP and fluorescence quenching with GTP [62,63], cyclophane 4 exhibits a different dependence of the fluorimetric response on the type of nucleotide. Namely, a fluorescence enhancement occurs upon binding of pyrimidine nucleotides TMP, TTP, and CMP, whereas an effective quenching of the fluorescence results from association with purine nucleotides AMP, ATP, and dGMP. This observation may be explained by the specific pH- and structure-dependent emission properties of the cyclophane 4. Firstly, the amino functionalities of the linker units quench the emission of such cyclophanes by a PET reaction (see above) [62], which readily explains the low emission at the applied pH of 7.2. More importantly, cyclophane 4 exhibits two different emission maxima: a fluorescence maximum at $\lambda = 429$ nm in the unbound state and a blue-shifted one around $\lambda = 384$ nm upon complexation of the nucleotides. As it has been observed already with similar aminoalkyl-linked cyclophanes that these compounds tend to form emitting excimers [63], it is proposed that the red-shifted emission of 4 also originates from an intramolecular excimer formation between the two styrylpyridine units (Scheme 3). This proposal is in agreement with the excimer formation of resembling azastilbene-type derivatives, which is accompanied by a red shift of the emission maximum [77–80]. Upon binding of the pyrimidine nucleobases with the cyclophane 4, the emission increases as a result of the formation of the host-guest complexes, presumably because the complexation of the nucleotide involves hydrogen bonding with the amino functionalities [81,82], which in turn suppresses the PET quenching of the photoexcited fluorophore and leads to increased fluorescence intensity. In addition, the accommodation of the nucleotide in the cavity of the cyclophane also inhibits the excimer formation so only the blue-shifted monomer emission is detected. In contrast, the binding of purine nucleobases leads to emission quenching of cyclophane **4**. This fluorescence quenching of cyclophane **4** by purine nucleotides may be explained by a different binding mode of the purine nucleotides ATP, AMP, and dGMP, as compared with one of pyrimidine nucleotides, which leads to a fluorescence enhancement upon formation of the cyclophane-nucleotide complex [83–85]. At the same time, it cannot be excluded that the purine nucleotides bind in a similar mode as the pyrimidine nucleotides and that the fluorescence quenching by ATP, AMP, and dGMP is just the result of a stronger quenching efficiency of the purine bases. Accordingly, the latter have a much lower reduction potential than the pyrimidine bases [86,87] and can, therefore, induce an efficient fluorescence quenching by a photoinduced electron transfer reaction with the excited styrylpyridine.

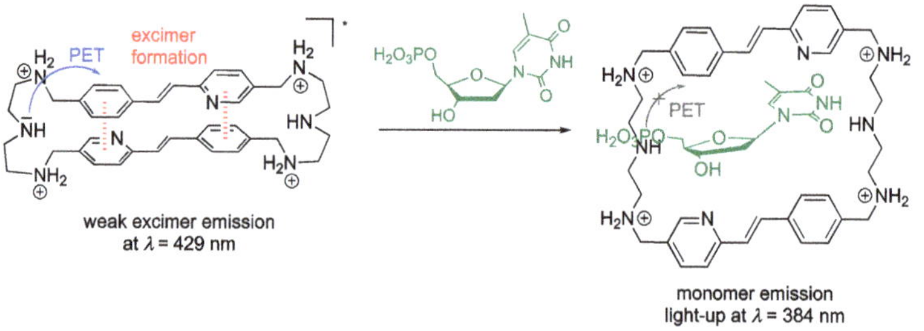

Scheme 3. Formation of excimer complex of cyclophane **4** and proposed binding mode of pyrimidine nucleotide TMP (green) with **4** (* indicates the excited state).

To the best of our knowledge, this is the first reported cyclophane-based fluorescent probe that can discriminate between purine and pyrimidine nucleobases based on a clear light-up effect induced by the latter. Nevertheless, a resembling anthracene-based derivative bearing two imidazolium-containing alkyl chains is known to show these properties [43]. Because of the significant light-up effect of **4** upon binding to TMP and TTP, cyclophane **4** may be employed as a fluorescent probe for the detection of thymine-based nucleotides. Notably, the detection of nucleotides is accomplished under physiological conditions at pH 7.2, rendering cyclophane **4** also interesting for biological applications. For comparison, only a few examples of cyclophanes have been explicitly reported that enable the detection of nucleotides at neutral pH [54,88], so there is still a demand to develop such recognition systems for nucleotides, that is, as the one reported herein, which operate in a physiological pH range.

4. Conclusions

The spectroscopic investigation of the nucleotide-binding properties of the cyclophane **4** revealed that purine bases AMP, ATP, and dGMP are binding upon fluorescence quenching, whereas in contrast, with pyrimidine bases TMP, TTP, and CMP, a clear, distinguishable fluorescence light-up effect was observed. Overall, we have demonstrated that the styrylpyridine unit is a useful and complementary fluorophore for the development of selective nucleotide-targeting fluorescent probes based on alkylamino-linked cyclophanes, especially considering the observation that this probe operates at the physiological pH range. Therefore, further studies of the particular binding modes as well as systematic variations of the substitution pattern, should enable the development of efficient chemical sensors for bioanalytical applications.

5. Materials and Methods

The commercially available chemicals (Alfa, Merck, Fluorochem, or BLDpharm) were of reagent grade and used without further purification. Nucleotides ATP (adenosine-5′-

triphosphate disodium salt) and CMP (cytidine-5′-monophosphate disodium salt) were purchased from Feinbiochemika (Heidelberg, Germany), and nucleotides TMP (thymidine-5′-monophosphate disodium salt hydrate), TTP (thymidine-5′-triphosphate tetrasodium salt), AMP (adenosine-5′-monophosphate sodium salt) and dGMP (2′-deoxyguanosine-5′-monophosphate sodium salt hydrate) were purchased from Sigma-Aldrich (St. Louis, MO, USA). ^{1}H NMR spectra were recorded with a JEOL ECZ 500 (^{1}H: 500 MHz and ^{13}C: 125 MHz) and a Varian VNMR S600 (^{1}H: 600 MHz and ^{13}C: 150 MHz) at $T = 25$ °C. The ^{1}H NMR and ^{13}C{1H} NMR spectra were referenced to an internal standard in CDCl$_3$ [TMS: $\delta(^1\text{H}) = 0.00$ ppm, $\delta(^{13}\text{C}) = 0.00$ ppm]. Structures were assigned with additional information from gCOSY, gHSQC, and gHMBC experiments, and the spectra were processed with the software MestreNova. The mass spectra were recorded with a Finnigan LCQ Deca (driving current: 6 kV, collision gas: argon, capillary temperature: 200 °C, support gas: nitrogen) and an Orbitrap mass spectrometer Thermo Fisher Exactive (driving current: 3.5 kV, capillary temperature: 300 °C, capillary voltage: 45 V, injection rate: 5 µL/min, scanning range: 150–750 m/z, and resolution: ultra-high) and processed with the software Xcalibur. The CHNS analysis data were determined in-house with a HEKAtech EuroEA combustion analyzer. The melting points were measured with a melting point apparatus BÜCHI 545 (Büchi, Flawil, CH) and are uncorrected. The absorption spectra were recorded on a Varian Cary 100 Bio absorption spectrometer with Hellma quartz glass cuvettes 110-QS (layer thickness $d = 10$ mm). The emission spectra were recorded on a Varian Cary Eclipse fluorescence spectrometer with Hellma quartz glass cuvettes 115 FQS (layer thickness $d = 10$ mm). All measurements were recorded at $T = 20$ °C as adjusted with a thermostat if not stated otherwise. The sample solutions in the titration experiments were mixed with a reaction vessel shaker Top-Mix 11118 (Fisher Bioblock Scientific). E-Pure water was obtained with an ultrapure water system D 4632-33 (Wilhelm Werner GmbH, Leverkusen, D) with filters D 0835, D 0803, and D 5027 (2×).

Synthesis of (12E,25E)-1^1,3,6,9,14^1,16,19,22-Octaazapentacyclo-1,14(3,6)-dipyridina-11,24(1,4)-dibenzenacyclo-hexacosaphane-12,25-diene (4)

Under an argon gas atmosphere, a solution of 2 (100 mg, 420 µmol) in CH$_2$Cl$_2$ (15 mL) and MeCN (55 mL) was added dropwise to a solution of ethylentriamine (45.4 µL, 43.5 mg, 420 µmol) in MeCN (30 mL) at room temperature, and the mixture was stirred for 4 d at room temperature. Approximately half the volume of the solvent was removed under reduced pressure, and the precipitated solid was filtered off, washed with MeCN (2 × 10 mL), dried under reduced pressure (0.5 mbar, 1 h), and suspended in a mixture of CH$_2$Cl$_2$ (5 mL) and MeOH (2.5 mL). NaBH$_4$ (100 mg, 2.66 mmol) was added, and the mixture was stirred at room temperature for 3 h under an argon gas atmosphere. The solvent was removed under reduced pressure, and the remaining residue was dissolved in aqueous NaOH (20 mL, 1.0 M) and extracted with CHCl$_3$ (3 × 20 mL). The combined organic layers were dried with K$_2$CO$_3$ and filtered, and the solvent was removed under reduced pressure. The crude product was dissolved in CHCl$_3$ (5 mL), precipitated with hexane (25 mL), filtered off and recrystallized from toluene to give the product 4 as light yellow amorphous solid (60 mg, 97 µmol, 23%); mp 184–187 °C.-^{1}H NMR (500 MHz, CDCl$_3$): $\delta = 1.74$ (br s, 6H, 6 × NH), 2.76–2.82 (m, 8H, 4 × CH$_2$, γ-H, δ-H, fornumbering, see Scheme 1), 2.83–2.92 (m, 8H, 4 × CH$_2$, β-H, ε-H), 3.77 (s, 8H, 4 × CH$_2$, α-H, ζ-H), 7.11, 7.12 (2 × d, $^3J = 16$ Hz, 2H, 1′-H), 7.16, 7.23 (2 × d, $^3J = 8$ Hz, 2H, 5-H), 7.26, 7.28 (2 × d, $^3J = 8$ Hz, 4H, 2″-H, 4″-H), 7.43, 7.45 (2 × d, $^3J = 8$ Hz, 4H, 1″-H, 5″-H), 7.52–7.57 (m, 2H, 4-H), 7.57, 7.60 (2 × d, $^3J = 16$ Hz, 2H, 2′-H), 8.52 (s, 2H, 2-H).-^{13}C NMR (125 MHz, CDCl$_3$): $\delta = 48.3$, 2 × 48.4 (4 × C, Cγ, Cδ), 48.8, 2 × 48.9 (4 × C, Cβ, Cε), 51.0 (2 × C, Cζ), 53.6 (2 × C, Cα), 121.8, 121.9 (2 × C, C5), 127.1, 127.2 (4 × C, C1″, C5″), 127.4, 127.5 (2 × C, C1′), 2 × 128.4 (4 × C, C2″, C4″), 132.0, 132.1 (2 × C, C2′), 2 × 134.2 (2 × C, C3), 2 × 135.4 (2 × C, C6″), 136.2, 136.3 (2 × C, C4), 140.7, 140.8 (2 × C, C3″), 2 × 149.4 (2 × C, C2), 154.5, 154.6 (2 × C, C6).-MS (ESI$^+$): m/z (%) = 617 (100) [M + H]$^+$.-El. Anal. for C$_{38}$H$_{48}$N$_8$ × H$_2$O calc. (%): C 71.89, H 7.94, N 17.65, found: C 72.11, H 7.68, N 17.07.

Supplementary Materials: The following supporting information can be downloaded at: https://www.mdpi.com/article/10.3390/chemistry5020082/s1, Synthesis of **1** and **2** [71]; determination of fluorescence quantum yields [72]; Figure S1: Plot of fluorescence intensity at selected wavelength versus concentration of nucleotides; Figure S2: ^{1}H-NMR spectrum of **1**; Figure S3: ^{13}C-NMR spectrum of **1**; Figure S4: ^{1}H-NMR spectrum of **2**; Figure S5: ^{13}C-NMR spectrum of **2**; Figure S6: ^{1}H-NMR spectrum of **4**; Figure S7: ^{13}C-NMR spectrum of **4**; Table S1: Limit of detection [89,90].

Author Contributions: Conceptualization, J.S. and H.I.; methodology, J.S. and H.I.; experimental work (synthesis, analysis, documentation), J.S.; experimental work (synthesis) J.F.M.H.; experimental work (calculation of binding constants), D.V.B.; writing—original draft preparation, J.S.; writing—review and editing, H.I.; project administration, H.I.; funding acquisition, H.I. All authors have read and agreed to the published version of the manuscript.

Funding: Generous funding by the *Deutsche Forschungsgemeinschaft* (Ih24/15-1) and the University of Siegen is gratefully acknowledged.

Data Availability Statement: Data are available from the authors.

Acknowledgments: We thank Christoph Dohmen for the photographic documentation.

Conflicts of Interest: The authors declare no conflict of interest.

Sample Availability: Samples are not available from the authors.

References

1. Lane, A.N.; Fan, T.W.-M. Regulation of mammalian nucleotide metabolism and biosynthesis. *Nucleic Acids Res.* **2015**, *43*, 2466–2485. [CrossRef] [PubMed]
2. Berdis, A. Nucleobase-modified nucleosides and nucleotides: Applications in biochemistry, synthetic biology, and drug discovery. *Front. Chem.* **2022**, *10*, 1051525. [CrossRef] [PubMed]
3. Hirsch, A.K.H.; Fischer, F.R.; Diederich, F. Phosphate recognition in structural biology. *Angew. Chem. Int. Ed.* **2007**, *46*, 338–352. [CrossRef]
4. Illes, P.; Klotz, K.-N.; Lohse, M.J. Signaling by extracellular nucleotides and nucleosides. *Naunyn Schmiedebergs Arch. Pharmacol.* **2000**, *362*, 295–298. [CrossRef]
5. Florea, M.; Nau, W.M. Implementation of anion-receptor macrocycles in supramolecular tandem assays for enzymes involving nucleotides as substrates, products, and cofactors. *Org. Biomol. Chem.* **2010**, *8*, 1033–1039. [CrossRef]
6. Ojida, A.; Takashima, I.; Kohira, T.; Nonaka, H.; Hamachi, I. Turn-On Fluorescence Sensing of Nucleoside Polyphosphates Using a Xanthene-Based Zn(II) Complex Chemosensor. *J. Am. Chem. Soc.* **2008**, *130*, 12095–12101. [CrossRef] [PubMed]
7. Malojčić, G.; Piantanida, I.; Marinić, M.; Žinić, M.; Marjanović, M.; Kralj, M.; Pavelić, K.; Schneider, H.-J. A novel bis-phenanthridine triamine with pH controlled binding to nucleotides and nucleic acids. *Org. Biomol. Chem.* **2005**, *3*, 4373–4381. [CrossRef]
8. Sakamoto, T.; Ojida, A.; Hamachi, I. Molecular recognition, fluorescence sensing, and biological assay of phosphate anion derivatives using artificial Zn(ii)–Dpa complexes. *Chem. Commun.* **2009**, *2*, 141–152. [CrossRef]
9. Hewitt, S.H.; Ali, R.; Mailhot, R.; Antonen, C.R.; Dodson, C.A.; Butler, S.J. A simple, robust, universal assay for real-time enzyme monitoring by signalling changes in nucleoside phosphate anion concentration using a europium(iii)-based anion receptor. *Chem. Sci.* **2019**, *10*, 5373–5381. [CrossRef]
10. Sessler, J.L.; Gale, P.; Cho, W.-S. *Anion Receptor Chemistry*; The Royal Society of Chemistry: London, UK, 2006.
11. Wu, Q.; Lei, Q.; Zhong, H.-C.; Ren, T.-B.; Sun, Y.; Zhang, X.-B.; Yuan, L. Fluorophore-based host-guest assembly complexes for imaging and therapy. *Chem. Commun.* **2023**, *59*, 3024–3039. [CrossRef]
12. Niu, H.; Liu, J.; O'Connor, H.M.; Gunnlaugsson, T.; James, T.D.; Zhang, H. Photoinduced electron transfer (PeT) based fluorescent probes for cellular imaging and disease therapy. *Chem. Soc. Rev.* **2023**, *52*, 232–2357. [CrossRef] [PubMed]
13. Klymchenko, A.S. Fluorescent Probes for Lipid Membranes: From the Cell Surface to Organelles. *Acc. Chem. Res.* **2023**, *56*, 1–12. [CrossRef] [PubMed]
14. Luo, C.; Zhang, Q.; Sun, S.; Li, H.; Xu, Y. Research progress of auxiliary groups in improving the performance of fluorescent probes. *Chem. Commun.* **2023**, *59*, 2199–2207. [CrossRef] [PubMed]
15. Fang, H.; Chen, Y.; Jiang, Z.; He, W.; Guo, Z. Fluorescent Probes for Biological Species and Microenvironments: From Rational Design to Bioimaging Applications. *Acc. Chem. Res.* **2023**, *56*, 258–269. [CrossRef] [PubMed]
16. Neto, B.A.D.; Correa, J.R.; Spencer, J. Fluorescent Benzothiadiazole Derivatives as Fluorescence Imaging Dyes: A Decade of New Generation Probes. *Chem. Eur. J.* **2022**, *28*, e202103262. [CrossRef]
17. Manna, S.K.; Mondal, S.; Jana, B.; Samanta, K. Recent advances in tin ion detection using fluorometric and colorimetric chemosensors. *New J. Chem.* **2022**, *46*, 7309–7328. [CrossRef]

18. Krämer, J.; Kang, R.; Grimm, L.M.; de Cola, L.; Picchetti, P.; Biedermann, F. Molecular Probes, Chemosensors, and Nanosensors for Optical Detection of Biorelevant Molecules and Ions in Aqueous Media and Biofluids. *Chem. Rev.* **2022**, *122*, 3459–3636. [CrossRef]
19. Niko, Y.; Klymchenko, A.S. Emerging solvatochromic push-pull dyes for monitoring the lipid order of biomembranes in live cells. *J. Biochem.* **2021**, *170*, 163–174. [CrossRef]
20. Klymchenko, A.S. Solvatochromic and Fluorogenic Dyes as Environment-Sensitive Probes: Design and Biological Applications. *Acc. Chem. Res.* **2017**, *50*, 366–375. [CrossRef]
21. Hargrove, A.E.; Nieto, S.; Zhang, T.; Sessler, J.L.; Anslyn, E.V. Artificial Receptors for the Recognition of Phosphorylated Molecules. *Chem. Rev.* **2011**, *111*, 6603–6782. [CrossRef]
22. Kataev, E.A.; Shumilova, T.A.; Fiedler, B.; Anacker, T.; Friedrich, J. Understanding Stacking Interactions between an Aromatic Ring and Nucleobases in Aqueous Solution: Experimental and Theoretical Study. *J. Org. Chem.* **2016**, *81*, 6505–6514. [CrossRef] [PubMed]
23. Suzuki, Y.; Masuko, M.; Hashimoto, T.; Hayashita, T. Selective ATP recognition by boronic acid-appended cyclodextrin and a fluorescent probe supramolecular complex in water. *New J. Chem.* **2023**, *47*, 7035–7040. [CrossRef]
24. Kuchelmeister, H.Y.; Schmuck, C. Molecular Recognition of Nucleotides. In *Designing Receptors for the Next Generation of Biosensors*; Piletsky, S.A., Whitcombe, M.J., Eds.; Springer: Berlin/Heidelberg, Germany, 2013; pp. 53–65. ISBN 978-3-642-32329-4.
25. Agafontsev, A.M.; Ravi, A.; Shumilova, T.A.; Oshchepkov, A.S.; Kataev, E.A. Molecular Receptors for Recognition and Sensing of Nucleotides. *Chem. Eur. J.* **2019**, *25*, 2684–2694. [CrossRef] [PubMed]
26. Li, W.; Gong, X.; Fan, X.; Yin, S.; Su, D.; Zhang, X.; Yuan, L. Recent advances in molecular fluorescent probes for organic phosphate biomolecules recognition. *Chin. Chem. Lett.* **2019**, *30*, 1775–1790. [CrossRef]
27. Yue, Y.; Huo, F.; Cheng, F.; Zhu, X.; Mafireyi, T.; Strongin, R.M.; Yin, C. Functional synthetic probes for selective targeting and multi-analyte detection and imaging. *Chem. Soc. Rev.* **2019**, *48*, 4155–4177. [CrossRef] [PubMed]
28. Fontecilla-Camps, J.C. The Complex Roles of Adenosine Triphosphate in Bioenergetics. *ChemBioChem* **2022**, *23*, e202200064. [CrossRef] [PubMed]
29. Pontes, M.H.; Sevostyanova, A.; Groisman, E.A. When Too Much ATP Is Bad for Protein Synthesis. *J. Mol. Biol.* **2015**, *427*, 2586–2594. [CrossRef]
30. Vultaggio-Poma, V.; Sarti, A.C.; Di Virgilio, F. Extracellular ATP: A Feasible Target for Cancer Therapy. *Cells* **2020**, *9*, 2496. [CrossRef]
31. Khojastehnezhad, A.; Taghavi, F.; Yaghoobi, E.; Ramezani, M.; Alibolandi, M.; Abnous, K.; Taghdisi, S.M. Recent achievements and advances in optical and electrochemical aptasensing detection of ATP based on quantum dots. *Talanta* **2021**, *235*, 122753. [CrossRef]
32. Wu, Y.; Wen, J.; Li, H.; Sun, S.; Xu, Y. Fluorescent probes for recognition of ATP. *Chin. Chem. Lett.* **2017**, *28*, 1916–1924. [CrossRef]
33. Kumar, P.; Pachisia, S.; Gupta, R. Turn-on detection of assorted phosphates by luminescent chemosensors. *Inorg. Chem. Front.* **2021**, *8*, 3587–3607. [CrossRef]
34. Jun, Y.W.; Sarkar, S.; Kim, K.H.; Ahn, K.H. Molecular Probes for Fluorescence Imaging of ATP in Cells and Tissues. *ChemPhotoChem* **2019**, *3*, 214–219. [CrossRef]
35. Huang, B.; Liang, B.; Zhang, R.; Xing, D. Molecule fluorescent probes for adenosine triphosphate imaging in cancer cells and in vivo. *Coord. Chem. Rev.* **2022**, *452*, 214302. [CrossRef]
36. Butler, S.J.; Jolliffe, K.A. Anion Receptors for the Discrimination of ATP and ADP in Biological Media. *ChemPlusChem* **2021**, *86*, 59–70. [CrossRef] [PubMed]
37. Zhou, X.; Shang, L. Recent Advances in Nanomaterial-based Luminescent ATP Sensors. *Curr. Anal. Chem.* **2022**, *18*, 677–688. [CrossRef]
38. Bazzicalupi, C.; Bencini, A.; Giorgi, C.; Valtancoli, B.; Lippolis, V.; Perra, A. Exploring the Binding Ability of Polyammonium Hosts for Anionic Substrates: Selective Size-Dependent Recognition of Different Phosphate Anions by Bis-macrocyclic Receptors. *Inorg. Chem.* **2011**, *50*, 7202–7216. [CrossRef]
39. Yoo, S.; Kim, S.; Eom, M.S.; Kang, S.; Lim, S.-H.; Han, M.S. Development of a highly sensitive colorimetric thymidine triphosphate chemosensor using gold nanoparticles and the p-xylyl-bis(Hg2+-cyclen) complex: Improved selectivity by metal ion tuning. *Tetrahedron Lett.* **2016**, *57*, 4484–4487. [CrossRef]
40. Zhao, X.J.; He, R.X.; Li, Y.F. A terbium(III)-organic framework for highly selective sensing of cytidine triphosphate. *Analyst* **2012**, *137*, 5190–5192. [CrossRef]
41. Gupta, A.K.; Dhir, A.; Pradeep, C.P. Ratiometric Detection of Adenosine-5′-triphosphate (ATP) and Cytidine-5′-triphosphate (CTP) with a Fluorescent Spider-Like Receptor in Water. *Eur. J. Org. Chem.* **2015**, *2015*, 122–129. [CrossRef]
42. Morozov, B.S.; Oshchepkov, A.S.; Klemt, I.; Agafontsev, A.M.; Krishna, S.; Hampel, F.; Xu, H.-G.; Mokhir, A.; Guldi, D.; Kataev, E. Supramolecular Recognition of Cytidine Phosphate in Nucleotides and RNA Sequences. *JACS Au* **2023**, *3*, 964–977. [CrossRef]
43. Kim, H.N.; Moon, J.H.; Kim, S.K.; Kwon, J.Y.; Jang, Y.J.; Lee, J.Y.; Yoon, J. Fluorescent sensing of triphosphate nucleotides via anthracene derivatives. *J. Org. Chem.* **2011**, *76*, 3805–3811. [CrossRef] [PubMed]
44. Roy, I.; David, A.H.G.; Das, P.J.; Pe, D.J.; Stoddart, J.F. Fluorescent cyclophanes and their applications. *Chem. Soc. Rev.* **2022**, *51*, 5557–5605. [CrossRef] [PubMed]

45. Tay, H.M.; Beer, P. Optical sensing of anions by macrocyclic and interlocked hosts. *Org. Biomol. Chem.* **2021**, *19*, 4652–4677. [CrossRef]
46. Wang, D.-X.; Wang, M.-X. Exploring Anion-π Interactions and Their Applications in Supramolecular Chemistry. *Acc. Chem. Res.* **2020**, *53*, 1364–1380. [CrossRef] [PubMed]
47. Xiong, S.; He, Q. Photoresponsive macrocycles for selective binding and release of sulfate. *Chem. Commun.* **2021**, *57*, 13514–13517. [CrossRef] [PubMed]
48. Lichosyt, D.; Dydio, P.; Jurczak, J. Azulene-Based Macrocyclic Receptors for Recognition and Sensing of Phosphate Anions. *Chem. Eur. J.* **2016**, *22*, 17673–17680. [CrossRef]
49. Katayev, E.A.; Myshkovskaya, E.N.; Boev, N.V.; Khrustalev, V.N. Anion binding by pyrrole–pyridine-based macrocyclic polyamides. *Supramol. Chem.* **2008**, *20*, 619–624. [CrossRef]
50. Flood, A.H. Creating molecular macrocycles for anion recognition. *Beilstein J. Org. Chem.* **2016**, *12*, 611–627. [CrossRef]
51. Evans, N.H.; Beer, P.D. Advances in Anion Supramolecular Chemistry: From Recognition to Chemical Applications. *Angew. Chem. Int. Ed.* **2014**, *53*, 11716–11754. [CrossRef]
52. Sarkar, S.; Ballester, P.; Spektor, M.; Kataev, E.A. Micromolar Affinity and Higher: Synthetic Host-Guest Complexes with High Stabilities. *Angew. Chem. Int. Ed.* **2022**, e202214705. [CrossRef]
53. Anda, C.; Angeles Martínez, M.; Llobet, A. A Systematic Evaluation of Molecular Recognition Phenomena: Part 5. Selective Binding of Tripolyphosphate and ATP to Isomeric Hexaazamacrocyclic Ligands Containing Xylylic Spacers. *Supramol. Chem.* **2005**, *17*, 257–266. [CrossRef]
54. Zhang, H.; Cheng, L.; Nian, H.; Du, J.; Chen, T.; Cao, L. Adaptive chirality of achiral tetraphenylethene-based tetracationic cyclophanes with dual responses of fluorescence and circular dichroism in water. *Chem. Commun.* **2021**, *57*, 3135–3138. [CrossRef] [PubMed]
55. Dhaenens, M.; Lehn, J.-M.; Vigneron, J.-P. Molecular Recognition of Nucleosides, Nucleotides and Anionic Planar Substrates by a Water-soluble Bis-intercaland-type Receptor Molecule. *J. Chem. Soc. Perkin Trans. 2* **1993**, *7*, 1379–1381. [CrossRef]
56. Neelakandan, P.P.; Nandajan, P.C.; Subymol, B.; Ramaiah, D. Study of cavity size and nature of bridging units on recognition of nucleotides by cyclophanes. *Org. Biomol. Chem.* **2011**, *9*, 1021–1029. [CrossRef] [PubMed]
57. van Eker, D.; Samanta, S.K.; Davis, A.P. Aqueous recognition of purine and pyrimidine bases by an anthracene-based macrocyclic receptor. *Chem. Commun.* **2020**, *56*, 9268–9271. [CrossRef] [PubMed]
58. Neelakandan, P.P.; Hariharan, M.; Ramaiah, D. A supramolecular ON-OFF-ON fluorescence assay for selective recognition of GTP. *J. Am. Chem. Soc.* **2006**, *128*, 11334–11335. [CrossRef]
59. Moreno-Corral, R.; Lara, K.O. Complexation Studies of Nucleotides by Tetrandrine Derivatives Bearing Anthraquinone and Acridine Groups. *Supramol. Chem.* **2008**, *20*, 427–435. [CrossRef]
60. Rhaman, M.M.; Powell, D.R.; Hossain, M.A. Supramolecular Assembly of Uridine Monophosphate (UMP) and Thymidine Monophosphate (TMP) with a Dinuclear Copper(II) Receptor. *ACS Omega* **2017**, *2*, 7803–7811. [CrossRef]
61. Ramaiah, D.; Neelakandan, P.P.; Nair, A.K.; Avirah, R.R. Functional cyclophanes: Promising hosts for optical biomolecular recognition. *Chem. Soc. Rev.* **2010**, *39*, 4158–4168. [CrossRef]
62. Agafontsev, A.M.; Shumilova, T.A.; Rüffer, T.; Lang, H.; Kataev, E.A. Anthracene-Based Cyclophanes with Selective Fluorescent Responses for TTP and GTP: Insights into Recognition and Sensing Mechanisms. *Chem. Eur. J.* **2019**, *25*, 3541–3549. [CrossRef]
63. Agafontsev, A.M.; Shumilova, T.A.; Oshchepkov, A.S.; Hampel, F.; Kataev, E.A. Ratiometric Detection of ATP by Fluorescent Cyclophanes with Bellows-Type Sensing Mechanism. *Chem. Eur. J.* **2020**, *26*, 9991–9997. [CrossRef] [PubMed]
64. Agafontsev, A.M.; Oshchepkov, A.S.; Shumilova, T.A.; Kataev, E.A. Binding and Sensing Properties of a Hybrid Naphthalimide-Pyrene Aza-Cyclophane towards Nucleotides in an Aqueous Solution. *Molecules* **2021**, *26*, 980. [CrossRef] [PubMed]
65. Kataev, E.A. Converting pH probes into "turn-on" fluorescent receptors for anions. *Chem. Commun.* **2023**, *59*, 1717–1727. [CrossRef] [PubMed]
66. Granzhan, A.; Kotera, N.; Teulade-Fichou, M.-P. Finding needles in a basestack: Recognition of mismatched base pairs in DNA by small molecules. *Chem. Soc. Rev.* **2014**, *43*, 3630–3665. [CrossRef] [PubMed]
67. Granzhan, A.; Largy, E.; Saettel, N.; Teulade-Fichou, M.-P. Macrocyclic DNA-mismatch-binding ligands: Structural determinants of selectivity. *Chem. Eur. J.* **2010**, *16*, 878–889. [CrossRef]
68. Kotera, N.; Granzhan, A.; Teulade-Fichou, M.-P. Comparative study of affinity and selectivity of ligands targeting abasic and mismatch sites in DNA using a fluorescence-melting assay. *Biochimie* **2016**, *128–129*, 133–137. [CrossRef]
69. Schlosser, J.; Ihmels, H. Ligands for Abasic-Site containing DNA and their Use as Fluorescent Probes. *Curr. Org. Synth.* **2023**, *20*, 96–113. [CrossRef]
70. Yaragorla, S.; Singh, G.; Dada, R. C(sp3)–H functionalization of methyl azaarenes: A calcium-catalyzed facile synthesis of (E)-2-styryl azaarenes and 2-aryl-1,3-bisazaarenes. *Tetrahedron Lett.* **2015**, *56*, 5924–5929. [CrossRef]
71. Seo, J.; Park, S.-R.; Kim, M.; Suh, M.C.; Lee, J. The role of electron-transporting Benzo[f]quinoline unit as an electron acceptor of new bipolar hosts for green PHOLEDs. *Dyes Pigm.* **2019**, *162*, 959–966. [CrossRef]
72. Brouwer, A.M. Standards for photoluminescence quantum yield measurements in solution (IUPAC Technical Report). *Pure Appl. Chem.* **2011**, *83*, 2213–2228. [CrossRef]

73. Buettelmann, B.; Alanine, A.; Bourson, A.; Gill, R.; Heitz, M.-P.; Mutel, V.; Pinard, E.; Trube, G.; Wyler, R. 2-Styryl-pyridines and 2-(3,4-Dihydro-naphthalen-2-yl)pyridines as Potent NR1/2B Subtype Selective NMDA Receptor Antagonists. *Chimia* **2004**, *58*, 630. [CrossRef]
74. Mittapalli, R.R.; Namashivaya, S.S.R.; Oshchepkov, A.S.; Kuczyńska, E.; Kataev, E.A. Design of anion-selective PET probes based on azacryptands: The effect of pH on binding and fluorescence properties. *Chem. Commun.* **2017**, *53*, 4822–4825. [CrossRef]
75. Beggiato, G.; Favaro, G.; Mazzucato, U. Acid-base equilibria of dipyridylethylenes studied by absorption and fluorescence spectrometry. *J. Heterocycl. Chem.* **1970**, *7*, 583–587. [CrossRef]
76. Zhang, C.; Li, M.; Liang, W.; Zhang, G.; Fan, L.; Yao, Q.; Shuang, S.; Dong, C. Substituent Effect on the Properties of pH Fluorescence Probes Containing Pyridine Group. *ChemistrySelect* **2019**, *4*, 5735–5739. [CrossRef]
77. Budyka, M.F.; Fedulova, J.A.; Gavrishova, T.N.; Li, V.M.; Potashova, N.I.; Tovstun, S.A. 2+2 Photocycloaddition in a bichromophoric dyad: Photochemical concerted forward reaction following Woodward-Hoffmann rules and photoinduced stepwise reverse reaction of the ring opening via predissociation. *Phys. Chem. Chem. Phys.* **2022**, *24*, 24137–24145. [CrossRef]
78. Chen, D.; Zhong, C.; Zhao, Y.; Nan, L.; Liu, Y.; Qin, J. A two-dimensional molecule with a large conjugation degree: Synthesis, two-photon absorption and charge transport ability. *J. Mater. Chem. C* **2017**, *5*, 5199–5206. [CrossRef]
79. Budyka, M.F.; Potasheva, N.I.; Gavrishova, T.N.; Li, V.M. Photoisomerization and [2 + 2] photocycloaddition in bichromophoric styrylbenzoquinoline dyads with o-xylene bridge group. *High Energy Chem.* **2017**, *51*, 201–208. [CrossRef]
80. Budyka, M.F.; Gavrishova, T.N.; Li, V.M.; Potashova, N.I.; Fedulova, J.A. Emissive and reactive excimers in a covalently-linked supramolecular multi-chromophoric system with a balanced rigid-flexible structure. *Spectrochim. Acta Part A Mol. Biomol. Spectrosc.* **2022**, *267*, 120565. [CrossRef] [PubMed]
81. Tripathy, M.; Subuddhi, U.; Patel, S. A styrylpyridinium dye as chromogenic and fluorogenic dual mode chemosensor for selective detection of mercuric ion: Application in bacterial cell imaging and molecular logic gate. *Dyes Pigm.* **2020**, *174*, 108054. [CrossRef]
82. Sivakumar, R.; Lee, N.Y. Paper-Based Fluorescence Chemosensors for Metal Ion Detection in Biological and Environmental Samples. *BioChip J.* **2021**, *15*, 216–232. [CrossRef]
83. Bazany-Rodríguez, I.J.; Salomón-Flores, M.K.; Bautista-Renedo, J.M.; González-Rivas, N.; Dorazco-González, A. Chemosensing of Guanosine Triphosphate Based on a Fluorescent Dinuclear Zn(II)-Dipicolylamine Complex in Water. *Inorg. Chem.* **2020**, *59*, 7739–7751. [CrossRef] [PubMed]
84. Viviano-Posadas, A.O.; Romero-Mendoza, U.; Bazany-Rodríguez, I.J.; Velázquez-Castillo, R.V.; Martínez-Otero, D.; Bautista-Renedo, J.M.; González-Rivas, N.; Galindo-Murillo, R.; Salomón-Flores, M.K.; Dorazco-González, A. Efficient fluorescent recognition of ATP/GTP by a water-soluble bisquinolinium pyridine-2,6-dicarboxamide compound. Crystal structures, spectroscopic studies and interaction mode with DNA. *RSC Adv.* **2022**, *12*, 27826–27838. [CrossRef] [PubMed]
85. Dorazco-González, A.; Alamo, M.F.; Godoy-Alcántar, C.; Höpfl, H.; Yatsimirsky, A.K. Fluorescent anion sensing by bisquinolinium pyridine-2,6-dicarboxamide receptors in water. *RSC Adv.* **2014**, *4*, 455–466. [CrossRef]
86. Steenken, S.; Jovanovic, S.V. How Easily Oxidizable Is DNA? One-Electron Reduction Potentials of Adenosine and Guanosine Radicals in Aqueous Solution. *J. Am. Chem. Soc.* **1997**, *119*, 617–618. [CrossRef]
87. Seidel, C.A.M.; Schulz, A.; Sauer, M.H.M. Nucleobase-Specific Quenching of Fluorescent Dyes. 1. Nucleobase One-Electron Redox Potentials and Their Correlation with Static and Dynamic Quenching Efficiencies. *J. Phys. Chem.* **1996**, *100*, 5541–5553. [CrossRef]
88. Hu, P.; Yang, S.; Feng, G. Discrimination of adenine nucleotides and pyrophosphate in water by a zinc complex of an anthracene-based cyclophane. *Org. Biomol. Chem.* **2014**, *12*, 3701–3706. [CrossRef] [PubMed]
89. Lohani, C.R.; Kim, J.-M.; Chung, S.-Y.; Yoon, J.; Lee, K.-H. Colorimetric and fluorescent sensing of pyrophosphate in 100% aqueous solution by a system comprised of rhodamine B compound and Al_3^+ complex. *Analyst* **2010**, *135*, 2079–2084. [CrossRef]
90. MacDougall, D.; Crummett, W.B. Guidelines for data acquisition and data quality evaluation in environmental chemistry. *Anal. Chem.* **1980**, *52*, 2242–2249. [CrossRef]

Disclaimer/Publisher's Note: The statements, opinions and data contained in all publications are solely those of the individual author(s) and contributor(s) and not of MDPI and/or the editor(s). MDPI and/or the editor(s) disclaim responsibility for any injury to people or property resulting from any ideas, methods, instructions or products referred to in the content.

Article

Modification of the Bridging Unit in Luminescent Pt(II) Complexes Bearing CˆN*N and CˆN*NˆC Ligands

Stefan Buss [1,2], María Victoria Cappellari [1,2], Alexander Hepp [1], Jutta Kösters [1] and Cristian A. Strassert [1,2,*]

[1] Institut für Anorganische und Analytische Chemie, Westfälische Wilhelms-Universität Münster, Corrensstraße 28/30, 48149 Münster, Germany
[2] CeNTech, CiMIC, SoN, Westfälische Wilhelms-Universität Münster, Heisenbergstraße 11, 48149 Münster, Germany
* Correspondence: ca.s@wwu.de

Abstract: In this work, we explored the synthesis and characterization of Pt(II) complexes bearing different tri- and tetradentate luminophores acting as CˆN*N- and CˆN*NˆC-chelators. Thus, we investigated diverse substitution patterns in order to improve their processability and assessed the effects of structural variations on their excited state properties. Hence, a detailed analysis of the different synthetic pathways is presented; the photophysical properties were studied by using steady-state and time-resolved photoluminescence spectroscopy. We determined the absorption and emission spectra, the photoluminescence efficiencies, and the excited state lifetimes of the complexes in fluid solutions at room temperature and frozen glassy matrices at 77 K. Finally, a structure–property relationship was established, showing that the decoration of the bridging unit on the tridentate luminophores only marginally affects the excited state properties, whereas the double cyclometallation related to the tetradentate chelator prolongs the excited state lifetime and increases the photoluminescence quantum yield.

Keywords: soluble triplet emitters; synthesis of Pt(II) complexes; (time-resolved) photoluminescence spectroscopy

Citation: Buss, S.; Cappellari, M.V.; Hepp, A.; Kösters, J.; Strassert, C.A. Modification of the Bridging Unit in Luminescent Pt(II) Complexes Bearing CˆN*N and CˆN*NˆC Ligands. *Chemistry* 2023, 5, 1243–1255. https://doi.org/10.3390/chemistry5020084

Academic Editor: Catherine Housecroft

Received: 1 April 2023
Revised: 8 May 2023
Accepted: 8 May 2023
Published: 15 May 2023

Copyright: © 2023 by the authors. Licensee MDPI, Basel, Switzerland. This article is an open access article distributed under the terms and conditions of the Creative Commons Attribution (CC BY) license (https://creativecommons.org/licenses/by/4.0/).

1. Introduction

In recent years, triplet emitters have received increasing attention due to their range of different applications relying on their balance between solubility and tendency towards aggregation [1–6]. Among organometallic compounds, Pt(II) complexes represent a special case where the intrinsic high ligand field splitting (LFS) and strong spin-orbit coupling (SOC) on the metal center lead to desirable photophysical properties, such as long excited state lifetimes and high photoluminescence quantum yields [7–13]. In addition, the d^8 configuration of the Pt(II) center leads to square planar coordination geometries, leaving the d_z^2 orbitals available for intermolecular coupling of the metal atoms. Hence, the aggregation properties of these complexes can be exploited in the context of supramolecular approaches by harnessing the unique characteristics of triplet states delocalized over dimeric species [14–16]. These concepts have been used for OLED fabrication [17–22], photocatalysis [7,23–26], bioimaging [27–29], and sensing technologies [30–33], among others.

As mentioned above, a high LFS plays an important role, as it increases the energy of dissociative metal-centered (MC) excited states to prevent thermally activated deactivation processes involving radiationless pathways that shorten the excited state lifetimes while reducing the efficiency of more desirable mechanisms [34]. In this regard, phenide-based σ-donors (cyclometalated aryl-functions) are widely used to increase the LFS due to their negative charge [35]. In addition, the rigidity of the chromophoric ligand contributes to the overall performance; hence, complexes bearing tri- and tetradentate chelators are preferred [34]. While tetradentate ligands lead to highly stable complexes and show an overall better performance, complexes with tridentate ligands can show tunable properties depending on the monodentate co-ligand, which can be adjusted depending on the intended outcome [14–16,36].

Huo et al. reported on studies focused on the optimization of the coordination environment derived from tri- and tetradentate ligands for Pt(II). Hence, they synthesized complexes with a combination of five- and six-membered metalacycles (including bis-cyclometalation for the tetradentate ligands) displaying interesting photophysical properties (Figure 1) [37,38]. Inspired by the work of Huo et al., we explored different alternative concepts. For instance, we decorated the cyclometalating unit with fluorine atoms to adjust the emission wavelength as well as the aggregation behavior (Figure 1) [39].

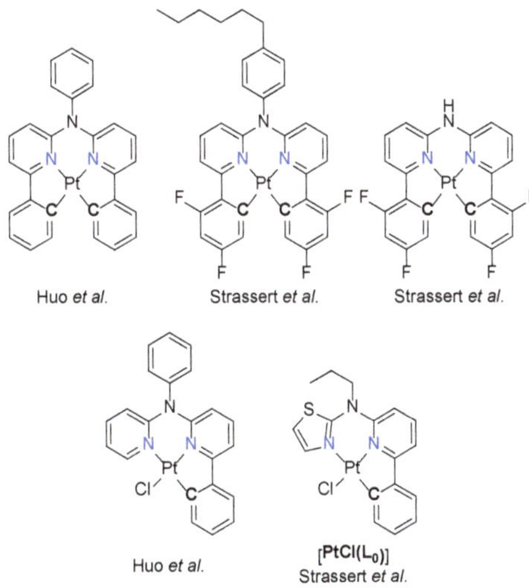

Figure 1. Selected Pt(II) complexes from previous studies. **Top**: C^N*N^C-based complexes with phenyl-amine (left) [37], 4-hexylphenyl-amine (center) [39], and secondary amine (right) [40] bridges. **Bottom**: N*N^C complexes with phenyl-amine (left) and [38] propyl-amine connectors (right; [PtCl(L$_0$)]) [41].

In 2019, we reported the synthesis of a bis-cyclometalated Pt(II) complex bearing a tetradentate ligand with a secondary amine bridge (Figure 1) [40]. Additionally, in 2020, the synthesis of Pt(II) complexes with C^N*N luminophores based on thiazol moieties was reported (Figure 1) [41]. In the present work, we aimed at an improvement of their rather low solubility while maintaining the planar coordination environment, and we designed a synthetic route to introduce a secondary amine that is subsequently functionalized by an alkylation step. Since the range of commercially available alkyl-bromides is wide, the alkylation reaction could provide a versatile tool to enable a vast range of decoration patterns. In this report, the impact on the chemical and photophysical properties of the resulting complexes was also investigated.

2. Experimental Section

General information about experimental procedures including instrumental and synthetic methods, structural characterization of the ligand precursors and the complexes, as well as photophysical measurements are provided in the Supplementary Materials.

Materials: All chemicals were used as purchased from commercially available sources. For the photophysical measurements, spectroscopic-grade solvents (Uvasol®) were used.

Synthesis: The detailed synthetic procedures and analytical data are provided in the Supplementary Materials. Each new compound was characterized by ^{1}H, ^{13}C, and

2D nuclear magnetic resonance spectroscopies (NMR, see Figures S1–S106) as well as mass spectrometry (EM-ESI-MS or MALDI-MS). The metal complexes were further analyzed by steady-state and time-resolved photoluminescence spectroscopy.

X-ray diffractometry: Suitable single crystals for X-ray diffraction measurements were obtained by slowly evaporating the solvent from a saturated DCM solution or by the diffusion of cyclohexane into such a solution; in the case of [PtCl(L$_4$)], the complex was crystallized from diethylether. The full set of data is given in the Supplementary Materials (Tables S1–S4; Figures S107–S114), as well as in the CCDC database (CCDC-Nr. 2252944 (I$_1$), 2252945 (A$_4$), 2252950 (A$_3$), 2252953 ([PtL$_6$]), 2252954 ([PtCl(L$_4$)]), 2252967 ([PtCl(L$_2$)])). The graphical representation of the molecular structures was realized by using the Mercury software package from CCDC [42].

3. Results and Discussion

With the selected precursors (A$_5$, A$_6$), we were able to establish a three-step synthesis route, based on our previous report [41], yielding complexes bearing tri- and tetradentate ligands.

3.1. Tridentate Coordination

The first step in the synthesis of the tridentate ligands involves the alkylation of the secondary amine A$_5$ by a previously described procedure [41]. Due to the tautomeric equilibrium involving the 2-amino thiazole unit, this reaction produces both an amine (A$_x$) and an imine (I$_x$) [41,43,44]. The observed ratios of A/I yields suggest that the formation of imines is favored when bulkier alkylbromides are employed. The molecular structures in the single crystals of I$_1$ and A$_3$ are depicted in Figure 2; both show a coplanar orientation of the heterocycles (for further information, see Figures S107–S108 and Table S1 in the Supplementary Materials). The two isomers can be easily distinguished by the $^3J_{HH}$-coupling constant in the ^{1}H-NMR spectrum, as the amine has a value of $^3J_{HH}$ = 3.6 Hz, whereas the imine reaches $^3J_{HH}$ = 4.8 Hz. Interestingly, the coordination of the thiazole on the metal center leads to a constant of around $^3J_{HH}$ = 4.1 Hz (*vide infra*). The formation of the two species was one of the main drawbacks of these reactions, due to the loss of significant amounts of potential products. Therefore, we explored the post-functionalization of the secondary amino group by acylation. By dissolving A$_5$ in hot valeric anhydride, the formation of the amide A$_4$ was achieved without the formation of the imine-like tautomer; the moderate yields are due to the incomplete conversion of the precursor. The amide A$_4$ is a crystalline white solid, its molecular structure in the single crystal is shown in Figure 2 (see also Figure S109 in the Supplementary Materials, as well as Table S1), and it is clear that the thiazolyl moiety is coplanar with the amide, whereas the pyridine is bent out of plane. Subsequent attempts to reduce the amide to yield an amine failed with various reagents (LiAH, NaBH$_4$, BH$_3$·THF, SiEt$_3$H), as only the secondary amine A$_5$ was obtained. Nonetheless, with the four amine precursors in hand, the versatile Suzuki–Miyaura coupling reactions with suitable boronic acids led to the four tridentate ligand precursors (L$_x$H). For the alkyl-substituted precursors (L$_1$H-L$_3$H), the final cyclometallation step was carried out using well-established reaction conditions (i.e., glacial acetic acid as a solvent at reflux paired with K$_2$[PtCl$_4$]), which successfully yielded the complexes [PtCl(L$_x$)]. The low yield of [PtCl(L$_1$)] can be attributed to the worse solubility and higher retention time during column chromatographic purification. In the case of [PtCl(L$_3$)], the low yield seems to originate from the acid-mediated lability of the benzylic group [45]. On the other hand, the acylated ligand precursor L$_4$H was unstable under the cyclometallation conditions and only the complex [PtCl(L$_5$)] was formed, as the amide was cleaved towards the complex with a secondary amine. In order to obtain the desired amide-substituted complex [PtCl(L$_4$)], milder conditions were used (i.e., MeCN/H$_2$O at reflux). The synthesis scheme is summarized in Figure 2 [46]. Regarding the solubility of these compounds, compared to the already published propyl-substituted complex [41], the methyl-substituted species ([PtCl(L$_1$)]) and the free secondary amine ([PtCl(L$_5$)]) presented poorer solubility; quite surprisingly, the 3,3-dimethyl-butyl complex ([PtCl(L$_2$)]) is comparable to the already-reported compound. The last two complexes ([PtCl(L₃)], [PtCl(L₄)]) showed better solubility than the other exemplars; unfortunately, they correspond to the least stable ligands under the explored reaction conditions.

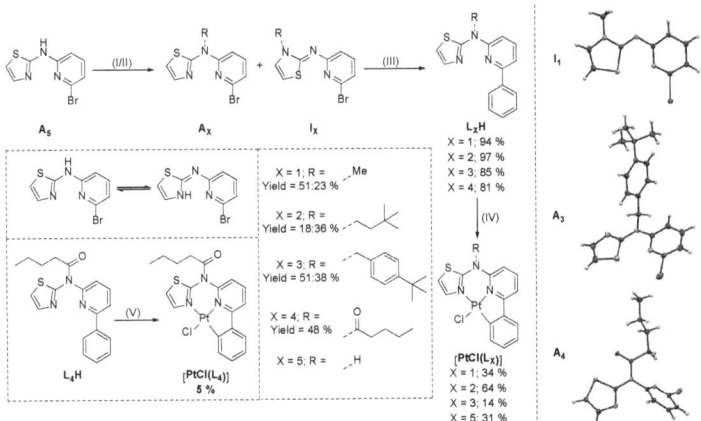

Figure 2. Left—Reaction schemes for the synthesis of the complexes with tridentate ligands. (I) Alkylation: bromoalkane, Cs₂CO₃, THF, reflux, 16 h, 18–51% yield. (II) Acylation: valeric anhydride, 170 °C, 3 h, 48%. (III) Suzuki–Miyaura cross-coupling: phenyl-boronic acid, K₂CO₃, [Pd(PPh₃)₄], THF, H₂O, reflux, 16 h, 81–97% yield. (IV) Cyclometallation: K₂[PtCl₄], ⁿBu₄NCl, glacial acetic acid, reflux, 16 h, 14–64% yield. (V) Cyclometallation: K₂[PtCl₄], MeCN/H₂O, reflux, 16 h, 5% yield. Right—Molecular structure in single crystals of **I₁** (top), **A₃** (center), and **A₄** (bottom). Displacement ellipsoids are shown at 50% probability.

We were able to obtain the molecular structures of [PtCl(L₂)] and [PtCl(L₄)] in single crystals by X-ray diffractometry, as shown in Figures 3 and 4, respectively (further details are found in the Supplementary Materials, Figures S110–S113 and Tables S2 and S3). The crystal structure of [PtCl(L₂)] corresponds to the $P2_1/c$ space group and confirms the square planar coordination environment with a chlorido unit as the fourth ligand. The overall coordination geometry is practically square planar, in agreement with the previously reported propyl-substituted complex [PtCl(L₀)]. Also in the present study, the formation of head-to-tail dimers is apparent. However, in this case, the 3D packing arises from the interactions between the π system and the hydrogen atoms, rather than being a chain evolving from stacked dimers in one direction supported by H-π interactions [41].

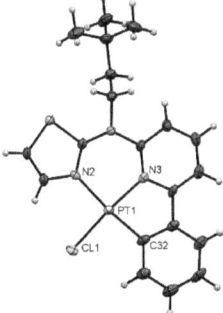

Figure 3. Molecular structure in the single crystal of [PtCl(L₂)]. Displacement ellipsoids are shown at 50% probability.

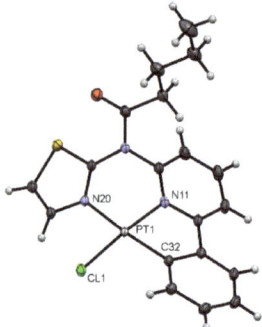

Figure 4. Molecular structure in the single crystal of [**PtCl(L₄)**]. Displacement ellipsoids are shown at 50% probability.

The crystal structure of the complex [**PtCl(L₄)**] involves three distinguishable conformers of the monomeric units, which crystallize in the $P\bar{1}$ space group. The coordination of the Pt(II) center in the case of [**PtCl(L₄)**] (i.e., one tridentate chromophore and one chlorido co-ligand) shows that the forced planarity of the amide bridge leads to an increased sterical demand for the chain and results in the bending of the chelating ligand. To release the sterical strain, the phenylpyridine and the thiazole moiety bend out of the coordination plane, which is visible in the higher deviation of the bond angle (N20-Pt1-C32 = 168.36°; for [**PtCl(L₂)**] = 175.97°) from the optimal 180° coordination geometry. The main differences between the three conformers are the placement and orientation of the alkyl group. The main intermolecular interactions for the 3D structure are the π–π interactions from the phenylpyrindine luminophores, combined with van der Waals interactions. Overall, no significant Pt-Pt coupling can be traced. The higher sterical strain, if compared with the alkyl-substituted analogs, in combination with the intrinsic reactivity of the amide in the presence of acids, may lead to the lability of the ligand during the cyclometalating reaction.

3.2. Tetradentate Coordination

The remarkable photophysical properties of Pt(II) complexes with C^N*N^C-type ligands are attributed to the large LFS and rigidity of the coordination environment [38]. For these luminophoric chelators, the Buchwald–Hartwig reaction limits the substitution pattern to aryl-amines and restricts the potential of substituents to enhance the processability of the complexes. In this sense, hexyl-phenyl substituents were necessary to attain meaningful solubility in organic solvents [39,40].

We therefore developed a synthetic approach similar to the one used for the herein-described tridentate thiazole-based compounds, but with a proper adaptation for tetradentate bis-cyclometalating ligand precursors. In this way, the Buchwald–Hartwig cross-coupling is avoided while giving access to alkyl-substitution on the bridging nitrogen atom. The di(bromo-pyridine)amine precursor is known from the literature [40,47] and can be modified as in the case of **A₅**. The alkylation can be achieved by following the procedure reported in the literature, using an alkylbromide and NaH in DMF at room temperature, thus yielding **A₆** [47]. The second step again involves a Suzuki–Miyaura cross-coupling towards the tetradentate ligand precursor **L₆H₂**. Unfortunately, the incomplete conversion leads to a product mixture of the di-substituted **L₆H₂** and the mono-substituted **L₆BrH** species. This explains the relatively low yields; nonetheless, the remaining bromine atom could represent a versatile intermediate where a second functionalization step could provide an asymmetric tetradentate chelator with interesting photophysical properties [48]. In the final step, the cyclometalation yielded the complex [**Pt(L₆)**]; due to the double cyclometalation, the reaction time needs to be adjusted to ensure a reasonable yield (from 16 h to 48 h; the synthesis is summarized in Figure 5). From pure DCM, we were able to obtain suitable crystals for X-ray diffractometry (the molecular structure in the crystal is shown in Figure 6; more details can be found in the Supplementary Materials, see Figure S114 and Table S4). The crystal structure shows a packing resembling [PtCl(L₂)], as both crystalize in a $P2_1/c$ space group, forming head-to-tail-dimers where the 3D packing results from interactions between the π systems and the hydrogen atoms.

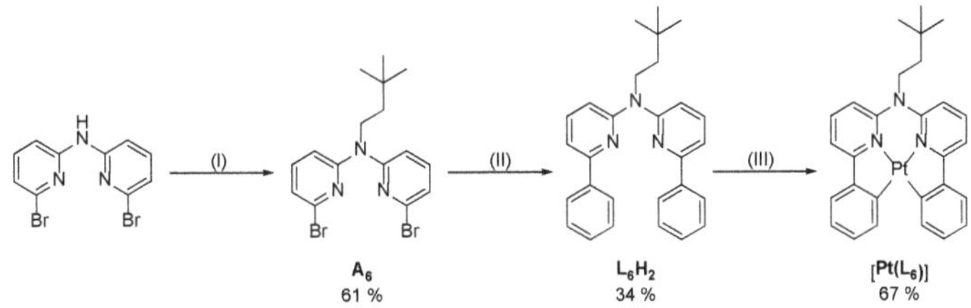

Figure 5. Synthetic route towards complexes with tetradentate ligands. (I) Alkylation: 1-bromo-3,3-dimethyl-butane, NaH, DMF, room temperature, 16 h, 61% yield. (II) Suzuki–Miyaura cross-coupling: phenyl-boronic acid, K₂CO₃, [Pd(PPh₃)₄], THF, H₂O, reflux, 16 h, 34% yield. (III) Cyclometallation: K₂[PtCl₄], ⁿBu₄NCl, glacial acetic acid, reflux, 16 h, 67% yield.

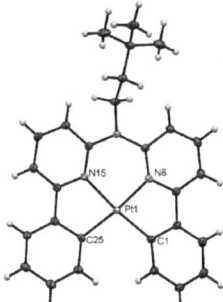

Figure 6. Molecular structure in the single crystal of [Pt(L₆)]. Displacement ellipsoids are shown at 50% probability.

3.3. Photophysics

The photophysical properties of all the complexes are summarized in Table 1; the absorption and photoluminescence spectra in fluid solutions at 298 K and in frozen glassy matrices at 77 K are shown in Figure 7 (as representative examples including [PtCl(L₅)] and [PtCl(L₆)]; the complete set of spectra, photoluminescence decay plots, as well as the uncertainties and the multiexponential lifetime components are found in the Supplementary Materials, see Table S5 and Figures S115–S137).

The assignment of the absorption bands by comparison with related compounds [37–39] indicates that the higher-energy bands with strong absorption coefficients below 350 nm correspond to transitions into $^1\pi\pi^*$ configurations (i.e., with ligand-centered character). The lower-energy bands around 375 nm and above can be generally assigned to transitions into mixed charge-transfer states. The UV/Vis spectra (Figure S115) for [PtCl(L₁₋₃)] and [PtCl(L₅)] strongly resemble the reported profile of [PtCl(L₀)], suggesting that the alkyl groups have no significant effect on the optical transitions. However, for the amide-substituted complex [PtCl(L₄)], a small red shift ($\Delta\lambda \approx 8$ nm) is observable, indicating that the electron-withdrawing effect of the amide substituent affects the energy of the excited electronic states. For complex [Pt(L₆)], the bands have distinct features: they are more intense and a low-energy band with an intense absorption at λ_{abs} = 404 nm becomes visible, suggesting a higher charge-transfer character in the excited state. The complexes bearing tridentate ligands and a chlorido co-ligand exhibit very weak luminescence at room temperature, due to the low LFS caused by the monodentate moiety that acts as an efficient π-donor. This leads to emission spectra with poor signal-to-noise ratios for **[PtCl(L4)]** while impairing the measurement of reliable photoluminescence data at room temperature. As a result of the bis-cyclometalation with an intrinsically higher LFS and rigidity from the tetradentate chromophore, **[Pt(L6)]** shows the best performance.

Table 1. Selected photophysical data for the complexes. A complete set of data is provided in the Supplementary Materials (see Table S5).

Complex	λ_{abs} / nm (ε / 10^3 M^{-1} cm^{-1})	Medium (T / K)	λ_{em} / nm	τ_{av} [a]	$\Phi_L \pm 0.02 / \pm 0.05$ [b]
[PtCl(L0)] [40]	246 (12.5), 266 (20.1), 278 (18.1), 316 (8.1), 348 (8.3), 370 (5.2)	DCM, Ar (298)	496	19.1 ns	<0.02
		Glassy matrix (77)	487	31.8 µs	0.98
[PtCl(L1)]	266 (17.8), 279 (17.4), 314 (6.7), 348 (6.6), 370 (3.7)	DCM, Ar (298)	495	14.58 ns	<0.02
		Glassy matrix (77)	483	23.52 µs	0.98
[PtCl(L2)]	266 (19.9), 280 (17.4), 291 (13.5), 315 (8.2), 349 (8.3), 371 (5.0)	DCM, Ar (298)	495	15.94 ns	<0.02
		Glassy matrix (77)	487	32.86 µs	0.98
[PtCl(L3)]	267 (30.1), 267 (27.7), 316 (10.9), 348 (11.7), 368 (6.7)	DCM, Ar (298)	496	25.7 ns	<0.02
		Glassy matrix (77)	484	24.879 µs	0.98
[PtCl(L4)]	273 (17.7), 293 (15.0), 345 (5.4), 378 (3.3)	DCM, Ar (298)	504	n.d.	n.d.
		Glassy matrix (77)	487	12.42 µs	0.98
[PtCl(L5)]	262 (22.6), 276 (20.9), 288 (16.1), 313 (9.0), 346 (8.6), 370 (5.1)	DCM, Ar (298)	493	0.2695 µs	<0.02
		Glassy matrix (77)	488	45.7 µs	0.98
[Pt(L6)]	274 (43.9), 290 (33.0), 318 (23.5), 334 (22.1), 366 (16.8), 404 (7.2)	DCM, Ar (298)	510	4.2035 µs	0.54
		Glassy matrix (77)	500	11.436 µs	0.97

(a) λ_{exc} = 376 nm, expressed as amplitude-weighted averaged lifetimes according to the suggestions from the relevant literature [49] (the single exponential components, relative amplitudes, and uncertainties are listed in the Supplementary Materials, see Table S5). (b) The uncertainty for the glassy matrix is estimated as ± 0.05 due to the measurement setup.

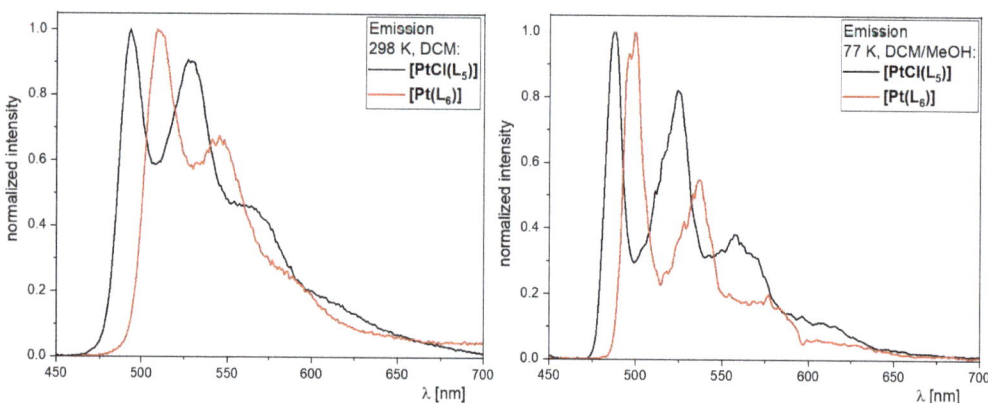

Figure 7. Selected normalized photoluminescence spectra (λ_{exc} ≈ 370 nm) for **[PtCl(L5)]** (black) and **[Pt(L6)]** (red). Measured in Ar-purged fluid DCM at 298 K (left) or in frozen glassy matrices (DCM/MeOH 1:1) at 77 K (right).

As the emission maxima peak at around λ_{em} = 495 nm, with a typical vibrational progression, excited state lifetimes (τ_{av}) in the ns range, and photoluminescence quantum yields (Φ_L) below the detection limit of our equipment at room temperature, the complexes **[PtCl(L1-5)]** reproduce the already-reported properties observed for **[PtCl(L0)]**. For **[PtCl(L0-4)]**, the deactivation of the luminescent triplet state is too fast to be affected by oxygen. Only **[PtCl(L5)]** shows a longer lifetime ($\tau_{av(Ar)}$ = 0.2695 µs), but it is also not affected by the presence of oxygen. Upon switching from fluid solution at RT to a glassy matrix at 77 K (DCM:MeOH = 1:1), we observed a blue shift ($\Delta\lambda$ ≈ 10 nm) of the emission maxima, due to the loss of solvent stabilization resulting in an enhanced ligand-centered character for the excited state. The long excited state lifetimes also point to a primarily ligand-centered character for the emissive states (τ_{av} = 23.5 µs–45.7 µs); only the amide-substituted complex [**PtCl(L$_4$)**] shows a faster decay of τ_{av} = 12.42 µs. We assume that the bent coordination plane observed in the molecular structure causes a faster deactivation of the excited state. At 77 K, all complexes show Φ_L close to unity, meaning that the fast relaxation at room temperature can be (mostly) attributed to the radiationless deactivation via thermally accessible metal-centered states; in fact, our previous report showed that this can be overcome by increasing the LFS with a suitable co-ligand, e.g., a cyanido unit acting as a π acceptor. Nonetheless, the nearly invariant photophysical properties suggest that the substitution pattern at the bridging amine group has a negligible effect on the excited state character and the concomitant deactivation rates.

The complex with the tetradentate ligand, [**Pt(L$_6$)**], shows a completely different behavior at room temperature compared to the tridentate chelation pattern. The emission maximum is red-shifted ($\Delta\lambda \approx 15$ nm) and the vibrational progression is less pronounced. Moreover, the excited state of this complex show a strong sensitivity to the presence of oxygen with a prolonged lifetime (τ_{av} = 0.2354 µs to τ_{av} = 4.2035 µs) and enhanced photoluminescence intensity (Φ_L = 0.02 to Φ_L = 0.53) upon deoxygenation of the solution. This distinct behavior can be attributed to the higher LFS, in combination with a higher degree of metal-to-ligand charge-transfer character in the excited state resulting from the exchange of the thiazol and chlorido σ-donors by a second phenylpyridine arm. Still, in the glassy matrix at 77 K, a blue shift ($\Delta\lambda \approx 10$ nm) was observed compared to room temperature. Due to the lack of thermally accessibly metal-centered states at 77 K, the Φ_L increases to almost unity with τ_{av} = 11.436 µs. These observations and values are in line with other reports [38,40,50].

4. Conclusions

Five coordination compounds bearing tridentate cyclometallated C^N*N ligands and one Pt(II) complex with a tetradentate bis-cyclometalated C^N*N^C luminophore were synthesized and characterized. In summary, we were able to modify the substitution patterns and solubilities without affecting the excited state properties. The complexes with a tridentate ligand show emission from mostly ligand-centered states, leading to fast radiationless deactivation via thermally accessible metal-centered states due to the low LFS resulting from the chlorido co-ligand; however, upon cooling down to 77 K, unitary quantum yields were achieved. On the other hand, the complex [**Pt(L$_6$)**] with the tetradentate chelator displays higher efficiencies (Φ_L = 0.53) and longer lifetimes in fluid solutions at room temperature, mostly due to a higher rigidity and ligand field splitting resulting from cyclometallation and enhanced metal participation in the excited state, which in turn hampers radiationless deactivation pathways and improves the phosphorescence rate, respectively.

The alkylation step constitutes a bottleneck due to the intrinsic amine–imine tautomerism related to the 2-amino-thiazol moiety. However, the attempted acylation followed by the reduction of the amide to the alkylamine failed due to the relatively high lability of the acylated species. In the future, milder reducing will be explored, such as diborane [51,52], as the product should be stable under acidic workup conditions. If successful, the tautomerism-related limitation could be overcome. The synthetic strategy will facilitate future efforts involving tailored substitution patterns for applications such as bioimaging. In this sense, special attention will be devoted to the complex [**PtCl(L$_5$)**]. Hence, the secondary amine bridge can be modified to avoid harsh reaction conditions linked to cyclometallation.

Supplementary Materials: The following supporting information can be downloaded at: https://www.mdpi.com/article/10.3390/chemistry5020084/s1, Figure S1: ^{1}H-NMR spectrum (400 MHz, DCM-d$_2$) of A$_1$. Figure S2: ^{13}C{^{1}H}-NMR spectrum (101 MHz, DCM-d$_2$) of A$_1$. Figure S3: ^{1}H/^{1}H-COSY-NMR spectrum (400 MHz/400 MHz, DCM-d$_2$) of A$_1$. Figure S4: ^{1}H/^{13}C-gHSQC-NMR spectrum (400

MHz/101 MHz, DCM-d$_2$) of A$_1$. Figure S5: ^{1}H/^{13}C-gHMBC-NMR spectrum (400 MHz/101 MHz, DCM-d$_2$) of A$_1$. Figure S6: ^{1}H-NMR spectrum (500 MHz, DCM-d$_2$) of I$_1$. Figure S7: ^{13}C{^{1}H}-NMR spectrum (101 MHz, DCM-d$_2$) of I$_1$. Figure S8: ^{1}H/^{1}H-COSY-NMR spectrum (400 MHz/400 MHz, DCM-d$_2$) of I$_1$. Figure S9: ^{1}H/^{13}C-gHSQC-NMR spectrum (400 MHz/101 MHz, DCM-d$_2$) of I$_1$. Figure S10: ^{1}H/^{13}C-gHMBC-NMR spectrum (400 MHz/101 MHz, DCM-d$_2$) of I$_1$. Figure S11: ^{1}H-NMR spectrum (500 MHz, DCM-d$_2$) of A$_2$. Figure S12: ^{13}C{^{1}H}-NMR spectrum (126 MHz, DCM-d$_2$) of A$_2$. Figure S13: ^{1}H/^{1}H-COSY-NMR spectrum (500 MHz/500 MHz, DCM-d$_2$) of A$_2$. Figure S14: ^{1}H/^{13}C-gHSQC-NMR spectrum (500 MHz/126 MHz, DCM-d$_2$) of A$_2$. Figure S15: ^{1}H/^{13}C-gHMBC-NMR spectrum (500 MHz/126 MHz, DCM-d$_2$) of A$_2$. Figure S16: ^{1}H-NMR spectrum (400 MHz, DCM-d$_2$) of I$_2$. Figure S17: ^{13}C{^{1}H}-NMR spectrum (101 MHz, DCM-d$_2$) of I$_2$. Figure S18: ^{1}H/^{1}H-COSY-NMR spectrum (400 MHz/400 MHz, DCM-d$_2$) of I$_2$. Figure S19: ^{1}H/^{13}C-gHSQC-NMR spectrum (400 MHz/101 MHz, DCM-d$_2$) of I$_2$. Figure S20: ^{1}H/^{13}C-gHMBC-NMR spectrum (400 MHz/101 MHz, DCM-d$_2$) of I$_2$. Figure S21: ^{1}H-NMR spectrum (400 MHz, DCM-d$_2$) of A$_3$. Figure S22: ^{13}C{^{1}H}-NMR spectrum (101 MHz, DCM-d$_2$) of A$_3$. Figure S23: ^{1}H/^{1}H-COSY-NMR spectrum (400 MHz/400 MHz, DCM-d$_2$) of A$_3$. Figure S24: ^{1}H/^{13}C-gHSQC-NMR spectrum (400 MHz/101 MHz, DCM-d$_2$) of A$_3$. Figure S25: ^{1}H/^{13}C-gHMBC-NMR spectrum (400 MHz/101 MHz, DCM-d$_2$) of A$_3$. Figure S26: ^{1}H-NMR spectrum (400 MHz, DCM-d$_2$) of I$_3$. Figure S27: ^{13}C{^{1}H}-NMR spectrum (101 MHz, DCM-d$_2$) of I$_3$. Figure S28: H/^{1}H-COSY-NMR spectrum (400 MHz/400 MHz, DCM-d$_2$) of I$_3$. Figure S29: ^{1}H/^{13}C-gHSQC-NMR spectrum (400 MHz/101 MHz, DCM-d$_2$) of I$_3$. Figure S30: ^{1}H/^{13}C-gHMBC-NMR spectrum (400 MHz/101 MHz, DCM-d$_2$) of I$_3$. Figure S31: ^{1}H-NMR spectrum (400 MHz, DCM-d$_2$) of A$_4$. Figure S32: ^{13}C{^{1}H}-NMR spectrum (101 MHz, DCM-d$_2$) of A$_4$. Figure S33: ^{1}H/^{1}H-COSY-NMR spectrum (400 MHz/400 MHz, DCM-d$_2$) of A$_4$. Figure S34: ^{1}H/^{13}C-gHSQC-NMR spectrum (400 MHz/101 MHz, DCM-d$_2$) of A$_4$. Figure S35: ^{1}H/^{13}C-gHMBC-NMR spectrum (400 MHz/101 MHz, DCM-d$_2$) of A$_4$. Figure S36: ^{1}H-NMR spectrum (500 MHz, DCM-d$_2$) of L$_1$H. Figure S37: ^{13}C{^{1}H}-NMR spectrum (126 MHz, DCM-d$_2$) of L$_1$H. Figure S38: ^{1}H/^{1}H-COSY-NMR spectrum (500 MHz/500 MHz, DCM-d$_2$) of L$_1$H. Figure S39: ^{1}H/^{13}C-gHSQC-NMR spectrum (500 MHz/126 MHz, DCM-d$_2$) of L$_1$H. Figure S40: ^{1}H/^{13}C-gHMBC-NMR spectrum (500 MHz/126 MHz, DCM-d$_2$) of L$_1$H. Figure S41: ^{1}H-NMR spectrum (500 MHz, DCM-d$_2$) of L$_2$H. Figure S42: ^{13}C{^{1}H}-NMR spectrum (101 MHz, DCM-d$_2$) of L$_2$H. Figure S43: ^{1}H/^{1}H-COSY-NMR spectrum (400 MHz/400 MHz, DCM-d$_2$) of L$_2$H. Figure S44: ^{1}H/^{13}C-gHSQC-NMR spectrum (400 MHz/101 MHz, DCM-d$_2$) of L$_2$H. Figure S45: ^{1}H/^{13}C-gHMBC-NMR spectrum (400 MHz/101 MHz, DCM-d$_2$) of L$_2$H. Figure S46: ^{1}H-NMR spectrum (400 MHz, DCM-d$_2$) of L$_3$H. Figure S47: ^{13}C{^{1}H}-NMR spectrum (101 MHz, DCM-d$_2$) of L$_3$H. Figure S48: ^{1}H/^{1}H-COSY-NMR spectrum (400 MHz/400 MHz, DCM-d$_2$) of L$_3$H. Figure S49: ^{1}H/^{13}C-gHSQC-NMR spectrum (400 MHz/101 MHz, DCM-d$_2$) of L$_3$H. Figure S50: ^{1}H/^{13}C-gHMBC-NMR spectrum (400 MHz/101 MHz, DCM-d$_2$) of L$_3$H. Figure S51: ^{1}H-NMR spectrum (500 MHz, CDCl$_3$) of L$_4$H. Figure S52: ^{13}C{^{1}H}-NMR spectrum (126 MHz, CDCl$_3$) of L$_4$H. Figure S53: ^{1}H/^{1}H-COSY-NMR spectrum (500 MHz/500 MHz, CDCl$_3$) of L$_4$H. Figure S54: ^{1}H/^{13}C-gHSQC-NMR spectrum (500 MHz/126 MHz, CDCl$_3$) of L$_4$H. Figure S55: ^{1}H/^{13}C-gHMBC-NMR spectrum (500 MHz/126 MHz, CDCl$_3$) of L$_4$H. Figure S56: ^{1}H-NMR spectrum (500 MHz, DMSO-d$_6$) of [PtCl(L$_1$)]. Figure S57: ^{13}C{^{1}H}-NMR spectrum (126 MHz, DMSO-d$_6$) of [PtCl(L$_1$)]. Figure S58: ^{195}Pt-NMR spectrum (107 MHz, DMSO-d$_6$) of [PtCl(L$_1$)]. Figure S59: ^{1}H/^{1}H-COSY-NMR spectrum (500 MHz/500 MHz, DMSO-d$_6$) of [PtCl(L$_1$)]. Figure S60: ^{1}H/^{13}C-gHSQC-NMR spectrum (500 MHz/126 MHz, DMSO-d$_6$) of [PtCl(L$_1$)]. Figure S61: ^{1}H/^{13}C-gHMBC-NMR spectrum (500 MHz/126 MHz, DMSO-d$_6$) of [PtCl(L$_1$)]. Figure S62: ^{1}H-NMR spectrum (400 MHz, DCM-d$_2$/MeOD-d$_4$) of [PtCl(L$_2$)]. Figure S63: ^{13}C{^{1}H}-NMR spectrum (101 MHz, DCM-d$_2$/MeOD-d$_4$) of [PtCl(L$_2$)]. Figure S64: ^{195}Pt-NMR spectrum (107 MHz, DCM-d$_2$/MeOD-d$_4$) of [PtCl(L$_2$)]. Figure S65: ^{1}H/^{1}H-COSY-NMR spectrum (400 MHz/400 MHz, DCM-d$_2$/MeOD-d$_4$) of [PtCl(L$_2$)]. Figure S66: ^{1}H/^{13}C-gHSQC-NMR spectrum (400 MHz/101 MHz, DCM-d$_2$/MeOD-d$_4$) of [PtCl(L$_2$)]. Figure S67: ^{1}H/^{13}C-gHMBC-NMR spectrum (400 MHz/101 MHz, DCM-d$_2$/MeOD-d$_4$) of [PtCl(L$_2$)]. Figure S68: ^{1}H-NMR spectrum (500 MHz, DCM-d$_2$) of [PtCl(L$_3$)]. Figure S69: ^{13}C{^{1}H}-NMR spectrum (101 MHz, DCM-d$_2$) of [PtCl(L$_3$)]. Figure S70: ^{195}Pt-NMR spectrum (107 MHz, DCM-d$_2$) of [PtCl(L$_3$)]. Figure S71: ^{1}H/^{1}H-COSY-NMR spectrum (400 MHz/400 MHz, DCM-d$_2$) of [PtCl(L$_3$)]. Figure S72: ^{1}H/^{13}C-gHSQC-NMR spectrum (400 MHz/101 MHz, DCM-d$_2$) of [PtCl(L$_3$)]. Figure S73: ^{1}H/^{13}C-gHMBC-NMR spectrum (400 MHz/101 MHz, DCM-d$_2$) of [PtCl(L$_3$)]. Figure S74: ^{1}H-NMR spectrum (400 MHz, DCM-d$_2$) of [PtCl(L$_4$)]. Figure S75: ^{13}C{^{1}H}-NMR spectrum (101 MHz, DCM-d$_2$) of [PtCl(L$_4$)]. Figure S76: ^{195}Pt-NMR spectrum (86 MHz, DCM-d$_2$) of [PtCl(L$_4$)]. Figure S77: ^{1}H/^{1}H-COSY-NMR spectrum (400 MHz/400 MHz, DCM-d$_2$) of [PtCl(L$_4$)]. Figure S78: ^{1}H/^{13}C-gHSQC-NMR spectrum (400 MHz/101 MHz, DCM-d$_2$) of [PtCl(L$_4$)]. Figure S79: ^{1}H/^{13}C-gHMBC-NMR spectrum (400 MHz/101 MHz, DCM-d$_2$) of [PtCl(L$_4$)]. Figure S80: ^{1}H-NMR spectrum (400 MHz, DMSO-d$_6$) of [PtCl(L$_5$)]. Figure S81: ^{13}C{^{1}H}-NMR spectrum (101 MHz, DMSO-d$_6$) of [PtCl(L$_5$)]. Figure S82: ^{195}Pt-NMR spectrum (86 MHz, DMSO-d$_6$) of [PtCl(L$_5$)]. Figure S83: ^{1}H/^{1}H-

COSY-NMR spectrum (400 MHz/400 MHz, DMSO-d_6) of [PtCl(L_5)]. Figure S84: ^{1}H/^{13}C-gHSQC-NMR spectrum (400 MHz/101 MHz, DMSO-d_6) of [PtCl(L_5)]. Figure S85: ^{1}H/^{13}C-gHMBC-NMR spectrum (400 MHz/101 MHz, DMSO-d_6) of [PtCl(L_5)]. Figure S86: ^{1}H-NMR spectrum (400 MHz, DCM-d_2) of A_6. Figure S87: ^{13}C{^{1}H}-NMR spectrum (101 MHz, DCM-d_2) of A_6. Figure S88: ^{1}H/^{1}H-COSY-NMR spectrum (400 MHz/400 MHz, DCM-d_2) of A_6. Figure S89: ^{1}H/^{13}C-gHSQC-NMR spectrum (400 MHz/101 MHz, DCM-d_2) of A_6. Figure S90: ^{1}H/^{13}C-gHMBC-NMR spectrum (400 MHz/101 MHz, DCM-d_2) of A_6. Figure S91: ^{1}H-NMR spectrum (500 MHz, CDCl$_3$) of L_6H_2. Figure S92: ^{13}C{^{1}H}-NMR spectrum (126 MHz, CDCl$_3$) of L_6H_2. Figure S93: ^{1}H/^{1}H-COSY-NMR spectrum (500 MHz/500 MHz, CDCl$_3$) of L_6H_2. Figure S94: ^{1}H/^{13}C-gHSQC-NMR spectrum (500 MHz/126 MHz, CDCl$_3$) of L_6H_2. Figure S95: ^{1}H/^{13}C-gHMBC-NMR spectrum (500 MHz/126 MHz, CDCl$_3$) of L_6H_2. Figure S96: ^{1}H-NMR spectrum (400 MHz, DCM-d_2) of L_6BrH. Figure S97: ^{13}C{^{1}H}-NMR spectrum (101 MHz, DCM-d_2) of L_6BrH. Figure S98: ^{1}H/^{1}H-COSY-NMR spectrum (400 MHz/400 MHz, DCM-d_2) of L_6BrH. Figure S99: ^{1}H/^{13}C-gHSQC-NMR spectrum (400 MHz/101 MHz, DCM-d_2) of L_6BrH. Figure S100: ^{1}H/^{13}C-gHMBC-NMR spectrum (400 MHz/101 MHz, DCM-d_2) of L_6BrH. Figure S101: ^{1}H-NMR spectrum (500 MHz, DCM-d_2) of [PtL_6]. Figure S102: ^{13}C{^{1}H}-NMR spectrum (101 MHz, DCM-d_2) of [PtL_6]. Figure S103: ^{195}Pt-NMR spectrum (86 MHz, DCM-d_2) of [PtL_6]. Figure S104: ^{1}H/^{1}H-COSY-NMR spectrum (400 MHz/400 MHz, DCM-d_2) of [PtL_6]. Figure S105: ^{1}H/^{13}C-gHSQC-NMR spectrum (400 MHz/101 MHz, DCM-d_2) of [PtL_6]. Figure S106: ^{1}H/^{13}C-gHMBC-NMR spectrum (400 MHz/101 MHz, DCM-d_2) of [PtL_6]. Figure S107: Molecular structure of compound I_1 in a single crystal and display of the packing. Figure S108: Molecular structure of compound A_3 in a single crystal and display of the packing. Figure S109: Molecular structure of compound A_4 in a single crystal and display of the packing. Figure S110: Molecular structure of compound [PtCl(L_2)] in a single crystal and display of the packing. Figure S111: Molecular structures of compound [PtCl(L_4)] in a single crystal and the display of the distortion of the coordination plane for all three different molecules. Figure S112: Asymmetric cell of the crystal structure of [PtCl(L_4)]. Figure S113: Display of the crystal packing of [PtCl(L_4)]. Figure S114: Molecular structure of compound [Pt(L_6)] in a single crystal and display of the packing. Figure S115: Molar absorption coefficient of [PtCl(L_1)], [PtCl(L_2)], [PtCl(L_3)], [PtCl(L_4)], [PtCl(L_5)] and [Pt(L_6)]. Figure S116: Excitation and emission spectra of [PtCl(L_1)] at 298 K in fluid DCM and at 77 K in a frozen glassy DCM/MeOH matrix. Figure S117: Left: Raw (experimental) time-resolved photoluminescence decay of [PtCl(L_1)] in fluid DCM at 298 K (air-equilibrated), including the residuals. Figure S118: Left: Raw (experimental) time-resolved photoluminescence decay of [PtCl(L_1)] in fluid DCM at 298 K (Ar-purged), including the residuals. Figure S119: Left: Raw (experimental) time-resolved photoluminescence decay of [PtCl(L_1)] in a frozen glassy DCM/MeOH matrix at 77 K, including the residuals. Figure S120: Excitation and emission spectra of [PtCl(L_2)] at 298 K in fluid DCM as a solid and at 77 K in a frozen glassy DCM/MeOH matrix. Figure S121: Left: Raw (experimental) time-resolved photoluminescence decay of [PtCl(L_2)] in fluid DCM at 298 K (air-equilibrated), including the residuals. Figure S122: Left: Raw (experimental) time-resolved photoluminescence decay of [PtCl(L_2)] in fluid DCM at 298 K (Ar-purged), including the residuals. Figure S123: Left: Raw (experimental) time-resolved photoluminescence decay of [PtCl(L_2)] in a frozen glassy DCM/MeOH matrix at 77 K, including the residuals. Figure S124: Excitation and emission spectra of [PtCl(L_3)] at 298 K in fluid DCM, at 77 K in a frozen glassy DCM/MeOH matrix and as a solid. Figure S125: Left: Raw (experimental) time-resolved photoluminescence decay of [PtCl(L_3)] in fluid DCM at 298 K (air-equilibrated), including the residuals and IRF. Figure S126: Left: Raw (experimental) time-resolved photoluminescence decay of [PtCl(L_3)] in fluid DCM at 298 K (Ar-purged), including the residuals and IRF. Figure S127: Left: Raw (experimental) time-resolved photoluminescence decay of [PtCl(L_3)] in a frozen glassy DCM/MeOH matrix at 77 K, including the residuals. Figure S128: Excitation and emission spectra of [PtCl(L_4)] at 298 K in fluid DCM and at 77 K in a frozen glassy DCM/MeOH matrix. Figure S129: Left: Raw (experimental) time-resolved photoluminescence decay of [PtCl(L_4)] in a frozen glassy DCM/MeOH matrix at 77 K, including the residuals. Figure S130: Excitation and emission spectra of [PtCl(L_5)] at 298 K in fluid DCM, at 77 K in a frozen glassy DCM/MeOH matrix and as a solid. Figure S131: Left: Raw (experimental) time-resolved photoluminescence decay of [PtCl(L_5)] in fluid DCM at 298 K (air-equilibrated), including the residuals. Figure S132: Left: Raw (experimental) time-resolved photoluminescence decay of [PtCl(L_5)] in fluid DCM at 298 K (Ar-purged), including the residuals. Figure S133: Left: Raw (experimental) time-resolved photoluminescence decay of [PtCl(L_5)] in a frozen glassy DCM/MeOH matrix at 77 K, including the residuals. Figure S134: Excitation and emission spectra of [Pt(L_6)] at 298 K in fluid DCM, at 77 K in a frozen glassy DCM/MeOH matrix and as a solid. Figure S135: Left: Raw (experimental) time-resolved photoluminescence decay of [Pt(L_6)] in fluid DCM at 298 K (air-equilibrated), including the residuals. Figure S136: Left: Raw (experimental) time-resolved photoluminescence decay of [Pt(L_6)] in fluid DCM at 298 K (Ar-purged), including the residuals. Figure S137: Left: Raw (experimental) time-resolved photoluminescence decay of [Pt(L_6)] in a frozen glassy DCM/MeOH matrix at 77 K, including the residuals. Table S1: Parameters and data from the single crystal measurements. Table S2: Selected bond lengths and angles for [PtCl(L$_2$)]. Table S3: Selected bond lengths and angles for [PtCl(L$_4$)]. Table S4: Selected bond lengths and angles for [Pt(L$_6$)]. Table S5: Complete emission data and Φ_L, as well as exited state lifetime data for each complex in DCM at 298 K and in frozen glassy matrix of DCM/MeOH at 77 K. References [41,47] are cited in the Supplementary Materials.

Author Contributions: Conceptualization, S.B. and C.A.S.; validation, M.V.C., S.B. and C.A.S.; formal analysis, S.B.; NMR-measurements, A.H.; NMR-structure analysis, A.H. and S.B., X-ray diffractometry and structure solution, J.K.; photophysical investigation, M.V.C. and S.B.; resources, C.A.S.; data curation, S.B. and M.V.C.; writing—original draft preparation, S.B.; writing—review and editing, M.V.C. and C.A.S.; visualization, S.B.; supervision, C.A.S.; project administration, C.A.S.; funding acquisition, C.A.S. All authors have read and agreed to the published version of the manuscript.

Funding: C.A.S. gratefully acknowledges funding from the Deutsche Forschungsgemeinschaft (DFG, German Research Foundation)—Project-ID 433682494—SFB 1459, as well as Project STR 1186/6-2 within the Priority Programm 2102 "Light-controlled reactivity of metal complexes". C.A.S. gratefully acknowledges the generous financial support for the acquisition of an "Integrated Confocal Luminescence Spectrometer with Spatiotemporal Resolution and Multiphoton Excitation" (DFG/Land NRW: INST 211/915-1 FUGG; DFG EXC1003: "Berufungsmittel").

Data Availability Statement: CCDC 2252944, 2252945, 2252950, 2252953, 2252954, 2252967 contain the supplementary crystallographic data for this paper. These data can be obtained free of charge via www.ccdc.cam.ac.uk/data_request/cif, or by emailing data_request@ccdc.cam.ac.uk, or by contacting The Cambridge Crystallographic Data Center, 12 Union Road, Cambridge CB2 1EZ, UK; fax: +44 1223 226033.

Acknowledgments: S.B. thanks Christiane Terlinde for the help in the synthesis of the [**Pt(L$_6$)**] complex.

Conflicts of Interest: The authors declare no conflict of interest.

References

1. Wunschel, K.R.; Ohnesorge, W.E. Luminescence of Iridium(II) Chelates with 2,2′-bipyridine and with 1,10-phenanthroline. *J. Am. Chem. Soc.* **1967**, *89*, 2777–2778. [CrossRef]
2. Baldo, M.A.; Lamansky, S.; Thompson, M.E.; Forrest, S.R. Very High-Efficiency Green Organic Light-Emitting Devices Based on Electrophosphorescence. *Appl. Phys. Lett.* **1999**, *75*, 4–6. [CrossRef]
3. Cheung, T.-C.; Cheung, K.-K.; Peng, S.-M.; Che, C.-M. Photoluminescent Cyclometallated Diplatinum(II,II) Complexes: Photophysical Properties and Crystal Structures of [PtL(PPh3)]ClO4 and [Pt2L2(μ-dppm)][ClO4]2(HL = 6-phenyl-2,2′-bipyridine, dppm = Ph2PCH2PPh2). *J. Chem. Soc., Dalton Trans.* **1996**, 1645–1651. [CrossRef]
4. Wong, Y.S.; Tang, M.C.; Ng, M.; Yam, V.W.W. Toward the Design of Phosphorescent Emitters of Cyclometalated Earth Abundant Nickel(II) and Their Supramolecular Study. *J. Am. Chem. Soc.* **2020**, *142*, 7638–7646. [CrossRef] [PubMed]
5. Williams, J.A.G.; Beeby, A.; Davies, E.S.; Weinstein, J.A.; Wilson, C. An Alternative Rout to Highly Luminescent Platinum(II) Complexes: Cyclometalation with N^C^N-Coordinating Dipyridylbenzene Ligands. *Inorg. Chem.* **2003**, *42*, 8609–8611. [CrossRef]
6. Otto, S.; Grabolle, M.; Förster, C.; Kreitner, C.; Resch-Genger, U.; Heinze, U. [Cr(ddpd)2]3+: A Molecular, Water-Soluble, Highly NIR-Emissive Ruby Analogue. *Angew. Chem. Int. Ed.* **2015**, *54*, 11572–11576. [CrossRef]
7. Chow, P.-K.; Cheng, G.; Tong, G.S.M.; To, W.-P.; Kwong, W.-L.; Low, K.-H.; Kwok, C.-C.; Ma, C.; Che, C.-M. Luminescent Pincer Platinum(II) Complexes with Emission Quantum Yields up to Almost Unity: Photophysics, Photoreductive C-C Bond Formation, and Materials Applications. *Angew. Chem. Int. Ed.* **2015**, *54*, 2084–2089. [CrossRef]
8. Sanning, J.; Ewen, P.; Stegemann, L.; Schmidt, J.; Daniliuc, C.G.; Koch, T.; Doltsinis, N.L.; Wegner, D.; Strassert, C.A. Scanning-Tunneling-Spectroscopy-Directed Design of Tailored Deep-Blue Emitters. *Angew. Chem. Int. Ed.* **2015**, *54*, 786–791. [CrossRef]
9. Rossi, E.; Colombo, A.; Dragonetti, C.; Roberto, D.; Ugo, R.; Valore, A.; Falciola, L.; Brulatti, P.; Cocchi, M.; Williams, J.A.G. Novel N^C^N-Cyclometallated Platinum Complexes with Acetylide Co-Ligands as Efficient Phosphors for OLEDs. *J. Mater. Chem.* **2012**, *22*, 10650–10655. [CrossRef]
10. Kayano, T.; Takayasu, S.; Sato, K.; Shinozaki, K. Luminescence Color Tuning of PtII Complexes and a Kinetic Study of Trimer Formation in the Photoexcited State. *Chem. Eur. J.* **2014**, *20*, 16583–16589. [CrossRef]
11. Cebrián, C.; Mauro, M. Recent Advances in Phosphorescent Platinum Complexes for Organic Light-Emitting Diodes, Beilstein. *J. Org. Chem.* **2018**, *14*, 1459–1481.
12. Cheng, G.; Kwak, Y.; To, W.P.; Lam, T.L.; Tong, G.S.M.; Sit, M.K.; Gong, S.; Choi, B.; Choi, W.I.; Yang, C.; et al. High-Efficiency Solution-Processed Organic Light-Emitting Diodes with Tetradentate Platinum(II) Emitters. *ACS Appl. Mater. Interfaces* **2019**, *11*, 45161–45170. [CrossRef] [PubMed]
13. Zhang, Y.; Wang, Y.; Song, J.; Qu, J.; Li, B.; Zhu, W.; Wong, W.-Y. Near-Infrared Emitting Materials via Harvesting Triplet Excitons: Molecular Design, Properties, and Application in Organic Light Emitting Diodes. *Adv. Opt. Mater.* **2018**, *6*, 1800466. [CrossRef]

14. Sanning, J.; Stegemann, L.; Ewen, P.R.; Schwermann, C.; Daniliuc, C.G.; Zhang, D.; Lin, N.; Duan, L.; Wegner, D.; Doltsinis, N.L.; et al. Colour-Tunable Asymmetric Cyclometalated Pt(II) Complexes and STM-Assisted Stability Assessment of Ancillary Ligands for OLEDs. *J. Mater. Chem. C* **2016**, *4*, 2560–2565. [CrossRef]
15. Koshevoy, I.O.; Krause, M.; Klein, A. Non-Covalent Intramolecular Interactions through Ligand-Design Promoting Efficient Luminescence from Transition Metal Complexes. *Coord. Chem. Rev.* **2020**, *405*, 213094. [CrossRef]
16. Ravotto, L.; Ceroni, P. Aggregation induced phosphorescence of metal complexes: From principles to applications. *Coord. Chem. Rev.* **2017**, *346*, 62–76. [CrossRef]
17. Nisic, F.; Colombo, A.; Dragonetti, C.; Roberto, D.; Valore, A.; Malicka, J.M.; Cocchi, M.; Freeman, G.R.; Williams, J.A.G. Platinum(II) Complexes with Cyclometallated 5-π-Delocalized-Donor-1,3-di(2-pyridyl)benzene Ligands as Efficient Phosphors for NIROLEDs. *J. Mater. Chem. C* **2014**, *2*, 1791–1800. [CrossRef]
18. Tam, A.Y.-Y.; Tsang, D.P.-K.; Chan, M.-Y.; Zhu, N.; Yam, V.W.-W. A Luminescent Cyclometalated Platinum(II) Complex and its Green Organic Light Emitting Device with High Device Performance. *Chem. Commun.* **2011**, *47*, 3383–3385. [CrossRef]
19. Lu, W.; Mi, B.-X.; Chan, M.C.W.; Hui, Z.; Zhu, N.; Lee, S.-T.; Che, C.-M. [(C^N^N)Pt(C≡C)nR] (HC^N^N = 6-aryl-2,2′-bipyridine, n = 1-4, R = aryl, SiMe3) as a New Class of Light Emitting Materials and their Applications in Electrophosphorescent Devices. *Chem. Commun.* **2002**, 206–207. [CrossRef]
20. Mao, M.; Peng, J.; Lam, T.-L.; Ang, W.-H.; Li, H.; Cheng, G.; Che, C.-M. High-performance organic light-emitting diodes with low-efficiency roll-off using bulky tetradentate [Pt(O^N^C^N)] emitters. *J. Mater. Chem. C* **2019**, *7*, 7230–7236. [CrossRef]
21. Kalinowski, J.; Fattori, V.; Cocchi, M.; Williams, J.A.G. Light-emitting devices based on organometallic platinum complexes as emitters. *Coord. Chem. Rev.* **2011**, *255*, 2401–2425. [CrossRef]
22. Yersin, H.; Rausch, A.F.; Czerwieniec, R.; Hofbeck, T.; Fischer, T. The triplet state of organo-transition metal compounds. Triplet harvesting and singlet harvesting for efficient OLEDs. *Coord. Chem. Rev.* **2011**, *255*, 2622–2652. [CrossRef]
23. Zhong, J.-J.; Meng, Q.-Y.; Wang, G.-X.; Liu, Q.; Chen, B.; Feng, K.; Tung, C.-H.; Wu, L.-Z. A Highly Efficient and Selective Aerobic Cross-Dehydrogenative-Coupling Reaction Photocatalyzed by a Platinum(II) Terpyridyl Complex. *Chem. Eur. J.* **2013**, *19*, 6443–6450. [CrossRef] [PubMed]
24. Mori, K.; Yamashita, H. Metal Complexes Supported on Solid Matrices for Visible-Light-Driven Molecular Transformations. *Chem. Eur. J.* **2016**, *22*, 11122–11137. [CrossRef] [PubMed]
25. Parasram, M.; Gevorgyan, V. Visible light-induced transition metal-catalyzed transformations: Beyond conventional photosensitizers. *Chem. Soc. Rev.* **2017**, *46*, 6227–6240. [CrossRef] [PubMed]
26. Choi, W.J.; Choi, S.; Ohkubo, K.; Fukuzumi, S.; Cho, E.J.; You, Y. Mechanisms and applications of cyclometalated Pt(II) complexes in photoredox catalytic trifluoromethylation. *Chem. Sci.* **2015**, *6*, 1454–1464. [CrossRef]
27. Wu, P.; Wong, E.L.-M.; Ma, D.-L.; Tong, G.S.-M.; Ng, K.-M.; Che, C.-M. Cyclometalated Platinum(II) Complexes as Highly Sensitive Luminescent Switch-On Probes for Practical Application in Protein Staining and Cell Imaging. *Chem. Eur. J.* **2009**, *15*, 3652–3656. [CrossRef]
28. Chung, C.Y.-S.; Li, S.P.-Y.; Louie, M.-W.; Lo, K.K.-W.; Yam, V.W.-W. Induced Self-Assembly and Disassembly of Water-Soluble Alkynylplatinum(II) Terpyridyl Complexes with "Switchable" NearInfrared (NIR) Emission Modulated by Metal-Metal Interaction over Physiological pH: Demonstration of pH-Responsive NIR Luminescent Probes in Cell-Imaging Studies. *Chem. Sci.* **2013**, *4*, 2453–2462. [CrossRef]
29. Baggaley, E.; Botchway, S.W.; Haycock, J.W.; Morris, H.; Sazanovich, I.V.; Williams, J.A.G.; Weinstein, J.A. Long-Lived Metal Complexes open up Microsecond Lifetime Imaging Microscopy under Multiphoton Excitation: From FLIM to PLIM and beyond. *Chem. Sci.* **2014**, *5*, 879–886. [CrossRef]
30. Zhang, K.Y.; Yu, Q.; Wei, H.; Liu, S.; Zhao, Q.; Huang, W. Long-Lived Emissive Probes for Time-Resolved Photoluminescence Bioimaging and Biosensing. *Chem. Rev.* **2018**, *118*, 1770–1839. [CrossRef]
31. Guerchais, V.; Fillaut, J.-L. Sensory Luminescent Iridium(III) and Platinum(II) Complexes for Cation Recognition. *Coord. Chem. Rev.* **2011**, *255*, 2448–2457. [CrossRef]
32. Ma, D.-L.; Ma, V.P.-Y.; Chan, D.S.-H.; Leung, K.-H.; He, H.-Z.; Leung, C.-H. Recent Advances in Luminescent Heavy Metal Complexes for Sensing. *Coord. Chem. Rev.* **2012**, *256*, 3087–3113. [CrossRef]
33. Liu, L.; Wang, X.; Hussain, F.; Zeng, C.; Wang, B.; Li, Z.; Kozin, I.; Wang, S. Multiresponsive Tetradentate Phosphorescent Metal Complexes as Highly Sensitive and Robust Luminescent Oxygen Sensors: Pd(II) Versus Pt(II) and 1,2,3-Triazolyl Versus 1,2,4-Triazolyl. *ACS Appl. Mater. Interfaces* **2019**, *11*, 12666–12674. [CrossRef] [PubMed]
34. Li, K.; Tong, G.S.M.; Wan, Q.; Cheng, G.; Tong, W.-Y.; Ang, W.-H.; Kwong, W.-L.; Che, C.-M. Highly Phosphorescent Platinum(II) Emitters: Photophysics, Materials and Biological Application. *Chem. Sci.* **2016**, *7*, 1653–1673. [CrossRef]
35. Williams, J.A.G. Photochemistry and Photophysics of Coordination Compounds II. *Top. Curr. Chem.* **2007**, *281*, 205–268.
36. Wu, W.; Huang, D.; Zhao, J. Tridentate Cyclometalated Platinum(II) Complexes with Strong Absorption of Visible Light and Long-Lived Triplet Excited States as Photosensitizers for Triplet–Triplet Annihilation Upconversion. *Dyes Pigm.* **2013**, *96*, 220–231. [CrossRef]
37. Ravindranathan, D.; Vezzu, D.A.K.; Bartolotti, L.; Boyle, P.D.; Hou, S. Improvement in Phosphorescence Efficiency through Tuning of Coordination Geometry of Tridentate Cyclometalated Platinum(II) Complexes. *Inorg. Chem.* **2010**, *49*, 8922–8928. [CrossRef]

38. Vezzu, D.A.K.; Deaton, J.C.; Jones, J.S.; Bartolotti, L.; Harris, C.F.; Marchetti, A.P.; Kondakova, M.; Pike, R.D.; Huo, S. Highly Luminescent Tetradentate Bis-Cyclometalated Platinum Complexes: Design, Synthesis, Structure, Photophysics, and Electroluminescence Application. *Inorg. Chem.* **2010**, *49*, 5107–5119. [CrossRef]
39. Wilde, S.; Ma, D.; Koch, T.; Bakker, A.; Gonzalez-Abradelo, D.; Stegemann, L.; Daniliuc, C.G.; Fuchs, H.; Gao, H.; Doltsinis, N.L.; et al. Toward Tunable Electroluminescent Devices by Correlating Function and Submolecular Structure in 3D Crystals, 2D-Confined Monolayers, and Dimers. *ACS Appl. Mater. Interfaces* **2018**, *10*, 22460–22473. [CrossRef]
40. Ren, J.; Cnudde, M.; Brünink, D.; Buss, S.; Daniliuc, C.G.; Liu, L.; Fuchs, H.; Strassert, C.A.; Gao, H.-Y.; Doltsinis, N.L. On-Surface Reactive Planarization of Pt(II) Complexes. *Angew. Chem. Int. Ed.* **2019**, *58*, 15396–15400. [CrossRef]
41. Knedel, T.-O.; Buss, S.; Maisuls, I.; Daniliuc, C.G.; Schlüsener, C.; Brandt, P.; Weingart, O.; Vollrath, A.; Janiak, C.; Strassert, C.A. Encapsulation of Phosphorescent Pt(II) Complexes in Zn-Based Metal–Organic Frameworks toward Oxygen-Sensing Porous Materials. *Inorg. Chem.* **2020**, *59*, 7252–7264. [CrossRef] [PubMed]
42. Macrae, C.F.; Sovago, I.; Cottrell, S.J.; Galek, P.T.A.; McCabe, P.; Pidcock, E.; Platings, M.; Shields, G.P.; Stevens, J.S.; Towler, M.; et al. Mercury 4.0: From Visualization to Analysis, Design and Prediction. *J. Appl. Cryst.* **2020**, *53*, 226–235. [CrossRef] [PubMed]
43. Schnürch, M.; Waldner, B.; Hilber, K.; Mihovilovic, M.D. Synthesis of 5-arylated N-arylthiazole-2-amines as Potential Skeletal Muscle Cell Differentiation Promoters. *Bioorg. Med. Chem. Lett.* **2011**, *21*, 2149–2154. [CrossRef] [PubMed]
44. Dao-Huy, T.; Waldner, B.; Wimmer, L.; Schnürch, M.; Mihovilovic, M.D. Synthesis of endo- and exo-N-Protected 5-Arylated 2-Aminothiazols through Direct Arylation: An Efficient Route to Cell Differentiation Accelerators. *Eur. J. Org. Chem.* **2015**, 4765–4771. [CrossRef]
45. Sintenis, F. Beiträge zur Kenntniss der Benzyläther. *Liebigs Ann. Chem.* **1872**, *161*, 329–346. [CrossRef]
46. Lai, S.-W.; Cheung, T.-C.; Chan, M.C.W.; Cheung, K.-K.; Peng, S.-M.; Che, C.-M. Luminescent Mononuclear and Binuclear Cyclometalated Palladium(II) Complexes of 6-Phenyl-2,2′-Bipyridines: Spectroscopic and Structural Comparisons with Platinum(II) Analogues 1,2. *Inorg. Chem.* **2000**, *39*, 255–262. [CrossRef]
47. Zhang, E.-X.; Wang, D.-X.; Huang, Z.-T.; Wang, M.-X. Synthesis of (NH)m(NMe)4−m-Bridged Calix [4]pyridines and the Effect of NH Bridge on Structure and Properties. *J. Org. Chem.* **2009**, *74*, 8595–8603. [CrossRef]
48. Solomatina, A.I.; Galenko, E.E.; Kozina, D.O.; Kalinichev, A.A.; Baigildin, V.A.; Prudovskaya, N.A.; Shakirova, J.R.; Khlebnikov, A.F.; Prosev, V.V.; Evarestov, R.A.; et al. Nonsymmetric [Pt(C^N*N'^C')] Complexes: Aggregation-Induced Emission in the Solid State and in Nanoparticles Tuned by Ligand Structure. *Chem. Eur. J.* **2022**, *28*, e202202207. [CrossRef]
49. Sillen, A.; Engelborghs, Y. The Correct Use of "Average" Fluorescence Parameters. *Photochem. Photobiol.* **1998**, *67*, 475–486. [CrossRef]
50. Wilde, S.; Mittelberg, L.; Daniliuc, C.G.; Koch, T.; Doltsinis, N.L.; Strassert, C.A. Studie über den Einfluss des Fluorierungsgrades an einem tetradentaten C^N*N^C-Luminophor auf die photophysikalischen Eigenschaften seiner Platin(II)-Komplexe und deren Aggregation. *Z. Naturforsch.* **2018**, *73*, 849–863. [CrossRef]
51. Strassert, C.A.; Dicelio, L.E.; Awruch, J. Reduction of an Amido Zinc(II) Phthalocyanine by Diborane. *Synthesis* **2006**, *5*, 799–802. [CrossRef]
52. Strassert, C.A.; Awruch, J. Conversion of Phthalimides to Isoinddolines by Diborane. *Monatsh. Chem.* **2006**, *137*, 1499–1503. [CrossRef]

Disclaimer/Publisher's Note: The statements, opinions and data contained in all publications are solely those of the individual author(s) and contributor(s) and not of MDPI and/or the editor(s). MDPI and/or the editor(s) disclaim responsibility for any injury to people or property resulting from any ideas, methods, instructions or products referred to in the content.

Article

Improved Synthesis and Coordination Behavior of 1H-1,2,3-Triazole-4,5-dithiolates (tazdt^{2-}) with Ni^{II}, Pd^{II}, Pt^{II} and Co^{III}

Nils Pardemann [1], Alexander Villinger [1] and Wolfram W. Seidel [1,2,*]

1 Institut für Chemie, Universität Rostock, Albert-Einstein-Straße 3a, 18059 Rostock, Germany
2 Leibniz-Institut für Katalyse e.V., Albert-Einstein-Straße 29a, 18059 Rostock, Germany
* Correspondence: wolfram.seidel@uni-rostock.de

Abstract: A new synthetic route to 1H-1,2,3-triazole-4,5-dithiols (tazdtH$_2$) as ligands for the coordination of Ni^{II}, Pd^{II}, Pt^{II} and Co^{III} via the dithiolate unit is presented. Different N-protective groups were introduced with the corresponding azide via a click-like copper-catalyzed azide-alkyne [3 + 2] cycloaddition (CuAAC) and fully characterized by NMR spectroscopy. Possible isomers were isolated and an alternative synthetic route was investigated and discussed. After removal of the benzyl protective groups on sulfur by in situ-generated sodium naphthalide, complexes at the [(dppe)M] (M = Ni, Pd, Pt), [(PPh$_3$)$_2$Pt] and [(η^5-C$_5$H$_5$)Co] moieties were prepared and structurally characterized by XRD analysis. In this process, the by-products **11** and **12** as monothiolate derivatives were isolated and structurally characterized as well. With regioselective coordination via the dithiolate unit, the electronic influence of different metals or protective groups at N was investigated and compared spectroscopically by means of UV/Vis spectroscopy and cyclic voltammetry. Complex [(η^5-C$_5$H$_5$)Co(**5c**)] (**10**), is subject to a dimerization equilibrium, which was investigated by temperature-dependent NMR and UV/Vis spectroscopy (solution and solid-state). The thermodynamic parameters of the monomer/dimer equilibrium were derived.

Keywords: dithiolene complex; 1,2,3-triazole ligands; click chemistry; CuAAC; thiol protective groups

1. Introduction

The award of the Nobel Prize to Sharpless, Meldal and Bertozzi in 2022 represents an accolade for click chemistry as a powerful synthetic method [1]. The concept of click chemistry was established as early as 2001 and describes a rapid and precise synthesis of molecules following the example of nature. The advantages of the method are high atomic efficiency, very few by-products and high yields while only the use of cheap and simple chemicals and short reaction time are needed [2]. Classically, click chemistry often includes Diels–Alder reactions, addition reactions on carbon–carbon double bonds, and especially copper-catalyzed Huisgen cycloaddition, which can be used for the synthesis of 1H-1,2,3-triazoles [3–5]. Sharpless and coworkers presented first protocols for the [3 + 2] cycloaddition of azides with terminal alkynes under Cu-catalyzed reaction conditions [5]. A [3 + 2] cycloaddition between azides and acetylenes are not regioselective [6–8]. Two regioisomeres with a substituent in 4- or 5-position are formed. Only in the case of electrophilically activated acetylenes is high regioselectivity possible [5,9,10]. The copper-catalyzed azide-alkyne [3 + 2] cycloaddition (CuAAC) opens a way for the regioselective synthesis of triazoles. In addition to various alkyl and aryl substituents, donors such as phosphanes, amines, sulfur and seleniums could be introduced into the 1H-1,2,3-triazole system as well [11–16]. Introduction of thiol groups at both 4- and 5-position of the triazole would result in a new ligand with five potential coordination sites in the form of the dithiolene unit and the N atoms. Both through the aromatic properties of the 1H-1,2,3-triazole ring and through the specific electronic situation of the dithiolene unit, the 1H-1,2,3-triazole-4,5-dithiolate (tazdt^{2-}) could serve as a versatile bridging ligand between several metal centers. In particular, the electronic properties appear potentially interesting due to the non-innocent character of the dithiolene unit [17–19]. In contrast to many other triazoles, a synthesis of 1H-1,2,3-triazole-4,5-dithiols by means of a click-like copper-catalyzed azide-alkyne [3 + 2] cycloaddition is not known to the best of our knowledge. So far, synthesis of 1H-1,2,3-triazole-4,5-disulfides was reported in a Ru-catalyzed [3 + 2] cycloaddition of an azide and a bis(alkylsulfanyl) acetylene at high temperatures under inert gas atmosphere [16,20]. Alternatively, this synthesis can be carried out with [(NHC)CuI] (NHC = 1-benzyl-3-n-butyl-1H-benz[d] imidazolylidene) as catalyst. The latter is easier to use, but the yields are lower compared with the Ru-based catalyst. In addition, 1H-1,2,3-triazoles have been synthesized in an Ir-catalyzed [3 + 2] cycloaddition of internal mono(alkylsulfanyl)alkynes with an azide [21]. Herein, we present a substantially improved synthesis of 1H-1,2,3-triazole-4,5-disulfides under CuAAC click conditions using the terminal benzylsulfanylacetylene. Pitfalls of the reductive removal of S-protective benzyl groups are identified by isolation of respective thiolato complexes. Finally, we describe coordination of the corresponding dithiols to group 10 metals and Co^{III}. The influence of the metal and the N-protective groups at the triazole on the electronic properties will be discussed.

2. Materials and Methods

2.1. Chemical Reagents and Instruments

Materials, details on physical measurements, X-ray determination data, original NMR and IR spectra of all products and preparative procedures as well as spectroscopic data of the only organic products (**1–4**) are provided in the ESI.

2.2. Synthetic Protocols

2.2.1. General Synthesis of **5**

A solution of **2a–c** (1 mmol) in THF (50 mL) was treated with sodium (5 mmol) and naphthalene (2.5 mmol). The red-brown suspension was stirred overnight, then cooled to 0 °C. MeOH (10 mL) was added and the mixture was stirred until gas evolution ceased. For purification, the solution was dried in vacuo, taken up in H$_2$O (40 mL) and washed three times with Et$_2$O (10 mL aliquots). The aqueous fraction was filtered over celite in a G3 frit and subsequently acidified with aqueous HCl (pH = 3–4), leading to the formation of a beige precipitate. The suspension was extracted four times with CH$_2$Cl$_2$ (aliquots of 10 mL). The organic fraction was dried over Na$_2$SO$_4$, filtered and dried in vacuo to isolate **5** as crude products. According to NMR, the samples are not analytically but sufficiently pure for successful complex synthesis. Potential by-products were characterized in the form of stable complexes as well (see compounds **11** and **12**).

H$_2$-**5a**, 1.049 g (2.42 mmol) **2a**, 0.284 g (12.35 mmol) sodium, 0.777 g (6.06 mmol) naphthalene: yield 0.174 g (28%, crude product). ^{1}H NMR (CDCl$_3$, δ, ppm, 300 MHz, 298 K): 7.25–7.22 (m, 2 H, H-o-(4-MOB)), 6.92–6.89 (m, 2 H, H-m-(4-MOB)), 5.45 (s, 2 H, NCH$_2$), 3.81 (s, 3 H, CH$_3$) + additional by-product signals. IR (CH$_2$Cl$_2$, $\tilde{\nu}$, cm^{-1}): 3686 (m), 2978 (s), 2873 (s), 2362 (w), 1604 (m), 1510 (m), 1384 (m), 1274 (s), 1110 (s), 763 (s), 697 (s).

H$_2$-**5b**, 0.649 g (1.401 mmol) **2b**, 0.165 g (7.174 mmol) sodium, 0.834 g (6.507 mmol) naphthalene: yield 0.323 g (40%, crude product). ^{1}H NMR (CD$_2$Cl$_2$, δ, ppm, 300 MHz, 298 K): 7.20–7.17 (m, 1 H, H-(2,4-dMOB)), 6.51–6.48 (m, 2 H, H-(2,4-dMOB)), 5.56 (s, 2 H, NCH$_2$), 3.82 (s, 3 H, CH$_3$), 3.79 (s, 3 H, CH$_3$) + additional by-product signals. IR (CH$_2$Cl$_2$, $\tilde{\nu}$, cm^{-1}): 2994 (s), 2890 (s), 2824 (s), 2496 (w), 1618 (s), 1509 (s), 1465 (m), 1300 (s), 1210 (s), 1157 (s), 1067 (s), 1038 (s), 901 (s), 840 (m), 697 (m).

H$_2$-**5c**, 0.632 g (1.53 mmol) **2c**, 0.173 g (7.52 mmol) sodium, 0.496 g (3.87 mmol) naphthalene: yield 0.356 g (76%, crude product). ^{1}H NMR (THF-D$_8$, δ, ppm, 300 MHz, 298 K): 4.57–4.31 (m, 2 H, NCH$_2$), 1.32–1.18 (m, 2 H, CH$_2$TMS), 0.07 (s, 9 H, CH$_3$-TMS) + additional by-product signals.

2.2.2. General Synthesis of the Metal Complexes 6 and 7

In a 50 mL flask 1 equivalent [(dppe)MCl$_2$] (M = Ni, Pd) was suspended in 15 mL H$_2$O. A solution of 1.1 equivalents **5b** in CH$_2$Cl$_2$ (25 mL) and 3 equivalents KOH were subsequently added. In the two-phase system, a color change from red to green (Ni) or colorless to violet (Pd) was observed in the lower phase. The reaction system was stirred for 3 days at room temperature. To purify the product, the aqueous phase was removed and the organic fraction washed four times with H$_2$O (15 mL), dried over Na$_2$SO$_4$ and filtered and the solvent was removed in vacuo. A column chromatographic purification was carried out with a CH$_2$Cl$_2$/MeOH solvent mixture (20/1) as a mobile phase. Suitable crystals for X-ray structure analysis were obtained from a CH$_2$Cl$_2$ solution by slow diffusion of *n*-pentane.

[(dppe)Ni(**5b**)] (**6**), 0.115 g (0.22 mmol) [(dppe)NiCl$_2$], 0.077 g (1.37 mmol) KOH, 0.068 g (approx. 0.24 mmol) **5b**: yield, 0.086 g (54%). Anal. Calcd. for C$_{37}$H$_{35}$N$_3$NiO$_2$P$_2$S$_2$: C, 60.18; H, 4.78; N, 5.69; S, 8.68%. Found: C, 59.79; H, 4.71; N, 5.77; S, 8.57%. ^{1}H NMR (CDCl$_3$, δ, ppm, 300 MHz, 298 K): 7.83–7.75 (m, 8 H, *H*-Ph), 7.52–7.43 (m, 12 H, *H*-Ph), 6.93 (d, $^3J_{H,H}$ = 8.3 Hz, 1 H, *H-o*-(2,4-dMOB)), 6.32 (d, $J_{H,H}$ = 2.4 Hz, 1 H, *H-m'*-(2,4-dMOB)), 6.28 (dd, $^3J_{H,H}$ = 8.3 Hz, $J_{H,H}$ = 2.4 Hz, 1 H, *H-m*-(2,4-dMOB)), 5.20 (s, 2 H, NCH$_2$), 3.73 (s, CH$_3$), 3.61 (s, CH$_3$), 2.40–2.22 (m, 4 H, CH$_2$-dppe). ^{13}C NMR (CDCl$_3$, δ, ppm, 75 MHz, 298 K): 160.3 (s, *C*-(2,4-dMOB)), 158.0 (s, *C*-(2,4-dMOB)), 156.7 (s, *C*-tazdt), 145.5 (d, $^3J_{C,P}$ = 17.3 Hz, *C*-tazdt), 133.6 (dd, $^3J_{C,P}$ = 10.6 Hz, $J_{C,P}$ = 2.2 Hz, *C*-Ph), 131.6 (s, *C*-Ph), 130.3 (s, *C*-(2,4-dMOB)), 129.1 (dd, $^1J_{C,P}$ = 46.9 Hz, $^3J_{C,P}$ = 16.1 Hz, *C*-Ph), 129.0 (dd, $^2J_{C,P}$ = 10.7 Hz, $J_{C,P}$ = 2.8 Hz, *C*-Ph), 117.5 (s, *C*-(2,4-dMOB)), 104.2 (s, *C*-(2,4-dMOB)), 98.3 (s, *C*-(2,4-dMOB)), 55.4 (s, CH$_3$), 55.4 (s, CH$_3$), 45.5 (s, NCH$_2$), 27.5–26.8 (m, CH$_2$-dppe). ^{31}P NMR (CDCl$_3$, δ, ppm, 122 MHz, 298 K): 60.5 (d, $^2J_{P,P}$ = 47.9 Hz, *P*-dppe), 58.7 (d, $^2J_{P,P}$ = 47.9 Hz, *P*-dppe). IR (CH$_2$Cl$_2$, $\tilde{v}$, cm^{-1}): 2963 (w), 1614 (m), 1508 (m), 1437 (m), 1261 (s), 1208 (m), 1103 (m), 739 (s), 691 (m), 532 (m).

[(dppe)Pd(**5b**)] (**7**), 0.161 g (0.28 mmol) [(dppe)PdCl$_2$], 0.055 g (0.98 mmol) KOH, 0.084 g (approx. 0.30 mmol) **5b**: yield 0.097 g (44%). ^{1}H NMR (DMF-D$_7$, δ, ppm, 300 MHz, 298 K): 8.03–7.87 (m, 8 H, *H*-Ph), 7.66–7.61 (m, 12 H, *H*-Ph), 6.88 (d, $^3J_{H,H}$ = 8.4 Hz, 1 H, *H-o*-(2,4-dMOB)), 6.60 (d, $J_{H,H}$ = 2.3 Hz, 1 H, *H-m'*-(2,4-dMOB)), 6.49 (dd, $^3J_{H,H}$ = 8.4 Hz, $J_{H,H}$ = 2.3 Hz, 1 H, *H-m*-(2,4-dMOB)), 5.17 (s, 2 H, NCH$_2$), 3.83 (s, 6 H, CH$_3$), 3.03–2.86 (m, 4 H, CH$_2$-dppe). ^{13}C NMR (DMF-D$_7$, δ, ppm, 75 MHz, 298 K): 160.9 (s, *C*-(2,4-dMOB)), 158.0 (s, *C*-(2,4-dMOB)), 154.5 (dd, $^3J_{C,P}$ = 10.8 Hz, $J_{C,P}$ = 3.9 Hz, *C*-tazdt), 143.3 (dd, $^3J_{C,P}$ = 12.3 Hz, $J_{C,P}$ = 3.7 Hz, *C*-tazdt), 133.9 (dd, $^3J_{C,P}$ = 11.4 Hz, $J_{C,P}$ = 7.0 Hz, *C*-Ph), 132.1 (d, $J_{C,P}$ = 2.4 Hz, *C*-Ph), 130.7–129.9 (m, *C*-Ph), 129.4 (s, *C*-(2,4-dMOB)), 129.3 (dd, $^2J_{C,P}$ = 10.7 Hz, $J_{C,P}$ = 7.3 Hz, *C*-Ph), 117.3 (s, *C*-(2,4-dMOB)), 104.8 (s, *C*-(2,4-dMOB)), 98.4 (s, *C*-(2,4-dMOB)), 55.6 (s, CH$_3$), 55.3 (s, CH$_3$), 44.6 (s, NCH$_2$), 28.2–27.5 (m, CH$_2$-dppe). ^{31}P NMR (DMF-D$_7$, δ, ppm, 122 MHz, 298 K): 58.4 (d, $^3J_{P,P}$ = 18.0 Hz, *P*-dppe), 56.1 (d, $^3J_{P,P}$ = 18.0 Hz, *P*-dppe). MS (ESI-TOF, 9:1 MeOH:H$_2$O with 0.1% HCOOH, *m/z*): 786 (M + H$^+$). IR (CH$_2$Cl$_2$, $\tilde{v}$, cm^{-1}): 3043 (w), 1647 (m), 1437 (m), 1259 (s), 739 (s), 705 (s).

2.2.3. Synthesis of [(dppe)Pt(5b)] (8)

In a 50 mL Schlenk flask [(dppe)PtCl$_2$] (0.088 g, 1.326 mmol) was dissolved in MeOH (10 mL). A solution of **5b** (0.042 g, approx. 1.484 mmol) and KOH (0.017 g, 0.303 mmol) in MeOH (10 mL) was added. The yellow suspension was diluted with CH$_2$Cl$_2$ (15 mL) and stirred for 3 days at room temperature. After drying in vacuo the purification was carried out chromatographically with a CH$_2$Cl$_2$/MeOH solvent mixture (20/1) as mobile phase. Crystals suitable for X-ray structural analysis were obtained from a saturated CH$_2$Cl$_2$ solution with *n*-pentane: yield 0.088 g (75%). ^{1}H NMR (CD$_2$Cl$_2$, δ, ppm, 300 MHz, 298 K): 7.85–7.75 (m, 8 H, *H*-Ph), 7.52–7.49 (m, 12 H, *H*-Ph), 6.87 (d, $^3J_{H,H}$ = 8.3 Hz, 1 H, *H-o*-(2,4-dMOB)), 6.40 (d, $J_{H,H}$ = 2.4 Hz, 1 H, *H-m'*-(2,4-dMOB)), 6.33 (dd, $^3J_{H,H}$ = 8.3 Hz, $J_{H,H}$ = 2.4 Hz, 1 H, *H-m*-(2,4-dMOB)), 5.18 (s, 2 H, NCH$_2$), 3.75 (s, 3 H, CH$_3$), 3.70 (s, 3 H, CH$_3$), 2.51–2.45 (m, 4 H, CH$_2$-dppe). ^{13}C NMR (CD$_2$Cl$_2$, δ, ppm, 75 MHz, 298 K): 161.0 (s, *C*-(2,4-dMOB)), 158.3 (s, *C*-(2,4-dMOB)), 134.0 (dd, $^2J_{C,P}$ = 11.0 Hz, $J_{C,P}$ = 1.1 Hz, *C*-Ph), 132.3–132.2 (m, *C*-Ph), 130.0 (s, *C*-Ph), 129.2 (dd, $^2J_{C,P}$ = 11.0 Hz, $J_{C,P}$ = 2.4 Hz, *C*-Ph), 117.3 (s, *C*-(2,4-dMOB)), 104.5 (s, *C*-(2,4-dMOB)), 98.6 (s, *C*-(2,4-dMOB)), 55.8 (s, CH$_3$),

55.7 (s, CH$_3$), 45.8 (s, NCH$_2$), 29.4–28.6 (m, CH$_2$-dppe). ^{31}P NMR (CD$_2$Cl$_2$, δ, ppm, 122 MHz, 298 K): 45.7 (d, $^3J_{P,P}$ = 10.3 Hz, P-dppe, Pt-satellites: dd, $^1J_{P,Pt}$ = 2854.1 Hz, $^3J_{P,Pt}$ = 10.3 Hz), 45.4 (d, $^3J_{P,P}$ = 10.3 Hz, P-dppe, Pt-satellites: dd, $^1J_{P,Pt}$ = 2782.8 Hz, $^3J_{P,Pt}$ = 10.3 Hz). MS (ESI-TOF, 9:1 MeOH:H$_2$O with 0.1% HCOOH, m/z): 874.1354 (M + H$^+$). IR (CH$_2$Cl$_2$, $\tilde{\nu}$, cm^{-1}): 3049 (w), 1437 (m), 1269 (s), 1105 (w), 748 (s), 721 (s), 533 (m).

2.2.4. General Synthesis of **9**

In a 50 mL Schlenk flask 1 equivalent [(PPh$_3$)$_2$PtCl$_2$] was suspended in MeOH (10 mL). A solution of 1.1 equivalents **5a–c** and 3 equivalents NaOMe in MeOH (10 mL) and CH$_2$Cl$_2$ (10 mL) was added. After stirring for 3 days at room temperature, the clear yellow solution was dried in vacuo and purified by column chromatography with CH$_2$Cl$_2$/MeOH solvent mixture (20/1) as mobile phase. Crystals suitable for X-ray structural analysis were obtained from a saturated CH$_2$Cl$_2$ solution with n-pentane.

[(PPh$_3$)$_2$Pt(**5a**)] (**9a**) and (**12**), 0.207 g (0.26 mmol) [(PPh$_3$)$_2$PtCl$_2$], 0.034 g (0.63 mmol) NaOMe, 0.074 g (approx. 0.30 mmol) **5a**: yield 0.088 g (76%, **9a**). **12** could be isolated from the first fraction of the same chromatography. ^{1}H NMR (CD$_2$Cl$_2$, δ, ppm, 500 MHz, 298 K): 7.50–7.47 (m, 13 H, H-Ph), 7.38–7.35 (m, 4 H, H-Ph), 7.24–7.18 (m, 13 H, H-Ph), 7.10 (d, $^3J_{H,H}$ = 8.7 Hz, 2 H, H-o-(4-MOB)), 6.76 (d, $^3J_{H,H}$ = 8.7 Hz, 2 H, H-m-(4-MOB)), 4.99 (s, 2 H, NCH$_2$), 3.78 (s, 3 H, CH$_3$). ^{13}C NMR (CD$_2$Cl$_2$, δ, ppm, 125 MHz, 298 K): 159.7 (s, C-(4-MOB)), 154.7 (d, $^3J_{C,P}$ = 10.8 Hz, C-tazdt), 142.7 (d, $^3J_{C,P}$ = 12.0 Hz, C-tazdt), 135.4 (dd, $^3J_{C,P}$ = 8.3 Hz, $J_{C,P}$ = 1.9 Hz, C-Ph), 131.2 (dd, $J_{C,P}$ = 3.6 Hz, $J_{C,P}$ = 1.9 Hz, C-Ph), 130.4 (s, C-(4-MOB)) 130.3 (dd, $^1J_{C,P}$ = 59.0 Hz, $^3J_{C,P}$ = 9.2 Hz, C-Ph), 128.1 (dd, $^2J_{C,P}$ = 11.1 Hz, $J_{C,P}$ = 4.5 Hz, C-Ph), 114.0 (s, C-(4-MOB)), 55.6 (s, CH$_3$), 51.0 (s, CH$_2$). ^{31}P NMR (CD$_2$Cl$_2$, δ, ppm, 202 MHz, 298 K): 17.2 (d, $^3J_{P,P}$ = 21.0 Hz, P-dppe, Pt-satellites: dd, $^1J_{P,Pt}$ = 2914.9 Hz, $^3J_{P,Pt}$ = 20.8 Hz), 17.1 (d, $^3J_{P,P}$ = 21.0 Hz, P-dppe, Pt-satellites: dd, $^1J_{P,Pt}$ = 2943.3 Hz, $^3J_{P,Pt}$ = 20.8 Hz). IR (CH$_2$Cl$_2$, $\tilde{\nu}$, cm^{-1}): 1436 (m), 1259 (s), 1094 (m), 738 (s), 708 (s), 525 (m).

[(PPh$_3$)$_2$Pt(**5b**)] (**9b**), 0.192 g (0.24 mmol) [(PPh$_3$)$_2$PtCl$_2$], 0.046 g (0.85 mmol) NaOMe, 0.069 g (approx. 0.24 mmol) **5b**: yield 0.103 g (42%). Anal. Calcd. for C$_{47}$H$_{41}$N$_3$O$_2$P$_2$PtS$_2$: C, 56.39; H, 4.13; N, 4.20; S, 6.41%. Found: C, 56.66; H, 4.27; N, 4.27; S, 6.63%. ^{1}H NMR (CD$_2$Cl$_2$, δ, ppm, 500 MHz, 298 K): 7.53–7.43 (m, 12 H, H-Ph), 7.38–7.34 (m, 6 H, H-Ph), 7.23–7.18 (m, 12 H, H-Ph), 6.86 (dd, $^3J_{H,H}$ = 7.98 Hz, $J_{H,H}$ = 0.56 Hz, 1 H, H-o-(2,4-dMOB)), 6.36 (t, $J_{H,H}$ = 2.41 Hz, 1 H, H-m-(2,4-dMOB)), 6.33 (d, $J_{H,H}$ = 2.41 Hz, 1 H, H-m'-(2,4-dMOB)), 5.03 (s, 2 H, NCH$_2$), 3.78 (s, 3 H, CH$_3$), 3.65 (s, 3 H, CH$_3$). ^{13}C NMR (CD$_2$Cl$_2$, δ, ppm, 125 MHz, 298 K): 161.0 (s, C-(2,4-dMOB)), 158.5 (s, C-(2,4-dMOB)), 154.2 (dd, $^3J_{C,P}$ = 14.2 Hz, $J_{C,P}$ = 3.2 Hz, C-tazdt), 143.3 (dd, $^3J_{C,P}$ = 16.0 Hz, $J_{C,P}$ = 3.6 Hz, C-tazdt), 135.4 (t, $J_{C,P}$ = 10.8 Hz, C-Ph), 131.2 (dd, $J_{C,P}$ = 12.3 Hz, $J_{C,P}$ = 2.4 Hz, C-Ph), 130.8 (s, C-(2,4-dMOB)), 130.4 (ddd, $^1J_{C,P}$ = 56.4 Hz, $^3J_{C,P}$ = 29.3 Hz, $J_{C,P}$ = 1.7 Hz, C-Ph), 128.1 (d, $^2J_{C,P}$ = 11.1 Hz, C-Ph), 117.0 (s, C-(2,4-dMOB)), 104.5 (s, C-(2,4-dMOB)), 98.5 (s, C-(2,4-dMOB)), 55.8 (s, CH$_3$), 55.7 (s, CH$_3$), 45.5 (s, NCH$_2$). ^{31}P NMR (CD$_2$Cl$_2$, δ, ppm, 202 MHz, 298 K): 17.7 (d, $^3J_{P,P}$ = 21.0 Hz, P-dppe, Pt-satellites: dd, $^1J_{P,Pt}$ = 2996.7 Hz, $^3J_{P,Pt}$ = 20.8 Hz), 16.7 (d, $^3J_{P,P}$ = 21.0 Hz, P-dppe, Pt-satellites: dd, $^1J_{P,Pt}$ = 2835.2 Hz, $^3J_{P,Pt}$ = 20.8 Hz). IR (CH$_2$Cl$_2$, $\tilde{\nu}$, cm^{-1}): 3055 (m), 1437 (m), 1268 (s), 1094 (w), 738 (s), 710 (s), 526 (m).

[(PPh$_3$)$_2$Pt(**5c**)] (**9c**), 0.260 g (0.33 mmol) [(PPh$_3$)$_2$PtCl$_2$], 0.067 g (1.24 mmol) NaOMe, 0.080 g (approx. 0.34 mmol) **5c**: yield 0.238 g (80%). Anal. Calcd. for C$_{43}$H$_{43}$N$_3$P$_2$PtS$_2$Si: C, 54.30; H, 4.56; N, 4.42; S, 6.74%. Found: C, 54.37; H, 4.39; N, 4.29; S, 6.43%. ^{1}H NMR (CD$_2$Cl$_2$, δ, ppm, 500 MHz, 298 K): 7.53–7.47 (m, 12 H, H-Ph), 7.37–7.34 (m, 6 H, H-Ph), 7.23–7.19 (m, 12 H, H-Ph), 3.96–3.92 (m, 2 H, NCH$_2$), 1.05–1.01 (m, 2 H, CH$_2$TMS), −0.06 (s, 9 H, CH$_3$-TMS). ^{13}C NMR (CD$_2$Cl$_2$, δ, ppm, 125 MHz, 298 K): 154.5 (dd, $^3J_{C,P}$ = 13.7 Hz, $J_{C,P}$ = 3.1 Hz, C-tazdt), 142.1 (dd, $^3J_{C,P}$ = 15.9 Hz, $J_{C,P}$ = 3.0 Hz, C-tazdt), 135.4 (dd, $J_{C,P}$ = 10.7 Hz, $J_{C,P}$ = 6.2 Hz, C-Ph), 131.2 (s, C-Ph), 130.6 (s, C-Ph), 128.1 (dd, $^2J_{C,P}$ = 10.7 Hz, $J_{C,P}$ = 5.5 Hz, C-Ph), 44.4 (s, NCH$_2$), 17.8 (s, CH$_2$TMS), -1.8 (s, CH$_3$-TMS). ^{31}P NMR (CD$_2$Cl$_2$, δ, ppm, 202 MHz, 298 K): 17.4 (d, $^3J_{P,P}$ = 21.0 Hz, P-dppe, Pt-satellites: dd, $^1J_{P,Pt}$ = 2861.0 Hz, $^3J_{P,Pt}$ = 20.9 Hz), 16.9 (d, $^3J_{P,P}$ = 21.0 Hz, P-dppe, Pt-satellites: dd, $^1J_{P,Pt}$ = 2988.1 Hz,

$^3J_{P,Pt}$ = 20.9 Hz). ^{29}Si-NMR (CD$_2$Cl$_2$, δ, ppm, 99 MHz, 298 K): 0.6–0.1 (m, *Si*-TMS). IR (CH$_2$Cl$_2$, $\tilde{\nu}$, cm^{-1}): 3056 (w), 2967 (w), 1437 (m), 1259 (s), 1094 (m), 724 (s), 526 (m).

2.2.5. Synthesis of **10**

In a 50 mL Schlenk flask, **5c** (0.081 g, approx. 0.35 mmol) were dissolved in THF (50 mL). Next, [(η^5-C$_5$H$_5$)Co(CO)I$_2$] (0.142 g, 0.35 mmol) and NEt$_3$ (0.11 mL, 0.76 mmol) were added to the solution. The blue solution was stirred for 4 days at room temperature. The purification was carried out chromatographically with a CH$_2$Cl$_2$/MeOH solvent mixture (20/1) as mobile phase. Crystals suitable for X-ray structural analysis were obtained from a saturated CH$_2$Cl$_2$ solution with *n*-pentane: yield 0.053 g (43%). ^{1}H NMR (dimer, CDCl$_3$, δ, ppm, 500 MHz, 298 K): 4.80 (s, 5 H, *H*-Cp), 4.42–4.30 (m, 2 H, NC*H*$_2$), 1.49–1.33 (m, 2 H, C*H*$_2$TMS), 0.16 (s, 9 H, C*H*$_3$-TMS). ^{13}C NMR (dimer, CDCl$_3$, δ, ppm, 125 MHz, 298 K): 156.0 (*C*-tazdt), 151.2 (*C*-tazdt), 88.7 (*C*-Cp), 45.7 (*N*CH$_2$), 18.1 (*C*H$_2$TMS), −1.6 (*C*H$_3$-TMS). ^{29}Si NMR (dimer, CDCl$_3$, δ, ppm, 99 MHz, 298 K): 1.1–0.4 (m, *Si*-TMS). MS (ESI-TOF, 9:1 MeOH:H$_2$O with 0.1% HCOOH, *m/z*): 356 (*M* + H$^+$), 710 (*M*$_2$). IR (CH$_2$Cl$_2$, $\tilde{\nu}$, cm^{-1}): 2968 (w), 2879 (w), 1483 (w), 1267 (s), 846 (s), 748 (s), 708 (s), 558 (m).

2.2.6. Synthesis of **11**

In a 50 mL Schlenk flask, a solution of **5a** (0.103 g, approx. 0.41 mmol) in THF (30 mL) was treated with [(η^5-C$_5$H$_5$)Co(CO)I$_2$] (0.165 g, 0.41 mmol) and NEt$_3$ (0.12 mL, 0.90 mmol). The blue solution was stirred for 5 days at room temperature. The purification was carried out chromatographically with a CH$_2$Cl$_2$/MeOH solvent mixture (20/1). Compound **11** was isolated from the first blue fraction. Crystals suitable for X-ray structural analysis were obtained from a saturated CH$_2$Cl$_2$ solution with *n*-pentane. Yield: 0.008 g (1%).

3. Results and Discussion

3.1. Ligand Synthesis

In contrast to the [3 + 2] cyclization reaction using bis(sulfanyl)acetylene described in a recent publication, mono(sulfanyl)ethyne was used to check whether an insertion of the second benzylsulfanyl group is more advantageous at the cyclized triazole than at the alkyne [16]. The synthesis of the sulfur-substituted triazole derivatives **1a–g** was carried out by a CuAAC reaction with an azide bearing the N-protective groups 4-methoxybenzyl (4-MOB), 2,4-dimethoxybenzyl (2,4-dMOB), 3,4-dimethoxybenzyl (3,4-dMOB), 2-(trimethylsilyl)ethyl (TMS-C$_2$H$_4$), 2,6-dimethylphenyl (Xy), benzyl (Bn) or 2-picolyl (2-Pic) and benzylsulfanylacetylene (Scheme 1, Table 1). Simply, CuSO$_4$ · 5 H$_2$O was used here as the catalyst system, which was reacted in situ with sodium ascorbate (NaAsc) to obtain the catalytically active Cu^I (Scheme 1) [5,10,22,23].

After purification by column chromatography, the N-protected 1*H*-1,2,3-triazole-4-monosulfides were isolated in yields of 36% to 97% (Table 1) and were characterized by NMR spectroscopy. It should be noted that the regioselective cyclization led exclusively to the 4-sulfido derivative, which is in accord with observations of Meldal and Sharpless. [5,24] The introduction of the second sulfur substituent is carried out analogously to synthesis of bis(benzylsulfanyl)acetylene described in the literature. [25] For this purpose, the corresponding triazoles **1a–g** were deprotonated with *n*-butyllithium at −78 °C, reacted with elemental sulfur and subsequently trapped with benzyl bromide (Scheme 1). After purification by column chromatography, the corresponding triazoles **2a–e** were isolated in yields between 38% and 89% (Table 1).

In the ^{1}H NMR spectra, 2 new signals were observed at a chemical shift between 3.55 ppm and 3.79 ppm for the CH$_2$ protons of the introduced benzyl group, while the triazole proton of **1a–g** between 7.05 ppm and 7.68 ppm had disappeared (Figures S32–S46). In the case of compound **1g**, the introduction of sulfur at 5-position failed.

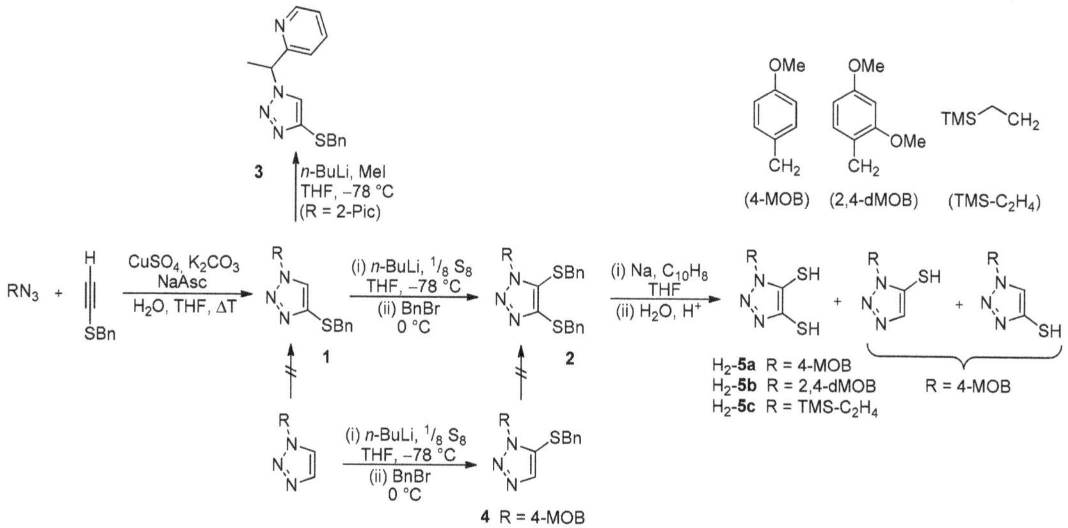

Scheme 1. CuAAC reaction to build 4-benzylsulfanyl-1H-1,2,3-triazole, subsequent introduction of a second sulfide group and reductive removal of the S-benzyl groups to form the free dithiol derivatives.

Table 1. List of N-protective groups and respective yields with regard to Scheme 1 (The letters in column 1 refer to the different N-R triazole derivatives in Scheme 1).

	R		1	2
a	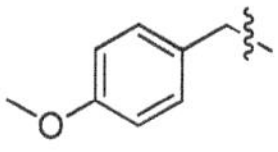	4-MOB	97%	76%
b	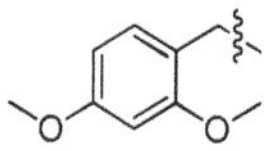	2,4-dMOB	83%	89%
c	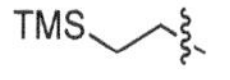	TMS-C$_2$H$_4$	93%	48%
d	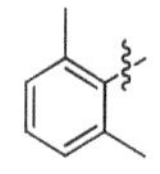	Xy	49%	38%
e	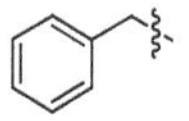	Bn	80%	65%
f	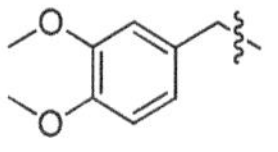	3,4-dMOB	36%	76%
g	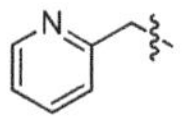	2-Pic	91%	-

Due to the electron-withdrawing pyridine substituent in the 2-picolyl protective group, the acidity of the methylene proton is higher than that of the triazole proton. Accordingly, deprotonation and subsequent methylation with MeI occurs at the N-2-picolyl group to give **3**, as can be observed by the doublet ^{1}H NMR signal at 1.88 ppm for the methyl group attached to the N-protective group (Figure S49). Also in a [3 + 2] cycloaddition of bis(benzylsulfanyl)acetylene and 2-picolyl azide with CuSO$_4$/NaAsc as catalyst **2a** was not isolated. A terminal acetylene is necessary for an end-on coordination of the Cu^I to catalyze the [3 + 2] cycloaddition [10].

Nevertheless, this new two-step method for the generation of a disulfide unit on the 1H-1,2,3-triazole shows clear advantages in comparison with the synthesis described in the literature. Thus, sensitive and expensive catalyst systems [(NHC)CuI] and [(η^5-C$_5$Me$_5$)(cod)RuCl] are dispensable [16]. Moreover, anaerobic and anhydrous conditions are not necessary in the first reaction steps and the overall yields are higher. While Schallenberg et al. achieved a yield of 39% with the benzyl group, a yield of 65% was realized with the new route [16]. Accordingly, it was also investigated whether the disulfide unit can be introduced stepwise into a 1,2,3-triazole by the direct method. For this purpose, the unsubstituted 1-(4-methoxybenzyl)-1H-1,2,3-triazole was deprotonated with n-butyllithium and subsequently reacted with elemental sulfur and benzyl bromide for alkylation (Scheme 1). After chromatographic purification, the ^{1}H NMR spectrum of the isolated product **4** revealed a methylene singlet at 3.67 ppm and a triazole proton at 7.48 ppm, indicating introduction of the sulfur in 5- instead of 4-position (Figure S52). Interestingly, a preference for the 5-substituted derivatives was also observed by Fokin et al. by ruthenium-catalyzed [3 + 2] cycloadditions of terminal alkynes with azides [26–28]. The regioselective deprotonation can be explained by the greater stabilization of the carbanion in 5-position due to resonance (Figure 1). Consistently, a subsequent introduction of the second sulfur substituent at 4-position by the same procedure proved unsuccessful. Respective attempts always led to the recovery of the starting material, which can be attributed to a lack of resonance stabilization in the carbanion.

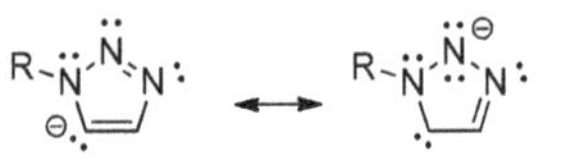

Figure 1. Mesomeric structures after deprotonation.

To enable coordination via dithiolene unit, the benzyl protective groups on sulfur must be removed. Due to having the best yields, compounds **2a–c** were used for coordination experiments. As we previously reported, this could readily be achieved by reductive removal with elemental sodium in presence of naphthalene in THF [16]. After an acidic work-up, the corresponding dithiols **5a–c** were isolated as yellow oils in reasonable yields (Scheme 1). The samples are not analytically but sufficiently pure for coordination experiments (*vide infra*).

3.2. Synthesis of Metal Complexes

Coordination experiments with 1H-1,2,3-triazoles-4,5-dithiols were performed with particular attention to the regioselective dithiolate over N-coordination. The dithiols H$_2$-**5a–c** were reacted with the first-row and group-10 transition metals Co^{III}, Ni^{II}, Pd^{II} and Pt^{II}. The Co^{III} complex **10** was synthesized by reacting the ligand H$_2$-**5c** with [(η^5-C$_5$H$_5$)Co(CO)I$_2$] in THF in presence of NEt$_3$ (Scheme 2). The reaction progress could be observed by a decrease of the CO band in IR spectroscopy and the reaction solution turning blue.

In contrast to the free dithiol H$_2$-**5c**, the corresponding complex could be purified by flash chromatography, such that a dark purple compound was isolated and identified as the Co-complex **10**. Further, the dppe-complexes **6** and **7** with group-10 metals were obtained either by reaction of H$_2$-**5b** in a two-phase system (CH$_2$Cl$_2$/H$_2$O) with KOH and the precursors [(dppe)MCl$_2$] {M = Ni, Pd; dppe = 1,2-bis(diphenylphosphino)ethane} or with [(dppe)PtCl$_2$] and [(PPh$_3$)$_2$PtCl$_2$], respectively, in MeOH using NaOMe as a base. After aqueous work-up and chromatographic purification, a green Ni compound (**6**), a reddish Pd compound (**7**) and yellow Pt compounds (**8** and **9a–c**) were isolated.

In addition to the main products, by-products were surprisingly isolated from the reaction mixtures with the crude dithiol H$_2$-**5a** and corresponding metal precursors (Scheme 3). From the reaction with [(η^5-C$_5$H$_5$)Co(CO)I$_2$], a tetranuclear complex **11** and from the reaction with [(PPh$_3$)$_2$PtCl$_2$] the by-product **12** were isolated and crystallized.

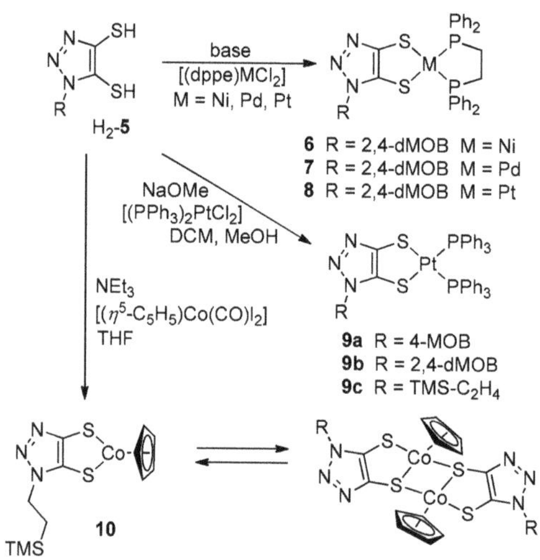

Scheme 2. Coordination of 5^{2-} at Ni^{II} (**6**), Pd^{II} (**7**), Pt^{II} (**8**) and Co^{III} (**10**) (base = KOH or NaOMe).

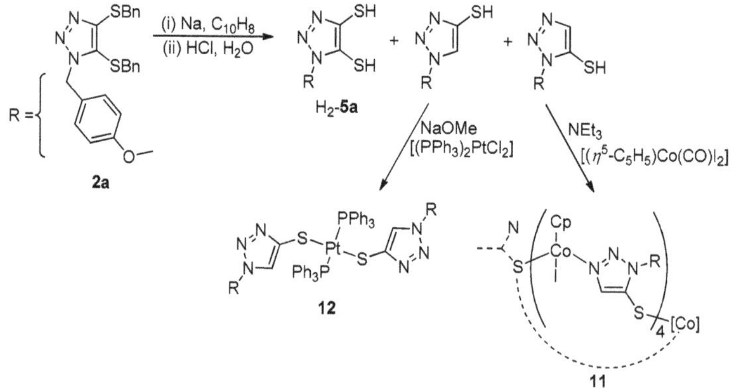

Scheme 3. Coordination to by-products **11** and **12**.

3.3. Molecular Structure of the Complexes

The molecular structures of all complexes **6–12** were determined by single-crystal XRD analysis (Figures 2, 3, S4 and S5). With the exception of the complexes **11** and **12**, which are by-products, all complexes exhibited an exclusive dithiolato coordination. The molecular structures of the group-10 metals showed the expected square planar geometry, including a planar dithiolate unit. The deviation from the SCCS planarity fell between 1.0(5)° and 3.1(3)°, which is very much in accordance with the values described in the literature [29]. Table 2 lists selected bond lengths and angles. In comparison to classical dithiolene complexes, a larger obtuse bite angle and, related to that, somewhat longer metal–sulfur bonds are evident [30–33]. The former follows the geometric requirements of a five-membered backbone ring, in which a regular internal angle leads to a formal C–C–S angle of 126°. In addition, comparison of the metric parameters in compounds **9a** and **9b** does not show any influence by the protective group on nitrogen in the bonding situation at the dithiolate unit.

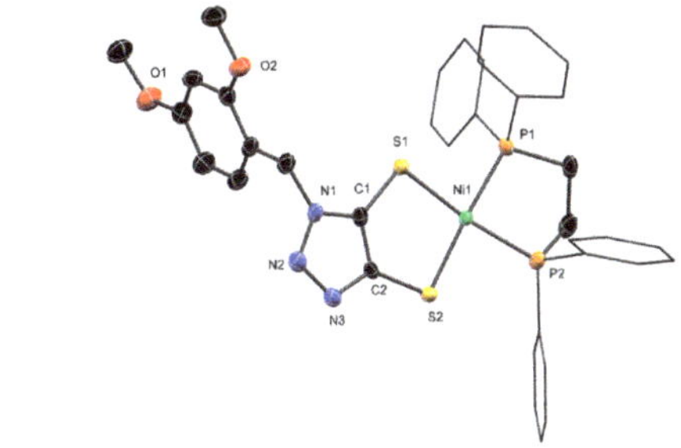

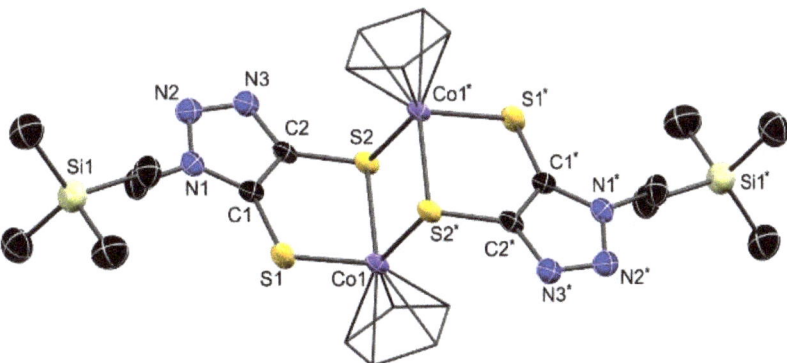

Figure 2. Molecular structure of **6–8**, **11** and **12** in the crystal with ellipsoids set at 50% probability. Hydrogen atoms have been omitted for clarity and phenyl or 4-methoxybenzyl (**11**) substituents are displayed as wireframe.

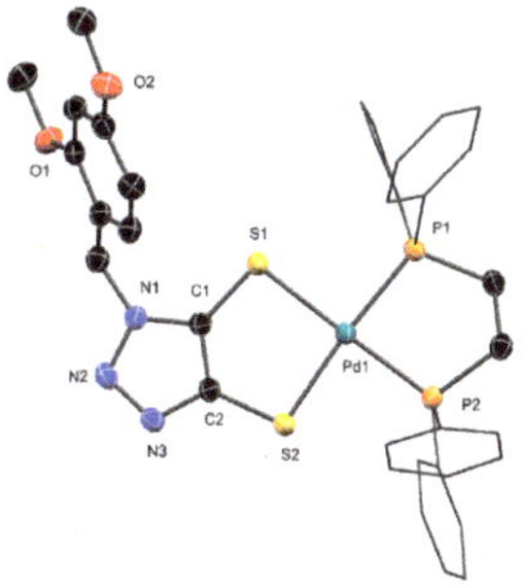

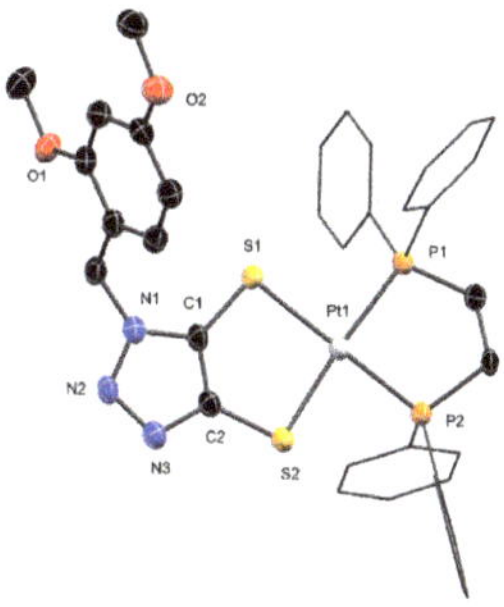

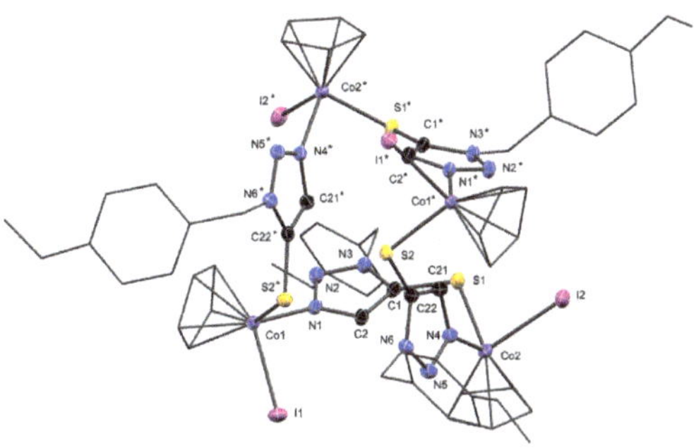

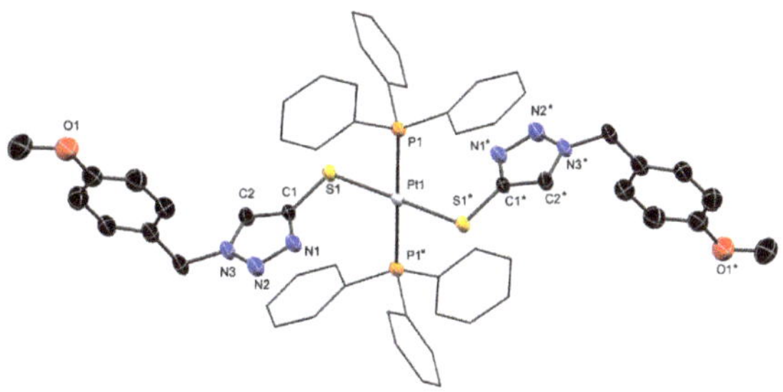

Figure 3. Molecular structure of the dimer **10** in the crystal with ellipsoids set at 50% probability. Hydrogen atoms have been omitted for clarity and η^5-C_5H_5 rings are displayed as wireframe.

Table 2. Comparison of essential bond lengths [Å] and bite angles [°].

	C–S	C–S	C–C	M–S1	M–S2/M–S2*	S1–M–S2
6	1.726(4)	1.747(4)	1.368(6)	2.199(1)	2.187(1)	95.80(4)
[(dppe)Ni(tazdt-Bn)] [16]	1.719(3)	1.750(3)	1.370(4)	2.1982(8)	2.1925(8)	96.09(3)
7	1.725(5)	1.748(6)	1.381(5)	2.354(2)	2.334(1)	92.99(5)
[(dppe)Pd(tazdt-Bn)] [16]	1.7333(15)	1.7400(17)	1.377(2)	2.3475(4)	2.3397(4)	92.90(1)
8	1.741(6)	1.736(4)	1.369(5)	2.349(1)	2.335(1)	92.15(4)
[(dppe)Pt(dmit)] [34]	1.710(11)	1.716(11)	1.366(16)	2.315(3)	2.308(3)	90.0(1)
[(dppe)Pt(dddt)] [35]	-	-	-	2.3157(13)	2.3235(15)	88.25(5)
9a	1.724(3)	1.739(3)	1.373(3)	2.3344(7)	2.3487(7)	90.82(2)
9b	1.725(5)	1.752(3)	1.369(6)	2.3536(9)	2.336(1)	91.08(4)
[(PPh$_3$)$_2$Pt(dmit)] [36]	1.722(4)	1.750(4)	1.349(6)	2.3319(11)	2.3192(11)	89.22(4)
10	1.718(5)	1.751(7)	1.380(8)	2.278(2)	2.271(1)/2.269(1)	93.81(5)
[(η^5-C$_5$H$_5$)Co(Cl$_3$bdt)] [33]	1.734(11)	1.765(11)	1.384(18)	2.211(2)	2.214(3)/2.270(3)	89.61(12)
[(η^5-C$_5$H$_5$)Co(bdt)] [32]	1.757(4)	1.783(3)	1.382	2.246(1)	2.230(1)/2.272(1)	89.73(4)
11	1.739(2)	1.739(2)	1.739(2)/1.379(3)	2.2601(7)	2.2721(6)	-
12		1.747(3)	1.379(4)	2.3274(7)		-

dddt = 5,6-dihydro-1,4-dithiin-2,3-dithiolate; dmit = 1,3-dithiole-2-thione-4,5-dithiolate.

Moreover, when replacing the metal center from Ni^{II} (**6**) to Pd^{II} (**7**) or Pt^{II} (**8**), the dithiolate moiety does not show significant differences in the bond lengths C1–C2 with 1.368(6) Å to 1.381(5) Å or C1–S1 and C2–S2, which are between 1.725(5) Å and 1.748(6) Å. On the other hand, the M–S bond lengths show a distinct elongation by going from Ni^{II} to Pd^{II} and Pt^{II}, which is essentially related to the increasing size of the metal atom. However, the bond lengths Pd–S in **7** {2.354(2) Å and 2.334(1) Å} and Pt–S in **8** {2.349(1) Å and 2.335(1) Å} are virtually equal. This effect is well-known and is attributed to the relativistic effect of the Pt atom and the resulting shrinking of the d orbitals [34].

The molecular structure of **10** in the solid state reveals a dimerization, in which not only is the Co^{III} center coordinated by one dithiolate unit, but a third sulfur atom of a neighboring dithiolate moiety is bound to cobalt and vice versa. The observed dimerization to (**10**)$_2$ can be rationalized by fulfilling the 18 valence electron rule. On the other hand, the monomer constitutes a 16 valence electron complex, which is less stable but more readily solvated due to the free coordination site. Such dimerization equilibria are regularly observed in related [(η^5-C$_5$H$_5$)Co(dithiolene)] complexes [32,33,37–40].

If the η^5-C$_5$H$_5$ ring is considered as occupying a single coordination site, the Co^{III} centers show a τ-parameter of 0.76, which is close to $\tau = 1$ of a tetrahedron [41]. The bond length M–S2* of 2.269(1) Å is comparable to that of M–S1 {2.278(2) Å} and M–S2 {2.271(1) Å}. In studies on the compounds [CpCo(Cl$_3$bdt)]$_2$ and [CpCo(bdt)]$_2$ (bdt = benzene-1,2-dithiolate), the Co–S bond lengths fall between 2.211(2) Å and 2.246(1) Å and are again slightly shorter than the bond lengths determined in **10**. [32,33] Accordingly, as described in the literature, the distance between the Co centers between 3.212(6) Å and 3.2893(4) Å is slightly shorter than the distance determined in **10** with 3.3055(9) Å. None correspond to a direct Co–Co bond of 2.32 Å. [21].

A by-product of the reaction of H$_2$-**5a** with [(η^5-C$_5$H$_5$)Co(CO)I$_2$] was isolated after chromatography and crystallization. The crystal structure of **11** undisclosed an unexpected tetranuclear complex, in which the Co^{III} ions are linked in a cyclic fashion by N-4-methoxybenzyl-1,2,3-tiazole-5-thiolate ligands (Figure 2). Herein, each Co^{III} is coordinated by a thiolate of one triazole and by a nitrogen atom in the third position of another. The coordination sphere of each Co^{III} center is saturated by one iodide and one η^5-C$_5$H$_5$ ligand. This structural motif uncovered the loss of one thiolate substituent at 4-position of the 1,2,3-triazole ligand.

Likewise, the triazole ligands in the by-product **12** do not contain a dithiolate unit. Instead, the two triazole ligands in **12**, next to two trans-standing triphenylphosphine ligands, are coordinated via one remaining thiolate in 4-position in a quadratic planar geometry around a Pt^{II} center. A comparison of complex **12** with **9a** with respect to the influence of cis/trans configuration is interesting, because the ligands are highly similar. The *trans* arrangement leads to longer Pt–P1/P1* bonds (2.3220(8) Å) in **12** compared to the *cis* arrangement in **9a** with Pt–P1/P2: 2.2853(7) Å/2.2944(7) Å, which reflects some symbiotic π-bonding effect in **9a**. The successful isolation of low-yield by-products **11** and **12** indicate limitation of side reactions in the reductive removal of the thiol protective groups. Remarkably, the cleavage of the whole benzylthiolate is possible both at 4- and 5-position.

3.4. NMR Spectroscopy of Metal Complexes

The phosphine ligands in the complexes **6–8** and **9a–c** are valuable probes for the electronic situation of the metal, which can be investigated by ^{31}P NMR spectroscopy. The Ni complex **6** as well as the Pd compound **7** show two doublets at chemical shifts of 58.7/60.5 ppm, and 56.1/58.4 ppm, respectively. The observed doublets result from the C_1 symmetry and the related chemical non-equivalence of the phosphorus atoms. Consistently, a slightly smaller coordination chemical shift $\Delta\delta$ of the Pd-dppe signals is combined with a lower $^{31}P/^{31}P$ coupling constant of 18.0 Hz. The Ni-dppe complex **6** shows a substantially larger coupling constant of 47.9 Hz. The doublet signals for the corresponding Pt^{II} compound **8** were detected at 45.4 ppm and 45.7 ppm, with a coupling constant of 10.5 Hz confirming the trend $J_{P,P}(Ni) > J_{P,P}(Pd) > J_{P,P}(Pt)$ and $\delta(Ni) > \delta(Pd) > \delta(Pt)$. Related observations were already reported for [(dppe)M(mnt)] (mnt = maleonitriledithiolate) serving as a selected example [29].

With the change of the ligand dppe to PPh_3 in compounds **9a–c**, two doublets are observed at the chemical shift between 16.7 ppm and 17.7 ppm. In addition to the $^{31}P/^{31}P$ coupling (J_{PP} = 21.0 Hz), $^{31}P/^{195}Pt$ coupling constants between 2861 Hz and 2998 Hz are observed (Table 3), which are in good agreement with other dithiolene-Pt compounds [31,42,43].

Table 3. Chemical shifts in ^{31}P NMR spectroscopy and the respective coupling constants.

	M	δ [ppm]	$J_{P,P}$ [Hz]	$J_{P,Pt}$ [Hz]
6	Ni	60.5/58.7	47.9	-
7	Pd	58.4/56.1	18.0	-
8	Pt	45.7/45.4	10.5	2778/2760
9a	Pt	17.2/17.1	21.0	2915/2943
9b	Pt	17.7/16.7	21.0	2862/2998
9c	Pt	17.4/16.9	21.0	2861/2988

Here, the PPh_3 is particularly well-suited for observing changes in the electronic situation of the complex by means of ^{31}P-NMR spectroscopy [42]. The individual N-protective group in **9a–c** exerts only a minor influence on the $^{31}P/^{195}Pt$ coupling constant. However, the slightly differing trans effect of the asymmetric dithiolate on the phosphines is reflected in the variance of the $^{31}P/^{193}Pt$ coupling constant, spanning ΔJ range from 28 Hz (**9a**) to 136 Hz (**9b**).

3.5. Electronic Structure Elucidation

The different electronic situation in compounds **6–8** is revealed by UV/Vis spectroscopy and cyclic voltammetry. Figure 4 shows the UV/Vis spectra of compounds **6**, **7** and **8**. In the visible range between 400 and 700 nm characteristic absorption bands at 409 nm (**8**), 523 nm (**7**) and 602 nm (**6**) are observed, which are responsible for the characteristic color of the compounds: green (**6**), red (**7**) and yellow (**8**). According to TD-DFT calculations, the underlying excitation can be assigned to a dithiolate-π to metal-d transition. Hence, the trend **6** > **7** > **8** in λ reflect the increasing ligand field splitting in the order Ni, Pd, Pt. Consistently, in cyclic voltammetry, a reduction process requires lower potentials for heavier metals. The Ni compound **6** shows a reversible Ni^{II}/Ni^{I} reduction with a half-step potential of -1.79 V, while an irreversible signal at a potentials of -2.14 V and -2.60 V, respectively, are observed for complexes **7** and **8**. DFT calculations on related Ni and Pd dppe complexes of N-2,6-dimethylphenyltriazole-4,5-dithiolate and the corresponding anions resulted that the reversible reduction $Ni^{II,I}$ is based on a substantial distortion to tetrahedral, which is not relevant for Pd and Pt. Accordingly, the calculated ΔG value for the reduction are higher for Pd and Pt compared with Ni. [16].

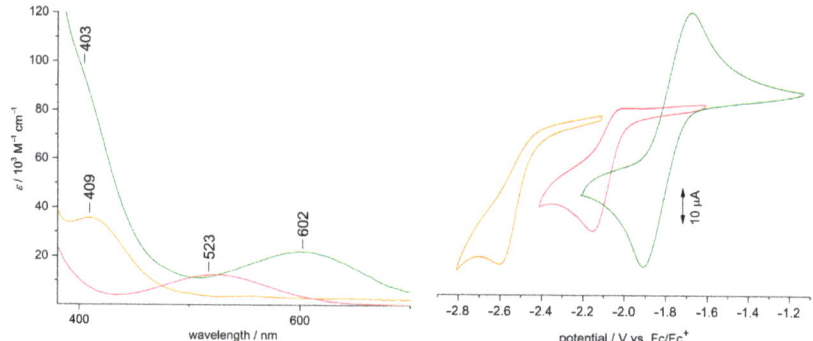

Figure 4. UV/Vis spectra (CH_2Cl_2, **left**) and cyclic voltammetry (CH_2Cl_2 or DMF, **right**) of the compounds **6** (green), **7** (red) and **8** (yellow).

3.6. Investigation of Dimerization

The dimerization of complex **10** to form **(10)₂** found in the solid state could be of great interest for the assembly of coordination polymers on multiple N-coordinated triazole ligands at one metal ion. Therefore, the dimerization equilibrium in solution was investigated by ^{1}H NMR and UV/Vis spectrometry as well as cyclic voltammetry. Variable temperature ^{1}H NMR demonstrated that at concentrations of about 0.02 mol/L, the dimer at 4.79 ppm prevails (Figure 5, right), while the monomer is detected at 5.48 ppm. A dimerization constant K_D of 290 L/mol was determined at 25 °C and a Van't-Hoff plot of K_D at decreasing temperatures resulted a ΔH value of -10.63 kcal/mol and ΔS of -23.6 cal/mol·K (Figure S89). In contrast, in UV/Vis spectroscopy at about 2×10^{-4} mol/L in CH_2Cl_2 the monomer is dominant. The violet crystals yielded a dark blue solution. Two absorption bands, at 485 nm and 619 nm, respectively, were observed in the visible range. For the solid state, reflectance UV/Vis spectroscopy was carried out (Figure 5, left). The absorption bands at 351 nm and 510 nm apparently belong to the dimer **(10)₂**. Accordingly, the strongest absorption band at 619 nm is assigned to a dithiolate-π to Co^{III} charge transfer in the monomer **10**. Compared to the complex $[(\eta^5\text{-}C_5H_5)Co(bdt)]$ ($\lambda = 566$ nm), the band is bathochromicly shifted by 1500 cm^{-1}. [44] This difference can be attributed to the stronger dithiolate character in 1H-1,2,3-triazole-4,5-dithiolate ligands compared with the benzene-1,2-dithiolate, which shows a stronger conjugation to the aromatic system due to better electronegativity matching. Comparable charge transfer bands were reported for many other semi-sandwich complexes with a cobalt dithiolene ligand. [45–47] As expected, the equilibrium between the monomer and the dimer can be influenced by changing the temperature between 0 °C and 40 °C. An increased temperature results in an increased concentration of the monomer at 619 nm.

The cyclic voltammograms of **10** were measured at a concentration range, at which the dimer **(10)₂** is the main species (Figure 6). The signal at a potential $E_{1/2}$ of -0.99 V for the Co^{III}/Co^{II} redox couple exhibits quasi-reversible features. The peak difference increases from 370 mV at a scan rate of 100 mV/s to 520 mV at 300 mV/s, which supports a weakly coupled two-electron process for **(10)₂**. In addition, irreversible oxidation at about +0.8 V causes the appearance of a new signal at slightly higher potential compared with the original Co^{II}/Co^{III} couple. This can reasonably be assigned to the monomer, because, being easier to reduce, the 16 valence electron monomer **10** should exhibit a higher potential. Apparently, one-electron oxidation leads to a release of the monomer **10**.

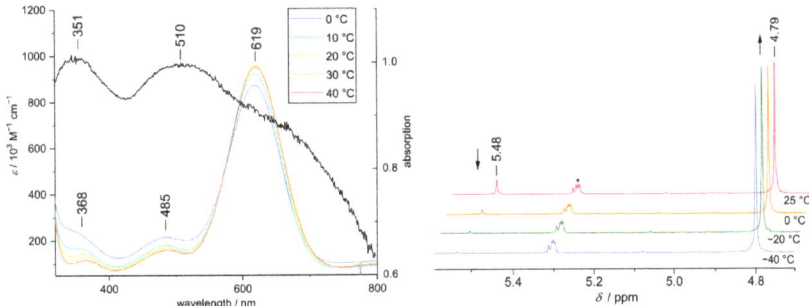

Figure 5. Temperature dependent spectra of **10**/(**10**)$_2$: UV/Vis spectra in CH$_2$Cl$_2$ solution and solid state (**left**) and ^{1}H NMR spectra in CD$_2$Cl$_2$ (*) (**right**).

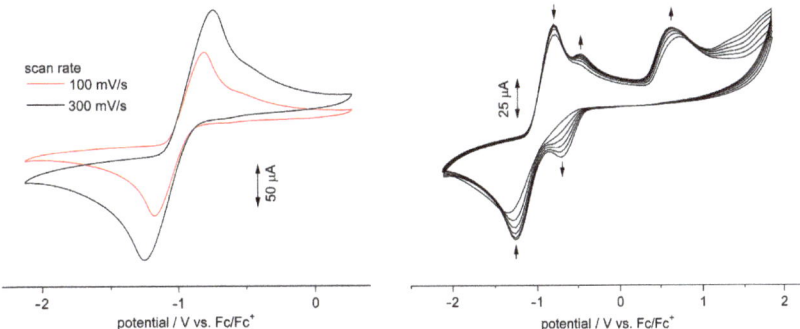

Figure 6. Cyclic voltammograms of the compound **10** in CH$_2$Cl$_2$ at different scan rates (**left**) and changes in the course of multiple potential scans (**right**).

4. Conclusions

In this publication, a new synthetic route for the assembly of 1H-1,2,3-triazole-4,5-dithiolenes was presented, which made use of click chemistry. Instead of complicated, expensive and sensitive catalysts, very high yields of the mono-substituted triazole sulfides **1** could be achieved using CuSO$_4$ in CuAAC. The second sulfur substituent could be introduced by facile deprotonation of the triazole ring and subsequent reaction with sulfur and benzyl bromide, yielding the triazole disulfides **2**. Nevertheless, this new synthetic method for the generation of a dithiolene unit at the 1H-1,2,3-triazole shows clear advantages in comparison with the synthesis described in the literature. [16] In addition, all attempts at a direct introduction of both sulfide substituents into the prototype 1H-1,2,3-triazole led exclusively to the monosulfide isomers **4**. Subsequent reductive removal of the S-protective groups with sodium in THF in the presence of naphthalene yielded the desired dithiol derivatives. However, by-products indicating a competing removal of the whole benzyl thiolate at either 4-or 5-position, respectively, were isolated in form of Co^{III} and Pt^{II} complexes (**11** and **12**). In coordination experiments with the dithiols, several complexes with Ni^{II}, Pd^{II}, Pt^{II} and Co^{III} could be isolated and fully characterized. It was shown that dithiolate coordination dominates the coordination behavior. Neither the coordinated metal (**6**, **7**, **8**) nor the protective group at the nitrogen atom of the triazole (**9a–c**) have a strong effect on the electronic situation at the dithiolate unit. With coordination of the [(η^5-C$_5$H$_5$)Co] moiety, a 16 valence electron Co^{III} center could be introduced at the dithiolate unit giving complex **10**. Instead of a conceivable coordination of a triazole N atom, this complex showed a dimerization via dual μ-sulfur coordination in the solid state. By means of a temperature-dependent NMR and UV/Vis spectroscopic measurements completed by cyclic voltammetry, the thermodynamic parameters of the monomer–dimer equilibrium were determined.

Supplementary Materials: The following supporting information can be downloaded at: https://www.mdpi.com/article/10.3390/chemistry5020086/s1, Tables S1–S4: Crystallographic details for **1d, 1g, 2a, 6–9b** and **10–12**; Figure S1: Molecular structure of **1d** in the crystal; Figure S2: Molecular structure of **1g** in the crystal; Figure S3: Molecular structure of **2a** in the crystal; Figure S4: Molecular structure of **9a** in the crystal; Figure S5: Molecular structure of **9b** in the crystal; Materials, Measurements and Synthese of organic products (**1–4**); Figure S6: ^{1}H NMR spectrum (300 MHz) of 2,4-dimethoxybenzyl azide in CDCl$_3$ at 298 K; Figure S7: IR spectroscopy of 2,4-dimethoxybenzyl azide in THF; Figure S8: ^{1}H NMR spectrum (300 MHz) of 2-(trimethylsilyl)ethyl azide with traces of n-hexane in CDCl$_3$ at 298 K; Figure S9: ^{13}C NMR spectrum (75 MHz) of 2-(trimethylsilyl)ethyl azide with traces of n-hexane in CDCl$_3$ at 298 K; Figure S10: ^{29}Si NMR spectrum (60 MHz) of 2-(trimethylsilyl)ethyl azide in CDCl$_3$ at 298 K; Figure S11: IR spectroscopy of 2-(trimethylsilyl)ethyl azide in Et$_2$O with traces of DMF; Figure S12: ^{1}H NMR spectrum (500 MHz) of **1a** in CDCl$_3$ at 298 K; Figure S13: ^{13}C NMR spectrum (125 MHz) of **1a** in CDCl$_3$ at 298 K; Figure S14: IR spectroscopy of **1a** in CH$_2$Cl$_2$; Figure S15: ^{1}H NMR spectrum (300 MHz) of **1b** in acetone-D$_6$ at 298 K; Figure S16: ^{13}C NMR spectrum (75 MHz) of **1b** in CDCl$_3$ at 298 K; Figure S17: IR spectroscopy of **1b** in CH$_2$Cl$_2$; Figure S18: ^{1}H NMR spectrum (300 MHz) of **1c** in CDCl$_3$ at 298 K; Figure S19: ^{13}C NMR spectrum (75 MHz) of **1c** in CDCl$_3$ at 298 K; Figure S20: ^{29}Si NMR spectrum (60 MHz) of **1c** in CDCl$_3$ at 298 K; Figure S21: ^{1}H NMR spectrum (300 MHz) of **1d** in CDCl$_3$ at 298 K; Figure S22: ^{13}C NMR spectrum (75 MHz) of **1d** in CDCl$_3$ at 298 K; Figure S23: IR spectroscopy of **1d** in CH$_2$Cl$_2$; Figure S24: ^{1}H NMR spectrum (500 MHz) of **1e** in CDCl$_3$ at 298 K; Figure S25: ^{13}C NMR spectrum (75 MHz) of **1e** in CDCl$_3$ at 298 K; Figure S26: IR spectroscopy of **1e** in CH$_2$Cl$_2$; Figure S27: ^{1}H NMR spectrum (300 MHz) of **1f** in acetone-D$_6$ at 298 K; Figure S28: ^{13}C NMR spectrum (75 MHz) of **1f** in acetone-D$_6$ at 298 K; Figure S29: IR spectroscopy of **1f** in CH$_2$Cl$_2$; Figure S30: ^{1}H NMR spectrum (300 MHz) of **1g** in CDCl$_3$ at 298 K; Figure S31: ^{13}C NMR spectrum (75 MHz) of **1g** in CDCl$_3$ at 298 K; Figure S32: ^{1}H NMR spectrum (250 MHz) of **2a** with traces of EtOAc in CDCl$_3$ at 298 K; Figure S33: ^{13}C NMR spectrum (75 MHz) of **2a** in CDCl$_3$ at 298 K; Figure S34: IR spectroscopy of **2a** in CH$_2$Cl$_2$; Figure S35: ^{1}H NMR spectrum (300 MHz) of **2b** in CDCl$_3$ at 298 K; Figure S36: ^{13}C NMR spectrum (75 MHz) of **2b** in CDCl$_3$ at 298 K; Figure S37: IR spectroscopy of **2b** in CH$_2$Cl$_2$; Figure S38: ^{1}H NMR spectrum (300 MHz) of **2c** in CDCl$_3$ at 298 K; Figure S39: ^{13}C NMR spectrum (75 MHz) of **2c** in CDCl$_3$ at 298 K; Figure S40: ^{29}Si NMR spectrum (60 MHz) of **2c** in CDCl$_3$ at 298 K; Figure S41: IR spectroscopy of **2c** in CH$_2$Cl$_2$; Figure S42: ^{1}H NMR spectrum (300 MHz) of **2d** in CDCl$_3$ at 298 K; Figure S43: IR spectroscopy of **2d** in CH$_2$Cl$_2$; Figure S44: ^{1}H NMR spectrum (500 MHz) of **2e** in CD$_2$Cl$_2$ at 298 K; Figure S45: IR spectroscopy of **2e** in CH$_2$Cl$_2$; Figure S46: ^{1}H NMR spectrum (500 MHz) of **2f** in CDCl$_3$ at 298 K; Figure S47: ^{13}C NMR spectrum (126 MHz) of **2f** in CDCl$_3$ at 298 K; Figure S48: IR spectroscopy of **2f** in CH$_2$Cl$_2$; Figure S49: ^{1}H NMR spectrum (300 MHz) of **3** in CDCl$_3$ at 298 K; Figure S50: ^{13}C NMR spectrum (75 MHz) of **3** in CDCl$_3$ at 298 K; Figure S51: IR spectroscopy of **3** in CH$_2$Cl$_2$; Figure S52: ^{1}H NMR spectrum (300 MHz) of **4** in CDCl$_3$ at 298 K; Figure S53: ^{13}C NMR spectrum (75 MHz) of **4** in CDCl$_3$ at 298 K; Figure S54: IR spectroscopy of **4** in CH$_2$Cl$_2$; Figure S55: ^{1}H NMR spectrum (300 MHz) of **5a** in CDCl$_3$ at 298 K; Figure S56: IR spectroscopy of **5a** in CH$_2$Cl$_2$; Figure S57: ^{1}H NMR spectrum (300 MHz) of **5b** in CD$_2$Cl$_2$ at 298 K; Figure S58: IR spectroscopy of **5b** in CH$_2$Cl$_2$; Figure S59: ^{1}H NMR spectrum (300 MHz) of **5c** in THF-D$_8$ at 298 K; Figure S60: ^{1}H NMR spectrum (300 MHz) of **6** with traces of CH$_2$Cl$_2$ in CDCl$_3$ at 298 K; Figure S61: ^{13}C NMR spectrum (75 MHz) of **6** in CDCl$_3$ at 298 K; Figure S62: ^{31}P NMR spectrum (122 MHz) of **6** in CDCl$_3$ at 298 K; Figure S63: IR spectroscopy of **6** in CH$_2$Cl$_2$; Figure S64: ^{1}H NMR spectrum (300 MHz) of **7** with traces of CH$_2$Cl$_2$ and CH$_3$OH in DMF-D$_7$ at 298 K; Figure S65: ^{13}C NMR spectrum (75 MHz) of **7** in DMF-D$_7$ at 298 K; Figure S66: ^{1}H NMR spectrum (122 MHz) of **7** in DMF-D$_7$ at 298 K; Figure S67: IR spectroscopy of **7** in CH$_2$Cl$_2$; Figure S68: ^{1}H NMR spectrum (300 MHz) of **8** in CD$_2$Cl$_2$ at 298 K; Figure S69: ^{13}C NMR spectrum (75 MHz) of **8** in CD$_2$Cl$_2$ at 298 K; Figure S70: ^{31}P NMR spectrum (122 MHz) of **8** in CD$_2$Cl$_2$ at 298 K; Figure S71: IR spectroscopy of **8** in CH$_2$Cl$_2$; Figure S72: ^{1}H NMR spectrum (500 MHz) of **9a** with traces of CH$_2$Cl$_2$ in CD$_2$Cl$_2$ at 298 K; Figure S73: ^{13}C NMR spectrum (125 MHz) of **9a** in CD$_2$Cl$_2$ at 298 K; Figure S74: ^{31}P NMR spectrum (202 MHz) of **9a** in CD$_2$Cl$_2$ at 298 K; Figure S75: IR spectroscopy of **9a** in CH$_2$Cl$_2$; Figure S76: ^{1}H NMR spectrum (500 MHz) of **9b** with traces of CH$_2$Cl$_2$ in CD$_2$Cl$_2$ at 298 K; Figure S77: ^{13}C NMR spectrum

(125 MHz) of **9b** in CD$_2$Cl$_2$ at 298 K; Figure S78: ^{31}P NMR spectrum (202 MHz) of **9b** in CD$_2$Cl$_2$ at 298 K; Figure S79: IR spectroscopy of **9b** in CH$_2$Cl$_2$; Figure S80: ^{1}H NMR spectrum (500 MHz) of **9c** with traces of CH$_2$Cl$_2$ in CD$_2$Cl$_2$ at 298 K; Figure S81: ^{13}C NMR spectrum (125 MHz) of **9c** in CD$_2$Cl$_2$ at 298 K; Figure S82: ^{29}Si NMR spectrum (99 MHz) of **9c** in CD$_2$Cl$_2$ at 298 K; Figure S83: ^{31}P NMR spectrum (202 MHz) of **9c** in CD$_2$Cl$_2$ at 298 K; Figure S84: IR spectroscopy of **9c** in CH$_2$Cl$_2$; Figure S85: ^{1}H NMR spectrum (500 MHz) of **10** in CDCl$_3$ at 298 K; Figure S86: ^{13}C NMR spectrum (125 MHz) of **10** in CDCl$_3$ at 298 K; Figure S87: ^{29}Si NMR spectrum (99 MHz) of **10** in CDCl$_3$ at 298 K; Figure S88: IR spectroscopy of **10** in CH$_2$Cl$_2$; Figure S89: Van't-Hoff-plot of the monomer-dimer equilibrium **10**/(**10**)$_2$. References [25,48–63] are cited in Supplementary Materials.

Author Contributions: Investigation, visualization, writing and review & editing N.P. and W.W.S.; X-ray diffractometry and structure solution, A.V.; conceptualization and supervision W.W.S. All authors have read and agreed to the published version of the manuscript.

Funding: This research received no external funding.

Data Availability Statement: CCDC 225417-2254527 contain the supplementary crystallographic data for this paper. These data can be obtained free of charge via www.ccdc.cam.ac.uk/data_request/cif (accessed on 10 April 2023), or by emailing data_request@ccdc.cam.ac.uk, or by contacting The Cambridge Crystallographic Data Center, 12 Union Road, Cambridge CB2 1EZ, UK, fax: +44-1223-226033.

Conflicts of Interest: The authors declare no conflict of interest.

References

1. Peplow, M. Click and bioorthogonal chemistry win 2022 Nobel Prize in Chemistry. *Chem. Eng. News* **2022**, *36*, 3.
2. Kolb, H.C.; Finn, M.G.; Sharpless, K.B. Click Chemistry: Diverse Chemical Function from a Few Good Reactions. *Angew. Chem. Int. Ed.* **2001**, *40*, 2004–2021. [CrossRef]
3. Lewis, W.G.; Green, L.G.; Grynszpan, F.; Radić, Z.; Carlier, P.R.; Taylor, P.; Finn, M.G.; Sharpless, K.B. Click Chemistry In Situ: Acetylcholinesterase as a Reaction Vessel for the Selective Assembly of a Femtomolar Inhibitor from an Array of Building Blocks. *Angew. Chem. Int. Ed.* **2002**, *41*, 1053–1057. [CrossRef]
4. Ramachary, D.B.; Barbas, C.F. Towards Organo-Click Chemistry: Development of Organocatalytic Multicomponent Reactions Through Combinations of Aldol, Wittig, Knoevenagel, Michael, Diels-Alder and Huisgen Cycloaddition Reactions. *Chem. Eur. J.* **2004**, *10*, 5323–5331. [CrossRef]
5. Rostovtsev, V.V.; Green, L.G.; Fokin, V.V.; Sharpless, K.B. A Stepwise Huisgen Cycloaddition Process: Copper(I)-Catalyzed Regioselective "Ligation" of Azides and Terminal Alkynes. *Angew. Chem. Int. Ed.* **2002**, *41*, 2596–2599. [CrossRef]
6. Clarke, D.; Mares, R.W.; McNab, H. Preparation and pyrolisis of 1-(pyrazol-5-yl)-1,2,3-triazoles and related compounds. *J. Chem. Soc. Parkin Trans. 1* **1997**, 1799–1804. [CrossRef]
7. Huisgen, R. 1,3-Dipolar Cycloadditions. Past and Future. *Angew. Chem. Int. Ed. Engl.* **1963**, *2*, 565–598. [CrossRef]
8. Qin, A.; Jim, C.K.W.; Lu, W.; Lam, J.W.Y.; Häussler, M.; Dong, Y.; Sung, H.H.Y.; Williams, I.D.; Wong, G.K.L.; Tang, B.Z. Click Polymerization: Facile Synthesis of Functional Poly(aroyltriazole)s by Metal-Free, Regioselective 1,3-Dipolar Polycycloaddition. *Macromolecules* **2007**, *40*, 2308–2317. [CrossRef]
9. Jasiński, R. Nitroacetylene as dipolarophile in 2 + 3 cycloaddition reactions with allenyl-type three-atom components: DFT computational study. *Monatsh. Chem.* **2015**, *146*, 591–599. [CrossRef]
10. Hein, J.E.; Fokin, V.V. Copper-catalyzed azide-alkyne cycloaddition (CuAAC) and beyond: New reactivity of copper(I) acetylides. *Chem. Soc. Rev.* **2010**, *39*, 1302–1315. [CrossRef]
11. Beerhues, J.; Aberhan, H.; Streit, T.-N.; Sarkar, B. Probing Electronic Properties of Triazolylidenes through Mesoionic Selones, Triazolium Salts, and Ir-Carbonyl-Triazolylidene Complexes. *Organometallics* **2020**, *39*, 4557–4564. [CrossRef]
12. Tran, H.-V.; Haghdoost, M.M.; Poulet, S.; Tcherkawsky, P.; Castonguay, A. Exploiting *exo* and *endo* furan-maleimide Diels-Alder linkages for the functionalization of organoruthenium complexes. *Dalton Trans.* **2022**, *51*, 2214–2218. [CrossRef]
13. Crowley, J.D.; Bandeen, P.H.; Hanton, L.R. A one pot multi-component CuAAC "click" approach to bidentate and tridentate pyridyl-1,2,3-triazole ligands: Synthesis, X-ray structures and copper(II) and silver(I) complexes. *Polyhedron* **2010**, *29*, 70–83. [CrossRef]
14. Struthers, H.; Mindt, T.L.; Schibli, R. Metal chelating systems synthesized using the copper(I) catalyzed azide-alkyne cycloaddition. *Dalton Trans.* **2010**, *39*, 675–696. [CrossRef]
15. Haas, A.; Krächter, H.-U. Darstellung und Reaktionen Trifluormethylchalkogenyl-substituierter Alkine. *Chem. Ber.* **1988**, *121*, 1833–1840. [CrossRef]
16. Schallenberg, D.; Pardemann, N.; Villinger, A.; Seidel, W.W. Synthesis and coordination behaviour of 1H-1,2,3-triazole-4,5-dithiolates. *Dalton Trans.* **2022**, *51*, 13681–13691. [CrossRef]

17. Szilagyi, R.K.; Lim, B.S.; Glaser, T.; Holm, R.H.; Hedman, B.; Hodgson, K.O.; Solomon, E.I. Description of the Ground State Wave Functions of Ni Dithiolenes Using Sulfur K-edge X-ray Absorption Spectroscopy. *J. Am. Chem. Soc.* **2003**, *125*, 9158–9169. [CrossRef]
18. Lim, B.S.; Fomitchev, D.V.; Holm, R.H. Nickel Dithiolenes Revisited: Structures and Electron Distribution from Density Functional Theory for the Three-Member Electron-Transfer Series $[Ni(S_2C_2Me_2)_2]^{0,1-,2-}$. *Inorg. Chem.* **2001**, *40*, 4257–4262. [CrossRef]
19. Aragoni, M.C.; Caltagirone, C.; Lippolis, V.; Podda, E.; Slawin, A.M.Z.; Woollins, J.D.; Pintus, A.; Arca, M. Diradical Character of Neutral Heteroleptic Bis(1,2-dithiolene) Metal Complexes: Case Study of [Pd(Me$_2$timdt)(mnt)] (Me$_2$timdt = 1,3-Dimethyl-2,4,5-trithioxoimidazolidine; mnt^{2-} = 1,2-Dicyano-1,2-ethylenedithiolate). *Inorg. Chem.* **2020**, *59*, 17385–17401. [CrossRef]
20. Schallenberg, D. *Der Aufbau Polynuklearer Komplexe auf der Basis N-Heterozyklischer Dithiolene*; Universität Rostock: Rostock, Germany, 2013; pp. 1–49.
21. Ding, S.; Jia, G.; Sun, J. Iridium-Catalyzed Intermolecular Azide-Alkyne Cycloaddition of Internal Thioalkynes under Mild Conditions. *Angew. Chem. Int. Ed.* **2014**, *53*, 1877–1880. [CrossRef]
22. Haldón, E.; Nicasio, M.C.; Pérez, P.J. Copper-catalysed azide-alkyne cycloadditions (CuAAC): An update. *Org. Biomol. Chem.* **2015**, *13*, 9528–9550. [CrossRef] [PubMed]
23. Kappe, C.O.; Van der Eycken, E. Click chemistry under non-classical reaction conditions. *Chem. Soc. Rev.* **2010**, *39*, 1280–1290. [CrossRef] [PubMed]
24. Tornøe, C.W.; Christensen, C.; Meldal, M. Peptidotriazoles on Solid Phase: [1,2,3]-Triazoles by Regiospecific Copper(I)-Catalyzed 1,3-Dipolar Cycloadditions of Terminal Alkynes to Azides. *J. Org. Chem.* **2002**, *67*, 3057–3064. [CrossRef] [PubMed]
25. Seidel, W.W.; Meel, M.J.; Schaffrath, M.; Pape, T. In Pursuit of an Acetylenedithiolate Synthesis. *Eur. J. Org. Chem.* **2007**, *2007*, 3526–3532. [CrossRef]
26. Rasmussen, L.K.; Boren, B.C.; Fokin, V.V. Ruthenium-Catalyzed Cycloaddition of Aryl Azides and Alkynes. *Org. Lett.* **2007**, *9*, 5337–5339. [CrossRef]
27. Zhang, L.; Chen, X.; Xue, P.; Sun, H.H.Y.; Williams, I.D.; Sharpless, K.B.; Fokin, V.V.; Jia, G. Ruthenium-Catalyzed Cycloaddition of Alkynes and Organic Azides. *J. Am. Chem. Soc.* **2005**, *127*, 15998–15999. [CrossRef]
28. Boren, B.C.; Narayan, S.; Rasmussen, L.K.; Zhang, L.; Zhao, H.; Lin, Z.; Jia, G.; Fokin, V.V. Ruthenium-Catalyzed Azide-Alkyne Cycloaddition: Scope and Mechanism. *J. Am. Chem. Soc.* **2008**, *130*, 8923–8930. [CrossRef]
29. Landis, K.G.; Hunter, A.D.; Wagner, T.R.; Curtin, L.S.; Filler, F.L.; Jansen-Varnum, S.A. The synthesis and characterisation of Ni, Pd and Pt maleonitriledithiolate complexes: X-ray crystal structure of the isomorphous Ni, Pd and Pt (Ph$_2$PCH$_2$CH$_2$PPh$_2$)M(maleonitriledithiolate) congeners. *Inorg. Chim. Acta* **1998**, *282*, 155–162. [CrossRef]
30. Wrixon, J.D.; Hayward, J.J.; Raza, O.; Rawson, J.M. Oxidative addition chemistry of tetrathiocines: Synthesis, structures and properties of group 10 dithiolate complexes. *Dalton Trans.* **2014**, *43*, 2134–2139. [CrossRef]
31. Wrixon, J.D.; Ahmed, Z.S.; Anwar, M.U.; Beldjoudi, Y.; Hamidouche, N.; Hayward, J.J.; Rawson, J.M. Oxidative addition of *bis*-(dimethoxybenzo)-1,2,5,6-tetrathiocins to Pt(PPh$_3$)$_4$: Synthesis and structures of mono- and di-metallic platinum dithiolate complexes, (dmobdt)Pt(PPh$_3$)$_2$ and [(dmobdt)Pt(PPh$_3$)]$_2$. *Polyhedron* **2016**, *108*, 115–121. [CrossRef]
32. Miller, E.J.; Brill, T.B.; Rheingold, A.L.; Fultz, W.C. A Reversible Chemical Reaction in a Single Crystal. The Dimerization of (η^5-C$_5$H$_5$)Co(S$_2$C$_6$H$_4$). *J. Am. Chem. Soc.* **1983**, *105*, 7580–7584. [CrossRef]
33. Nomura, M.; Sasao, T.; Hashimoto, T.; Sugiyama, T.; Kajitani, M. Structures and electrochemistry of monomeric and dimeric CpCo(dithiolene) complexes with substituted benzene-1,2-dithiolate ligand. *Inorg. Chim. Acta* **2010**, *363*, 3647–3653. [CrossRef]
34. Vicente, R.; Ribas, J.; Solans, X.; Font-Alaba, M.; Mari, A.; de Loth, P.; Cassoux, P. Electrochemical, EPR, and crystal structure studies on mixed-ligand 4,5-Dimercapto-1,2-dithia-2-thione phosphine complexes of nickel, palladium and platinum, M(dmit)(dppe) and Pt(dmit)(PPh$_3$)$_2$. *Inorg. Chim. Acta* **1987**, *132*, 229–236. [CrossRef]
35. Lee, S.-K.; Shin, K.-S.; Noh, D.-Y.; Jeannin, O.; Barrière, F.; Bergamini, J.-F.; Fourmigué, M. Redox Multifunctionality in a Series of Pt^{II} Dithiolene Complexes of a Tetrathiafulvalene-Based Diphosphine Ligand. *Chem. Asian J.* **2010**, *5*, 169–176. [CrossRef]
36. Keefer, C.E.; Purrington, S.T.; Bereman, R.D.; Knight, B.W.; Bedgood, D.R., Jr.; Boyle, P.D. The synthesis and characterization of platinum monodithiolene complexes containing 1,3-dithiole-2-oxo-4,5-dithiolate (dmid^{2-}) and 1,3-dithiole-2-thione-4,5-dithiolate (dmit^{2-}). *Inorg. Chim. Acta* **1998**, *282*, 200–208. [CrossRef]
37. Habe, S.; Yamada, T.; Nankawa, T.; Mizutani, J.; Murata, M.; Nishihara, H. Synthesis, Structure, and Dissociation Equilibrium of Co(η^5-C$_5$H$_5$)(Se$_2$C$_6$H$_4$)$_2$, a Novel Metalladiselenolene Complex. *Inorg. Chem.* **2003**, *42*, 1952–1955. [CrossRef]
38. Nomura, M.; Fourmigué, M. Isostructural diamagnetic cobalt(III) and paramegnetic nickel(III) dithiolene complexes with an extended benzdithiolate core [CpM^{III}(bdtodt)] (M = Co and Ni). *J. Organomet. Chem.* **2007**, *692*, 2491–2499. [CrossRef]
39. Nomura, M.; Kondo, S.; Yamashita, S.; Suzuki, E.; Toyota, Y.; Alea, G.V.; Janairo, G.C.; Fujita-Takayama, C.; Sugiyama, T.; Kajitani, M. Sulfur-rich CpCo(dithiolene) complexes: Isostructural or non-isostructural couples of CpCo(III) with CpNi(III) dithiolene complexes. *J. Organomet. Chem.* **2010**, *695*, 2366–2375. [CrossRef]
40. Watanabe, L.K.; Ahmed, Z.S.; Hayward, J.J.; Heyer, E.; Macdonald, C.; Rawson, J.M. Oxidative addition of 1,2,5,6-Tetrathiocins to Co(I): A Re-Examination of Crown Ether Functionalized Benzene Dithiolate Cobalt(III) Complexes. *Organometallics* **2022**, *41*, 226–234. [CrossRef]
41. Yang, L.; Powell, D.R.; Houser, R.P. Structural variation in copper(I) complexes with pyridylmethylamide ligands: Structural analysis with a new four-coordinate geometry index, τ_4. *Dalton Trans.* **2007**, 955–964. [CrossRef]

42. Keefer, C.E.; Bereman, R.D.; Purrington, S.T.; Knight, B.W.; Boyle, P.D. The ^{195}Pt NMR of L$_2$Pt(1,2-dithiolene) Complexes. *Inorg. Chem.* **1999**, *38*, 2294–2302. [CrossRef]
43. Kimura, T.; Nakahodo, T.; Fujihara, H. Preparation and reactivity of benzo-1,2-dichalcogenete derivatives and their bis(triphenylphosphine)platinum complexes. *Heteroat. Chem.* **2018**, *29*, e21472. [CrossRef]
44. Takeo, A.; Yoshihiro, W.; Akiko, M.; Toru, K.; Hirobumi, U.; Masatsugu, K.; Kunio, S.; Akira, S. Photochemical Reduction of (η^5-Cyclopentadienyl)(1,2-disubstituted 1,2-ethylenedichalcogenolato)cobalt(III) and (η^5-Cyclopentadienyl)(1,2-benzenedithiolato)cobalt(III) Complexes in the Presence of Triethanolamine. *Bull. Chem. Soc. Jpn.* **1992**, *65*, 1047–1051. [CrossRef]
45. Matsuo, Y.; Ogumi, K.; Maruyama, M.; Nakagawa, T. Divergent Synthesis and Tuning of the Electronic Structures of Cobalt–Dithiolene–Fullerene Complexes for Organic Solar Cells. *Organometallics* **2014**, *33*, 659–664. [CrossRef]
46. Murata, M.; Habe, S.; Araki, S.; Namiki, K.; Yamada, T.; Nakagawa, N.; Nankawa, T.; Nihei, M.; Mizutani, J.; Kurihara, M. Synthesis of Herometal Cluster Complexes by the Reaction of Cobaltadichalcogenolato Complexes with Groups 6 and 8 Metal Carbonyls. *Inorg. Chem.* **2006**, *45*, 1108–1116. [CrossRef]
47. Tsukada, S.; Kondo, M.; Sato, H.; Gunji, T. Fine electronic state tuning of cobaltadithiolene complexes by substituent groups on the benzene ring. *Polyhedron* **2016**, *117*, 265–272. [CrossRef]
48. Sheldrick, G.M. *SHELXS-2013, Program for Solution of Crystal Structure*; Universität Göttingen: Göttingen, Germany, 2013.
49. Sheldrick, G.M. *SHELXL-2013. Program for Refinement of Crystal Structure*; Universität Göttingen: Göttingen, Germany, 2013.
50. Sheldrick, G.M. SHELXT–Integrated space-group and crystal-structure determination. *Acta Crystallogr. A Found. Adv.* **2015**, *71*, 3–8. [CrossRef]
51. Kitano, H.; Choi, J.-H.; Ueda, A.; Ito, H.; Hagihara, S.; Kan, T.; Kawagishi, H.; Itami, K. Discovery of Plant Growth Stimulants by C-H Arylation of 2-Azahypoxanthine. *Org. Lett.* **2018**, *20*, 5684–5687. [CrossRef]
52. Tran, T.K.; Bricaud, Q.; Ocafrain, M.; Blanchard, P.; Roncali, J.; Lenfant, S.; Godey, S.; Vuillaume, D.; Rondeau, D. Thiolate chemistry: A powerful and versatile synthetic tool for immobilization/functionalization of oligothiophenes on a gold surface. *Chem. Eur. J.* **2011**, *17*, 5628–5640. [CrossRef]
53. Matake, R.; Niwa, Y.; Matsubara, H. Phase-vanishing method with acetylene evolution and its utilization in several organic syntheses. *Org. Lett.* **2015**, *17*, 2354–2357. [CrossRef]
54. Spencer, L.P.; Altwer, R.; Wei, P.; Gelmini, L.; Gauld, J.; Stephan, D.W. Pyridine– and Imidazole–Phosphinimine Bidentate Ligand Complexes: Considerations for Ethylene Oligomerization Catalysts. *Organometallics* **2003**, *22*, 3841–3854. [CrossRef]
55. Zanato, C.; Cascio, M.G.; Lazzari, P.; Pertwee, R.; Testa, A.; Zanda, M. Tricyclic Fused Pyrazoles with a 'Click' 1,2,3-Triazole Substituent in Position 3 Are Nanomolar CB$_1$ Receptor Ligands. *Synthesis* **2015**, *47*, 817–826. [CrossRef]
56. Howell, S.J.; Spencer, N.; Philp, D. Recognition-mediated regiocontrol of a dipolar cycloaddition reaction. *Tetrahedron* **2001**, *57*, 4945–4954. [CrossRef]
57. Wang, T.; Wang, C.; Zhou, S.; Xu, J.; Jiang, W.; Tan, L.; Fu, J. Nanovalves-Based Bacteria-Triggered, Self-Defensive Antibacterial Coating: Using Combination Therapy, Dual Stimuli-Responsiveness, and Multiple Release Modes for Treatment of Implant-Associated Infections. *Chem. Mater.* **2017**, *29*, 8325–8337. [CrossRef]
58. Kumar, A.S.; Ghule, V.D.; Subrahmanyam, S.; Sahoo, A.K. Synthesis of thermally stable energetic 1,2,3-triazole derivatives. *Chem. Eur. J.* **2013**, *19*, 509–518. [CrossRef]
59. Standley, E.A.; Smith, S.J.; Müller, P.; Jamison, T.F. A Broadly Applicable Strategy for Entry into Homogeneous Nickel(0) Catalysts from Air-Stable Nickel(II) Complexes. *Organometallics* **2014**, *33*, 2012–2018. [CrossRef]
60. Lobana, T.S.; Bawa, G.; Hundal, G.; Butcher, R.J.; Castineiras, A. The Influence of Substituents at C 2 Carbon of Thiosemicarbazones on the Bonding Pattern of Bis(diphenylphopshano)alkanes in Palladium(II) Complexes. *Z. Anorg. Allg. Chem.* **2009**, *635*, 1447–1453. [CrossRef]
61. Li, X.; Zha, M.-Q.; Gao, S.-Y.; Low, P.-J.; Wu, Y.-Z.; Gan, N.; Cao, R. Synthesis, photoluminescence, catalysis and multilayer film assembly of an ethynylpyridine platinum compound. *CrystEngComm* **2011**, *13*, 920–926. [CrossRef]
62. Ramos-Lima, F.J.; Quiroga, A.G.; Pérez, J.M.; Font-Bardia, M. Synthesis and Characterization of New Transplatinum Complexes Containing Phosphane Groups—Cytotoxic Studies in Cisplatin-Resistant Cells. *Eur. J. Inorg. Chem.* **2003**, *2003*, 1591–1598. [CrossRef]
63. Shapley, J.R. *Inorganic Syntheses*; Wiley-Interscience: Hoboken, NJ, USA, 2004; Volume 34.

Disclaimer/Publisher's Note: The statements, opinions and data contained in all publications are solely those of the individual author(s) and contributor(s) and not of MDPI and/or the editor(s). MDPI and/or the editor(s) disclaim responsibility for any injury to people or property resulting from any ideas, methods, instructions or products referred to in the content.

Article

On the Redox Equilibrium of TPP/TPPO Containing Cu(I) and Cu(II) Complexes

Stephanie L. Faber, Nesrin I. Dilmen and Sabine Becker *

Fachbereich Chemie, RPTU Kaiserslautern-Landau, Erwin-Schroedinger-Str. 54, 67663 Kaiserslautern, Germany
* Correspondence: sabine.becker@chem.rptu.de; Tel.: +49-(0)631-2055964

Abstract: Copper(II) clusters of the type [Cu$^{II}_4$OCl$_6$L$_4$] (L = ligand or solvent) are a well-studied example of inverse coordination compounds. In the past, they have been studied because of their structural, magnetic, and spectroscopic features. They have long been believed to be redox-inactive compounds, but recent findings indicate a complex chemical equilibrium with diverse mononuclear as well as multinuclear copper(I) and copper(II) compounds. Furthermore, depending on the ligand system, such cluster compounds have proven to be versatile catalysts, e.g., in the oxidation of cyclohexane to adipic acid. This report covers a systematic study of the formation of [Cu$^{II}_4$OCl$_6$(TPP)$_4$] and [Cu$^{II}_4$OCl$_6$(TPPO)$_4$] (TPP = triphenylphosphine, PPh$_3$; TPPO = triphenylphosphine oxide, O=PPh$_3$) as well as the redox equilibrium of these compounds with mononuclear copper(I) and copper(II) complexes such as [Cu^{II}Cl$_2$(TPPO)$_2$], [{Cu^{II}Cl$_2$}$_2$(TPPO)$_2$], [{Cu^{II}Cl$_2$}$_3$(TPPO)$_2$], [{Cu$^{II}_4$Cl$_4$}(TPP)$_4$], and [Cu^{I}Cl(TPP)$_n$] (n = 1–3).

Keywords: coordination chemistry; copper clusters; inverse coordination; μ_4-oxido copper clusters; TPP; TPPO

1. Introduction

Copper(II) clusters of the type [Cu$^{II}_4$OCl$_6$L$_4$] (L = ligand or solvent) are a well-studied example of inverse coordination compounds [1–3]. The first example of these compounds was reported for [Cu$^{II}_4$OCl$_6$(TPPO)$_4$] (TPPO = triphenylphosphine oxide) by Bertrand et al. in 1966 [4]. In the past, such compounds mainly have been investigated regarding their magnetic properties [4–8]. Although these clusters were initially not considered to be of interest concerning their reactivity, recent findings indicate their catalytic activity, for example, in the conversion of cyclohexane to adipic acid [9–13]. However, studies of the reactivity of such compounds can be challenged by the complex chemical equilibrium through which they are linked to other mononuclear and multinuclear compounds [14–16]. With this in mind, we were particularly interested in [Cu$^{II}_4$OCl$_6$(TPP)$_4$] (TPP = triphenylphosphine) and [Cu$^{II}_4$OCl$_6$(TPPO)$_4$]. To date, it is unknown if these compounds are also a part of such a complex chemical equilibrium. Yet, it is known that the redox reaction of simple Cu(II) salts with phosphines leads to Cu(I) and TPPO [17–19]. This redox reaction already occurs if traces of H$_2$O and/or O$_2$ are present and leads to a number of mononuclear and dinuclear copper complexes. E.g., Makáňová et al. described the mononuclear Cu(I) complexes [Cu^{I}Cl(TPP)$_n$] (n = 1, 2, 3) as well as the Cu(II) complex [Cu^{II}Cl$_2$(TPPO)$_n$] (n = 2, 4) as products of the reaction of CuCl$_2$ with TPP in acetone under inert conditions. Interestingly, [Cu$^{II}_4$OCl$_6$(TPPO)$_4$] was also found in this mixture, which already hints at a complex chemical equilibrium of mononuclear and multinuclear compounds [17]. Both TPP and TPPO are widely used ligands in coordination chemistry. TPP and its derivatives are mainly studied for the luminescent properties of their copper(I) complexes [20–24], which makes the chemical equilibrium with other copper compounds of interest.

In general, the preference of Cu(I) for TPP and Cu(II) for TPPO (as predicted by the HSAB concept) results in a high number of Cu(II)-TPPO and Cu(I)-TPP complexes. In

Citation: Faber, S.L.; Dilmen, N.I.; Becker, S. On the Redox Equilibrium of TPP/TPPO Containing Cu(I) and Cu(II) Complexes. *Chemistry* 2023, 5, 1288–1301. https://doi.org/10.3390/chemistry5020087

Academic Editors: Christoph Janiak, Sascha Rohn and Georg Manolikakes

Received: 14 April 2023
Revised: 11 May 2023
Accepted: 13 May 2023
Published: 17 May 2023

Copyright: © 2023 by the authors. Licensee MDPI, Basel, Switzerland. This article is an open access article distributed under the terms and conditions of the Creative Commons Attribution (CC BY) license (https://creativecommons.org/licenses/by/4.0/).

contrast, few Cu(II)-phosphine complexes are known. In addition to [Cu$^{II}_4$OCl$_6$(TPP)$_4$] [25], one exception is [Cu^{II}(hfac)$_2$PR$_3$] with the chelating ligand hfac = hexafluoroacetylacetonate [26].

In this report, a systematic study concerning the formation of [Cu$^{II}_4$OCl$_6$(TPP)$_4$] and [Cu$^{II}_4$OCl$_6$(TPPO)$_4$] is presented. Furthermore, the chemical equilibrium with other Cu(I) and Cu(II) compounds is investigated. Therefore, three synthetic procedures that are known to yield [Cu$^{II}_4$OCl$_6$L$_4$] (L = ligand with amine donor or solvent) have been investigated with regard to the possible formation of [Cu$^{II}_4$OCl$_6$(TPPO)$_4$] and [Cu$^{II}_4$OCl$_6$(TPP)$_4$]. Our results establish the complex redox equilibrium of mononuclear and multinuclear Cu(I) and Cu(II) compounds and expands the already known set of TPP- and TPPO-containing compounds that are part of this equilibrium.

2. Materials and Methods

Preparation of [Cu$^{II}_4$OCl$_6$(MeOH)$_4$]. [Cu$_4^{II}$OCl$_6$(MeOH)$_4$] was prepared according to procedures described in the literature and under inert conditions (under N$_2$ atmosphere, dried solvents) [25]. CuCl$_2$ (6.72 g, 50.0 mmol) and CuO (1.43 mg, 18.0 mmol) were suspended in absolute MeOH (20 mL) and heated to reflux for 5 h. Then, insoluble solid was filtered off the hot reaction mixture. The solvent of the filtrate was removed, which yielded an ochre-colored solid that was dried in vacuum. Yield: 2.47 g (4.03 mmol, 24%).

Preparation of [Cu$^{II}_4$OCl$_6$(CH$_3$CN)$_4$]. Synthesis was carried out under inert conditions. CuCl$_2$·2H$_2$O (0.60 g, 3.52 mmol, 1.0 eq.), CuCl$_2$ (3.20 g, 23.8 mmol, 6.9 eq.), and sodium tert-butylate (0.75 g, 7.50 mmol, 2.1 eq.) were suspended in absolute CH$_3$CN (20 mL). After heating to reflux for 2 h, the precipitate was filtered off the hot suspension and discarded. The red filtrate was stored at 4.5 °C. Based on previous experience (yield ca. 90%) the concentration was estimated to be 0.34 mol·L^{-1}.

Preparation of [Cu$^{II}_4$OCl$_6$(TPP)$_4$] (**1**). Preparation of **1** was carried out according to procedures described in the literature [25]. Synthesis was carried out under inert conditions. [Cu$^{II}_4$OCl$_6$(MeOH)$_4$] (350 mg, 0.57 mmol, 1 eq.) was dissolved in absolute diethyl ether (50 mL). A solution of TPP (681 mg, 2.60 mmol, 4.6 eq.) in diethyl ether (10 mL) was added dropwise, which yielded a black-brownish precipitate. This solid was filtered off, washed with diethyl ether (10 mL), dried under vacuum, and stored under inert conditions. Yield: 439 mg (0.28 mmol, 50%).

Preparation of **1-crude**. CuCl$_2$ (2.22 g, 16.5 mmol, 2.8 eq.), CuO (0.47 g, 5.90 mmol, 1 eq.), and TPP (5.90 g, 22.5 mmol, 4.1 eq.) were suspended in dry methanol (15 mL) under inert conditions and heated to reflux for 3 h. After hot filtration, the filter cake was discarded. Removal of the solvent yielded a dark green solid.

Preparation of **1a–1c**. A fixed amount of CuCl$_2$·2H$_2$O (100 mg, 0.57 mmol, 1 eq.) was dissolved in acetone (3 mL) under atmospheric conditions and a varying amount of TPP (0.5–4 eq.) in acetone (2–4 mL) was slowly added. After a few minutes, a precipitate formed, which was filtered, washed with a small amount of acetone (6 mL), and dried under vacuum. [Cu^{I}Cl(TPP)] (**1a**): Addition of 0.5 eq. TPP (78 mg, 0.29 mmol). Yield: 13 mg (0.035 mmol, 12%). Alternatively: addition of 1 eq. TPP (153 mg, 0.58 mmol). Yield: 77 mg (0.21 mmol, 37%). [Cu^{I}Cl(TPP)$_2$] (**1b**): addition of 2 eq. TPP (308 mg, 1.17 mmol). Yield: 235 mg (0.36 mmol, 63%). [Cu^{I}Cl(TPP)$_3$] (**1c**): addition of 4 eq. TPP (613 mg, 2.35 mmol). Yield: 435 mg of **1c** that contained 1.5 eq. of acetone (0.48 mmol, 84%, corrected yield for pure **1c**: 78%).

Preparation of [Cu$^{II}_4$OCl$_6$(TPPO)$_4$] (**2**). Recrystallization of **2-crude** from acetone via a diethyl ether diffusion yielded red crystals of **2**. The yield was not determined.

Preparation of **2-crude**. CuCl$_2$ (1.11 g, 8.24 mmol, 2.8 eq.), CuO (0.24 g, 2.98 mmol, 1 eq.), and TPPO (3.15 g, 11.3 mmol, 3.8 eq.) were suspended in dry methanol (10 mL) under inert conditions and heated to reflux for 3 h. After filtration of the hot suspension, the solvent was removed from the dark green filtrate, which yielded an ochre-colored solid (2.7 g).

Preparation of [Cu^{II}Cl$_2$(TPPO)$_2$] (**2a**). Recrystallization of **1-crude** from acetone via a diethylether diffusion yielded yellow crystals of **2a**. The yield was not determined.

Preparation of [Cu^{II}Cl$_2$(TPPO)$_2$]·[{Cu^{II}Cl$_2$}$_2$(TPPO)$_2$] (**2a·2b**). We obtained **2a·2b** by recrystallization of **2-crude** via diethyl ether diffusion from CH$_3$CN under atmospheric conditions. Yield: A few green crystals.

Preparation of [{Cu^{II}Cl$_2$}$_3$(TPPO)$_2$] (**2c**). Recrystallization of **2-crude** from a mixture of DCM:MeCN (8:1) via diethyl ether diffusion at 4.8 °C yielded a few green crystals of **2c**.

Preparation of [{Cu$^I_4$Cl$_4$}(TPP)$_4$] (**3**). Recrystallization of **1-crude** from acetone via a diethyl ether diffusion under inert conditions yielded colorless crystals of **3**. Yield: 34.0 mg.

3. Results

There are three synthetic routes that are known to yield [Cu$^{II}_4$OCl$_6$L$_4$] (L = ligand): (a) template reactions of the *solvento* cluster [Cu$^{II}_4$OCl$_6$(MeOH)$_4$] (MeOH = methanol) with the corresponding ligand L [5,25]; (b) the reaction of CuO, CuCl$_2$, and L in a dry and refluxing solvent, e.g., methanol [25]; and (c) the reaction of CuCl$_2$·2H$_2$O with L in methanol under atmospheric conditions [14,15]. With the exception of the preparation of **1** via a template synthesis, none of these synthetic protocols has been tested for L = TPP/TPPO before. In the following, the results of all three procedures are described for L = TPP/TPPO.

3.1. Template Reactions of [Cu$^{II}_4$OCl$_6$(MeOH)$_4$] with TPP/TPPO

The literature describes the preparation of [Cu$^{II}_4$OCl$_6$(TPP)$_4$] (**1**) under inert conditions via the template reaction of the *solvento* cluster [Cu$^{II}_4$OCl$_6$(MeOH)$_4$] with TPP [5,25]. Following this procedure, we were able to obtain **1** from the reaction of the *solvento* cluster [Cu$^{II}_4$OCl$_6$(MeOH)$_4$] with TPP; however, the choice of solvent (and possibly remaining traces of H$_2$O) was of great importance for the success of this synthesis: According to the literature, when [Cu$^{II}_4$OCl$_6$(MeOH)$_4$] was suspended in diethyl ether and TPP (dissolved in diethyl ether) was added, **1** precipitated as amorphous black solid and could be isolated. We additionally repeated this procedure in methanol and, in contrast, when mixing [Cu$^{II}_4$OCl$_6$(MeOH)$_4$] with TPP in methanol, the solution turned dark for a few seconds and then decolorized to form a colorless precipitate, a behavior that has not been described as yet. This precipitate turned out to be a mixture of diverse [Cu^{I}Cl(TPP)n] (n = 1: **1a**, n = 2: **1b**, n = 3: **1c**) complexes. These complexes have been described previously by Makáňová et al.; however, they had obtained them from the reaction of pure CuCl$_2$ with TPP in dry acetone [17]. Similarly, **1c** has been described by Jardine et al., who obtained **1c** by heating CuCl$_2$·2H$_2$O and TPP in ethanol to reflux [27].

We characterized **1** via IR spectroscopy and elemental analysis (see Supplementary Materials). Despite much effort, it was not possible to obtain crystals suitable for X-ray diffraction analysis. Instead, if the diethyl ether solution was stored for a longer time period, colorless crystals formed that were isolated and determined to the already known cubane cluster [{Cu$^I_4$Cl$_4$}(TPP)$_4$] (**3**, Figure 1). Similarly, as for **1a–1c**, the formation of **3** has already been described for the reaction of CuCl$_2$·2H$_2$O with TPP (1:1.5) in hot ethanol [27]. Thus, even though this reactivity has not been observed for clusters of the type [Cu$^{II}_4$OCl$_6$L$_4$], the reduction of Cu(II) to Cu(I) fits with previously described observations in the literature, which describe the reduction of Cu(II) to Cu(I) by an excess of TPP [17–19].

The spectroscopic and spectrometric characteristics of **3** are in accordance with previously reported characteristics and are listed in the Supplementary Materials. Interestingly, few red crystals of **2** also occurred as side product.

To obtain pure **2**, a similar procedure as for the preparation of **1** was applied. Therefore, TPPO was added to a solution of [Cu$^{II}_4$OCl$_6$(CH$_3$CN)$_4$] in CH$_3$CN. However, and despite the solvent used, it was not possible via this attempt to isolate pure **2**. Instead, this attempt led to [Cu$^{II}_4$OCl$_6$(CH$_3$CN)$_{4-n}$(TPPO)$_n$], a mixture of heteroleptic and homoleptic μ_4-oxido clusters with varying amounts of CH$_3$CN/TPPO ligands, which will be reported elsewhere.

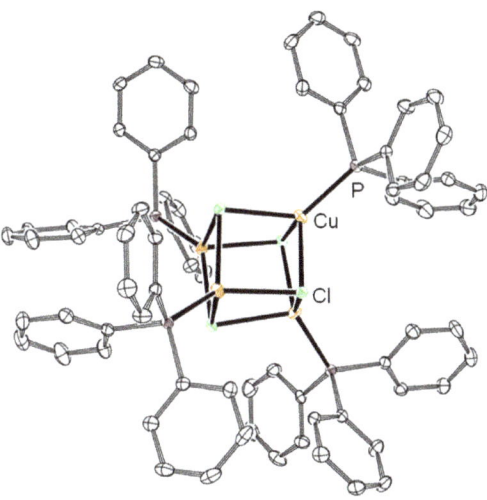

Figure 1. Molecular structure of **3**. Thermal ellipsoids at 50%. Hydrogen atoms omitted for clarity.

3.2. Reaction of CuO, CuCl$_2$, and TPP/TPPO

Another common synthetic route for the preparation of [Cu$^{II}_4$OCl$_6$L$_4$] starts from the simple Cu(II) salts CuO and CuCl$_2$. Here, CuO, CuCl$_2$, and the ligand are suspended in a dry solvent, e.g., methanol, and are heated to reflux [25]. To date, the preparation of either one of **1** or **2** has not been tested via this procedure. To obtain **2**, we mixed CuO, CuCl$_2$, and TPPO in dry methanol and heated it to reflux for 3 h. We obtained an ochre-colored crude product (**2-crude**), from which several Cu(II) complexes could be obtained. For this, we recrystallized **2-crude** from diverse solvents via diethyl ether diffusion. Depending on the solvent used (e.g., acetone, dichloromethane, CH$_3$CN or CHCl$_3$), we obtained different complexes that are described below. When using acetone, red cubic crystals of **2** (Figure 2) were obtained.

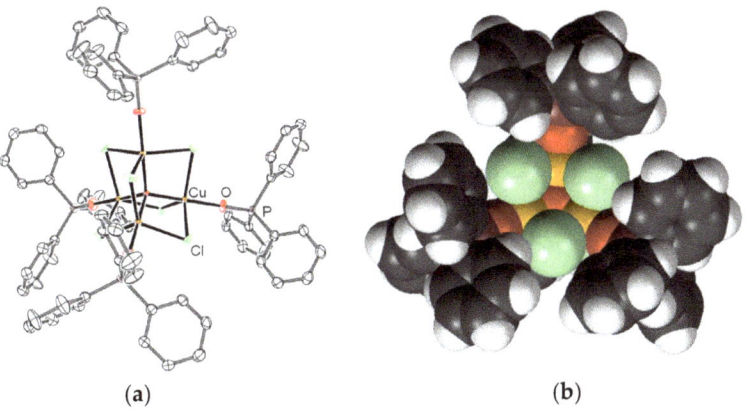

Figure 2. (**a**) Molecular structure of **2**. Thermal ellipsoids set at 50%; (**b**) space-filling model of **2**.

Crystallization of **2** occurred in the cubic space group $F\overline{4}3c$ and was first described in 1966 by Bertrand and Kelley, who obtained **2** by recrystallization of [Cu^{II}Cl$_2$(TPP)$_2$] from methyl isobutyl ketone [4]. The central oxide is inversely coordinated by four Cu(II) centers, yielding a tetrahedron, in which the four Cu(II) centers occupy the corners. Each Cu(II) center is coordinated by three additional Cl$^-$ anions in the equatorial plane and one TPPO ligand in the axial position, thus generating a trigonal bipyramidal coordination geometry for each Cu(II) center. The parameters determined are in good agreement with parameters already published (see Supplementary Materials) and are not discussed herein. Even though such μ_4-oxido copper clusters are known to be sensitive to moisture [14,25,28,29], **2** can be stored under atmospheric conditions for weeks without being hydrolyzed. In fact, we even observed the formation of **2** under air when mixing $CuCl_2 \cdot 2H_2O$ with TPP (vide infra). One reason for this behavior could be the shielding of the central oxide by the rather bulky TPPO ligands; however, as highlighted by the space-filling model of **2** (Figure 2b), the central oxide remains accessible. Accordingly, the shielding effect by TPPO cannot be the main reason for the comparatively high stability against moisture.

Interestingly, when storing the crystallization solution (acetone) for a couple of weeks, the yellow-colored and already known complex $[Cu^{II}Cl_2(TPPO)_2]$ (**2a**, Figure 3) slowly co-crystallized. Instead of the preferred square-planar coordination geometry for Cu(II) ions, a slightly elongated tetrahedron is observed in **2a**. Depending on the severity of tetragonal distortion, two isomers, α and β, with distinct spectroscopic features, can be distinguished [30,31]. The yellow-colored α-$[Cu^{II}Cl_2(TPPO)_2]$, which also is described within this report, is thermodynamically more stable.

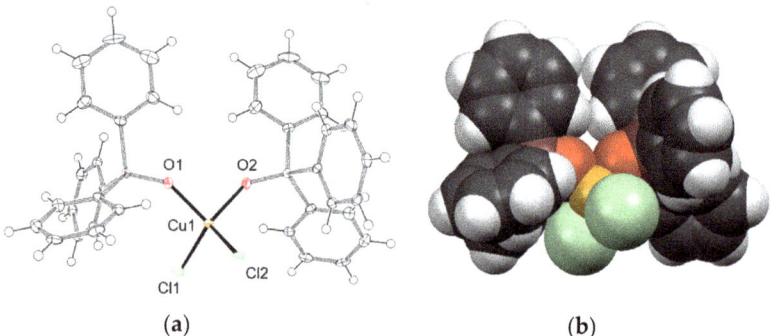

Figure 3. (a) Molecular structure of **2a**. Thermal ellipsoids set at 50%; (b) space-filling model of **2a**. Chosen bond distances: Cu1-O1: 1.9627(19) Å, Cu1-O2: 1.9711(19) Å, Cu1-Cl1: 2.1827(8) Å, Cu1-Cl2: 2.1870(8) Å.

Goodgame and Cotton first obtained **2a** in 1961 and studied its configuration [32]. They obtained **2a** by the reaction of $CuCl_2 \cdot H_2O$ and TPPO in ethanol with subsequent recrystallization from butanone. Additionally, **2a** has been prepared from the reaction of $CuCl_2$ and TPP in an ethanol/acetone mixture at room temperature [17]. The spectroscopic features of **2a** are in accordance with those described in the literature and are listed in the SI.

When using a mixture of CH_2Cl_2 or $CHCl_3$ and CH_3CN (ratio 4:1) for the recrystallization of **2-crude**, again, **2a** was obtained. By increasing the amount of CH_3CN in the CH_2Cl_2/CH_3CN solvent mixture, we obtained two similar and as yet unknown compounds, which only differ in the number of $\{CuCl_2\}$ units. Recrystallization of **2-crude** from pure CH_3CN led to green needles of $[Cu^{II}Cl_2(TPPO)_2] \cdot [\{Cu^{II}Cl_2\}_2(TPPO)_2]$ (**2a·2b**). In contrast, a mixture of CH_2Cl_2/CH_3CN (8:1) yielded green needles of $[\{Cu^{II}Cl_2\}_3(TPPO)_2]$ (**2c**, Figure 4).

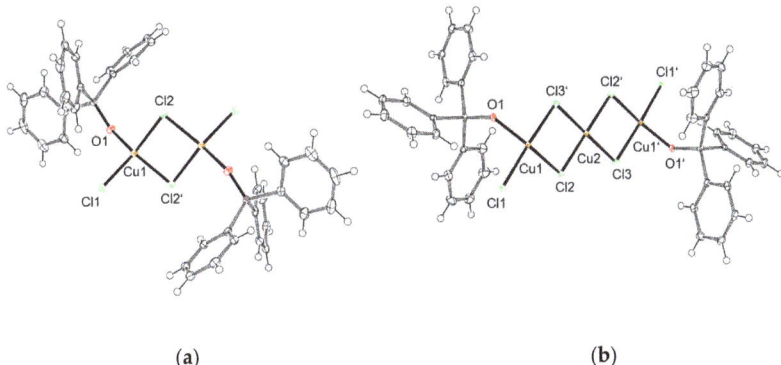

(a)   (b)

Figure 4. Molecular structures of **2b** (**a**) and **2c** (**b**). Thermal ellipsoids set at 50%. Chosen bond distances in **2b**: Cu1-O1: 1.9133(19) Å, Cu1-Cl1: 2.1977(8) Å, Cu1-Cl2: 2.2684(8) Å, Cu1-Cl2′: 2.2877(8) Å, Cu1-Cu1′: 3.310(1) Å. Chosen bond distances in **2c**: Cu1-Cl1: 2.2618(6) Å, Cu1-Cl2: 2.3269(6) Å, Cu1-Cl3′: 2.3288(7) Å, Cu1-Cl2′: 2.5490(7) Å, Cu2-Cl2: 2.2535(6) Å, Cu2-Cl3′: 2.2501(6) Å, Cu2-Cl3: 2.2501(6) Å, Cu2-Cl2′: 2.2535(6) Å, Cu1-Cu2: 3.371(1) Å.

Compound **2b** forms individual complex units that pack in a 1:1 ratio with molecules of **2a** in the crystal structure. Both Cu(II) centers show a distorted square-planar coordination geometry, which also is observed for **2a**; however, in **2a**, the distortion is much more pronounced, as indicated by the τ_4 value of 0.59, which even suggests a rather distorted tetrahedral geometry [33]. In **2b**, the τ_4 value is 0.36 and, thus, much closer to the ideal square-planar geometry ($\tau_4 = 0$) than to a tetrahedron ($\tau_4 = 1$). As a consequence of the dinuclearity of **2b**, the Cu-Cl bond distances in the central {Cu_2Cl_2} diamond are elongated by appr. 3.6% in comparison to the terminal Cu-Cl bond distances. The latter are comparable to the Cu-Cl bond distances in **2a**. Additionally, the Cu-O bonds in **2b** are shorter by appr. 2.8% than in **2a**. In contrast to **2a** and **2b**, **2c** forms zigzag-like chains, in which the [{$Cu^{II}Cl_2$}$_3$(TPPO)$_2$] units are linked via the terminal copper centers of the {Cu_3Cl_6} units (Figure 5). As a result, the terminal Cu(II) centers display a trigonal-bipyramidal coordination geometry with two Cl$^-$ anions and the oxygen atom of TPPO in the equatorial plane. The central Cu(II) center shows a square-planar geometry and lies in the same plane as the other two Cu(II) centers. In general, copper(II) halides are known to form chain-like structures with repeating {CuX_2} units; e.g., for $CuCl_2$ and $CuBr_2$, researchers involved in this study have previously reported similar compounds with CH_3CN instead of TPPO as ligands [15,16].

To complete our investigations, we repeated the synthetic experiment with TPP instead of TPPO. Thus, instead of mixing CuO and $CuCl_2$ with TPPO, we added TPP and heated the suspension in dry methanol to reflux for 3 h. After filtration and removal of the solvent from the filtrate, a green-colored solid (**1-crude**) was obtained. Recrystallization of **1-crude** from acetone yielded **2a** as the main product. The formation of **2a** can be explained by the redox reaction of the redox pairs Cu(II)/Cu(I) and TPP/TPPO; however, the origin of oxygen is not certain. It is known that traces of H_2O in dry solvents leads to the oxidation of TPP to TPPO [17], but at the same time, CuO could have been the oxidant and origin of the oxygen atom. In accordance with the oxidation of TPP to TPPO, the reduction of Cu(II) to Cu(I) was observed: Additionally to **2a**, colorless crystals of CuCl and the new compound CuCl·CH_3CN (Figure 6) were found in the sample. In contrast to pure CuCl, CuCl·CH_3CN crystallized as a zigzag ladder structure of {Cu_2Cl_2} diamonds, in which the copper(I) ions are coordinated by three chloride ions. The additional CH_3CN ligand completes the tetrahedral coordination geometry. The Cu-Cl bond distances are 2.390(6) Å and thus, slightly elongated in comparison to those of pure CuCl (d(Cu-Cl): 2.339(6) Å). The structure is isostructural with the previously reported structure of CuBr·CH_3CN [15].

Interestingly, similar structures containing TPP or triethylphosphine (PEt$_3$) are known; however, [{Cu^{I}Br}$_4$(TPP)$_4$], [{Cu^{I}I}$_4$(TPP)$_4$], and [{Cu^{I}Br}$_4$(PEt$_3$)$_4$] are described as tetramers with a step configuration and not as coordination polymer [18,34–36]. A similar ladder-like structure with phosphine is only known for tetraphosphine ligands that yield octanuclear complexes with CuX (X = Cl, Br, I) [37].

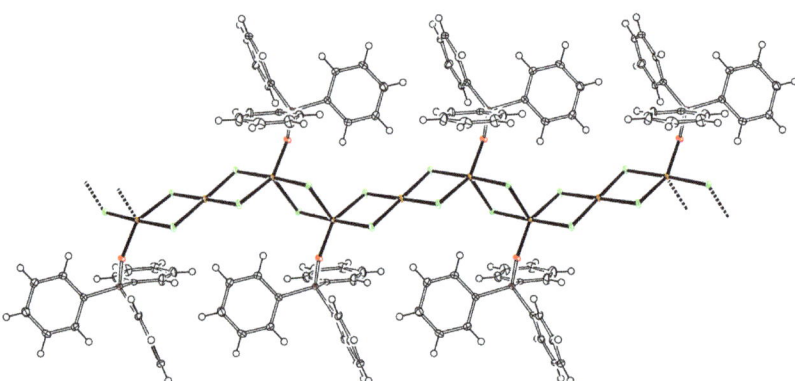

Figure 5. Zigzag-like chain that is formed by [{Cu^{II}Cl$_2$}$_3$(TPPO)$_2$] units in **2c**. Thermal ellipsoids at 50%.

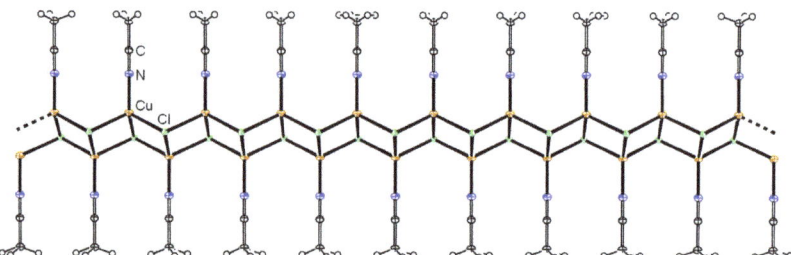

Figure 6. Molecular structure of CuCl·CH$_3$CN. Thermal ellipsoids set at 50%.

Interestingly, when refluxing CuO, CuCl$_2$, and TPP in acetone instead of methanol, a gray-colored [Cu^{I}Cl(TPP)]$_n$ (n = 1, 2, 3) precipitate also appeared. After separation from the liquid, **3** could be crystallized from the filtrate via diethyl ether diffusion. In general, it is known that even at room temperature, simple Cu(II) and Cu(I) complexes are in a chemical equilibrium with multinuclear species [14,15]. Here, the formation of the main products can be influenced by the stoichiometric ratio of the reactants [14,15]. A similar observation was described by Makáňová et al., who investigated the reactivity of CuCl$_2$ with TPP in acetone with the ratios 1:1 and 1:4 [17]. When using equimolar amounts of CuCl$_2$ and TPP, the Cu(I) compounds [Cu^{I}Cl(TPP)] (**1a**), [Cu^{I}Cl(TPP)$_2$] (**1b**), and [Cu^{I}Cl(TPP)$_3$] (**1c**) were obtained. However, when applying an excess of TPP, the Cu(II) compounds [Cu^{II}Cl$_2$(TPPO)$_2$] (**2a**), [Cu^{II}Cl$_2$(TPPO)$_4$]·2H$_2$O, and [Cu^{II}$_4$OCl$_6$(TPPO)$_4$] (**2**) were obtained. In contrast, in our studies, we could observe a mixture of Cu(II) and Cu(I) compounds without varying the stoichiometric ratios but varying the reaction conditions/starting materials.

3.3. Reaction of CuCl$_2$·2H$_2$O and TPP/TPPO under Atmospheric Conditions

It is known that depending on the ligand, [Cu^{II}$_4$OCl$_6$L$_4$] can be prepared by simply mixing L with CuCl$_2$·2H$_2$O in methanol [14,15]. Here, it is important to carry out the reaction under atmospheric conditions to provide an oxygen source. Additionally, the stoichiometric ratio of the reactants governs the reaction outcome because of the complex chemical equilibrium of $[Cu^{II}_4OCl_6(L)_4]$ with simple copper complexes [14,15]. As described above, the outcome of the reaction between $CuCl_2$ and TPP under inert conditions can be influenced by the stoichiometric ratio of the reactants as well [17]. Thus, to complete our studies, we carried out similar reactions to Makáňová et al. In contrast to Makáňová et al., we used non-dry copper(II) chloride, carried out the reactions under atmospheric conditions, and expanded the range of the stoichiometric ratio by the inclusion of the ratios (Cu:TPP) 0.5:1 and 1:2. Therefore, we added TPP in various ratios to a solution of $CuCl_2 \cdot 2H_2O$ in acetone at room temperature. Table 1 summarizes the products that could be isolated from these reaction mixtures.

Table 1. Stoichiometric ratios and products isolated from the reaction of $CuCl_2 \cdot 2H_2O$ with TPP under atmospheric conditions in acetone.

Entry	Cu(II):TPP	Isolated Product	Yield
1	1:0.5	$[Cu^{II}_4OCl_6(TPPO)_4]$ (2)	Few crystals
		$[Cu^ICl(TPP)]$ (1a)	12%
2	1:1	$[Cu^ICl(TPP)]$ (1a)	37%
3	1:2	$[Cu^ICl(TPP)_2]$ (1b)	63%
4	1:4	$[Cu^ICl(TPP)_3]$ (1c)	78%

Even though our reaction setup differed, the results mainly agree with those described by Makáňová et al. The already described trend that an excess of TPP reduces Cu(II) to Cu(I) could be confirmed, as well [17,18]. When $CuCl_2 \cdot 2H_2O$ was added in excess (Table 1, entry 1), red crystals of **2** occurred as the main product. Additionally, small amounts of $[Cu^ICl(TPP)]$ (**1a**) were obtained as gray precipitate. When using equimolar amounts of $CuCl_2 \cdot 2H_2O$ and TPP (Table 1, entry 2), **1a** was obtained as the main product. In contrast, when using a slight excess of TPP (Table 1, entry 3), a colorless solid precipitated that was identified as $[Cu^ICl(TPP)_2]$ (**1b**) formed. When the amount of TPP was further increased (Table 1, entry 4), $[Cu^ICl(TPP)_3]$ (**1c**) was obtained as colorless precipitate. The remaining filtrate was left for crystallization. Via diethyl ether diffusion, colorless crystals of **1c** could be obtained that were suitable for X-ray diffraction analysis (Figure 7). As known from previous analysis [38,39], the Cu(I) center displays a tetrahedral coordination geometry. Co-crystallization of **1c** occurs with acetone solvent molecules, whose presence corresponds to the traces of acetone in the IR spectrum (see Supplementary Materials) and minor deviations of the elemental analysis (see Supplementary Materials). In general, minor deviations in the elemental analysis results of **1a**–**1c** are caused by remaining solvent molecules.

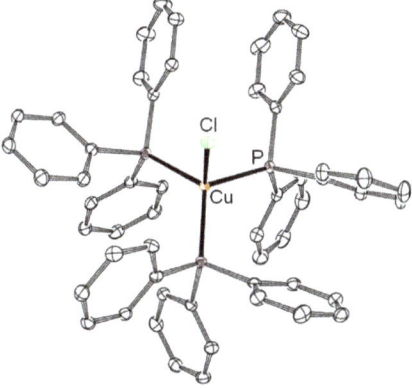

Figure 7. Molecular structure of **1c**. Thermal ellipsoids at 50%. Hydrogen atoms and solvent molecules omitted for clarity.

To complete our investigations, we repeated the experiments described above by adding TPPO instead of TPP to the $CuCl_2 \cdot 2H_2O$ solution. Notably, it was only possible to isolate a product when excess of TPPO (1:4) was added. Here, yellow crystals of **2a** and red crystals of **2** appeared. In all other cases, CuCl crystallized as the only product even though the solution colors changed from an initial slightly blue color to brown or orange (ratio 1:4). A change of the solvent to either methanol or ethanol did not lead to any isolable complexes. Even **2** and **2a** could not be isolated from the alcoholic mixtures.

4. Discussion

To date, only the reactivity of simple copper salts towards TPP/TPPO had been investigated: it is known that under inert conditions, an excess of TPP reduces $CuCl_2$ to form $[Cu^I Cl(TPP)_n]$ (n = 1: **1a**, n = 2: **1b**, n = 3: **1c**). Depending on the stoichiometric ratio of the reactants, $[Cu^{II}_4 OCl_6(TPPO)_4]$ (**2**) and $[Cu^{II} Cl_2(TPPO)_2]$ (**2a**) were found as by-products [17,18]. However, it is not known to what extent μ_4-oxido copper clusters $[Cu^{II}_4 OCl_6 L_4]$ behave similarly to simple copper salts and, for example, form multinuclear Cu(I) compounds during the reaction with TPP. Further, it is unknown to what extent multinuclear Cu(I) and Cu(II) species are in equilibrium with the aforementioned, largely mononuclear complexes and, thus, behave analogously to μ_4-oxido copper clusters with amine ligands known from the literature [14–16]. Therefore, building on previous results, we investigated the formation of $[Cu^{II}_4 OCl_6(TPP)_4]$ and $[Cu^{II}_4 OCl_6(TPPO)_4]$ as well as the reactivity of $[Cu^{II}_4 OCl_6 L_4]$ (L = MeOH, CH_3CN) towards TPP/TPPO. With these experiments, we sought to elucidate the complex redox equilibrium of TPP/TPPO containing Cu(I) and Cu(II) compounds and thus, to complete the set of already known compounds that form this equilibrium. For this purpose, we have chosen three synthetic routes for our investigations, which are known to generally lead to the formation of μ_4-oxido copper clusters with amine donors. To date, only the first approach had been described in the literature as a synthetic route for the preparation of **1**.

In our experiments, we could obtain a huge variety of mononuclear as well as multinuclear copper(II) and copper(I) compounds; these included three new compounds (**2b**, **2c**, $CuCl \cdot CH_3CN$) that could be crystallographically characterized. Furthermore, due to the systematic variation of the reaction conditions, we could show that the reaction between copper(II) chloride and TPP/TPPO yields a complex chemical redox equilibrium of Cu(I)/TPP and Cu(II)/TPPO compounds, which, in contrast to previous studies, does not only include **2** and mononuclear compounds but also multinuclear compounds such as **1**, **2b**, **2c**, and **3**. Furthermore, we could show that the solvent clusters $[Cu^{II}_4 OCl_6(solv)_4]$ (solv = methanol, CH_3CN) show a similar reactivity towards TPP as that of $CuCl_2$. Depending on the crystallization conditions, individual compounds can be crystallized, as described below.

The template synthesis of $[Cu^{II}_4 OCl_6(solv)_4]$ (solv = methanol, CH_3CN) yielded **1**, **2**, and **3**. Using acetone instead of methanol and CH_3CN as solvent, pure **3** could be crystallized, thereby confirming our assumption that multinuclear Cu(I) compounds are also involved in the chemical equilibrium. Additionally, $[Cu^I Cl(TPP)_n]$ (n = 1, 2, 3) was obtained from the reaction of $[Cu^{II}_4 OCl_6(MeOH)_4]$ with TPP in methanol, which, as yet, had only been reported for the reaction of $CuCl_2$ and TPP. It is known that the redox instability of such copper(II) phosphine complexes formed from $CuCl_2$ and TPP results in their decomposition to form Cu(I) complexes [19,25,27], which goes in hand with the reduction of Cu(II) to Cu(I) by an excess of TPP at room temperature [17]. The oxidized product of this reaction is TPPO, which is known to be formed if traces of H_2O are in the reaction solution [17]. The driving force of this reaction is the formation of the P=O bond [40] that is one of the strongest double bonds with 575 kJmol^{-1} bond energy [18,41]. However, in general, the autoxidation of arylphosphines is slow because of the high stability of the intermediate phosphoranyl radical $PH_3\dot{P}O_2R$ [42].

In our experiments, the trend to form Cu(I) compounds was even more pronounced in the reaction of CuO, $CuCl_2$, and TPP under inert conditions at 60 °C, from which **2a** was obtained as the main product as well as the copper(I) compounds CuCl, CuCl·CH$_3$CN (from CH$_3$CN), and **3** (from acetone); of these Cu(I) compounds, CuCl·CH$_3$CN could be crystallographically characterized for the first time. When carrying out the same reaction with TPPO instead of TPP, a variety of copper(II) compounds (**2**, **2a**, **2b**, and **2c**) were obtained, which extends the known set of CuCl$_2$-TPPO complexes by **2b** and **2c** and highlights the tendency of CuCl$_2$ to form multinuclear complexes that are bridged via chloride ions.

The reaction of CuCl$_2$·2H$_2$O and TPP under atmospheric conditions and at room temperature yielded diverse copper(I) complexes **1a–1c**, which already are known to be formed due to the reaction of CuCl$_2$ and TPP under inert conditions. As to be expected, the number of TPP ligands increased with increasing excess of TPP. Additionally, **2** formed over a long standing time in the reaction mixture with CuCl$_2$·2H$_2$O in excess. Similar studies with TPPO instead of TPP only yielded CuCl and **2a**, whereby the reaction yielding Cu(I) under these conditions is not known. The results of all experiments are presented in Table 2. In general, at least concerning the copper(I) complexes **1a-1c**, indeed, the reaction outcome can be influenced by the stoichiometric ratio of the starting compounds. However, with regard to the copper(II) compounds as well as the multinuclear compounds **1**, **2**, and **3**, the choice of solvent for the preparation as well as for the recrystallization governs the reaction outcome.

Table 2. Overview of compounds obtained depending on the reaction conditions.

Entry	Starting Compound	Ligand Added	Products
1	[Cu$^{II}_4$OCl$_6$(MeOH)$_4$]	TPP	[Cu^{I}Cl(TPP)$_n$] (n = 1–3, **1a–1c**) [Cu$^{II}_4$OCl$_6$(TPP)$_4$] (**1**) [Cu$^{II}_4$OCl$_6$(TPPO)$_4$] (**2**) [{Cu$^I_4$Cl$_4$}(TPP)$_4$] (**3**)
2	[Cu$^{II}_4$OCl$_6$(CH$_3$CN)$_4$]	TPPO	[Cu$^{II}_4$OCl$_6$(CH$_3$CN)$_{n-4}$(TPPO)$_n$] [Cu^{II}Cl$_2$(TPPO)$_2$] (**2a**)
3	CuO, CuCl$_2$	TPP	[{Cu$^I_4$Cl$_4$}(TPP)$_4$] (**3**) Cu^{I}Cl·CH$_3$CN
4	CuO, CuCl$_2$	TPPO	[Cu$^{II}_4$OCl$_6$(TPPO)$_4$] (**2**) [Cu^{II}Cl$_2$(TPPO)$_2$] (**2a**) [{Cu^{II}Cl$_2$}$_2$(TPPO)$_2$] (**2b**) [{Cu^{II}Cl$_2$}$_3$(TPPO)$_2$] (**2c**)
5	CuCl$_2$·2H$_2$O	TPP	[Cu$^{II}_4$OCl$_6$(TPPO)$_4$] (**2**) [Cu^{I}Cl(TPP)] (**1a**) [Cu^{I}Cl(TPP)$_2$] (**1b**) [Cu^{I}Cl(TPP)$_3$] (**1c**)
6	CuCl$_2$·2H$_2$O	TPPO	[Cu$^{II}_4$OCl$_6$(TPPO)$_4$] (**2**) [Cu^{II}Cl$_2$(TPPO)$_2$] (**2a**) CuCl

We characterized the compounds prepared via IR spectroscopy, melting points, elemental analysis, and, if possible, X-ray diffraction analysis (see Supplementary Materials). Even though only few IR data have been reported to date [25,43,44], IR spectroscopy is a suitable method for the differentiation of the compounds named above. In general, all compounds show similar vibration bands; however, they can be differentiated by the characteristic νCu$^{II}_4$O, νCu$^I_4$Cl$_4$, νC-P, and νP-O in the IR spectra. Compounds **1**, **2**, and **3** display characteristic vibrations of the metal core at 540 cm^{-1}, 572 cm^{-1}, and 542 cm^{-1}, respectively, whereby the values for **1** and **2** correspond well to those described in the literature (see Supplementary Materials Figure S15) [4,17,18,25]. In the free TPPO ligand, νC-P (542 cm^{-1}) is blue-shifted in comparison to the free TPP ligand (513 cm^{-1}). Hence, νC-P of the TPPO-containing compounds **2** (536 cm^{-1}) and **2a** (544 cm^{-1}) falls into the range of the metal core vibrations of **1** and **3** (see Supplementary Materials, Figure S16).

However, the confusion of these bands, and hence compounds, can be excluded by comparison with the range 1300 cm^{-1}–1050 cm^{-1}, where only the TPPO-containing compounds **2** and **2a** show the characteristic νP-O (**2**: 1198 cm^{-1}, **2a**: 1143 cm^{-1}, free TPPO: 1190 cm^{-1}, see Supplementary Materials, Figure S17). Furthermore, **1**, **2**, and **3** show 3 characteristic absorption bands for δC-H$_{arom.}$ (*out of plane*) in the range 775 cm^{-1}–650 cm^{-1}, whereas **1a–1c** only show 2 characteristic bands (see Supplementary Materials, Figure S18). Again, the TPPO-containing compounds **2** and **2a** as well as free TPPO show three absorption bands (see Supplementary Materials, Figure S19); thus, a clear differentiation of **1**, **1a–1c**, **2**, **2a**, and **3** based on IR spectroscopy alone is only possible via the appearance/absence of νP-O.

Compounds **1a-1c** can be differentiated by the absorptions in the fingerprint region (570–475 cm^{-1}) of the IR spectra. The spectra of **1b** and **1c** show vibrations similar to free TPP (517 cm^{-1}, 503 cm^{-1}, and 494 cm^{-1}, TPP: 513 cm^{-1}, 500 cm^{-1}, and 494 cm^{-1}), with the vibration at lowest energy only pronounced as a shoulder. An additional absorption band is observed at 527 cm^{-1}. The IR spectrum of **1a**, however, shows only 2 absorption bands, at 542 cm^{-1} and 501 cm^{-1} in this range (see Supplementary Materials Figure S20). Further differences for these compounds are observed in the region 1250 cm^{-1}–1050 cm^{-1}. Even though all three compounds show similar absorption bands to the free TPP ligand, the shift and intensity of the vibrations are characteristic for **1a**, **1b**, and **1c** (see Supplementary Materials, Figure S21). The IR spectrum of **1c** shows 3 bands of similar intensity at 1221 cm^{-1}, 1184 cm^{-1}, and 1156 cm^{-1}. In contrast, the intensity of the absorption band at 1221 cm^{-1} is much lower in **1a** and **1b**. Furthermore, the spectrum of **1a** shows an additional absorption band at 1121 cm^{-1} that is not observed either for **1b** nor **1c**. Unfortunately, because of the small amounts obtained of **2b** and **2c**, it was not possible to record IR data.

5. Conclusions

In this paper, the formation of [Cu$^{II}_4$OCl$_6$(TPP)$_4$] and [Cu$^{II}_4$OCl$_6$(TPPO)$_4$] as well as the reactivity of [Cu$^{II}_4$OCl$_6$(solv)$_4$] (solv = solvent) towards TPP and TPPO was investigated. In addition, the reactivity of CuCl$_2$ (·2H$_2$O) towards TPP and TPPO was more extensively reinvestigated. Therefore, the reaction parameters as well as the copper-containing starting compounds were systematically varied. In doing so, previous reports, according to which TPP reduces Cu(II) to Cu(I) to yield [Cu^{I}Cl(TPP)$_n$ (n = 1–3)] (**1a-c**) [17], were also confirmed and could be transferred to the reaction of [Cu$^{II}_4$OCl$_6$(solv)$_4$] with TPP and of CuCl$_2$·2H$_2$O with TPP under atmospheric conditions. In general, our investigations allowed us to show that the previously described mononuclear complexes **1a–1c**, [Cu^{II}Cl$_2$(TPPO)$_4$]·2H$_2$O, and [Cu^{II}Cl$_2$(TPPO)$_2$] are in a chemical redox equilibrium with several multinuclear compounds such as [Cu$^{II}_4$OCl$_6$(TPP)$_4$] (**1**), [Cu$^{II}_4$OCl$_6$(TPPO)$_4$] (**2**), [{Cu$^I_4$Cl$_4$}(TPP)$_4$] (**3**), [{Cu^{II}Cl$_2$}$_2$(TPPO)$_2$] (**2b**), and [{Cu^{II}Cl$_2$}$_3$(TPPO)$_2$] (**2c**). Additionally, we identified and crystallographically characterized the three compounds **2b**, **2c**, and CuCl·CH$_3$CN for the first time. Taking together our results with a previous study concerning the solution equilibria of tertiary phosphine complexes of CuCl [45], it can be assumed that in solution, all compounds are in equilibrium and are obtained as a complex solid mixture when removing the solvent. The individual compounds can selectively be crystallized from this mixture by varying the solvent for recrystallization. This variation of the solvent and subsequent fractional crystallization could, accordingly, prove to be more suitable for reaction control and product retention than the previous methods, purely based on variation of the stoichiometric ratio of the reactants.

Supplementary Materials: The following supporting information can be downloaded at https://www.mdpi.com/article/10.3390/chemistry5020087/s1: Figures S1–S7: Molecular structure of **1c**, **2**, **2a**, **2b**, **2c**, **3** and **CuCl·CH$_3$CN** respectfully; Figures S8–S21: IR spectra, Tables S1–S14: Crystal data and structure refinement and selected bond lengths [Å] and angles [°] for **1c**, **2**, **2a**, **2a·2b**, **2c**, **3** and **CuCl·CH3CN** respectfully; Table S15: Elemental analysis of **1**, **1a**, **1b**, **1c**, **2**, **2a** and **3**; Table S16: Melting points of **1a**, **1b**, **1c**, **2**, and **2a**. References [46–52] are cited in the Supplementary Materials.

Author Contributions: Conceptualization, S.L.F. and S.B.; validation, S.L.F., N.I.D. and S.B.; investigation, S.L.F. and N.I.D.; writing—original draft preparation, S.L.F.; writing—review and editing, S.B.; visualization, S.L.F. and S.B.; supervision, S.B.; funding acquisition, S.B. All authors have read and agreed to the published version of the manuscript.

Funding: This research was funded by the SFB/TRR 88 "3MET–Cooperative effects in homo- and heterometallic complexes".

Institutional Review Board Statement: Not applicable.

Informed Consent Statement: Not applicable.

Data Availability Statement: Additional data (IR spectra, melting points, elemental analyses reports) are presented in the Supplementary Materials. The cif files of all crystal structures have been deposited with the Cambridge Structural Database (CSD). All determined parameters such as bond distances, U_{ij} components, etc., can be retrieved free of charge from the CSD (CCDC and CSD numbers: 2255472-2255478).

Acknowledgments: The authors thank Jonathan Becker (JLU Gießen) for the crystal measurement.

Conflicts of Interest: The authors declare no conflict of interest.

References

1. Haiduc, I. ReviewInverse coordination–An emerging new chemical concept. Oxygen and other chalcogens as coodination centers. *Coord. Chem. Rev.* **2017**, *338*, 1. [CrossRef]
2. Melník, M.; Kabešová, M.; Koman, M.; Macáškova, L.; Holloway, C.E. Copper(II) coordination compounds: Classification and analysis of crystallographic and structural data IV. Trimeric and oligomeric compounds. *J. Coord. Chem.* **1999**, *48*, 271. [CrossRef]
3. Melník, M.; Koman, M.; Ondrejovič, G. Tetramers Cu_4 $(\mu_4$-$O)(\eta$-$X)_6$ (L_4): Analysis of structural data. *Coord. Chem. Rev.* **2011**, *255*, 1581. [CrossRef]
4. Bertrand, J.A.; Kelley, J.A. Five-coordinate complexes. II. [1] Trigonal Bipyramidal Copper(II) in a Metal Atom Cluster. *J. Am. Chem. Soc.* **1966**, *88*, 4746. [CrossRef]
5. Bertrand, J.A. Five-coordinate complexes. III. Structure and properties of. μ_4-oxohexa-μ-chlorotetrakis {(triphenylphosphine oxide)copper (II)}[1]. *Inorg. Chem.* **1967**, *6*, 495. [CrossRef]
6. Lines, M.E.; Ginsberg, A.P.; Martin, R.L.; Sherwood, R.C. Magnetic Exchange in Transition Metal Complexes. VII. Spin Coupling in μ_4-Oxohexa-μ-halotetrakis [(Triphenylphosphine Oxide or Pyridine) Copper (II)]: Evidence for Antisymmetric Exchange. *J. Chem. Phys.* **1972**, *57*, 1. [CrossRef]
7. Wong, H.; Dieck, H.T.; O'Connor, C.J.; Sinn, E. Magnetic exchange interactions in tetranuclear copper(II) complexes. Effect of ligand electronegativity. *J. Chem. Soc. Dalton Trans.* **1980**, *5*, 786. [CrossRef]
8. Reim, J.; Griesar, K.; Haase, W.; Krebs, B. Structure and magnetism of novel tetranuclear μ_4-oxo-bridged copper(II) complexes. *J. Chem. Soc. Dalton Trans.* **1995**, *16*, 2649. [CrossRef]
9. Kirillova, M.V.; Kozlov, Y.N.; Shul'pina, L.S.; Lyakin, O.Y.; Kirillov, A.M.; Talsi, E.P.; Pombeiro, A.J.L.; Shul'pin, G.B. Remarkably fast oxidation of alkanes by hydrogen peroxide catalyzed by a tetracopper(II) triethanolaminate complex: Promoting effects of acid co-catalysts and water, kinetic and mechanistic features. *J. Catal.* **2009**, *268*, 26. [CrossRef]
10. Sun, H.; Harms, K.; Sundermeyer, J. Aerobic oxidation of 2,3,6-Trimethylphenol to Trimethyl-1,4-benzoquinone with Copper(II) Chloride as Catalyst in Ionic Liquid and Structure of the Active Species. *J. Am. Chem. Soc.* **2004**, *126*, 9550. [CrossRef]
11. Kirillov, A.M.; Kopylovich, M.N.; Kirillova, M.V.; Haukka, M.; Da Silva, M.F.C.G.; Pombeiro, A.J.L. Mild Peroxidative Oxidation of Cyclohexane Catalyzed by Mono-, Di-, Tri-, Tetra-and Polynuclear Copper Triethanolamine Complexes. *Angew. Chem.* **2005**, *117*, 4419. [CrossRef]
12. Kirillov, A.M.; Kopylovich, M.N.; Kirillova, M.V.; Haukka, M.; Da Silva, M.F.C.G.; Pombeiro, A.J.L. Mild Peroxidative Oxidation of Cyclohexane Catalyzed by Mono-, Di-, Tri-, Tetra-and Polynuclear Copper Triethanolamine Complexes. *Angew. Chem. Int. Ed.* **2005**, *44*, 4345. [CrossRef] [PubMed]
13. Gawlig, C.; Schindler, S.; Becker, S. One-Pot Conversion of Cyclohexane to Adipic Acid Using a μ_4-Oxido-Copper Cluster as Catalyst Together with Hydrogen Peroxide. *Eur. J. Inorg. Chem.* **2020**, *2020*, 248. [CrossRef]
14. Löw, S.; Becker, J.; Würtele, C.; Miska, A.; Kleeberg, C.; Behrens, U.; Walter, O.; Schindler, S. Reactions of Copper(II) Chloride in Solution: Facile Formation of Tetranuclear Copper Clusters and Other Complexes That Are Relevant in Catalytic Redox Processes. *Chem. Eur. J.* **2013**, *19*, 5342. [CrossRef]
15. Becker, S.; Behrens, U.; Schindler, S. Investigations Concerning [$Cu_4OX_6L_4$] Cluster Formation of Copper(II) Chloride with Amine Ligands Related to Benzylamine. *Eur. J. Inorg. Chem.* **2015**, *14*, 2437. [CrossRef]
16. Becker, S.; Dürr, M.; Miska, A.; Becker, J.; Gawlig, C.; Behrens, U.; Ivanović-Burmazović, I.; Schindler, S. Copper Chloride Catalysis: Do μ_4-Oxido Copper Clusters Play a Significant Role? *Inorg. Chem.* **2016**, *55*, 3759. [CrossRef]
17. Makáňová, D.; Ondrejovič, G.; Gažo, J. Reaktionen des Kupfer(II)-chlorids mit Triphenylphosphin Isolierung und Identifizierung einiger Oxidoreduktionsprodukte. *Chem. Zvesti* **1973**, *27*, 4.

18. Berners-Price, S.J.; Sadler, P.J. *Bioinorganic Chemistry*; Springer: Berlin/Heidelberg, Germany, 1988; Volume 70, p. 27. [CrossRef]
19. Cahours, A.; Gal, H. Untersuchungen über neue Derivate des Triäthylphosphins. *Ann. Chem. Pharm.* **1870**, *155*, 355. [CrossRef]
20. Boden, P.; Di Martino-Fumo, P.; Busch, J.M.; Rehak, F.R.; Steiger, S.; Fuhr, O.; Nieger, M.; Volz, D.; Klopper, W.; Bräse, S.; et al. Cover Feature: Investigation of Luminescent Triplet States in Tetranuclear Cu^I Complexes: Thermochromism and Structural Characterization. *Chem. Eur. J.* **2021**, *27*, 5309. [CrossRef]
21. Lapprand, A.; Dutartre, M.; Khiri, N.; Levert, E.; Fortin, D.; Rousselin, Y.; Soldera, A.; Jugé, S.; Harvey, P.D. Luminescent P-chirogenic copper clusters. *Inorg. Chem.* **2013**, *52*, 7958. [CrossRef]
22. Perruchas, S.; Le Goff, X.F.; Maron, S.; Maurin, I.; Guillen, F.; Garcia, A.; Gacoin, T.; Boilot, J.-P. Mechanochromic and Thermochromic Luminescence of a Copper Iodide Cluster. *J. Am. Chem. Soc.* **2010**, *132*, 10967. [CrossRef] [PubMed]
23. Perruchas, S.; Tard, C.; Le Goff, X.F.; Fargues, A.; Garcia, A.; Kahlal, S.; Saillard, J.-Y.; Gacoin, T.; Boilot, J.-P. Thermochromic luminescence of Copper Iodide Clusters: The Case of Phosphine Ligands. *Inorg. Chem.* **2011**, *50*, 10682. [CrossRef] [PubMed]
24. Ford, P.C.; Cariati, E.; Bourassa, J. Photoluminescence Properties of Multinuclear Copper(I) Compounds. *Chem. Rev.* **1999**, *99*, 3625. [CrossRef] [PubMed]
25. Dieck, H.T.; Brehm, H. Vierkernige Komplexe mit trigonal-bipyramidal koordiniertem Kupfer (II), II. Synthese und Substitution an der axialen Position. *Chem. Ber.* **1969**, *102*, 3577. [CrossRef]
26. Zelonka, R.A.; Baird, M.C. Five-coordinate Tertiary Phosphine and Arsine Adducts of Copper(hexafluoroacetylacetonate)$_2$. *Can. J. Chem.* **1972**, *50*, 1269. [CrossRef]
27. Jardine, F.H.; Rule, L.; Vohra, A.G. The chemistry of copper (I) complexes. Part I. Halogeno-complexes. *J. Chem. Soc. A* **1970**, 238–240. [CrossRef]
28. Bock, H.; Dieck, H.T.; Pyttlik, H.; Schnöller, M. Über Kupfer-Komplexe Cu$_4$O(Amin)$_4$(Hal)$_6$ mit tetraedrisch koodiniertem Sauerstoff. *Z. Anorg. Allg. Chem.* **1968**, *357*, 54. [CrossRef]
29. Dieck, H.T. Tetranuclear complexes of Trigonal-bipyramidal Copper(II). III. Electronic and Infrared Spectra. *Inorg. Chim. Acta* **1973**, *7*, 397. [CrossRef]
30. Ondrejovič, G.; Melník, M.; Makáňová, D.; Gažo, J. Darstellung und Eigenschaften zweier isomerer Formen des Dichlorobis[triphenylphosphinoxid—Kupfer(II)]-Komplexes. *Monatsh. Chem.* **1977**, *108*, 1047. [CrossRef]
31. Bertrand, J.A.; Kalyanaraman, A.R. The Crystal and Molecular Structure of Dichlorobis(Triphenylphosphine Oxide)Copper(II). *Inorg. Chim. Acta* **1971**, *5*, 341. [CrossRef]
32. Goodgame, D.M.L.; Cotton, F.A. 445. Phosphine Oxide Complexes. Part IV. Tetrahedral, Planar, and Binuclear Complexes of Copper(II) with Phosphine Oxides, and some Arsine Oxide Analogues. *J. Chem. Soc.* **1961**, 2298–2305. [CrossRef]
33. Yang, L.; Powell, D.R.; Houser, R.P. Structural variation in copper (I) complexes with pyridylmethylamide ligands: Structural analysis with a new four-coordinate geometry index, τ_4. *Dalton Trans.* **2007**, *9*, 955. [CrossRef]
34. Churchill, M.R.; DeBoer, B.G.; Donovan, D.J. Molecules with an M$_4$X$_4$ core. IV. Crystallographic Detection of a Step Configuration for the Cu$_4$I$_4$ Core in Tetrameric Triphenylphosphinecopper (I) Iodide, [PPh$_3$CuI]$_4$. *Inorg. Chem.* **1975**, *14*, 617. [CrossRef]
35. Churchill, M.R.; Kalra, K.L. Molecules with an M$_4$X$_4$ core. II. X-ray Crystallographic Determination of the Molecular Structure of Tetrameric Triphenylphosphinecopper(I) Bromide in Crystalline [PPH$_3$CuBr]$_4$·2 CHCl3. *Inorg. Chem.* **1974**, *13*, 1427. [CrossRef]
36. Churchill, M.R.; Kalra, K.L. Crystallographic Studies on Sulfur Dioxide Insertion compounds. V.^{1}Elucidation of the Molecular Geometry of cis-μ-carbonyl-μ-(sulfur dioxide)-bis (π-cyclopentadienylcarbonyliron), cis-(π.-C$_5$H$_5$)$_2$Fe$_2$ (CO)$_3$ (SO$_2$). *Inorg. Chem.* **1973**, *12*, 1650. [CrossRef]
37. Takemura, Y.; Nakajima, T.; Tanase, T. Interconversion between ladder-type octanuclear and linear tetranuclear copper(I) complexes supported by tetraphosphine ligands. *Dalton Trans.* **2009**, *46*, 10231. [CrossRef]
38. Costa, G.; Pellizer, G.; Rubessa, F. Complex of copper-methyl with triphenylphosphine and other copper(I) complexes with triphenylphosphine. *J. Inorg. Nucl. Chem.* **1964**, *26*, 961. [CrossRef]
39. Albano, V.G.; Bellon, P.L.; Ciani, G.; Manassero, M. Crystal and Molecular Structure of Monoclinic Di-μ-chloro-tris- (triphenylphosphine) dicopper(I) Cu$_2$Cl$_2$(PPh$_3$)$_3$. *J. Chem. Soc., Dalton Trans.* **1972**, 171–175. [CrossRef]
40. Hartley, S.B.; Holmes, W.S.; Jacques, J.K.; Mole, M.F.; McCoubrey, J.C. Thermochemical properties of phosphorus compounds. *Q. Rev. Chem. Soc.* **1963**, *17*, 204. [CrossRef]
41. Clayden, J.; Greeves, N.; Warren, S.G. *Organische Chemie*; Springer: Berlin/Heidelberg, Germany, 2017. [CrossRef]
42. Ogata, Y.; Yamashita, M. Kinetics of the Autoxidation of Trimethyl Phosphite, Methyl Diphenyl-phosphinite, and Triphenylphosphine. *J. Chem. Soc. Perkin Trans.* **1972**, *2*, 730. [CrossRef]
43. Costa, G.; Reisenhofer, E.; Stefani, L. Complexes of copper(I) with triphenylphosphine. *J. Inorg. Nucl. Chem.* **1965**, *27*, 2581. [CrossRef]
44. Favarin, L.R.V.; Rosa, P.P.; Pizzuti, L.; Machulek, A.; Caires, A.R.L.; Bezerra, L.S.; Pinto, L.M.C.; Maia, G.; Gatto, C.C.; Back, D.F.; et al. Synthesis and Structural Characterization of New Heteroleptic Copper(I) Complexes Based on Mixed Phosphine/Thiocarbamoyl-pyrazoline Ligands. *Polyhedron* **2017**, *121*, 185. [CrossRef]
45. Fife, D.J.; Moore, W.M.; Morse, K.W. Solution Equilibria of Tertiary Phosphine Complexes of Copper(I) Halides. *Inorg. Chem.* **1984**, *23*, 1684. [CrossRef]
46. Krause, L.; Herbst-Irmer, R.; Sheldrick, G.M.; Stalke, D. Comparison of silver and molybdenum microfocus X-ray sources for single-crystal structure determination. *J. Appl. Crystallogr.* **2015**, *48*, 3–10. [CrossRef]
47. *APEX2*, version 2; Bruker AXS, Inc.: Madison, WI, USA, 2007.

48. *XPREP*, Bruker AXS, Inc.: Madison, WI, USA, 2014.
49. Sheldrick, G.M. SHELXT-Integrated space-group and crystal-structure determination. *Acta Crystallogr. A Found. Adv.* **2015**, *71*, 3–8. [CrossRef]
50. Sheldrick, G.M. Crystal structure refinement with SHELXL. *Acta Crystallogr. C Struct. Chem.* **2015**, *71*, 3–8. [CrossRef]
51. Hübschle, C.B.; Sheldrick, G.M.; Dittrich, B. ShelXle: A Qt graphical user interface for SHELXL. *J. Appl. Crystallogr.* **2011**, *44*, 1281–1284. [CrossRef]
52. *PLATON*, Multipurpose Crystallographic Tool, Spek, A.L.; University of Utrecht: Utrecht, The Netherlands, 2008.

Disclaimer/Publisher's Note: The statements, opinions and data contained in all publications are solely those of the individual author(s) and contributor(s) and not of MDPI and/or the editor(s). MDPI and/or the editor(s) disclaim responsibility for any injury to people or property resulting from any ideas, methods, instructions or products referred to in the content.

Article

Reactivity of Rare-Earth Oxides in Anhydrous Imidazolium Acetate Ionic Liquids

Sameera Shah [1,2], Tobias Pietsch [1,2], Maria Annette Herz [1], Franziska Jach [1,2] and Michael Ruck [1,2,*]

[1] Faculty of Chemistry and Food Chemistry, Technische Universität Dresden, 01062 Dresden, Germany
[2] Max Planck Institute for Chemical Physics of Solids, Nöthnitzer Straße 40, 01187 Dresden, Germany
* Correspondence: michael.ruck@tu-dresden.de

Abstract: Rare-earth metal sesquioxides (RE_2O_3) are stable compounds that require high activation energies in solid-state reactions or strong acids for dissolution in aqueous media. Alternatively, dissolution and downstream chemistry of RE_2O_3 have been achieved with ionic liquids (ILs), but typically with additional water. In contrast, the anhydrous IL 1-butyl-3-methylimidazolium acetate [BMIm][OAc] dissolves RE_2O_3 for RE = La–Ho and forms homoleptic dinuclear metal complexes that crystallize as $[BMIm]_2[RE_2(OAc)_8]$ salts. Chloride ions promote the dissolution without being included in the compounds. Since the lattice energy of RE_2O_3 increases with decreasing size of the RE^{3+} cation, Ho_2O_3 dissolves very slowly, while the sesquioxides with even smaller cations appear to be inert under the applied conditions. The Sm and Eu complex salts show blue and red photoluminescence and Van Vleck paramagnetism. The proton source for the dissolution is the imidazolium cation. Abstraction of the acidic proton at the C^2-atom yields an N-heterocyclic carbene (imidazole-2-ylidene). The IL can be regenerated by subsequent reaction with acetic acid. In the overall process, RE_2O_3 is dissolved by anhydrous acetic acid, a reaction that does not proceed directly.

Keywords: anhydrous complexes; dinuclear complexes; ionic liquids; photoluminescence; rare-earth oxides; regeneration; Van Vleck paramagnetism

Citation: Shah, S.; Pietsch, T.; Herz, M.A.; Jach, F.; Ruck, M. Reactivity of Rare-Earth Oxides in Anhydrous Imidazolium Acetate Ionic Liquids. *Chemistry* 2023, 5, 1378–1394. https://doi.org/10.3390/chemistry5020094

Academic Editors: Christoph Janiak, Sascha Rohn and Georg Manolikakes

Received: 31 March 2023
Revised: 19 May 2023
Accepted: 22 May 2023
Published: 2 June 2023

Copyright: © 2023 by the authors. Licensee MDPI, Basel, Switzerland. This article is an open access article distributed under the terms and conditions of the Creative Commons Attribution (CC BY) license (https://creativecommons.org/licenses/by/4.0/).

1. Introduction

Rare earths, such as monazite, bastnasite, loparite and laterite clays, are the natural source of rare-earth elements and their compounds, which are essential, e.g., for luminophores, magnets, catalysts, superconductors, biomedical diagnosis, therapy, and environmental chemistry [1–3]. Although rare-earth elements are not as rare as their name suggests, their extraction is demanding, which has an impact on both price and environment. In the first production step, the phosphate mineral monazite is dissolved either in concentrated H_2SO_4 at temperatures between 150 and 200 °C (acid process) or in 70% NaOH (basic process) at 140 to 150 °C using autoclaves. After this activation step, an elaborate procedure for the separation of the different metal species follows. Nowadays, the recycling of industrial and electronic waste has also become an important source of rare-earth metals. In many of these processes, rare-earth metal oxides occur as intermediate or final products. For any downstream chemistry, high temperatures or aggressive chemicals must again activate these stable oxides.

With the goal of establishing a more sustainable process, much research has been done in recent years on the chemistry of rare-earth compounds in ionic liquids (ILs) [4–9]. In most investigations, the focus lies on the development of new liquid extraction and separation techniques. Unfortunately, most ILs successfully tested for this purpose have severe weaknesses. First-generation ILs, such as $[BMIm]Cl \cdot nAlCl_3$ ($[BMIm]^+$ = 1-n-butyl-3-methylimidazolium), are sensitive to air and moisture, which strongly limits their usefulness [10]. Super-acidic ILs combined with a stream of chlorine gas also dissolve rare-earth oxides [11]; however, using a highly toxic gas does not achieve the aim of green chemistry. Nevertheless, there are also some existing reports on the direct dissolution of rare-earth oxides in ILs [12–17]. Water-stable ILs containing Brønsted acidic groups, in particular the very effective IL [Hbet][NTf$_2$] ([Hbet]$^+$ = betainium, [NTf$_2$]$^-$ = bis(trifluoromethylsulfonyl)imide), are much more suitable. Its excellent dissolution power for metal oxides is based on the acidity of the [Hbet]$^+$ cation and the complexation capacity of the conjugate base (bet) [18]. The weakly coordinating anion [NTf$_2$]$^-$, however, is expensive and ecologically questionable [19]. Moreover, these IL-based approaches typically involve water as a co-solvent. However, water narrows the electrochemical window and can have a detrimental effect on the subsequent chemistry.

In this work, we show that the anhydrous IL 1-butyl-3-methylimidazolium acetate [BMIm][OAc] can dissolve rare-earth oxides. Acetate, [OAc]$^-$, is much cheaper than [NTf$_2$]$^-$ and ecologically harmless. Moreover, it has a complexing effect. Similarly, the nitrogen atoms in the positions 1 and 3 of the imidazole ring can form N-donor coordination complexes [20]. The proton at the C^2-position of the ring is acidic and can be abstracted by a base. This generates an N-heterocyclic carbene (imidazole-2-ylidene), which can be functional in organo-catalytic reactions [21,22]. Imidazolium acetate ILs have been used to dissolve biomass, e.g., cellulose, and to store CO_2 [23–26].

It is known that traces of chloride, which is a strong nucleophile in most ILs, facilitate the dissolution of metal oxides [27,28] but also of covalently bonded substances such as red phosphorus [29–31]. We have also applied this here and succeeded in establishing an effective and inexpensive method for the chemical activation of RE_2O_3 (RE = La–Ho, except the radioactive Pm) at moderate temperature, including the recycling of the IL. In addition, we crystallized anhydrous, homoleptic, dinuclear rare-earth metal complexes from these solutions and studied the magnetic and luminescence properties of the Eu and Sm complexes, which were previously synthesized from expensive anhydrous rare-earth metal trihalides [32].

2. Materials and Methods

Chemicals. The ionic liquids 1-butyl-3-methylimidazolium acetate [BMIm][OAc] (98%), 1-butyl-2,3-dimethylimidazolium chloride [BDMIm]Cl, and 1-butyl-3-methylimidazolium chloride [BMIm]Cl (99%) were purchased from iolitec (ionic liquids technology GmbH (Heilbronn, Germany). Potassium acetate (99.9%) was purchased from Merck (Darmstadt, Germany). [BDMIm][OAc] was synthesized from [BDMIm]Cl and potassium acetate in ethanol following a literature procedure for the production of [BMIm][OAc] [33]. All ILs were dried at 110 °C under dynamic vacuum (Schlenk line) overnight and transferred to the glove box before use. HPLC grade ethanol and acetonitrile were obtained from Merck (Darmstadt, Germany). Acetic acid (100%) was obtained from Carl Roth (Karlsruhe, Germany). Ce_2O_3 was synthesized by heating CeO_2 and Ce to 1500 °C for 24 h in a sealed tantalum ampule [34]. The other rare-earth oxides (99.9%) were obtained from Fluka Chemie GmbH (Buchs, Switzerland).

Synthesis of [BDMIm][OAc]. [BDMIm][OAc] was synthesized by mixing 20 mmol (3.75 g) of [BDMIm]Cl in 10 mL anhydrous ethanol with 50 mL of an ethanolic solution of 4.8 g potassium acetate under vigorous stirring at room temperature [33]. KCl precipitated quickly, but the solution remained in the stirring bath overnight at room temperature for a complete reaction. KCl was removed via vacuum filtration, and the resultant filtrate was dried in a rotatory evaporator at 70 °C under reduced pressure. Post-precipitation of potassium acetate required additional vacuum filtration. The resulting pale yellow liquid was dried under dynamic vacuum (Schlenk line) at 110 °C overnight.

Synthesis of [BMIm]$_2$[RE$_2$(OAc)$_8$]. For RE = La–Eu, 1 mmol of RE_2O_3 powder was mixed with 10 mmol of [BMIm][OAc] and 1 mmol of [BMIm]Cl in a round bottom flask with magnetic stirring at 175 °C. For RE = Gd–Ho, the amount of [BMIm]Cl was increased to 5 mmol and 175 °C was applied for 48 h. After cooling the solutions to room temperature, platelet-shaped crystals of the complex salts [BMIm]$_2$[RE$_2$(OAc)$_8$]

precipitated within few days. The crystals were separated by centrifugation and washed several times with acetonitrile.

Regeneration of [BMIm][OAc]. After separation of the [BMIm]$_2$[RE$_2$(OAc)$_8$] precipitate, the reacted IL contained a large portion of imidazole-2-ylidene, residuals of the dissolved complex as well as acetonitrile from the washing procedure. The IL was regenerated by mixing with a 2 mol L^{-1} solution of acetic acid in acetonitrile at room temperature under inert conditions. A white precipitate formed, which was centrifuged, and the supernatant liquid phase was pipetted off. Acetonitrile and excess acetic acid were evaporated off from the organic phase, first at 70 °C using a rotatory evaporator for 3 h and then at 100 °C on a Schlenk line under dynamic vacuum overnight.

Powder X-ray Diffraction. PXRD data were collected on an Empyrean diffractometer PANalytical (Malvern, Worcestershire, UK) equipped with a curved Ge(111)-monochromator in Bragg-Brentano geometry at 296 K using Cu-Kα_1 radiation (λ = 1.54056 Å). The data were collected in reflection mode in the range $5° \leq 2\theta \leq 90°$, with a step width of $\Delta(2\theta) = 0.01°$.

Single-Crystal Structure Determination. Single-crystal X-ray diffraction data were collected on a four-circle Kappa APEX II CCD diffractometer (Bruker, Madison, WI, USA) with a graphite(002)-monochromator and a CCD detector using Mo-Kα radiation (λ = 0.71073 Å) under flowing nitrogen gas at T = 100 (2) K. The data were corrected for background, Lorentz factor, and polarization factor using the APEX III software. Empirical multi-scan absorption correction was applied [35,36]. The structures were solved using SHELXT in OLEX2 [37,38]. SHELXL was used for refinement against F^2 [39,40]. Anisotropic displacement parameters were refined for all non-hydrogen atoms. H atoms were refined without constraints. The structure was visualized with the Diamond software Version 4.6.8 (Klaus Brandenburg, Bonn, Germany) [41].

NMR Spectroscopy. The samples were dissolved in deuterated DMSO-d6 and filled in an NMR tube. Measurements were performed on a Bruker Avance (Leipzig, Germany) Neo WB 300 MHz NMR spectrometer at resonance frequencies of 300 MHz for ^{1}H and 75.5 MHz for ^{13}C spectra. The chemical shifts were referenced externally according to tetramethylsilane (TMS) and were calibrated with 10% ethylbenzene in CDCl$_3$ as external reference. The signals in the spectra were assigned with the NMR Predictor from ACD/Labs [42].

UV-Vis Spectroscopy. The sample was a powder consisting of 10 mg of the complex salt and 100 mg of BaSO$_4$. The spectra were recorded with a solid-state UV-Vis spectrophotometer Varian-4000 (Varian BV, Middelburg, The Netherland).

Emission Spectroscopy. Photoluminescence of powder samples was measured on a spectrofluorometer Fluoro Max-4r (Horiba scientific, Piscataway, NJ, USA). The light source was a xenon arc lamp with photomultiplier R928P detector.

IR Spectroscopy. FTIR spectra of liquid samples soon after dissolution (no visible precipitates) as well as a crystal of [BMIm]$_2$[Eu$_2$(OAc)$_8$] were measured on a Bruker Vertex 70 FTIR spectrophotometer (Bruker Optics GmbH, Ettlingen, Germany) with attenuated total reflection (ATR) accessory in the range from 500 to 4000 cm^{-1}, with 32 scans per measurement. The data were analyzed using the OPUS 6.5 software (Bruker, Ettlingen, Germany) [43].

Thermogravimetric Analysis (TGA) and Differential Scanning Calorimeter (DSC). The thermal stability of several complex salts was investigated with TGA using an STA 409 Luxx (Netzsch, Selb, Germany). The TGA experiments were carried out in the temperature range from 25 to 500 °C, with a heating rate of 5 K min^{-1} in an argon stream. The thermal effects upon heating with 5 K min^{-1} in an argon stream were determined using a DSC1 (Mettler Toledo GmbH, Giessen, Germany).

Magnetic Measurement. The magnetic susceptibility of the Sm and Eu complex salts were measured with helium-cooled mini-CFMS (Cryogenic, The Vale London, UK) equipped with a vibrating-sample magnetometer (VSM, The Vale London, UK) in the temperature range from 300 K to 2 K.

3. Results and Discussion

3.1. Dissolution of RE_2O_3

In orienting experiments, we tested the dissolution of the comparatively inexpensive La_2O_3 in seven phosphonium- or imidazolium-based ILs at 175 °C in closed flasks under argon (Table S1). Only in the case of [BMIm][OAc] was no solid residue found after 48 h. Droplets of water condensed in the upper, colder part of the flask. After cooling the solution to room temperature, the complex compound $[BMIm]_2[La_2(OAc)_8]$ crystallized within some days (see below). The only hydrogen source for the formation of water in this system is the acidic proton at the C^2 atom of the imidazolium cation. In fact, La_2O_3 did not dissolve in 1-butyl-2,3-dimethyl imidazolium acetate, [BDMIm][OAc], which is methylated at the C^2-position and thus has no acidic proton. The reaction is summarized in Equation (1):

$$La_2O_3 + 8\,[BMIm][OAc] \rightarrow [BMIm]_2[La_2(OAc)_8] + 3\,H_2O + 6\,BMIm\text{-2-ylidene} \quad (1)$$

The deprotonated [BMIm]$^+$ cation is a N-heterocyclic carbene (imidazole-2-ylidene). This is sensitive to air at the reaction temperature, which leads to decomposition when the reaction is performed in a system open to the atmosphere. However, at room temperature the reactivity of the carbene is strongly reduced, allowing handling of the solution and separation of the solid product in air.

It is known that the basic acetate ion can abstract the acidic proton from the [BMIm]$^+$ cation and form acetic acid [44,45]. Remarkably, La_2O_3 did not dissolve in neat acetic acid (100%) at 175 °C under argon. However, when La_2O_3 was mixed with [BDMIm][OAc] and acetic acid, a vigorous reaction occurred already at room temperature, resulting in a white microcrystalline precipitate of a lanthanum complex that was soluble in water, partially soluble in ethanol, and insoluble in acetone (for PXRD see Figure S1). Obviously, the cation of the IL is necessary for the dissolution; through this, there is enough free acetate in the solution for the formation of the acetate complex, and a stable salt can be precipitated.

According to Equation (1), the molar ratio of RE_2O_3 and [BMIm][OAc] must be at least 1:8. Since the IL is not only a reactant but also the solvent, an excess is advisable to avoid overly high viscosity. To still obtain fairly concentrated solutions, we kept the ratio of 1:10 for the subsequent experiments with RE_2O_3. In view of the catalytic effect of chloride on the dissolution of many metal oxides in ILs, we also repeated the dissolution experiments with admixtures of [BMIm]Cl in diverse molar ratios as the chloride source. A summary of the dissolution experiments performed can be found in Table S2.

Achieving complete dissolution below 150 °C proved difficult, as the viscosity increased with decreasing temperature, which affected the diffusion and wetting behavior of the IL. In addition, below 120 °C, abstraction of the proton from the substituted imidazolium cations was hindered [44].

The lighter RE_2O_3 (RE = La–Nd) dissolved completely, also without the addition of chloride. However, chloride shortened the reaction time and the quality of the crystals was improved (Figure S2), although chloride is not a component of the complex salt. For RE = Dy–Ho, the dissolution took much more time, especially without chloride. RE_2O_3 with RE = Er–Lu, Sc, or Y did not react with [BMIm][OAc], even when the reaction time, the temperature, or the concentration of the IL was increased or chloride had been added (Figure 1). Instead, decomposition of the IL was observed.

Such a dissolution behavior is well known and understood. In the series of RE^{3+} cations, the effective ionic radius (for coordination number 9) decreases from 1.216 Å (La^{3+}) to 1.032 Å (Lu^{3+}) [46]. The smaller the RE^{3+} cation, the shorter the cation–anion distance, the higher the lattice energy, and the poorer the solubility of the RE_2O_3. In the underlying case, the boundary is between 1.08 and 1.07 Å, with Ho (1.072 Å) showing poor solubility and Y (1.075 Å) and Er (1.062 Å) apparently none. CeO_2, which has a much high lattice energy due to the higher cation charge, did not dissolve at all.

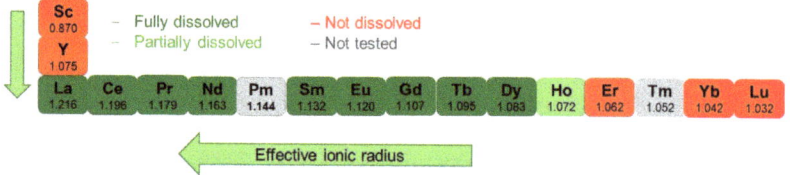

Figure 1. Qualitative solubility of RE_2O_3 in [BMIm][OAc] or mixtures of [BMIm][OAc] with [BMIm]Cl. The effective ionic radii of the RE^{3+} cations for CN 9 are given in Å [46].

This behavior can be used for the separation of rare-earth elements. The four equimolar mixtures La_2O_3/Sc_2O_3, La_2O_3/Lu_2O_3, Eu_2O_3/Y_2O_3, and Eu_2O_3/CeO_2 were treated at 175 °C in an IL consisting of [BMIm][OAc] and [BMIm]Cl (molar ratio of 10:1) for few hours. The unreacted solid was separated from the liquid, washed three times with acetonitrile, weighted, and analyzed with PXRD. In all cases, the diffraction patterns showed no evidence of the soluble oxide (Figure S3), and the weight of the recovered insoluble oxide was within 1% of the original. From the separated solutions, the $[BMIm]_2[RE_2(OAc)_8]$ salts (RE = La, Eu) precipitated.

3.2. Regeneration of the Reacted Ionic Liquid

Under prolonged thermal stress, the dialkylimidazolium cations of the corresponding acetate ILs decompose into alkylated anions and imidazole derivatives [25,47–49] (see also Section 3.4). The decomposition is the reverse reaction of the quaternization (reversed Menshutkin-like reaction). Further decomposition products result from the deprotonation of the C^2-atom of the imidazolium ring by strong nucleophiles [50], as the resulting N-heterocyclic carbene reacts either with the IL or with other decomposition products [51]. However, the carbene formation is reversible, and the IL can be regenerated by the addition of acid. Irreversible decomposition of the IL can be largely avoided by heating at moderate temperature and for only short periods of time.

The regeneration of used [BMIm][OAc] (Figure S4) was done by reaction with a solution of acetic acid in acetonitrile at room temperature under inert conditions. After evaporation of the acetonitrile and excess acetic acid, the recovered IL was analyzed spectroscopically. The FTIR spectra of neat and regenerated IL are almost identical (Figure S5). In the ^{13}C and ^1H NMR spectra (Figure 2), no residual acetic acid and acetonitrile was found in the regenerated IL. Generally, the proton signal of acetic acid lies between 10 and 15 ppm, and that of acetonitrile near 2 ppm [52,53]. The decomposition products methyl acetate and 1-butylimidazole were also not detected. While the ester was likely cleaved in the treatment, the alkylimidazole likely precipitated as a coordination compound (white precipitate, see Materials and Methods).

No solid residue remained after combustion of the regenerated IL in air. The ^{13}C NMR spectra of regenerated and neat IL showed no difference as long as the freshly regenerated IL was stored under inert conditions (sample A). After some time in air (sample B), the ^1H NMR spectrum of the regenerated IL showed a broad resonance signal at 5 ppm, which can be attributed to H_2O, and also an up-field shift of the protons at C^2, C^4, and C^5. The same effect was observed when water was added to the freshly regenerated IL (Figure S6) as has been reported previously [54,55]. The new signals marked with asterisks in the spectra of (B) indicate a decomposition product. This is probably due to the small amount of carbene that is formed by the internal acid–base equilibrium between $[BMIm]^+$ and $[OAc]^-$ (Section 3.4).

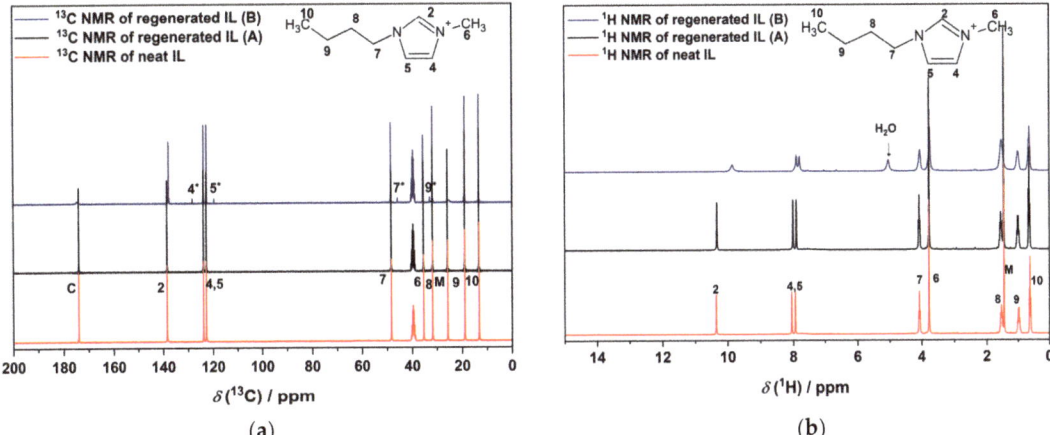

Figure 2. (a) ^{13}C and (b) ^{1}H NMR spectra of neat [BMIm][OAc] and the regenerated IL. The regenerated IL was stored under inert conditions (A) or in air (B). The signals of [BMIm]$^+$ are assigned to the atoms at the marked positions in the structures above [26]. M and C represent the methyl and carboxylic groups of acetate. The asterisk sign * indicates the decomposition product in regenerated IL in air (B). Deuterated DMSO-d6 was used for diluting all samples.

3.3. [BMIm]$_2$[RE$_2$(OAc)$_8$] Complex Salts

After one to four days at room temperature, salts of homoleptic dinuclear rare-earth metal complexes [BMIm]$_2$[RE$_2$(OAc)$_8$] (RE = La–Pr, Nd, Sm–Ho) crystallized from the RE_2O_3 solutions in [BMIm][OAc] (with or without chloride). The yields were between 25 and 84% (Table S2). The powder X-ray diffractogram of [BMIm]$_2$[Eu$_2$(OAc)$_8$] confirms a single-phase product (Figure S7).

Unlike many other complexes obtained from RE_2O_3 and ILs, these are anhydrous compounds. Up to now, such anhydrous complexes were made from anhydrous chlorides, nitrates, or acetates, i.e., starting materials that must be synthesized and are much more expensive than the oxides [56–58].

The washed crystals of [BMIm]$_2$[RE$_2$(OAc)$_8$] are air-stable (thick) platelets and appear colorless, except the light-green Pr salt. In the case of the Pr and Dy compounds, the platelet-shaped crystals transformed into thin needles within a day when stored in the reacted IL in air at 10 °C (Figure S8). The powder diffractogram and the IR spectra of the platelets and the needles clearly differ (Figures S9 and S10). Subsequently, the needles turned into yellowish translucent flakes that decompose into dark brown microscale needles within several days.

The structure of plate-like crystals was determined with single-crystal X-ray diffraction (Table S3). As the sum formula suggests, the compounds consist of [BMIm]$^+$ cations and negatively charged [RE_2(OAc)$_8$]$^{2-}$ complexes. Nine oxygen atoms of acetate anions coordinate each RE^{3+} cation in the shape of a distorted tricapped trigonal prism (Figure 3). The RE–O bond length develops according to the effective radii of the RE^{3+} cations (Table S4). The carboxylate groups of the acetate anions coordinate in three different modes (Figure 4). The two RE^{3+} anions are connected through four carboxylate groups.

The salts [BMIm]$_2$[RE$_2$(OAc)$_8$] adopt two different types of crystal structure, depending on the size of the rare-earth (RE) cation. Both structure types have triclinic symmetry (space group type $P\bar{1}$) and contain two formula units in the unit cell; however, their lattice translations are very different. Bond lengths (Table S5) and the basic arrangements of the molecules are similar in the two structure types, but they differ in translational and molecular point symmetry. Figure S11 shows analogous subcells of the two structure types, reduced to the RE^{3+} cations. Their positions relative to the locations of the inversion centers differ in the two cases. Translational pseudosymmetry is observed (B-centering for RE = La–Pr and A-centering for RE = Nd–Ho). Yet, there are clear superstructure reflections that corroborate the primitive Bravais lattices.

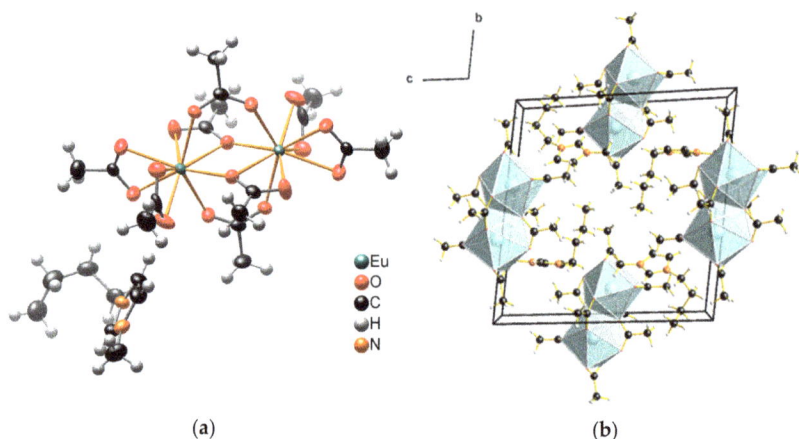

(a) (b)

Figure 3. (a) Molecular and (b) crystal structure of [BMIm]$_2$[Eu$_2$(OAc)$_8$]. The ellipsoids (in (a)) enclose a space in which with 90% probability the electron density of the atoms can be found at 100 K.

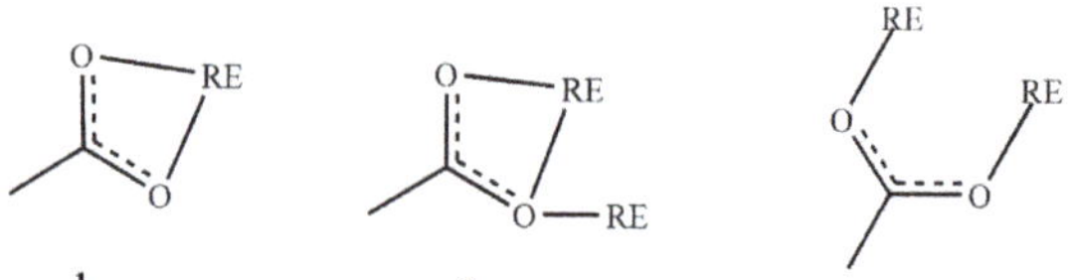

Figure 4. Coordination modes of acetate ions in the [BMIm]$_2$[RE$_2$(OAc)$_8$] complexes. (**1**) Chelating (η^2 mode), (**2**) chelating, bridging (μ_2–η^2: η^1 mode), (**3**) bridging (μ_2–η^1: η^1 mode). η represents the number of coordinating ligand atoms to the metal atom, and μ represents the number of metal atoms that are coordinated by the same ligand.

For RE = La–Pr, the anionic [RE$_2$(OAc)$_8$]$^{2-}$ complex is non-centrosymmetric, with two crystallographically independent RE atoms. However, it has a pseudo-center of inversion. The two crystallographically distinct [BMIm]$^+$ cations show a combined disorder. The majority position of one cation is connected to the minority position of the other cation by the pseudo-inversion. The fact that there are two positions for the [BMIm]$^+$ cation that can be alternatively occupied without significant impact on the position of the anions suggests that the complexes form a packing that includes larger cavities than needed for the organic cations.

In the second structure type, which is adopted by the compounds RE = Nd–Ho, the [RE$_2$(OAc)$_8$]$^{2-}$ complexes are truly centrosymmetric. The packing is slightly different, and the [BMIm]$^+$ cations are ordered. Thus, it could be argued that this packing is more efficient than the first; however, there is no unusual reduction in the volume of the unit cell when changing the structure type from the Pr to the Nd compound.

3.4. IR and NMR Spectra

Nockemann et al. showed that dinuclear rare-earth betaine complexes [RE$_2$(bet)$_8$(H$_2$O)$_2$]$^{6+}$ can dissociate into mononuclear units in ILs [59]. It could therefore be argued that, also in the underlying case, the dissolved species might be mononuclear. Acetates have been studied extensively with vibrational spectroscopy. Among the 15 active infrared frequencies in acetate ions, the symmetric and asymmetric stretching frequencies of the carboxylic group were analyzed using FTIR spectroscopy. The spectra of various RE_2O_3 dissolved in the IL as well as of crystalline $[BMIm]_2[Eu_2(OAc)_8]$ show additional bands along with shifted and split bands of the neat IL (Figure 5).

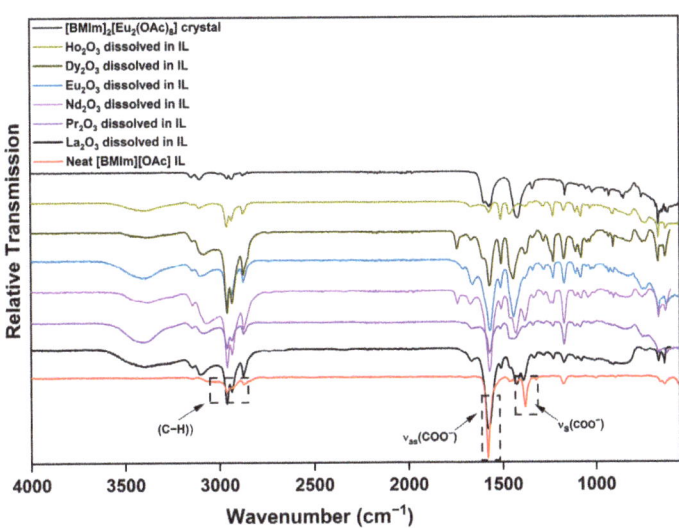

Figure 5. FTIR spectra of the $[BMIm]_2[Eu_2(OAc)_8]$ salt, diverse solutions of RE_2O_3 in [BMIm][OAc], and the neat IL.

The C–H stretching frequencies of the organic cation in neat [BMIm][OAc] lie at about 3000 cm^{-1}, e.g., $\nu_{as}(CH_2) = 2952$, $\nu_{as}(CH_3) = 2932$, and $\nu_s(CH_3)$ and $\nu_s(CH_2) = 2873$ cm^{-1}. The asymmetric ν_{as} and symmetric ν_s stretching frequencies of the carboxylic group are found at 1575 and 1385 cm^{-1} [60]. For the solutions and the complex salt, the signals of the carboxylic group are slightly shifted and split or broadened according to the diverse coordination modes of the acetate ion. Nakamoto et al. showed that the nature of the acetate coordination is related to the difference of $\nu_{as}-\nu_s$ [61]. Since the diverse signals cannot be reliably assigned to the different coordination modes, further evaluation is not possible. Nonetheless, the differences in the FTIR spectra suggest that the complexes in solution can differ from the crystallized one [62]. The broad bands at about 3400 cm^{-1} in the spectra of the solutions indicate the water formed during the dissolution of the oxides. When the system was kept open during the reaction, this led to the decomposition of the IL, as shown by the FTIR spectrum in Figure S12.

Using ^{13}C and ^1H NMR spectroscopy (Figure 6), we compared three samples (Figure S4): the unreacted IL, a freshly prepared solution of La_2O_3 in [BMIm][OAc] from which no precipitation had occurred, and the supernatant of a solution from which the complex salt had precipitated and which had been stored in air for several months. In addition, a dark solution containing decomposition products of [BMIm][OAc] after heating La_2O_3 in the IL for 4 days at 175 °C was investigated (Figure S13).

In the ^{13}C and ^1H NMR spectra of the freshly reacted solution, methyl acetate and 1-butylimidazole are identified as decomposition products known from the literature [25,63,64]. ^{13}C NMR-DEPT-135 spectroscopy (DEPT = Distortionless Enhancement by Polarization Transfer; Figure S14) supports the assignment of the signals. The ^{13}C NMR signal of the carboxylic group at 175 ppm is broadened and drastically reduced. Since the signal of the methyl group of acetate is still intense, we attribute this effect to the interaction with the La^{3+} cations. A closer look into the ^1H NMR spectrum shows also the presence of water at 5.7 ppm. However, no sign of the formation of BMIm-2-

ylidene or its reaction products could be found, although it must have formed at least intermediately. Although dimerization according to the Wanzlick equilibrium is not expected for this type of N-heterocyclic carbene [65,66], it cannot be excluded that the high reaction temperature allows for it.

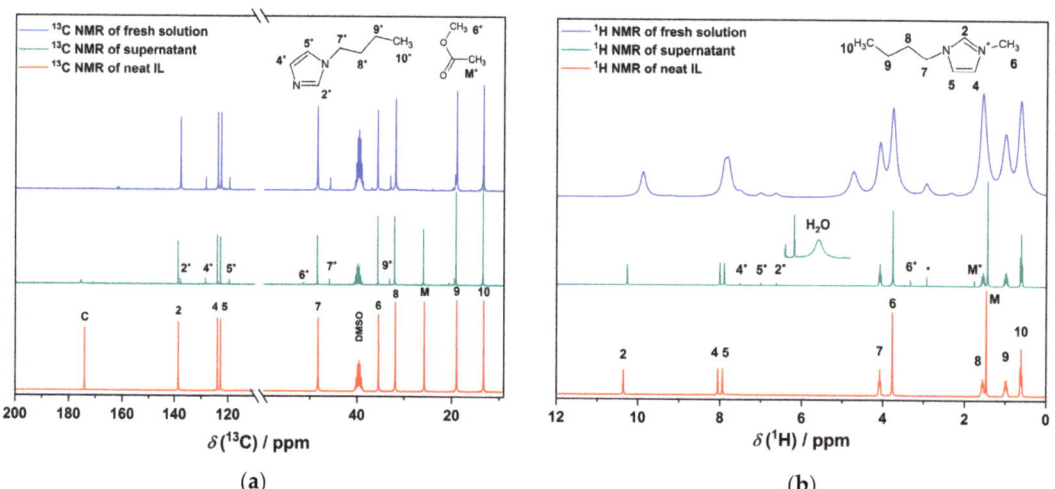

Figure 6. (a) ^{13}C and (b) ^{1}H NMR spectra of neat [BMIm][OAc], a freshly prepared solution of La$_2$O$_3$ in [BMIm][OAc] from which no precipitation had occurred, and the supernatant of a solution from which the complex salt had precipitated and which had been stored in air for several months. The signals of [BMIm]$^+$ are assigned to the atoms at the marked positions in the structures above [26]. The asterisk sign * indicates the decomposition product in the reacted IL. M* and 6* represent the methyl group of methyl acetate (decomposition product of the IL). Perdeuterated DMSO-d6 was used for diluting all samples (^{13}C signal around 40 ppm).

The ^{1}H NMR spectrum of the aged supernatant shows only weak and broadened resonances, while the ^{13}C NMR signals are still sharp. We attribute this to a higher water content that has accumulated in the hygroscopic IL during the long period of exposure to the air. With a higher water content, the water signals shift to 4.7 ppm. The influence of the water concentration on the NMR signal is shown in Figure S6. Remarkably, the ^{13}C NMR signals of the [BMIm]$^+$ cation are still visible, but no corresponding (carbon-based) anion can be identified.

3.5. Stability of the Complex Salts

The [BMIm]$_2$[RE$_2$(OAc)$_8$] complexes were obtained as single-phase products, allowing their further chemical and physical characterization. Since no decomposition was observed by visual inspections over a period of ten months, the washed crystals can be considered stable under ambient conditions. In contrast, unwashed samples are degraded to an unknown product (Figure S15).

The thermal stability in argon atmosphere of washed [BMIm]$_2$[RE$_2$(OAc)$_8$] crystals (RE = Pr, Sm, Eu, Dy) was investigated with a differential scanning calorimeter (DSC) and thermogravimetric analysis (TGA) in the temperature range from 25 to 500 °C. All samples showed a two-step decomposition (Figure 7a), with a uniform onset temperature of 250 °C for first weight loss of 54.3%. Remarkably, the decomposition of imidazolium-based ILs was reported to start at 216 °C in argon atmosphere, resulting in complete evaporation of all fragments at 300 °C [67]. In a second step starting at about 330° C and up to the maximum temperature of 500 °C, the mass was reduced further by 8% of the initial mass. The residual of 37.7% of the initial mass matches the calculated mass of $Eu_2O_2CO_3$ (37.5%), whose formation under similar conditions has been described previously [62]. The PXRD patterns of the solid residual shows a poorly crystalline material, which could be $Eu_2O_2CO_3$ (Figure S16). The DSC thermograms of the $[BMIm]_2[Eu_2(OAc)_8]$ complex salt shows a sharp endothermic transition at 167 °C, which indicates melting of the compound, and a broad endothermic effect starting above 220 °C, which is associated with its decomposition (Figure 7b).

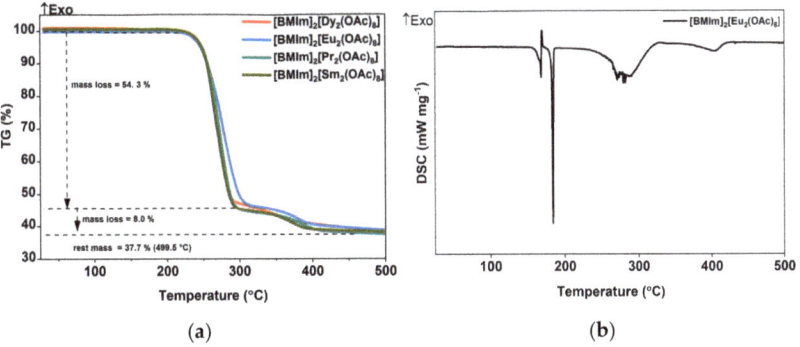

Figure 7. (a) TG and (b) DSC analysis of crystal structure of $[BMIm]_2[Eu_2(OAc)_8]$ during heating in a stream of argon.

3.6. Photoluminescence of the Eu and Sm Salts

Isolation of the homoleptic dinuclear rare-earth complexes as single-phase materials allowed us to characterize their optical properties. The UV-Vis spectra of $[BMIm]_2[Eu_2(OAc)_8]$ and $[BMIm]_2[Sm_2(OAc)_8]$ showed broad absorption bands at 214 and 261 nm (Figure 8). This absorptions correspond to $\pi \to \pi^*$ transitions of the IL cation [68]. The absorption bands observed above 350 nm were assigned to the absorption bands of the respective rare-earth metal cations.

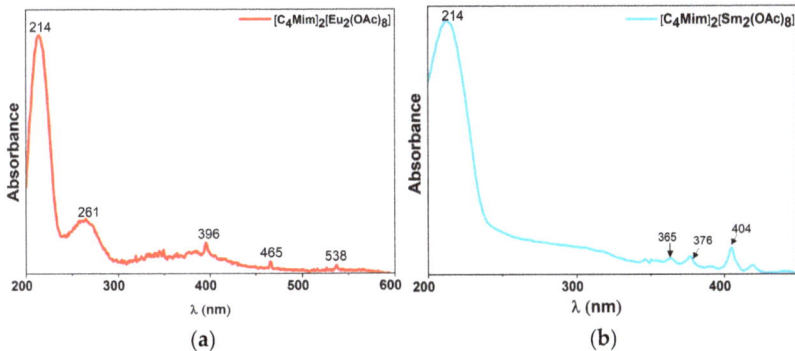

Figure 8. Solid-state UV-Vis spectra of (a) $[BMIm]_2[Eu_2(OAc)_8]$ and (b) $[BMIm]_2[Sm_2(OAc)_8]$.

Under a UV lamp, $[BMIm]_2[Eu_2(OAc)_8]$ emits red light (Figure 9), $[BMIm]_2[Tb_2(OAc)_8]$ pale-green light (Figure S17), and $[BMIm]_2[Sm_2(OAc)_8]$ blue light (Figure 9). Rare-earth metal ions can show strong luminescence due to transitions involving their 4f orbitals. A narrow emission band is characteristic of metalorganic complexes [69], in which the organic ligand absorbs energy and the intramolecular transfer of this energy from π^* or n^* to the next closest 4f level occurs (antenna effect). The intensity of the photoluminescence depends on the energy gap between the singlet S1 and the triplet state of the ligand, and the band gap between the triplet T1 and the 4f resonance states of the metal cation. Geometry and electronic features of the rare-earth metal complex influence the emission of spectra [7,70]. The coordination of polar molecules reduces the lifetime of the luminescence and the quantum yields. The depopulation of excited states takes place through the vibronic coupling of X-H oscillators (X = N, C, O) [71].

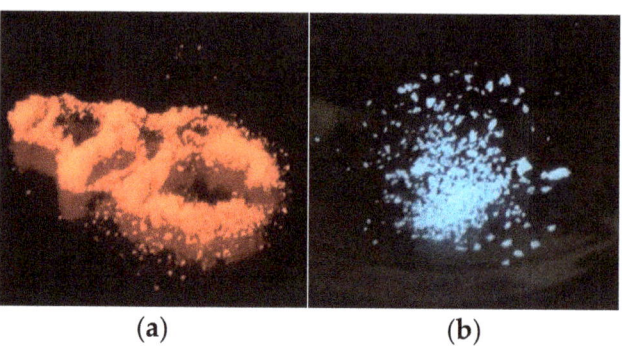

Figure 9. (a) [BMIm]$_2$[Eu$_2$(OAc)$_8$] and (b) [BMIm]$_2$[Sm$_2$(OAc)$_8$] under UV radiation of 312 nm.

The excitation and emission spectra were measured at room temperature for [BMIm]$_2$[Eu$_2$(OAc)$_8$] and [BMIm]$_2$[Sm$_2$(OAc)$_8$] (Figure 10). The excitation spectrum of the Eu compound was monitored at 615 nm, that is, the emission of Eu^{3+} ($^5D_0 \to {}^7F_2$). A series of intense and narrow lines was observed (in nm): 361, 367 ($^7F_0 \to {}^5D_4$), 374 ($^7F_{0/1} \to {}^5G_j$), 381, 385 ($^7F_{0/1} \to {}^5L_7, {}^5G_j$), 396 ($^7F_0 \to {}^5L_6$), 416 ($^7F_0 \to {}^5D_3$), and 465 ($^7F_0 \to {}^5D_2$). The most prominent excitations are at 396 and 465 nm, which is similar to the excitation spectrum of Eu(DCA)$_3$ [5]. Using the excitation at 396 nm, the emission spectrum showed lines at (in nm): 580 ($^5D_0 \to {}^7F_0$), 592 ($^5D_0 \to {}^7F_1$), 615 ($^5D_0 \to {}^7F_2$), and 651 ($^5D_0 \to {}^7F_3$), while the ($^5D_0 \to {}^7F_4$) transition band composed of two stark sublevels at 687 nm and 697 nm under the electric field (crystal field) were observed [72–74]. The very sharp $^5D_0 \to {}^7F_2$ transition is more intense than $^5D_0 \to {}^7F_1$. This correlates with a Eu^{3+} cation in a distorted tricapped trigonal prism [5].

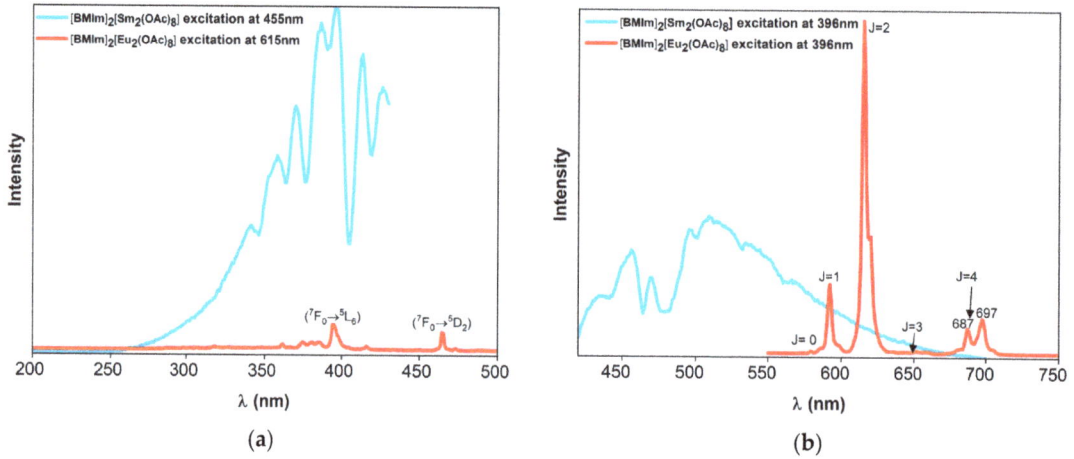

Figure 10. (a) Excitation and (b) emission spectra of [BMIm]$_2$[Eu$_2$(OAc)$_8$] and [BMIm]$_2$[Sm$_2$(OAc)$_8$].

In the case of the Sm complex salt, the excitation spectra were measured at 455 nm. The entire excitation spectrum is composed of multiple broad bands, ranging from 270 to 420 nm, most likely due to inefficient energy transfer from the ligand to the Sm^{3+} cation. The maximum excitation was observed at 396 nm. The emission lines are also broad, with overlapping bands ranging from 400 to 700 nm. Commonly, Sm^{3+} emits orange-red light, but in $[BMIm]_2[Sm_2(OAc)_8]$, the emission band is in the blue region, which cannot be explained without further investigations. Energy transfer from the imidazolium cation, which shows fluorescence between 250 and 400 nm [68], may be involved.

3.7. Magnetism of the Eu and Sm Salts

The rare-earth metal ions show magnetism mainly due to the interelectron repulsion and spin-orbit coupling. Strong spin-orbit coupling can also affect the degenerate state of the lanthanides and split them into $^{2S+1}L_J$ multiplets that further split into Stark levels through the crystal field. The magnetic property of the rare-earth ions thus depends on the splitting and thermal population of the resulting states. The magnetic moment of $[BMIm]_2[Eu_2(OAc)_8]$ and $[BMIm]_2[Sm_2(OAc)_8]$ complexes were measured from 2 K to 300 K under a DC magnetic field. The temperature dependence of the molar susceptibility χ_m (Figure 11) indicates ordinary Langevin paramagnetism for $[BMIm]_2[Sm_2(OAc)_8]$. In contrast, for $[BMIm]_2[Eu_2(OAc)_8]$, an increase of the magnetic susceptibility with decreasing temperature to a plateau below 100 K was observed. This behavior is consistent with Van Vleck paramagnetism [75], commonly observed for Eu^{3+}-containing compounds. The increase in χ_m below 15 K could be explained by traces of Eu^{2+} ions, as has been observed in dinuclear europium complexes before [76].

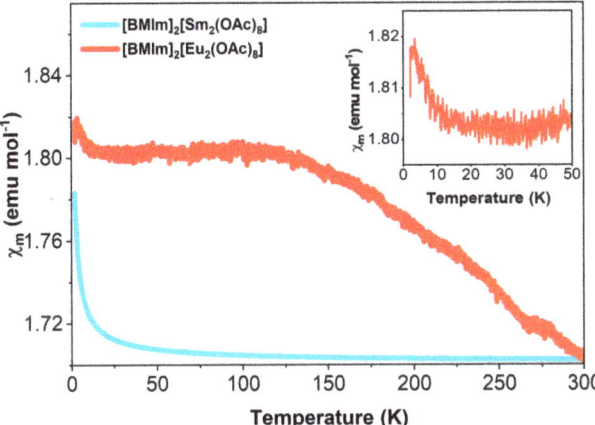

Figure 11. Temperature dependence of the molar susceptibility of $[BMIm]_2[Eu_2(OAc)_8]$ and $[BMIm]_2[Sm_2(OAc)_8]$.

Generally, only the electronic ground states are occupied at low and room temperatures in the free rare-earth metal ions due to the larger-than-k_BT energy splitting between the multiplets [77]. However, the Eu^{3+} ($4f^6$) and Sm^{3+} ($4f^5$) complexes are exceptions, as they show temperature-independent van Vleck contributions to the susceptibility. The indicative sign for this behavior is a significant curvature in the inverse susceptibility [78], which is observed for both samples (Figures S18 and S19). Due to a small energetic difference between the ground and the first excited state, the latter is substantially populated at higher temperatures [79,80]. At 300 K, $\chi_m T$ is 2.37 emu K mol^{-1} for $[BMIm]_2[Eu_2(OAc)_8]$ and 1.24 emu K mol^{-1} for $[BMIm]_2[Sm_2(OAc)_8]$. $\chi_m T$ decreases continuously with decreasing temperature, which is caused by the depopulation of excited states. At low temperatures, a rigid linear change is still observed, where the product $\chi_m T$ remains nonzero due to the thermal population of the first excited state [79]. The effective magnetic moment per Eu^{3+} ion of 3.02 μ_B at 0.1 T in the high-temperature region decreases to 0.15 μ_B without a field using the molar susceptibility at 300 K. The effective magnetic moment per Sm^{3+} ion is 2.40 μ_B at 0.1 T. Similar values were found in the literature [58,79,81].

4. Conclusions

A straightforward approach was used for the study of anhydrous dissolution of RE_2O_3 in an imidazolium acetate ionic liquid. The fact that the use of water can be avoided opens up new possibilities for downstream chemistry. The proton source for the dissolution is the imidazolium cation. Abstraction of the acidic proton at the C^2-atom yields an N-heterocyclic carbene (imidazole-2-ylidene). The used IL can be regenerated by reaction with acetic acid. Chloride promotes the dissolution, especially for the heavier RE_2O_3 (RE = Gd–Ho), although it is not included in the precipitating salts [BMIm]$_2$[RE_2(OAc)$_8$] (RE = La–Ho). The latter contain anhydrous, homoleptic, dinuclear complexes. The samarium and europium complex salts show blue and red photoluminescence, respectively. They are paramagnetic, with substantial van Vleck contributions.

Supplementary Materials: The following supporting information can be downloaded at: https://www.mdpi.com/article/10.3390/chemistry5020094/s1, Table S1. Results of reactions of 1 mmol La$_2$O$_3$ with 10 mmol of various imidazolium or phosphonium-based ILs. Table S2. Reaction parameters of RE_2O_3 used in this work. Figure S1. PXRD pattern of the white microcrystalline precipitate formed by the reaction of La$_2$O$_3$ with [BDMIm][OAc] and acetic acid at room temperature together with the patterns of some reference compounds. Figure S2. Crystals of [BMIm]$_2$[La$_2$(OAc)$_8$] formed in the absence or presence of chloride. Figure S3. PXRD diffractogram of the solid residuals after selective solution of one component from the mixtures La$_2$O$_3$/Sc$_2$O$_3$, La$_2$O$_3$/Lu$_2$O$_3$, Eu$_2$O$_3$/Y$_2$O$_3$, and Eu$_2$O$_3$/CeO$_2$. Figure S4. Pure [BMIm][OAc], regenerated IL, La$_2$O$_3$ dissolved in the IL, and decomposed IL. Figure S5. FTIR spectra of neat and regenerated [BMIm][OAc]. Figure S6. ^{1}H NMR spectra of regenerated [BMIm][OAc] with different water content. Figure S7. Simulated and measured powder X-ray diffractogram of [BMIm]$_2$[Eu$_2$(OAc)$_8$]. Figure S8. Transformation of [BMIm]$_2$[Dy$_2$(OAc)$_8$]. Figure S9. PXRD diffractogram measured at 100 K of [BMIm]$_2$[Dy$_2$(OAc)$_8$] crystals grown at room temperature or at 5 °C. Figure S10. FTIR spectra of the IL and [BMIm]$_2$[Dy$_2$(OAc)$_8$] crystals grown at room temperature or at 5 °C. Table S3. Crystallographic data for [BMIm]$_2$[RE_2(OAc)$_8$] salts at 100(2) K. Table S4. Metal–metal and metal–oxygen bond distances of all complexes studied in this work. Table S5. Selected interatomic distances in the Eu and La compounds, which belong to different structure types. Figure S11. Analogous sections of the structures of [BMIm]$_2$[Pr$_2$(OAc)$_8$] and [BMIm]$_2$[Ho$_2$(OAc)$_8$]. Figure S12. FTIR spectra of neat [BMIm][OAc], a fresh solution of La$_2$O$_3$ in the IL, and a solution with decomposed IL. Figure S13. ^{13}C and ^{1}H NMR spectra of neat [BMIm][OAc], of a freshly prepared solution of La$_2$O$_3$ in the IL, and of a solution with largely decomposed IL. Figure S14. ^{13}C NMR spectrum and DEPT 135 ^{13}C NMR spectrum of a freshly prepared La$_2$O$_3$ solution in [BMIm][OAc]. Figure S15. Degradation of [BMIm]$_2$[Dy$_2$(OAc)$_8$] crystals in the reaction mixture after one month in air. Figure S16. PXRD diffractogram of [BMIm]$_2$[Eu$_2$(OAc)$_8$] samples after DSC or after TG. Figure S17. Luminescence of [BMIm]$_2$[Tb$_2$(OAc)$_8$] under UV light of 312 nm. Figure S18. $\chi_m T$ and χ_m^{-1} plots for [BMIm]$_2$[Sm$_2$(OAc)$_8$] and [BMIm]$_2$[Eu$_2$(OAc)$_8$]. Figure S19. Field-dependent magnetization of [BMIm]$_2$[Sm$_2$(OAc)$_8$] and [BMIm]$_2$[Eu$_2$(OAc)$_8$].

Author Contributions: Conceptualization, M.R. and S.S.; methodology, S.S. and T.P.; validation, M.R.; investigation, S.S., T.P. and F.J.; resources, M.R.; writing—original draft preparation, S.S.; writing—review and editing, S.S., M.A.H., T.P. and M.R.; visualization, S.S. and T.P.; supervision, M.R. All authors have read and agreed to the published version of the manuscript.

Funding: This research received no external funding.

Data Availability Statement: The data presented in this study are available in the supplementary materials and available on request from the corresponding author. Crystal structure data are available at the CCDC database with the following depository numbers: [BMIm]$_2$[La$_2$(OAc)$_8$] 2252467, [BMIm]$_2$[Pr$_2$(OAc)$_8$] 2252287, [BMIm]$_2$[Nd$_2$(OAc)$_8$] 2252437, [BMIm]$_2$[Sm$_2$(OAc)$_8$] 2252450, [BMIm]$_2$[Eu$_2$(OAc)$_8$] 2252260, [BMIm]$_2$[Ho$_2$(OAc)$_8$] 2252271, and [BMIm]$_2$[Tb$_2$(OAc)$_8$] 2252456.

Acknowledgments: The authors thank P. Höhn (MPI CPfS, Dresden) for the synthesis of Ce$_2$O$_3$. I. Kuhnert, F. Pabst, and E. Carrillo Aravena (all TU Dresden) are acknowledged for experimental support and discussions. We are grateful to A. Eychmüller and T. O. Starzynski (TU Dresden) for the photoluminescence measurement.

Conflicts of Interest: The authors declare no conflict of interest. The funders had no role in the design of the study; in the collection, analyses, or interpretation of data; in the writing of the manuscript; or in the decision to publish the results.

References

1. Yan, B. *Photofunctional Rare Earth Hybrid Materials*, 1st ed.; Springer Series in Materials Science; Springer: Singapore, 2017; pp. 1–261. [CrossRef]
2. Peng, J.B.; Kong, X.J.; Ren, Y.P.; Long, L.-S.; Huang, R.-B.; Zheng, L.-S. Trigonal Bipyramidal Dy$_5$ Cluster Exhibiting Slow Magnetic Relaxation. *Inorg. Chem.* **2012**, *51*, 2186–2190. [CrossRef] [PubMed]
3. Nockemann, P.; Beurer, E.; Driesen, K.; Van Deun, R.; Van Hecke, K.; Van Meervelt, L.; Binnemans, K. Photostability of a Highly Luminescent Europium β-Diketonate Complex in Imidazolium Ionic Liquids. *Chem. Commun.* **2005**, *34*, 4354–4356. [CrossRef] [PubMed]
4. Billard, I.; Gaillard, C.; Hennig, C. Dissolution of UO$_2$, UO$_3$ and of Some Lanthanide Oxides in BumimTf$_2$N: Effect of Acid and Water and Formation of UO$_2$(NO$_3$)$^{3-}$. *Dalton Trans.* **2007**, *37*, 4214–4221. [CrossRef]
5. Tang, S.-F.; Smetana, V.; Mishra, M.K.; Kelley, S.P.; Renier, O.; Rogers, R.D.; Mudring, A.-V. Forcing Dicyanamide Coordination to f-Elements by Dissolution in Dicyanamide-Based Ionic Liquids. *Inorg. Chem.* **2020**, *59*, 7227–7237. [CrossRef]
6. Tang, S.; Babai, A.; Mudring, A.-V. Europium-Based Ionic Liquids as Luminescent Soft Materials. *Angew. Chem. Int. Ed.* **2008**, *47*, 7631–7634. [CrossRef]
7. Prodius, D.; Mudring, A.V. Rare Earth Metal-Containing Ionic Liquids. *Coord. Chem. Rev.* **2018**, *363*, 1–16. [CrossRef]
8. Huang, Y.; Wang, D.; Duan, D.; Liu, J.; Cao, Y.; Peng, W. A Novel Dissolution and Synchronous Extraction of Rare Earth Elements from Bastnaesite by a Functionalized Ionic Liquid [Hbet][Tf$_2$N]. *Minerals* **2022**, *12*, 1592. [CrossRef]
9. Khodakarami, M.; Alagha, L. Separation and Recovery of Rare Earth Elements Using Novel Ammonium-Based Task-Specific Ionic Liquids with Bidentate and Tridentate O-Donor Functional Groups. *Sep. Purif. Technol.* **2020**, *232*, 115952. [CrossRef]
10. Brown, L.C.; Hogg, J.M.; Swadźba-Kwaśny, M. Lewis Acidic Ionic Liquids. *Top. Curr. Chem.* **2017**, *375*, 78. [CrossRef]
11. Yao, A.; Qu, F.; Liu, Y.; Qu, G.; Lin, H.; Hu, S.; Wang, X.; Chu, T. Ionic Liquids with Polychloride Anions as Effective Oxidants for the Dissolution of UO$_2$. *Dalton Trans.* **2019**, *48*, 16249–16257. [CrossRef]
12. Shahr El-Din, A.M.; Rizk, H.E.; Borai, E.H.; El Afifi, E.S.M. Selective Separation and Purification of Cerium(III) from Concentrate Liquor Associated with Monazite Processing by Cationic Exchange Resin as Adsorbent. *Chem. Pap.* **2023**, *77*, 2525–2538. [CrossRef]
13. Liu, T.; Ivanov, A.S. N-Oxide Ligands for Selective Separations of Lanthanides: Insights from Computation. *RSC Adv.* **2023**, *13*, 764–769. [CrossRef]
14. Fan, F.-L.; Qin, Z.; Cao, S.-W.; Tan, C.-M.; Huang, Q.-G.; Chen, D.-S.; Wang, J.-R.; Yin, X.-J.; Xu, C.; Feng, X.-G. Highly Efficient and Selective Dissolution Separation of Fission Products by an Ionic Liquid [Hbet][Tf$_2$N]: A New Approach to Spent Nuclear Fuel Recycling. *J. Inorg. Chem.* **2019**, *58*, 603–609. [CrossRef]
15. Mohapatra, P.K.; Mahanty, B. Direct Dissolution of Metal Oxides in Ionic Liquids as a Smart Strategy for Separation: Current Status and Prospective. *Sep. Sci. Technol.* **2022**, *57*, 2792–2823. [CrossRef]
16. Entezari-Zarandi, A.; Larachi, F. Selective Dissolution of Rare-Earth Element Carbonates in Deep Eutectic Solvents. *J. Rare Earths* **2019**, *37*, 528–533. [CrossRef]
17. Li, X.; Li, Z.; Orefice, M.; Binnemans, K. Metal Recovery from Spent Samarium-Cobalt Magnets Using a Trichloride Ionic Liquid. *ACS Sustain. Chem. Eng.* **2019**, *7*, 2578–2584. [CrossRef]
18. Nockemann, P.; Thijs, B.; Parac-Vogt, T.N.; Van Hecke, K.; Van Meervelt, L.; Tinant, B.; Hartenbach, I.; Schleid, T.; Vu, T.N.; Minh, T.N.; et al. Carboxyl-Functionalized Task-Specific Ionic Liquids for Solubilizing Metal Oxides. *Inorg. Chem.* **2008**, *47*, 9987–9999. [CrossRef]
19. Quintana, A.A.; Sztapka, A.M.; Ebinuma, V.D.C.S.; Agatemor, C. Enabling Sustainable Chemistry with Ionic Liquids and Deep Eutectic Solvents: A Fad or the Future? *Angew. Chem. Int. Ed.* **2022**, *61*, e202205609. [CrossRef]
20. Kelley, S.P.; Smetana, V.; Emerson, S.D.; Mudring, A.V.; Rogers, R.D. Benchtop Access to Anhydrous Actinide N-Donor Coordination Complexes Using Ionic Liquids. *Chem. Commun.* **2020**, *56*, 4232–4235. [CrossRef]

21. Shah, S.; Pietsch, T.; Ruck, M. N-Heterocyclic Carbene-Mediated Oxidation of Copper (I) in an Imidazolium Ionic Liquid. *Z. Naturforsch. B* **2023**, *78*, 105–112. [CrossRef]
22. Flanigan, D.M.; Romanov-Michailidis, F.; White, N.A.; Rovis, T. Organocatalytic Reactions Enabled by N-Heterocyclic Carbenes. *Chem. Rev.* **2015**, *115*, 9307–9387. [CrossRef] [PubMed]
23. Liu, C.; Li, Y.; Hou, Y. Basicity Characterization of Imidazolyl Ionic Liquids and Their Application for Biomass Dissolution. *Int. J. Chem. Eng.* **2018**, *2018*, 7501659. [CrossRef]
24. Xu, A.; Guo, O.; Xu, R. Understanding the Dissolution of Cellulose in 1-Butyl-3-Methylimidazolium Acetate+DMAc Solvent. *Int. J. Biol. Macromol.* **2015**, *81*, 1000–1004. [CrossRef]
25. Liebner, F.; Patel, I.; Ebner, G.; Becker, E.; Horix, M.; Potthast, A.; Rosenau, T. Thermal Aging of 1-Alkyl-3-Methylimidazolium Ionic Liquids and Its Effect on Dissolved Cellulose. *Holzforschung* **2010**, *64*, 161–166. [CrossRef]
26. Besnard, M.; Cabaço, M.I.; Vaca Chávez, F.; Pinaud, N.; Sebastião, P.J.; Coutinho, J.A.P.; Mascetti, J.; Danten, Y. CO_2 in 1-Butyl-3-Methylimidazolium Acetate. 2. NMR Investigation of Chemical Reactions. *J. Phys. Chem. A* **2012**, *116*, 4890–4901. [CrossRef] [PubMed]
27. Richter, J.; Ruck, M. Dissolution of Metal Oxides in Task-Specific Ionic Liquid. *RSC Adv.* **2019**, *9*, 29699–29710. [CrossRef]
28. Orefice, M.; Binnemans, K.; Vander Hoogerstraete, T. Metal Coordination in the High-Temperature Leaching of Roasted NdFeB Magnets with the Ionic Liquid Betainium Bis(Trifluoromethylsulfonyl)Imide. *RSC Adv.* **2018**, *8*, 9299–9310. [CrossRef]
29. Lê Anh, M.; Wolff, A.; Kaiser, M.; Yogendra, S.; Weigand, J.J.; Pallmann, J.; Brunner, E.; Ruck, M.; Doert, T. Mechanistic Exploration of the Copper(I) Phosphide Synthesis in Phosphonium-Based and Phosphorus-Free Ionic Liquids. *Dalton Trans.* **2017**, *46*, 15004–15011. [CrossRef]
30. Wolff, A.; Pallmann, J.; Boucher, R.; Weiz, A.; Brunner, E.; Doert, T.; Ruck, M. Resource-Efficient High-Yield Ionothermal Synthesis of Microcrystalline $Cu_{3-x}P$. *Inorg. Chem.* **2016**, *55*, 8844–8851. [CrossRef]
31. Wolff, A.; Pallmann, J.; Brunner, E.; Doert, T.; Ruck, M. On the Anion Exchange of PX_3 (X = Cl, Br, I) in Ionic Liquids Comprising Halide Anions. *Z. Anorg. Allg. Chem.* **2017**, *643*, 20–24. [CrossRef]
32. Getsis, A.; Balke, B.; Felser, C.; Mudring, A.V. Dysprosium-Based Ionic Liquid Crystals: Thermal, Structural, Photo- and Magnetophysical Properties. *Cryst. Growth Des.* **2009**, *9*, 4429–4437. [CrossRef]
33. Abu-Eishah, S.I.; Elsuccary, S.A.A.; Al-Attar, T.H.; Khanji, A.A.; Butt, H.P.; Mohamed, N.M. Production of 1-Butyl-3-Methylimidazolium Acetate [Bmim][Ac] Using 1-Butyl-3-Methylimidazolium Chloride [Bmim]Cl and Silver Acetate: A Kinetic Study. In *Ionic Liquids—Thermophysical Properties and Applications*, 2nd ed.; Murshed, S.M.S., Ed.; IntechOpen Limited: London, UK, 2021; pp. 1–21. [CrossRef]
34. Hamm, C.M.; Alff, L.; Albert, B. Synthesis of Microcrystalline Ce_2O_3 and Formation of Solid Solutions between Cerium and Lanthanum Oxides. *Z. Anorg. Allg. Chem.* **2014**, *640*, 1050–1053. [CrossRef]
35. Sheldrick, G.M. *SADABS, Area-Detector Absorption Correction*; Bruker AXS Inc.: Madison, WI, USA, 2016.
36. Evans, P.R. An Introduction to Data Reduction: Space-Group Determination, Scaling and Intensity Statistics. *Acta Crystallogr. Sect. D* **2011**, *67*, 282–292. [CrossRef]
37. Sheldrick, G.M. A short history of SHELX. *Acta Crystallogr. Sect. A* **2008**, *64*, 112–122. [CrossRef]
38. OlexSys. *OLEX2*, version 1.2; OlexSys Ltd.: Durham, UK, 2014.
39. Sheldrick, G.M.; Schneider, T.R. SHELXL: High-Resolution Refinement. *Methods Enzymol.* **1997**, *277*, 319–343. [CrossRef]
40. Sheldrick, G.M. Crystal Structure Refinement with SHELXL. *Acta Crystallogr. Sect. C Struct. Chem.* **2015**, *71*, 3–8. [CrossRef]
41. Brandenburg, K. *DIAMOND 4, Crystal and Molecular Structure Visualization*, Version 4.6.8; Crystal Impact GbR: Bonn, Germany, 2022.
42. Advanced Chemistry Development, Inc. *ACchD/C+H NM Predictor and DB 2017.1.3*; Advanced Chemistry Development, Inc.: Toronto, ON, Canada, 2017; Available online: www.acdlabs.com (accessed on 10 July 2020).
43. *OPUS*, Version 6.5; Bruker Optik GmbH: Ettlingen, Germany, 2004.
44. Chiarotto, I.; Feroci, M.; Inesi, A. First Direct Evidence of N-Heterocyclic Carbene in BMIm Acetate Ionic Liquids. An Electrochemical and Chemical Study on the Role of Temperature. *New J. Chem.* **2017**, *41*, 7840–7843. [CrossRef]
45. Rodríguez, H.; Gurau, G.; Holbrey, J.D.; Rogers, R.D. Reaction of Elemental Chalcogens with Imidazolium Acetates to Yield Imidazole-2-Chalcogenones: Direct Evidence for Ionic Liquids as Proto-Carbenes. *Chem. Commun.* **2011**, *47*, 3222–3224. [CrossRef]
46. Shannon, R.D.; Prewitt, C.T. Effective Ionic Radii in Oxides and Fluorides. *Acta Crystallogr. Sect. B* **1969**, *25*, 925–946. [CrossRef]
47. Wendler, F.; Todi, L.N.; Meister, F. Thermostability of Imidazolium Ionic Liquids as Direct Solvents for Cellulose. *Thermochim. Acta* **2012**, *528*, 76–84. [CrossRef]
48. Kroon, M.C.; Buijs, W.; Peters, C.J.; Witkamp, G.J. Quantum Chemical Aided Prediction of the Thermal Decomposition Mechanisms and Temperatures of Ionic Liquids. *Thermochim. Acta* **2007**, *465*, 40–47. [CrossRef]
49. De Vos, N.; Maton, C.; Stevens, C.V. Electrochemical Stability of Ionic Liquids: General Influences and Degradation Mechanisms. *ChemElectroChem* **2014**, *1*, 1258–1270. [CrossRef]
50. Kirchner, B.; Blasius, J.; Alizadeh, V.; Gansäuer, A.; Hollóczki, O. Chemistry Dissolved in Ionic Liquids. A Theoretical Perspective. *J. Phys. Chem. B* **2022**, *126*, 766–777. [CrossRef] [PubMed]
51. Clough, M.T.; Geyer, K.; Hunt, P.A.; Mertes, J.; Welton, T. Thermal Decomposition of Carboxylate Ionic Liquids: Trends and Mechanisms. *Phys. Chem. Chem. Phys.* **2013**, *15*, 20480–20495. [CrossRef]

52. Fulmer, G.R.; Miller, A.J.M.; Sherden, N.H.; Gottlieb, H.E.; Nudelman, A.; Stoltz, B.M.; Bercaw, J.E.; Goldberg, K.I. NMR Chemical Shifts of Trace Impurities: Common Laboratory Solvents, Organics, and Gases in Deuterated Solvents Relevant to the Organometallic Chemist. *Organometallics* **2010**, *29*, 2176–2179. [CrossRef]
53. Socha, O.; Dračínský, M. Dimerization of Acetic Acid in the Gas Phase—NMR Experiments and Quantum-Chemical Calculations. *Molecules* **2020**, *25*, 2150. [CrossRef]
54. Zhang, Q.G.; Wang, N.N.; Yu, Z.W. The Hydrogen Bonding Interactions between the Ionic Liquid 1-Ethyl-3-Methylimidazolium Ethyl Sulfate and Water. *J. Phys. Chem. B* **2010**, *114*, 4747–4754. [CrossRef]
55. Troshenkova, S.V.; Sashina, E.S.; Novoselov, N.P.; Arndt, K.F.; Jankowsky, S. Structure of Ionic Liquids on the Basis of Imidazole and Their Mixtures with Water. *Russ. J. Gen. Chem.* **2010**, *80*, 106–111. [CrossRef]
56. Moriguchi, T.; Kawata, H.; Jalli, V. Design, Synthesis, Crystal Structure and Photoluminescence Properties of Four New Europium(III) Complexes with Fluorinated β-Diketone Ligand. *Cryst. Struct. Theory Appl.* **2021**, *10*, 1–13. [CrossRef]
57. Zheng, X.; Wang, M.; Li, Q. Synthesis and Luminescent Properties of Europium Complexes Covalently Bonded to Hybrid Materials Based on MCM-41 and Poly(Ionic Liquids). *Materials* **2018**, *11*, 677. [CrossRef]
58. Andruh, M.; Bakalbassis, E.; Kahn, O.; Trombe, J.C.; Porcher, P. Structure and Spectroscopic and Magnetic Properties of Rare Earth Metal(III) Derivatives with the 2-Formy 1-4-Methyl-6-(A-(2-Pyridylethyl)Formimidoyl) Phenol Ligand. *Inorg. Chem.* **1993**, *32*, 1616–1622. [CrossRef]
59. Nockemann, P.; Thijs, B.; Lunstroot, K.; Parac-Vogt, T.N.; Görller-Walrand, C.; Binnemans, K.; Van Hecke, K.; Van Meervelt, L.; Nikitenko, S.; Daniels, J.; et al. Speciation of Rare-Earth Metal Complexes in Ionic Liquids: A Multiple-Technique Approach. *Chem. Eur. J.* **2009**, *15*, 1449–1461. [CrossRef]
60. Marekha, B.A.; Bria, M.; Moreau, M.; De Waele, I.; Miannay, F.-A.; Smortsova, Y.; Takamuku, T.; Kalugin, O.N.; Kiselev, M.; Idrissi, A. Intermolecular Interactions in Mixtures of 1-n-Butyl-3-Methylimidazolium Acetate and Water: Insights from IR, Raman, NMR Spectroscopy and Quantum Chemistry Calculations. *J. Mol. Liq.* **2015**, *210*, 227–237. [CrossRef]
61. Nakamoto, K.; Fujita, J.; Tanaka, S.; Kobayashi, M. Infrared Spectra of Metallic Complexes. IV. Comparison of the Infrared Spectra of Unidentate and Bidentate Metallic Complexes. *J. Am. Chem. Soc.* **1957**, *79*, 4904–4908. [CrossRef]
62. Patil, K.C.; Chandrashekhar, G.V.; George, M.V.; Rao, C.N.R. Infrared Spectra and Thermal Decompositions of Metal Acetates and Dicarboxylates. *Can. J. Chem.* **1968**, *46*, 257–265. [CrossRef]
63. Trujillo-Rodríguez, M.J.; Anderson, J.L. In Situ Generation of Hydrophobic Magnetic Ionic Liquids in Stir Bar Dispersive Liquid-Liquid Microextraction Coupled with Headspace Gas Chromatography. *Talanta* **2019**, *196*, 420–428. [CrossRef]
64. Radzymińska-Lenarcik, E. Search for the Possibility of Utilizing the Differences in Complex-Forming Capacities of Alkylimidazoles for Selective Extraction of Some Metal Ions from Aqueous Solutions. *Pol. J. Chem. Technol.* **2008**, *10*, 73–78. [CrossRef]
65. Kirmse, W. The beginning of N-heterocyclic carbenes. *Angew. Chem. Int. Ed.* **2010**, *49*, 8798–8801. [CrossRef]
66. Böhm, V.P.W.; Herrmann, W.A. The Wanzlick Equilibrium. *Angew. Chem.* **2000**, *39*, 4036–4038. [CrossRef]
67. Efimova, A.; Hubrig, G.; Schmidt, P. Thermal Stability and Crystallization Behavior of Imidazolium Halide Ionic Liquids. *Thermochim. Acta* **2013**, *573*, 162–169. [CrossRef]
68. Paul, A.; Samanta, A. Optical Absorption and Fluorescence Studies on Imidazolium Ionic Liquids Comprising the Bis(Trifluoromethanesulphonyl)Imide Anion. *J. Chem. Sci.* **2006**, *118*, 335–340. [CrossRef]
69. Tilley, R.J.D. Lanthanoid Ion Color. In *Encyclopedia of Color Science and Technology*; Shamey, R., Ed.; Springer: Berlin/Heidelberg, Germany, 2020; pp. 1–10. [CrossRef]
70. Pereira, C.C.L.; Carretas, J.M.; Monteiro, B.; Leal, J.P. Luminescent Ln-Ionic Liquids beyond Europium. *Molecules* **2021**, *26*, 4834. [CrossRef] [PubMed]
71. Mudring, A.V.; Babai, A.; Arenz, S.; Giernoth, R.; Binnemans, K.; Driesen, K.; Nockemann, P. Strong Luminescence of Rare Earth Compounds in Ionic Liquids: Luminescent Properties of Lanthanide(III) Iodides in the Ionic Liquid 1-Dodecyl-3-Methylimidazolium Bis(Trifluoromethanesulfonyl)Imide. *J. Alloys Compd.* **2006**, *418*, 204–208. [CrossRef]
72. Binnemans, K. Interpretation of Europium(III) Spectra. *Coord. Chem. Rev.* **2015**, *295*, 1–45. [CrossRef]
73. Wada, S.; Kitagawa, Y.; Nakanishi, T.; Gon, M.; Tanaka, K.; Fushimi, K.; Chujo, Y.; Hasegawa, Y. Electronic Chirality Inversion of Lanthanide Complex Induced by Achiral Molecules. *Sci. Rep.* **2018**, *8*, 16395. [CrossRef]
74. Durham, D.A.; Frost, G.H.; Hart, F.A. Lanthanide Complexes-VIII Tris(2,2′6′,2″-Terpyridine)Lanthanide(III)Perchlorates: Fluorescence and Structure. *J. Inorg. Nucl. Chem.* **1969**, *31*, 833–838. [CrossRef]
75. Takikawa, Y.; Ebisu, S.; Nagata, S. Van Vleck Paramagnetism of the Trivalent Eu Ions. *J. Phys. Chem. Solids* **2010**, *71*, 1592–1598. [CrossRef]
76. Matsuda, Y.; Izawa, K.; Vekhter, I. Nodal Structure of Unconventional Superconductors Probed by Angle Resolved Thermal Transport Measurements. *J. Phys. Condens. Matter* **2006**, *18*, R705–R752. [CrossRef]
77. Ferenc, W.; Cristóvão, B.; Sarzynski, J. Magnetic, Thermal and Spectroscopic Properties of Lanthanide(III) 2-(4-Chlorophenoxy) Acetates, Ln($C_8H_6ClO_3$)$_3 \cdot nH_2O$. *J. Serbian Chem. Soc.* **2013**, *78*, 1335–1349. [CrossRef]
78. Mugiraneza, S.; Hallas, A.M. Tutorial: A Beginner's Guide to Interpreting Magnetic Susceptibility Data with the Curie-Weiss Law. *Commun. Phys.* **2022**, *5*, 95. [CrossRef]
79. Farger, P.; Leuvrey, C.; Gallart, M.; Gilliot, P.; Rogez, G.; Rocha, J.; Ananias, D.; Rabu, P.; Delahaye, E. Magnetic and Luminescent Coordination Networks Based on Imidazolium Salts and Lanthanides for Sensitive Ratiometric Thermometry. *Beilstein J. Nanotechnol.* **2018**, *9*, 2775–2787. [CrossRef]

80. Fomina, I.G.; Dobrokhotova, Z.V.; Ilyukhin, A.B.; Aleksandrov, G.G.; Kazak, V.O.; Gehman, A.E.; Efimov, N.N.; Bogomyakov, A.S.; Zavorotny, Y.S.; Gerasimova, V.I.; et al. Binuclear Samarium(III) Pivalates with Chelating N-Donors: Synthesis, Structure, Thermal Behavior, Magnetic and Luminescent Properties. *Polyhedron* **2013**, *65*, 152–160. [CrossRef]
81. Wang, X.J.; Cen, Z.M.; Ni, Q.L.; Jiang, X.F.; Lian, H.C.; Gui, L.C.; Zuo, H.H.; Wang, Z.Y. Synthesis, Structures and Properties of Functional 2-D Lanthanide Coordination Polymers [Ln$_2$(Dpa)$_2$(C$_2$O$_4$)$_2$(H$_2$O)$_2$]$_n$ (Dpa = 2,2'-(2-Methylbenzimidazolium-1,3-Diyl)Diacetate, C$_2$O$_4^{2-}$ = Oxalate, Ln = Nd, Eu, Gd, Tb). *Cryst. Growth Des.* **2010**, *10*, 2960–2968. [CrossRef]

Disclaimer/Publisher's Note: The statements, opinions and data contained in all publications are solely those of the individual author(s) and contributor(s) and not of MDPI and/or the editor(s). MDPI and/or the editor(s) disclaim responsibility for any injury to people or property resulting from any ideas, methods, instructions or products referred to in the content.

Communication

Suzuki–Miyaura Reaction in the Presence of *N*-Acetylcysteamine Thioesters Enables Rapid Synthesis of Biomimetic Polyketide Thioester Surrogates for Biosynthetic Studies

Sebastian Derra, Luca Schlotte and Frank Hahn *

Department of Chemistry, Faculty Biology, Chemistry and Earth Sciences, Universität Bayreuth, 95447 Bayreuth, Germany; sebastian1.derra@uni-bayreuth.de (S.D.)
* Correspondence: frank.hahn@uni-bayreuth.de

Abstract: Biomimetic *N*-acetylcysteamine thioesters are essential for the study of polyketide synthases, non-ribosomal peptide synthetases and fatty acid synthases. The chemistry for their preparation is, however, limited by their specific functionalization and their susceptibility to undesired side reactions. Here we report a method for the rapid preparation of *N*-acetylcysteamine (SNAC) 7-hydroxy-2-enethioates, which are suitable for the study of various enzymatic domains of megasynthase enzymes. The method is based on a one-pot sequence of hydroboration and the Suzuki–Miyaura reaction. The optimization of the reaction conditions made it possible to suppress potential side reactions and to introduce the highly functionalized SNAC methacrylate unit in a high yield. The versatility of the sequence was demonstrated by the synthesis of the complex polyketide-SNAC thioesters **12** and **33**. Brown crotylation followed by the hydroboration to Suzuki–Miyaura reaction sequence enabled the introduction of the target motif in significantly fewer steps and with a higher overall yield and stereoselectivity than previously described approaches. This is the first report of a transition-metal-catalyzed cross-coupling reaction in the presence of an SNAC thioester.

Keywords: Suzuki–Miyaura reaction; biomimetic thioesters; polyketide synthases; enzymes; cyclases

Citation: Derra, S.; Schlotte, L.; Hahn, F. Suzuki–Miyaura Reaction in the Presence of *N*-Acetylcysteamine Thioesters Enables Rapid Synthesis of Biomimetic Polyketide Thioester Surrogates for Biosynthetic Studies. *Chemistry* **2023**, *5*, 1407–1418. https://doi.org/10.3390/chemistry5020096

Academic Editors: Christoph Janiak, Sascha Rohn and Georg Manolikakes

Received: 28 April 2023
Revised: 2 June 2023
Accepted: 2 June 2023
Published: 8 June 2023

Copyright: © 2023 by the authors. Licensee MDPI, Basel, Switzerland. This article is an open access article distributed under the terms and conditions of the Creative Commons Attribution (CC BY) license (https:// creativecommons.org/licenses/by/ 4.0/).

1. Introduction

Thioesters are an important functional group in many biosynthetic systems. They often serve to link biosynthetic acyl intermediates to carrier thiols, which can be free molecules such as coenzyme A (CoA) or proteins. The important systems working with protein-bound metabolites are so-called megasynthase enzymes, such as fatty acid synthases, polyketide synthases (PKS) and non-ribosomal peptide synthetases (NRPS) and their hybrids [1,2]. They are responsible for the formation of polyketide and peptide natural products, including some of the most important small-molecule drugs in clinical use, such as erythromycin, rapamycin or epothilone. The availability of suitable substrate surrogates is essential for the functional study of such biosynthetic systems (Figure 1A). *N*-Acetylcysteamine (SNAC) thioesters are of particular importance for this as they effectively mimic the protein attachment of the substrate via the 4′-phosphopantetheine arm and thus allow simplified studies with active enzymes (Figure 1B) [3–10].

Acylated SNACs contain an acetamide and a thioester as conserved reactive functional groups, which afford them problematic properties (Figure 1C) [7,11]. The thioester can undergo side reactions with external or internal nucleophiles, resulting in the irreversible loss of substance. Due to its polarity, the acetamide can cause problems during substance purification and can, as a nucleophilic/protic group, cause undesired side reactions. The functionalization distance between the acetamide and thioester carries the risk that they act as a chelate ligand and interact with metals. The synthesis of the SNAC thioester surrogates of late-stage biosynthetic intermediates is as challenging as the synthesis of natural products of similar structural complexity but, for the above-mentioned reasons, has the challenge of an additional problematic functional group. A useful strategy to overcome this problem would be to introduce the SNAC moiety at a late stage of synthesis along with a larger fraction of the polyketide moiety.

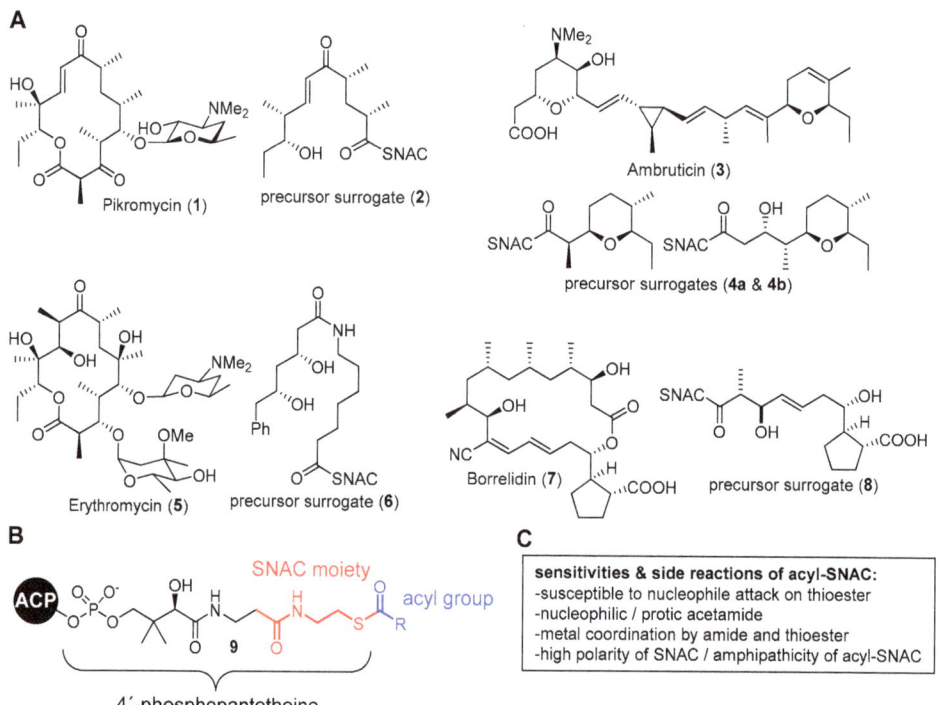

Figure 1. (A) Structure of the 4′-phosphopantetheine prosthetic group of acyl carrier proteins and the partial structure that is mimicked by SNAC. (B) Susceptibilities and side reactions of acyl-SNACs. (C) Prominent natural products and the structures of biomimetic SNAC thioesters that have been used for their biosynthesis studies.

Improving the specific methodology for the synthesis of complex polyketide–SNAC thioesters is therefore of great interest to the biosynthetic research community. Transition metal-mediated reactions are well suited to late-stage attachment in the convergent synthesis of complex biosynthetic thioester surrogates, but have only very rarely been described in the presence of SNAC thioesters. To the best of our knowledge, the literature currently only contains a report about olefin cross-metathesis between SNAC–acrylates and hydroxyolefins catalyzed by the second-generation Grubbs catalyst [12].

The Suzuki–Miyaura reaction (SMR) is a highly versatile Pd-catalyzed cross-coupling reaction. It allows couplings between halides and non-toxic boronic acid derivatives under relatively mild conditions (Figure 2) [13,14]. In addition to sp^2–sp^2 bond formations, it is now possible to carry out couplings between sp^2 and sp^3 centers, as well as between two sp^3 centers. Two aspects of the SMR could be problematic when applied to SNAC thioesters. On the one hand, the use of a base is necessary to accelerate the essential group transfer from the boronic acid to the Pd during the catalytic cycle (step 3). Moreover, Pd can also be inserted into the C–S bond of the thioester instead of the C–halide bond (step 1) [15]. This reactivity is so pronounced that it forms the basis of the Liebeskind–Srogl reaction, a modification of the SMR for the direct synthesis of ketones from thioesters [16,17].

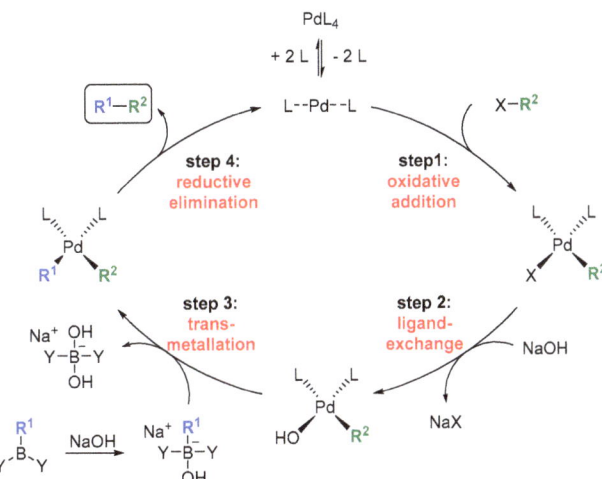

Figure 2. Mechanism of the SMR.

Among the diverse enzymatic PKS domains, cyclases that form saturated oxygen heterocycles via intramolecular oxa-Michael addition (IMOMA) stand out due to the synthetical value of this transformation (Figure 3A) [18,19]. It has been shown that they catalyze a ring formation with exceptional stereoselectivity and therefore represent a potential new type of biocatalyst [20–29]. For the study of such enzymes, SNAC-7-hydroxy-2-enethioates are required as substrate surrogates. The synthetic methodology used for the selective installation of this structural motif is, however, not well developed, making the generation of precursor libraries a difficult task. The multi-step routes described in the literature are either highly elaborate, are not stereoselective or lack flexibility, and are therefore narrow in their applicability [20–23]. For example, the synthesis of the SNAC surrogate of **10** in stereochemical pure form was accomplished in eight steps and required multiple purification procedures [21,22]. Furthermore, a lack of convergence makes it necessary to carry out the largest part of this sequence using different starter building blocks to access derivatives with variations in the eastern part of the molecule. Other reported routes are shorter, but also less flexible due to the choice of larger starting building blocks or the introduction reaction chosen for the SNAC thioester. Olefin cross-metathesis, for example, is only possible with SNAC–acrylthioates and not with SNAC–methacrylthioates. Therefore, we set out to develop a flexible, straightforward and broadly applicable method for the preparation of SNAC-7-hydroxy-2-enethioates.

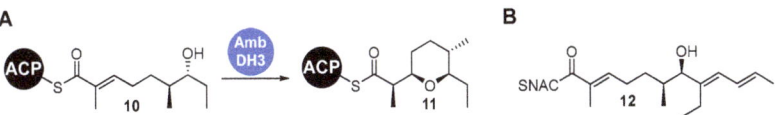

Figure 3. (**A**) IMOMA cyclases catalyze the intramolecular oxa-Michael addition to oxygen heterocycles. The natural reaction of AmbDH3 is shown as an example. (**B**) Structure of the target compound required for our biosynthetic studies.

As a solution, we turned to a sequence of hydroboration and SMR to assemble the backbone and directly introduce the SNAC moiety. The specific challenge was to effectively perform the SMR in the presence of the SNAC thioester, which has not been achieved before to the best of our knowledge. The versatility of the method should be shown on the example of the synthesis of **12** (Figure 3B). This compound was, on the one hand, specifically required for our enzymatic studies on new IMOMA cyclases (Figure 3A). On the other hand, it represents a particularly challenging substrate during whose preparation various detrimental side reactions occur; it is thus a reasonable benchmark.

2. Materials and Methods

2.1. General Methods and Materials

All chemicals and solvents were obtained from Abcr (Karlsruhe, Germany), Acros Organics (Geel, Belgium), BLD Pharm (Kaiserslautern, Germany), Carbolution (St. Ingbert, Germany), Eurisotop (Saarbrücken, Germany), Fluorochem (Hadfield, UK), Grüssing (Filsum, Germany), Roth (Karlsruhe, Germany), Sigma-Aldrich (Schnelldorf, Germany), TCI (Zwijndrecht, Belgium) Thermo Fisher Chemical (Schwerte, Germany), and VWR (Rednor, DE, USA) and were, unless otherwise stated, used without further purification. Dry solvents were obtained from Acros Organics. All reactions were performed under argon gas using dry solvents and reagents. Light-sensitive substances were handled in brown glass- or aluminum-foil-wrapped flasks. The reactions were monitored via TLC using Alugram SilG/UV254 TLC foils from Macherey-Nagel (Düren, Germany). The substances were detected using UV light and a $KMnO_4$ stain (1.50 g of $KMnO_4$, 10.0 g of K_2CO_3, 2.50 mL of 5% NaOH, 200 mL of H_2O). Products were purified via flash chromatography on SiO_2 (Macherey-Nagel MN Kieselgel 60, 40–63 μm). Semi-preparative HPLC was performed using a Waters HPLC (600 controller, 2487 Dual wavelength absorbance detector) and a C18-SP stationary phase (H_2O:MeCN = 95:5 {5 min}, Gradient H_2O:MeCN 95:5 → 5:95 {20 min}, H_2O:MeCN = 5:95 {5 min}, 20 mL/min).

All NMR spectra were recorded using Bruker Avance III HD 500 (Rheinstetten, Germany) with the residual solvent signal as an internal standard: $CDCl_3$ 7.26 ppm for ^1H and 77.16 ppm for ^{13}C.[2] Signal multiplicities are stated, using the following abbreviations: s = singlet, d = doublet, t = triplet, q = quartet, m = multiplet and br = broad. For ^{13}C-NMR, the following abbreviations were used: q = quarternary, t = tertiary, s = secondary and p = primary. The chemical shifts are reported as values of the δ-scale in [ppm] and the coupling constants J in [Hz]. Signal assignments were made with 2D NMR spectra (COSY, HSQC, HMBC, NOESY). High-resolution mass spectra (HRMS) were obtained using a Thermo Fisher Scientific Q Exactive (Orbitrap) mass spectrometer. Optical rotation was recorded on a Jasco P-1020 polarimeter (10 cm cell) from Portman Instruments (Biel-Benken, Switzerland) using the sodium D line (589 nm). The given value of $[\alpha]^D$ represents the average of 50 individual measurements and is stated as deg·mL·g^{-1}·dm^{-1}. Elemental analyses were carried out using a 2400 CHN elemental analyzer from Perkin-Elmer (Waltham, MA, USA).

2.2. General Procedures

2.2.1. General Procedure 1: STEGLICH Esterification for SNAC Thioester Formation

A solution of carboxylic acid (**GP2**, 1.0 eq.) and *N*-acetylcysteamine (**GP1**, 1.5 eq.) in CH_2Cl_2 (0.2 M) was cooled to 0 °C. Subsequently, DMAP (0.1 eq.) and EDC*HCl (1.5 eq.) were added. After warming to room temperature, the solution was stirred for 2 h, before diluting with saturated aqueous NH_4Cl solution. The resulting phases were separated and the aqueous one was extracted three times with CH_2Cl_2. The combined organic phases were washed with brine, dried over $MgSO_4$ and filtrated. The solvent was removed in vacuo and the crude thioester (**GP3**) was purified via flash chromatography.

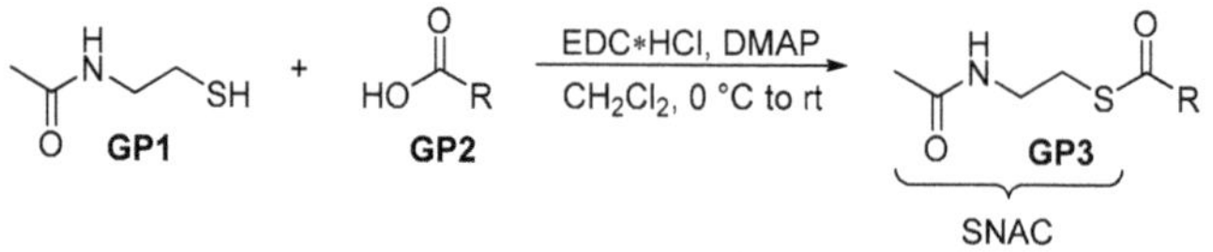

Scheme 1. General procedure for thioesterification.

2.2.2. General Procedure 2: Protection of Hydroxyls as Silylethers

To a stirred solution of secondary alcohol (**GP4**, 1.0 eq.) in DMF (1 M), silylchloride (1.5 eq.) and imidazole (2.5 eq.) were added. After stirring the mixture at room temperature overnight, pentane and water were added. The resulting phases were separated and the aqueous one was extracted three times with pentane. Subsequently, the combined organic phases were washed with brine, dried over $MgSO_4$ and filtrated. The solvent was removed in vacuo and the crude silylether (**GP5**) was purified via filtration over a short plug of silica (pentane).

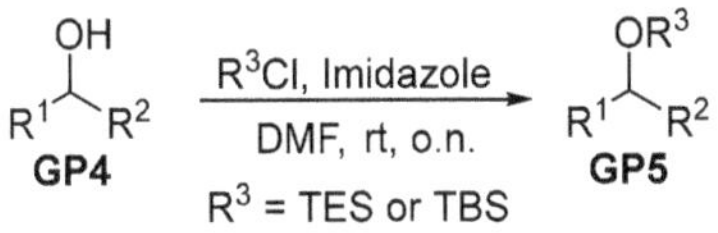

Scheme 2. General procedure for silylether protection.

2.2.3. General Procedure 3: Hydroboration-SUZUKI-MIYAURI Reaction

A solution of terminal alkene (**GP7**, 1.5 eq.) in freshly degassed THF (1 M) was cooled to 0 °C and 9-BBN (1.5 eq., 0.5 M in THF) was added dropwise. The mixture was stirred overnight while being allowed to slowly warm to room temperature. Subsequently, freshly degassed DMF (total 0.2 M), thioester vinylhalogenide (**GP6**, 1.0 eq.), $PdCl_2$ (dppf) (5 mol%), $AsPh_3$ (5 mol%) and Cs_2CO_3 (2.0 eq.) were added, and the suspension was heated to 50 °C. After the complete consumption of the starting material (TLC), EtOAc was added, and the mixture was transferred to a separating funnel containing aqueous LiCl solution (10% wt). After the separation and extraction of the aqueous phase using EtOAc (3×), the combined organics were washed with brine and dried over $MgSO_4$. After concentration in vacuo, the crude product (**GP8**) was purified via flash chromatography.

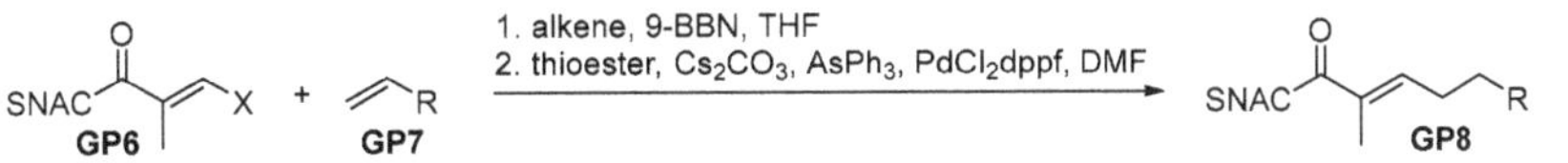

Scheme 3. General procedure for SMR.

2.2.4. General Procedure 4: Deprotection

The silylether (**GP9**, 1.0 eq.) was dissolved in HF-containing stock solution (70% HF*pyridine/pyridine/THF (1:2:8)) at 0 °C. After warming to room temperature and the complete consumption of the starting material according to TLC, the saturated aqueous $NaHCO_3$ solution was added dropwise until no more formation of CO_2 was observed. Subsequently, the phases were separated and the aqueous one was extracted three times using EtOAc. The combined organics were washed with brine and dried over $MgSO_4$. After concentration in vacuo, the crude product (**GP7**) was purified via semi-preparative HPLC.

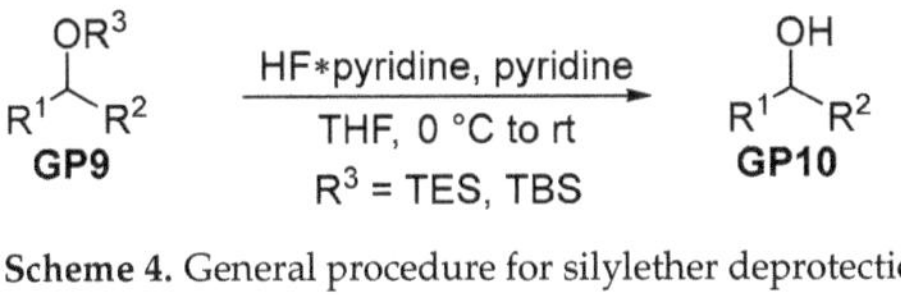

Scheme 4. General procedure for silylether deprotection.

3. Results

Thioester-halides are rare substrates in SMRs. The literature, however, contains an example of the coupling reactions of simple 4-bromothiophenols with 4-tolyl-boronic acid, in which the bulky acyl unit of the thioesters served as a protecting group for the thiol [15]. We used the reported conditions for the synthesis of an ethylenoate that was sensitive to hydrolysis and a base described by Suzuki et al. as the starting point for our studies [30]. These were carried out using the SNAC (E)-3-bromo-2-methylprop-2-enethioate **13a** and the OTBS-protected 3-hydroxyolefin **14** (1.0 equiv. of **13a**, 1.1 equiv. of **14**, 1.1 equiv. 9-BBN, 5mol% $PdCl_2$(dppf) (dppf: 1,1′-bis(diphenylphosphino)ferrocene) and 2.0 equiv. K_2CO_3) (Schemes 1 and 2). Although a basic coupling reactivity was observed, the yields of the reactions varied hardly reproducibly over a wide range and showed a strong dependence on even small variations in the amounts of thioester, alkene, borane and Pd catalyst. This suggests that several side reactions might proceed at rates similar to the desired pathway. We therefore carried out a systematic optimization study (Table 1, see Supplementary Materials pages 3–7).

Table 1. Optimization of the conditions for the coupling of the SNAC thioester halides **13a/13b** and TBS-protected olefin **14**.

Entry	X	Base	Additive	Temperature [°C]	Isolated Yield [%]
a	Br	K_2CO_3	-	50	54
b	Br	K_2CO_3	P(o-furyl)$_3$	50	23
c	Br	K_2CO_3	$AsPh_3$	50	55
d	Br	Cs_2CO_3	-	50	55
e	Br	K_2CO_3	-	20	13
f	I	K_2CO_3	-	50	55
g	I	Cs_2CO_3	$AsPh_3$	50	34
h	I	Cs_2CO_3	$AsPh_3$	65	-
i	I	Cs_2CO_3	$AsPh_3$	20	78

General reaction conditions: 1. **14** (1.5 eq., 1 M in THF), 9-BBN (1.5 eq., 0.5 M in THF), 0 °C to 20 °C, o.n.; 2. DMF (total 0.2 M), **13** (1.0 eq.), base (2.0 eq.), $PdCl_2$dppf (5 mol%), additive (5 mol%), reaction control via TLC. Reaction scale: 90–100 µmol.

For this, we varied the individual reaction parameters. Since we assumed that the side reactions of the 3-bromoacryl thioate **13a** were a particular problem, we worked with an excess of 1.5 equiv. of alkene **14** and 9-BBN. Different thioester halides (Br and I), bases (K_2CO_3 and Cs_2CO_3), additives (P(o-furyl)$_3$ and $AsPh_3$) and temperatures (20 °C, 50 °C and 65 °C) were tested. All reactions were carried out on a scale of 90–100 µmol of **13a/13b** and compared based on the isolated yield after column chromatography. The yields in the basic experiment with an excess of 1.5 equiv. of **14** and 9-BBN (entry a) were fortunately stable upon repetition in a range slightly above 50%. The variation in the individual parameters did not lead to a marked increase in this value, whereas the addition of P(o-furyl)$_3$ and the decrease in the reaction temperature even significantly reduced the yield (entries b and e). Fortunately, the combined change in several parameters led to a significantly improved result (entry i). Using 3-iodoacryl thioate **13b**, Cs_2CO_3, $AsPh_3$ and carrying out the reaction at room temperature gave **15** a yield of 78%. The TBS deprotection of **16** achieved a 52% yield using PPTS under conditions that we identified as successful in the synthesis of other SNAC-7-hydroxy-2-enethioates [21–23].

Side products that could not be isolated and fully analyzed were regularly detected in the low-yielding hydroboration SMRs in Table 1. According to TLC, these were highly polar compounds whose migration behavior suggests that they were derived from SNAC. We assume that a major part of this is the homocoupling product of the thioester acrylates **13a/13b** and the 2-(N-acetamidyl)-ethylketone resulting after the C–S bond insertion of the Pd, a side reaction described previously for low-functionalized thioesters [15]. The yield improvement observed in the optimization study is consistent with the suppression of these side reactions. Cs_2CO_3 is much more soluble in DMF than K_2CO_3, leading to a much higher effective concentration of carbonate. This should significantly improve the activation of the *ate* complex for alkyl group transfer to the Pd (step 3 in Figure 2) and accelerate the heterocoupling reaction. The iodoacrylate is more reactive towards Pd insertion than the bromoacrylate, thus favoring this productive reaction (step 1) compared to the insertion of Pd into the C–S bond. This selectivity is expected to be even more pronounced at room temperature than at 50 °C. The addition of AsPh$_3$ supports these effects by accelerating both the formation of the active Pd(0) from the Pd(II) species and transmetallation, due to its lower σ-donor effect than PPh$_3$ [31,32].

We now turn to the coupling between the thioesters **13a/13b** and the olefins **17a** and **17b**, which resemble the sensitive 5-hydroxy-tri-1,3,7-ene in target molecule **12** (Table 2, see SI pages 7–13). Their higher degree of functionalization makes them more susceptible to side reactions during the introduction and removal of the protecting group and during the coupling cascade. In addition to screening the same thioester halides (**13a/13b**) as in Table 2, a broader panel of bases (Cs$_2$CO$_3$, K$_2$CO$_3$ and K$_3$PO$_4$) and hydroxyl protection groups (TBS and TES) on the olefinic coupling partner were examined. Due to the superiority of the previous optimization, only AsPh$_3$ was applied as an additive and only 20 °C and 50 °C were tested as reaction temperatures.

Table 2. Optimization of the conditions for the coupling between **13a/13b** and protected trienes **17a/17b**, varying protecting group, halogenide, base, additives and temperature.

Entry	X	PG	Base	Additive	Temperature [° C]	Isolated Yield [%]
a	Br	TBS	2 eq. K$_2$CO$_3$	-	50	27
b	Br	TBS	3 eq. K$_3$PO$_4$	-	50	17
c	Br	TES	2 eq. K$_2$CO$_3$	-	50	25
d	Br	TES	3 eq. K$_3$PO$_4$	-	50	12
e	Br	TES	2 eq. K$_2$CO$_3$	-	20	15
f	I	TES	2 eq. K$_2$CO$_3$	-	50	49
g	Br	TES	2 eq. Cs$_2$CO$_3$	-	50	80
h	Br	TES	2 eq. K$_2$CO$_3$	AsPh$_3$	50	77
i	I	TES	2 eq. K$_2$CO$_3$	-	20	63
j	I	TES	2 eq. Cs$_2$CO$_3$	AsPh$_3$	20	87

Reaction conditions: 1. **17** (1.5 eq., 1 M in THF), 9-BBN (1.5 eq., 0.5 M in THF), 0 °C to 20 °C, o.n.; 2. DMF (total 0.2 M), **13** (1.0 eq.), base (2.0 eq.), PdCl$_2$dppf (5 mol%), additive (5 mol%), reaction control via TLC.

Compared to the basic experiments (entries a and c, Table 2), a decrease in the yield of **18** was observed when the reaction was carried out at 20 °C instead of 50 °C, or when K$_3$PO$_4$ was employed as a base (entries a–e). In contrast to the experiments using the simpler coupling partner **14** (Table 1), the change in one or two reaction parameters led to an increase in the yield of up to 80% (entries f–i). When these measures were combined and the reaction was carried out at 20 °C, a further increase to an 87% yield of **18b** was achieved (entry j, Scheme 3). The TES group was expected to be more easily removable than the TBS group (vide infra). As the former demonstrated stability under the conditions tested, and as both protecting groups gave similar yields in the comparable entries a–d, the optimization in entries f–j was conducted using the TES protection group. The results summarized in Table 2 are in agreement with those observed in Table 1 and confirm the conclusions/interpretations drawn from them.

Numerous side reactions were conceivable during the deprotection of the silyl ethers **18a** and **18b**, such as eliminations, intramolecular oxa-Michael additions or interferences with the thioester. The slightly acidic conditions of the standard deprotection protocol with PPTS (see Table 1, step 2) resulted in the elimination of the alcohol/silylether (entry a, Table 3, see SI page 14). With TBAF, the formation of the desired product was also not observed in any case. No reaction of the TBS ether **18a** was found after 1 h at 0 °C (entry b). Decomposition occurred for the TBS ether **18b** after overnight reaction at 20 °C and for the TES ether after only 1 h at 0 °C (entries c and d). Standard HF*pyridine treatment also resulted in decomposition (entry e). A successful procedure was finally adopted from a protocol previously reported by Carreira et al., which relied on using a premixed stock solution of HF*pyridine in THF supplemented with additional pyridine at 0 °C [33]. Deprotection was successful for both silylethers and led to the attainment of the desired alcohol **19** in pure form after column chromatography (entries f and g, Scheme 4). The reactions were continuously monitored by TLC and stopped before noticeable decomposition occurred. The yield for the TES ether was significantly better than that of the TBS ether, suggesting that the former is the preferable protecting group for the synthesis of **12**.

Table 3. Testing conditions for silylether deprotection.

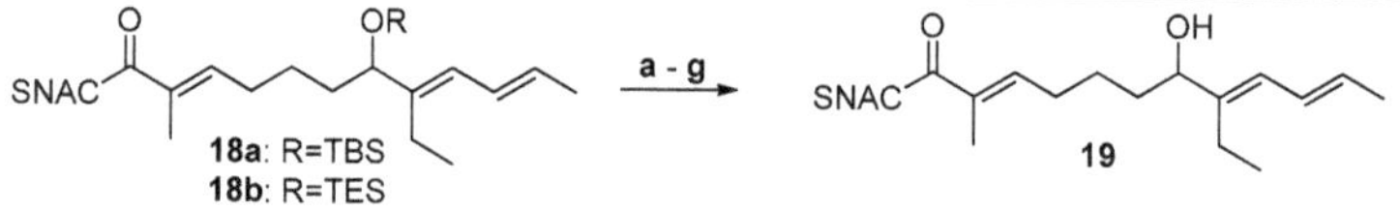

Entry	PG	Reagent	Conditions	Result
a	TBS	PPTS	DMSO, 50 °C, o.n.	Decomposition
b	TBS	TBAF	THF, 0 °C, 1 h	No reaction
c	TBS	TBAF	THF, 0 → 20 °C, o.n.	Decomposition
d	TES	TBAF	THF, 0 °C, 1 h	Decomposition
e	TBS	HF*pyridine	THF, 0 °C, 3 h	Decomposition
f	TBS	HF*pyridine, pyridine	THF, 0 → 20 °C, 3 h	51%
g	TES	HF*pyridine, pyridine	THF, 0 → 20 °C, 3 h	81%

The reaction sequences to **12** were carried out starting from aldehyde **20** (see Supplementary Materials pages 14–16). Brown crotylation first afforded the highly sensitive hydroxytriene **21** in a yield of 71% and an 86% e.e., which was immediately transformed into the isolatable TBS and TES ethers **22a** and **22b**, with yields of 91% and 84% (Scheme 5). This was followed by the established one-pot two-step cascade of hydroboration and SMR to give **28a** and **28b**, which were deprotected under the optimized conditions to give 7-hydroxy-2-ene-SNAC thioate **12** in overall yields of 16–60% (Table 4, see Supplementary Materials pages 17–19). These results confirm, on the one hand, that TES is the preferable protecting group compared to TBS as it leads to higher yields in both the coupling and deprotection reaction (entries a and b). On the other hand, they show the positive effect of optimizing the SMR conditions, which led to an improvement in the yield from 51% to 74% in the coupling step (entries b and c).

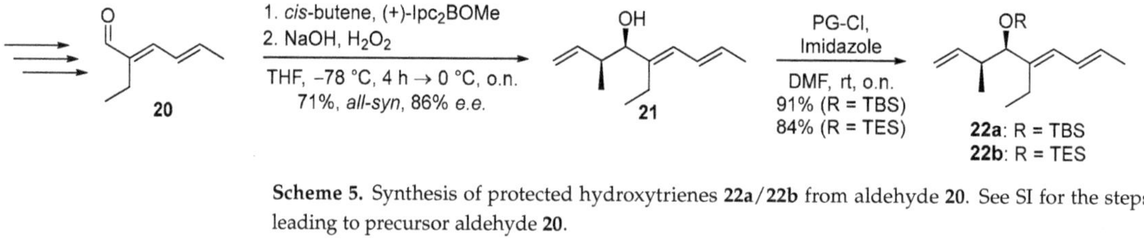

Scheme 5. Synthesis of protected hydroxytrienes **22a**/**22b** from aldehyde **20**. See SI for the steps leading to precursor aldehyde **20**.

To illustrate the broader synthetic utility of the synthesis of chiral 7-hydroxy-2-ene-SNAC thioates, we additionally applied the method to the synthesis of octaketide **33** (Scheme 6, see SI pages 20–24). In comparison to **12**, this compound exhibits a highly hydrophobic heptyl chain, an additional chiral secondary alcohol and a relative *anti*-configuration at the vicinal stereocenters at C-6 and C-7. Olefin **31** was obtained from aldehyde **29** via Brown crotylation and TBS protection. Despite the presence of two sterically demanding TBS groups, the hydroboration to SMR sequence proceeded similarly well, as in the synthesis of **28b**, giving **32** a yield of 79% starting from **31**. The deprotection of **32** was achieved in a 93% yield under the conditions optimized for the synthesis of the sensitive allylic alcohols **12** and **19**. These conditions thus also provide an advantage for the deprotection of SNAC 2-enethioate silyl ethers devoid of critical (poly)enes, as is evident from the comparison made with the deprotection of **15** to **16** that was carried out using PPTS and that led to a yield of only 52%.

Table 4. Two-pot three-step reaction sequence for thioester **12** starting from **13a/13b** and **22a/22b**. The coupling step is formulated under the assumption that the SMR proceeds via a Pd(0)/Pd(II) mechanism.

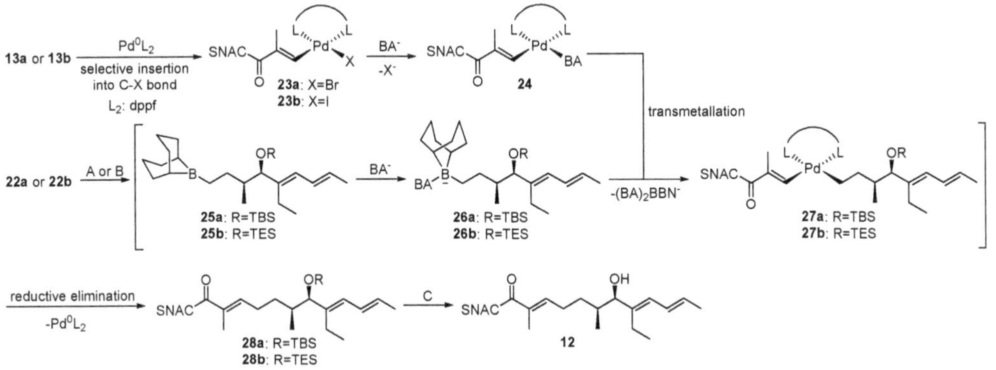

Entry	X	PG	Conditions	Coupling Yield [%]	Deprotection Yield [%]	Overall Yield [%]
a	Br	TBS	A	30	53	16
b	Br	TES	A	51	86	44
c	I	TES	B	74	81	60

Reaction conditions: A. **22** (1.0 eq., 1 M in THF), 9-BBN (1.0 eq., 0.5 M in THF), 0 °C to 20 °C, o.n.; 2. DMF (total 0.2 M), **13** (1.5 eq.), K_2CO_3 (2.0 eq.), $PdCl_2$dppf (5 mol%), 50 °C, reaction control via TLC; B. **22** (1.5 eq., 1 M in THF), 9-BBN (1.5 eq., 0.5 M in THF), 0 °C to 20 °C, o.n.; 2. DMF (total 0.2 M), **13** (1.0 eq.), Cs_2CO_3 (2.0 eq.), $PdCl_2$dppf (5 mol%), $AsPh_3$ (5 mol%), 20 °C, reaction control via TLC; C. **22** (10.0 mg, 1.0 eq.), 110 μL of HF-containing stock solution (1 part HF*pyridine, 2 parts pyridine, 8 parts THF).

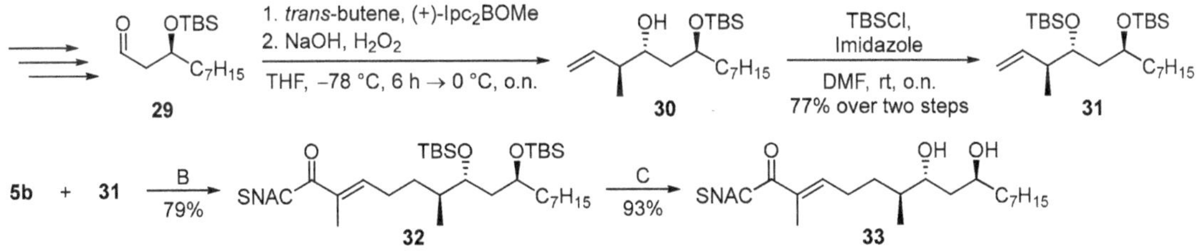

Scheme 6. Synthesis of chiral hydroxythioate **33**. See SI for the steps leading to precursor aldehyde **29**.

4. Discussion

In summary, a useful method for the rapid assembly of SNAC hydroxyenethioates that makes use of a cascade of hydroboration to SMR, followed by optimized silylether deprotection, was developed. The four SNAC hydroxyenethioates **12**, **16**, **19** and **33** were synthesized using this strategy. The chiral **12** and **33** were obtained in four synthetic operations using the aldehydes **20** and **29**, respectively, with overall yields of 36% and 57% and high stereoisomeric purity. This represents a significant improvement over previously described routes to similar compounds, which either required significantly more steps or gave lower overall yields (eight steps, 10% overall yield for the SNAC thioester analog of **10** starting from propionaldehyde) [21]. Other routes gave SNAC 7-hydroxy-2-ene thioates in five steps from TBS-protected 1,5-hexanediol with a total yield of 23% [20]. The latter, however, only provided access to racemic products, which were also not branched in the α-position. It did also not offer the flexibility in backbone installation that the presented method does.

The SMR-based coupling method presented here is compatible with the presence of SNAC thioesters and can be used in the future for the flexible and efficient preparation of substrate surrogates for studies of IMOMA cyclases and other enzymatic megasynthase domains that act on similar functionalization patterns as those present at C-1–C-6 in **12, 16, 19** and **33**. The method should also be of interest for the synthesis of precursors of non-enzymatic IMOMA reactions. It has been shown for chemically catalyzed IMOMA reactions that *cis*-THP stereoselectivity can be more reliably achieved using enethioates rather than enoates, meaning that the former are attractive precursors.

Supplementary Materials: The following supporting information can be downloaded at: https://www.mdpi.com/article/10.3390/chemistry5020096/s1, The Supplementary Materials contain detailed synthetic procedures and analytical data including ^{1}H and ^{13}C NMR spectra in Figures S1–S50. References [34–37] are cited in Supplementary Materials.

Author Contributions: Conceptualization, S.D. and F.H.; investigation, S.D. and L.S.; resources, F.H.; writing—original draft preparation, F.H.; writing—review and editing, S.D., L.S. and F.H.; supervision, F.H.; project administration, F.H.; funding acquisition, F.H. All authors have read and agreed to the published version of the manuscript.

Funding: This research was funded by the Deutsche Forschungsgemeinschaft (DFG), grant numbers HA 5841/5-1 and HA 5841/7-1.

Data Availability Statement: The data presented in this study are available on request from the corresponding author (Frank Hahn).

Acknowledgments: We thank Central Analytics of the Department of Chemistry, as well as the North Bavarian NMR Centre (NBNC) at the University of Bayreuth.

Conflicts of Interest: The authors declare no conflict of interest.

References

1. Weissman, K.J.; Müller, R. Protein–Protein Interactions in Multienzyme Megasynthetases. *ChemBioChem* **2008**, *9*, 826–848. [CrossRef]
2. Grininger, M. Enzymology of Assembly Line Synthesis by Modular Polyketide Synthases. *Nat. Chem. Biol.* **2023**, *19*, 401–415. [CrossRef] [PubMed]
3. Ge, H.-M.; Huang, T.; Rudolf, J.D.; Lohman, J.R.; Huang, S.-X.; Guo, X.; Shen, B. Enediyne Polyketide Synthases Stereoselectively Reduce the β-Ketoacyl Intermediates to β-d-Hydroxyacyl Intermediates in Enediyne Core Biosynthesis. *Org. Lett.* **2014**, *16*, 3958–3961. [CrossRef] [PubMed]
4. Sahner, J.H.; Sucipto, H.; Wenzel, S.C.; Groh, M.; Hartmann, R.W.; Müller, R. Advanced Mutasynthesis Studies on the Natural α-Pyrone Antibiotic Myxopyronin from *Myxococcus Fulvus*. *ChemBioChem* **2015**, *16*, 946–953. [CrossRef] [PubMed]
5. Pinto, A.; Wang, M.; Horsman, M.; Boddy, C.N. 6-Deoxyerythronolide B Synthase Thioesterase-Catalyzed Macrocyclization Is Highly Stereoselective. *Org. Lett.* **2012**, *14*, 2278–2281. [CrossRef] [PubMed]
6. Hansen, D.A.; Rath, C.M.; Eisman, E.B.; Narayan, A.R.H.; Kittendorf, J.D.; Mortison, J.D.; Yoon, Y.J.; Sherman, D.H. Biocatalytic Synthesis of Pikromycin, Methymycin, Neomethymycin, Novamethymycin, and Ketomethymycin. *J. Am. Chem. Soc.* **2013**, *135*, 11232–11238. [CrossRef]
7. Franke, J.; Hertweck, C. Biomimetic Thioesters as Probes for Enzymatic Assembly Lines: Synthesis, Applications, and Challenges. *Cell Chem. Biol.* **2016**, *23*, 1179–1192. [CrossRef]
8. Hahn, F.; Kandziora, N.; Friedrich, S.; Leadlay, P.F. Synthesis of Complex Intermediates for the Study of a Dehydratase from Borrelidin Biosynthesis. *Beilstein J. Org. Chem.* **2014**, *10*, 634–640. [CrossRef]
9. Berkhan, G.; Merten, C.; Holec, C.; Hahn, F. The Interplay between a Multifunctional Dehydratase Domain and a C-Methyltransferase Effects Olefin Shift in Ambruticin Biosynthesis. *Angew. Chem. Int. Ed.* **2016**, *55*, 13589–13592. [CrossRef]

10. Schröder, M.; Roß, T.; Hemmerling, F.; Hahn, F. Studying a Bottleneck of Multimodular Polyketide Synthase Processing: The Polyketide Structure-Dependent Performance of Ketoreductase Domains. *ACS Chem. Biol.* **2022**, *17*, 1030–1037. [CrossRef]
11. Wunderlich, J.; Roß, T.; Schröder, M.; Hahn, F. Step-Economic Synthesis of Biomimetic β-Ketopolyene Thioesters and Demonstration of Their Usefulness in Enzymatic Biosynthesis Studies. *Org. Lett.* **2020**, *22*, 4955–4959. [CrossRef]
12. Sundaram, S.; Kim, H.J.; Bauer, R.; Thongkongkaew, T.; Heine, D.; Hertweck, C. On-Line Polyketide Cyclization into Diverse Medium-Sized Lactones by a Specialized Ketosynthase Domain. *Angew. Chem. Int. Ed.* **2018**, *57*, 11223–11227. [CrossRef]
13. Hooshmand, S.E.; Heidari, B.; Sedghi, R.; Varma, R.S. Recent Advances in the Suzuki–Miyaura Cross-Coupling Reaction Using Efficient Catalysts in Eco-Friendly Media. *Green Chem.* **2019**, *21*, 381–405. [CrossRef]
14. Miyaura, N.; Suzuki, A. Palladium-Catalyzed Cross-Coupling Reactions of Organoboron Compounds. *Chem. Rev.* **1995**, *95*, 2457–2483. [CrossRef]
15. Zeysing, B.; Gosch, C.; Terfort, A. Protecting Groups for Thiols Suitable for Suzuki Conditions. *Org. Lett.* **2000**, *2*, 1843–1845. [CrossRef]
16. Liebeskind, L.S.; Srogl, J. Thiol Ester−Boronic Acid Coupling. A Mechanistically Unprecedented and General Ketone Synthesis. *J. Am. Chem. Soc.* **2000**, *122*, 11260–11261. [CrossRef]
17. Cheng, H.-G.; Chen, H.; Liu, Y.; Zhou, Q. The Liebeskind–Srogl Cross-Coupling Reaction and Its Synthetic Applications. *Asian J. Org. Chem.* **2018**, *7*, 490–508. [CrossRef]
18. Meng, S.; Tang, G.-L.; Pan, H.-X. Enzymatic Formation of Oxygen-Containing Heterocycles in Natural Product Biosynthesis. *ChemBioChem* **2018**, *19*, 2002–2022. [CrossRef]
19. Hemmerling, F.; Hahn, F. Biosynthesis of Oxygen- and Nitrogen-Containing Heterocycles in Polyketides. *Beilstein J. Org. Chem.* **2016**, *12*, 1512–1550. [CrossRef] [PubMed]
20. Pöplau, P.; Frank, S.; Morinaka, B.I.; Piel, J. An Enzymatic Domain for the Formation of Cyclic Ethers in Complex Polyketides. *Angew. Chem. Int. Ed.* **2013**, *52*, 13215–13218. [CrossRef]
21. Berkhan, G.; Hahn, F. A Dehydratase Domain in Ambruticin Biosynthesis Displays Additional Activity as a Pyran-Forming Cyclase. *Angew. Chem. Int. Ed.* **2014**, *53*, 14240–14244. [CrossRef] [PubMed]
22. Hollmann, T.; Berkhan, G.; Wagner, L.; Sung, K.H.; Kolb, S.; Geise, H.; Hahn, F. Biocatalysts from Biosynthetic Pathways: Enabling Stereoselective, Enzymatic Cycloether Formation on a Gram Scale. *ACS Catal.* **2020**, *10*, 4973–4982. [CrossRef]
23. Wagner, L.; Stang, J.; Derra, S.; Hollmann, T.; Hahn, F. Towards Understanding Oxygen Heterocycle-Forming Biocatalysts: A Selectivity Study of the Pyran Synthase PedPS7. *Org. Biomol. Chem.* **2022**, *20*, 9645–9649. [CrossRef]
24. Sung, K.H.; Berkhan, G.; Hollmann, T.; Wagner, L.; Blankenfeldt, W.; Hahn, F. Insights into the Dual Activity of a Bifunctional Dehydratase-Cyclase Domain. *Angew. Chem. Int. Ed.* **2018**, *57*, 343–347. [CrossRef]
25. Wagner, L.; Roß, T.; Hollmann, T.; Hahn, F. Cross-Linking of a Polyketide Synthase Domain Leads to a Recyclable Biocatalyst for Chiral Oxygen Heterocycle Synthesis. *RSC Adv.* **2021**, *11*, 20248–20251. [CrossRef]
26. Wagner, D.T.; Zhang, Z.; Meoded, R.A.; Cepeda, A.J.; Piel, J.; Keatinge-Clay, A.T. Structural and Functional Studies of a Pyran Synthase Domain from a Trans-Acyltransferase Assembly Line. *ACS Chem. Biol.* **2018**, *13*, 975–983. [CrossRef] [PubMed]
27. Ueoka, R.; Uria, A.R.; Reiter, S.; Mori, T.; Karbaum, P.; Peters, E.E.; Helfrich, E.J.N.; Morinaka, B.I.; Gugger, M.; Takeyama, H.; et al. Metabolic and Evolutionary Origin of Actin-Binding Polyketides from Diverse Organisms. *Nat. Chem. Biol.* **2015**, *11*, 705–712. [CrossRef]
28. Luhavaya, H.; Dias, M.V.B.; Williams, S.R.; Hong, H.; de Oliveira, L.G.; Leadlay, P.F. Enzymology of Pyran Ring A Formation in Salinomycin Biosynthesis. *Angew. Chem. Int. Ed.* **2015**, *54*, 13622–13625. [CrossRef]
29. Woo, A.J.; Strohl, W.R.; Priestley, N.D. Nonactin Biosynthesis: The Product of NonS Catalyzes the Formation of the Furan Ring of Nonactic Acid. *Antimicrob. Agents Chemother.* **1999**, *43*, 1662–1668. [CrossRef] [PubMed]
30. Miyaura, N.; Ishiyama, T.; Sasaki, H.; Ishikawa, M.; Sato, M.; Suzuki, A. Palladium-Catalyzed Inter- and Intramolecular Cross-Coupling Reactions of B-Alkyl-9-Borabicyclo [3.3.1] Nonane Derivatives with 1-Halo-1-Alkenes or Haloarenes. Syntheses of Functionalized Alkenes, Arenes, and Cycloalkenes via a Hydroboration-Coupling Sequence. *J. Am. Chem. Soc.* **1989**, *111*, 314–321. [CrossRef]
31. Farina, V.; Krishnan, B. Large Rate Accelerations in the Stille Reaction with Tri-2-Furylphosphine and Triphenylarsine as Palladium Ligands: Mechanistic and Synthetic Implications. *J. Am. Chem. Soc.* **1991**, *113*, 9585–9595. [CrossRef]
32. Chishiro, A.; Konishi, M.; Inaba, R.; Yumura, T.; Imoto, H.; Naka, K. Tertiary Arsine Ligands for the Stille Coupling Reaction. *Dalton Trans.* **2021**, *51*, 95–103. [CrossRef] [PubMed]
33. Carreira, E.M.; Du Bois, J. (+)-Zaragozic Acid C: Synthesis and Related Studies. *J. Am. Chem. Soc.* **1995**, *117*, 8106–8125. [CrossRef]
34. Li, Y.; Zhang, W.; Zhang, H.; Tian, W.; Wu, L.; Wang, S.; Zheng, M.; Zhang, J.; Sun, C.; Deng, Z.; et al. Structural Basis of a Broadly Selective Acyltransferase from the Polyketide Synthase of Splenocin. *Angew. Chem.* **2018**, *130*, 5925–5929. [CrossRef]
35. Campbell, N.E.; Sammis, G.M. Single-Electron/Pericyclic Cascade for the Synthesis of Dienes. *Angew. Chem. Int. Ed.* **2014**, *53*, 6228–6231. [CrossRef]

36. Whittaker, A.M.; Lalic, G. Monophasic Catalytic System for the Selective Semireduction of Alkynes. *Org. Lett.* **2013**, *15*, 1112–1115. [CrossRef]
37. Baldwin, J.E.; Moloney, M.; Parsons, A.F. An intramolecular cobalt cyclisation for the construction of substituted pyrrolidines. *Tetrahedron* **1992**, *48*, 9373–9384. [CrossRef]

Disclaimer/Publisher's Note: The statements, opinions and data contained in all publications are solely those of the individual author(s) and contributor(s) and not of MDPI and/or the editor(s). MDPI and/or the editor(s) disclaim responsibility for any injury to people or property resulting from any ideas, methods, instructions or products referred to in the content.

Article

Unraveling the Synthesis of SbCl(C₃N₆H₄): A Metal-Melaminate Obtained through Deprotonation of Melamine with Antimony(III)Chloride

Elaheh Bayat, Markus Ströbele and Hans-Jürgen Meyer *

Section for Solid State and Theoretical Inorganic Chemistry, Institute of Inorganic Chemistry, University of Tübingen, Auf der Morgenstelle 18, 72076 Tübingen, Germany
* Correspondence: juergen.meyer@uni-tuebingen.de

Abstract: The discovery of melamine by Justus von Liebig was fundamental for the development of several fields of chemistry. The vast majority of compounds with melamine or melamine derivatives appear as adducts. Herein, we focus on the development of novel compounds containing anionic melamine species, namely the melaminates. For this purpose, we analyze the reaction of SbCl$_3$ with melamine by differential scanning calorimetry (DSC). The whole study includes the synthesis and characterization of three antimony compounds that are obtained during the deprotonation process of melamine to melaminate with the reaction sequence from SbCl$_4$(C$_9$N$_{18}$H$_{19}$) (**1**) via (SbCl$_4$(C$_6$N$_{12}$H$_{13}$))$_2$ (**2**) to SbCl(C$_3$N$_6$H$_4$) (**3**). Compounds are characterized by single-crystal X-ray diffraction (SXRD), powder X-ray diffraction (PXRD), and infrared spectroscopy (IR). The results give an insight into the mechanism of deprotonation of melamine, with the replacement of one, two, or eventually three hydrogen atoms from the three amino groups of melamine. The structure of (**3**) suggests that metal melaminates are likely to form supramolecular structures or metal-organic frameworks (MOFs).

Keywords: melaminate; antimony; melamine; melaminium; deprotonation; crystal structures

1. Introduction

In the 19th century, the foundation of amine-substituted s-triazine derivatives was laid for the first time by Liebig and Gmelin [1–3] with the synthesis of melamine, melam, melem and their condensation product called melon. Melamine (1,3,5-triazine-2,4,6-triamine) is the simplest and most intensively studied C/N/H compound synthesized from potassium thiocyanate and ammonium chloride for the first time (1834) by Liebig [1,4]. However, it can also be easily achieved with trimerization of cyanamide (CN$_2$H$_2$), while today, industrial productions take place from urea in tons [5–7]. Melamine has a relatively high-melting point for an organic compound and undergoes condensation reactions on heating. The condensation products melem, melam, and melon (Figure 1) of this ancestry compound have been studied extensively using various spectroscopy techniques [8,9]. For a long period, thermal condensation was not fully understood due to the chemical inertness and low solubility of these products [10]. In 1959, May conducted a study on the pyrolysis of melamine at temperatures between 200 °C and 500 °C [11]. Afterward, this process was investigated by several scientists, particularly by Schnick and Lotsch [12,13]. The temperature-programmed XRD (TPXRD) was used in the temperature range between 25 °C and 660 °C to clarify the exact temperatures of formation of condensed products, which shows that the sublimation temperature of melamine is approximately 360–370 °C at atmospheric pressure. However, TPXRD in semi-closed systems shows that the X-ray reflections of melamine disappear at 296 °C, and then melamine forms melam and melem, which are stable up to 379 °C [13]. Pure melem, which consists of internally hydrogen-bonded heptazine molecules, can be obtained at 379 °C and is stable up to 500 °C. The polymeric carbon nitride material melon is also achieved with further heating [13].

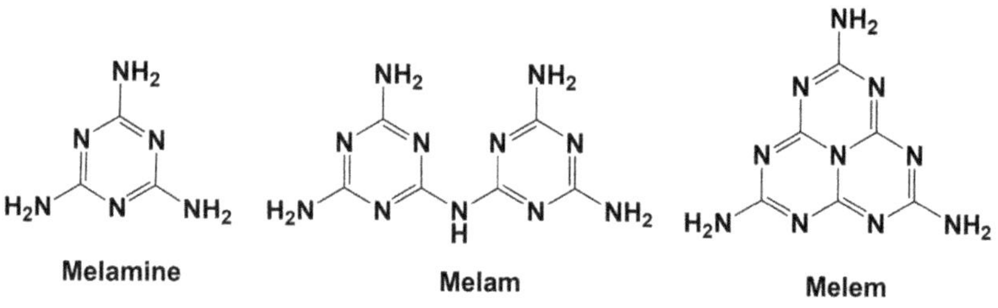

Figure 1. Molecular structures of melamine ($C_3N_6H_6$), melam ($C_6N_{11}H_9$) and melem ($C_6N_{10}H_6$).

In addition to the many applications melamine has, such as surface coating [14], flame redundancy [14–17], and heavy metal removal [18,19], it has some unique characteristics which make it a relevant research topic up to this day. The most important potential of melamine is its ability to create a metal-organic framework (MOF) [20] or porous-organic framework (POFs) [21] by the formation of metal melaminates.

Justus von Liebig's discovery of melamine was essential in the progress of C/N/H chemistry. Most melamine-containing compounds and their derivatives are found as adducts. Cationic C/N/H ions are present in various molecular compounds, including melamine, melam, and melem. These ions are formed by protonating the ring nitrogen atoms, which are more basic than the terminal amino groups. The most common cations are monoprotonated, but di- or trications have also been observed. More research into the chemistry of these substances led to the discovery of melaminium [22–26], melamium, and melemium salts. By far, the majority of salts were produced by melamine, including melaminium sulfate [27], melaminium nitrate, melaminium phosphates [16,28], melaminium chloride [29], organic slats of phthalates [30], benzoates, or citrates [31], and many inorganic salts containing complex anions [32,33]. On the other hand, a small number of melamium salts have been studied, such as melamium bromide and iodide [26]. Recently, melemium salts, namely melemium sulfate, triple protonated melemium methylsulfonate, and melemium perchlorate, are also discovered [34,35]. Melamine was also reported to coordinate with metal halides to form organic-inorganic hybrid copper halides such as $Cu_2Br_2(C_3N_6H_7)]_n$, $[Cu_3Cl_3(C_3N_6H_7)]_n$ [36], the silver complex $[Ag(C_3N_6H_6)(H_2O)(NO_3)]_n$, and the mercury compound $(C_3N_6H_7)(C_3N_6H_6)HgCl_3$ [37], which have biochemical applications and nonlinear optical properties [38,39].

Regarding the basic property of melamine, protonation is easy, and a great variety of such compounds, either theoretically or experimentally, have been investigated [25,26,40,41]. A new class of chemistry related to melaminates (deprotonated melamine) has received less attention until now; however, it is very important from either a chemistry or application perspective. The coordination behavior of molecules such as guanidine or melamine, capable of forming extended hydrogen bonds, can be changed by deprotonation [36,42]. Thus, it is a promising strategy for synthesizing interconnected supramolecular structures or MOFs. Despite the challenge which arises from the rigidity of its heterocyclic structure, the affinity of ring-N atoms to act as H-bond acceptors, and the steric hindrance of neighboring amino groups [35], the deprotonation of melamine seems to be plausible since guanidine (a stronger base) has already been deprotonated twice [43]. Franklin pioneered the work on anionic melaminate by synthesizing two compounds of $K(C_3N_6H_5)\cdot NH_3$ and $K_3(C_3N_6H_3)$ [44,45] in liquid ammonia. However, these compounds were only characterized using elemental analyses, and no crystallographic structure information was provided. Later, Dronskowski and coworkers confirmed the presence of the two ammonia adducts, $K(C_3N_6H_5)\cdot NH_3$ and $Rb(C_3N_6H_5)\cdot \frac{1}{2}NH_3$ by single-crystal X-ray diffraction. Ammonia-free $K_3(C_3N_6H_3)$ has been assigned by its characteristic infrared bands, being compared with calculated bands from density-functional theory (DFT) [42]. There was no further research reported on these classes of compounds and their properties until the discovery and identification of the copper melaminate $Cu_3(C_3N_6H_3)$ with a layered framework structure by Meyer & coworkers [20].

In this work, the step-wise deprotonation of melamine in a solid state has been studied by thermal analysis. Herein, antimony (III) chloride is used for the deprotonation of melamine due to its low melting point of 73.4 °C [46]. The recorded reaction sequence shows three compounds that were prepared and later characterized by powder X-ray diffraction (PXRD), single-crystal diffraction, and IR measurements. The structure of $SbCl(C_3N_6H_4)$ suggests the potential of synthesizing interconnected supramolecular structures or metal-organic frameworks (MOFs).

2. Materials and Methods

2.1. Materials

The starting materials, melamine (2,4,6-triamino-1,3,5-triazine, purchased from Sigma-Aldrich, 99%), and antimony(III)chloride (Sigma-Aldrich, 99%), ammonium chloride (Sigma-Aldrich, 99.99%) were used without further purification. The reaction mixtures were prepared under an argon atmosphere in a glovebox with moisture and oxygen levels below 1 ppm and transferred into homemade silica tubing (inner diameter 13 mm and 7 mm) and sealed under vacuum. The reactions were carried out in Simon–Müller and Carbolite chamber furnaces.

2.1.1. Synthesis

Synthesis of $SbCl_4(C_9N_{18}H_{19})$ (1):

Precursors were pestled in an agate mortar with a 1:4 molar ratio of antimony(III)chloride and melamine. A mixture of antimony(III) chloride and melamine with a total mass of ≈200.0 mg was transferred into a homemade silica ampule and sealed therein under vacuum. The ampoule was placed into a Simon–Müller furnace and heated to 200 °C for 20 h with a heating rate of 2 °C/min and cooling ramp of 0.5 °C/min (Figure S1). The reaction produced a white color product crystallized on the top of the ampule (>90% yield w.r.t Sb). A temperature gradient seemed to play an essential role in the separation of the product (1).

The solubility of compound (1) has been investigated in acetonitrile, THF, DCM, ethanol, methanol, and water. The PXRD measurements showed the decomposition of this product to unknown phases.

Synthesis of $(SbCl_4(C_6N_{12}H_{13}))_2$ (2):

Similar to the previous preparation, the mixture of antimony (III)chloride and melamine was mixed in a 1:2 molar ratio (total mass of ≈200.0 mg) and heated to 200 °C for 20 h with a heating and cooling rate of 2 °C/min (Figure S1). The product was X-ray amorphous powder and contained transparent single crystals of (2) (10% w.r.t Sb).

Synthesis of $SbCl(C_3N_6H_4)$ (3):

The structure of (3) was obtained from both (1:2 and 1:4) ratios of antimony(III)chloride and melamine by heating the 1:2 ratio at 250 °C for 20 h, or by heating the 1:4 ratio at 280 °C for 20 h (Figure S1). The beige color product was isolated in 50% yield w.r.t Sb.

The solubility of compound (3) has been studied in many solvents. The powder was soaked for one hour in acetonitrile, THF, DMF, DCM, ethanol, methanol, water, and diluted acetic acid. Subsequent PXRD measurements were undertaken. The results showed that compound (3) remains stable in acetonitrile, THF, DMF, DCM, ethanol, and methanol. However, in water and DMSO, compound (3) decomposes and forms Sb_2O_3 and an unknown phase, respectively. The schematic synthesis of all three compounds is presented in Figure S1.

2.1.2. X-ray Powder Diffraction

The X-ray diffraction of prepared powders was recorded with a powder diffractometer (STOE Darmstadt, STADIP, Ge-monochromator) using Cu-K$_{\alpha 1}$ (λ = 1.540598 Å) radiation in the range of 5 < 2θ < 120°. Match3! Software [47] was used to compare the patterns with patterns of the corresponding crystal structures.

2.1.3. Single-Crystal X-ray Diffraction

Single Crystals of (**1**), (**2**), and (**3**) were selected and placed on a single-crystal X-ray diffractometer (Rigaku XtaLab Synergy-S) with Cu-K$_\alpha$ radiation (λ = 1.54184 Å) and a mirror monochromator at 150 or 220 K. Crystal structures were solved by direct methods (SHELXT) [48], followed by full-matrix least-squares structure refinements (SHELXL-2014) [49]. The absorption correction of X-ray intensities was performed with numerical methods using the CrysAlisPro 1.171.41.92a software (Rigaku Oxford Diffraction). Hydrogen atoms were found in the difference map and refined therefrom isotropically.

2.1.4. Thermoanalytic Studies

Differential scanning calorimetry (DSC) was carried out using a DSC 204 F1 Phoenix (Fa. Netzsch, Selb, Germany). The starting materials were enclosed under Ar in a glovebox into gold-plated (5 µm) steel autoclaves with a volume of 100 µL (Bächler Feintech AG in Hölstein, Switzerland). The reactions of SbCl$_3$ with melamine were analyzed for different ratios between room temperature and 500 °C at a heating and cooling rate of 2 °C/min.

2.1.5. Infrared Spectra

The infrared (IR) spectra of samples were recorded with a Bruker VERTEX 70 FT-IR spectrometer within the spectra range of 400–4000 cm^{-1}. Tablets of KBr were used as a background.

3. Results and Discussion

3.1. Thermoanalytic Studies

Thermal analyses based on DSC and DTA have been shown to be highly insightful regarding the examination of reaction sequences [50,51] and for comprehensive studies of binary or ternary systems [52], especially when combined with PXRD studies. Following this method, the formation or decomposition of a crystalline species is usually indicated by a thermal event, and the newly formed species is characterized by X-ray diffraction techniques.

The differential scanning calorimetric (DSC) measurements of 1:2 and 1:4 molar mixtures of antimony chloride and melamine are shown in Figure 2, with heating and cooling rates of 2 °C/min. The DSC patterns display a small exothermic peak at around 70 °C, which can be attributed to the melting point of antimony(III)chloride. Figure 2a,b show multiple exothermic effects between 200 °C and 300 °C. The resolution of thermal events in this region is rather poor and cannot be significantly improved by changing the heating ramp. For example, we have explored different heating rates throughout. At lower heating rates, the signals were smeared out and were not as sharp as the signals shown in Figure 2a,b, so the resolution was worse. The presented heating rate is the optimized heating rate with respect to the signal-to-noise ratio. Moreover, the effects are slightly different for different ratios of starting materials, with lower reaction temperatures in the presence of more melamine.

Powder XRD patterns were recorded on samples obtained under conditions given in the DSC experiments, being interrupted at certain temperatures, especially in the temperature region between 200 °C and 280 °C. Compound (**1**) was already formed at 200 °C from a 1:4 ratio of starting materials, and compound (**2**) was observed in the XRD pattern from a 1:2 ratio at the same temperature. Compound (**3**) was identified at around 280 °C from a 1:4 ratio of starting materials or, alternatively, at 250 °C from a 1:2 ratio. The endothermic peaks at 370 °C indicate the melting/decomposition of (excess) melamine, which appears sharper for the 1:4 ratio due to the larger amount of melamine. This assignment is confirmed by a DSC of melamine (Figure S2), which shows a similar endothermic peak with a slight shift at 361 °C. At slightly higher temperatures, compound (**3**) is decomposed, which is followed by a strong exothermic peak at 400 °C and 417 °C for the 1:4 and a 1:2 ratio, respectively.

These intense exothermic peaks led to the formation of a phase with a yellow color (**4**) that looked glassy under the microscope. This was further studied by means of IR spectroscopy (see the relevant section).

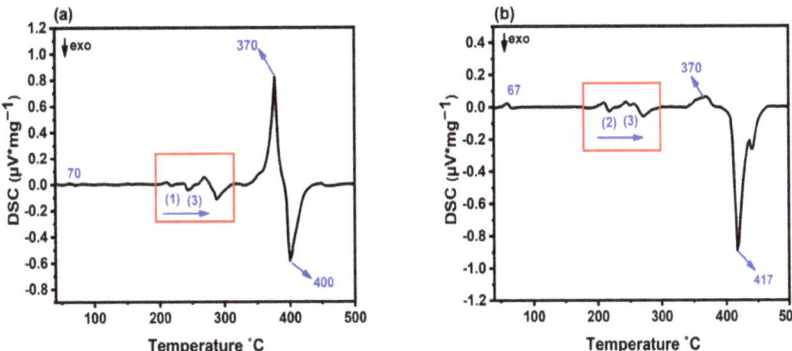

Figure 2. (a) DSC of the reaction of SbCl$_3$ and melamine in a ratio of 1:4, (b) DTA of the reaction of SbCl$_3$ and melamine in a ratio of 1:2.

From this study, we note that reactions in the given system proceed very quickly, almost simultaneously, making the assignment of compounds and their preparations challenging. This is due to the high reactivity of reaction partners.

3.2. Crystal Structures

Crystal structures of all three compounds (**1**), (**2**), and (**3**) were solved and refined based on single-crystal X-ray diffraction data with triclinic ($P\bar{1}$) and monoclinic ($P2_1/c$ and $P2_1/n$) space groups, respectively, with crystallographic details summarized in Table 1 and relevant distances given in Table 2. The asymmetric unit of each compound is shown in Figure S3.

The crystal structure of (**1**) is composed of one deprotonated melamine, two protonated melamine and a single chloride ion besides SbCl$_3$ to make up SbCl$_4$(C$_3$N$_6$H$_5$)(C$_3$N$_6$H$_7$)$_2$. The crystal structure contains a sequence of three distinct layers stacked on top of each other (along *b*); one of them is displayed in Figure 3. Stacking behavior is most common for melamine-based structures.

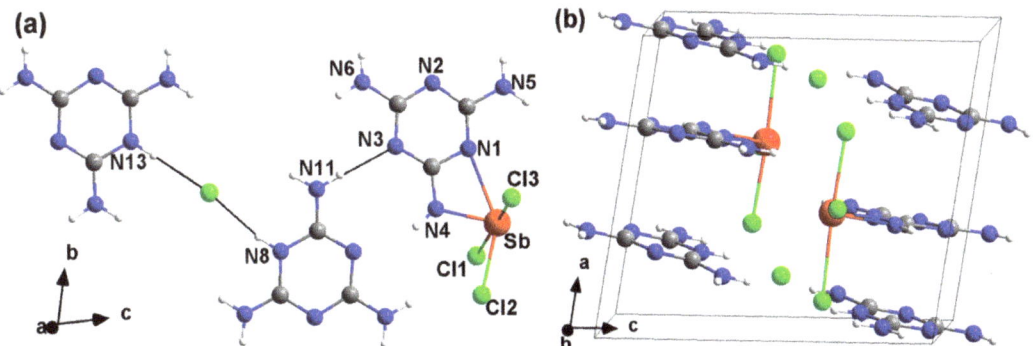

Figure 3. (a) Constituents of one layer in the structure of (**1**) as SbCl$_4$(C$_3$N$_6$H$_5$)(C$_3$N$_6$H$_7$)$_2$, and (b) a perspective view of the unit cell of (**1**) along the *b*-axis, with the color code: N: blue, C: gray, H: white, Cl: green, Sb: red).

Table 1. Crystallographic details of the crystal structure refinement of compounds (1), (2), and (3).

Compound	(1)	(2)	(3)
CCDC code	2201273	2210244	2213381
Formula weight	642.97	1033.67	281.32
Temperature/K	220.0(1)	150.0(1)	150.0(1)
Wavelength/Å	1.54184	1.54184	1.54184
Space group	$P\bar{1}$	$P2_1/c$	$P2_1/n$
a/Å	9.5878(5)	13.2780(2)	5.3562(2)
b/Å	10.5395(3)	10.6878(1)	10.5432(3)
c/Å	11.4338(5)	24.0953(2)	12.5618(4)
α/°	74.011(3)	90	90
β/°	79.122(4)	105.860(1)	93.710(3)
γ/°	85.602(3)	90	90
Volume/Å^3	1090.37(8)	3289.26(7)	707.90(4)
Z	2	4	4
R_{int}	0.0364	0.0485	0.0312
Goodness-of-fit on F^2	1.074	1.044	1.044
wR_2 (all data)	0.0660	0.0607	0.0267
wR_2	0.0643	0.0603	0.0264
Final R indices (all data)	0.0339	0.0251	0.0120
R_1	0.0278	0.0243	0.0110
$\theta_{Max.}/°$	4.365	3.460	5.483
$\theta_{Min.}/°$	66.585	66.583	70.067
μ/mm^{-1}	14.93	19.478	33.932
$\Delta\rho_{Max.}$/e·Å^{-3}	0.508	2.491	0.322
$\Delta\rho_{Min.}$/e·Å^{-3}	−0.593	−0.597	−0.451
Completeness/%	97.3	100	99.8

Table 2. Selected interatomic distances (pm) of compounds (1), (2), and (3).

Compound (1)			Compound (2)			Compound (3)		
Atom	Atom	Length/pm	Atom	Atom	Length/pm	Atom	Atom	Length/pm
Sb1	Cl1	276.5(8)	Sb1	Cl3	284.7(6)	Sb1	N2	241.5(1)
Sb1	N1	253.6(3)	Sb1	Cl4	260.6(0)	Sb1	N6	208.6(8)
Sb1	Cl2	247.0(0)	Sb1	Cl5	248.7(2)	Sb1	N4	204.4(6)
Sb1	Cl3	256.8(2)	Sb1	Cl1	240.2(1)	Sb1	Cl1	254.9(3)
Sb1	N4	204.7(3)	Sb2	N1	256.1(3)			
			Sb2	N4	204.7(3)			
			Sb2	Cl6	279.2(7)			
			Sb2	Cl7	247.2(8)			
			Sb2	Cl8	251.1(1)			

The SbCl$_3$ entity in the structure, with its lone pair, is well known from several crystal structures having average Sb-Cl distances of 260.1 pm [53]. The antimony is connected with an exocyclic nitrogen atom of the melaminate ion (C$_3$H$_5$N$_6$)$^-$ via Sb-N4 (204.7(3) pm) and an obviously weaker interaction via Sb-N1 (253.6(3) pm). The constituents in each layer in (1) are interconnected by a network of hydrogen bonds (Figure 3a). An isolated Cl$^-$ ion in the structure is interconnected by hydrogen bridges at d$_{H-Cl}$ = 216.3 pm and 225.6 pm, consistent with the corresponding value in melaminium chloride d$_{H-Cl}$ = 239.7 pm [29].

The crystal structure of (2) comprises one deprotonated melamine, three protonated melaminium ions, an SbCl$_3$ unit and an (SbCl$_5$)$^{2-}$ ion to make up (SbCl$_4$)$_2$(C$_3$N$_6$H$_5$)(C$_3$N$_6$H$_7$)$_3$ displayed in Figure 4. The average Sb-Cl distances in SbCl$_3$ are 259.3 pm, and those of SbCl$_5$ are 262.1 pm and 257.8 pm, supporting the presence of Sb^{3+} throughout. Antimony in SbCl$_3$ is interconnected with the melaminate ion (C$_3$H$_5$N$_6$)$^-$ via Sb-N4 (204.4(6) pm) and an obviously weaker interaction via Sb-N1 (256.1(3) pm). Again, the crystal structure features a layered arrangement and hydrogen bridging within layers.

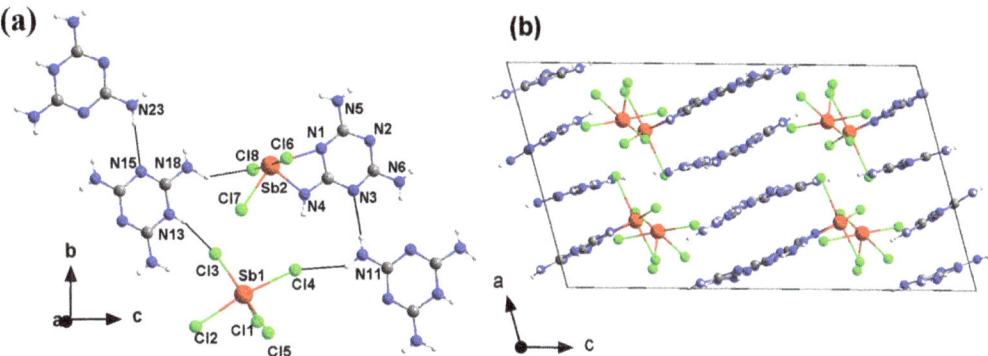

Figure 4. (a) Section of the crystal structure of (2) as $(SbCl_4)_2(C_3N_6H_5)(C_3N_6H_7)_3$ projected on the bc-plane, and (b) a perspective view along the b-axis with the color code: N: blue, C: gray, H: white, Cl: green, Sb: red).

The crystal structure of (3) features the presence of $(SbCl)^{2+}$ and the melaminate ion $(C_3N_6H_4)^{2-}$ in $SbCl(C_3N_6H_4)$. Unlike the two previous systems, this structure can be described as an infinite chain structure due to the bridging connectivity of the divalent melaminate anion, all shown in Figure 5. The $(SbCl)^{2+}$ (d_{Sb-Cl} = 254.9(3) pm) is interconnected via exocyclic nitrogen atoms of two melaminate ions via Sb-N4 (204.4(6) pm) and Sb-N6 (208.6(8) pm) interactions and an obviously weaker interaction via Sb-N2 (241.5(1) pm).

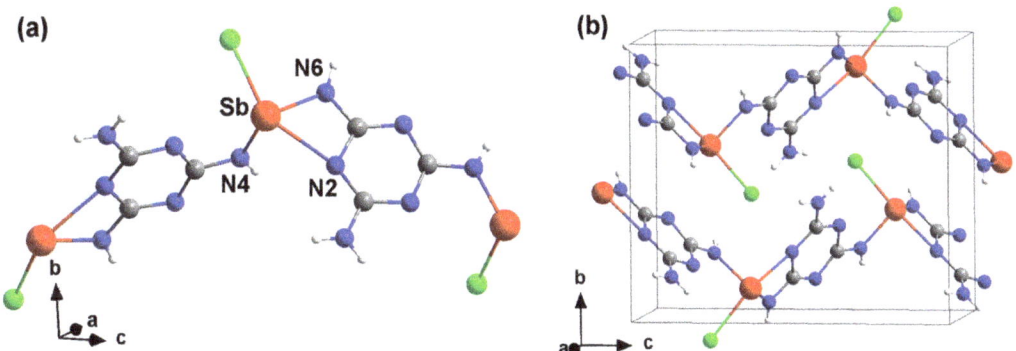

Figure 5. (a) Section of a chain section of the crystal structure of (3) and (b) a perspective view of the unit cell roughly along the a-axis with the color code: N: blue, C: gray, H: white, Cl: green, Sb: red.

The stacking sequences of layers are often dominated by the preference that the N atom of the triazine ring in one layer is alternating with a C atom of the triazine ring in the next layer, which is a characteristic feature in copper melaminate [20] and metal cyanurates as well [54,55]. This is achieved by rotating or shifting C_3N_3 units in adjacent layers relative to each other. However, this is not apparent in the structure of compounds (1–3). Layered arrangements of C_3N_3 units are quite clearly visible in compounds (1) and (2) but not in compound (3) (Figures S3–S6). Hence there is the possibility of π-π interactions between C_3N_3 units in (1) and (2). Such interactions can play a crucial role in the stabilization of parallel and antiparallel ring architectures in the crystal structure. The centroid-to-centroid distance at which C_3N_3 rings may be considered representative of π-π stacking interactions is 360–390 pm in compounds (1) and (2). This distance increases to 560–590 pm in (3) which might present no π-π interactions between layers in this structure.

The range from 357–393 pm was previously reported for several compounds [38,56,57]; for example, in a zinc(II) complex containing melamine (392.8 pm) [56]. In many other studies of copper halide complexes (357.2–389.2 pm), we can see the stacking behavior of twisted melamine rings, which represents the π-π interactions [58].

3.3. X-ray Powder Diffraction and Infrared Spectroscopy

The reaction products were investigated by PXRD, and the XRD patterns of (1), and (3) are provided in Figures S7 and S8. Therein, the recorded data are compared with the calculated patterns obtained from the structure refinement based on single-crystal data. Compound (2) was obtained in low yield; therefore, no powder pattern of this intermediate could be recorded. This compound was always found in the presence of (3) or melamine at higher and lower temperatures, respectively. The powder pattern of compound (3) in Figure S8 shows some unidentified diffraction peaks.

3.4. Infrared Spectroscopy (IR)

The IR spectrum of (1), (3) and (4) has been compared to that of melamine and melaminium chloride, as presented in Figure 6. Table S1 lists the frequencies associated with each vibrational mode of these molecules, along with the corresponding bond assignments. Three IR absorption bands, indicative of the asymmetric and symmetric stretching of -NH_2 groups of melamine, can be found in the 3500–3300 cm^{-1} range of the melamine spectrum [29,59]. These vibrations can overlap with the -NH^+ in melaminium chloride, and due to coupling, the peak is broadened [60]. The characteristic bands of -NH_2 groups and -NH^+ are also seen in the spectrum of compound (1) at 3462, 3357, and 3433 cm^{-1}. We can see that the first peak (3462 cm^{-1}) in compound (1) is shifted to lower wavenumbers when compared to melamine. This shift may be due to the presence of protonated melamine units in (1). In fact, the presence of hydrogen bonding would shift -NH_2 IR bands to lower wavenumbers, as the hydrogen bond would weaken the NH_2 bond and lower its vibrational frequency [61]. However, due to coupling with the N-H ··· Cl stretching mode or the presence of heavier atoms (Cl, Sb) in compound (1), the second peak (3357 cm^{-1}) is shifted to slightly higher wavenumbers and also broadened [29]. Similarly, infrared spectra for compounds (3) and (4) indicate that the asymmetric and symmetric vibrations of -NH_2 overlap with those of -NH^+ in both compounds, as evidenced by the disappearance of the first peak (3471 cm^{-1} for melamine) and the broadening of the other two peaks. The bending mode bands for melaminium chloride and compound (1) are substantially higher than those for melamine (1652 cm^{-1}) at 1722, 1676, 1649 cm^{-1} and 1679, 1656, and 1612 cm^{-1}, respectively. This is explained by the fact that melaminium chloride and (1) have fewer intermolecular interactions than melamine. In compounds (3) and (4), bending modes are split into multiple bands, showing that -NH_2 groups in these compounds have different vibrational frequencies due to the presence of neighboring atoms and molecular interactions in these structures. The region below 1500 cm^{-1} is related to C-N, and C=N ring stretching modes, C-N side group stretching, N-H rocking, and triazine ring breath and bending vibrations, which are listed with detailed numbers in Table S1. The exact position of these peaks can depend on various factors, such as the substitution pattern of the triazine ring and the nature of the surrounding chemical environment, which agrees well with the slight shifts in each region for compounds (1), (3), and (4). The strong split-band at 800 cm^{-1}, which is brought on by the sextant-bend of both the triazine and heptazine rings, provides additional evidence that compound (4) is still either a heptazine- or triazine-based compound (IR cannot differentiate between triazine and heptazine) [13,22,62], whereas the yellow emission color of the compound under ultraviolet radiation rather indicates a heptazine based compound.

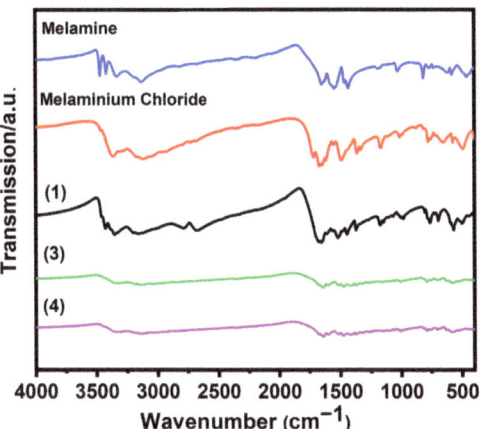

Figure 6. FTIR spectrum of melamine, melaminium chloride powder, compared with compounds (**1**), (**3**), (**4**).

4. Conclusions

The development of metal melaminates is just at its beginning. A preparative concept for the development of melaminates was recently established for $Cu_3(C_3H_3N_3)$ based on the reaction of CuCl with melamine. The same concept is employed in this study for the reaction of $SbCl_3$ with melamine. Thermal studies (DSC) reveal a narrow sequence of thermal events, or rather intertwining reactions that reveal new compounds, following the sequence (**1**), (**2**) and (**3**) with increasing temperature.

Indeed, the final product of the given reaction cascade is compound (**3**), observed via compounds (**1**) and (**2**). For a better description of the reaction sequence of compounds, we use the abbreviation *Mel* for melamine, with $Mel^{(n-)}$ for melaminate and $Mel^{(+)}$ for melaminium. The overall reaction representing the formation of compound (**3**) can be described as follows:

$$SbCl_3 + Mel \rightarrow SbClMel^{(2-)} + 2\ HCl\uparrow$$

The formation of HCl in this reaction can be equivalent to melaminium chloride ($Mel^{(+)}Cl$), which is, in fact, present in compound (**1**) but is lost at elevated temperatures through sublimation, which indeed has been reported as a side-phase for the corresponding reaction of CuCl and melamine [20]. This reaction scheme with metal halide and melamine is indeed a useful way to develop metal melaminates. However, reactions with melamine are intrinsically difficult due to the high reactivity and condensation behavior of melamine.

The reaction of $SbCl_3$ with excess melamine passes through some intermediate reaction stages with the formations of melamine derivatives (Mel^-, Mel^+) that are successively lost with increasing temperature from (**1**) to (**2**) and finally (**3**). Compound (**1**) is best described as $SbCl_4Mel^{(-)}(Mel^{(+)})_2$ containing three melamine species per antimony atom, and compound (**2**) is given as $(SbCl_4)_2Mel^{(-)}(Mel^{(+)})_3$ and contains only two melamine derivatives per antimony atom until only one melaminate is left in (**3**). The formation of the expectable compound $SbMel^{(3-)}$ is not observed.

Supplementary Materials: The following supporting information can be downloaded at: https://www.mdpi.com/article/10.3390/chemistry5020099/s1, Reference [29] is also cited in the supplementary materials.

Author Contributions: Conceptualization, supervision, funding acquisition, review and editing, H.-J.M.; Synthesis, PXRD and IR, writing, E.B.; X-ray diffraction refinements and structure solutions, M.S. All authors have read and agreed to the published version of the manuscript.

Funding: This research received no external funding.

Data Availability Statement: Data is contained within the article and Supplementary Materials.

Acknowledgments: A sincere thank you to Mike Healey Smith for his diligent English revisions and proofreading of this article.

Conflicts of Interest: The authors declare no conflict of interest.

References

1. Liebig, J. Über einige Stickstoff-Verbindungen. *Ann. Pharm. Fr.* **1834**, *10*, 1–47. [CrossRef]
2. Gmelin, L. Über einige Verbindungen des Melon's. *Ann. Pharm. Fr.* **1835**, *15*, 252–258. [CrossRef]
3. Liebig, J. Über die Constitution der Mellonverbindungen. *Justus Liebigs Ann. Chem.* **1855**, *95*, 257–282. [CrossRef]
4. Finkel'shtein, A.; Boitsov, E. The molecular structure of 1,3,5-triazine and its derivatives. *Russ. Chem. Rev.* **1962**, *31*, 712. [CrossRef]
5. Crews, G.M.; Ripperger, W.; Kersebohm, D.B.; Güthner, T.; Mertschenk, B. Melamine and guanamines. In *Ullmann's Encyclopedia of Industrial Chemistry*; Wiley: Weinheim, Germany, 2001. [CrossRef]
6. Keßler, F.K. Structure and Reactivity of S-Triazine-Based Compounds in C/N/H Chemistry. Ph.D. Thesis, Ludwig Maximilian University of Munich, Munich, Germany, 2019.
7. Klason, P. Über Melamverbindungen. *J. Prakt. Chem.* **1886**, *33*, 285–289. [CrossRef]
8. Finkel'shtein, A.I.; Spiridonova, N.Y.V. Chemical properties and molecular structure of derivatives of sym-heptazine [1,3,4,6,7,9,9b-heptaazaphenalene, tri-1,3,5-triazine]. *Russ. Chem. Rev.* **1964**, *33*, 400. [CrossRef]
9. Jürgens, B.; Irran, E.; Senker, J.; Kroll, P.; Müller, H.; Schnick, W. Melem (2,5,8-Triamino-tri-s-triazine), an Important Intermediate during Condensation of Melamine Rings to Graphitic Carbon Nitride: Synthesis, Structure Determination by X-ray Powder Diffractometry, Solid-State NMR, and Theoretical Studies. *J. Am. Chem. Soc.* **2003**, *125*, 10288–10300. [CrossRef]
10. Kroke, E.; Schwarz, M. Novel group 14 nitrides. *Coord. Chem. Rev.* **2004**, *248*, 493–532. [CrossRef]
11. May, H. Pyrolysis of melamine. *J. Appl. Chem.* **1959**, *9*, 340–344. [CrossRef]
12. Sattler, A.; Pagano, S.; Zeuner, M.; Zurawski, A.; Gunzelmann, D.; Senker, J.; Müller-Buschbaum, K.; Schnick, W. Melamine–melem adduct phases: Investigating the thermal condensation of melamine. *Chem.-Eur. J.* **2009**, *15*, 13161–13170. [CrossRef]
13. Lotsch, B.V.; Schnick, W. New light on an old story: Formation of melam during thermal condensation of melamine. *Chem.-Eur. J.* **2007**, *13*, 4956–4968. [CrossRef] [PubMed]
14. Weil, E.D. Fire-Protective and Flame-Retardant Coatings—A State-of-the-Art Review. *J. Fire Sci.* **2011**, *29*, 259–296. [CrossRef]
15. Schartel, B.; Weiß, A.; Mohr, F.; Kleemeier, M.; Hartwig, A.; Braun, U. Flame retarded epoxy resins by adding layered silicate in combination with the conventional protection-layer-building flame retardants melamine borate and ammonium polyphosphate. *J. Appl. Polym. Sci.* **2010**, *118*, 1134–1143. [CrossRef]
16. Yang, H.; Song, L.; Tai, Q.; Wang, X.; Yu, B.; Yuan, Y.; Hu, Y.; Yuen, R.K.K. Comparative study on the flame retarded efficiency of melamine phosphate, melamine phosphite and melamine hypophosphite on poly(butylene succinate) composites. *Polym. Degrad. Stab.* **2014**, *105*, 248–256. [CrossRef]
17. Weil, E.D.; Levchik, S.V. Flame Retardants in Commercial Use or Development for Textiles. *J. Fire Sci.* **2008**, *26*, 243–281. [CrossRef]
18. Yin, N.; Wang, K.; Xia, Y.a.; Li, Z. Novel melamine modified metal-organic frameworks for remarkably high removal of heavy metal Pb(II). *Desalination* **2018**, *430*, 120–127. [CrossRef]
19. Cao, Y.; Huang, J.; Li, Y.; Qiu, S.; Liu, J.; Khasanov, A.; Khan, M.A.; Young, D.P.; Peng, F.; Cao, D.; et al. One-pot melamine derived nitrogen doped magnetic carbon nanoadsorbents with enhanced chromium removal. *Carbon* **2016**, *109*, 640–649. [CrossRef]
20. Kallenbach, P.; Bayat, E.; Ströbele, M.; Romao, C.P.; Meyer, H.-J. Tricopper Melaminate, a Metal-Organic Framework Containing Dehydrogenated Melamine and Cu-Cu Bonding. *Inorg. Chem.* **2021**, *60*, 16303–16307. [CrossRef]
21. Pareek, K.; Rohan, R.; Cheng, H. Polymeric organo–magnesium complex for room temperature hydrogen physisorption. *RSC Adv.* **2015**, *5*, 10886–10891. [CrossRef]
22. Kessler, F.K.; Schuhbeck, A.M.; Schnick, W. Melamium Thiocyanate Melam, a Melamium Salt with Disordered Anion Sites. *Z. Anorg. Allg. Chem.* **2019**, *645*, 840–847. [CrossRef]
23. Sattler, A.; Schnick, W. Preparation and Structure of Melemium Melem Perchlorate $HC_6N_7(NH_2)_3ClO_4 \cdot C_6N_7(NH_2)_3$. *Z. Anorg. Allg. Chem.* **2008**, *634*, 457–460. [CrossRef]
24. Sattler, A.; Schnick, W. Melemium Hydrogensulfate $H_3C_6N_7(NH_2)_3(HSO_4)_3$—The First Triple Protonation of Melem. *Z. Anorg. Allg. Chem.* **2010**, *636*, 2589–2594. [CrossRef]
25. Vella-Zarb, L.; Braga, D.; Guy Orpen, A.; Baisch, U. The influence of hydrogen bonding on the planar arrangement of melamine in crystal structures of its solvates, cocrystals and salts. *CrystEngComm* **2014**, *16*, 8147–8159. [CrossRef]
26. Kessler, F.K.; Koller, T.J.; Schnick, W. Synthesis and Structure of Melemium Bromide $C_6N_{11}H_{10}Br$ and Melemium Iodide $C_6N_{11}H_{10}I$. *Z. Anorg. Allg. Chem.* **2018**, *644*, 186–192. [CrossRef]
27. Braml, J.; Perpétuo, G.J. Bis (melaminium) sulfate dihydrate. *Acta Crystallogr. Sect. C Cryst. Struct. Commun.* **2001**, *57*, 1431–1433. [CrossRef]
28. Volfkovi, S.I.; Feldmann, W.W.; Kozmina, M.L. Über Kondensierte Phosphate des Melamins. *Z. Anorg. Allg. Chem.* **1979**, *457*, 20–30. [CrossRef]
29. Janczak, J.; Perpétuo, G.J. Melaminium chloride hemihydrate. *Acta Crystallogr. Sect. C Cryst. Struct. Commun.* **2001**, *57*, 1120–1122. [CrossRef] [PubMed]

30. Janczak, J.; Perpétuo, G.J. Melaminium phthalate. *Acta Crystallogr. Sect. C Cryst. Struct. Commun.* **2001**, *57*, 123–125. [CrossRef]
31. Marchewka, M.; Pietraszko, A. Structure and spectra of melaminium citrate. *J. Phys. Chem. Solids* **2003**, *64*, 2169–2181. [CrossRef]
32. Colombo, A.; Menabue, L.; Motori, A.; Pellacani, G.C.; Porzio, W.; Sandrolini, F.; Willett, R. Crystal structure and spectroscopic, magnetic, and electrical properties of a copper (II) dimer, melaminium hexachlorodicuprate, exhibiting a new stacking interaction. *Inorg. Chem.* **1985**, *24*, 2900–2905. [CrossRef]
33. Kroenke, W.J.; Fackler, J.P., Jr.; Mazany, A.M. Structure and bonding of melaminium. beta-octamolybdate. *Inorg. Chem.* **1983**, *22*, 2412–2416. [CrossRef]
34. Sattler, A.; Seyfarth, L.; Senker, J.; Schnick, W. Synthesen, Kristallstrukturen und spektroskopische Eigenschaften des Melem-Adduktes $C_6N_7(NH_2)_3 \cdot H_3PO_4$ sowie der Melemium-Salze $(H_2C_6N_7(NH_2)_3)SO_4 \cdot 2H_2O$ und $(HC_6N_7(NH_2)_3)ClO_4 \cdot H_2O$. *Z. Anorg. Allg. Chem.* **2005**, *631*, 2545–2554. [CrossRef]
35. Sattler, A.; Schönberger, S.; Schnick, W. Melemium Methylsulfonates $HC_6N_7(NH_2)_3H_2C_6N_7(NH_2)_3(SO_3Me)_3 \cdot H_2O$ and $H_2C_6N_7(NH_2)_3(SO_3Me)_2 \cdot H_2O$. *Z. Anorg. Allg. Chem.* **2010**, *636*, 475–482. [CrossRef]
36. Zhang, L.; Zhang, J.; Li, Z.-J.; Cheng, J.-K.; Yin, P.-X.; Yao, Y.-G. New Coordination Motifs of Melamine Directed by N−H···X (X=Cl or Br) Hydrogen Bonds. *Inorg. Chem.* **2007**, *46*, 5838–5840. [CrossRef]
37. Rana, A.; Bera, M.; Chowdhuri, D.S.; Hazari, D.; Jana, S.K.; Zangrando, E.; Dalai, S. 3D Coordination Network of Ag(I) Ions with μ_3-Bridging Melamine Ligands. *J. Inorg. Organomet. Polym. Mater.* **2012**, *22*, 360–368. [CrossRef]
38. Bai, Z.; Lee, J.; Kim, H.; Hu, C.L.; Ok, K.M. Unveiling the Superior Optical Properties of Novel Melamine-Based Nonlinear Optical Material with Strong Second-Harmonic Generation and Giant Optical Anisotropy. *Small*, **2023**; in press. [CrossRef] [PubMed]
39. Liu, L.; Wu, Y.; Ma, L.; Fan, G.; Gao, W.; Wang, W.; Ma, X. A new melamine-based Cu(I) coordination polymer with an excellent photocatalytic activity, therapeutic and nursing effects on the blood glucose regulation. *J. Struct. Chem.* **2022**, *63*, 302–309. [CrossRef]
40. Braml, N.E.; Sattler, A.; Schnick, W. Formation of melamium adducts by pyrolysis of thiourea or melamine/NH_4Cl mixtures. *Chem.-Eur. J.* **2012**, *18*, 1811–1819. [CrossRef] [PubMed]
41. Mukherjee, S.; Ren, J. Gas-phase acid-base properties of melamine and cyanuric acid. *J. Am. Soc. Mass Spectrom.* **2010**, *21*, 1720–1729. [CrossRef]
42. Gorne, A.L.; Scholz, T.; Kobertz, D.; Dronskowski, R. Deprotonating Melamine to Gain Highly Interconnected Materials: Melaminate Salts of Potassium and Rubidium. *Inorg. Chem.* **2021**, *60*, 15069–15077. [CrossRef]
43. Gorne, A.L.; George, P.; van Leusen, J.; Duck, G.; Jacobs, P.; Chogondahalli Muniraju, N.K.; Dronskowski, R. Ammonothermal Synthesis, Crystal Structure, and Properties of the Ytterbium(II) and Ytterbium(III) Amides and the First Two Rare-Earth-Metal Guanidinates, $YbC(NH)_3$ and $Yb(CN_3H_4)_3$. *Inorg. Chem.* **2016**, *55*, 6161–6168. [CrossRef]
44. Franklin, E.C. The ammono carbonic acids. *J. Am. Chem. Soc.* **1922**, *44*, 486–509. [CrossRef]
45. Schnick, W.; Huppertz, H. Darstellung, Kristallstruktur und Eigenschaften von Kaliumhydrogencyanamid. *Z. Anorg. Allg. Chem.* **1995**, *621*, 1703–1707. [CrossRef]
46. Haynes, W.M. *CRC Handbook of Chemistry and Physic*, 95th ed.; CRC Press LLC: Boca Raton, FL, USA, 2016.
47. Putz, H.; Brandenburg, K. *Match!—Phase Analysis Using Powder Diffraction*, Version 3.15.x; Crystal Impact: Bonn, Germany, 2023.
48. Sheldrick, G.M. *SHELXS-97 and SHELXL-97, Program for Crystal Structure Solution and Refinement*; University of Göttingen: Göttingen, Germany, 1997.
49. Dolomanov, O.; Bourhis, L.; Gildea, R.; Howard, J.; Puschmann, H. OLEX2: A complete structure solution, refinement and analysis program. *J. Appl. Crystallogr.* **2009**, *42*, 339–341. [CrossRef]
50. Mos, A.; Castro, C.; Indris, S.; Ströbele, M.; Fink, R.F.; Meyer, H.-J. From WCl_6 to WCl_2: Properties of Intermediate Fe−W−Cl Phases. *Inorg. Chem.* **2015**, *54*, 9826–9832. [CrossRef]
51. Ströbele, M.; Mos, A.; Meyer, H.-J. Cluster harvesting by successive reduction of a metal halide with a nonconventional reduction agent: A benefit for the exploration of metal-rich halide systems. *Inorg. Chem.* **2013**, *52*, 6951–6956. [CrossRef] [PubMed]
52. Ströbele, M.; Meyer, H.-J. Pandora's box of binary tungsten iodides. *Dalton Trans.* **2019**, *48*, 1547–1561. [CrossRef]
53. Lizarazo-Jaimes, E.H.; Reis, P.G.; Bezerra, F.M.; Rodrigues, B.L.; Monte-Neto, R.L.; Melo, M.N.; Frezard, F.; Demicheli, C. Complexes of different nitrogen donor heterocyclic ligands with $SbCl_3$ and $PhSbCl_2$ as potential antileishmanial agents against Sb(III)-sensitive and -resistant parasites. *J. Inorg. Biochem.* **2014**, *132*, 30–36. [CrossRef]
54. Kalmutzki, M.; Ströbele, M.; Bettinger, H.F.; Meyer, H.-J. Development of Metal Cyanurates: The Example of Barium Cyanurate (BCY). *Eur. J. Inorg. Chem.* **2014**, *2014*, 2536–2543. [CrossRef]
55. Kalmutzki, M.; Ströbele, M.; Enseling, D.; Jüstel, T.; Meyer, H.-J. Synthesis, Structure, and Luminescence of Rare Earth Cyanurates. *Eur. J. Inorg. Chem.* **2015**, *2015*, 134–140. [CrossRef]
56. Salah, T.; Mhadhbi, N.; Ben Ahmed, A.; Hamdi, B.; Krayem, N.; Loukil, M.; Guesmi, A.; Khezami, L.; Houas, A.; Ben Hamadi, N. Physico-Chemical Characterization, DFT Modeling and Biological Activities of a New Zn(II) Complex Containing Melamine as a Template. *Crystals* **2023**, *13*, 746. [CrossRef]
57. Li, F.; Xu, H.; Xu, X.; Cang, H.; Xu, J.; Chen, S. Supramolecular salts assembled by melamine and two organic hydroxyl acids: Synthesis, structure, hydrogen bonds, and luminescent property. *CrystEngComm* **2021**, *23*, 2235–2248. [CrossRef]
58. Mitra, M.; Hossain, A.; Manna, P.; Choudhury, S.R.; Kaenket, S.; Helliwell, M.; Bauzá, A.; Frontera, A.; Mukhopadhyay, S. Melamine-mediated self-assembly of a Cu(II)–methylmalonate complex assisted by $\pi+-\pi+$ and anti-electrostatic H-bonding interactions. *J. Coord. Chem.* **2017**, *70*, 463–474. [CrossRef]

59. Araar, H.; Benounis, M.; Direm, A.; Touati, A.; Atailia, S.; Barhoumi, H.; Jaffrezic-Renault, N. A new thin film modified glassy carbon electrode based on melaminium chloride pentachlorocuprate (II) for selective determination of nitrate in water. *Mon. Chem.* **2019**, *150*, 1737–1744. [CrossRef]
60. Hesse, M.; Meier, H.; Zeeh, B. *Spektroskopische Methoden in der Organischen Chemie*; Georg Thieme: Stuttgart, Germany, 2005.
61. Pavia, D.L.; Lampman, G.M.; Kriz, G.S.; Vyvyan, J.A. *Introduction to Spectroscopy*; Cengage Learning: Boston, MA, USA, 2014; pp. 14–101.
62. Lotsch, B.V.; Döblinger, M.; Sehnert, J.; Seyfarth, L.; Senker, J.; Oeckler, O.; Schnick, W. Unmasking melon by a complementary approach employing electron diffraction, solid-state NMR spectroscopy, and theoretical calculations—Structural characterization of a carbon nitride polymer. *Chem.-Eur. J.* **2007**, *13*, 4969–4980. [CrossRef]

Disclaimer/Publisher's Note: The statements, opinions and data contained in all publications are solely those of the individual author(s) and contributor(s) and not of MDPI and/or the editor(s). MDPI and/or the editor(s) disclaim responsibility for any injury to people or property resulting from any ideas, methods, instructions or products referred to in the content.

Article

Syntheses, Crystal and Electronic Structures of Rhodium and Iridium Pyridine Di-Imine Complexes with O- and S-Donor Ligands: (Hydroxido, Methoxido and Thiolato)

Michel Stephan, Max Völker, Matthias Schreyer and Peter Burger *

Institute of Inorganic and Applied Chemistry, Department of Chemistry, University of Hamburg, Martin-Luther-King-Platz 6, 20146 Hamburg, Germany; michel.stephan@uni-hamburg.de (M.S.); max.johannes.voelker@uni-hamburg.de (M.V.); matthias.schreyer@gmx.de (M.S.)
* Correspondence: burger@chemie.uni-hamburg.de; Tel.: +49-40-42838-3662

Abstract: The syntheses of new neutral square-planar pyridine di-imine rhodium and iridium complexes with O- and S-donor (OH, OR, SH, SMe and SPh) ligands along with analogous cationic compounds are reported. Their crystal and electronic structures are investigated in detail with a focus on the non-innocence/innocence of the PDI ligand. The oxidation states of the metal centers were analyzed by a variety of experimental (XPS and XAS) and theoretical (LOBA, EOS and OSLO) methods. The $d\pi$-$p\pi$ interaction between the metal centers and the π-donor ligands was investigated by theoretical methods and revealed the partial multiple-bond character of the M-O,S bonds. Experimental support is provided by a sizable barrier for the rotation about the Ir-S bond in the methyl thiolato complex and confirmed by DFT and LNO-CCSD(T) calculations. This was corroborated by the high Ir-O and Ir-S bond dissociation enthalpies calculated at the PNO-CCSD(T) level.

Keywords: metal–ligand $d\pi$-$p\pi$-interaction; Ir-O,S bond dissociation enthalpies; DFT; P/LNO-CCSD(T) calculations; oxidation state analysis; metal–ligand charge transfer; non-innocent ligand; pyridine di-imine ligand

Citation: Stephan, M.; Völker, M.; Schreyer, M.; Burger, P. Syntheses, Crystal and Electronic Structures of Rhodium and Iridium Pyridine Di-Imine Complexes with O- and S-Donor Ligands: (Hydroxido, Methoxido and Thiolato). *Chemistry* 2023, 5, 1961–1989. https://doi.org/10.3390/chemistry5030133

Academic Editor: Spyros P. Perlepes

Received: 19 July 2023
Revised: 23 August 2023
Accepted: 1 September 2023
Published: 5 September 2023

Copyright: © 2023 by the authors. Licensee MDPI, Basel, Switzerland. This article is an open access article distributed under the terms and conditions of the Creative Commons Attribution (CC BY) license (https://creativecommons.org/licenses/by/4.0/).

1. Foreword

Contributing to the special issue commemorating the 150-year legacy of Justus von Liebig is a great honor for us. Liebig's contributions to all fields of chemistry established him as a true general chemist. Although not as widely known, Liebig had a remarkable friendship and collaboration with Friedrich Wöhler, and they shared a common interest in inorganic chemistry [1]. Together with Auguste Laurent and Friedrich Wöhler, Liebig developed the "Radikaltheorie" to explain the structure of organic molecules [2], and his invention of elemental analysis made it possible to determine their constitution [3].

The correspondence author of this publication is an inorganic chemist who obtained his PhD under the guidance of Hans-Herbert Brintzinger at the University of Konstanz. Brintzinger himself was a PhD student in Basel under the mentorship of Hans Erlenmeyer, who was part of the Erlenmeyer line of chemists. Emil Erlenmeyer, in turn, received his PhD from Justus Liebig, which instilled a deep respect for Liebig's contributions to the field of chemistry in all of us [4].

In this publication, we will summarize our research on the synthesis, structure, and reactivity of rhodium and iridium pyridine di-imine (PDI) complexes with alkoxide and thiolate ligands. We combine theory and experiment, as exemplified by Liebig, to address the reactivity of these complexes, revealing new insights into their structure and reactivity. We hope that this publication will contribute to the ongoing legacy of Justus von Liebig and inspire further research in the fields of inorganic and general chemistry.

2. Introduction

Low-valent, late-transition-metal complexes with π-donor ligands, e.g., anionic amido, alkoxido, thiolato and fluorido, -NR$_2$, -OR, -SR and -F groups present a special class of compounds. Their study is primarily motivated by their anticipated unique properties and reactivities based on the HSAB principle for these mismatched soft–hard metal–ligand couples. Furthermore, 4-electron-2-orbitaldestabilizing dπ-pπ orbital interactions between occupied d-orbitals on the metal center and ligand-based lone pairs with π-symmetry might lead to a weakening of the bonds and hence higher reactivity. The chemistry of these types of complexes was summarized in several review articles [5–7]; the unique bonding properties and strengths of the M-OR, M-SR and M-F bonds were also addressed by several authors [8–12]. A particular focus was placed on alkoxido ligands with β-hydrogen atoms, e.g., the methoxido group. Initiated by a β-hydrogen elimination step, these complexes were frequently found to be rather unstable [11,13].

Our group investigates square-planar complexes with pyridine di-imine NNN-donors, of which we and others have demonstrated the ability to behave as non-innocent ligands and reasonable π-acceptors [14–20]. Their unique steric and electronic properties enable the stabilization of highly reactive compounds, such as rhodium and iridium methyl and terminal nitrido complexes, capable of intra- and intermolecular C-H, H-H, Si-H and even C-C activation in ferrocene [21–25]. For the synthesis of these methyl and nitrido complexes, using methoxido ligands as starting materials played a crucial role. We noted that the particular thermal stability of the rhodium and iridium OMe unit, which is resistant to β-hydride elimination even at elevated temperatures, was due to push–pull π-interactions between the oxygen π-donor and the PDI π-acceptor [20,26].

In this paper, we report the synthesis, spectroscopic and crystallographic characterization of rhodium and iridium PDI complexes with O- and S-donors. Theoretical methods are employed to provide insight into their electronic structure with a focus on M-O,S dπ-pπ interactions.

3. Materials and Methods

3.1. Syntheses

3.1.1. Synthesis of Ligand 2

Tridentate pyridine di-imine NNN-donor ligands are ubiquitous and employed in main group and transition metal complexes, including both, mono- and dinuclear systems [27–31]. They are commonly prepared by the Schiff-base condensation of a 2,6-diketo or dialdehyde pyridine with two equivalents of a desired aniline or amine derivative [27,28]. The most common precursor is 2,6-diacetyl pyridine, which was also mostly employed by us [27,28]. The ketimine methyl groups in these ligands are rather acidic, and the complexes can be deprotonated under basic conditions [32]. We therefore switched to the corresponding phenyl and 4-t-butyl-phenyl imine-substituted alternatives and selected N-aryl groups with 2,6-di-isopropyl substituent groups to increase the solubility of the complexes in non-polar solvents. The synthesis of 2,6-dibenzoyl pyridine starting material and corresponding PDI ligand is reported in the literature [33]. For the preparation of the 4-t-butyl phenyl analogue, the required diketo pyridine derivative **1** was obtained in analytically pure form at 43% yield from the reaction of 2 equiv. 4-tert-butylphenyllithium [34] with N2,N2,N6,N6-tetramethylpyridine-2,6-dicarboxamide [35] (see Supplementary Materials). Condensation with 2,6-di-isopropyl aniline under acid-catalyzed Dean–Stark conditions in toluene then provided the new ligand **2** at 78% yield as yellow crystals (details see Supplementary Materials).

3.1.2. Synthesis of the Rhodium and Iridium Chlorido Complexes

For the syntheses of the rhodium and iridium chlorido complexes, we followed our previously established route [20,22,26] by the reaction of the di-μ-chlorido bridged tetraethylene dimetal precursor [(Rh,Ir(μ-Cl(ethylene)$_2$)$_2$] with $\frac{1}{2}$ equiv. of the PDI ligand in THF at RT (Scheme 1). The green rhodium and iridium chlorido complexes **4** and **5** were thus obtained in moderate to good yields (70–90%).

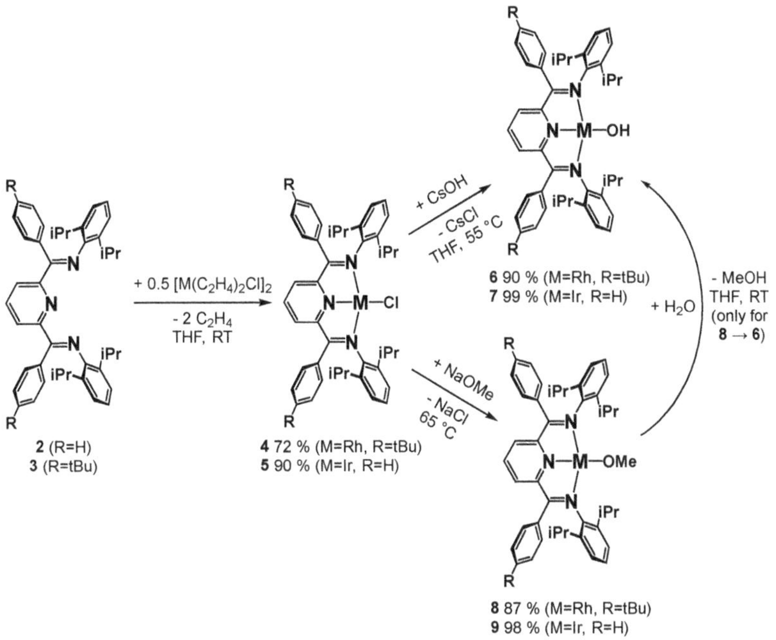

Scheme 1. Synthesis of the chlorido, hydroxido and methoxido complexes.

The observed sharp signals in the ^{1}H and ^{13}C NMR spectra signaled diamagnetic complexes as expected for these d^8-configured square-planar systems. The presence of doublets and triplets for the homotopic meta and para protons of the pyridine ring in the ^{1}H NMR spectra were consistent with the (time-averaged) C_{2v}-symmetry of the chlorido complexes, which was further corroborated by the expected number of ^{13}C NMR resonances (see Supplementary Materials).

3.1.3. Synthesis of the Rhodium and Iridium Hydroxido and Methoxido Complexes

The hydroxido (**6, 7**) and methoxido complexes (**8, 9**) were prepared following our previously established route by salt metathesis of the corresponding chlorido compounds with excess cesium hydroxide or sodium methoxide in THF (Scheme 1) [20,26]. The green hydroxido and methoxido complexes were readily isolated by extraction with toluene due to their higher solubility. Upon crystallization, the products were available in moderate to excellent yields. The methoxido complexes are highly water-sensitive, which can be employed to obtain the hydroxido compounds by reaction with water in quantitative yield (see Supplementary Materials). The constitution of these complexes was unambiguously established by elemental analysis and/or X-ray crystallography. While the X-ray crystal structures revealed C_s-symmetry of the planar N_3-M-O-Me,H core units with bent sp^2-hybridized oxygen atoms, the NMR spectra clearly evidenced time-averaged C_{2v}-symmetry in solution. This is exemplified by the observation of sharp signals for the pyridine para and meta protons, which appeared as a triplet and doublet in a 1:2 integration ratio for complexes **6–9**. Further support was provided by the fully assigned ^{13}C NMR spectra of the hydroxido and methoxido compounds, which revealed three resonances for the ortho, meta and para pyridine carbon atoms and one signal for the carbon atom of the ketimine unit. This is consistent with our previous results for the analogous complexes with methyl-substituted ketimines, for which time-averaged C_{2v}-symmetry was also revealed through

^{1}H and ^{13}C NMR spectra [20,26]. The compounds are readily soluble in aromatic solvents, pentane and ethers, and readily decompose in dichloromethane to chlorido complexes.

3.1.4. Synthesis of Cationic Rhodium and Iridium PDI Complexes

The cationic complexes **10–13** were prepared in good to excellent yields by protonation of the hydroxido or methoxido complexes **7–9** using acids with non- or weakly coordinating counterions (Scheme 2). Depending on the conditions and starting material, either the cationic aqua (**12**), methanol (**11**) or THF (**10, 13**) square-planar diamagnetic complexes were obtained. The ^{1}H NMR coordination chemical shifts of the latter ligands in solution along with sharp resonances and X-ray crystallography in the solid state revealed that the counterions were not coordinated in any case. In dichloromethane, the cationic complexes were almost instantaneously converted to the neutral chlorido compounds **4** or **5**; in fluorobenzene, however, they were stable and dissolved well. In THF solution, the solvent occasionally polymerizes, which can be prevented by the use of the related cyclic ether, 2-methyl tetrahydrofuran.

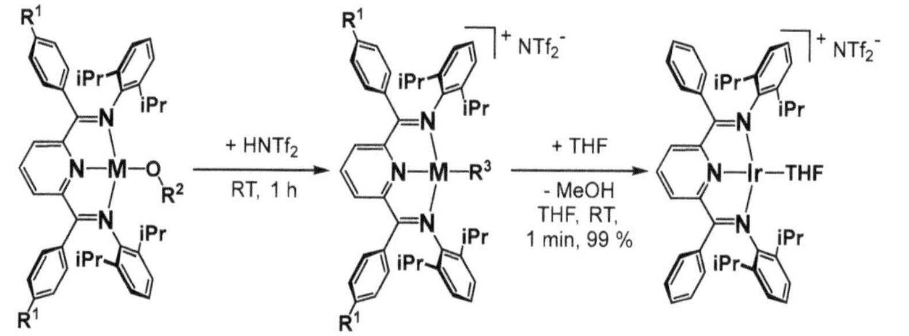

7 (M=Ir, R^1=H, R^2=H)
8 (M=Rh, R^1=tBu, R^2=Me)
9 (M=Ir, R^1=H, R^2=Me)

10 100 % (M=Rh, R$_1$=tBu, R^3=THF)
11 99 % (M=Ir, R^1=H, R$_3$=MeOH)
12 93 % (M=Ir, R^1=H, R^3=H$_2$O)

Scheme 2. Synthesis of the cationic complexes.

3.1.5. Synthesis of the Iridium Thiolato Complexes

For the synthesis of the iridium thiolato compounds **14–16**, we followed the established routes reported in the literature [36,37]. For the methyl and phenyl sulfido compounds **14** and **15**, salt metathesis with NaSR (R = Me, Ph) in the chlorido complexes was employed (Scheme 3). The hydrosulfido compound **16**, PDI-Ir-SH, was obtained by the reaction of a commercial H$_2$S solution in THF with the methoxido complex **9**, which was related to the aforementioned reaction of **8** with the weaker acid water (Scheme 2). Most noticeable is the violet color of these compounds, which deviates strongly from the green (green–brown) color observed for all the other square-planar Rh(I) and Ir(I) PDI complexes previously studied by our group. All of the new sulfido complexes were obtained as diamagnetic crystalline solids in moderate to excellent yields. The proposed constitution based on NMR spectroscopy was unambiguously confirmed by X-ray crystallography (vide infra). As will be discussed below, the observed idealized C$_s$-symmetrical structure of the PDI-Ir-SMe core unit in the crystal structure of **14** can also be witnessed in solution by temperature-dependent ^{1}H NMR spectroscopy (vide infra).

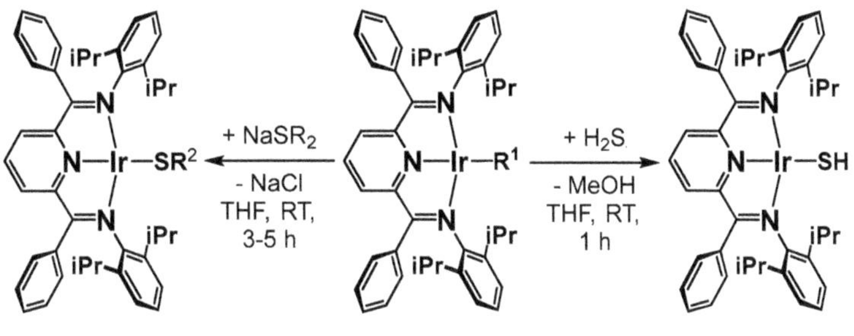

Scheme 3. Synthesis of the methyl, phenyl thiolato and sulfido complexes.

3.1.6. Synthesis of the Iridium Methyl Complex 18

For comparison, we also synthesized and crystallographically characterized the PDI iridium methyl complex **18**, which carries methyl rather than aryl groups at the ketimine carbon atom. This diamagnetic compound was prepared from the chlorido complex by reaction with dimethyl zinc.

3.2. Methods
3.2.1. Theoretical Methods
DFT Calculations

For the geometry optimizations of the ground and transition states, we employed DFT calculations with the PBE functional [38], including dispersion corrections by Grimme's D3 method [39] with Becke–Johnson damping (D3BJ) [40]. Def2-TZVP basis sets were employed for all atoms [41]. For rhodium and iridium, def2-ECP pseudopotentials were used (Ir: ECP-60-MWB, Rh: ECP-28-MWB) [42]. The RI-DFT method [43] was used with the corresponding RIJ-auxiliary basis [44]. For the PW6B95 hybrid functional [45], semi-numeric exchange ($senex keyword in Turbomole) was employed [46]. Solvation effects were included within the COSMO formalism [47] using a dielectric constant of $\varepsilon = 7.6$ for THF. The geometries were fully optimized without geometry or symmetry constraints; minima were confirmed by the absence of imaginary frequencies in the calculations of the analytic second derivatives; for the transition states, only one imaginary frequency was observed. Transition state optimizations were carried out with Kästners DL-FIND optimizer [48] implemented in TCL-Chemshell 3.7 [49] starting from the transition state geometries obtained from linear transit searches. IRC calculations were carried out to confirm that the transition states connected the starting material and product. The calculations were carried out with version 7.7.1 of the parallelized Turbomole program package [50] on our local 32-core and 96-core machines equipped with 512 GB and 3 TB RAM, respectively, and the two 40-core nodes (1 TB RAM) of the "Hummel" computing cluster of the University of Hamburg computing center (RRZ) employing fast SSD/NVME scratch disk space.

Local Coupled Cluster Calculations

For the closed-shell systems, local natural orbital LNO-CCSD(T) calculations [51] were carried out with the freely available MRCC (2022) program package using the default thresholds (lcorthr = normal) [52]. Geometries optimized at the PBE-D3BJ/def2-TZVP level and def2-TZVPP basis sets in combination with complementary def2-TZVPP/C auxiliary correlation basis [53,54] sets and def2-TZVP pseudopotentials [42] were employed. For reference, density-fitted Hartree–Fock calculations def2-QZVPP/JK auxiliary basis were used [55]. The solvation correction was obtained from the energy differences of two single-point calculations at the PBE-D3BJ(COSMO($\varepsilon = 7.6$)/def2-TZVP) and PBE-D3BJ/def2-TZVP (gas phase) levels. Back-corrections for the LNO-CCSD(T) energies to free enthalpies (ΔG_{298}) were carried out with thermochemical data obtained from the DFT calculations at the PBE-D3BJ/def2-TZVP level. A value of 1.011 was taken as the scaling factor from Truhlars database (ver. 5.0) [56]. The values for the T1 and D1 diagnostics were typically T1 ≈ 0.015 and D1 ≈ 0.15, signaling single reference cases.

The evaluation of bond dissociation enthalpies requires local couple cluster calculations of open-shell (S = 1/2) systems. Since this feature is currently not available in MRCC, we switched to PNO-U(R)-CCSD(T1) calculations [57] with Molpro version 2022.3 with the domopt = tight setting [58]. The SO-SCI SCF optimization scheme was employed to converge to the ground state of the Hartree–Fock reference wave function. Enthalpy corrections of the thermochemical data were provided by the "freeh" program of the Turbomole package from the analytical second derivatives obtained at the UPBE-D3BJ/def2-TZVP level. A value of 1.011 was taken as the scaling factor from Truhlars database (ver. 5.0). For single atoms (H, Cl), a value of 5/2 RT was used. The default def2-TZVP pseudopotentials were employed for rhodium (ECP-28MWB) and iridium (ECP-60MWB). For all atoms, the corresponding def2-MP2FIT auxiliary density-fitting basis was used [54]. The values for the T1 and D1 diagnostics were typically T1 ≈ 0.015 and D1 ≈ 0.15, signaling single reference cases, which was further corroborated by negligible spin contamination (<S**2> ≈ 0.75) of the Hartree–Fock reference wave functions of the S = 1/2 radicals.

Charge Transfer and Electron Decomposition Analysis

ALMO-EDA & ALMO-CTA: Energy decomposition analysis (EDA) [59] and charge transfer analysis (CTA) [60] are methods for examining interactions within a chemical system involving fragments [59]. It allows for the decomposition of the overall interaction energy into distinct contributions, e.g., permanent electrostatics, Pauli repulsion, dispersion, and charge transfer. EDA and CTA approaches are applied with great success in theoretical chemical investigations of electronic structures. A recent variant is ALMO-EDA [61] implemented in the Qchem program package [62], which builds on absolutely localized molecular orbitals and enables the analysis of covalently bonded fragments. For our study, we were particularly interested in the charge transfer to and from the PDI ligand fragment, which was analyzed with regard to the energetics (in kJ/mol) and transferred electrons (in me$^-$). The most important contributions were identified by the corresponding localized orbitals called complementary occupied virtual pairs (COVPs). ALMO-EDA/CTA calculations were performed with Qchem 6.0 for converged WB97X-D [63] Kohn–Sham wave functions (def2-TZVP basis and def2-ECP pseudopotentials for rhodium and iridium).

QTAIM analysis: The topology analysis method, a component of quantum theory of atoms in molecules (QTAIM), studies electron density gradients (ρ) with a focus on critical points (CPs), where these gradients are zero ($\nabla \rho(CP) = 0$) [64]. CPs are further characterized by negative eigenvalues ($\lambda_1, \lambda_2, \lambda_3$) of the Hessian matrix. We explored bond critical points (termed (3, −1)) with two negative eigenvalues, located on bond paths connecting local maxima in all directions ((3, −3) critical points). The latter aligned closely with nuclear positions. The size of $\nabla \rho(BCP)$ and the sign of $\nabla^2 \rho(BCP)$ distinguish interaction types (covalent, ionic, and van der Waals). Bond ellipticity (ε), defined as $\varepsilon = \lambda_1/\lambda_2 - 1$ ($|\lambda_1| \geq |\lambda_2|$), reveals bonding types. For $\varepsilon = 0$, like cylindrically symmetrical distributions, examples include the single bond in ethane and the triple bond in acetylene. ε serves as a measure for π double bonding, reaching 0.45 for ethylene. QTAIM defines atom basins, within which electronic density integration provides the atomic charge. For the charge transfer analysis, the aggregated charges of the atoms of the PDI ligand fragment were used. The QTAIM calculations were performed with the MultiWfn (ver. 3.8) program package [65]. The required wfx file was obtained from a Turbomole-generated Molden file and subsequent conversion by Molden2aim [66]. Both basis sets with ECPs (def2-TZVP) and the x2c-TZVPall all-electron basis sets in combination with the X2C relativistic all-electron approach were tested and gave essentially identical results.

NBO analysis: Natural bond analysis (NBO) is a computational method to analyze the electronic structure and chemical bonding within molecules [67]. It focuses on identifying natural orbitals and Lewis-like orbitals by computing bonding orbitals characterized by the highest electron density. NBO helps reveal the nature of chemical bonds, lone pairs, and charge transfer interactions, providing valuable insights into a molecule's electronic characteristics and bonding patterns. The NBO 7.0 calculations [68] were performed with a FILE47 input file generated by an Orca 5.04 [69] DFT PBE-D3BJ/def2-TZVP single-point calculation using geometries optimized by Turbomole. Molecular orbital plots (NBO and ALMO-EDA(COVP)) were generated from cube files with Chemcraft ver. 1.80 [70].

Oxidation State Analysis

For the determination of the (non-physical observable) oxidation states of the metal centers, three different theoretical methods—(i) the localized orbital bond analysis (LOBA) [71], (ii) the effective oxidation states (EOS) [72,73] and (iii) the oxidation state localized orbitals (OSLO) analyses—were employed [74]. The LOBA method was previously promoted by Head-Gordon et al., but displays deficiencies for systems with strong electron delocalization [71]. It was superseded by the OSLO methodology by the same group, which helped to solve difficult cases [74]. The OSLO method was recently benchmarked with the effective oxidation state (EOS) analysis [73] developed by Salvador et al., which indicated that both are robust methods to assign oxidation states based on the defined fragments [74]. For the oxidation state analyses, Head-Gordon's WB97X-D functional was employed using either Qchem 6.018 [62] or Gaussian 16 rev. C01 19 [75] with tight DFT grids (def2-TZVP basis and def2-ECP pseudopotentials for rhodium and iridium). Oxidation states based on the LOBA and OSLO method were calculated with a prelease version of Q-Chem 6.1 with either Loewdin or Mulliken charges. The APOST-3D program of Pedro Salvador Sedano [76] was used to perform the calculations of the oxidation states by the EOS method using a Gaussian 16 formatted checkpoint file. QTAIM and TFVC charges were employed.

Bond Order Analysis

The AOMIX program package ver. 6.92 was utilized to decompose the Wiberg/Mayer bond orders into their symmetry components [77–79]. The required Molden input file was prepared by Turbomole with the tm2aomix program for a PBE-D3BJ/def2-TZVP(def2-ECP) calculation of an optimized geometry at that level.

Wieghardt's (Geometric) Analysis of Bonding Metrics

Based on the analysis of a large number of X-ray crystal structures of main group and transition metal PDI complexes, Wieghardt et al. devised a diagnostic tool to assign the reduction state of the PDI ligand, ranging from neutral PDI to PDI^{-4}, based on its bonding metrics [16]. For this method, the C-N and C-C distances in the pyridine and ketimine units ($r(C_{py}-N_{py}$, $r(C_{im}-N_{im})$ and the exocyclic C-C bond ($rC_{py}-C_{im}$) between the pyridine and ketimine carbon atoms are condensed into a single parameter, $\Delta_{geo} = r_{avg}(C_{py}-C_{im}) - (r_{avg}(C_{py}-N_{py}) + r_{avg}(C_{im}-N_{im}))/2$ (avg: averaged). Combining the analyzed crystallographic data with other spectroscopic or theoretical findings, Wieghardt et al. defined ranges for Δ_{geo} corresponding to the reduction state of the PDI ligand. This method relies on the fact that the occupation of π* orbitals in the reduced PDI ligand leads to shortening of the exocyclic C-C bonds and lengthening of the C-N bonds in the pyridine and ketimine groups. It should be noted that Wieghardt primarily utilized 3d metal systems to calibrate the Δ_{geo} parameter. In comparison to their 4d and 5d congeners, 3d transition metal complexes exhibit a significantly larger variation in the bonding parameter for reduced PDI ligands.

Local Vibrational Mode Analysis

Cremer and Kraka et al. developed the local mode vibrational theory, which allows for the derivation of local vibrational force constants based on the Hessian matrix of a QM calculation [80–82]. The calculations were performed with the LMODEA(F90) program kindly provided by Elfi Kraka [83]. The required Hessians (and dipole gradients) were obtained from Turbomole (aoforce) calculations of the analytical second derivatives.

4. Results and Discussion

4.1. X-ray Crystal Structures

We were able to determine X-ray crystal structures for most of the compounds reported herein. The most important structural parameters are compiled in Table 1, which includes reference data for related square-planar complexes for comparison. The molecular structures are presented in Figure 1.

Table 1. Selected averaged bond distances and angles determined by X-ray crystallography with DFT-optimized values (PBE-D3BJ/def2-TZVP, Rh, Ir, def2-ECP) in parentheses and Wieghardt's Δ_{geo} parameter.

Cpd./Parameter	M: Σ of Angles [°]	N_{imine}-M-N_{imine} [°]	M-X-R [°]	r_{avg}((PDI)Ir-X) [Å]	r_{avg} (M-N_{imine}) [Å]	r_{avg}(M-$N_{pyridine}$) [Å]	r_{avg} (C_{im}-N_{im}) [Å]	Δ_{geo} [Å] [a]
IrSPh 15	359.89 (360.10)	157.41 (157.31)	122.38 (124.03)	2.251 (2.252)	2.032 (2.030)	1.914 (1.929)	1.332 (1.349)	0.083 (0.072)
IrSH 16	359.89 (360.0)	157.90 (158.53)	(103.54)	2.269 (2.262)	2.019 (2.010)	1.913 (1.922)	1.338 (1.348)	0.082 (0.078)
IrSMe 14	360.04 (359.99)	157.49 (157.63)	118.76 (118.82)	2.241 (2.243)	2.017 (2.020)	1.920 (1.931)	1.340 (1.350)	0.078 (0.074)
IrOMe 9	359.95 (360.01)	158.43 (159.21)	130.50 (129.96)	1.968 (1.954)	2.022 (2.014)	1.893 (1.896)	1.337 (1.348)	0.082 (0.071)
^{Me}IrOMe 17	359.98	159.01	134.24	1.951	2.00	1.870	1.349	0.063
RhOMe [d] 8	359.96 (360.01)	158.69 (158.73)	134.01 (130.39)	1.942 (1.949)	2.035 (2.028)	1.889 (1.897)	1.327 (1.338)	0.100 (0.088)
IrOH [b] 7	359.98 (360.0)	159.29 (159.95)	(111.16)	2.027 (1.964)	2.00 (2.005)	1.876 (1.892)	1.335 (1.348)	0.054 (0.074)
RhOH 6	359.99 (359.99)	158.76 (159.64)	(110.47)	2.012 (1.957)	2.010 (2.015)	1.888 (1.892)	1.325 (1.338)	0.104 (0.095)
IrCl [c] 5	359.87 (359.99)	158.61 (159.88)		2.256 (2.309)	2.015 (2.008)	1.910 (1.898)	1.333 (1.343)	0.088 (0.087)
RhCl 4	359.95 (359.99)	157.03 (159.60)		2.303 (2.309)	2.075 (2.017)	1.883 (1.897)	1.319 (1.333)	0.113 (0.102)
IrMe 18	359.97 (359.85)	157.54 (157.46)		2.115 (2.086)	2.003 (2.008)	1.933 (1.949)	1.333 (1.340)	0.091 (0.088)
Ir(MeOH)$^+$ 11	359.96 (359.99)	159.56 (160.30)		2.097 (2.130)	2.023 (2.012)	1.891 (1.884)	1.338 (1.336)	0.081 (0.113)
Ir(H$_2$O)$^{+}$ [d] 12	360.00 (360.01)	160.26 (161.05)		2.090 (2.137)	2.026 (2.002)	1.873 (1.879)	1.315 (1.335)	0.132 (0.106)
Ir(THF)$^+$ 13	360.18 (360.04)	158.55 (159.59)		2.111 (2.141)	2.033 (2.026)	1.883 (1.889)	1.321 (1.336)	0.108 (0.099)
Rh(THF)$^+$ 10	360.14 (360.02)	158.62 (159.3)		2.181 (2.156)	2.026 (2.036)	1.901 (1.887)	1.308 (1.328)	0.123 (0.111)
IrN 19	360.1 (360.15)	150.86 (149.71)		1.647 (1.699)	2.010 (2.031)	2.003 (2.036)	1.329 (1.347)	0.065 (0.054)
IrNO 20	360.01 (360.01)	156.55 (155.74)	176.36 (179.83)	1.751 (1.783)	1.983 (2.017)	1.886 (1.932)	1.333 (1.354)	0.037 (0.037)
IrNH$_2$ 21	360.0 (360.0)	159.51 (158.57)		1.925 (1.938)	2.002 (2.005)	1.886 (1.904)	1.363 (1.346)	0.036 (0.063)
IrCO$^+$ 22	360.0 (360.0)	157.08 (156.82)	180.0 (180.0)	1.846 (1.873)	2.022 (2.028)	1.980 (1.990)	1.308 (1.321)	0.155 (0.135)

[a] $\Delta_{geo} = r_{avg}(C_{py} - C_{im}) - (r_{avg}(C_{py} - N_{py}) + r_{avg}(C_{im} - N_{im}))/2$; [b] structure of lower quality; [c] disorder not resolved; [d] M and $N_{py,para}$ on special positions (C_2-axis).

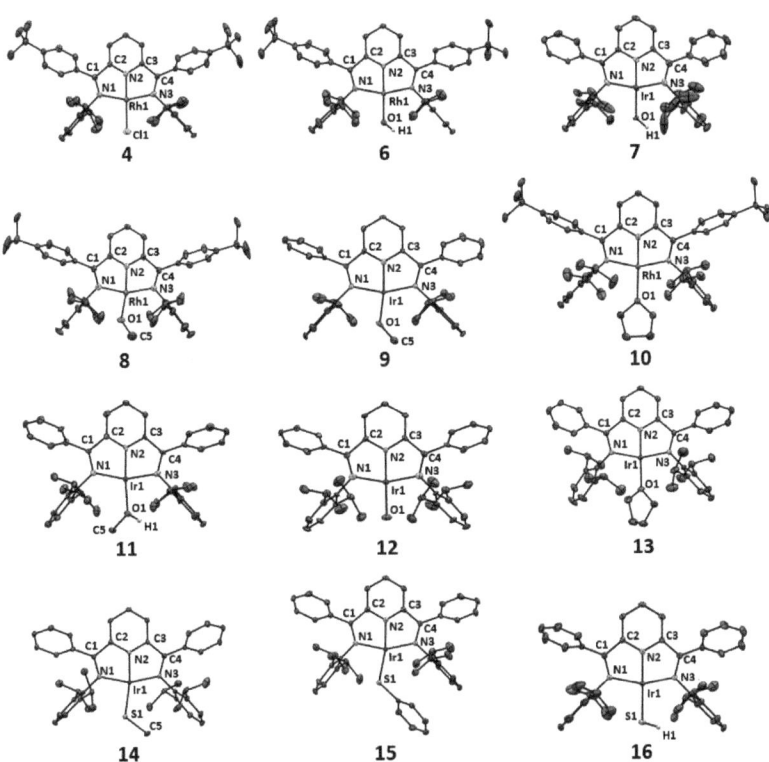

Figure 1. Ortep plots of the complexes reported herein with anisotropic displacement parameters shown at the 50% probability level. Hydrogen atoms except for the OH and SH units, counterions and solvent molecules are omitted for clarity.

The sum of the angles around the metal centers was approximately 360° for all compounds, revealing planar coordination geometries. As previously observed for these types of four-coordinate Rh and Ir PDI complexes, the angles between the trans imine nitrogen atoms and the metal centers fell within a narrow range of 157.4–160.3°, which deviated from the ideal 180° in square-planar geometry due to the short imine C-N double bonds [20]. This deviation has significant implications for the interactions of the d_{xz} and d_{yz} orbitals and the π^*-system of the PDI ligand and is particularly relevant when comparing with PNP–pincer ligands. Notably, the metal–N_{imine} bond distances showed only a negligible variation, with an average of 2.02 ± 0.01 Å. Similarly, the $N_{pyridine}$-metal bond lengths fell within a narrow range of 1.88 to 1.92 Å, despite the variation of the trans influences of the ligands investigated (as compiled in Table 1). The rigidity of these metal PDI geometric parameters was only broken in the carbonyl and nitrido complexes, which exerted the strongest trans influence, leading to Ir-$N_{pyridine}$ distances of 1.980 Å and 2.003 Å. This required a reduction of the N_{imine}-Ir-N_{imine} angle to 150.86° in the nitrido complex to maintain a planar structure. In conclusion, all the complexes studied in this work can be assigned to a pseudo-square-planar coordination geometry. For the sake of clarity, further on we will refer to this geometry as square-planar.

The forthcoming analysis of the electronic structures of the complexes described herein will focus on two aspects: (i) the non-innocence of the PDI ligand and the oxidation states of the metal centers, and (ii) the characterization of the Ir-X bond.

4.2. Electronic Structure of the Complexes

4.2.1. Non-Innocence of the PDI Ligand

The non-innocence of PDI ligands is well-established and was mostly studied for 3d transition metal systems [14–18]. We have previously analyzed the situation of our square-planar 4d and 5d PDI transition metal complexes for a wide variety of additional fourth ligands ranging from strong π-donors to π-acceptors, i.e., nitrido and carbonyl groups [20]. For these investigations, experimental XPS, XAS and ^{13}C-NMR data were combined with theory (DFT and CASSCF) including a localized orbital analysis (LOBA) to assess the oxidation states of the metal centers [19,20,23]. In contrast to their 3d congeners, in most cases the metal PDI interaction in the Rh and Ir complexes is best described by back-donation to the π*-acceptor orbital of the PDI ligand. This can be traced to the better metal–ligand dπ–π* overlap of the larger and more diffuse 4d and 5d orbitals. Nevertheless, it should be noted that the transition between the limiting back-donation and biradical scenarios in non-innocent and innocent PDI ligands is continuous [17]. We identified clear examples with non-innocent, doubly and singly reduced PDI ligands in the nitrido and nitrosyl iridium complexes ((PDI^{2-})Ir(III)(N^-), (PDI^{2-})Ir(I)(NO^+)) and neutral carbonyl and dinitrogen compounds ((PDI^-)Ir(I)(CO), (PDI^-)Rh(I)(N_2)). On the other hand, neutral PDI ligands were established for cationic complexes, while the situation for complexes with N and O π-donors (NRH: R = H, SiR_3 and Ph; OR: R = H, Me) was borderline in terms of non-innocent PDI ligands. In the course of the analysis of the hydrogenolysis experiments described herein, we made one more attempt to investigate the electronic situation of the PDI ligands and included new diagnostic tools based on experimental bonding metrics and theory.

To analyze the innocence or non-innocence of the PDI ligand by theoretical methods, we utilized small parent model complexes instead of those with methyl or aryl substituents of the ketimine units. The Absolutely Localized Molecular Orbitals Energy Decomposition Analysis (ALMO-EDA) in Q-Chem [61,84] was employed to assess the charge transfer between the PDI ligand and the M-X units in the (PDI)M-X complexes. Two fragments were defined: the PDI ligand and the M-X units, and their interaction energies and charge transfer were analyzed [61,84]. A further attempt to determine the charge transfer between the PDI ligand and the IrX unit was made by investigating the aggregated QTAIM charges of the PDI ligand atoms (Table 2). Additionally, the oxidation states of the metal centers were assessed by three different theoretical methods: (i) the localized orbital bond analysis (LOBA) [71], (ii) the effective oxidation states (EOS) [72,73] and (iii) the oxidation state localized orbital (OSLO) analyses [74]. Where available, we also included complementary experimental XPS and XAS data. Furthermore, the Wieghardt et al. diagnostic geometrical parameter Δ_{geo} for the PDI ligands was tabulated for the PDI ligand derived from X-ray crystal structure data along with those from DFT-optimized geometries for the full system in parentheses [16]. To illustrate the variation for the different complexes, we also compiled the experimental (X-ray) and DFT-calculated C-N bond distance between the pyridine and ketimine carbon atoms (rC_{py}-C_{im}). Finally, the local force constants for this bond in the corresponding model complex obtained by local vibrational mode analysis with the LMODEA program package [80] are compiled in Table 2.

Table 2. Analysis of the charge of the PDI ligand, (ALMO-EDA) electron transfer from to the PDI ligand, oxidation states of the metal centers (XPS/XAS, LOBA, EOS, OSLO) and Wieghardt's Δ_{geo}-parameter (theoretical value in parentheses). Δ_{CT} equals $\Delta_{CT} = CT(PDI \rightarrow MX) - CT(PMX \rightarrow PDI)$ (values in charge units ΔQ of 1/1000 of an electron: me$^-$). The non-/innocence of the PDI ligand is indicated by the color-coding (orange: non-innocent; yellow: borderline; green: innocent); for details see text.

	ALMO-EDA		QTAIM		Oxidation State Analysis				Wieghardt Analysis		Local Force
Cpd. M-X/Method	E_{CT}/CT PDI $\rightarrow$ MX [kJ/mol]/ [me$^-$]	E_{CT}/CT MX $\rightarrow$ PDI [kJ/mol] /[me$^-$]	Δ_{CT} [me$^-$]	SPDI Charge [me$^-$]	XPS /XAS	LOBA[c] M	EOS M, X, PDI	OSLO M, X, PDI	Δ_{geo} Parameter [Å]	$r(C_{py}C_{im})$ X-ray (calc.) [Å]	ν $(C_{py}C_{im})$ [mdyn/Å]
IrNO	−383/148	−151/208	−60	−627	+1	+1	+1, +1, −2	+5, −1, −4 *	0.037 (0.037)	1.377 (1.403)	6.103
IrN	−348/132	−128/157	−25	−530	+3	+3	+3, −1, −2 [b]	+3, −1, −2	0.065 (0.054)	1.424 (1.421)	5.758
IrNH$_2$	−357/126	−169/180	−54	−607	+1	+3	+1, −1, 0 [a]	+1, −1, 0 *	0.036 (0.063)	1.392 (1.415)	5.536
IrOMe	−368/132	−153/150	−18	−487		+3	+1, −1, 0	1, −1, 0 *	0.082 (0.071)	1.444 (1.440)	5.420
IrMe	−342/120	−152/154	−34	−469		+1	+1, −1, 0	1, 1, −2 *	0.091 (0.088)	1.445 (1.445)	5.170
IrSMe	−365/135	−135/128	7	−464		+1	+1, −1, 0	1, −1, 0 *	0.078 (0.074)	1.439 (1.441)	5.426
RhOMe	−251/102	−111/102	0	−376		+1	+1, −1, 0	1, −1, 0 *	0.100 (0.088)	1.451 (1.447)	5.300
IrCl	−390/146	−135/117	29	−362	+1	+1	+1, −1, 0	+1, −1, 0 *	0.088 (0.087)	1.438 (1.450)	5.238
RhCl	−273/115	−96/78	37	−253		+1	+1, −1, 0	+1, −1, 0 *	0.113 (0.102)	1.459 (1.456)	5.098
IrTHF$^+$	−403/156	−114/91	65	−8		+1	1, 0, 0	1, 0, 0	0.108 (0.099)	1.456 (1.456)	5.190
RhTHF$^+$	−282/125	−81/60	65	+82		+1	1, 0, 0	1, 0, 0	0.123 (0.111)	1.458 (1.461)	5.075
IrCO$^+$	−406/159	−67/42	117	+302		+1	+1, 0, 0	1, 0, 0	0.155 (0.135)	1.481 (1.472)	4.590

[a] +3, −1, −2 (TFVC scheme); [b] QTAIM scheme; +1, +1, −2 with Mulliken scheme; +1, +1, −2 (L); [c] PBE0 functional; * branching option was used.

4.2.2. ALMO-EDA

The ALMO-EDA was facilitated by the inspection of the complementary occupied/virtual pairs of orbitals (COVPs) displaying the most significant contributions between pairs of the interacting fragments. The COVPs for the ligand → metal and metal → ligand interactions are exemplified for the methyl thiolato model complex shown in Figure 2.

Inspection of the COVPs for the PDI → IrSMe charge transfer shown in Figure 2 reveals the expected σ-donation of the ketimine and pyridine N-donors to the metal center. The corresponding π-back-donation to the PDI ligand was reflected in the IrSMe → PDI COVPs. This interaction was weaker and involved the d_{xz} and d_{yz} orbitals with the approximately identical energies of −48.3 and −45.0 kJ/mol, respectively. It deserves special mention that there was also a weaker back-donation from the d_{xy} orbital to a ligand-based orbital. Energetically, the binding was dominated overall by donation, which can be readily seen from the aggregated charge transfer energies of −365 kJ/mol for E_{CT}(PDI → IrSMe) and −135 kJ/mol for E_{CT}(IrSMe → PDI) (Table 2). This was contrasted by systems with essentially no charge transfer between the fragments, as in the methyl thiolato model complex (PDI → IrSMe: 135 me$^-$ vs. IrS → PDI 128 me$^-$). After this more detailed analysis of the methyl thiolato model complex, we will turn to the data for the other complexes compiled in Table 2.

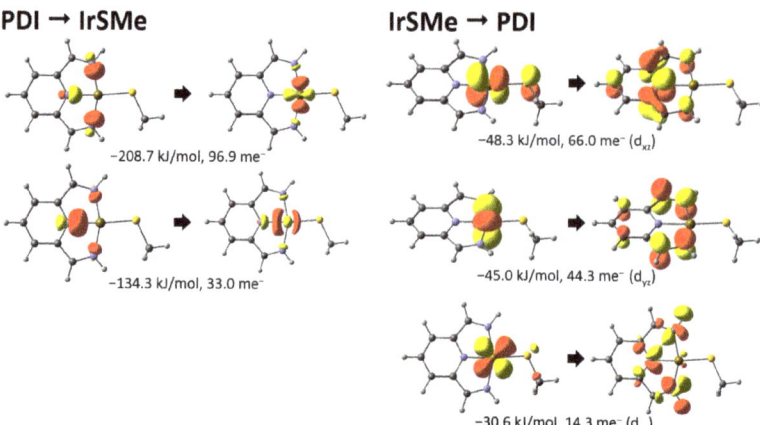

Figure 2. Most important COVPs for the methyl thiolato model complex with energy and charge transfer contributions.

Firstly, the ALMO-EDA data clearly established a substantially weaker donation of the PDI ligands in rhodium complexes (E_{CT}(PDI → RhX) < 300 kJ/mol) compared to the values for the iridium systems, which ranged from ca. 350 to 400 kJ/mol. This was also observed for the metal to PDI π-back-donation and is readily explained by stronger bonding in the heavier 5d iridium metal centers due to relativistic effects; it is also noted for the M-X bond strengths discussed below (Table 3).

With regard to the charge transfer, the PDI → MX donation showed only a small variation (Ir: 120–159 me$^-$; Rh: 100–125 me$^-$); as expected, smaller values were observed for better donors X and larger ones for the cationic systems. This is in contrast to the charge transfer related to back-donation (MX → PDI), which displayed a strong dependence on the MX unit, ranging from 208 me$^-$ for the nitrosyl ligand to 42 me$^-$ for the IrCO$^+$ fragment in the cationic carbonyl complex.

Overall, the direction of charge transfer was best analyzed by the difference Δ_{CT} = CT(PDI → MX) − CT(MX → PDI). For the strong donor ligands, X = NO, NH$_2$, N, Me and OMe; negative values for Δ_{CT} indicate a net charge transfer in the direction of the PDI ligand in the iridium systems. The values of Δ_{CT} = 7 me$^-$ and 0 me$^-$ for the iridium methyl thiolato and rhodium methoxido complex (Δ_{CT} = 0 me$^-$), on the other hand, signal a balanced charged transfer between the PDI and M-SMe,OMe fragments. Finally, the positive values for Δ_{CT} for the residual chlorido and cationic complexes revealed a charge depletion of the PDI ligands with regard to the MX units. This trend of Δ_{CT} was also reflected in the aggregated QTAIM charges of the PDI ligand, which were most strongly negative in the iridium nitrosyl system (-627 me$^-$) and even increased to positive values for the cationic systems, i.e., $+82$ and $+302$ me$^-$ for the rhodium THF and iridium carbonyl complexes. It seems obvious to correlate strongly negative values for Δ_{CT} and the aggregated PDI ligand charge with the non-innocence of the PDI ligand; accordingly, clear-cut examples were the amido and nitrosyl complexes.

Table 3. Binding mode/bond analysis of M-O/S-bonds in (PDI)M-X-R complexes (M = Rh, Ir; X = O, S; R = H, Me, Ph).

Property/Bond/Complex			Rh-OMe	Ir-OH	Ir-OMe	Ir-SH	Ir-SMe	Ir-SPh	Ir-NH$_2$
			Full Systems						
Bond distance (M-X) [Å]		exp.[a]	1.942	2.027	1.968	2.269	2.241	2.251	1.926
		DFT[b]	1.949	1.964	1.954	2.262	2.243	2.252	1.929
Bond angle (M-X-R) [°]		exp.[a]	134.01	n/a[c]	130.37	n/a[c]	118.76	122.38	
		DFT[b]	130.39	111.16	130.27	103.54	118.82	124.03	
Bond Order Analysis		Wiberg BO	0.7179	0.8547	0.7679	1.0574	1.1407	1.0702	
		Wiberg/Löwdin BO	0.967	1.0615	0.9953	1.3923	1.4068	1.3636	
		Fuzzy BO	1.442	1.4573	1.4115	1.5550	1.5938	1.6791	
QTAIM Analysis		ρ at BCP(M-X) [au]	0.1228	0.1358	0.1364	0.1178	0.1225	0.1191	
		$\nabla^2\rho$ BCP(M-X) [au]	0.5853	0.5573	0.5866	0.1878	0.1856	0.1760	
		ellipticity at BCP(M-X)	0.1471	0.2055	0.1819	0.3345	0.1961	0.1551	
Local Force Constant k(M-X) [mdyn/Å] [d]			2.543	2.999	3.000	2.083	2.187	2.015	
Bond Dissociation Energy CCSD(T)/def2-TZVPP [kcal/mol] [e]			69.28	87.90	76.36	79.08	74.36	71.03	
			small model complexes						
			0.729	0.872	0.819	1.098	1.165	1.106	0.975
C$_s$-symmetry: Wiberg Bond Order a' σ (and π d$_{xy}$) a'' π d$_{xz}$			0.50	0.60	0.56	0.78	0.82	0.80	0.62
			0.24	0.26	0.26	0.32	0.34	0.32	0.36
C$_{2v}$-symmetry: Wiberg Bond Order a$_1$ σ									0.56
b$_1$ π d$_{xy}$									0.06
b$_2$ π d$_{xz}$									0.36

[a] det. by X-ray crystallography; [b] geometry-optimized (PBE-DB3J/def2-TZVP/def2-ECP); [c] hydrogen position could not be determined; [d] local force constant. [e] For comparison: BDE((PDI)Rh-Cl): 87.77; BDE((PDI)Ir-Cl): 91.15; BDE((PDI)Ir-Me): 55.64; all values in kcal/mol.

4.2.3. Oxidation State Analysis

In order to further judge this assumption, we performed an analysis of the oxidation state of the metal centers and ligands using the LOBA, EOS and OSLO methodologies. For some of the compounds described herein, we previously employed a LOBA for the metal centers, which was complemented by an oxidation state assignment based on experimental XPS and XAS data (Table 2). For our study, we employed the PDI ligand; metal centers and the residual fourth ligand X were defined as fragments. The results are compiled in Table 2. For the OSLO method, the branching option in the QChem program package was employed for borderline cases. For the EOS analysis, different population schemes, e.g., Mulliken and TFVC, were tested, which occasionally provided different results, as denoted in Table 2.

We color-coded Table 2 to indicate the oxidation state of the metal centers and the reduction state of the PDI ligand. A green color indicates an Rh(I) and Ir(I) oxidation state and innocent neutral PDI ligands. This correlates well with the results of the ALMO-EDA and is the case for all complexes with a positive value for Δ_{CT}, i.e., for all systems with a net charge flow from the PDI ligand to the MX units. For the iridium chlorido complex, this assignment agrees with the results of the XPS measurements. Note that the methyl thiolato complex and rhodium methoxido complexes PDI-Rh-OMe belong to this group.

The Ir methoxido compound, PDI-Ir-OMe, on the other hand, is part of the borderline cases, which are color-coded in yellow. The latter also includes the methyl and amido iridium complex PDI-Ir-Me,NH$_2$, for which all negative values were calculated for Δ_{CT}. Both the OSLO and EOS methods (with the QTAIM scheme) indicated iridium(I) oxidation states and neutral PDI ligands for these systems. For the amido complex, this assignment agrees with our findings based on the experimental XAS and XPS data. As previously noted by Head-Gordon et al., the LOBA method seemed to be less reliable for the delocalized electronic structures of the PDI systems and provided oxidation states of +III for the methoxido and amido compound [74]. It deserves special mention that a doubly reduced PDI^{2-} ligand was derived by the OSLO method, which required a cationic methyl group to form a neutral complex ([(PDI^{2-})(Ir^{+1})(Me^{1+})]). Furthermore, different oxidation states were obtained for the amido system, when the EOS (TFVC) scheme was employed. In this case, the complex was described with an Ir(III) center, and negatively charged NH$_2^-$ and PDI^{2-} ligands, i.e., an [(PDI^{2-})(Ir^{+3})(NH$_2^{1-}$)] electronic structure. This description comes close to the results of the ALMO-EDA, which displayed the largest amount of charge transfer to the PDI ligand (CT(IrNH$_2 \to$ PDI) = 180 me$^-$) and the overall strongest interaction energy E$_{CT}$(IrNH$_2 \to$ PDI) = 169 kJ/mol, as well as the second largest aggregated QTAIM charge of the PDI ligand (627 me$^-$). Overall, it should be emphasized that the concept of oxidation states definitely reaches its limit for these highly delocalized systems.

The final group highlighted in orange in Table 2 consists of the nitrosyl and nitrido complexes (X = NO and N), The OSLO and EOS analysis clearly revealed reduced non-innocent PDI ligands for all these systems. With the EOS method, doubly reduced PDI^{2-} ligands and Ir(I) and Ir(III) metal centers were derived for the nitrosyl and nitrido complexes, which is in agreement with previous assignments based on XPS and XAS measurements. For the NO and nitrido ligands, a positive NO$^+$ and negative N$^-$ charge was calculated, which combine to form (PDI^{2-})(NO^{1+})Ir$^{(1+)}$ and (PDI^{2-})(N^{1-})Ir$^{(3+)}$. This is in contrast to the results of the OSLO analysis, which revealed a negative nitrosyl group (NO$^-$) and a quadruply reduced PDI^{-4} ligand, i.e., (PDI^{4-})(NO^{1-})Ir$^{(5+)}$, in contradiction to the XPS data. Overall, these results are consistent with the results of the ALMO-EDA and the total QTAIM charges for the PDI ligand, which showed a large accumulation of charge on the latter.

4.2.4. Bonding Metrics and Local Mode Vibrational Analysis

Furthermore, we considered Wieghardt's diagnostic parameter Δ_{geo} of the PDI ligand, which allows for assigning the reduction state of the PDI ligand, ranging from PDI to PDI^{-4}, based on the value of Δ_{geo}. As the rC$_{py}$-C$_{im}$ bond length is expected to correlate with the bond order, we also looked into the local vibrational force constant of this C-C bond by aid of the LMODEA tool (Table 2) [81–83].

The Wieghardt et al. method is based on crystallographic data. While this geometric approach has great applicability, limitations arise when dealing with poor or distorted crystal structures. We therefore also calculated the Δ_{geo} parameter for DFT-optimized geometries (Tables 1 and 2). In agreement with the results of our analysis presented in Sections 4.2.2 and 4.2.3, small Δ_{geo} parameters in the range of 0.037–0.067 Å were observed for the doubly reduced PDI ligands in the (PDI)IrX complexes for X = NO, N (orange color in Table 2). Consistently, these compounds also displayed short rC$_{py}$C$_{im}$ bond distances, which reached the range of C-C double bonds in the nitrosyl complex (rC$_{py}$C$_{im}$ = 1.377 Å). This was also reflected in the high values for the corresponding local vibrational force constants of these complexes, which lies above 6 mdyn/Å for the nitrosyl compound.

The local vibrational force constant of the rC$_{py}$C$_{im}$ bond for the members of the second group with innocent PDI ligands and strong back-donation (M = Ir, X = NH$_2$, OMe, Me, color encoded in yellow) was also high, indicating a clear double-bond character. Unexpectedly, the experimental value of 0.036 Å for the Δ_{geo} parameter in the iridium amido complex reached the one for the nitrosyl complex. This might be explained by the rather poor X-ray crystal structure of the amido complex, which is supported by a comparison of the Δ_{geo} parameters of 0.037 Å (X = NO) and 0.063 Å (X = NH$_2$) for the DFT-optimized geometries of the full system. Combining the results of all the diagnostic tools, it becomes apparent that the amido system is borderline. The PDI ligand experienced a very strong charge transfer from the IrNH$_2$ unit, but may yet be considered neutral/innocent. The iridium methoxido and methyl complexes displayed larger values of 0.083 and 0.091 Å, consistent with a sizable but reduced charge transfer to the PDI ligand. Overall, these data are therefore consistent with the assignment to the yellow-coded entries in Table 2 for complexes with PDI ligands on the brink of non-innocence.

The last category encoded in green in Table 2 consists of complexes with innocent PDI ligands. This is most clear-cut for the cationic complexes, displaying Δ_{geo} parameters of >0.1 Å for the unambiguously innocent PDI ligands, which was most pronounced for the (PDI)Ir-CO complex with the CO π-acceptor ligand (Δ_{geo} = 0.155 Å). This cationic carbonyl complex also displayed the longest C_{py}-C_{im} bond distance ($rC_{py}C_{im}$ = 1.481 Å) and the smallest local vibrational constant (4.59 mdyn/Å) of all the compounds. On the other end of this group, complexes with SMe and OMe π-donor ligands were found. For the iridium methyl thiolato complex **14**, the values of 0.078 Å and 1.439 Å for Δ_{geo} and $rC_{py}C_{im}$ as well as the corresponding large local vibrational constants of 5.426 and 4.59 mdyn/Å, respectively, reached or surpassed those of the yellow-encoded entries in Table 2. This clearly signals substantial π back-donation to the PDI ligand, which is only partially reflected in the results of the ALMO-EDA (vide supra).

In conclusion, a nearly seamless transition can be observed from the cationic iridium carbonyl complex, characterized by minimal π back-bonding and an innocent, neutral PDI ligand, to the nitrosyl complex featuring a non-innocent, doubly reduced PDI^{2-} ligand.

4.2.5. Characterization of the M-X Bond

In our initial paper on PDI complexes over two decades ago, we emphasized the remarkably short Ir-O bond length of 1.949(4) Å in a methoxido iridium complex [26]. At that time, this bond distance was the shortest among d^8-configured square-planar iridium complexes containing hydroxido, alkoxido, or aryloxido ligands. We hypothesized that this observation could be attributed, at least partially, to the weaker trans influence of the pyridine unit in the PDI complexes compared to related L$_3$Ir-OR (R = H, alkyl, aryl) compounds, which predominantly featured CO or phosphine ligands in the trans position.

One intriguing characteristic of the aforementioned iridium and methoxido PDI complexes is their remarkable thermal stability. When heated in C$_6$D$_6$ for several days, these compounds exhibited minimal decomposition, with primarily unreacted starting material remaining. This starkly contrasted the facile β-hydrogen elimination process observed in square-planar rhodium and iridium complexes in the presence of phosphine donors, as reported in the literature [6,13]. Relevant theoretical and kinetic studies were also reported for nickel and palladium methoxido PCP–pincer systems [85].

DFT calculations showed that the β-hydrogen elimination leading to formaldehyde and the corresponding hydrido complex is highly thermodynamically unfavorable by more than 30 kcal/mol [26]. This exceptional stability of the M-O bond was attributed to favorable π-donation of the O-donor rather than a destabilizing 4-electron 2-orbital destabilization of the occupied d$_{xz}$ orbital and the p-orbital of the oxygen atom, i.e., the well-known dπ-pπ repulsion between electron-rich late-transition-metal centers and π-donor ligands [7,10,12,86]. This is due to an empty PDI π* orbital and results in a rotational barrier for the rotation around the M-O bond. In our previous DFT calculations, we estimated an activation barrier of approximately 12 kcal/mol for this rotation. Despite our attempts, (cf below) we have not yet been able to experimentally confirm this hypothesis.

Hence, when we started to look into the chemistry of the corresponding thiolato complexes, we revisited the analysis of the dπ-pπ interaction including more recent theoretical diagnostics, e.g., NBO [67] and QTAIM [64] analysis. Furthermore, as described below, we were able to present experimental evidence supporting the existence of a rotational barrier in the methyl thiolato complex.

We will begin by comparing the Rh-O, Ir-O and Ir-S bond lengths with values reported in the literature. Regarding the M-O bond distances, it is worth noting that the value of 2.027 Å observed in the iridium hydroxido complex **7** falls outside the narrow range of 1.94–1.97 Å (Table 1). We attribute this deviation to certain problems encountered during the X-ray crystal structure determination, as detailed in the Supplementary Materials. This conclusion is supported by the shorter bond distance of 1.965 Å observed in the previously studied analogous rhodium hydroxido complex **6** and the value of 1.964 Å obtained from the DFT-optimized geometry. By comparing our PDI complexes to the average of 2.05 ± 0.05 Å reported for Rh/Ir-O bonds in the current Cambridge Structural Database (CSD) for square-planar L_3M-OR complexes (M = Rh, Ir, R = H, alkyl, aryl), it becomes clear that the M-OR bonds in our PDI complexes continue to exhibit exceptionally short M-O distances within this molecular class. For further information on the CSD search, please refer to the Supplementary Materials. It is worth mentioning that the Schneider group observed another instance of a short M-O bond length, measuring 1.935 Å, in a (PNP)Ir(III) hydroxido complex [87].

The Ir-S bond distances in our PDI Ir sulfido complexes, as described herein, range from 2.241 Å to 2.269 Å. These values are extremely well-reproduced in the DFT-optimized geometries (Table 3). The analysis of rhodium and iridium sulfur bond distances in square-planar complexes reported in the CSD revealed substantially longer bonds averaging at 2.37 ± 0.04 Å (details see Supplementary Materials). There is only one example reported by Braun et al. for a (POP)Rh-SH pincer complex, which displayed an Rh-S bond distance of 2.286 Å [37], coming close to the very short values observed in the compounds described herein.

Taking into consideration the moderate to weak trans influence of the pyridine ligand [88,89], the remarkably short Ir-S bond distances in the PDI complexes suggest the presence of strong Ir-S bonds with a multiple-ligand bond character. This was confirmed by calculating the Wiberg and Fuzzy bond orders (BOs), which were found to be greater than 1 in the C_s-symmetrical model complexes listed in Table 3. To further analyze the bond orders, the AOMIX program package was utilized to decompose the Wiberg/Mayer bond orders into their symmetry components for the corresponding small model complexes [79].

In C_s-symmetry, the contributions of orbitals with a″ symmetry can be clearly assigned to Ir-S π-bonding interactions involving the iridium d_{xz} and sulfur p_z orbitals. On the other hand, MOs with a′ symmetry are related to a combination of σ- and π-bonding involving the $d_{x^2-y^2}/d_{z^2}$ and d_{xy} orbitals, respectively (using D_{4h}-symmetry notation). Examining Table 3, it is evident that π-bonding plays a significant role, contributing values exceeding 0.3 to the Wiberg/Mayer bond order, which corresponds to approximately 40%. For comparison, a similar analysis was performed for the previously isolated C_{2v}-symmetrical PDI iridium amido complex [22], allowing for a further decomposition of contributions into their σ- and π-components. This calculation revealed a comparable contribution of the MOs with d_{xz} participation to the bond order, while those involving d_{xy} orbitals exhibited a substantial reduction. Additionally, while smaller Mayer bond orders were calculated for the M-O bonds, partial π-bonding was also identified through the symmetry decomposition in the metal methoxido and hydroxido complexes (Table 3).

The π-character of the M-O,S bonds was confirmed by the results of the QTAIM (Quantum Theory of Atoms in Molecules) analysis [64], which identified bond critical points (BCP(3, −1)) for all M-O and M-S bonds. The corresponding bond ellipticities deviated significantly from zero, indicating non-rotational symmetry of the electron density at the BCP and thus confirming the presence of π-bonding [64]. The electron densities at the bond critical points of the M-O and M-S bonds were substantial (ranging from 0.118 to 0.136 e/Å^3). Combined with the positive values for the Laplacian $\nabla \rho^2$, this suggests an ionic character for the M-X bonds [90], which aligns with the electronegativity differences between the metal centers and the chalcogenides (EN (Pauling scale): Rh: 2.28; Ir: 2.20; O: 3.44; S: 2.58).

We have also calculated the bond dissociation energies at the PNO-CCSD(T)/def2-TZVPP//PBE-D3BJ/TZVP level of theory. In this regard, a previous work by Ess et al. focusing on the binding of late transition metals (Ru, Rh, Ir, Pt) to heteroatoms deserves special mention [8]. They employed DFT and canonical CCSD(T) calculations with a small basis set (6–31 G(d,p)) due to computational and methodological limitations at that time. However, their study successfully reproduced the experimentally established correlation between M-X and H-X bond energies [91]. The values of the bond dissociation energies for the (PDI)M-O and (PDI)M-S ranged from ca. 70–88 kcal/mol, suggesting rather strong bonds (Table 3). As expected, the M-O bond in the methoxido compounds was ca. 7 kcal/mol higher in the iridium complex compared to the 4d rhodium congener, which we attribute to relativistic effects. Furthermore, in agreement with previously established correlations for H-X with M-X bond dissociation enthalpies [91] (e.g., BDE(H-OH) = 118.9 kcal/mol, BDE(H-SH) = 91.29 kcal/mol), weaker bonds were observed in the thiolato complexes (Ir-OH: 87.9 vs. Ir-SH: 79.08 kcal/mol).

Additional support for the π-bonding nature of the M-O and M-S bonds was provided by NBO (Natural Bond Orbital) analysis of the methoxido and methyl thiolato iridium model compounds. This analysis revealed clear sp^2 hybridization of the oxygen and sulfur atoms. In the case of the methyl thiolato complex, four doubly occupied lone pairs (LPs) were assigned to the d_{xy}, d_{yz}, d_{z^2}, and d_{xz} orbitals of the d^8-configured iridium(I) center (see Supplementary Materials). However, the situation is more complex for the methoxido compound. The leading NBO structure involves only three LPs for the Ir center, representing the d_{xy}, d_{yz}, and d_{z^2} orbitals. The d_{xz} orbital is engaged in a π-bonding interaction with the pyridine nitrogen orbital, with approximately equal contributions from the nitrogen and iridium atoms (Figure 3).

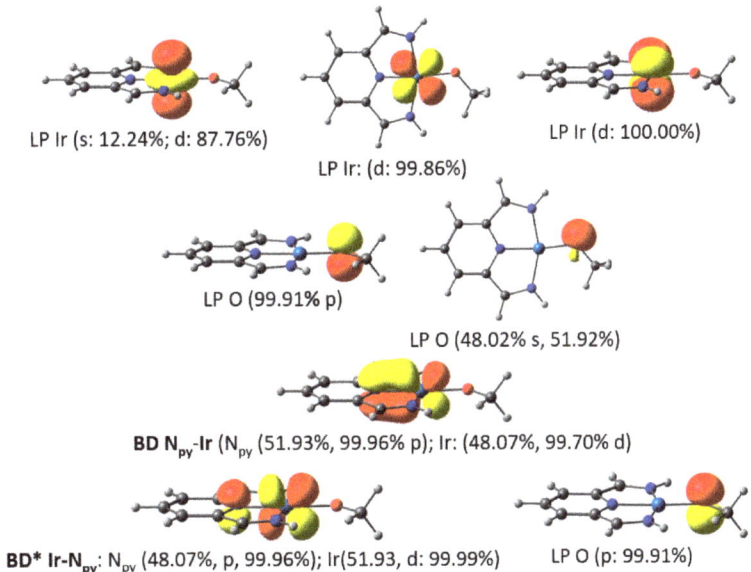

Figure 3. NBOs related to the Ir-O bonding in the iridium methoxido model complex with population numbers.

The NBO analysis further revealed an Ir-O σ-bond and two additional lone pairs of the oxygen atom in the methoxido complex. One of these lone pairs corresponded to the sp^2 orbital located in the square plane, while the other represented a pure p$_z$ orbital. The latter exhibited a strong second-order perturbation interaction of 31.3 kcal/mol and contributed to the antibonding combination of the NBO associated with the aforementioned $N_{pyridine}$-Ir(d_{xz}) π-bond (Figure 3).

4.3. Experimental Evidence for M-X π-Bonding–M-X-R Rotational Barrier

Due to the partial π-bond character of the M-XR (X = O, S) bond, a rotational barrier for the M-X bond could be envisaged. In the solid state, the alkoxido and hydroxido as well as the thiolato compounds display an idealized C_s-symmetry with bent sp²-hybri-dized M-O,S-R units located in the square plane (Table 1 and Figure 1). In solution, however, the vT ¹H and ¹³C NMR spectra are consistent with a time-averaged C_{2v}-symmetry of the complexes for all but the methyl thiolato compound **14**. Of particular diagnostic use in this regard are the protons of the pyridine ring, which exhibit a triplet for the para proton and a doublet for the (time-averaged) homotopic (or enantiotopic) meta protons with an integration ratio of 1:2. For the methyl thiolato complex **14**, we observed temperature-dependent ¹H spectra, with a splitting of the meta pyridine protons into two separate diastereotopic resonances (doublets) at <263 K, consistent with the idealized C_s-symmetrical structure found in the solid state and in the DFT-optimized geometry (Figure 1 and Supplementary Materials). Upon warming, these signals broadened, coalesced at 318 K in the 400 MHz spectrum and turned into a doublet at elevated temperatures (Figure 4 left and Supplementary Materials).

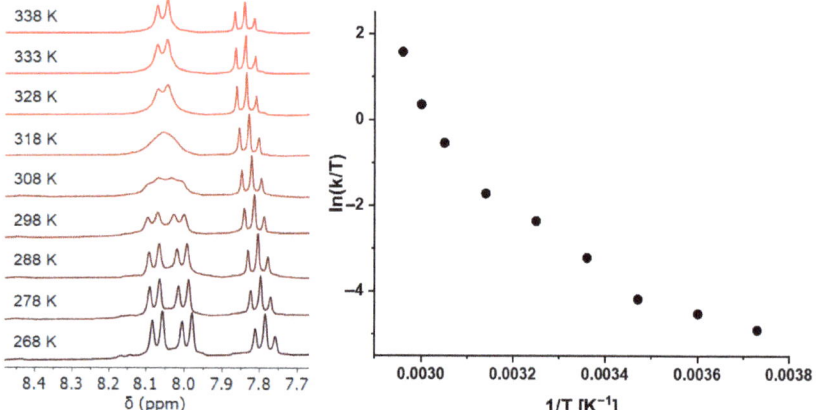

Figure 4. Pyridine section of the vT ¹H NMR spectra of the iridium methyl thiolato complex **14** (**left**) and Eyring plot for the rate constants derived from line shape analysis of the ¹H NMR data (**right**).

This reversible dynamic process was modeled by line shape analysis (LA) of the vT ¹H NMR spectra in toluene in the temperature range of 268–338 K (details see Supplementary Materials). It deserves special mention that the derived barrier at the coalescence temperature of $\Delta G^{\#}$ = 15.7 kcal/mol was consistent with the estimate by the Gutowski–Holm equation, which gave $\Delta G_{318}^{\#}$ = 16.2 kcal/mol. With the fitted rate constants of the LA, we attempted to obtain the activation parameters from an Eyring plot (Figure 4 right).

Much to our surprise, rather than a linear dependence of ln(k/T) on the reciprocal temperature, we clearly obtained a curved line, which is an indication of (at least) two independent processes with different activation parameters [92].

Besides rotation about the M-X bond shown below in Scheme 4 (item a), other processes (items b–e) were considered to explain the observed time-averaged spectra.

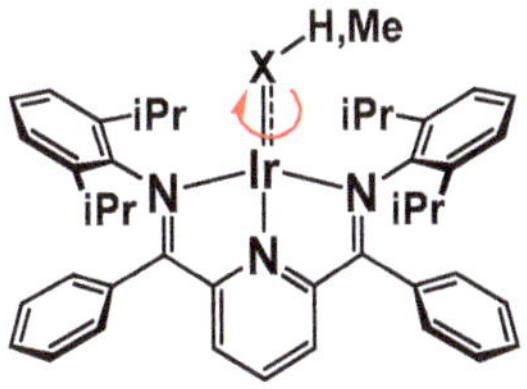

Scheme 4. a. Rotation about the Ir-X bond.

This further includes:

b. R = H, Me, Ph: ionization and recombination of the ions (PDI)M-XR ⇌ (PDI)M$^+$ + XR$^-$
c. R = H: sequence of α-H elimination (IrOH, IrSH) and the microscopic reverse 1,2-H shift (less likely for R = Me and Ph) (PDI)M-X-H ⇌ (PDI)M(H) = X
d. "windshield wiper" process with a C$_{2v}$-symmetrical transition state and sp-hybridized oxygen or sulfur atoms
e. R = Me: β-H elimination, (rotation about of the form/thioaldehyde unit) and reinsertion (PDI)M-X-H ⇌ (PDI)M(H)(η^2-CH$_2$X) (⇌ rotation about (PDI)M(H)-(η^{2}CH$_2$X) ⇌ (PDI)M-X-H)

The values for the rotational barriers about the M-X bonds (item a) were estimated from the corresponding transition states and are compiled in Table 4 for the iridium methoxido and thiolato complexes.

Table 4. Rotational barriers calculated at the DFT and LNO-CCSD(T) levels according to Scheme 4 (item a) and bond orders and distances in the ground and transition states.

Method/Complex	TS(IrOMe)	TS(IrSH)	TS(IrSMe)	TS(IrSMe)THF
ΔE$^\#$ [kcal/mol] DFT(PW6B95-D3BJ(def2-TZVP))/ PBE-D3BJ(def2-TZVP)	9.06	13.70	15.43	15.87
ΔE$^\#$ [kcal/mol] LNO-CCSD(T)(def2-TZVPP)/ PBE-D3B-J(def2-TZVP)	8.76	14.29	15.02	17.22
ΔG$^\#_{298}$ [kcal/mol] LNO-CCSD(T)(def2-TZVPP)/ PBE-D3B-J(def2-TZVP)	8.15	16.08	16.68	18.07
Wiberg Bond Order(GS)/ distance (Ir-X) [Å]	0.768/1.954	1.057/2.264	1.141/2.243	1.141/2.243
Wiberg Bond Order(TS)/ distance (Ir-X) [Å]	0.630/1.980	0.877/2.379	0.884/2.343	0.930/2.325

First of all, the good agreement between the DFT and LNO-CCSD(T) calculations deserves special mention. There was only a very small rotational barrier of ca. 9 kcal/mol for the iridium methoxido complex **9**. The larger activation energies of ca. 14 (SH) and 15 kcal/mol (SMe) were derived for the thiolato complexes **14** and **16**. It is assumed that in contrast to the methyl thiolato complex **14**, the smaller rotational barrier of 14 kcal/mol for the terminal hydrosulfido ligand in **16** may be just slightly too small to manifest itself in the vT ^{1}H NMR spectra. The transition state for the methyl thiolato compound **14** depicted in Figure 5 (left) clearly shows that the S-methyl group is no longer located in the square plane, but rather oriented perpendicular to it. This leads to an idealized C$_s$-symmetry, which gives rise to enantiotopic meta-pyridine protons in the ^{1}H NMR spectrum. Inspection of Table 4 reveals that the Ir-X bonds are elongated in the transition state, which is accompanied by a sizable reduction of the corresponding bond orders. We anticipate that this is a consequence of the missing dπ-pπ interaction in the transition state.

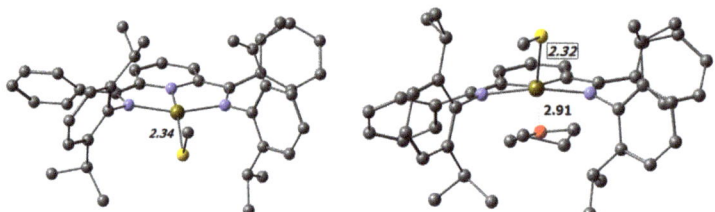

Figure 5. Transition states for rotation about the Ir-S bond. **Left**: without the contribution of the solvent; **Right**: with the contribution of THF. Selected distances are shown in Å. (C: •, N: •, Ir: •, O: •, S: •).

In the search for another process contributing to the non-linear Eyring plot, we identified an additional five-coordinate transition state, which involved coordination of the THF solvent to the iridium center in the transition state (Figure 5 right).

The calculated activation energy of 15.87 kcal/mol (DFT) is only slightly larger than the barrier in the four-coordinate complex (15.43 kcal/mol). Although the energy difference between these two transition states comes out slightly larger in the LNO-CCSD(T) calculation, we anticipate that this process is certainly competitive and might hence serve as a good explanation for the curved Eyring plot.

b. Ionization and recombination

For the ionization process, we calculated the reaction energy ΔE of the full iridium methyl thiolato complex **14** in THF according to Equation (1) at the DFT PBE-D3BJ(def2-QZVPPD)/COSMO (ε = 7.6) level.

$$(PDI)Ir-SMe + THF \rightleftharpoons (PDI)Ir-THF^+ + SMe^-, \Delta E_R = +45.5 \text{ kcal/mol} \quad (1)$$

The COSMO dielectric continuum solvation model might not fully and adequately describe the true picture; therefore, the calculated value of +45.5 kcal/mol has to be considered an upper limit. It should be noted, however, that ionization is uphill by 28.7 kcal/mol even for solvation with water (ε = 80). Therefore, we anticipate that the reaction energy is too endothermic to compete with other processes. This is also in agreement with our previous molecular conductivity measurements of a PDI iridium complex bearing a significantly better ionizable triflato ligand, which is only partially ionized in THF solution [26].

c. α-H Elimination and the microscopic reverse step

The α-H elimination and its reverse process, the [1,2] H-shift, is well-established for alkyl ligands [93], but for oxo and sulfido ligands this reaction is less common [86,94]. We calculated the thermodynamics for the formation of the hydrido, oxo/sulfido complex by α-H elimination in the hydroxido and hydrosulfido complexes. The DFT-optimized geometries of the α-H elimination products revealed slightly distorted square-pyramidal structures, which are shown in Figure 6. It deserves special mention that the Ir-S bond distance was only marginally shortened in the product (2.22 vs. 2.24 Å), while the Ir-O bond length dropped by 0.14 Å.

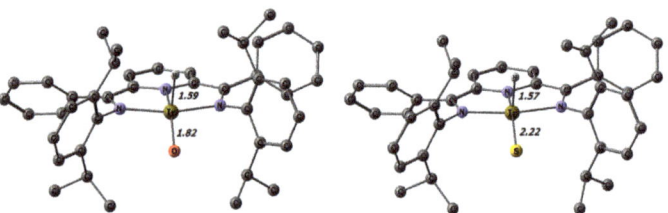

Figure 6. α-H elimination products of **7** and **16** with selected distances in Å.

Both products shown in Figure 6 are substantially energetically uphill from their corresponding hydroxido and hydrosulfido educts. The values of +50.9 kcal/mol for the oxido and +31.4 kcal/mol for the sulfido complex calculated at the DFT level with the PW6B95-D3BJ functional (def2-TZVP basis) clearly rule out that the α-H elimination step plays a role in the observed dynamic behavior.

d. "Windshield" wiper process

The "windshield wiper" process refers to the in-plane movement of the O,S-R bond group (R = H, Me, Ph) from one molecule side to the other. This process was modeled for the unsubstituted methyl thiolato complex; the corresponding Walsh diagram is presented in Figure 7.

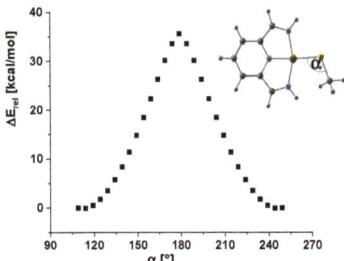

Figure 7. Walsh diagram for the windshield wiper process in the methyl thiolato iridium model complex.

The windshield wiper process exhibited an activation energy of approximately 40 kcal/mol. The transition state adopted a C_{2v}-symmetrical structure, in which the hydrogen atoms of the methyl group were disregarded. For the complete system, an even higher barrier is expected; consequently, it is concluded that this process is not involved in the observed dynamic behavior.

e. β-H Elimination and reinsertion

For the methyl-substituted methoxido and methyl thiolato complexes, ß-hydrogen elimination leading to C_s-symmetrical hydrido η^2-formaldehyde and thioformaldehyde complexes can be envisaged. This process was studied by DFT and LNO-CCSD(T) calculations; the geometry-optimized structures of the transition states are depicted in Figure 8.

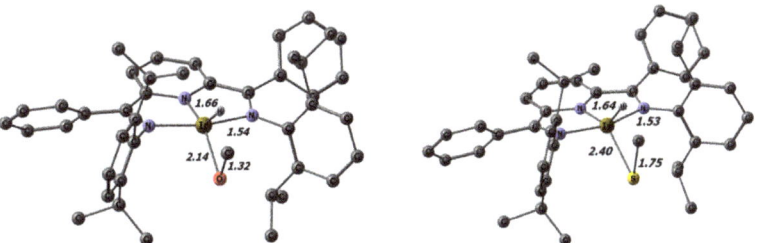

Figure 8. Transition states for ß-hydrogen elimination in the methoxido and methyl thiolato iridium complexes **9** and **14** with selected distances in Å.

The activation energies calculated at the DFT PW6B95-D3BJ/def2-TZVP level for the iridium methoxido and methyl thiolato complexes amounted to 16.7 and 23.4 kcal/mol, respectively. These barriers are approximately 7 kcal/mol higher than those for the rotation about the M-O and M-S bonds. Therefore, the β-hydrogen elimination processes are not considered to be competitive with regard to the observed dynamic behavior. However, it has to be emphasized that these β-hydrogen elimination barriers are quite low. Nevertheless, further elimination of the thio-/aldehyde to yield the hydrido complex was not observed. This contrasts with the results of other late-transition-metal alkoxide complexes reported in the literature. Frequently, these compounds exhibit relatively low thermal stability, which is due to β-hydrogen elimination and the consecutive elimination of the aldehyde or ketone [9,13,85,95,96]. For the PDI complexes, this difference can be rationalized by the fact that extrusion of the aldehyde/thioaldehyde is highly thermodynamically unfavorable (Figure 9). At the LNO-CCSD(T) level, this step is 25.58 (X = O) and 50.05 (X = S) kcal/mol uphill (Table 5).

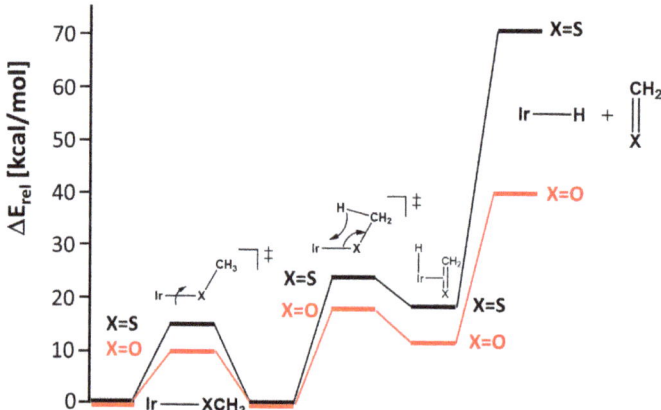

Figure 9. Calculated energy profile at the LNO-CCSD(T)/def2-TZVPP//DFT(PBE-D3B)/def2-TZVP level for the rotation about the Ir-X bond (X = O,S) in complexes **9** and **14,** and ß-hydrogen elimination with consecutive extrusion of thio-/formaldehyde.

Table 5. Energy levels for ß-hydrogen elimination and thio-/formaldehyde extrusion in the iridium methoxido and methyl thiolato complexes **9** and **14**.

STEP/Method	$\Delta E^{\#}_{rel}$ [kcal/mol] DFT(PW6B95-D3BJ (def2-TZVP))		$\Delta E^{\#}$ [kcal/mol] LNO-CCSD(T) (def2-TZVPP)	
	X = O	X = S	X = O	X = S
Complex				
TS(ß-H elimination)	17.91	24.21	18.27	22.97
intermediate	11.82	18.48	11.71	16.28
product	39.94	70.80	37.29	66.33
product–intermed.	28.12	52.32	25.58	50.05

The reverse association process of formaldehyde is essentially barrierless, as can be recognized from the inspection of the linear transit for this process shown in Figure 10.

Finally, we investigated the isomerization pathway from the methoxido to the hydroxymethylene complex according to Equation (2).

$$(PDI)Ir\text{-}OCH_3 \leftrightarrows (PDI)Ir\text{-}CH_2OH \quad (2)$$

This type of transformation was reported by Wayland et al. for a rhodium porphyrin system and was found to be energetically in favor of the hydroxymethylene complex by ca. 6 kcal/mol [97]. We studied this reaction by DFT and LNO-CCSD(T) calculations, which are summarized in Figure 11.

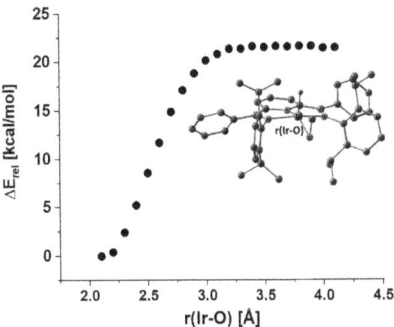

Figure 10. Energy profile for extrusion or association of formaldehyde in the iridium hydrido η²-formaldehyde complex at the r2scan-3c DFT level (def2-mTZVPP basis; ECP: def2-ECP).

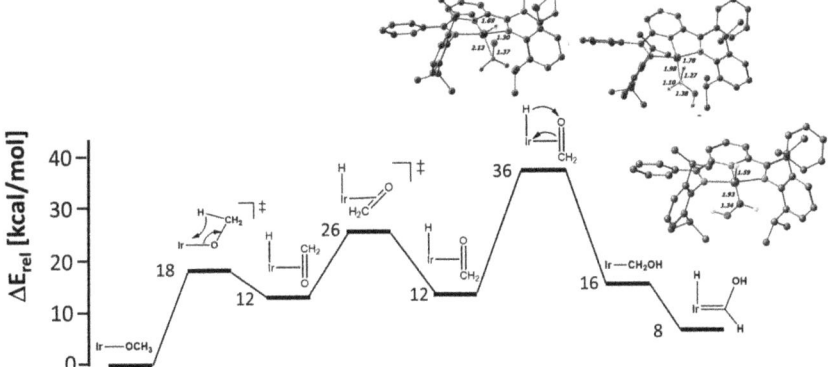

Figure 11. Energy profile at the LNO-CCSD(T)/def2-TZVPP//DFT(PBE-D3BJ/def2-TZVP) level for the transformation of the iridium methoxido complex **9** to the hydroxymethylene and hydroxycarbene congeners. Selected distances in Å are shown.

In contrast to the porphyrin system, the formation of the hydroxymethylene system is significantly uphill for both the Rh (+24.2 kcal/mol) and Ir (18.8 kcal/mol) complexes. We take this as one more hint that the dπ-pπ interaction in the methoxido complexes is stabilizing. We noted two more interesting features in these calculations. First of all, there is an apparent α–H–agostic interaction in the hydroxymethylene complexes, as indicated by the short M···H-CHOH interactions of 1.87 (M = Rh) and 1.78 Å (M = Ir) with concomitant elongated C-H bonds of 1.19 and 1.27 Å (see Figure 11 for M = Ir and ESI for M = Rh). These C-H bonds are on the brink of α-elimination to the corresponding hydrido hydroxycarbene complexes. For iridium, the carbene compound lies only 7.7 kcal/mol above the methoxido complex, while the rhodium carbene system is significantly less stable (+17.7 kcal/mol). As expected for the multiple-bond character in the hydroxycarbene complexes, the M-C bonds are shortened with respect to the hydroxymethylene complexes (Ir: 0.05 Å, Rh: 0.08 Å) (Figure 11 and ESI).

Finally, this section can be concluded by saying that the (PDI)M-O,S-R complexes (M = Rh, Ir, R = H, Me, Ph) display significant thermal stability, which can be attributed to partially stabilizing dπ-pπ interactions. Complexes with methoxido and methyl thiolato ligands are particularly noteworthy. Despite their small barriers for ß-hydrogen elimination, they are stabilized by the thermodynamically highly unfavorable extrusion of the thio-/aldehyde in the consecutive step. Finally, it can therefore be concluded that only the rotational process with or without solvent contribution is involved in the dynamic process.

5. Conclusions

In this article, we analyzed the bonding situation in square-planar PDI rhodium and iridium complexes with a particular focus on O,S π-donors. The PDI ligand serves as an electron acceptor and assists in reducing the destabilizing electron-2 orbital orbital dπ-pπ interactions between the metal centers and the O,S–heteroatom ligands. Consequently, a partial Ir-O,S multiple-bond character can be derived, which is also reflected in the observed sizable Ir-S rotational barrier and the large calculated homolytic BDEs. We are currently in the final stages of our investigation on the hydrogenation of the hydroxido, methoxido and thiolato complexes and will provide a comprehensive report in the near future.

Supplementary Materials: The following supporting information can be downloaded at: https://www.mdpi.com/article/10.3390/chemistry5030133/s1, Figure S1: NMR assignment for the respective protons and carbon atoms; Figure S2: ^{1}H NMR spectra of 2 in dibromomethane-d_2 between 296 K and 353 K; Figure S3: ^{1}H NMR spectra of 14 in toluene-d8 between 268 K and 338 K; Figure S4: ATR-IR spectra of 6 (Rh-OH) and the deutero isotopologue **6D** (RH-OD); Figure S5: Ortep diagram of the molecular structure of **1**. Hydrogen atoms are omitted for clarity, ellipsoids are shown at the 50% probability level; Figure S6: Ortep diagram of the molecular structure of **2**. Hydrogen atoms and solvent molecules are omitted for clarity, ellipsoids are shown at the 50% probability level; Figure S7: Ortep diagram of the molecular structure of **4**. Hydrogen atoms and solvent molecules are omitted for clarity, ellipsoids are shown at the 50% probability level; Figure S8: Ortep diagram of the molecular structure of **6**. Hydrogen atoms with exception of the OH group and solvent molecules are omitted for clarity, ellipsoids are shown at the 50% probability level; Figure S9: Ortep diagram of the molecular structure of **7**. Hydrogen atoms with exception of the OH group and solvent molecules are omitted for clarity, ellipsoids are shown at the 50% probability level; Figure S10: Ortep diagram of the molecular structure of **8**. Hydrogen atoms and solvent molecules are omitted for clarity, ellipsoids are shown at the 50% probability level; Figure S11: Ortep diagram of the molecular structure of **9**. Hydrogen atoms and solvent molecules are omitted for clarity, ellipsoids are shown at the 50% probability level; Figure S12: Ortep diagram of the molecular structure of **10**. Hydrogen atoms, solvent molecules and the counter ion are omitted for clarity, ellipsoids are shown at the 50% probability level; Figure S13: Ortep diagram of the molecular structure of **11**. Hydrogen atoms except of the OH group, solvent molecules and the counter ion are omitted for clarity, ellipsoids are shown at the 50% probability level; Figure S14: Ortep diagram of the molecular structure of **12**. Hydrogen atoms, solvent molecules and the counter ion are omitted for clarity, ellipsoids are shown at the 50% probability level; Figure S15: Ortep diagram of the molecular structure of **13**. Hydrogen atoms, solvent molecules and the counter ion are omitted for clarity, ellipsoids are shown at the 50% probability level; Figure S16: Ortep diagram of the molecular structure of **14**. Hydrogen atoms and solvent molecules are omitted for clarity, ellipsoids are shown at the 50% probability level; Figure S17: Ortep diagram of the molecular structure of **15**. Hydrogen atoms and solvent molecules are omitted for clarity, ellipsoids are shown at the 50% probability level; Figure S18: Ortep diagram of the molecular structure of **16**. Hydrogen atoms with exception of the SH group and solvent molecules are omitted for clarity, ellipsoids are shown at the 50% probability level; Figure S19: Ortep diagram of the molecular structure of **18**. Hydrogen atoms and solvent molecules are omitted for clarity, ellipsoids are shown at the 50% probability level; Table S1: Selected averaged bond distances and angles determined by X-ray crystallography with DFT-optimized values (PBE-D3BJ/def2-TZVP, Rh, Ir, def2-ECP) in parentheses and ^{13}C NMR data related to the PDI unit; Table S2: Summary of the crystal data and structure refinement of **1**, **2** and **4**; Table S3: Summary of the crystal data and structure refinement for the reported complexes **6**, **7** and **8**; Table S4: Summary of the crystal data and structure refinement for the reported complexes **9**, **10** and **11**; Table S5: Summary of the crystal data and structure refinement for the reported complexes **12**, **13** and **14**; Table S6: Summary of the crystal data and structure refinement for the reported complexes **15**, **16** and **18**; Figure S20: Rh,Ir-O Bond Distances and employed structures by CCSD code; Figure S21: Rh,Ir-S bond distances and employed structures by CCSD code; Figure S22: NBOs of (PDI)Ir-SMe model complex: LPs of the Ir and S atoms. References [26,33–35,48,49,51,62,65,68,69,75–78,81–83,98–108] are cited in the Supplementary Materials.

Author Contributions: Conceptualization, P.B.; methodology, P.B. and M.S. (Michel Stephan); validation, M.S. (Michel Stephan), M.V. and P.B.; investigation, P.B., M.S. (Michel Stephan), M.V. and M.S. (Matthias Schreyer); resources, P.B. (University of Hamburg); data curation, M.S. (Michel Stephan) and P.B.; writing—original draft preparation, P.B.; writing—review and editing, P.B., M.S. (Michel Stephan) and M.V. visualization, P.B. and M.S. (Michel Stephan); supervision, P.B.; project administration, P.B.; funding acquisition, P.B. All authors have read and agreed to the published version of the manuscript.

Funding: This research received funding by the University of Hamburg.

Data Availability Statement: The X-ray crystallographic data were submitted to the Cambridge Crystallographic Database and can be accessed via the deposition numbers: CCD 2282018-2282033.

Acknowledgments: We are indebted to Pedro Salvador Sedano for generously sharing the APOST-3D code and for his invaluable assistance in conducting the EOS analysis. We are grateful to QChem for providing a pre-release of QChem rel. 6.1, which allowed us to make full use of the branching option of the OSLO method, and would also like to acknowledge the support and assistance of Abdul Aaldossary. We would like to thank Marc Prosenc for discussions on the NBO analysis and assistance with the X-ray crystal structure refinement of the twinned crystal of complex **9**.

Conflicts of Interest: The authors declare no conflict of interest.

References

1. Esteban, S. Liebig–Wöhler Controversy and the Concept of Isomerism. *J. Chem. Educ.* **2008**, *85*, 1201–1203. [CrossRef]
2. Liebig, J. Ueber Laurent's Theorie der organischen Verbindungen. *Ann. Phar.* **1838**, *25*, 1–31. [CrossRef]
3. Liebig, J. Ueber einen neuen Apparat zur Analyse organischer Körper, und über die Zusammensetzung einiger organischen Substanzen. *Ann. Phys. Chem.* **1831**, *97*, 1–43. [CrossRef]
4. Academic Tree. Available online: https://academictree.org/chemistry/tree.php?pid=747282 (accessed on 26 June 2023).
5. Martínez-Prieto, L.M.; Cámpora, J. Nickel and Palladium Complexes with Reactive σ-Metal-Oxygen Covalent Bonds. *Isr. J. Chem.* **2020**, *60*, 373–393. [CrossRef]
6. Groysman, S.; Grass, A. Alkoxide Ligands. In *Comprehensive Coordination Chemistry III*; Elsevier: Amsterdam, The Netherlands, 2021; pp. 158–177.
7. Caulton, K.G. The Influence of π-Stabilized Unsaturation and Filled/Filled Repulsions in Transition Metal Chemistry. *New J. Chem.* **1994**, *18*, 25–41.
8. Devarajan, D.; Gunnoe, T.B.; Ess, D.H. Theory of late-transition-metal alkyl and heteroatom bonding: Analysis of Pt, Ru, Ir, and Rh complexes. *Inorg. Chem.* **2012**, *51*, 6710–6718. [CrossRef]
9. Yuwen, J.; Jiao, Y.; Brennessel, W.W.; Jones, W.D. Determination of Rhodium-Alkoxide Bond Strengths in Tp'Rh(PMe$_3$)(OR)H. *Inorg. Chem.* **2016**, *55*, 9482–9491. [CrossRef] [PubMed]
10. Holland, P.L.; Andersen, R.A.; Bergman, R.G. Application of the E-CApproach to Understanding the Bond Energies Thermodynamics of Late-Metal Amido, Aryloxo and Alkoxo Complexes: An Alternative to pπ/dπ Repulsion. *Comments Inorg. Chem.* **1999**, *21*, 115–129. [CrossRef]
11. Martinez-Prieto, L.M.; Palma, P.; Alvarez, E.; Campora, J. Nickel Pincer Complexes with Frequent Aliphatic Alkoxo Ligands [((iPr)PCP)Ni-OR] (R = Et, nBu, iPr, 2-hydroxyethyl). An Assessment of the Hydrolytic Stability of Nickel and Palladium Alkoxides. *Inorg. Chem.* **2017**, *56*, 13086–13099. [CrossRef]
12. Ashby, M.T.; Enemark, J.H.; Lichtenberger, D.L. Destabilizing dπ-pπ orbital interactions and the alkylation reactions of iron(II)-thiolate complexes. *Inorg. Chem.* **2002**, *27*, 191–197. [CrossRef]
13. Mann, G.; Hartwig, J.F. Palladium Alkoxides: Potential Intermediacy in Catalytic Amination, Reductive Elimination of Ethers, and Catalytic Etheration. Comments on Alcohol Elimination from Ir(III). *J. Am. Chem. Soc.* **1996**, *118*, 13109–13110. [CrossRef]
14. Bart, S.C.; Chlopek, K.; Bill, E.; Bouwkamp, M.W.; Lobkovsky, E.; Neese, F.; Wieghardt, K.; Chirik, P.J. Electronic structure of bis(imino)pyridine iron dichloride, monochloride, and neutral ligand complexes: A combined structural, spectroscopic, and computational study. *J. Am. Chem. Soc.* **2006**, *128*, 13901–13912. [CrossRef]
15. Knijnenburg, Q.; Gambarotta, S.; Budzelaar, P.H. Ligand-centred reactivity in diiminepyridine complexes. *Dalton Trans.* **2006**, *46*, 5442–5448. [CrossRef]
16. Römelt, C.; Weyhermüller, T.; Wieghardt, K. Structural characteristics of redox-active pyridine-1,6-diimine complexes: Electronic structures and ligand oxidation levels. *Coord. Chem. Rev.* **2019**, *380*, 287–317. [CrossRef]
17. Zhu, D.; Budzelaar, P.H.M. A Measure for σ-Donor and π-Acceptor Properties of Diiminepyridine-Type Ligands. *Organometallics* **2008**, *27*, 2699–2705. [CrossRef]
18. Zhu, D.; Thapa, I.; Korobkov, I.; Gambarotta, S.; Budzelaar, P.H. Redox-active ligands and organic radical chemistry. *Inorg. Chem.* **2011**, *50*, 9879–9887. [CrossRef] [PubMed]
19. Angersbach-Bludau, F.; Schulz, C.; Schöffel, J.; Burger, P. Syntheses and electronic structures of mu-nitrido bridged pyridine, diimine iridium complexes. *Chem. Commun.* **2014**, *50*, 8735–8738. [CrossRef] [PubMed]

20. Sieh, D.; Schlimm, M.; Andernach, L.; Angersbach, F.; Nückel, S.; Schöffel, J.; Šušnjar, N.; Burger, P. Metal–Ligand Electron Transfer in 4d and 5d Group 9 Transition Metal Complexes with Pyridine, Diimine Ligands. *Eur. J. Inorg. Chem.* **2012**, *2012*, 444–462. [CrossRef]
21. Schiller, C.; Sieh, D.; Lindenmaier, N.; Stephan, M.; Junker, N.; Reijerse, E.; Granovsky, A.A.; Burger, P. Cleavage of an Aromatic C–C Bond in Ferrocene by Insertion of an Iridium Nitrido Nitrogen Atom. *J. Am. Chem. Soc.* **2023**, *145*, 11392–11401. [CrossRef]
22. Schöffel, J.; Rogachev, A.Y.; DeBeer George, S.; Burger, P. Isolation and hydrogenation of a complex with a terminal iridium-nitrido bond. *Angew. Chem. Int. Ed.* **2009**, *48*, 4734–4738. [CrossRef] [PubMed]
23. Schöffel, J.; Šušnjar, N.; Nückel, S.; Sieh, D.; Burger, P. 4D vs. 5D—Reactivity and Fate of Terminal Nitrido Complexes of Rhodium and Iridium. *Eur. J. Inorg. Chem.* **2010**, *2010*, 4911–4915. [CrossRef]
24. Sieh, D.; Burger, P. Si-H activation in an iridium nitrido complex—A mechanistic and theoretical study. *J. Am. Chem. Soc.* **2013**, *135*, 3971–3982. [CrossRef]
25. Sieh, D.; Schöffel, J.; Burger, P. Synthesis of a chloro protected iridium nitrido complex. *Dalton Trans.* **2011**, *40*, 9512–9524. [CrossRef] [PubMed]
26. Nückel, S.; Burger, P. Transition Metal Complexes with Sterically Demanding Ligands, 3.1 Synthetic Access to Square-Planar Terdentate Pyridine–Diimine Rhodium(I) and Iridium(I) Methyl Complexes: Successful Detour via Reactive Triflate and Methoxide Complexes. *Organometallics* **2001**, *20*, 4345–4359. [CrossRef]
27. Gibson, V.C.; Redshaw, C.; Solan, G.A. Bis(imino)pyridines: Surprisingly reactive ligands and a gateway to new families of catalysts. *Chem. Rev.* **2007**, *107*, 1745–1776. [CrossRef]
28. Flisak, Z.; Sun, W.-H. Progression of Diiminopyridines: From Single Application to Catalytic Versatility. *ACS Catal.* **2015**, *5*, 4713–4724. [CrossRef]
29. Reinhart, E.D.; Jordan, R.F. Synthesis and Ethylene Reactivity of Dinuclear Iron and Cobalt Complexes Supported by Macrocyclic Bis(pyridine-diimine) Ligands Containing o-Terphenyl Linkers. *Organometallics* **2020**, *39*, 2392–2404. [CrossRef]
30. Stephan, M.; Dammann, W.; Burger, P. Synthesis and reactivity of dinuclear copper(I) pyridine diimine complexes. *Dalton Trans.* **2022**, *51*, 13396–13404. [CrossRef]
31. Dammann, W.; Buban, T.; Schiller, C.; Burger, P. Dinuclear tethered pyridine, diimine complexes. *Dalton Trans.* **2018**, *47*, 12105–12117. [CrossRef]
32. McTavish, S.; Britovsek, G.J.P.; Smit, T.M.; Gibson, V.C.; White, A.J.P.; Williams, D.J. Iron-based ethylene polymerization catalysts supported by bis(imino)pyridine ligands: Derivatization via deprotonation/alkylation at the ketimine methyl position. *J. Mol. Catal. A Chem.* **2007**, *261*, 293–300. [CrossRef]
33. Lukmantara, A.Y.; Kalinowski, D.S.; Kumar, N.; Richardson, D.R. Synthesis and biological evaluation of 2-benzoylpyridine thiosemicarbazones in a dimeric system: Structure-activity relationship studies on their anti-proliferative and iron chelation efficacy. *J. Inorg. Biochem.* **2014**, *141*, 43–54. [CrossRef]
34. Zhang, Z.; Wang, Q.; Jiang, H.; Chen, A.; Zou, C. Bidentate Pyridyl-Amido Hafnium Catalysts for Copolymerization of Ethylene with 1-Octene and Norbornene. *Eur. J. Inorg. Chem.* **2022**, *26*, e202200725. [CrossRef]
35. Okamoto, I.; Terashima, M.; Masu, H.; Nabeta, M.; Ono, K.; Morita, N.; Katagiri, K.; Azumaya, I.; Tamura, O. Acid-induced conformational alteration of cis-preferential aromatic amides bearing N-methyl-N-(2-pyridyl) moiety. *Tetrahedron* **2011**, *67*, 8536–8543. [CrossRef]
36. Hartwig, J.F. *Organotransition Metal Chemistry: From Bonding to Catalysis*; University Science Books: Melville, NY, USA, 2010.
37. Wozniak, M.; Braun, T.; Ahrens, M.; Braun-Cula, B.; Wittwer, P.; Herrmann, R.; Laubenstein, R. Activation of SF6 at a Xantphos-Type Rhodium Complex. *Organometallics* **2018**, *37*, 821–828. [CrossRef]
38. Perdew, J.P.; Burke, K.; Ernzerhof, M. Generalized Gradient Approximation Made Simple. *Phys. Rev. Lett.* **1996**, *77*, 3865. [CrossRef] [PubMed]
39. Grimme, S.; Antony, J.; Ehrlich, S.; Krieg, H. A consistent and accurate ab initio parametrization of density functional dispersion correction (DFT-D) for the 94 elements H-Pu. *J. Chem. Phys.* **2010**, *132*, 154104. [CrossRef]
40. Grimme, S.; Ehrlich, S.; Goerigk, L. Effect of the damping function in dispersion corrected density functional theory. *J. Comput. Chem.* **2011**, *32*, 1456–1465. [CrossRef]
41. Weigend, F.; Ahlrichs, R. Balanced basis sets of split valence, triple zeta valence and quadruple zeta valence quality for H to Rn: Design and assessment of accuracy. *Phys. Chem. Chem. Phys.* **2005**, *7*, 3297–3305. [CrossRef]
42. Andrae, D.; Huermann, U.; Dolg, M.; Stoll, H.; Preu, H. Energy-adjustedab initio pseudopotentials for the second and third row transition elements. *Theor. Chim. Acta* **1990**, *77*, 123–141. [CrossRef]
43. Eichkorn, K.; Treutler, O.; Öhm, H.; Häser, M.; Ahlrichs, R. Auxiliary basis sets to approximate Coulomb potentials (Chem. Phys. Letters 240 (1995) 283–290). *Chem. Phys. Lett.* **1995**, *242*, 652–660. [CrossRef]
44. Weigend, F. Accurate Coulomb-fitting basis sets for H to Rn. *Phys. Chem. Chem. Phys.* **2006**, *8*, 1057–1065. [CrossRef]
45. Zhao, Y.; Truhlar, D.G. Design of density functionals that are broadly accurate for thermochemistry, thermochemical kinetics, and nonbonded interactions. *J. Phys. Chem. A* **2005**, *109*, 5656–5667. [CrossRef]
46. Holzer, C. An improved seminumerical Coulomb and exchange algorithm for properties and excited states in modern density functional theory. *J. Chem. Phys.* **2020**, *153*, 184115. [CrossRef] [PubMed]
47. Schäfer, A.; Klamt, A.; Sattel, D.; Lohrenz, J.C.W.; Eckert, F. COSMO Implementation in TURBOMOLE: Extension of an efficient quantum chemical code towards liquid systems. *Phys. Chem. Chem. Phys.* **2000**, *2*, 2187–2193. [CrossRef]

48. Kästner, J.; Carr, J.M.; Keal, T.W.; Thiel, W.; Wander, A.; Sherwood, P. DL-FIND: An open-source geometry optimizer for atomistic simulations. *J. Phys. Chem. A* **2009**, *113*, 11856–11865. [CrossRef]
49. ChemShell, a Computational Chemistry Shell. Available online: www.chemshell.org (accessed on 26 June 2023).
50. Franzke, Y.J.; Holzer, C.; Andersen, J.H.; Begusic, T.; Bruder, F.; Coriani, S.; Della Sala, F.; Fabiano, E.; Fedotov, D.A.; Furst, S.; et al. TURBOMOLE: Today and Tomorrow. *J. Chem. Theory Comput.* **2023**. [CrossRef] [PubMed]
51. Nagy, P.R.; Kallay, M. Approaching the Basis Set Limit of CCSD(T) Energies for Large Molecules with Local Natural Orbital Coupled-Cluster Methods. *J. Chem. Theory Comput.* **2019**, *15*, 5275–5298. [CrossRef] [PubMed]
52. Kallay, M.; Nagy, P.R.; Mester, D.; Rolik, Z.; Samu, G.; Csontos, J.; Csoka, J.; Szabo, P.B.; Gyevi-Nagy, L.; Hegely, B.; et al. The MRCC program system: Accurate quantum chemistry from water to proteins. *J. Chem. Phys.* **2020**, *152*, 074107. [CrossRef]
53. Hellweg, A.; Rappoport, D. Development of new auxiliary basis functions of the Karlsruhe segmented contracted basis sets including diffuse basis functions (def2-SVPD, def2-TZVPPD, and def2-QVPPD) for RI-MP2 and RI-CC calculations. *Phys. Chem. Chem. Phys.* **2015**, *17*, 1010–1017. [CrossRef]
54. Hellweg, A.; Hättig, C.; Höfener, S.; Klopper, W. Optimized accurate auxiliary basis sets for RI-MP2 and RI-CC2 calculations for the atoms Rb to Rn. *Theor. Chem. Acc.* **2007**, *117*, 587–597. [CrossRef]
55. Weigend, F. Hartree-Fock exchange fitting basis sets for H to Rn. *J. Comput. Chem.* **2008**, *29*, 167–175. [CrossRef]
56. Kanchanakungwankul, S.; JBao, J.L.; Zheng, J.; Alecu, I.M.; Lynch, B.J.; Zhao, Y.; Truhlar, D.G. Database of Frequency Scale Factors for Electronic Model Chemistries—Version 5. Available online: https://comp.chem.umn.edu/freqscale/ (accessed on 26 June 2023).
57. Molpro 2022.3; 2022. Available online: www.molpro.net (accessed on 26 June 2023).
58. Werner, H.J.; Knowles, P.J.; Manby, F.R.; Black, J.A.; Doll, K.; Hesselmann, A.; Kats, D.; Kohn, A.; Korona, T.; Kreplin, D.A.; et al. The Molpro quantum chemistry package. *J. Chem. Phys.* **2020**, *152*, 144107. [CrossRef] [PubMed]
59. Levine, D.S.; Head-Gordon, M. Energy decomposition analysis of single bonds within Kohn-Sham density functional theory. *Proc. Natl. Acad. Sci. USA* **2017**, *114*, 12649–12656. [CrossRef]
60. Khaliullin, R.Z.; Bell, A.T.; Head-Gordon, M. Analysis of charge transfer effects in molecular complexes based on absolutely localized molecular orbitals. *J. Chem. Phys.* **2008**, *128*, 184112. [CrossRef] [PubMed]
61. Mao, Y.; Loipersberger, M.; Horn, P.R.; Das, A.; Demerdash, O.; Levine, D.S.; Prasad Veccham, S.; Head-Gordon, T.; Head-Gordon, M. From Intermolecular Interaction Energies and Observable Shifts to Component Contributions and Back Again: A Tale of Variational Energy Decomposition Analysis. *Annu. Rev. Phys. Chem.* **2021**, *72*, 641–666. [CrossRef]
62. Epifanovsky, E.; Gilbert, A.T.B.; Feng, X.; Lee, J.; Mao, Y.; Mardirossian, N.; Pokhilko, P.; White, A.F.; Coons, M.P.; Dempwolff, A.L.; et al. Software for the frontiers of quantum chemistry: An overview of developments in the Q-Chem 5 package. *J. Chem. Phys.* **2021**, *155*, 084801. [CrossRef]
63. Chai, J.D.; Head-Gordon, M. Long-range corrected hybrid density functionals with damped atom-atom dispersion corrections. *Phys. Chem. Chem. Phys.* **2008**, *10*, 6615–6620. [CrossRef]
64. Matta, C.F.; Boyd, R.J. *The Quantum Theory of Atoms in Molecules*; Wiley-VCH: Weinheim, Germany, 2007.
65. Lu, T.; Chen, F. Multiwfn: A multifunctional wavefunction analyzer. *J. Comput. Chem.* **2012**, *33*, 580–592. [CrossRef]
66. Available online: https://github.com/zorkzou/Molden2AIM (accessed on 1 May 2023).
67. Weinhold, F.; Landis, C.R.; Glendening, E.D. What is NBO analysis and how is it useful? *Int. Rev. Phys. Chem.* **2016**, *35*, 399–440. [CrossRef]
68. Glendening, E.D.; Badenhoop, J.K.; Reed, A.E.; Carpenter, J.E.; Bohmann, J.A.; Morales, C.M.; Karafiloglou, P.; Landis, C.R.; Weinhold, F. *NBO 7.0*; Theoretical Chemistry Institute, University of Wisconsin: Madison, WI, USA, 2018.
69. Neese, F. Software Update: The ORCA Program System—Version 5.0. *WIREs Comput. Mol. Sci.* **2022**, *12*, e1606. [CrossRef]
70. Zhurko, G.A. Chemcraft Ver. 1.80. Chemcraft—Graphical Program for Visualization of Quantum Chemistry Computations. Ivanovo, Russia. 2005. Available online: https://chemcraftprog.com (accessed on 1 August 2023).
71. Thom, A.J.; Sundstrom, E.J.; Head-Gordon, M. LOBA: A localized orbital bonding analysis to calculate oxidation states, with application to a model water oxidation catalyst. *Phys. Chem. Chem. Phys.* **2009**, *11*, 11297–11304. [CrossRef] [PubMed]
72. Gimferrer, M.; Comas-Vila, G.; Salvador, P. Can We Safely Obtain Formal Oxidation States from Centroids of Localized Orbitals? *Molecules* **2020**, *25*, 234. [CrossRef] [PubMed]
73. Postils, V.; Delgado-Alonso, C.; Luis, J.M.; Salvador, P. An Objective Alternative to IUPAC's Approach To Assign Oxidation States. *Angew. Chem. Int. Ed.* **2018**, *57*, 10525–10529. [CrossRef]
74. Gimferrer, M.; Aldossary, A.; Salvador, P.; Head-Gordon, M. Oxidation State Localized Orbitals: A Method for Assigning Oxidation States Using Optimally Fragment-Localized Orbitals and a Fragment Orbital Localization Index. *J. Chem. Theory Comput.* **2022**, *18*, 309–322. [CrossRef] [PubMed]
75. Frisch, M.J.; Trucks, G.W.; Schlegel, H.B.; Scuseria, G.E.; Robb, M.A.; Cheeseman, J.R.; Scalmani, G.; Barone, V.; Petersson, G.A.; Nakatsuji, H.; et al. *Gaussian 16 Rev. C.01*; Gaussian, Inc.: Wallingford, CT, USA, 2016.
76. Salvador, P.; Ramos-Cordoba, E. *APOST-3D*; Universitat de Girona: Girona, Spain, 2012.
77. Gorelsky, S.I. AOMix: Program for Molecular Orbital Analysis, 6.92. 2018. Available online: https://www.sg-chem.net/aomix/ (accessed on 27 July 2023).
78. Gorelsky, S.I.; Lever, A.B.P. Electronic structure and spectra of ruthenium diimine complexes by density functional theory and INDO/S. Comparison of the two methods. *J. Organomet. Chem.* **2001**, *635*, 187–196. [CrossRef]

79. Gorelsky, S.I.; Basumallick, L.; Vura-Weis, J.; Sarangi, R.; Hodgson, K.O.; Hedman, B.; Fujisawa, K.; Solomon, E.I. Spectroscopic and DFT investigation of [MHB(3,5-iPr$_2$pz)$_3$(SC$_6$F$_5$)] (M = Mn, Fe, Co, Ni, Cu, and Zn) model complexes: Periodic trends in metal-thiolate bonding. *Inorg. Chem.* 2005, *44*, 4947–4960. [CrossRef]
80. Kraka, E.; Quintano, M.; La Force, H.W.; Antonio, J.J.; Freindorf, M. The Local Vibrational Mode Theory and Its Place in the Vibrational Spectroscopy Arena. *J. Phys. Chem. A* 2022, *126*, 8781–8798. [CrossRef]
81. Zou, W.L.; Tao, Y.W.; Freindorf, M.; Cremer, D.; Kraka, E. Local vibrational force constants—From the assessment of empirical force constants to the description of bonding in large systems. *Chem. Phys. Lett.* 2020, *748*, 137337. [CrossRef]
82. Kraka, E.; Zou, W.L.; Tao, Y.W. Decoding chemical information from vibrational spectroscopy data: Local vibrational mode theory. *Wiley Rev. Comput. Mol. Sci.* 2020, *10*, e1480. [CrossRef]
83. Kraka, E.; Zou, W.; Tao, Y. *LMODEA(F90)*; Version 2.0.0; Computational and Theoretical Chemistry Group (CATCO): SMU: Dallas, TX, USA, 2020.
84. Khaliullin, R.Z.; Cobar, E.A.; Lochan, R.C.; Bell, A.T.; Head-Gordon, M. Unravelling the origin of intermolecular interactions using absolutely localized molecular orbitals. *J. Phys. Chem. A* 2007, *111*, 8753–8765. [CrossRef]
85. Martinez-Prieto, L.M.; Avila, E.; Palma, P.; Alvarez, E.; Campora, J. β-Hydrogen Elimination Reactions of Nickel and Palladium Methoxides Stabilised by PCP Pincer Ligands. *Chemistry* 2015, *21*, 9833–9849. [CrossRef]
86. Mayer, J.; Nugent, W. *Metal-Ligand Multiple Bonds: The Chemistry of Transition Metal Complexes Containing Oxo, Nitrido, Imido, Alkylidene, or Alkylidyne Ligands*; John Wiley & Sons: Hoboken, NJ, USA, 1988.
87. Delony, D.; Kinauer, M.; Diefenbach, M.; Demeshko, S.; Würtele, C.; Holthausen, M.C.; Schneider, S. A Terminal Iridium Oxo Complex with a Triplet Ground State. *Angew. Chem. Int. Ed.* 2019, *58*, 10971–10974. [CrossRef]
88. Appleton, T.G.; Clark, H.C.; Manzer, L.E. The trans-influence: Its measurement and significance. *Coord. Chem. Rev.* 1973, *10*, 335–422. [CrossRef]
89. Sajith, P.K.; Suresh, C.H. Quantification of mutual trans influence of ligands in Pd(II) complexes: A combined approach using isodesmic reactions and AIM analysis. *Dalton Trans.* 2010, *39*, 815–822. [CrossRef] [PubMed]
90. Yang, H.; Boulet, P.; Record, M.-C. A rapid method for analyzing the chemical bond from energy densities calculations at the bond critical point. *Comput. Theor. Chem.* 2020, *1178*, 112784. [CrossRef]
91. Bryndza, H.E.; Fong, L.K.; Paciello, R.A.; Tam, W.; Bercaw, J.E. Relative metal-hydrogen, -oxygen, -nitrogen, and -carbon bond strengths for organoruthenium and organoplatinum compounds; equilibrium studies of Cp*(PMe$_3$)$_2$RuX and (DPPE)MePtX systems. *J. Am. Chem. Soc.* 2002, *109*, 1444–1456. [CrossRef]
92. Anslyn, E.V.; Dougherty, D.A. *Modern Physical Organic Chemistry*; University Science Books: Melville, NY, USA, 2005.
93. Luecke, H.F.; Arndtsen, B.A.; Burger, P.; Bergman, R.G. Synthesis of Fischer Carbene Complexes of Iridium by C−H Bond Activation of Methyl and Cyclic Ethers: Evidence for Reversible α-Hydrogen Migration. *J. Am. Chem. Soc.* 1996, *118*, 2517–2518. [CrossRef]
94. Mei, J.; Carsch, K.M.; Freitag, C.R.; Gunnoe, T.B.; Cundari, T.R. Variable pathways for oxygen atom insertion into metal-carbon bonds: The case of Cp*W(O)$_2$(CH$_2$SiMe$_3$). *J. Am. Chem. Soc.* 2013, *135*, 424–435. [CrossRef] [PubMed]
95. Blum, O.; Milstein, D. Direct Observation of OH Reductive Elimination from Ir^{III} Complexes. *Angew. Chem. Int. Ed.* 1995, *34*, 229–231. [CrossRef]
96. Milstein, D. Carbon-hydrogen vs. oxygen-hydrogen reductive elimination of methanol from a metal complex. Which is a more likely process? *J. Am. Chem. Soc.* 2002, *108*, 3525–3526. [CrossRef]
97. Li, S.; Cui, W.; Wayland, B.B. Competitive O-H and C-H oxidative addition of CH$_3$OH to rhodium(II) porphyrins. *Chem. Commun.* 2007, 4024–4025. [CrossRef] [PubMed]
98. Poater, A. Versatile deprotonated NHC: C,N-bridged dinuclear iridium and rhodium complexes. *Beilstein J. Org. Chem.* 2016, *12*, 117–124. [CrossRef]
99. Vaughan, B.A.; Webster-Gardiner, M.S.; Cundari, T.R.; Gunnoe, T.B. Organic chemistry. A rhodium catalyst for single-step styrene production from benzene and ethylene. *Science* 2015, *348*, 421–424. [CrossRef]
100. Kleigrewe, N.; Steffen, W.; Blomker, T.; Kehr, G.; Frohlich, R.; Wibbeling, B.; Erker, G.; Wasilke, J.C.; Wu, G.; Bazan, G.C. Chelate bis(imino)pyridine cobalt complexes: Synthesis, reduction, and evidence for the generation of ethene polymerization catalysts by Li+ cation activation. *J. Am. Chem. Soc.* 2005, *127*, 13955–13968. [CrossRef]
101. Wark, T.A.; Stephan, D.W. Early metal thiolato species as metalloligands in the formation of early/late heterobimetallic complexes: Syntheses and molecular structures of Cp2Ti(SMe)2, Cp2V(SMe)2, (Cp2Ti(μ-SMe)2)2Ni and (Ni(μ-SMe)2)6. *Organometallics* 1989, *8*, 2836–2843. [CrossRef]
102. Nishiyama, Y.; Katsuura, A.; Okamoto, Y.; Hamanaka, S. Bis(acyl) Diselenides as Convenient Acylating Reagents. *Chem. Lett.* 2002, *18*, 1825–1826. [CrossRef]
103. Sheldrick, G.M. A short history of SHELX. *Acta Crystallogr A* 2008, *64*, 112–122. [CrossRef]
104. Dolomanov, O.V.; Bourhis, L.J.; Gildea, R.J.; Howard, J.A.K.; Puschmann, H. OLEX2: A complete structure solution, refinement and analysis program. *J. Appl. Crystallogr.* 2009, *42*, 339–341. [CrossRef]
105. Balasubramani, S.G.; Chen, G.P.; Coriani, S.; Diedenhofen, M.; Frank, M.S.; Franzke, Y.J.; Furche, F.; Grotjahn, R.; Harding, M.E.; Hattig, C.; et al. TURBOMOLE: Modular program suite for ab initio quantum-chemical and condensed-matter simulations. *J. Chem. Phys.* 2020, *152*, 184107. [CrossRef]

106. Lefebvre, C.; Khartabil, H.; Boisson, J.C.; Contreras-Garcia, J.; Piquemal, J.P.; Henon, E. The Independent Gradient Model: A New Approach for Probing Strong and Weak Interactions in Molecules from Wave Function Calculations. *ChemPhysChem* **2018**, *19*, 724–735. [CrossRef]
107. Ma, Q.; Werner, H.J. Scalable Electron Correlation Methods. 7. Local Open-Shell Coupled-Cluster Methods Using Pair Natural Orbitals: PNO-RCCSD and PNO-UCCSD. *J. Chem. Theory Comput.* **2020**, *16*, 3135–3151. [CrossRef]
108. Neese, F.; Valeev, E.F. Revisiting the Atomic Natural Orbital Approach for Basis Sets: Robust Systematic Basis Sets for Explicitly Correlated and Conventional Correlated ab initio Methods? *J. Chem. Theory Comput.* **2011**, *7*, 33–43. [CrossRef]

Disclaimer/Publisher's Note: The statements, opinions and data contained in all publications are solely those of the individual author(s) and contributor(s) and not of MDPI and/or the editor(s). MDPI and/or the editor(s) disclaim responsibility for any injury to people or property resulting from any ideas, methods, instructions or products referred to in the content.

106. Lefebvre, C.; Khartabil, H.; Boisson, J.C.; Contreras-Garcia, J.; Piquemal, J.P.; Henon, E. The Independent Gradient Model: A New Approach for Probing Strong and Weak Interactions in Molecules from Wave Function Calculations. *ChemPhysChem* **2018**, *19*, 724–735. [CrossRef]
107. Ma, Q.; Werner, H.J. Scalable Electron Correlation Methods. 7. Local Open-Shell Coupled-Cluster Methods Using Pair Natural Orbitals: PNO-RCCSD and PNO-UCCSD. *J. Chem. Theory Comput.* **2020**, *16*, 3135–3151. [CrossRef]
108. Neese, F.; Valeev, E.F. Revisiting the Atomic Natural Orbital Approach for Basis Sets: Robust Systematic Basis Sets for Explicitly Correlated and Conventional Correlated ab initio Methods? *J. Chem. Theory Comput.* **2011**, *7*, 33–43. [CrossRef]

Disclaimer/Publisher's Note: The statements, opinions and data contained in all publications are solely those of the individual author(s) and contributor(s) and not of MDPI and/or the editor(s). MDPI and/or the editor(s) disclaim responsibility for any injury to people or property resulting from any ideas, methods, instructions or products referred to in the content.

MDPI
St. Alban-Anlage 66
4052 Basel
Switzerland
www.mdpi.com

Chemistry Editorial Office
E-mail: chemistry@mdpi.com
www.mdpi.com/journal/chemistry

Disclaimer/Publisher's Note: The statements, opinions and data contained in all publications are solely those of the individual author(s) and contributor(s) and not of MDPI and/or the editor(s). MDPI and/or the editor(s) disclaim responsibility for any injury to people or property resulting from any ideas, methods, instructions or products referred to in the content.

www.ingramcontent.com/pod-product-compliance
Lightning Source LLC
LaVergne TN
LVHW070225100526
838202LV00015B/2091